Advances in Experimental Medicine and Biology

Advances in Internal Medicine

Volume 1307

More information about this series at http://www.springer.com/series/13780

Md. Shahidul Islam

Editor

Diabetes: from Research to Clinical Practice

Volume 4

 Springer

Editor
Md. Shahidul Islam
Department of Clinical Science and Education
Karolinska Institutet
Stockholm, Sweden

Department of Emergency Care and Internal Medicine
Uppsala University Hospital, Uppsala University
Sweden

ISSN 0065-2598   ISSN 2214-8019   (electronic)
Advances in Experimental Medicine and Biology
ISSN 2367-0177   ISSN 2367-0185   (electronic)
Advances in Internal Medicine
ISBN 978-3-030-51091-6   ISBN 978-3-030-51089-3   (eBook)
https://doi.org/10.1007/978-3-030-51089-3

Dedicated to the living memory of Muhammad Ibrahim (c. 1911–1989)

Disclaimer

While every effort has been made to ensure that the contents of the book are up to date and correct, the authors, editors and the publisher assume no responsibility for eventual mistakes in the book, or eventual damages that might occur to the patients because of their treatment according to the information contained in the book. Clinicians are advised to follow approved local, national or international guidelines, and use their professional judgment to ensure correct and optimal treatment of their patients.

Contents

Adv Exp Med Biol - Advances in Internal Medicine (2020) 4: 1–5
https://doi.org/10.1007/5584_2020_553
© Springer Nature Switzerland AG 2020
Published online: 25 June 2020

Diabetes: From Research to Clinical Practice

Md. Shahidul Islam

Abstract

The number of people living with diabetes, the number of deaths attributable to it, and the cost of treating the disease and its complications are increasing exponentially. Centuries of research led to the discovery of insulin and other drugs based on pathophysiology from "the triumvirate to ominous octet". The agonists of the glucagon-like peptide-1 (GLP-1) receptor, and the inhibitors of the sodium-glucose transport protein 2 (SGLT2) are the new drugs that improve cardiovascular outcomes and provide renal protection, and they are being used increasingly for evidence-based treatment of type 2 diabetes. Bariatric surgery, when indicated, results in excellent weight- and metabolic-control, and in many instances even remission of diabetes. Technological advances like Flash glucose monitoring, continuous subcutaneous insulin infusion (CSII), and continuous glucose monitoring (CGM) have improved glycemic control, reduced episodes of severe hypoglycemia, and improved quality of life. For the treatment of diabetic macular edema intravitreal injection of several anti-VEGF agents are being used. Numerous people living in the middle- and low-income countries cannot afford the costs of care of diabetes. Institutions like the World Health Organization, the World Bank and the International Monetary Fund should roll out plans to convince the politicians to invest more in improving the diabetes care facilities.

Keywords

Anti-VEGF · Cardiovascular outcomes · Diabetes and GLP-1 receptor agonists · Diabetes and SGLT2 inhibitors · Diabetic keto-acidosis · Diabetic macular edema · Euglycemic diabetic keto-acidosis · Gestational diabetes · Hyperglycemic hyperosmolar comma · Hypoglycemia · Renal protection · Type 1 diabetes · Type 2 diabetes

M. S. Islam (✉)
Department of Clinical Science and Education, Södersjukhuset, Karolinska Institutet, Research Center, Stockholm, Sweden

Department of Emergency Care and Internal Medicine, Uppsala University Hospital, Uppsala, Sweden
e-mail: shahidul.islam@ki.se

In 2019, 463 million adults in the world were living with diabetes, and about half of them were unaware of their condition (Saeedi et al. 2019). About 80% of them were living in the middle- and low-income countries. Number of children and adolescents living with type 1 diabetes was 1.1 million (Patterson et al. 2019). The estimated global healthcare expenditure for diabetes in the world was 760 billion USD (Williams et al. 2020). In 2019, 2.4 million of deaths among adults (20–79 year) were attributed to diabetes (Saeedi et al. 2020).

Research in the past decades have increased our understanding of the different types of diabetes and their treatment. In addition to the type 1 diabetes, type 2 diabetes, and gestational diabetes, there are other forms of diabetes that include diabetes of cystic fibrosis, fibro-calculous pancreatopathy, drug-induced diabetes, and different monogenic diabetes syndromes. Some people with "type 1 diabetes" have no evidence of autoimmune β-cell damage. Some people with apparent "type 2 diabetes" have autoimmune diabetes (Latent Autoimmune Diabetes in Adults, LADA). They have an intermediate phenotype between type 1 diabetes and type 2 diabetes. These people can be identified by demonstrating anti-glutamic acid decarboxylase (GAD) antibody. Treatment of these people must be individualized, and in this respect measurement of C-peptide may help the clinicians in choosing the appropriate treatment.

When it comes to the drug treatment of diabetes, we are witnessing an era of more "personalized approach". Glucose-lowering drugs that are safe and that clearly improve cardiovascular outcomes are now available. Inhibitors of dipeptidyl peptidase-4 (DPP-4), agonists of the glucagon-like peptide-1 (GLP-1) receptor, and the inhibitors of the sodium-glucose transport protein 2 (SGLT2) represent major advances in the treatment of type 2 diabetes. GLP-1 receptor agonists and SGLT2 inhibitors reduce cardiovascular events and mortality by glycemic- and extra-glycemic effects. The use of these drugs is increasing, especially in the high-income countries. Metformin is no longer the first line drug, but it should be considered for overweight patients with type 2 diabetes who does not have cardiovascular disease or who have only moderate risk of cardiovascular disease. For patients with cardiovascular disease, an SGLT2 inhibitor, or a GLP-1 receptor agonist that offers cardiovascular protection, is recommended as the first line drug. GLP-1 receptor agonists are particularly suitable for overweight or obese people with type 2 diabetes. DPP-4 inhibitors, GLP-1 receptor agonists, and SGLT2 inhibitors are usually not associated with hypoglycemia. For many people, especially in the middle- and low-income countries, metformin and sulfonylurea still remain the first-line drugs because of their low prices (Mohan et al. 2020).

The history of discovery and clinical application of GLP-1 through decades of basic- and translational-researches is fascinating (Drucker et al. 2017). The list of beneficial effects of GLP-1 receptor agonists is long. These include stimulation of glucose-dependent insulin secretion, inhibition of glucagon secretion, delay of gastric emptying, activation of the "ileal brake", improvement of satiety, reduction of food intake, reduction of body weight, increase of natriuresis, reduction of blood pressure, and promotion of β-cell growth (Buteau 2008). The blood-pressure lowering effect of the GLP-1 receptor agonists is mediated partly by GLP-1 mediated increase in the secretion of the atrial natriuretic peptide and by the positive effects of GLP-1 on the endothelial cells (Helmstadter et al. 2020). The delaying effect on the gastric emptying is successively reduced when the short-acting GLP-1 receptor agonists are used for long time, or when the long-acting GLP-1 receptor agonists are used (Umapathysivam et al. 2014). Bariatric surgery, apart from causing anatomical changes, increases GLP-1-secreting L cells in the small intestine and increases secretion of the gastrointestinal satiety hormone peptide YY. These changes result in improvement of glycemia, remission of type 2 diabetes, and reduction of diabetes-associated morbidity and mortality.

In people with type 2 diabetes, GLP-1 receptor agonists reduce cardiovascular deaths, major adverse cardiac events, and hospitalization for heart failure. These drugs should be used whenever possible for the treatment of people with diabetes who have atherosclerotic cardiovascular diseases. These drugs should also be considered in people with type 2 diabetes who does not have established cardiovascular diseases, but have high risk factors for developing cardiovascular diseases (Buse et al. 2020).

It has taken about 178 years from the discovery of phlorizin in 1835 to the clinical use of a stable derivative of phlorizin as a specific inhibitor of sodium-glucose transport protein 2 (SGLT2) (Vick et al. 1973). This group of drugs should be used for the treatment of people with type 2 diabetes who have heart failure, specially heart failure with reduced ejection fraction, and in people with type 2 diabetes who have chronic kidney disease (Buse et al. 2020). They reduce major adverse cardiac events, cardiovascular deaths, hospitalization for heart failure, the rate of progression of the chronic kidney disease, and liver fats in people with nonalcoholic fatty liver disease (NAFLD) (Buse et al. 2020). Despite the risks of some side effects like urinary tract infections and euglycemic diabetic ketoacidosis, this group of drugs are being increasingly used, especially in the high-income countries.

In 2019, more than 20 million babies were born to mothers with diabetes during pregnancy (Yuen et al. 2019). Women must be screened for gestational diabetes but there is no consensus about the optimal evidence-based screening strategy, diagnostic methods and diagnostic thresholds. The new diagnostic criteria for gestational diabetes are based on the HAPO study that focused on the perinatal outcomes (Group HSCR 2009). Use of these criteria have increased the prevalence of gestational diabetes 6–11-fold (Behboudi-Gandevani et al. 2019). The benefits, and possible harms of reducing the threshold of the diagnostic criteria need careful evaluation. Gestational diabetes mellitus is treated by dietary modifications and exercise. For pharmacological treatment, insulin is the first choice. As an alternative to insulin, some women can be treated by metformin, but it is an off-label use. The sulfonylurea drug glyburide can also be used as second line drug in some selected patients. Both metformin and glyburide cross the placenta and their long-term safety remains unclear (American Diabetes Association 2020a).

Reduction of weight in people with type 2 diabetes who are obese, by lifestyle modification and optimal medical treatment is often difficult and sometimes impossible. Some of these people, when carefully selected, will benefit from bariatric surgical procedures (American Diabetes Association 2020b). Laparoscopic vertical sleeve gastrectomy, a newer low morbidity bariatric surgical procedure can result in sustained loss of excess bodyweight, and improvement or even resolution of diabetes (Gill et al. 2010). The decision to treat type 2 diabetes by bariatric surgery must be taken after serious consideration of patients' psychological and social situations, since these people will need long-term lifestyle support, monitoring of micronutrients, and some may need psychiatric support for adjustment to the changes after surgery.

Diabetic retinopathy affects about 80% of the people who have diabetes for >20 years. The main cause of blindness in diabetic retinopathy is diabetic macular edema. Imaging techniques including optical coherence tomography (OCT), OCT angiography, fluorescence angiography, ultra-widefield fluorescence angiography, are used for diagnosis, classification and follow up of diabetic macular edema. OCT, which is rapid, non-invasive, and accurate is the most widely used imaging technique in the clinical practice. Extensive research has increased our understanding about the role of angiogenesis and inflammation in the pathogenesis of diabetic macular edema. Intravitreal injection of anti-VEGF agents e.g. Bevacizumab, Ranibizumab and Aflibercept are recommended as first line treatment in diabetic macular edema (Schmidt-Erfurth et al. 2017).

Diabetes is the most common cause of chronic kidney disease and end-stage kidney disease in the world. The benefits of angiotensin converting enzyme inhibitors or angiotensin receptor blockers in patients who have diabetes and severely increased albuminuria is well established. In addition, the SGLT2 inhibitors should be used in people with type 2 diabetes who have eGFR >30 ml/min/1.73m^2. These drugs reduce the risks of death due to kidney disease, dialysis and kidney transplantation in people with type 2 diabetes (Neuen et al. 2019).

Diabetes increases the risk of non-alcoholic fatty liver disease (NAFLD), and its progressive phenotype nonalcoholic steatohepatitis (NASH). People with type 2 diabetes, especially those who are obese may be screened for NAFLD, and those at high risk of developing NASH should be refereed to hepatologists. Clinical prediction model based on routinely available clinical variables can be used for assessing the likelihood of NASH in people with diabetes and NAFLD. The degree of steatosis and fibrosis of the liver can be estimated by transient elastography. NAFLD and NASH in people with diabetes are treated by weight loss through lifestyle modification, and optimization of blood glucose control initially with metformin. The GLP-1 receptor agonist liraglutide leads to resolution of NASH and slows progression to fibrosis and it should be used in people who have diabetes and NAFLD or NASH (Armstrong et al. 2016). The insulin sensitizer pioglitazone also improves steatosis, inflammation and fibrosis but, it is not used so often because of the risk of weight gain and heart failure.

Hypoglycemia in people with or without diabetes is associated with increased morbidity and mortality. Fear of severe hypoglycemia limits the optimal control of glycemia and the quality of life. Prevention of hypoglycemia is an essential objective of diabetes management. Patient education, selection of appropriate insulin regime, dietary modifications, frequent monitoring of glucose, and use of automated insulin dose advisors, particularly for people with type 1 diabetes, can reduce the frequency of severe hypoglycemic episodes. Use of Flash glucose monitoring, continuous subcutaneous insulin infusion (CSII), continuous glucose monitoring (CGM) systems, and bionic pancreas can prevent severe hypoglycemic episodes, improve glycemic control and improve quality of life.

Diabetic ketoacidosis, hyperglycemic hyperosmolar state, and euglycemic diabetic ketoacidosis are life-threatening conditions. Prevention, immediate hospitalization, and guideline-based management of these conditions are essential to reduce mortality. Not surprisingly, the death rates from these acute conditions are higher in some of the middle- and low-income countries, where some people cannot obtain insulin or cannot afford the cost of insulin. It is essential that insulin is made available in all countries at affordable costs (Greene and Riggs 2015; The Lancet Diabetes E 2020).

Diabetes care is much more than prescribing new drugs or recommending new technologies. Delivering optimal diabetes care needs a dedicated team consisting of general practitioners, consultant diabetologists, diabetes specialist nurses, practice nurses, dietitians, diabetes educators, pharmacists, social workers, podiatrists, ophthalmologists, psychologists and other specialist consultants. People with diabetes must live their lives with meticulous attention to numerous details including what they eat, when they eat, how much they exercise, frequent glucose-monitoring, and multiple insulin injections every day. Attention to all these details and the fear of hypoglycemic episodes can make life stressful. It is essential to treat people with diabetes with warmth and empathy and provide emotional supports whenever they need it.

About 80% of the people with diabetes live in the middle- and low-income countries. The saddest reality is that many people living in these countries do not have access to diabetes care facilities of acceptable quality, and they cannot afford the costs of even the essential medicines for their treatment (Mohan et al. 2020; Manne-Goehler et al. 2019). The total healthcare expenditure for diabetes in the high-income countries is about 300 times more than in the low-income countries (Williams et al. 2020). While, mean annual health expenditure per person with diabetes in the USA is 11,915 USD, it is only about 64 USD in Bangladesh. This tragic inequality needs an innovative solution. International institutions like the World Health Organization, the World Bank and the International Monetary Fund should roll out plans to convince the politicians of these countries to invest in improvement of their diabetes care facilities so that people with diabetes can get improved care within the framework of universal health coverage (Moucheraud et al. 2019).

Acknowledgement Financial support was obtained from the Karolinska Institutet and the Uppsala County Council.

References

American Diabetes Association (2020a) 14. Management of diabetes in pregnancy: standards of medical care in diabetes-2020. Diabetes Care 43(Suppl 1):S183–SS92

American Diabetes Association (2020b) 8. Obesity management for the treatment of type 2 diabetes: standards of medical care in diabetes-2020. Diabetes Care 43 (Suppl 1):S89–S97

Armstrong MJ, Gaunt P, Aithal GP, Barton D, Hull D, Parker R et al (2016) Liraglutide safety and efficacy in patients with non-alcoholic steatohepatitis (LEAN): a multicentre, double-blind, randomised, placebo-controlled phase 2 study. Lancet 387(10019):679–690

Behboudi-Gandevani S, Amiri M, Bidhendi Yarandi R, Ramezani TF (2019) The impact of diagnostic criteria for gestational diabetes on its prevalence: a systematic review and meta-analysis. Diabetol Metab Syndr 11:11

Buse JB, Wexler DJ, Tsapas A, Rossing P, Mingrone G, Mathieu C et al (2020) 2019 update to: Management of Hyperglycemia in type 2 diabetes, 2018. A consensus report by the American Diabetes Association (ADA) and the European Association for the Study of diabetes (EASD). Diabetes Care 43(2):487–493

Buteau J (2008) GLP 1 receptor signaling: effects on pancreatic beta-cell proliferation and survival. Diabetes Metab 34(Suppl 2):S73–S77

Drucker DJ, Habener JF, Holst JJ (2017) Discovery, characterization, and clinical development of the glucagon-like peptides. J Clin Invest 127(12):4217–4227

Gill RS, Birch DW, Shi X, Sharma AM, Karmali S (2010) Sleeve gastrectomy and type 2 diabetes mellitus: a systematic review. Surg Obes Relat Dis 6(6):707–713

Greene JA, Riggs KR (2015) Why is there no generic insulin? Historical origins of a modern problem. N Engl J Med 372(12):1171–1175

Group HSCR (2009) Hyperglycemia and Adverse Pregnancy Outcome (HAPO) Study: associations with neonatal anthropometrics. Diabetes 58(2):453–459

Helmstadter J, Frenis K, Filippou K, Grill A, Dib M, Kalinovic S et al (2020) Endothelial GLP-1 (glucagon-like peptide-1) receptor mediates cardiovascular protection by Liraglutide in mice with experimental arterial hypertension. Arterioscler Thromb Vasc Biol 40(1):145–158

Manne-Goehler J, Geldsetzer P, Agoudavi K, Andall-Brereton G, Aryal KK, Bicaba BW et al (2019) Health system performance for people with diabetes in 28 low- and middle-income countries: a cross-sectional study of nationally representative surveys. PLoS Med 16(3):e1002751

Mohan V, Khunti K, Chan SP, Filho FF, Tran NQ, Ramaiya K et al (2020) Management of Type 2 diabetes in developing countries: balancing optimal glycaemic control and outcomes with affordability and accessibility to treatment. Diabetes Ther 11 (1):15–35

Moucheraud C, Lenz C, Latkovic M, Wirtz VJ (2019) The costs of diabetes treatment in low- and middle-income countries: a systematic review. BMJ Glob Health 4(1): e001258

Neuen BL, Young T, Heerspink HJL, Neal B, Perkovic V, Billot L et al (2019) SGLT2 inhibitors for the prevention of kidney failure in patients with type 2 diabetes: a systematic review and meta-analysis. Lancet Diabetes Endocrinol 7(11):845–854

Patterson CC, Karuranga S, Salpea P, Saeedi P, Dahlquist G, Soltesz G et al (2019) Worldwide estimates of incidence, prevalence and mortality of type 1 diabetes in children and adolescents: Results from the International Diabetes Federation Diabetes Atlas, 9th edition. Diabetes Res Clin Pract 157:107842

Saeedi P, Petersohn I, Salpea P, Malanda B, Karuranga S, Unwin N et al (2019) Global and regional diabetes prevalence estimates for 2019 and projections for 2030 and 2045: Results from the International Diabetes Federation Diabetes Atlas, 9(th) edition. Diabetes Res Clin Pract 157:107843

Saeedi P, Salpea P, Karuranga S, Petersohn I, Malanda B, Gregg EW et al (2020) Mortality attributable to diabetes in 20–79 years old adults, 2019 estimates: results from the International Diabetes Federation Diabetes Atlas, 9th edition. Diabetes Res Clin Pract 162:108086

Schmidt-Erfurth U, Garcia-Arumi J, Bandello F, Berg K, Chakravarthy U, Gerendas BS et al (2017) Guidelines for the management of diabetic macular edema by the European Society of Retina Specialists (EURETINA). Ophthalmologica 237(4):185–222

The Lancet Diabetes E (2020) Action on improving access to insulin. Lancet Diabetes Endocrinol 8(1):1

Umapathysivam MM, Lee MY, Jones KL, Annink CE, Cousins CE, Trahair LG et al (2014) Comparative effects of prolonged and intermittent stimulation of the glucagon-like peptide 1 receptor on gastric emptying and glycemia. Diabetes 63(2):785–790

Vick H, Diedrich DF, Baumann K (1973) Reevaluation of renal tubular glucose transport inhibition by phlorizin analogs. Am J Phys 224(3):552–557

Williams R, Karuranga S, Malanda B, Saeedi P, Basit A, Besancon S et al (2020) Global and regional estimates and projections of diabetes-related health expenditure: Results from the International Diabetes Federation Diabetes Atlas, 9th edition. Diabetes Res Clin Pract 162:108072

Yuen L, Saeedi P, Riaz M, Karuranga S, Divakar H, Levitt N et al (2019) Projections of the prevalence of hyperglycaemia in pregnancy in 2019 and beyond: Results from the International Diabetes Federation Diabetes Atlas, 9th edition. Diabetes Res Clin Pract 157:107841

Adv Exp Med Biol - Advances in Internal Medicine (2020) 4: 7–27
https://doi.org/10.1007/5584_2020_516
© Springer Nature Switzerland AG 2020
Published online: 22 March 2020

Glucose Lowering Treatment Modalities of Type 2 Diabetes Mellitus

Asena Gökçay Canpolat and Mustafa Şahin

Abstract

This chapter gives an overview of present knowledge and clinical aspects of antidiabetic drugs according to the recently available research evidence and clinical expertise.

Many agents are acting on eight groups of pathophysiological mechanisms, which is commonly called as "Ominous Octet" by DeFronzo. The muscle, liver and β-cell, the fat cell, gastrointestinal tract, α-cell, kidney, and brain play essential roles in the development of glucose intolerance in type 2 diabetic individuals (Defronzo, Diabetes 58:773–795, 2009).

A treatment paradigm shift is seen in the initiation of anti-hyperglycemic agents from old friends (meglitinides or sulphonylürea) to newer agents effecting on GLP-1 RA or SGLT-2 inhibitors. It is mostly about the other protective positive effects of thcsc agents for kidney, heart, etc. Although there are concerns for the long term safety profiles; they are used widely around the World. The delivery of patient-centered care, facilitating medication adherence, the importance of weight loss in obese patients, the importance of co-morbid conditions are the mainstays of selecting the optimal agent.

A. Gökçay Canpolat and M. Şahin (✉)
Department of Endocrinology and Metabolism, Ankara University School of Medicine, Ankara, Turkey
e-mail: mustafasahin@ankara.edu.tr; Mustafa.
Sahin@ankara.edu.tr; drsahinmustafa@yahoo.com

Keywords

Acarbose · Dipeptidyl peptidase 4 inhibitors · Glucagon-like peptide 1 receptor agonists · Glucose-lowering therapies · Meglitinides · Metformin · Sodium-glucose transporter two inhibitors · Sulphonylurea · Thiazolidinediones

1 Introduction

Diabetes mellitus (DM) is a chronic metabolic disease associated with the metabolism of carbohydrates, fats, and proteins due to insulin deficiency or insulin resistance. The treatment strategy of DM has to be based on the knowledge of its pathophysiology. The general goals of the treatment of DM are to provide glycemic control, avoid acute complications, prevent or delay the appearance of chronic complications of the disease, and thus to improve the quality of life. The management of type 2 DM also includes managing conditions associated with T2DM, such as obesity, hypertension, dyslipidemia, and cardiovascular disease.

Obesity, hypertension, and diabetes are increasingly common epidemics worldwide. What can be the reason for this incredible increase? Since a change in our genes can not explain this increase in such a short time, we can say that the cause is environmental and epigenetic. Western diet, endocrine disrupters,

Table 1 General features of major anti-hyperglycemic agents

	Hypo	Weight	Hba1c % reduction	Renal	Hepatic	GIS	Cardiac benefits/risks	Bone	Major side effects
Metformin	N	N	1.5	Contraindicated GFR <30 ml/min	Safe in mild to moderate CLD	Moderate	N	N	Gis
SU/Glinid	Moderate-Severe/Mild	Increase	1–2/1	Contraindicated GFR <30 ml/min	Avoided	N	N	N	Hypo
Agi	N	N	0.5–0.7	Contraindicated GFR <15 ml/min	Safe (except Child C)	Moderate	N	N	Gis
TZD	N	Increase	0.5–1.4	Contraindicated GFR <30 ml/min	Safe in Child A	N	Moderate risk	Moderate risk	Bone, heart, edema
DPP4 inh	N	N	0.5–1	Dose reduction (except lina)	Safe in Child A	N	Agent dependent	N	Skin, nasopharyngitis
SGLT-2 inh	N	Decrease	0.5–1	Dose reduction Contraindicated GFR <45 ml/min	Safe in Child A	N	Benefit	N/Minor risk	Genital infection
GLP-1RA	N	Decrease	1–1.5	Exenetide contraindicated GFR <30 ml/min	Safe	Moderate	Benefit	N	Gis

AGI alpha-glucosidase inhibitors, *CLD* Chronic Liver Disease, *DPP4 inh* Dipeptidyl peptidase 4 inhibitors, *GFR* Glomerular filtration rate, *GIS* Gastrointestinal, *GLP-1RA* Glucagon-like peptide 1 receptor agonists, *N* Neutral, *SGLT-2 inh* Sodium-glucose transporter 2 inhibitor, *SU* sulphonylurea, *TZD* Thiazolidinedione

processed food, changes in microbiota, technology itself, increased sugar consumption, inactivity, and stress are the leading causes of this outbreak. According to the 9th edition of new IDF atlas, 463 million adults are currently living with diabetes, and there will be 578 million adults with diabetes by 2030 and 700 million by 2045 (Saeedi et al. 2019).

The main elements of treatment are lifestyle modification and nutrition therapy, exercise, and medical therapies. Oral antidiabetic agents, insulin, and non-insulin injectable anti-hyperglycemic agents are mainstays for medical therapies (Table 1). Regarding the treatment approaches in recent years, national and international authorities published current treatment guidelines one after another (American Diabetes A 2019; Davies et al. 2018).

2 Pharmacotherapy for Type 2 Diabetes

The glucose-lowering agents target different pathophysiological pathways like insulin secretion, hepatic glucose production and utilization, insulin resistance, gastric emptying, satiety, and GLP-1 action (Fig. 1).

Many agents are acting on eight groups of pathophysiological mechanisms, which is commonly called as "Ominous Octet" by DeFronzo. The muscle, liver, and β-cell, the fat cell, gastrointestinal tract, α-cell, kidney, and brain play essential roles in the development of glucose intolerance in type 2 diabetic individuals on the mechanisms mentioned above.

2.1 Biguanides

2.1.1 Metformin

Since Phenformin was removed from markets due to deaths because of lactic acidosis in the '70s, metformin (1,1- dimethyl biguanide hydrochloride) is the only available biguanide in the world. It had been derived from Galega officinalis (French lilac) and used as the first-line agent

according to diabetes guidelines for approximately 60 years.

The mechanism of action of metformin is through activating AMP-activated Protein Kinase (AMPK), which is a nutrient sensor activated in states of low energy balance, activates uncoupling mitochondrial oxidative phosphorylation and increases cellular AMP levels. The resultant effect is inhibiting gluconeogenesis. Metformin also decreases intestinal glucose absorption, improves peripheral glucose uptake, lowers fasting plasma insulin levels, and increases insulin sensitivity (Wang et al. 2017). Another new hypothesis is to reduce intestinal bile acid resorption and lead to increase GLP-1 secretion, modulate the composition of the gut microbiota (Sansome et al. 2019).

Lifestyle intervention and metformin treatment may be useful in high-risk prediabetic patients (Moin 2019). Metformin also improves metabolic and cardiovascular risk factors, including osteoprotegerin and receptor activator of the nuclear factor–B ligand (RANKL) levels in prediabetes (Arslan et al. 2017).

The initial starting dose is generally 500 mg once or twice a day; then, the dosage is increased to 2000 mg/day. The reason for this increment is based on its gastrointestinal side effects with an incidence rate of 20–30%, such as nausea, vomiting, anorexia, diarrhea, and metallic taste. The scariest and serious adverse effect is lactic acidosis with an incidence rate of 1/30,000 and occurs in high-risk patients for this condition.

Metformin is contraindicated in patients with renal insufficiency (prominently Glomerular Filtration Rate (GFR) <30 ml/min) (Lee and Halter 2017). It is used precautiously in patients with a GFR <60 ml/min, and the dose is escalated. It is also contraindicated in clinical settings of hepatic insufficiency (liver cirrhosis), any form of acidosis, chronic heart failure, chronic lung diseases with severe hypoxemia, and alcohol abuse. Another general precaution to avoid lactic acidosis is to discontinue metformin 24–48 h before administration of high dose radiographic contrast material or general anesthesia. Elderly patients should be handled with care for metformin usage. B12 deficiency can be seen during

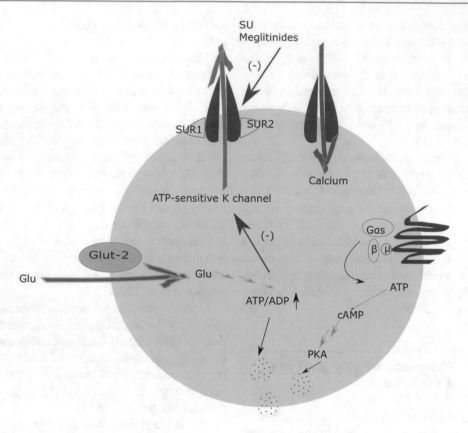

Fig. 1 Schematic mechanism of antidiabetic drugs

treatment and prevented by calcium coadministration (Bauman et al. 2000).

According to the recent guidelines, metformin is preferred as the initial therapy in individuals who do not achieve glycaemic targets despite diet and other lifestyle interventions. The oldest trials of metformin have demonstrated mean HbA$_{1c}$ reductions between 1% and 1.5% (DeFronzo and Goodman 1995; Garber et al. 1997). In head-to-head trials, metformin was shown to be equivalent to sulfonylureas, thiazolidinediones, glucagon-like peptide-1 (GLP-1) receptor agonists, and dipeptidyl peptidase-4 (DPP-4) inhibitors (Bennett et al. 2011; Russell-Jones et al. 2012).

There are some other potential indications for metformin because of its antitumor, anti-aging, cardiovascular protective, neuroprotective effects, and potential benefits for polycystic ovary syndrome (PCOS) (Wang et al. 2017).

2.2 Incretin Mimetics

Oral glucose administration has a more significant effect on insulin release than intravenous glucose administration, which is known as the "incretin effect (Nauck et al. 1986)." This effect is due to the secretion of some hormones from the gastrointestinal tract. Glucagon-like peptide-1 (GLP-1) and glucose-dependent insulinotropic polypeptide (GIP) are the major incretin hormones in humans. GIP is produced in the K-cells, and these cells are located predominantly in the duodenum.

In contrast, the L cells produce GLP-1, and they are located chiefly in the ileum and colon. They can also be found throughout the whole intestine. The incretins are cleaved by the enzyme dipeptidyl peptidase-4 (DPP-4). Because the incretin effect is found to be substantially reduced or absent in patients with T2DM, incretin-based therapies such as **DPP-4 inhibitors** or **GLP-1 receptor agonists** can be used for a therapeutic approach.

Their mechanism of action includes stimulation of glucose-dependent insulin release from the pancreatic islets, slowing gastric emptying, inhibition of post-meal glucagon release, and satiety. GLP-1 Rs have been detected in the central nervous system, including the hippocampus and hippocampal progenitor cells suggesting the existence of the gut-brain axis. GLP-1 is thought to provide a short and long term energy homeostasis by its regulating role on the appetite, food intake, and body weight (Kim and Egan 2008).

2.2.1 DPP4 Inhibitors

DPP-4 inhibitors are generally not considered as first-line therapy for T2DM and can be considered as add-on drug therapy for patients who are inadequately controlled with other first-line agents. This group consists of sitagliptin, saxagliptin, linagliptin, alogliptin, and vildagliptin. The group has nearly similar glycemic efficacy according to meta-analyses, although few head-to-head trials. The advantages of this group are they are weight-neutral and have a low risk of hypoglycemia.

Adverse effects of this group are generally well-tolerated such as headache, nasopharyngitis, upper respiratory tract infection, and gastrointestinal side effects (Amori et al. 2007).

The DPP4 inhibitors, except linagliptin, are excreted by the kidneys; therefore, dose adjustments are advisable for patients with renal dysfunction.

Vildagliptin and alogliptin have side effects of hepatic dysfunction, and in case of three times the upper limit of normal or higher liver enzyme elevations, drugs should be discontinued.

DPP4 enzyme is a pleiotropic enzyme and inactivates a large variety of hormones, peptides on the whole body. Possible crossover with other substrates of DPP-4, particularly in accordance with immune function, remains still a concern (Rohrborn et al. 2015).

Even though there is not sufficient data about the causal relationship between acute pancreatitis and DPP4 inhibitors, DPP-4 inhibitors should not be initiated in a patient with a history of pancreatitis or unexplained abdominal pain. They should be stopped in patients with persistent severe abdominal pain. Some reports also pointed out the increased risk of pancreatic cancer and neuro-endocrine tumors in sitagliptin, but there are lacking data (Pinto et al. 2018).

Another major concern about DPP4 inhibitors is skin reactions in recent years. Severe skin reactions, including Stevens-Johnson syndrome, have been reported for this group, especially with linagliptin and saxagliptin (Karagiannis et al. 2014).

Sitagliptin It can be either used as monotherapy or in combination therapy in patients who do not have adequate glycemic control with other agents. The usual dose is 100 mg per day with a reduction to 50 mg for moderate to severe renal insufficiency (GFR 30 to <45 mL/min) and 25 mg for severe renal insufficiency (GFR <30 mL/min). They have co-formulations with metformin in single tablets (50/500–850 or 1000 mg). Sitagliptin studies showed Hba1c reductions of approximately 0.6%.

Saxagliptin The usual dose of saxagliptin is 2.5 or 5 mg once daily, with the 2.5 mg dose recommended for patients with moderate to severe chronic kidney disease (glomerular filtration rate GFR ≤45 mL/min). It is metabolized through CYP 3A4/5, so 2.5 mg dose is suitable also for patients taking strong cytochrome P450 3A4/5 inhibitors such as ketoconazole. There are drug-specific concerns about increasing the hospitalization rates for heart failure for saxagliptin. In the case of known risk factors, including elevated baseline natriuretic peptides, prior heart failure, and chronic kidney disease, the risk for heart failure hospitalization under saxagliptin treatment should be considered (Scirica et al. 2013).

Linagliptin The optimal dose is 5 mg daily orally. Since it is primarily eliminated via the enterohepatic system, no dose adjustment is necessary for patients with renal or hepatic impairment.

Alogliptin The usual dose is 25 mg once daily. The dose reductions to 12.5 mg once daily are advised for patients with creatinine clearance between 30 and 60 mL/min and to 6.25 mg daily in patients with creatinine clearance <30 mL/min or undergoing dialysis. There is an increased rate of heart failure, like saxagliptin.

Vildagliptin The usual dose is 50 mg twice daily, while the dose is reduced to 50 mg once daily for moderate or severe renal impairment. No dose adjustment is necessary for patients with mild renal impairment.

GLP-1RA

DPP4 inhibitors rapidly metabolize GLP-1, and as a result, the half-life of GLP-1 is 1–2 min. Instead of the use of native peptide because of its shorter half-life, GLP-1 receptor agonists (GLP-1RA) with longer half-lives are manufactured for therapeutic use.

GLP-1RAs acts on stimulating glucose-dependent insulin secretion, inhibiting glucagon secretion, delaying gastric emptying, and lowering appetite (Nauck 2004).

Six GLP-1RA can be broadly grouped as xenopeptides (exenatide and lixisenatide), human GLP-1 analogs (liraglutide and semaglutide), and fusion peptides (albiglutide and dulaglutide). Liraglutide is administered once daily, but the other four once a week formulations (albiglutide, dulaglutide, once week exenatide, and semaglutide) provide longer durations of action.

Although they are injectable formulations such as insulin, they have comparatively more benefits like weight loss, less hypoglycemia, and in the setting of cardiovascular disease, benefits on myocardial infarction, stroke, and cardiovascular death. They are also associated with lipid and BP reductions (Raccah 2017).

The most common side effects of GLP-1 RA therapy are gastrointestinal symptoms, and they occur in as many as half of the patients (50% of patients). Semaglutide has the highest rates of nausea (11–40%), vomiting (4–13%), and diarrhea (9–17%), followed by dulaglutide, liraglutide, and exenatide. Liraglutide is slightly more likely to cause diarrhea than dulaglutide (Raccah 2017). The reactions are more frequent at higher doses of medications.

Injection site reactions are other side effects of this group.

There are safety concerns regarding the risk of medullary thyroid cancer (MTC), pancreatitis, pancreatic cancer, and cholelithiasis with these agents. But no significant differences were seen in pancreatitis, pancreatic cancer, or MTC between GLP-1 RA and placebo in blinded long-term cardiovascular outcome trials. The general approval is that GLP-1 RAs may increase the risk of cholelithiasis, but they do not lead an elevation in the risk of pancreatitis, pancreatic cancer, or medullary thyroid cancer. If the patients taking GLP-1 RA are suffering from unexplained persistent severe abdominal pain, the medication should be stopped in case of pancreatitis. Also, they should not be used in patients with a personal history of MTC or multiple endocrine neoplasia syndrome type 2.

Exenatide is renally cleared and is contraindicated in the clinical setting of advanced kidney failure (GFR <30 ml/min). No dose adjustment is required for liraglutide, albiglutide, or dulaglutide in CKD stages 2 and 3 (Davies et al. 2016).

2.2.2 Exenatide

Exendin-4 is a naturally occurring component of the saliva of the Gila monster (Heloderma suspectum) (Lee and Halter 2017). Exenatide is synthetic exendin-4, and the first GLP-1 RA approved for use by the Food and Drug Association (FDA) in 2005. It is injected twice daily

(5–10 µg) before meals subcutaneously or once a week for its extended-release formulation. It produces a reduction of 1% in HbA1C levels and a reduction in body weight with 4.5–9 kg/year. Cases of acute renal failure were reported due to exenatide therapy, especially in patients with preexisting kidney impairment and other risk factors for kidney disease.

2.2.3 Liraglutide

It has been engineered to be 97% homologous to native human GLP-1 by substituting arginine for lysine at position 34. Liraglutide is made by attaching a C-16 fatty acid (palmitic acid) with a glutamic acid spacer on the remaining lysine residue at position 26 of the peptide precursor according to product monograph. The 0.6 mg dose is a starting dose intended to reduce gastrointestinal symptoms. After 1 week at 0.6 mg per day, the dose should be increased to 1.2 and then 1.8 mg once daily to achieve maximum efficacy for glycemic control. The weight loss ranges from 0.5 kg to 2.7 kg. Also, liraglutide at a dose of 3 mg daily has approval for treatment of obesity.

No dose adjustment is required for patients with mild, moderate, or severe renal insufficiency.

2.2.4 Albiglutide

The half-life of albiglutide is 5 days, and the usual dose is 30 mg once a week by sc injection. It lowers Hba1c about 0.8%.

Atrial dysrhythmias such as atrial fibrillation were reported due to albiglutide treatment. Weight loss is less evident than exenatide or liraglutide.

2.2.5 Dulaglutide

The half-life of dulaglutide is 5 days, and the usual dosage is 0.75 mg once a week to a maximum dosage of 1.5 mg once a week. It has a reduction in body weight with 2–3 kg/year.

2.2.6 Lixisenatide

It is a modified exendin-5 molecule with a half-life of 2–4 h. The dose is initiated with 10 µg and titrated to a maximum dose of 20 µg per day. It lowers Hba1c by 0.7%.

Semaglutide Semaglutide (SEM) is the last approved selective long-acting GLP-1RA. It has a half-life of 1 week. It is produced from the basis of liraglutide by the main differences in the structure such as Ala to Aib substitution in position 8, a longer linker, and an increase in the length of the fatty diacid chain from C16 to C18. The affinity of semaglutide toward the GLP-1 R was reduced, and affinity with albumin was increased compared to liraglutide.

It is recommended to begin with a once a week sub-therapeutic dose of 0.25 mg, followed by an increase to 0.5 mg per week after 4 weeks. It can be further increased to 1 mg if required after at least 4 weeks. FDA has very recently approved oral semaglutide as the first and only oral GLP-1 RA. The tablet version is used once daily, containing 7 mg and 14 mg of semaglutide.

2.3 Sodium Glucose Co-Transporter Inhibitors

Kidneys have a significant role in normal glucose homeostasis by balancing the amount of glucose filtered from the plasma into the renal glomerular filtrate. The sodium-glucose co-transporter 2 (SGLT2), which is located in the proximal tubule of the kidney, is responsible for more than 90% reabsorption of filtered glucose. SGLT1 also plays a role in the intestinal absorption and the renal reabsorption of glucose but very negligible in healthy individuals. SGLT1 transporters have a more extensive-expression such as intestinal cells, myocardial, or vascular endothelial cells of the heart, central nervous system (Tsimihodimos et al. 2018). Because of the unknown long-term efficiency and safety of SGLT1 inhibitors, consequences of SGLT1 inhibition remain unreliable.

In patients with T2DM, plasma glucose levels exceed the glucose transport capacity, leading to glycosuria and this results in up-regulated SGLT genes, increased renal glucose reabsorption, and further hyperglycemia, thus making a vicious circle. Their mechanism of action is inhibition of SGLT-2, which results in the inhibition of renal glucose reabsorption without increasing the risk of hypoglycemia. They also reduce the renal glucose threshold and increase urinary glucose excretion. Besides, owing to its glucosuric effect, resultant reductions for HbA1C (by 0.5–1%), weight (−1.5 − −3.5 kg), and systolic BP (−3 − −5 mmHg) are observed. The main adverse effects are increased risk of genitourinary infections and slightly raised (3–8%) low-density lipoprotein cholesterol (LDL-C) levels (Hsia et al. 2017).

Euglycemic DKA diabetic ketoacidosis, limb amputations, increased risk of acute kidney injury, dehydration, orthostatic hypotension, fracture risk, and reduced bone mineral density (especially with canagliflozin) and Fournier's gangrene are other but less observed side effects of SGLT-2 inhibitors (Singh and Kumar 2018).

It is recommended to stop SGLT2 inhibitors 24 h before scheduled surgeries and anticipated metabolically stressful activities.

Canagliflozin, dapagliflozin, empagliflozin, ipragliflozin, ertugliflozin, and sotagliflozin are the components of the group.

Their cardioprotective benefits will be discussed in detail under the following headings.

Canagliflozin: The usual dose is 100 mg and up to 300 mg daily. It is contraindicated in patients with GFR <45 ml/min. Reduced bone density and increases in fractures are class effects, but upper limb fractures are much more observed with canagliflozin.

Dapagliflozin: The usual dose is 10 mg daily.

Empagliflozin: The usual dose is 10–25 mg daily.

Sotagliflozin: It has a unique property of its inhibition on both sodium-glucose co-transporter 1 (mild affinity) and 2 (with strong affinity). The recommended dose is 200 mg sotagliflozin once daily and approved by the European Medicines Agency recently.

2.4 Insulin Secretagogues

2.4.1 Sulfonylureas

They bind ATP sensitive potassium channels (KATP) of pancreatic beta cells. In the pancreatic β-cell, the KATP channel plays an essential role in coupling membrane excitability with glucose-stimulated insulin secretion. The resultant effect of the closure of the channel is the depolarization of the beta cells. Because of this depolarization, calcium shift is seen to the cell, and insulin secretion is promoted. K ATP channels consist of two subunits of the regulatory subunit SUR1 and the potassium channel subunit the inward-rectifier potassium ion channel (Kir6.2). The SUR1 regulatory subunit is encoded by the ATP-binding cassette, subfamily C, member 8 (ABCC8) gene, while the Kir6.2 subunit is encoded by the potassium inwardly-rectifying channel, subfamily J, member 11 (KCNJ11) gene (Haghvirdizadeh et al. 2015).

Sulfonylurea drugs promote insulin secretion by binding to the regulatory sulfonylurea receptor-1 (SUR1) subunit and inhibiting the KATP channel current. While SU closes the SUR1/Kir6.2 complex, diazoxide opens. Either inactivating mutations of SUR 1 or Kir6.2 lead to hyperinsulinemic hypoglycemia of infancy and activating mutations of them are related to neonatal Diabetes.

These drugs are mostly affected by T2DM individuals with a diabetes age of 5 years or less because of residual endogenous insulin production. Declining β-cell function, bad glycemic course besides with long-standing Diabetes are predisposing factors for sulfonylurea failure.

The first-generation sulfonylureas include tolbutamide, tolazamide chlorpropamide, and acetohexamide. They are very rarely used recently. The second-generation sulfonylureas include glyburide (also known as glibenclamide), glipizide, gliclazide, and glimepiride. This group is more potent than the first generation.

Most sulfonylureas are metabolized in the liver, primarily by the cytochrome P450 (CYP) 2C9 isoenzymes and excreted by the kidneys.

Some of the second generation SU are partly excreted from the bile.

The most common adverse effect of sulfonylureas is hypoglycemia. Sulfonylurea-induced hypoglycemia is more likely to occur with the longer-acting agents such as chlorpropamide and glibenclamide, especially in case of irregular eating habits and alcohol consumption.

Adverse effects are rare, and they include cholestatic jaundice, skin rash, hematologic toxicity (hemolytic anemia, thrombocytopenia, agranulocytosis), flushing (chlorpropamide) and hyponatremia (chlorpropamide).

To improve efficacy, medications in this class should be taken about 30 min before meals.

Sulfonylureas monotherapy lowers fasting plasma glucose by 20–40 mg/dL and reduces HbA1c by 1.0–2.0% (Thule and Umpierrez 2014). In the United Kingdom Prospective Diabetes Study (UKPDS), treatment with glyburide or chlorpropamide achieved an HbA1c <7% in 50% of patients at 3 years, and this number declined to 24% at 9 years. The mean percentage of patients per year with one or more episodes of any hypoglycemia was 17.7% for glyburide and 11.0% for chlorpropamide (Gangji et al. 2007). In contrast, the rate of severe hypoglycemia was ~0.5% per year, with both medications (UK Prospective Diabetes Study (UKPDS) Group 1998). The LEAD-2 study showed that glimepiride had a comparable mean HbA1c decrease of 1.0% with liraglutide when added-on to metformin over a 26-week week follow-up period (Nauck et al. 2009).

The University Group Diabetes Project (UGDP), and subsequent studies, reported an increased risk of coronary vascular mortality with sulfonylureas; however, extensive prospective randomized clinical studies, the UKPDS, ACCORD, ADOPT, DREAM, VADT, and RECORD did not support that result (Thule and Umpierrez 2014).

We describe the most commonly used second-generation SU;

Glyburide (Glibenclamide) The starting dose is one 5 mg tablet daily, and if the response is inadequate, the treatment can be raised in a stepwise fashion to 15 mg daily. Its unmetabolized compound has a half-life of 1–2 h but, its hypoglycemic effects lasts for nearly 24 h. It is contraindicated for the elderly, liver, and hepatic failure because of the particular risk for hypoglycemia.

Gliclazide This group has both conventional and extended duration preparations. The recommended dose of conventional form is initially 40–80 mg twice daily, with increases weekly to achieve adequate glycemic control. The long-acting form is initially 30–120 mg once daily, with a duration of 24 h. The formulation shows high bioavailability, and its absorption profile is unaffected by coadministration with food. It also has antioxidant properties that are independent of glycaemic control because of the free radical scavenging ability of its unique amino azabicyclo-octane ring.

Glipizide tablets of 5 and 10 mg and as an extended-release form in 2.5, 5 and 10 mg, the recommended dose initially is 5–10 mg daily, with increases based upon blood glucose and tolerance to a maximum of 15 mg (standard formulation) or 20 mg (extended-release formulation) once daily.

Glimepiride 1, 2, and 4 mg, the recommended dose initially is 1–2 mg once daily, with increases based upon blood glucose and tolerance to a maximum of 8 mg once daily.

For CKD, SU should be carefully given because of the hypoglycemia and prolonged effects of the class.

2.4.2 Meglitinide Analogs

The meglitinide analogs were first introduced in 1995. They also interact with the KATP channel, and the mechanism of action is the same as SU. The main differences from the SU group are comparatively shorter half-life, rapid onset of effect, and lower potency of hypoglycemia (Malaisse 2003). Because of the short half-life, they should be given immediately before meals. They significantly suppress meal-induced elevations in blood glucose concentrations and has advantages over SU for the possible occurrence of undesirable hypoglycemia. Another advantage of this group is their metabolism in the liver and secretion by bile. They are useful options for renal impairment and the elderly.

Repaglinide It is a carbamoylbenzoic acid derivative of the meglitinide class of insulin secretagogues. Although they have similar mechanisms of action, the molecular binding site of repaglinide is different from that of nateglinide. It is supplied as 0.5, 1, and 2 mg of blisters. Repaglinide has a half-life of 1 h. No dose adjustment necessary for chronic kidney disease (CKD) stage 3–4, and it can be initiated at 0.5 mg dose when GFR <40 mL/min/1.73m^2.

Mitiglinide It has approval for use in Japan. The primary role in therapy for mitiglinide is the treatment of elevated postprandial glucose in patients with T2DM because of selective action on the pancreatic β-cells and utility in patients with chronic kidney disease or at high risk of hypoglycemia.

Nateglinide It is a δ-phenylalanine derivative and approved in 2000. The binding sites are different from repaglinide, and the onset of inhibition of the KATP channel by nateglinide is more rapid than that by repaglinide. It has advantages for a shorter duration of action and a reduced risk of hypoglycemia compared with repaglinide. The recommended time for administration is 15 min before each meal with a maximum dosing of 120 mg per meal with nateglinide. It is initiated at low doses of 60 mg in the case of CKD stage 3–4 or kidney transplantation.

2.5 Peroxisome Proliferator-Activated Receptor (PPARγ) Agonists

The thiazolidinedione class of drugs is insulin sensitizers, which work through the activation of PPARγ. They lead an increase in GLUT 1 and 4 expressions, a decrease in free fatty acid levels,

a reduction of hepatic glucose output, an increase in differentiation from preadipocytes to adipocytes (Dubuisson et al. 2011).

Troglitazone was the first manufactured member of this group but was withdrawn from the market because of fatal hepatotoxicity. Rosiglitazone and pioglitazone are available agents of this class.

The liver metabolizes both rosiglitazone and pioglitazone with CYP 2C8 and CYP 2C8 and CYP 3A4, respectively. Because of the hepatic elimination, this class should be precautiously used in patients with liver disease and the case of high liver enzymes.

The dosage of rosiglitazone is 4–8 mg once a day, and pioglitazone is recommended at doses 15–45 mg once a day.

They have similar glucose and HbA1c lowering effects; however, they had differences in lipid metabolism and cardiovascular diseases. In a study, pioglitazone reduced the triglyceride levels (%9) and increased HDL-C (%15), whereas rosiglitazone increased total cholesterol, LDL-C, and triglycerides. LDL-C particle size is both improved with glitazones, but LDL-C particle number was improved only with pioglitazone (Deeg and Tan 2008).

Since this group was introduced to the market, glucose-lowering effects, being the only anti-hyperglycemic agent to reduce insulin resistance directly, potent HbA1C-lowering effects, low risk of hypoglycemia, lipid-lowering effects besides slowing effects on the progressive beta-cell deterioration and reduction in visceral fat promised a perfect drug option for Diabetes. But, side effects like weight gain, fluid retention, and increased risk of bone fractures limited their use. Because of the side effects of fluid retention, it is not recommended for the patients with Congestive Heart Failure NYHA stage 3–4 and graves orbitopathy. An association with bladder cancer and TZDs have been denied. Also, there are case reports about creatinine kinase elevation with rosiglitazone therapy (Sahin et al. 2005).

According to the cardiovascular safety issues in 2007, concurrent with the publication of Nissen et al., the Food and Drug Administration (FDA) issued a safety alert warning of a "potentially significant increase in the risk of heart attacks in patients taking the oral antidiabetic rosiglitazone (Nissen and Wolski 2007). The usage of this class shifted towards pioglitazone.

2.6 Alpha-Glucosidase Inhibitors

Alpha -Glucosidases, locate in the brush border of the small intestine, and they hydrolyze a-glucose residues to release a single a-glucose molecule.

Because of the structural similarity to disaccharides or oligosaccharides, they bind strongly to the enzyme and competitively inhibit the carbohydrate–glucosidase complexes. As a result of this competitive inhibition, the carbohydrate can not be absorbed in the intestine. There are four AGIs on the market: acarbose, miglitol, voglibose, and DNJ. The first three are called sugar-mimic compounds, and they are commercially available worldwide (Liu and Ma 2017). There are structural differences between acarbose and miglitol in their molecular masses and absorption. While acarbose has a molecular weight similar to a tetrasaccharide, miglitol is as small as glucose and absorbable. Miglitol is also not metabolized and excreted unchanged by the kidneys.

AGI improves postprandial glycemic excursions by delaying the absorption of dietary carbohydrates, so they have modest A1C-lowering effects and low risk for hypoglycemia.

Because of their gastrointestinal side effects (bloating, diarrhea, flatulence, abdominal pain, and constipation), their use has been limited. Slow titration of premeal doses may mitigate the side effects and facilitate tolerance to the AGIs. Acarbose and its metabolites, as well as miglitol, very probably accumulate in chronic kidney diseases. These agents should be used with caution in patients with CKD and contraindicated if GFR is <25 ml/min.

Acarbose is available as 50 and 100 mg tablets. They are taken three times daily with meals (immediately with the first mouthful of food) with a low starting dose and gradually increase the dose.

Miglitol is initiated with a dose of 25 mg and incrementally given at a dose of 50 mg three times a day.

Both acarbose and miglitol have demonstrated significantly higher efficacy versus placebo, and in most of the studies, AGIs showed more significant reductions in postprandial glucose levels. However, the changes in HbA1c and FPG appeared to be comparable between the oral agents and AGIs (Hedrington and Davis 2019). But when on-add therapy, the combination of acarbose and metformin resulted in higher reductions in HbA1c, fasting, and postprandial glucose levels (Jayaram et al. 2010).

2.7 Colesevelam

Colesevelam is a second-generation bile acid sequestrant that lowers glucose modestly and does not cause hypoglycemia. Observational information for the improvement of glycaemic control in patients with T2DM with colesevelam further provided the basis for the approval by the FDA in 2008 as an adjunct therapy for glycaemic control in adults with T2DM. Its use is associated with an increase in triglyceride levels and reductions of (15%) LDL-C. Gastrointestinal side effects, such as constipation and dyspepsia affect nearly 10% of patients, may contribute to limited use. The underlying mechanism for its anti-glycemic effects is not understood (Ooi and Loke 2012).

2.8 Bromocriptine

The quick-release formulation of dopamine receptor agonist bromocriptine has mild glucose-lowering properties. It is suggested that creating a circadian peak in central dopaminergic tone may improve insulin sensitivity. Since 1980 it was documented that bromocriptine exerts an inhibitory effect on hyperglycemia in type 2 diabetes (Lopez Vicchi et al. 2016). Bromocriptine also has a beneficial effect on plasma lipids (Luo et al. 1998).

Hba1c reductions are modest. Cardiovascular safety study (the Cylcoset trial) involving 3095 patients, demonstrated that the treatment group with bromocriptine QR showed less adverse severe effects and cardiovascular disease endpoints (Chamarthi et al. 2015).

It can cause frequent nausea and orthostasis as side effects.

2.8.1 Dual or Triple Therapy Strategies

Generally, treatment guidelines recommend a stepwise approach to intensification of therapies in patients with T2DM.

Another second or sometimes, third agents should be added on when monotherapy alone fails to achieve Hba1c targets suitable for an individual basis (Moon et al. 2017). Whatever anti-diabetic group, we decided to add to the monotherapy; the main point is to choose the best medication ever according to some issues. Second agents generally provide a reduction of approximately 0.62–1% in Hba1c independently from the agent (Phung et al. 2010). Besides this comparative effectiveness, their effects on body weight, plasma lipid levels, and their safety issues such as hypoglycemia or other adverse effects are the selection criteria for second/third agents (Bennett et al. 2011).

According to the American College of Endocrinology, dual or 2-drug therapy is recommended for A1c levels ≥7.5%, and the American Diabetes Association suggests dual treatment for levels ≥9% (Vaughan et al. 2017).

In a meta-analysis of the effect of antihyperglycemic agents added to metformin and a sulfonylurea on glycemic control and weight gain in type 2 diabetes, all drugs decreased hemoglobin A1c levels about equally when added to metformin and a sulfonylurea, without any differences. There were no distinct differences between antidiabetic groups as a third agent to add on metformin and SU combination. Insulin was the agent which was associated with more weight gain and hypoglycemia (Gross et al. 2011). Therefore, the decision making should be based on the patient's circumstances and the efficacy, side effects, risk of hypoglycemia, effect on body weight, patient preference, and co-morbidities.

Cost-effectiveness is another consideration for dual-triple therapies. The relatively higher cost of novel agents is an obstacle for local/national insurance agents (Cahn and Cefalu 2016).

3 The Role of Co-Morbidity and Special Conditions in the Selection of Antidiabetic Pharmacotherapy in Type-2 Diabetes

3.1 Chronic Kidney Disease

Chronic kidney disease (CKD) itself is associated with insulin resistance, and Diabetes is the leading cause of kidney failure. There is a vicious cycle between them. Antidiabetic agent choices for patients with Diabetes with CKD are limited because of the reduced GFR. It results in the accumulation of the drugs and prolongation of the side effects such as hypoglycemia. Although insulin therapy remains the mainstay of treatment with CKD, some agents can be used instead of insulin in the presence of CKD.

The HbA1C target that is associated with the best outcome in predialysis CKD patients has not been established. But the HbA1C target of approximately 7% is advised as the goal of therapy for predialysis CKD patients according to Kidney Disease Outcomes Quality Initiative (K/DOQI) and Kidney Disease: Improving Global Outcomes (KDIGO) guidelines.

The HbA1C target that is associated with the best outcome in dialysis patients has also not been established. Among dialysis patients, the goals should be decided according to the patient's co-morbidities, age, and the risk of hypoglycemia.

For the SU group, the agents that can be used for CKD stage 3–4, renal transplantation and dialysis are glipizide and gliclazide with no dose adjustment. Glimepiride is recommended with a dose of 1 mg but should be avoided in dialysis patients.

Clinical studies of acarbose have shown no change in renal function parameters. Because data in humans with impaired renal function in diabetic patients are lacking, the treatment with acarbose is not recommended, especially for patients with severe renal impairment (CrCl <25 mL/min).

No dose adjustment is necessary for repaglinide and nateglinide for CKD stage 3–4 and renal transplantation, but starting with a low dose for repaglinide (0.5 mg each meal) is recommended. No dose adjustment is necessary for repaglinide for dialysis. Nateglinide is prescribed 60 mg with each meal for CKD stage 3–4 but should be avoided for dialysis.

Metformin is contraindicated for GFR <30 mL/min. No dosage adjustment is necessary for GFR >45 ml/min, and monitoring renal function at least annually is recommended. For GFR of 30–45 mL/min, the use of metformin is not recommended for initiation of therapy. Still, if GFR falls between 30 and 45 mL/min during therapy, the cons and pros of continuing therapy should be considered. If continuing therapy is considered, a dosage reduction of 50% and monitoring of renal function every 3 months is recommended.

The half-lives of rosiglitazone and pioglitazone are similar in patients with ESRD and healthy individuals, so no dose adjustment is necessary for pioglitazone and rosiglitazone both for CKD Stage 3,4, or kidney transplant and dialysis.

Sitagliptin is renally eliminated mostly by excretion unchanged in the urine via active secretion and glomerular filtration (Bergman et al. 2007). Sitagliptin dose adjustment is recommended for patients with moderate to severe renal insufficiency and ESRD to doses of 25–50 mg.

Elimination of vildagliptin mainly involves renal excretion of unchanged drug. Vildagliptin is recommended to be used (50 mg/day) in half a dose of usual doses in patients with moderate to severe renal impairment.

Alogliptin dose is reduced by 50% (12.5 mg/day) when GFR is <50 and 30 mL/min and by 75% (6.25 mg/day) when GFR is below 30 mL/min. It is reduced by 75% (6.25 mg/day) for dialysis patients.

Renal excretion is a minor elimination pathway of linagliptin even at therapeutic doses, so dose adjustment in patients with renal impairment is not anticipated for linagliptin.

Exenatide and lixisenatide are not recommended in patients with GFR <30 mL/min. They should be cautiously applied for GFR >30 and <50 mL/min. No dose adjustment is necessary when GFR >50 mL/min.

No dose adjustments are necessary for dulaglutide and liraglutide for CKD stage 3–4 and renal transplantation and dialysis.

Although the cardiovascular outcome trials of SGLT2 inhibitors were not primarily designed for kidney outcomes, canagliflozin, empagliflozin, and dapagliflozin slowed the progression of kidney disease regardless of ASCVD, heart failure, or baseline kidney function.

3.2   Chronic Liver Disease or Liver Dysfunction

Management of diabetes in patients with CLD or Liver dysfunction is challenging because the liver is the primary site of metabolism for the antidiabetic agents and accompanying hypoalbuminemia, insulin resistance, lactic acidosis, and hypoglycemia are the significant risk factors for these patients.

For metformin, which is metabolized through the liver, it should be used with caution in patients with moderate CLD and avoided in hepatic failure for risk of lactic acidosis. It is beneficial in patients with non-alcoholic fatty liver disease. Insulin secretagogues also should be avoided in patients with T2DM and CLD, and lower doses should be preferred in Child-Pugh Class A and B (Gangopadhyay and Singh 2017).

Based on available evidence, repaglinide should be cautiously used in CLD patients, and nateglinide may be used in Child-Pugh Class A patients.

Pioglitazone should be used with caution in CLD patients, and it should be avoided in patients whose liver enzymes are >3 times the upper limits of normal (ULN) range. Pioglitazone may be used in Child-Pugh Class A patients (Gangopadhyay and Singh 2017).

AGIs can be used without dose modification in CLD patients, according to the available evidence. But they should not be used in Class C patients.

DPP-4 inhibitors except for vildagliptin can be used with caution without dose modification in Child-Pugh Class A patients. Vildagliptin should not be used in hepatic impairment and in patients whose liver enzymes are >3 times ULN range.

SGLT-2 inhibitors can be used with caution and lower doses in CLD patients. SGLT-2-inhibitors are safe in Child-Pugh Class A patients. Caution is advised, mainly due to the risk of dehydration and hypotension (Garcia-Compean et al. 2015).

There is limited information available about the safety and efficacy of GLP-1RAs in hepatic failure, but also, there is no dosage adjustment recommended for hepatic impairment for these agents.

3.3   Cardiovascular Disease and Chronic Heart Failure

The only glucose-lowering medication major cardiovascular (CV) outcome trials (CVOTs) in type 2 diabetes were from the University Group Diabetes Program and the UK Prospective Diabetes Study (UKPDS) before the twenty-first century. After the 2008 Food and Drug Administration (FDA) guidance and the CV safety concerns from heart failure (HF) events with pioglitazone and myocardial infarction with rosiglitazone, the FDA had mandated a requirement to achieve a maximum hazard of 1.30 in a CVOT with a composite endpoint of first MI, stroke or CV death event for cardiovascular reliability after licensing.

Among DPP4 inhibitors, for sitagliptin, The Trial Evaluating Cardiovascular Outcomes with Sitagliptin (TECOS) concluded that sitagliptin was not associated with an increased incidence of adverse CVO or heart failure admissions with a current diagnosis of CV disease. Also, the Saxagliptin Assessment of Vascular Outcomes Recorded in Patients with Diabetes Mellitus–Thrombolysis in Myocardial Infarction study (SAVOR-TIMI) declared that saxagliptin has no effect on the rate of ischemic events but increases the rate of hospitalization for heart failure. The CV safety study of alogliptin (The Examination of Cardiovascular Outcomes with Alogliptin versus Standard of Care- EXAMINE) showed that alogliptin did not increase the rates of major adverse CV incidents compared to placebo. A very recent CV safety study of linagliptin (the Cardiovascular Outcome Study of Linagliptin versus Glimepiride in patients with Type 2 Diabetes -CAROLINA) showed that the use of linagliptin compared with glimepiride over a median of 6.3 years resulted in a noninferior risk of a composite cardiovascular outcome among adults with relatively early type 2 diabetes and elevated cardiovascular risk (Rosenstock et al. 2019).

The first CVOT among GLP-1 RA was the Evaluation of Lixisenatide in Acute Coronary.

Syndrome (ELIXA) for lixisenatide, which showed that lixisenatide was not associated with significant CV effects in T2D patients with the recent acute coronary syndrome.

The Liraglutide Effect and Action in Diabetes: Evaluation of Cardiovascular Outcome Results (LEADER) trial, which assessed the CV of liraglutide, showed reduced CV events, all-cause mortality, and CV death. Liraglutide recently received FDA approval to reduce the risk of cardiovascular death, nonfatal myocardial infarction, and nonfatal stroke in adults with T2D and established cardiovascular disease. The following CVOT of GLP-1RA was the "Trial to Evaluate Cardiovascular, and Other Long-term Outcomes with Semaglutide in Subjects with Type 2 Diabetes" (SUSTAIN 6) demonstrated that semaglutide reduces CV death, nonfatal myocardial infarction and nonfatal stroke in diabetic patients at high risk of CV compared to placebo. The Exenatide Study of Cardiovascular Event Lowering (EXSCEL) reported that rates of death from CV causes, fatal or nonfatal MI, fatal or nonfatal stroke, hospitalization for HF, and hospitalization for acute coronary syndrome did not differ significantly between exenatide and placebo groups.

The first study of antidiabetic agents to show CV benefit was the Empagliflozin, Cardiovascular Outcomes, and Mortality in Type 2 Diabetes study (EMPAREG OUTCOME), which showed reduced CV mortality and all-cause mortality and reduced hospitalization for HF. Empagliflozin recently received FDA approval for the indication of lowering cardiac death in adults with T2D and established CVD.

Subsequently, treatment with canagliflozin showed significantly reduced risk of the combined cardiovascular outcomes of cardiovascular death, myocardial infarction, or nonfatal stroke, but increased risk of amputation (primarily at the level of the toe or metatarsal) in the CANVAS (Canagliflozin Cardiovascular Assessment Study in T2D patients who already had a high risk of CV disease.

The Multicenter Trial to Evaluate the Effect of Dapagliflozin on the Incidence of Cardiovascular Events (DECLARE-TIMI 58) was the longest and had the broadest participation of patients with T2DM and various multiple CV risk factors. The study declared a 27% reduction in hospitalization for heart failure but no significant difference in death from CV causes and no significant difference in all-cause mortality.

3.4   Cerebrovascular Disease

Diabetes mellitus (DM) is a significant risk factor for more severe stroke ischemic stroke and less favorable outcomes. Two large scale meta-analysis of 33 trials showed that intensive versus conventional glycemic control did not affect the incidence of stroke. Therefore, intensive glucose lowering strategies may not affect the risk of ischemic stroke. But agents may have effects on cerebrovascular disease.

From pieces of evidence in UKPDS, metformin treatment was showed to reduce the risk of ischemic stroke in overweight patients with newly diagnosed T2DM. In the same study, treatment with insulin did not affect the risk of ischemic stroke. In the PROspective pioglitAzone Clinical Trial In macroVascular Events (PROACTIVE), pioglitazone did not reduce the risk of ischemic stroke in the total study population but decreased the risk of recurrent stroke in patients with a history of ischemic stroke. In the Insulin Resistance Intervention after Stroke (IRIS) trial pioglitazone group had lower rates of ischemic stroke compared to placebo. Still, the risk of ischemic stroke did not differ between the pioglitazone and placebo groups. A metanalysis of pioglitazone in patients with a history of stroke or transient ischemic attack, the risk of recurrent stroke was reduced by 48% with pioglitazone.

According to EXAMINE, TECOS and SAVOR-TIMI 53 alogliptin, saxagliptin and sitagliptin did not affect the incidence of ischemic stroke compared with placebo.

In Sustain-6, the risk of ischemic stroke was decreased by 39% in patients who received semaglutide. Still, no reduction was seen at the risk of ischemic stroke for liraglutide in LEADER and lixisenatide in ELIXA.

Among SGLT-2 inhibitors, the rates of ischemic stroke were found higher in patients treated with empagliflozin, although fatal and recurrent strokes were not increased.

3.5   Gestational Diabetes Mellitus

Poor glycemic control in gestational diabetes mellitus (GDM) is associated with adverse maternal and neonatal outcomes. Increased red blood cell turnover can lower the normal A1C level in pregnancy, so the target of <6% HbA1C in pregnancy is offered. Insulin has traditionally been used as first-line therapy for GDM with lifestyle modifications. FDA approved insulin Detemir as long-acting, and insulin aspart, insulin lispro, and insulin glulisine as rapid-acting insulins are approved as analog insulins besides conventional regular insulin and Neutral Protamine Hagedorn (NPH) insulin.

Because glargine insulin has the potential affinity for insulin-like growth factor receptor (IGF-1R), glargine is still considered a category C drug by the FDA. The type of insulin, timing of administration, frequency, and dosing are based on pregnants' glycemic profile (Mukerji and Feig 2017).

While insulin has been the mainstay therapy for women with GDM for glycemic control, individual oral antidiabetic agents can also be used alternatively (Mukerji and Feig 2017).

Metformin crosses the placenta but can be considered as an alternative therapeutic option for the management of GDM because there is no evidence to suggest that this leads to fetal abnormalities in the short term. Metformin has been reported to decrease pregnancy-induced hypertension, LGA, neonatal hypoglycemia, and maternal weight gain (Finneran and Landon 2018). Still, long-term safety data are lacking (Sahin and Corapcioglu 2016).

There are no long-term data about fetal exposure to metformin. Also, metformin may reduce the levels of folate and vitamin B12 (Sahin et al. 2007).

The largest scaled comparison trial for metformin and insulin is Metformin in Gestational Diabetes Trial (MiG). There was a statistically significant difference in the rate of preterm delivery between the metformin group and the insulin group ($p = 0.02$). Still, the authors reported equivalent effects on glycemic control between study groups (Rowan et al. 2008). Besides, the offspring follow-up study (MiG TOFU) reported an increased mid-upper arm circumference, and biceps skin folds at 2 years of age (Rowan et al. 2011). A recent meta-analysis including 12 RCTs of metformin versus insulin for the treatment of GDM reported decreased rates of LGA,

macrosomia, NICU admission, neonatal hypoglycemia, and pregnancy-induced hypertension with metformin (Farrar et al. 2017).

Glyburide also has good efficacy, and it crosses the placenta. It is associated with increased rates of large-for-gestational-age (LGA) infants and neonatal hypoglycemia when compared with insulin according to the meta-analyses of glyburide (Finneran and Landon 2018). The largest trial of glyburide is the Langer study comparing insulin and glyburide therapy. There were no significant differences for LGA, macrosomia, birth weight, neonatal hypoglycemia, pulmonary complications, admission to the neonatal intensive care unit, and congenital anomalies between the two modalities (Langer et al. 2000).

The third oral antidiabetic agent, in GDM, was acarbose. In a small randomized study that compared glibenclamide versus acarbose, there was no evidence of a difference for any of their maternal or infant primary outcomes and neonatal hypoglycemia. However, gastrointestinal side effects were frequent in the acarbose group.

Sitagliptin is another medication investigated in pregnancy, but there was no assessment of placental transfer or obstetric outcomes reported. The glycemic parameters were found to be good when compared to the placebo (Sun et al. 2017).

Women with GDM should be screened for persistent diabetes at 6–12 weeks of the postpartum period because GDM has an increased risk of T2DM. According to The American College of Obstetricians and Gynecologists (ACOG), the screening should be planned for the first year and every 3 years in light of the risk factors (Committee on Practice B-O 2013).

In the postpartum period, the puerpera should be couraged to breastfeeding, and in the case of hyperglycemia, proper medical treatment should be advised. As well as suggesting to continue insulin as a treatment modality in lactation; other pharmaceutical agents such as metformin, glyburide, and glipizide (Glatstein et al. 2009). Both of the levels of these SU are found negligible in breast milk because of their extensive bindings to plasma proteins. Metformin is also detectable in low concentrations in milk, but it is considered to lead no adverse effects on breastfed infants (Anderson 2018).

3.6 Recommendations According to the Recent Society Guidelines

The European Association for the Study of Diabetes (EASD) and the American Diabetes Association (ADA) have produced an updated consensus statement on managing hyperglycemia in type 2 diabetes patients, including a range of recent trials of drug interventions (Buse et al. 2020). The new recommendations from the expert panel from both societies included the delivery of patient-centered care, facilitating medication adherence, the importance of weight loss in obese patients, the importance of medical nutrition therapy, and increased physical activity. Metformin is still the first-line recommended therapy for almost all patients with type 2 diabetes. The selection of anti-glycemic agents to metformin is based on patient preference and co-morbid conditions, including the presence of cardiovascular disease, heart failure, and kidney disease. For example, for patients with cardiovascular disease, an SGLT2 inhibitor or a GLP-1 RA with proven cardiovascular benefit is recommended, or for patients with CKD, congestive HF or atherosclerotic cardiovascular disease, an SGLT2 inhibitor with proven benefit should be considered.

Patient-centered decision making besides a healthy diet and exercise remain the principal of anti-glycaemic therapy. Starting therapy with metformin and selection of an additional second/third etc. glucose-lowering medication when it is necessary should be based on patient co-morbidities.

Metabolic surgery is recommended for T2DM with a BMI $\geq$ 40 or a BMI of 35–39.9 who do not achieve permanent weight loss and improvement in co-morbidities with medical therapies (Buse et al. 2020).

According to the recent ESC/EASD Guidelines, Metformin is no longer first-line therapy in patients with DM, but should now be considered in overweight patients with T2DM without CVD and at moderate CV risk. The SGLT2 inhibitors empagliflozin, canagliflozin, or dapagliflozin are recommended in patients with T2DM and CVD, or at very high/high CV risk, to reduce CV events. Empagliflozin is recommended in patients with T2DM and CVD to minimize the risk of death. The GLP-1RAs liraglutide, semaglutide, or dulaglutide are recommended in patients with T2DM and CVD, or very high/high CV risk, to reduce CV events. Liraglutide is recommended in patients with T2DM and CVD, or at high CV risk, to reduce the risk of death. SGLT2 inhibitors are recommended to lower the risk of HF hospitalization. GLP1-RAs and DPP4 inhibitors sitagliptin and linagliptin have a neutral effect on the risk of HF and may be considered, but the usage of saxagliptin and thiazolidinediones in HF is not recommended.

In anti-hyperglycemic treatment selection, as summarized in Fig. 2, glucose-lowering effects as well as side-effect profile, preference in specific patient groups, safety, tolerability, and cost issues should be considered.

3.6.1 Shared Decision Making with Patient Preference

Traditionally, treatment for diabetes, especially for type 2, is being given under evidence-based knowledge. The primary source of this evidence is mainly from clinical trials under optimal conditions. The main trouble of this actual knowledge is that can they point out the real-world circumstances and unique properties of patients? "Providing care that is respectful of and responsive to individual patient preferences, needs, and values and ensuring that patient values guide all clinical decisions" is first defined for patient-centered decision making according to ADA/EASD position statement (Inzucchi et al. 2015).

4 Conclusion

In conclusion with recent guidelines and current scientific data, individualizing glycemic targets

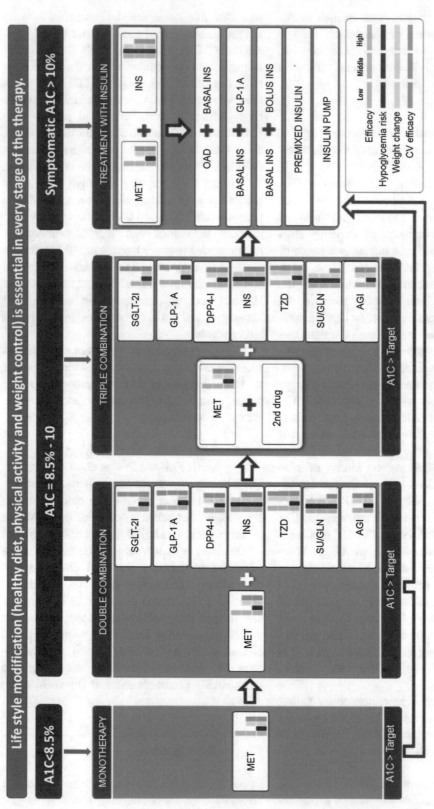

SEMT TYPE 2 DIABETES TREATMENT ALGORITHM[1,2]

[1]If A1C is >7% or above the individual target, then the treatment should be changed. [2]MET is the preferred monotherapy drug. If MET is contraindicated or there is intolerance for MET, treatment with another drug can be initiated.

A1C, glycated HbA1c; MET, metformin; SGLT2-I, sodium glucose co-transporter 2 inhibitor; GLP-1A, glucagon-like peptide-1 receptor agonist; DPP4-I, Dipeptidyl peptidase-4 inhibitor; INS, insulin; TZD, thiazolidinedione; SU, sulphonylurea; GLN, glinide; AGI, alpha-glucosidase inhibitor; CV, cardiovascular.

Fig. 2 Antiglycemic agent selection due to efficacy, Hba1c targets, and co-morbidities according to The Society of Endocrinology and Metabolism of Turkey

according to the patient's life expectancy, age, and comorbid situations such as obesity, cardiovascular, cerebrovascular, or chronic liver-renal diseases, consideration of adverse effects, and costs of each glucose-lowering medications should be major goals of diabetes therapy.

References

American Diabetes A (2019) 9. Pharmacologic approaches to glycemic treatment: standards of medical care in diabetes-2019. Diabetes Care 42:S90–S102

Amori RE, Lau J, Pittas AG (2007) Efficacy and safety of incretin therapy in type 2 diabetes: systematic review and meta-analysis. JAMA 298:194–206

Anderson PO (2018) Treating diabetes during breastfeeding. Breastfeed Med 13:237–239

Arslan MS, Tutal E, Sahin M et al (2017) Effect of lifestyle interventions with or without metformin therapy on serum levels of osteoprotegerin and receptor activator of nuclear factor kappa B ligand in patients with prediabetes. Endocrine 55:410–415

Bauman WA, Shaw S, Jayatilleke E, Spungen AM, Herbert V (2000) Increased intake of calcium reverses vitamin B12 malabsorption induced by metformin. Diabetes Care 23:1227–1231

Bennett WL, Maruthur NM, Singh S et al (2011) Comparative effectiveness and safety of medications for type 2 diabetes: an update including new drugs and 2-drug combinations. Ann Intern Med 154:602–613

Bergman AJ, Cote J, Yi B et al (2007) Effect of renal insufficiency on the pharmacokinetics of sitagliptin, a dipeptidyl peptidase-4 inhibitor. Diabetes Care 30:1862–1864

Buse JB, Wexler DJ, Tsapas A et al (2020) 2019 update to: management of hyperglycaemia in type 2 diabetes, 2018. A consensus report by the American Diabetes Association (ADA) and the European Association for the Study of Diabetes (EASD). Diabetologia 63:221–228

Cahn A, Cefalu WT (2016) Clinical considerations for use of initial combination therapy in type 2 diabetes. Diabetes Care 39(Suppl 2):S137–S145

Chamarthi B, Gaziano JM, Blonde L et al (2015) Timed bromocriptine-QR therapy reduces progression of cardiovascular disease and dysglycemia in subjects with well-controlled type 2 diabetes mellitus. J Diabetes Res 2015:157698

Committee on Practice B-O (2013) Practice bulletin no. 137: gestational diabetes mellitus. Obstet Gynecol 122:406–416

Davies M, Chatterjee S, Khunti K (2016) The treatment of type 2 diabetes in the presence of renal impairment: what we should know about newer therapies. Clin Pharmacol Adv Appl 8:61–81

Davies MJ, D'Alessio DA, Fradkin J et al (2018) Management of hyperglycaemia in type 2 diabetes, 2018. A consensus report by the American Diabetes Association (ADA) and the European Association for the Study of Diabetes (EASD). Diabetologia 61:2461–2498

Deeg MA, Tan MH (2008) Pioglitazone versus rosiglitazone: effects on lipids, lipoproteins, and apolipoproteins in head-to-head randomized clinical studies. PPAR Res 2008:520465

Defronzo RA (2009) Banting lecture. From the triumvirate to the ominous octet: a new paradigm for the treatment of type 2 diabetes mellitus. Diabetes 58:773–795

DeFronzo RA, Goodman AM (1995) Efficacy of metformin in patients with non-insulin-dependent diabetes mellitus. The Multicenter Metformin Study Group. N Engl J Med 333:541–549

Dubuisson O, Dhurandhar EJ, Krishnapuram R et al (2011) PPARgamma-independent increase in glucose uptake and adiponectin abundance in fat cells. Endocrinology 152:3648–3660

Farrar D, Simmonds M, Bryant M et al (2017) Treatments for gestational diabetes: a systematic review and meta-analysis. BMJ Open 7:e015557

Finneran MM, Landon MB (2018) Oral agents for the treatment of gestational diabetes. Curr Diab Rep 18:119

Gangji AS, Cukierman T, Gerstein HC, Goldsmith CH, Clase CM (2007) A systematic review and meta-analysis of hypoglycemia and cardiovascular events: a comparison of glyburide with other secretagogues and with insulin. Diabetes Care 30:389–394

Gangopadhyay KK, Singh P (2017) Consensus statement on dose modifications of antidiabetic agents in patients with hepatic impairment. Indian J Endocrinol Metab 21:341–354

Garber AJ, Duncan TG, Goodman AM, Mills DJ, Rohlf JL (1997) Efficacy of metformin in type II diabetes: results of a double-blind, placebo-controlled, dose-response trial. Am J Med 103:491–497

Garcia-Compean D, Gonzalez-Gonzalez JA, Lavalle-Gonzalez FJ, Gonzalez-Moreno EI, Maldonado-Garza HJ, Villarreal-Perez JZ (2015) The treatment of diabetes mellitus of patients with chronic liver disease. Ann Hepatol 14:780–788

Glatstein MM, Djokanovic N, Garcia-Bournissen F, Finkelstein Y, Koren G (2009) Use of hypoglycemic drugs during lactation. Can Fam Physician 55:371–373

Gross JL, Kramer CK, Leitao CB et al (2011) Effect of antihyperglycemic agents added to metformin and a sulfonylurea on glycemic control and weight gain in type 2 diabetes: a network meta-analysis. Ann Intern Med 154:672–679

Haghvirdizadeh P, Mohamed Z, Abdullah NA, Haghvirdizadeh P, Haerian MS, Haerian BS (2015) KCNJ11: genetic polymorphisms and risk of diabetes mellitus. J Diabetes Res 2015:908152

Hedrington MS, Davis SN (2019) Considerations when using alpha-glucosidase inhibitors in the treatment of type 2 diabetes. Expert Opin Pharmacother 20:2229–2235

Hsia DS, Grove O, Cefalu WT (2017) An update on sodium-glucose co-transporter-2 inhibitors for the treatment of diabetes mellitus. Curr Opin Endocrinol Diabetes Obes 24:73–79

Inzucchi SE, Bergenstal RM, Buse JB et al (2015) Management of hyperglycemia in type 2 diabetes, 2015: a patient-centered approach: update to a position statement of the American Diabetes Association and the European Association for the Study of Diabetes. Diabetes Care 38:140–149

Jayaram S, Hariharan RS, Madhavan R, Periyandavar I, Samra SS (2010) A prospective, parallel group, open-labeled, comparative, multi-centric, active controlled study to evaluate the safety, tolerability and benefits of fixed dose combination of acarbose and metformin versus metformin alone in type 2 diabetes. J Assoc Physicians India 58:679–682; 687

Karagiannis T, Boura P, Tsapas A (2014) Safety of dipeptidyl peptidase 4 inhibitors: a perspective review. Ther Adv Drug Saf 5:138–146

Kim W, Egan JM (2008) The role of incretins in glucose homeostasis and diabetes treatment. Pharmacol Rev 60:470–512

Langer O, Conway DL, Berkus MD, Xenakis EM, Gonzales O (2000) A comparison of glyburide and insulin in women with gestational diabetes mellitus. N Engl J Med 343:1134–1138

Lee PG, Halter JB (2017) The pathophysiology of hyperglycemia in older adults: clinical considerations. Diabetes Care 40:444–452

Liu Z, Ma S (2017) Recent advances in synthetic alpha-glucosidase inhibitors. ChemMedChem 12:819–829

Lopez Vicchi F, Luque GM, Brie B, Nogueira JP, Garcia Tornadu I, Becu-Villalobos D (2016) Dopaminergic drugs in type 2 diabetes and glucose homeostasis. Pharmacol Res 109:74–80

Luo S, Meier AH, Cincotta AH (1998) Bromocriptine reduces obesity, glucose intolerance and extracellular monoamine metabolite levels in the ventromedial hypothalamus of Syrian hamsters. Neuroendocrinology 68:1–10

Malaisse WJ (2003) Pharmacology of the meglitinide analogs: new treatment options for type 2 diabetes mellitus. Treat Endocrinol 2:401–414

Moin T (2019) Should adults with prediabetes be prescribed metformin to prevent diabetes mellitus? Yes: high-quality evidence supports metformin use in persons at high risk. Am Fam Physician 100:134–135

Moon MK, Hur KY, Ko SH et al (2017) Combination therapy of oral hypoglycemic agents in patients with type 2 diabetes mellitus. Korean J Intern Med 32:974–983

Mukerji G, Feig DS (2017) Pharmacological management of gestational diabetes mellitus. Drugs 77:1723–1732

Nauck MA (2004) Glucagon-like peptide 1 (GLP-1) in the treatment of diabetes. Horm Metab Res = Hormon- und Stoffwechselforschung = Hormones et metabolisme 36:852–858

Nauck MA, Homberger E, Siegel EG et al (1986) Incretin effects of increasing glucose loads in man calculated from venous insulin and C-peptide responses. J Clin Endocrinol Metab 63:492–498

Nauck M, Frid A, Hermansen K et al (2009) Efficacy and safety comparison of liraglutide, glimepiride, and placebo, all in combination with metformin, in type 2 diabetes: the LEAD (liraglutide effect and action in diabetes)-2 study. Diabetes Care 32:84–90

Nissen SE, Wolski K (2007) Effect of rosiglitazone on the risk of myocardial infarction and death from cardiovascular causes. N Engl J Med 356:2457–2471

Ooi CP, Loke SC (2012) Colesevelam for type 2 diabetes mellitus. Cochrane Database Syst Rev 12:CD009361

Phung OJ, Scholle JM, Talwar M, Coleman CI (2010) Effect of noninsulin antidiabetic drugs added to metformin therapy on glycemic control, weight gain, and hypoglycemia in type 2 diabetes. JAMA 303:1410–1418

Pinto LC, Rados DV, Barkan SS, Leitao CB, Gross JL (2018) Dipeptidyl peptidase-4 inhibitors, pancreatic cancer and acute pancreatitis: a meta-analysis with trial sequential analysis. Sci Rep 8:782

Raccah D (2017) Safety and tolerability of glucagon-like peptide-1 receptor agonists: unresolved and emerging issues. Expert Opin Drug Saf 16:227–236

Rohrborn D, Wronkowitz N, Eckel J (2015) DPP4 in diabetes. Front Immunol 6:386

Rosenstock J, Kahn SE, Johansen OE et al (2019) Effect of Linagliptin vs glimepiride on major adverse cardiovascular outcomes in patients with type 2 diabetes: the CAROLINA randomized clinical trial. JAMA 322 (21):1155–1166

Rowan JA, Hague WM, Gao W, Battin MR, Moore MP, Mi GTI (2008) Metformin versus insulin for the treatment of gestational diabetes. N Engl J Med 358:2003–2015

Rowan JA, Rush EC, Obolonkin V, Battin M, Wouldes T, Hague WM (2011) Metformin in gestational diabetes: the offspring follow-up (MiG TOFU): body composition at 2 years of age. Diabetes Care 34:2279–2284

Russell-Jones D, Cuddihy RM, Hanefeld M et al (2012) Efficacy and safety of exenatide once weekly versus metformin, pioglitazone, and sitagliptin used as monotherapy in drug-naive patients with type 2 diabetes (DURATION-4): a 26-week double-blind study. Diabetes Care 35:252–258

Saeedi P, Petersohn I, Salpea P et al (2019) Global and regional diabetes prevalence estimates for 2019 and projections for 2030 and 2045: results from the International Diabetes Federation Diabetes Atlas, 9th edition. Diabetes Res Clin Pract 157:107843

Sahin M, Corapcioglu D (2016) Metformin versus placebo in obese pregnant women without diabetes. N Engl J Med 374:2501

Sahin M, Bakiner O, Ertugrul D, Guvener ND (2005) Creatine kinase elevation in a patient taking rosiglitazone. Diabet Med 22:1624–1625

Sahin M, Tutuncu NB, Ertugrul D, Tanaci N, Guvener ND (2007) Effects of metformin or rosiglitazone on serum concentrations of homocysteine, folate, and vitamin B12 in patients with type 2 diabetes mellitus. J Diabetes Complicat 21:118–123

Sansome DJ, Xie C, Veedfald S, Horowitz M, Rayner CK, Wu T (2019) Mechanism of glucose-lowering by metformin in type 2 diabetes: role of bile acids. Diabetes Obes Metab 22(2):141–148

Scirica BM, Bhatt DL, Braunwald E et al (2013) Saxagliptin and cardiovascular outcomes in patients with type 2 diabetes mellitus. N Engl J Med 369:1317–1326

Singh M, Kumar A (2018) Risks associated with SGLT2 inhibitors: an overview. Curr Drug Saf 13:84–91

Sun X, Zhang Z, Ning H, Sun H, Ji X (2017) Sitagliptin down-regulates retinol-binding protein 4 and reduces insulin resistance in gestational diabetes mellitus: a randomized and double-blind trial. Metab Brain Dis 32:773–778

Thule PM, Umpierrez G (2014) Sulfonylureas: a new look at old therapy. Curr Diab Rep 14:473

Tsimihodimos V, Filippas-Ntekouan S, Elisaf M (2018) SGLT1 inhibition: pros and cons. Eur J Pharmacol 838:153–156

UK Prospective Diabetes Study (UKPDS) Group (1998) Intensive blood-glucose control with sulphonylureas or insulin compared with conventional treatment and risk of complications in patients with type 2 diabetes (UKPDS 33). Lancet 352:837–853

Vaughan EM, Johnston CA, Hyman DJ, Hernandez DC, Hemmige V, Foreyt JP (2017) Dual therapy appears superior to monotherapy for low-income individuals with newly diagnosed type 2 diabetes. J Prim Care Community Health 8:305–311

Wang YW, He SJ, Feng X et al (2017) Metformin: a review of its potential indications. Drug Des Devel Ther 11:2421–2429

Adv Exp Med Biol - Advances in Internal Medicine (2020) 4: 29–41
https://doi.org/10.1007/5584_2020_533
© Springer Nature Switzerland AG 2020
Published online: 19 May 2020

Latent Autoimmune Diabetes in Adults: A Review of Clinically Relevant Issues

Marta Hernández and Dídac Mauricio

Abstract

Latent autoimmune diabetes in adults (LADA) is still a poorly characterized entity. However, its prevalence may be higher than that of classical type 1 diabetes. Patients with LADA are often misclassified as type 2 diabetes. The underlying autoimmune process against β-cell has important consequences for the prognosis, comorbidities, treatment choices and even patient-reported outcomes with this diabetes subtype. However, there is still an important gap of knowledge in many areas of clinical relevance. We are herein focusing on the state of knowledge of relevant clinical issues than may help in the diagnosis and management of subjects with LADA.

M. Hernández
Department of Endocrinology & Nutrition, University Hospital Arnau de Vilanova, Lleida, Spain

Institut de Recerca Biomèdica de Lleida, Lleida, Spain

Department of Medicine, Universitat de Lleida (UdL), Lleida, Spain

D. Mauricio (✉)
Department of Endocrinology & Nutrition. Hospital de la Santa Creu i Sant Pau, Autonomous University of Barcelona, Barcelona, Spain

Centro de Investigación Biomédica en Red de Diabetes y Enfermedades Metabólicas Asociadas (CIBERDEM), Instituto de Salud Carlos III, Barcelona, Spain
e-mail: didacmauricio@gmail.com

Keywords

Latent autoimmune diabetes in adults · LADA · Diabetes mellitus · Diagnosis · Adults · Clinical management · Diabetes complications · Treatment · Glutamic decarboxylase autoantibodies (GADA) · Patient-reported outcomes · Quality of life

Abbreviations

LADA	latent autoimmune diabetes in adults
ICA	cytoplasmic islet cell autoantibodies
GADA	antibodies against glutamic acid decarboxylase
IA2A	antibodics against tyrosine phosphatase IA-2
ZnT8A	antibodies against zinc transporter 8
BMI	body mass index
SGLT2i	sodium-glucose cotransporter 2 inhibitors
QoL	quality of life

1 Latent Autoimmune Diabetes in Adults: A Review of Clinically Relevant Issues

Diabetes mellitus is a heterogeneous group of diseases. In the last few years, the scientific and clinical community has expressed a growing concern about the validity of its actual classification. Consistent with personalized medicine, a better understanding of the pathophysiological basis of a disease and its interaction with environmental factors will result in a better treatment for patients (Raz et al. 2013; Tuomi et al. 2014; Leslie et al. 2016; Schwartz et al. 2016; Skyler et al. 2017; McCarthy 2017; Ahlqvist et al. 2018).

Diabetes diagnosed during adulthood is usually classified as "type 2 diabetes mellitus", unless a clear insulinopenic clinical phenotype is present at diabetes onset, thus leading to the diagnosis of type 1 diabetes. However, a non-negligible number of subjects undergo a slowly-evolving autoimmune process that results in a clinical phenotype that initially fits with a clinical diagnosis of type 2 diabetes. This condition is usually known as latent autoimmune diabetes in adults; LADA carries important implications in terms of prognosis, development of comorbidities, treatment choices and even quality of life that differ from those of type 2 diabetes (Mauricio 2020). In addition, from the genetic point of view, LADA has been shown to be a mixture of type 1-like autoimmune genetic components and type 2-like metabolic genetic components (Mishra et al. 2017; Cousminer et al. 2018). These genotypes and its interaction with environmental factors result in a wide range of clinical phenotypes (Hjort et al. 2019).

The present review will focus on the relevant clinical aspects of this slow, progressive form of autoimmune diabetes in adults. We will use the term LADA, Latent Autoimmune Diabetes in Adults, to refer to this condition in this review.

2 Clinical Problem #1. What Is LADA?

The first reference to a subgroup of subjects with type 2 diabetes in whom autoimmunity was detected against the β-cell, but with preserved insulin secretion over time, dates back to 1977. Irvine et al. described the presence of islet cell anti-cytoplasm autoantibodies (ICA) in approximately 11% of patients with type 2 diabetes. In these ICA-positive patients, treatment with sulphonylurea failed earlier, and therefore insulin therapy was required much earlier than in the ICA-negative group (Irvine et al. 1977).

In the 1990s, Tuomi et al. and Zimmet et al. coined the acronym LADA to define a slowly progressive form of autoimmune diabetes that was initially controlled with oral drugs before requiring insulin treatment (Tuomi et al. 1993; Zimmet et al. 1994).

In 2005, the Diabetes Immunology Society proposed three criteria for the diagnosis of LADA (Fourlanos et al. 2005): (1) onset of diabetes after the age of 30, (2) the presence of circulating islet autoantibodies and (3) the lack of a requirement for insulin for at least 6 months after diagnosis. Of these three criteria, only the presence of circulating antibodies is an objective criterion, since subjects aged less than 30 years may present a slowly progressive autoimmune destruction of the β cells. Actually, the decision to start insulin treatment depends on criteria different from the pathophysiological process of the disease in many cases. Since then, the definition of LADA and a determination of whether it is a distinct clinical entity or one end of a spectrum of immune-mediated diabetes has generated a type of Byzantine debate in the scientific community (Gale 2005; Leslie and Pozzilli 2006; Stenström et al. 2005; Redondo 2013).

In 2019, the WHO updated the 1999 Classification of Diabetes Mellitus in an attempt to help health professionals choose appropriate treatments for their patients (World Health Organization 2019). This new classification extends beyond the dichotomous classification of type 1 and type 2 diabetes and incorporates a new category defined as "hybrid forms of diabetes". One of these hybrid forms of diabetes is "**slowly evolving, immune-mediated diabetes of adults**", formerly called LADA, and is described as similar to slowly evolving type 1 diabetes in adults that more often is characterized by features of metabolic syndrome, a single GAD autoantibody and greater β-cell function. Finally, in its traditional standards of care issued earlier this year, the American Diabetes Association acknowledged for the first time the existence of a debate around the clinical suitability of the term

LADA (American Diabetes Association 2020), although these patients remain without differentiation in the section on type 1 diabetes mellitus.

3 Clinical Problem #2. What Is the Best Method to Diagnose LADA?

The most prevalent autoantibodies detected in patients with slowly evolving immune-mediated diabetes of adults are directed against glutamic acid decarboxylase (GADA). These GADA are present in up to 90% of patients. Other autoantibodies, such as antibodies against tyrosine phosphatase IA-2 (IA2A) or zinc transporter 8 (ZnT8A), may also be positive in these subjects. Interestingly, although not yet available for its routine use in clinical practice, it has been shown that antibodies against the 257–760 domain of tyrosine phosphatase 2A are the marker with the highest sensitivity for detection of humoral IA-2 immunoreactivity in LADA (Tiberti et al. 2008). Unlike in patients with classic type 1 diabetes, patients with LADA are less frequently positive for more than one autoantibody at the time of the diabetes diagnosis (Lampasona et al. 2010).

GADA are the most prevalent and the most persistent autoantibodies detected during disease progression, and are the preferred immunological marker to diagnose LADA (Borg et al. 2002; Desai et al. 2007). Nevertheless, the limited data available from longitudinal studies show that GADA disappear in 59% of patients after 10 years of follow-up (Sørgjerd et al. 2012) and a pattern of a positive/negative fluctuation rate of 20% has been observed over 3 years (Turner et al. 1997; Huang et al. 2016).

4 Clinical Problem #3. How Prevalent Is LADA?

The prevalence of GADA ranges between 2% and 14% of subjects considered a priori as having type 2 diabetes. The data on the prevalence of GADA are variable, which probably reflects not only differences between populations but also the different designs of epidemiological studies published to date. Studies performed in populations recruited from primary care settings show lower rates than studies performed in secondary care settings. In general, the highest prevalence of positive β-cell autoimmunity in patients diagnosed with type 2 diabetes has been reported in northern European countries. In some populations, the overall number of subjects affected by this type of autoimmune diabetes is estimated to be higher than classic type 1 diabetes diagnosed according to the classical definition (Tuomi et al. 2014). We show the different studies that reported the frequency of positivity of diabetes-related autoantibodies among patients with type 2 diabetes in Table 1 to summarize the most relevant studies.

5 Clinical Problem #4. What Is the Clinical Phenotype of a Subject with LADA?

The majority of the studies assessing differences in phenotypes between subjects with LADA and type 2 diabetes share important selection biases, as they are conducted in reference secondary or tertiary centers, where more complex patients with more diabetes complications are treated (Zhou et al. 2013; Buzzetti et al. 2007; Hawa et al. 2009; Mollo et al. 2013; Li et al. 2019). As relevant information should be based on studies without recruitment bias, we will only summarize population-based studies with available phenotypic data.

In the Botnia study conducted in Finland (Tuomi et al. 1999), GADA-positive subjects presented lower C-peptide levels and insulin secretion than GADA-negative subjects. Accordingly, these patients were more frequently treated with insulin (30% vs. 12%), and insulin was initiated earlier after the diagnosis of diabetes. GADA-positive patients had a lower body mass index (BMI) and waist-to-hip ratio, lower systolic and diastolic blood pressure, lower triglyceride

Table 1 Positivity of autoimmune diabetes-related autoantibodies among patients with type 2 diabetes

Reference	Sample size (n)	Population	Frequency of Ab (+)	Population-based	New diabetes onset	Auto-antibody	Time to insulin criteria	Age criteria (years)
Xiang et al. (2018)	4671	China	6.6%	Yes	No	GADA, IA-2A, ZnT8A	–	>20
Stidsen et al. (2018)	5813	Denmark	2.8%*	Yes	Yes	GADA	–	>30
Wod et al. (2017)	4374	Denmark	7.4%**	No	"Relatively new"	GADA	–	>18
Maddaloni et al. (2015)	17,072	United Arab Emirates	2.6%	No	No	GADA, IA-2A	>6 months	30–70
Sachan et al. (2014)	618	Northern India	1.5%	No	No	GADA	>6 months	>30
Hawa et al. (2013)	6156	Europe	9.7%	No	<5 years	GADA, IA-2A, ZnT8A	>6 months	30–70
Zhou et al. (2013)	4880	China	5.9%	No (clinical-based, nationwide)	<1 year	GADA	>6 months	>30
Szepietowska et al. (2012)	212	Poland	8.9%	Yes	Yes	GADA, IAA	–	20–62
Qi et al. (2011)	498	Eastern China	9.2%	Yes	46% new onset	GADA	>6 months	>35
Park et al. (2011)	884	Republic of Korea	4.4%	No	<5 years	GADA, IA-2A, ZnT8A	>2 months	35–70
Maioli et al. (2010)	5568	Sardinia (Italy)	5%	No	<5 years	GADA	>8 months	35–70
Radtke et al. (2009)	1049	Norway	10.1%	Yes	No	GADA	>12 months ***	>20
Lee et al. (2009)	1270	Republic of Korea	5.1%	No	No	GADA, IA-2A	>6 months	>30
Davies et al. (2008)	387	South Wales	3.6%	Yes	Yes	GADA, IA-2A	>1 month	>18
Buzzetti et al. (2007)	4250	Italy	4.5%	No	<5 years	GADA, IA-2A	>6 months	–
Britten et al. (2007)	500	South Asian****	2.6%	No	No	GADA, IA-2A	–	31–89
Zinman et al. (2004)	4134	Canada, USA an Europe	4.2%	No	<3 years	GADA	>3 years	30–75
Takeda et al. (2002)	4980	Japan	3.8%	No	No	GADA	–	>20
Bosi et al. (1999)	193	Italy	2%	Yes	No	GADA, IA-2A	–	>40
Tuomi et al. (1999)	1122	Finland	9.3%	Yes	No	GADA	–	–
Turner et al. (1997)	3672	United Kingdom	11.7%	No	Yes	GADA, ICA	–	25–65

Ab (+): antibodies-positive. GADA: antibodies against glutamic acid decarboxylase. IA-2A: antibodies against tyrosine phosphatase IA-2. ZnT8A: antibodies against zinc transporter 8. ICA: cytoplasmic islet cell autoantibodies

(*) Percentage from people diagnosed of diabetes >18 years. (**) Fasting C-peptide >300 pmol/l. (***) Also included time to insulin <12 months if C-peptide >150 pmol/l. (****) South Asian population living in United Kingdom

levels and higher HDL-cholesterol concentrations.

In the HUNT study conducted in Norway (Radtke et al. 2009), subjects were stratified into groups according to the administration of insulin treatment and the data were compared. The proportion of patients with LADA who were treated with insulin was two times higher than patients with type 2 diabetes (40% vs. 22%). When comparing patients with LADA and type 2 diabetes, both of whom were not treated with insulin therapy, no differences were observed in adiposity, blood pressure or the lipid status. Conversely, patients with LADA treated with insulin therapy were leaner and presented a lower waist circumference and cholesterol-to-HDL cholesterol ratio than insulin-treated patients with type 2 diabetes. No differences in insulin doses or diabetes duration at the initiation of insulin therapy were observed.

Of particular interest are studies of population-based incident diabetes cases, since they are the most representative cases of this subtype of diabetes. According to Davies et al., GADA-positive patients with incident diabetes in South Wales showed a lower BMI and higher frequency of acute symptoms at diagnosis than GADA-negative patients (Davies et al. 2008). Szepietowska et al. analyzed incident cases of diabetes mellitus in Poland and showed that patients diagnosed with LADA were younger at the time of diagnosis and had a lower BMI and waist circumference than patients diagnosed with type 2 diabetes (Szepietowska et al. 2012).

Compared to patients with type 2 diabetes, patients with LADA show a higher prevalence of other organ-specific autoantibodies, such as anti-thyroid peroxidase, anti-transglutaminase, anti-gastric parietal cell or anti-21-hydroxylase antibodies. The prevalence is similar to that in patients with classic type 1 diabetes mellitus (Schloot et al. 2016) and appears to be higher in patients with high GADA titers (Zampetti et al. 2012). In the study of Zampetti et al. (Zampetti et al. 2012), 73.3% of subjects with a high GADA titer were positive for at least one or more antibodies, compared to 38.3% of patients with a low GADA titer.

We can conclude that LADA does not have a single phenotype. Instead, patients with LADA have an intermediate phenotype between patients with type 1 and type 2 diabetes in terms of adiposity, insulin secretion and insulin sensitivity. The phenotype of patients with a higher titer and greater number of islet cell auto-antibodies is more similar to that in patients with type 1 diabetes (Hawa et al. 2013; Zhou et al. 2013; Maioli et al. 2010; Buzzetti et al. 2007).

6 Clinical Problem #5. Chronic Diabetes Complications in Patients with LADA

Despite its important clinical relevance, the effect of chronic complications on patients with LADA is not well studied. To date, no relevant differences have been reported in the few studies that compared patients with classic type 1 diabetes, LADA, and type 2 diabetes.

Regarding microvascular complications, no differences were observed in the Botnia study between patients with LADA and type 2 diabetes, although a lower frequency of diabetic retinopathy was observed in patients with LADA than in patients with type 1 diabetes (Isomaa et al. 1999). Fewer subjects with LADA and a short diabetes duration are diagnosed with retinopathy and neuropathy, but these differences disappear at 5 years after the diagnosis (Lu et al. 2015). Conversely, no differences in diabetic retinopathy were reported between patients with LADA and patients with type 2 diabetes in another study of recent-onset diabetes mellitus (Martinell et al. 2016). In the few longitudinal studies comparing LADA and type 2 diabetes, either no differences between groups were reported (Hawa et al. 2014) or an increase in microalbuminuria was observed in patients with type 2 diabetes during follow-up (Myhill et al. 2008).

In a recent review, evidence was provided on how microvascular complications in LADA depend much more on diabetes duration compared to type 2 diabetes, probably because type 2 diabetes is diagnosed later in the history of the disease (Buzzetti et al. 2017). A faster worsening of glycemic control in LADA may be the main cause of the disappearance of the differences in diabetes microangiopathy between both groups of diabetes over time. A recent post-hoc analysis of the time course of microvascular complications in the UKPDS study is clearly in line with this hypothesis: at diabetes onset and during the first 9 years of follow-up LADA subjects had a 55% lower adjusted risk of microvascular complications than patients with type 2 diabetes. However, beyond 9 years, this situation was reversed, and subjects with LADA had a 33% greater adjusted risk for the microvascular composite outcome. This increase in microvascular complications in LADA can be mainly explained by their earlier worse glycemic control (Maddaloni et al. 2020). These results are very relevant, because the UKPDS study has the largest cohort of newly diagnosed LADA subjects (n = 564), with a follow-up of up to 30 years for clinical complications, evaluated in a randomized clinical trial.

A more severe neuropathy, particularly small fiber neuropathy, has been recently described in LADA compared to patients with type 2 diabetes, matched by age and diabetes duration (Alam et al. 2019). Although the data so far are very scarce, markers of bone turnover are decreased in LADA, as in type 1 and in type 2 diabetes (Napoli et al. 2018).

Regarding macrovascular complications, patients with LADA consistently show a better cardiovascular risk profile than patients with type 2 diabetes. Nevertheless, no differences in the prevalence and incidence of cardiovascular disease and cardiovascular mortality have been identified in prospective studies (Hawa et al. 2014; Myhill et al. 2008; Olsson et al. 2013). No differences were observed between patients with LADA and patients with type 1 diabetes in the only prospective study that has compared these 2 types of diabetes (Olsson et al. 2013). However, we have reported that asymptomatic patients with LADA have a higher frequency and burden of preclinical carotid atherosclerosis than patients with adult type 1 diabetes and type 2 diabetes (Hernández et al. 2017). Finally, the incidence of

cardiovascular disease was lower in the subgroup of autoantibody-positive patients with diabetes in the UKPDS study (11.2%, n = 567). However, after adjusting for confounders such as age or adiposity markers, the difference was no longer significant (Maddaloni et al. 2019).

To summarize, although few studies are available, on one side their findings suggest that there is at least a similar burden of macrovascular complications in subjects with LADA. This occurs despite a lower cardiovascular risk in subjects with LADA compared to those with type 2 diabetes. On the other hand, poorer long-term glycemic control of subjects with LADA, compared to their counterparts with type 2 diabetes, leads to their increased risk of microvascular complications (Mauricio 2020).

7 Clinical Problem #6. What Is the Best Strategy to Treat LADA?

The best treatment for patients with LADA should aim to achieve good glycemic control and preserve residual β-cell function, which decreases more rapidly in patients with LADA than in patients with type 2 diabetes (Hernandez et al. 2015; Pieralice and Pozzilli 2018). In the UKPDS cohort of patients who were not randomized to receive insulin, 56.3% of patients positive for antibodies received insulin treatment after 6 years of follow-up compared with only 4.5% of antibody-negative patients (Turner et al. 1997).

The problem is that the speed of β-cell failure is unpredictable at the diabetes diagnosis, and a significant percentage of patients with LADA will not need insulin throughout their lifetimes. A high titer and persistence of diabetes autoantibodies, a low C-peptide level and a phenotype that is more similar to type 1 diabetes are associated with a more rapid progression to insulinopenia (Sørgjerd et al. 2012; Mollo et al. 2013; Zampetti et al. 2014).

High-quality randomized controlled trials designed to discover the best treatment option for LADA are lacking. The most consistent result, still with a limited validity, is that sulphonylurea treatment achieves poorer glycemic control and accelerates β-cell failure compared to insulin treatment (Brophy et al. 2011).

Metformin is the drug of choice for the initial treatment of type 2 diabetes. No specific studies have examined the efficacy of metformin in treating LADA. Given the efficacy and safety of this drug, and the observation that these patients show an intermediate phenotype between type 1 and type 2 diabetes, metformin is usually used to treat LADA in clinical practice.

Thiazolidinediones are insulin-sensitizing drugs with anti-inflammatory and immunomodulatory effects. In a small study, the use of rosiglitazone combined with insulin appeared to preserve β-cell function better than sulphonylurea or insulin treatment alone (Yang et al. 2009). No studies have been conducted with pioglitazone, the only thiazolidinedione currently available.

Dipeptidyl peptidase-4 inhibitors exert a protective effect on β-cells in vitro and on animal models. Results of a post hoc analysis of a randomized clinical trials showed better β-cell parameters in patients treated with linagliptin (Johansen et al. 2014) or saxagliptin (Buzzetti et al. 2016) than in patients treated with glimepiride or placebo. A recent randomized trial did not report differences in glucagon-stimulated C-peptide levels when comparing the addition of insulin or sitagliptin to metformin in patients with LADA characterized by a short diabetes duration and preserved β-cell function (Hals et al. 2019).

Even fewer studies have analyzed GLP-1 receptor agonists. A post hoc analysis of three phase 3 randomized clinical trials of the efficacy of dulaglutide in patients with type 2 diabetes revealed similar hypoglycemic effects on GADA-positive and –negative patients (Pozzilli et al. 2018). However, the presence of pancreatic autoimmunity predicted a poorer response to GLP-1 receptor agonists (exenatide and liraglutide) in a prospective study designed to assess predictors of response to these drugs in patients with type 2 diabetes (Jones et al. 2016). The participants in the latter study had a longer diabetes duration and more severe insulinopenia, which may partially explain the poor response to

this class of drugs, similar to that in patients with type 1 diabetes.

No studies have examined the effects of sodium-glucose cotransporter 2 inhibitors (SGLT2i) on patients with LADA. β-cell failure and subsequent hyperglycemia are more frequently accelerated in patients with LADA than in patients with type 2 diabetes. This hyperglycemia may be masked by the intense glycosuria produced by SGLT2i, delaying insulin therapy and increasing the risk of diabetic ketoacidosis (Handelsman et al. 2016). Patients and clinicians should be aware of this possibility and implement educational measures when needed (Danne et al. 2019). Sotagliflozin and dapagliflozin have been recently approved by the European Medicines Agency for use in patients with type 1 diabetes, with a BMI >27 kg/m^2, and who have not achieved adequate glycemic control with insulin alone (Markham and Keam 2019). On the other hand, the U.S. Food and Drug Administration have not approved dapagliflozin, sotagliflozin and empagliflozin for use in patients with type 1 diabetes.

In the absence of current specific clinical guidelines for the treatment of LADA, the clinician must develop a personalized therapeutic approach. Measurements of random or fasting C-peptide levels, whenever possible, can help determine the appropriate therapy, although there is not strong evidence to support its use.

Thus, based on our clinical experience, we are hereby proposing that a patient with a phenotype more similar to subjects with type 2 diabetes and with preserved C-peptide levels can be treated according to the guidelines for type 2 diabetes. On the other hand, patients with a phenotype more similar to patients with type 1 diabetes and with low C-peptide levels will require insulin treatment, which ranges from a basal insulin treatment to a more complex insulin schedule, such as a bolus/basal regimen. In intermediate cases, the recommendations for the treatment of type 2 diabetes can be applied, after considering that the deterioration of the β-cell function may be more accelerated than usual and worsened by sulphonylurea treatment. We should also remember that hyperglycemia due to progressive β-cell failure can be masked by SGLT2i. This progressive insulinopenia should be evaluated periodically to prevent a delay in the initiation of insulin treatment as soon as it is needed.

Finally, we should consider that the treatment of all cardiovascular risk factors with an equal intensity to type 2 diabetes is also very important in subjects with LADA.

8 Clinical Problem #7. Patient-Reported Outcomes in LADA

Although diabetes guidelines recommend a regular assessment of quality of life, psychological well-being and diabetes-related distress in people with diabetes (American Diabetes Association 2018), a substantial knowledge gap exists in patient-reported outcomes in people with LADA. We assessed the diabetes-dependent quality of life (QoL) and treatment satisfaction in a small group of subjects with LADA and compared the results to those from patients with type 1 and type 2 diabetes (Granado-Casas et al. 2017). Patients with LADA had a lower diabetes-specific QoL and average weighted impact scores than patients with type 2 diabetes. The overall measure of treatment satisfaction was not different between patients with LADA, type 2 and type 1 diabetes, but the patients with LADA showed a poorer perception of the hyperglycemia

frequency than patients with type 2 diabetes and an increased perception of the frequency of hypoglycemia compared with patients with type 1 diabetes. To our knowledge, no additional studies are available examining this important issue.

9 Clinical Problem #8. Who Should Be Screened for LADA?

Currently, no guidelines are yet available to determine which patients should be screened for LADA. In a study designed to create a diagnostic tool for LADA, an age at diagnosis of less than 50 years, the presence of acute symptoms at diagnosis, a BMI less than 25 kg/m^2 and a history (personal or familial) of autoimmune diseases predicted GADA positivity (Fourlanos et al. 2006). Interestingly, Lynman et al. have recently developed and validated a model that integrates also clinical features (age, BMI) and different biomarkers (islet autoantibodies, type 1 diabetes genetic risk score) resulting in an accurate prediction of what they classify as type 1 diabetes with rapid insulin requirement in adults aged 18–50 years (Lynam et al. 2019).

As we have discussed in the text, there is a considerable gap in the knowledge of this chronic condition that is probably more prevalent than classical type 1 diabetes (Table 2). The clinical characterization of these patients in the primary care setting is lacking, and its clinical progression

Table 2 Gaps in current knowledge on LADA

Prevalence	Epidemiological population-based studies are needed, ideally in new-onset diabetes mellitus
Ethnicity	More data on populations of not-European ancestry are needed
Complications	More research addressing the occurrence and natural history of chronic diabetes-related complications
Patient-reported outcomes	Almost no evidence available on this matter. More insight is clearly needed
Treatment	Randomized controlled trials assessing which is the best treatment for subjects with LADA, not only focused on glycemic control, but also on the potential for preservation of insulin secretion
	Post-hoc analysis of RCT conducted in type 2 diabetes whenever GADA testing is feasible
Related autoinmune disorders	Research on the occurrence and clinical course of associated autoimmune diseases in different populations.
	New data will inform future screening policies

LADA latent autoimmune diabetes in adults, *RCT* randomized controlled trials, *GADA* antibodies against glutamic acid decarboxylase

is unpredictable. Researchers have not determined whether chronic vascular complications of diabetes equally affect patients with LADA, type 1 and type 2 diabetes. In addition, the best treatment options, both in terms of β-cell preservation and glycemic control, have not been identified, and consensus guidelines that will help clinicians in its treatment are not available.

Therefore, a reasonable approach appears to be to study the vast majority of patients diagnosed with diabetes by screening for GADA positivity. The cost of a single GADA measurement is currently approximately 5€ (or $7), and the titer of these antibodies only needs to be measured once throughout the life of the patient. Cost-effective studies on this approach have not been conducted, but it will probably be efficient if this strategy helps clinicians to choose better and more cost-effective treatments for a lifelong disease.

Additionally, knowledge of a potential autoimmune process underlying an alleged diagnosis of type 2 diabetes will allow the clinicians to assess other possible accompanying autoimmune diseases. We are providing the reader with highlights on the clinical management of patients with LADA in Table 3.

Finally, we propose that informing the patient of the nature of his/her diabetes will help him/her make consensual therapeutic decisions, despite the uncertainty that currently exists.

Table 3 Highlights on the clinical management of LADA

	Fact	Action
How to diagnose LADA	Determining GADA, the most sensitive marker. The cost is around 5€/$7	Test for GADA all newly diagnosed type 2 diabetes
		If not economically feasible, then assess newly diagnosed type 2 diabetes for clinical features; 2 or more of the following characteristics should lead to clinical suspicion of LADA: age <50 years, acute symptoms, BMI <25 kg/m², history of autoimmune diseases
Chronic diabetes complications	LADA subjects have a similar burden of micro- and macrovascular complications compared to type 1 and type 2 diabetes	Screen for diabetes complications from the onset of diabetes, following the same periodicity as in type 2 diabetes
Cardiovascular risk factors	Patients with LADA have a lower prevalence of other classical cardiovascular risk factors but a similar cardiovascular risk	Treat cardiovascular risk factors with the same intensity than in type 2 diabetes
Related autoimmune disorders	Patients with LADA have higher prevalence of other autoimmune disorders in the patient and his/her relatives	Be alert for the development of other autoimmune diseases
		Measure thyroid-stimulating hormone (TSH) levels with the same regularity than in type 1 diabetes
Hyperglycemia therapeutic approach	In LADA, there is a faster progression to insulinopenia leading to more severe hyperglycemia	Be ready to initiate insulin earlier than in type 2 diabetes
	Random C-peptide can help in treatment choice	Measure random C-peptide levels (better than fasting) when possible
	Sulphonylureas seem to accelerate beta-cell failure	Avoid sulphonylureas
	SGLT2i treatment can mask hyperglycemia due to insulin deficiency, and increase circulating levels of ketone bodies	SGLT2i prescription should be accompanied of educational measures

LADA latent autoimmune diabetes in adults, *GADA* antibodies against glutamic acid decarboxylase, *BMI* body mass index, *SGLT2i* sodium-glucose cotransporter 2 inhibitors

References

Ahlqvist E, Storm P, Käräjämäki A, Martinell M, Dorkhan M, Carlsson A, Vikman P, Prasad RB, Aly DM, Almgren P, Wessman Y, Shaat N, Spégel P, Mulder H, Lindholm E, Melander O, Hansson O, Malmqvist U, Lernmark Å, Lahti K, Forsén T, Tuomi T, Rosengren AH, Groop L (2018) Novel subgroups of adult-onset diabetes and their association with outcomes: a data-driven cluster analysis of six variables. Lancet Diabetes Endocrinol 6:361–369

Alam U, Jeziorska M, Petropoulos IN, Pritchard N, Edwards K, Dehghani C, Srinivasan S, Asghar O, Ferdousi M, Ponirakis G, Marshall A, Boulton AJM, Efron N, Malik RA (2019) Latent autoimmune diabetes of adulthood (LADA) is associated with small fibre neuropathy. Diabet Med 36(9):1118–1112

American Diabetes Association (2018) 4. Lifestyle management: standards of medical care in diabetes-2018. Diabetes Care 41:S38–S50

American Diabetes Association (2020) 2. Classification and diagnosis of diabetes: standards of medical care in diabetes-2020. Diabetes Care 43(Suppl 1):S14–S31

Borg H, Gottsäter A, Fernlund P, Sundkvist G (2002) A 12-year prospective study of the relationship between islet antibodies and beta-cell function at and after the diagnosis in patients with adult-onset diabetes. Diabetes 51:1754–1762

Bosi EP, Garancini MP, Poggiali F, Bonifacio E, Gallus G (1999) Low prevalence of islet autoimmunity in adult diabetes and low predictive value of islet autoantibodies in the general adult population of northern Italy. Diabetologia 42(7):840–844

Britten AC, Jones K, Törn C, Hillman M, Ekholm B, Kumar S, Barnett AH, Kelly MA (2007) Latent autoimmune diabetes in adults in a South Asian population of the U.K. Diabetes Care 30:3088–3090

Brophy S, Davies H, Mannan S, Brunt H, Williams R (2011) Interventions for latent autoimmune diabetes (LADA) in adults. Cochrane Database Syst Rev 9: CD006165

Buzzetti R, Di Pietro S, Giaccari A, Petrone A, Locatelli M, Suraci C, Capizzi M, Arpi ML, Bazzigaluppi E, Dotta F, Bosi E (2007) High titer of autoantibodies to GAD identifies a specific phenotype of adult-onset autoimmune diabetes. Diabetes Care 30:932–938

Buzzetti R, Pozzilli P, Frederich R, Iqbal N, Hirshberg B (2016) Saxagliptin improves glycaemic control and C-peptide secretion in latent autoimmune diabetes in adults (LADA). Diabetes Metab Res Rev 32:289–296

Buzzetti R, Zampetti S, Maddaloni E (2017) Adult-onset autoimmune diabetes: current knowledge and implications for management. Nat Rev Endocrinol 13 (11):674–686

Cousminer DL, Ahlqvist E, Mishra R, Andersen MK, Chesi A, Hawa MI, Davis A, Hodge KM, Bradfield JP, Zhou K et al (2018) First genome-wide association study of latent autoimmune diabetes in adults reveals novel insights linking immune and metabolic diabetes. Diabetes Care 41(11):2396–2403

Danne T, Garg S, Peters AL, Buse JB, Mathieu C, Pettus JH, Alexander CM, Battelino T, Ampudia-Blasco FJ, Bode BW, Cariou B, Close KL, Dandona P, Dutta S, Ferrannini E, Fourlanos S, Grunberger G, Heller SR, Henry RR, Kurian MJ, Kushner JA, Oron T, Parkin CG, Pieber TR, Rodbard HW, Schatz D, Skyler JS, Tamborlane WV, Yokote K, Phillip M (2019) International consensus on risk management of diabetic ketoacidosis in patients with type 1 diabetes treated with Sodium-Glucose Cotransporter (SGLT) inhibitors. Diabetes Care 42:1147–1154

Davies H, Brophy S, Fielding A, Bingley P, Chandler M, Hilldrup I, Brooks C, Williams R (2008) Latent autoimmune diabetes in adults (LADA) in South Wales: incidence and characterization. Diabet Med 25:1354–1357

Desai M, Cull CA, Horton VA, Christie MR, Bonifacio E, Lampasona V, Bingley PJ, Levy JC, Mackay IR, Zimmet P, Holman RR, Clark A (2007) GAD autoantibodies and epitope reactivities persist after diagnosis in latent autoimmune diabetes in adults but do not predict disease progression: UKPDS 77. Diabetologia 50(10):2052–2060

Fourlanos S, Dotta D, Greenbaum CJ, Palmer JP, Rolandsson O, Colman PG, Harrison LC (2005) Latent autoimmune diabetes in adults (LADA) should be less latent. Diabetologia 48:2206–2212

Fourlanos S, Perry C, Stein MS, Stankovich J, Harrison LC, Colman PG (2006) A clinical screening tool identifies autoimmune diabetes in adults. Diabetes Care 29:970–975

Gale EA (2005) Latent autoimmune diabetes in adults: a guide for the perplexed. Diabetologia 48:2195–2199

Granado-Casas M, Martínez-Alonso M, Alcubierre N, Ramírez-Morros A, Hernández M, Castelblanco E, Torres-Puiggros J, Mauricio D (2017) Decreased quality of life and treatment satisfaction in patients with latent autoimmune diabetes of the adult. PeerJ 5:e3928

Hals IK, Fiskvik Fleiner H, Reimers N, Astor MC, Filipsson K, Ma Z, Grill V, Björklund A (2019) Investigating optimal β-cell-preserving treatment in latent autoimmune diabetes in adults: results from a 21-month randomized trial. Diabetes Obes Metab 21:2219–2227

Handelsman Y, Henry RR, Bloomgarden ZT, Dagogo-Jack S, DeFronzo RA, Einhorn D, Ferrannini E, Fonseca VA, Garber AJ, Grunberger G, LeRoith D, Umpierrez GE, Weir MR (2016) American association of clinical endocrinologists and American College of Endocrinology position statement on the association of SGLT-2 inhibitors and diabetic ketoacidosis. Endocr Pract 22:753–762

Hawa MI, Thivolet C, Mauricio D, Alemanno I, Cipponeri E, Collier D, Hunter S, Buzzetti R, de Leiva A, Pozzilli P, Leslie RD, Action LADA Group (2009) Metabolic syndrome and autoimmune diabetes: action LADA 3. Diabetes Care 32(1):160–164

Hawa MI, Kolb H, Schloot N, Beyan H, Paschou SA, Buzzetti R, Mauricio D, De Leiva A, Yderstraede K, Beck-Neilsen H, Tuomilehto J, Sarti C, Thivolet C, Hadden D, Hunter S, Schernthaner G, Scherbaum WA, Williams R, Brophy S, Pozzilli P, Leslie RD, Action LADA Consortium (2013) Adult-onset autoimmune diabetes in Europe is prevalent with a broad clinical phenotype. Action LADA 7. Diabetes Care 36:908–913

Hawa MI, Buchan AP, Ola T, Wun CC, DeMicco DA, Bao W, Betteridge DJ, Durrington PN, Fuller JH, Neil HA, Colhoun H, Leslie RD, Hitman GA (2014) LADA and CARDS: a prospective study of clinical outcome in established adult-onset autoimmune diabetes. Diabetes Care 37:1643–1649

Hernandez M, Mollo A, Marsal JR, Esquerda A, Capel I, Puig-Domingo M, Pozzilli P, de Leiva A, Mauricio D, Action LADA Consortium (2015) Insulin secretion in patients with latent autoimmune diabetes (LADA): half way between type 1 and type 2 diabetes: action LADA 9. BMC Endocr Disord 15:1

Hernández M, López C, Real J, Valls J, Ortega-Martinez de Victoria E, Vázquez F, Rubinat E, Granado-Casas M, Alonso N, Molí T, Betriu A, Lecube A, Fernández E, Leslie RD, Mauricio D (2017) Preclinical carotid atherosclerosis in patients with latent autoimmune diabetes in adults (LADA), type 2 diabetes and classical type 1 diabetes. Cardiovasc Diabetol 16(1):94

Hjort R, Löfvenborg JE, Ahlqvist E, Alfredsson L, Andersson T, Grill V, Groop L, Sørgjerd EP, Tuomi T, Åsvold BO, Carlsson S (2019) Interaction between overweight and genotypes of HLA, TCF7L2, and FTO in relation to the risk of latent autoimmune diabetes in adults and type 2 diabetes. J Clin Endocrinol Metab 104(10):4815–4826

Huang G, Yin M, Xiang Y, Li X, Shen W, Luo S, Lin J, Xie Z, Zheng P, Zhou Z (2016) Persistence of glutamic acid decarboxylase antibody (GADA) is associated with clinical characteristics of latent autoimmune diabetes in adults: a prospective study with 3-year follow-up. Diabetes Metab Res Rev 32:615–622

Irvine WJ, Gray RS, McCallum CJ, Duncan LJP (1977) Clinical and pathogenic significance of pancreatic-islet-cell antibodies in diabetics treated with oral hypoglycaemic agents. Lancet 1:1025–1027

Isomaa B, Almgren P, Henricsson M, Taskinen MR, Tuomi T, Groop L, Sarelin L (1999) Chronic complications in patients with slowly progressing autoimmune type 1 diabetes (LADA). Diabetes Care 22:1347–1353

Johansen OE, Boehm BO, Grill V, Torjesen PA, Bhattacharya S, Patel S, Wetzel K, Woerle HJ (2014) C-peptide levels in latent autoimmune diabetes in adults treated with linagliptin versus glimepiride: exploratory results from a 2-year double-blind, randomized, controlled study. Diabetes Care 37:e11–e12

Jones AG, McDonald TJ, Shields BM, Hill AV, Hyde CJ, Knight BA, Hattersley AT, PRIBA Study Group (2016) Markers of β-cell failure predict poor glycemic response to GLP-1 receptor agonist therapy in type 2 diabetes. Diabetes Care 39(2):250–257

Lampasona V, Petrone A, Capizzi T, Spoletini M, di Pietro S, Songini M, Bonicchio S, Giorgino F, Bonifacio E, Bosi E, Buzzetti R (2010) Zinc transporter 8 antibodies complement GAD and IA-2 antibodies in the identification and characterization of adult-onset autoimmune diabetes Non Insulin Requiring Autoimmune Diabetes (NIRAD) 4. Diabetes Care 33:104–108

Lee SH, Kwon HS, Yoo SJ, Ahn YB, Yoon KH, Cha BY, Lee KW, Son HY (2009) Identifying latent autoimmune diabetes in adults in Korea: the role of C-peptide and metabolic syndrome. Diabetes Res Clin Pract 83(2):e62–e65

Leslie RD, Pozzilli WR (2006) Clinical review: type 1 diabetes and latent autoimmune diabetes in adults: one end of the rainbow. J Clin Endocrinol Metab 91:1654–1659

Leslie RD, Palmer J, Schloot NC, Lernmark A (2016) Diabetes at the crossroads: relevance of disease classification to pathophysiology and treatment. Diabetologia 59:13–20

Li X, Cao C, Tang X, Yan X, Zhou H, Liu J, Ji L, Yang X, Zhou Z (2019) Prevalence of metabolic syndrome and its determinants in newly-diagnosed adult-onset diabetes in China: a multi-center, cross-sectional survey. Front Endocrinol (Lausanne) 10:661. https://doi.org/10.3389/fendo.2019.00661

Lu J, Hou X, Zhang L, Hu C, Zhou J, Pang C, Pan X, Bao Y, Jia W (2015) Associations between clinical characteristics and chronic complications in latent autoimmune diabetes in adults and type 2 diabetes. Diabetes Metab Res Rev 31(4):411–420

Lynam A, McDonald T, Hill A, Dennis J, Oram R, Pearson E, Weedon M, Hattersley A, Owen K, Shields B, Jones A (2019) Development and validation of multivariable clinical diagnostic models to identify type 1 diabetes requiring rapid insulin therapy in adults aged 18-50 years. BMJ Open 9(9):e031586

Maddaloni E, Lessan N, Al Tikirti Buzzetti R, Pozzilli P, Barakat MT (2015) Latent autoinmune diabetes in adults in the United Arab Emirates: clinical features and factors related to insulin requirement. PLoS One 10(8):e0131837

Maddaloni E, Coleman RL, Pozzilli P, Holman RR (2019) Long-term risk of cardiovascular disease in individuals with latent autoimmune diabetes in adults (UKPDS 85). Diabetes Obes Metab 21:2115–2122

Maddaloni E, Coleman RL, Agbaje O, Buzzetti R, Holman RR (2020) Time-varying risk of microvascular complications in latent autoimmune diabetes of adulthood compared with type 2 diabetes in adults: a post-hoc analysis of the UK Prospective Diabetes Study 30-year follow-up data (UKPDS 86). Lancet Diabetes Endocrinol 8(3):206–215

Maioli M, Pes GM, Delitala G, Puddu L, Falorni A, Tolu F, Lampis R, Orrù V, Secchi G, Cicalò AM,

Floris R, Madau GF, Pilosu RM, Whalen M, Cucca F (2010) Number of autoantibodies and HLA genotype, more than high titers of glutamic acid decarboxylase autoantibodies, predict insulin dependence in latent autoimmune diabetes of adults. Eur J Endocrinol 163:541–549

Markham A, Keam SJ (2019) Sotagliflozin: first global approval. Drugs 79(9):1023–1029

Martinell M, Dorkhan M, Stålhammar J, Storm P, Groop L, Gustavsson C (2016) Prevalence and risk factors for diabetic retinopathy at diagnosis (DRAD) in patients recently diagnosed with type 2 diabetes (T2D) or latent autoimmune diabetes in the adult (LADA). J Diabetes Complicat 30(8):1456–1461

Mauricio D (2020) Latent autoimmune diabetes of adulthood: time to take action. Lancet Diabetes Endocrinol 8(3):177–179

McCarthy MI (2017) Painting a new picture of personalised medicine for diabetes. Diabetologia 60:793–799

Mishra R, Chesi A, Cousminer DL, Hawa MI, Bradfield JP, Hodge KM, Guy VC, Hakonarson H, Bone Mineral Density in Childhood Study, Mauricio D, Schloot NC, Yderstræde KB, Voight BF, Schwartz S, Boehm BO, Leslie RD, Grant SFA (2017) Relative contribution of type 1 and type 2 diabetes loci to the genetic etiology of adult-onset, non-insulin-requiring autoimmune diabetes. BMC Med 15(1):8

Mollo A, Hernandez M, Marsal JR, Esquerda A, Rius F, Blanco-Vaca F, Verdaguer J, Pozzilli P, de Leiva A (2013) Mauricio D; Action LADA 8. Latent autoimmune diabetes in adults is perched between type 1 and type 2: evidence from adults in one region of Spain. Diabetes Metab Res Rev 29:446–451

Myhill P, Davis WA, Bruce DG, Mackay IR, Zimmet P, Davis TM (2008) Chronic complications and mortality in community-based patients with latent autoimmune diabetes in adults: the Fremantle Diabetes Study. Diabet Med 25:1245–1250

Napoli N, Strollo R, Defeudis G, Leto G, Moretti C, Zampetti S, D'Onofrio L, Campagna G, Palermo A, Greto V, Manfrini S, Hawa MI, Leslie RD, Pozzilli P, Buzzetti R, NIRAD (NIRAD 10) and Action LADA Study Groups (2018) Serum Sclerostin and bone turnover in latent autoimmune diabetes in adults. J Clin Endocrinol Metab 103(5):1921–1928

Olsson L, Grill V, Midthjell K, Ahlbom A, Andersson T, Carlsson S (2013) Mortality in adult-onset autoimmune diabetes is associated with poor glycemic control: results from the HUNT Study. Diabetes Care 36 (12):3971–3978

Park Y, Hong S, Park L, Woo J, Baik S, Nam M, Lee K, Kim Y, KNDP Collaboratory Group (2011) LADA prevalence estimation and insulin dependency during follow-up. Diabetes Metab Res Rev 27:975–979

Pieralice S, Pozzilli P (2018) Latent autoimmune diabetes in adults: a review on clinical implications and management. Diabetes Metab J 42:451–464

Pozzilli P, Leslie RD, Peters AL, Buzzetti R, Shankar SS, Milicevic Z, Pavo I, Lebrec J, Martin S, Schloot NC (2018) Dulaglutide treatment results in effective glycaemic control in latent autoimmune diabetes in adults (LADA): a post-hoc analysis of the AWARD-2, −4 and −5 trials. Diabetes Obes Metab 20:1490–1498

Qi X, Sun J, Wang J, Wang PP, Xu Z, Murphy M, Jia J, Wang J, Xie Y, Xu W (2011) Prevalence and correlates of latent autoimmune diabetes in adults in Tianjin, China: a population-based cross-sectional study. Diabetes Care 34:66–70

Radtke MA, Midthjell K, Nilsen TI, Grill V (2009) Heterogeneity of patients with latent autoimmune diabetes in adults: linkage to autoimmunity is apparent only in those with perceived need for insulin treatment: results from the Nord-Trøndelag Health (HUNT) study. Diabetes Care 32:245–250

Raz I, Riddle MC, Rosenstock J, Buse JB, Inzucchi SE, Home PD, Del Prato S, Ferrannini E, Chan JC, Leiter LA, Leroith D, Defronzo R, Cefalu WT (2013) Personalized management of hyperglycemia in type 2 diabetes: reflections from a Diabetes Care Editors' Expert Forum. Diabetes Care 36(6):1779–1788

Redondo MJ (2013) LADA: time for a new definition. Diabetes 62:339–340

Sachan A, Zaidi G, Sahu RP, Agrawal S, Colman PG, Bhatia E (2014) Low prevalence of latent autoimmune diabetes in adults in northern India. Diabet Med 32 (6):810–813

Schloot NC, Pham MN, Hawa MI, Pozzilli P, Scherbaum WA, Schott M, Kolb H, Hunter S, Schernthaner G, Thivolet C, Seissler J, Leslie RD, Action LADA Group (2016) Inverse relationship between organ-specific autoantibodies and systemic immune mediators in type 1 diabetes and type 2 diabetes: Action LADA 11. Diabetes Care 39(11):1932–1939

Schwartz SS, Epstein S, Corkey BE, Grant SF, Gavin JR 3rd, Aguilar RB (2016) The time is right for a new classification system for diabetes: rationale and implications of the β-cell-centric classification schema. Diabetes Care 39:179–186

Skyler JS, Bakris GL, Bonifacio E, Darsow T, Eckel RH, Groop L, Groop PH, Handelsman Y, Insel RA, Mathieu C, McElvaine AT, Palmer JP, Pugliese A, Schatz DA, Sosenko JM, Wilding JP, Ratner RE (2017) Differentiation of diabetes by pathophysiology, natural history, and prognosis. Diabetes 66:241–255

Sørgjerd EP, Skorpen F, Kvaløy K, Midthjell K, Grill V (2012) Time dynamics of autoantibodies are coupled to phenotypes and add to the heterogeneity of autoimmune diabetes in adults: the HUNT study, Norway. Diabetologia 55:1310–1318

Stenström G, Gottsäter A, Bakhtadze E, Berger B, Sundkvist G (2005) Latent autoimmune diabetes in adults: definition, prevalence, beta-cell function, and treatment. Diabetes 54(Suppl 2):S68–S72

Stidsen JV, Henriksen JE, Olsen MH, Thomsen RW, Nielsen JS, Rungby J, Ulrichsen SP, Berencsi K,

Kahlert JA, Friborg SG, Brandslund I, Nielsen AA, Christiansen JS, Sørensen HT, Olesen TB, Beck-Nielsen H (2018) Pathophysiology-based phenotyping in type 2 diabetes: a clinical classification tool. Diabetes Metab Res Rev 34(5):e3005

Szepietowska B, Głębocka A, Puch U, Górska M, Szelachowska M (2012) Latent autoimmune diabetes in adults in a population-based cohort of Polish patients with newly diagnosed diabetes mellitus. Arch Med Sci 8:491–495

Takeda H, Kawasaki E, Shimizu I, Konoue E, Fujiyama M, Murao S, Tanaka K, Mori K, Tarumi Y, Seto I, Fujii Y, Kato K, Kondo S, Takada Y, Kitsuki N, Kaino Y, Kida K, Hashimoto N, Yamane Y, Yamawaki T, Onuma H, Nishimiya T, Osawa H, Saito Y, Makino H, Ehime Study (2002) Clinical, autoimmune, and genetic characteristics of adult-onset diabetic patients with GAD autoantibodies in Japan (Ehime Study). Diabetes Care 25(6):995–1001

Tiberti C, Giordano C, Locatelli M, Bosi E, Bottazzo GF, Buzzetti R, Cucinotta D, Galluzzo A, Falorni A, Dotta F (2008) Identification of tyrosine phosphatase 2 (256-760) construct as a new, sensitive marker for the detection of islet autoimmunity in type 2 diabetic patients: the non-insulin requiring autoimmune diabetes (NIRAD) study 2. Diabetes 57(5):1276–1283

Tuomi T, Groop LC, Zimmet PZ, Rowley MJ, Knowles W, Mackay IR (1993) Antibodies to glutamic acid decarboxylase reveal latent autoinmune diabetes mellitus in adults with a non-insulin-dependent onset of disease. Diabetes 42:359–362

Tuomi A, Carlsson H, Li H, Isomaa B, Miettinen A, Nilsson A, Nissén M, Ehrnström BO, Forsén B, Snickars B, Lahti K, Forsblom C, Saloranta C, Taskinen MR, Groop LC (1999) Clinical and genetic characteristics of type 2 diabetes with and without GAD antibodies. Diabetes 48:150–157

Tuomi T, Santoro N, Caprio S, Cai M, Weng J, Groop L (2014) The many faces of diabetes: a disease with increasing heterogeneity. Lancet 383:1084–1094

Turner R, Stratton I, Horton V, Manley S, Zimmet P, Mackay IR, Shattock M, Bottazzo GF, Holman R (1997) UKPDS 25: autoantibodies to islet-cell cytoplasm and glutamic acid decarboxylase for prediction of insulin requirement in type 2 diabetes. UK Prospective Diabetes Study Group. Lancet 350:1288–1293

Wod M, Yderstræde KB, Halekoh U, Beck-Nielsen H, Højlund K (2017) Metabolic risk profiles in diabetes stratified according to age at onset, islet autoimmunity and fasting C-peptide. Diabetes Res Clin Pract 134:62–71

World Health Organization (2019) Classification of diabetes mellitus. World Health Organization, Geneva. Licence: CC BY-NC-SA 3.0 IGO

Xiang Y, Huang G, Zhu Y, Zuo X, Liu X, Feng Q, Li X, Yang T, Lu J, Shan Z, Liu J, Tian H, Ji Q, Zhu D, Ge J, Lin L, Chen L, Guo X, Zhao Z, Li Q, Weng J, Jia W, Liu Z, Ji L, Yang W, Leslie RD, Zhou Z, China National Diabetes and Metabolic Disorders Study Group (2018) Identification of autoimmune type 1 diabetes and multiple organ-specific autoantibodies in adult-onset non-insulin-requiring diabetes in China: a population-based multicentre nationwide survey. Diabetes Obes Metab. https://doi.org/10.1111/dom.13595

Yang Z, Zhou Z, Li X, Huang G, Lin J (2009) Rosiglitazone preserves islet beta-cell function of adult-onset latent autoimmune diabetes in 3 years follow-up study. Diabetes Res Clin Pract 83:54–60

Zampetti S, Capizzi M, Spoletini M, Campagna G, Leto G, Cipolloni L, Tiberti C, Bosi E, Falorni A, Buzzetti R, NIRAD Study Group (2012) GADA titer-related risk for organ-specific autoimmunity in LADA subjects subdivided according to gender (NIRAD study 6). J Clin Endocrinol Metab 97(10):3759–3765

Zampetti S, Campagna G, Tiberti C, Songini M, Arpi ML, De Simone G, Cossu E, Cocco L, Osborn J, Bosi E, Giorgino F, Spoletini M, Buzzetti R, NIRAD Study Group (2014) High GADA titer increases the risk of insulin requirement in LADA patients: a 7-year follow-up (NIRAD study 7). Eur J Endocrinol 171(6):697–704

Zhou Z, Xiang Y, Ji L, Jia W, Ning G, Huang G, Yang L, Lin J, Liu Z, Hagopian WA, Leslie RD (2013) Frequency, immunogenetics, and clinical characteristics of latent autoimmune diabetes in China (LADA China study): a nationwide, multicenter, clinic-based cross-sectional study. Diabetes 62:543–550

Zimmet PZ, Tuomi T, Mackay IR, Rowley MJ, Knowles W, Cohen M, Lang DA (1994) Latent autoimmune diabetes mellitus in adults (LADA): the role of antibodies to glutamic acid decarboxylase in diagnosis and prediction of insulin dependency. Diabet Med 11:299–303

Zinman B, Kahn SE, Haffner, O'Neill MC, Heise MA, Freed MI, ADOPT Study Group (2004) Phenotypic characteristics of GAD antibody–positive recently diagnosed patients with type 2 diabetes in North America and Europe. Diabetes 53:3193–3200

Adv Exp Med Biol - Advances in Internal Medicine (2020) 4: 43–69
https://doi.org/10.1007/5584_2020_534
© Springer Nature Switzerland AG 2020
Published online: 14 May 2020

Hypoglycaemia

Muhammad Muneer

Abstract

In health hypoglycaemia is rare and occurs only in circumstances like extreme sports. Hypoglycaemia in type 1 Diabetes (T1D) and advanced type 2 Diabetes (T2D) are the result of interplay between absolute or relative insulin access and defective glucose counterregulation. The basic mechanism is, failure of decreasing insulin and failure of the compensatory increasing counterregulatory hormones at the background of falling blood glucose. Any person with Diabetes on anti-diabetic medication who behaves oddly in any way whatsoever is hypoglycaemic until proven otherwise. Hypoglycaemia can be a terrifying experience for a patient with Diabetes. By definition, hypoglycaemic symptoms are subjective and vary from person to person and even episode to episode in same person. Fear of iatrogenic hypoglycaemia is a major barrier in achieving optimum glycaemic control and quality of life which limits the reduction of diabetic complications. Diabetes patients with comorbidities especially with chronic renal failure, hepatic dysfunction, major limb amputation, terminal illness, cognitive dysfunction etc. are more vulnerable to hypoglycaemia. In most cases, prompt glucose intake reverts hypoglycaemia. Exogenous insulin in T1D and insulin treated advanced T2D have no control by pancreatic regulation. Moreover, failure of increase of glucagon and attenuated secretion in epinephrine causes the defective glucose counterregulation. In this comprehensive review, I will try to touch all related topics for better understanding of hypoglycaemia.

Keywords

T1D (Type 1 Diabetes) · T2D (Type 2 Diabetes) · IAH (Impaired Awareness of Hypoglycaemia) · HAAF (Hypoglycaemia Associated Autonomic Failure) · Autonomic, Neuroglycopenic · IHSG (International Hypoglycemia Study Group) · ADA (American Diabetes Association) · ES (The Endocrine Society) · UKHSG (UK Hypoglycaemia Study Group)

1 Introduction

Hypoglycaemia is the commonest acute complication in T1D and may also occur in T2D who are on insulin, sulphonylurea or glinide therapy (Balijepalli et al. 2017). It is a major physiological and psychological barrier to have targeted glycaemic control that might lead to the risk of diabetes complications and emotional morbidity to the patients and a serious concern to its caregivers as well. Monitoring hypoglycaemia

M. Muneer (✉)
Cardiff University, Cardiff, Heath Park, Cardiff, UK
e-mail: MuneerM1@cardiff.ac.uk;
genome2006@yahoo.ca

should be one of the key component of diabetes care in health and during sick days. Educating patient regarding its symptoms, causes, risk factors, prevention and treatment should be an integral part of each visit. Although intensive glycaemic control has been shown to reduce complications in diabetes, it also significantly increases the risk of hypoglycaemia (UKPDS Study Group 1998; The Diabetes Control and Complications Trial Research Group 1993). Severe hypoglycaemia can result in confusion, coma or seizures and requires external assistance for recovery (Yale et al. 2018). Severe hypoglycaemia puts patient at the risk of injury and threatens life (Seaquist et al. 2013). Frequency and severity of hypoglycaemia has an adverse effect on quality of life as well. Nocturnal hypoglycaemia, a fearful concern, because a sleeping patient cannot intervene and many episodes remain asymptomatic. Moreover, repeated exposure to nocturnal hypoglycaemia can blunt the counterregulatory defense mechanisms which makes serious clinical consequences. AO Whipple was first introduced a triad composed of symptoms of hypoglycaemia, a defined low sugar value and recovery of symptoms on glucose intake. This triad is justified in people without diabetes and in patients with endocrine tumours but not always the case in diabetes patients. Diabetes patients with autonomic neuropathy and IHA have no symptoms or minimal symptoms so the first point is not always justified. Moreover, in diabetes patients with poor control or recurrent hypoglycaemia cases, the cutoff value is different than that of good control cases (Whipple and Frantz 1935).

Iatrogenic hypoglycaemia is most common cause to be of concern. Skipping meals, unusual exercise and alcohol ingestion without food are common precipitating factors of hypoglycaemia in teen agers. Wrong dose at wrong time on wrong place insulin administration are the most common errors of insulin induced hypoglycaemia. Counter-regulatory mechanisms which normally protect against hypoglycaemia are impaired with increasing duration of diabetes. This further increases the risk of hypoglycaemia (Briscoe 2006). Prevention, early recognition and treatment are integral part of diabetes management plan. All international diabetes guidelines stress the need for adjustment of

individualized HbA1c targets for optimum control of hyperglycaemia without the potential risk of hypoglycaemia. It is important to remember that the primary cause of death in T1D under 40 years of age is an acute complication, especially hypoglycaemia. 'Dead in bed syndrome', where a previously well T1D is found sudden, unexplained night-time death in an undisturbed bed. This accounts for 5% of all deaths in long-term follow-up of T1D (Weston 2012). Most hypoglycaemic episodes happen at night and caregivers or families manage that. This is why there is less likely the patient to get education or adjustment of medication. Patient presents at emergency department with severe hypoglycaemia should be followed up. As there is strong link of increased mortality over the following year (Turchin et al. 2009).

2 Definition and Classification: Diabetic Hypoglycaemia

Hypoglycaemia is defined as a combination of a triad consisting of the development of autonomic or neuroglycopenic symptoms, a low plasma glucose level commonly below 3.9 mmol/L (70 mg/dL) in diabetes patients and the symptoms responding with carbohydrate intake (Yale et al. 2018). The glycaemic threshold for hypoglycaemic symptoms varies among individuals and in same person as well. It shifts to lower value with recent episode of hypoglycaemia and higher value with poorly controlled diabetes and infrequent hypoglycaemia. As a result, a single value of plasma glucose that defines hypoglycaemia is difficult to assign. So, ADA has defined hypoglycaemia in diabetes non-numerically as "All episodes of an abnormally low plasma glucose concentration that expose the individual to potential harm" (Heller 2016). However, the workgroup of the ADA and ES have defined an alert value of 3.9 mmol/L (70 mg/dL) or below that may be associated with potential harm (Seaquist et al. 2013).

IHSG recommends a glucose level < 3 mmol/L (<54 mg/dL) cut-off value be the clinically significant biochemical hypoglycaemia should be included in clinical trials of anti-diabetic medicines. Nevertheless, the IHSG thinks this value would enable the concerning authorities to intervene by effective measures with pharmacological, technological or educational ways. IHSG further elaborated that as a value <3 mmol/L (54 mg/dL) and < 2.8 mmol/L (<50 mg/dL) do not come under physiological conditions in non-diabetic persons so, they are unequivocally hypoglycaemic values (Heller et al. 2020a). We witnessed glucose value <2.8 mmol/L (50 mg/dL) caused excess mortality in ACCORD (Action to Control Cardiovascular Risk in Diabetes) study, in ORIGIN (Outcomes Reduction with an Initial Glargine Intervention) trial and in NICE-SUGAR (Normoglycaemia in Intensive Care Evaluation-Survival Using Glucose Algorithm Regulation) trial (Bonds et al. 2010; Mellbin et al. 2013; The NICE-SUGAR Study Investigators 2012).

Furthermore, a Joint Position Statement of the American Diabetes Association and the European Association for the Study of Diabetes defined a value less than 54 mg/dL (3 mmol/L) as serious, clinically important hypoglycaemia.

The workgroup of the ADA and the EASD proposed hypoglycaemia for clinical trials (Heller 2016) as follows:

Level 1: A glucose value of ≤ 3.9 mmol/L (70 mg/dL) is the alert value for hypoglycaemia needs to be reported routinely.
Level 2: A glucose value of ≤ 3 mmol/L (54 mg/dL) is the value to show sufficiently low value to denote serious, major, clinically important or clinically significant hypoglycaemia.
Level 3: A very low value to be denoted as severe hypoglycaemia defined by ADA which shows severe cognitive impairment requiring external assistance for recovery.

The level 2 of hypoglycaemia was recently validated with ADA definition in a study which did a post hoc analysis of SWITCH 1, SWITCH 2 and DEVOTE studies on T1D, T2D and T2D with high risk of MACE respectively and it concluded that the IHSG level 2 definition is valid to be uniformly adopted by international bodies to be used in future clinical trials (Heller et al. 2020b).

ADA and ES classified hypoglycaemia (Seaquist et al. 2013) as:

1 **Severe hypoglycaemia**
This is defined as hypoglycaemia causing significant cognitive impairment, which requires assistance from another person for recovery.
2. **Documented symptomatic hypoglycaemia**
An episode associated with typical symptoms of hypoglycaemia with a documented plasma glucose value of ≤ 3.9 mmol/L (≤ 70 mg/dL)
3. **Asymptomatic hypoglycaemia**
An episode with a documented plasma glucose value of ≤ 3.9 mmol/L (≤ 70 mg/dL) but without symptoms
4. **Probable symptomatic hypoglycaemia**
An episode with typical symptoms without measured plasma glucose level.
5. **Pseudo or relative hypoglycaemia**
An episode in which any of the typical symptoms of hypoglycaemia are experienced with a measured plasma glucose value above 3.9 mmol/L (>70 mg/dL) but is nearing this value.

Non-diabetic Hypoglycaemia This type is traditionally categorized as reactive or postprandial and postabsorptive or fasting hypoglycaemia but this classification is confusing to diagnose. It is because some causes can be present in both forms. So a useful approach based on clinical characteristics is more helpful. People who are apparently healthy, are having some specific causes and people looking ill also have some obvious causes. Drugs and alcohol can cause hypoglycaemia in both health and illness (Cryer et al. 2009b; Service 1999). The cause are described in (Table 3). The traditionally classified two types are summarised below:

1. **Reactive hypoglycaemia or postprandial hypoglycaemia**: It is mainly due to three causes- idiopathic, alimentary and congenital enzyme deficiencies. Alimentary form of reactive hypoglycaemia is most common and is due to upper gastrointestinal surgical procedures like gastrectomy, gastrojejunostomy, pyloroplasty, Bariatric surgery etc. (Hamdy et al. 2019; Desimone and Weinstock 2017). The surgical modifications especially Roux-en-Y bypass surgery cause rapid food entry in the intestine. This promotes

fast absorption of glucose from intestine after a meal provoking excessive insulin and GLP-1 secretion. Pancreatic islet nesidioblastosis also found responsible in some studies (Ritz et al. 2016; Service et al. 2005). It usually occurs within 4 h after a meal ingestion. Hereditary fructose intolerance and galactosemia are the form of inborn errors of metabolism that cause acute inhibition of hepatic glucose output after fructose or galactose is ingested. This happens due to deficiency of aldolase B (Hamdy et al. 2019).

2. **Fasting hypoglycaemia or postabsorptive hypoglycaemia**: Nesidoblastosis, a rare cause of fasting hypoglycaemia mostly found in infancy and extremely rare in adults. Diffuse budding of insulin secreting cells from pancreatic duct epithelium are found and microadenomas of beta-cells are also found in some cases. Inherited enzyme defects in ketogenesis and in fatty acid oxidation results fasting hypoglycaemia by restricting ketogenesis and fatty acid oxidation respectively. This deprives nonneural tissues to get energy during fasting or exercise which results high glucose uptake by nonneural tissues and subsequently hypoglycaemia (Hamdy et al. 2019).

Severity of Hypoglycaemia: Mild Where autonomic symptoms are present, and the person is able to self-treat. **Moderate**: Where autonomic and neuroglycopenic symptoms are present but the person can self-treat. **Severe**: Where a person needs assistance from another person. Unconsciousness may occur and plasma glucose is typically <2.8 mmol/L (<50 mg/dL).

3 Epidemiology

Addressing the epidemiology of hypoglycaemia is challenging because mild to moderate episodes are usually not recorded by patients. In studies CGMS rarely been used so asymptomatic episodes or moderate episodes go unrecorded. But the rates of severe hypoglycaemia can be reliably obtained as it needs another person's assistance. Moreover, studies were using different cut-off values. Recently, IHSG has put the clinical trial cut-off value so future studies will find the more accurate prevalence and incidence rates. UKHSG showed annual prevalence of severe hypoglycaemia was twice that of DCCT for T1D for less than 5 years and five times for T1D more than 15 years. It also showed the incidences are 3.3–16.37 per patient-year for T2D (UK Hypoglycaemia Study Group 2007). In another population based self-reported study in UK showed that frequency of non-severe hypoglycaemic episodes in T1D is 129.7 per year and 57.6 per year for insulin treated T2D (Frier et al. 2015).

There are very few real-world studies on the incidence of diabetes-related hypoglycaemia. In Canada a research team has conducted a study named InHypo-DM (In Hypo Diabetes Mellitus). It was an online questionnaire to T1D and T2D patients above 18 years comprising 552 responders who are on insulin and or secretagogues. It showed the incidence proportion and rate of non-severe events were higher among people with T1D versus T2D were 77% and 55.7 events per person-year vs 54% and 28.0 events per person-year respectively. Severe hypoglycaemia was reported by 41.8% of all respondents, at an average rate of 2.5 events per person-year. Over half, 65.2%, of the total respondents reported experiencing at least one event of non-severe or severe at an annualised crude incidence density of 35.1 events per person per year (Ratzki-Leewing et al. 2018). The mean incidence of mild hypoglycaemia in a European population has been reported at 1.8 episode per patient per week in T1D and 0.4–0.7 episode per patient per week in insulin treated type 2 diabetes, corresponding to annual incidence rate of 94 and 21–36, respectively (Östenson et al. 2013). Similarly, accurate reporting of severe hypoglycaemia can be influenced by amnesia and impaired awareness (Frier 2014).

EPIDIAR study comprising 1,070 T1D and 11,173 T2D patients evaluated during fasting of Ramadan showed that severe hypoglycaemia was significantly more during fasting of Ramadan and was 0.14 vs 0.03 episode/month for T1D and 0.03 vs 0.004 episode/month for T2D. On the other hand, in CREED study which was also done on

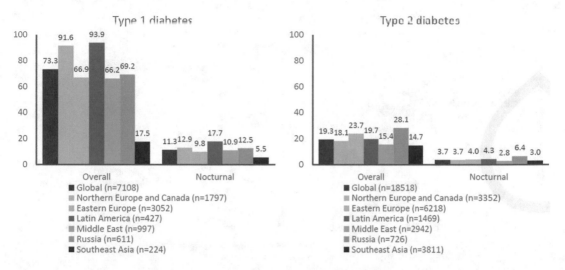

Fig. 1 IO HAT study showing T1D and T2D patients' total hypoglycaemic episodes of per patient per year in different study zones (Khunti et al. 2014a). (The figure is reproduced here with permission)

participants who were fasting during Ramadan, showed much less episodes of hypoglycaemia as healthcare professional were actively instructed Ramadan specific precautions to participants (Salti et al. 2004; Jabbar et al. 2017).

Intensive glycaemic control increases the risk of severe hypoglycaemia as seen in DCCT trial where 3 times more hypoglycaemia than conventional arm, in ACCORD study rate of treatment for hypoglycaemia was 3.1% in intensive arm and was just 1% in standard arm, in VADT annual rate of severe hypoglycaemia was 3.8% in intensive arm and 1.8% in standard arm (DCCT Research Group 1993; Seaquist et al. 2011; Saremi et al. 2016).

IO HAT study, an epidemiological, multicentre, multicountry study which comprised of 27,000 people with T1D and T2D from 24 countries to determine the frequency, predictive factors and consequences of hypoglycaemia. IO HAT study found that frequency of self-reported rates of severe hypoglycaemia in T2D on insulin were 5 times higher than previous reports which means the rates of severe hypoglycaemia in that group was 2.5 per person years (Khunti et al. 2014a) (Fig. 1).

An observational study to get better understanding of frequency of hypoglycaemia episodes reported in the real-world setting compared with finding in RCT. It showed that there are huge discrepancies in RCT. It observed patients with

severe hypoglycaemia, recurrent episodes of hypoglycaemia and patients with impaired awareness of hypoglycaemia were excluded. All these underestimates the true incidence of hypoglycaemia. So, it is prudent that consideration should be given to those under-represented patients with renal insufficiency, elderly or frail patients who are at risk of severe hypoglycaemia (Elliott et al. 2016) (Fig. 2).

4   Pathophysiology

4.1   Physiological Defence in Hypoglycaemia

The *'Rules of Metabolic Game'* which human being had to follow to ensure its survival. As humans evolved like hunter-gatherers and unlike humans of today. They used to consume irregular meals. Therefore, our body has evolved a mechanism to store food when it was in abundance in the form of glycogen in liver & skeletal muscles. Human body use these stores in the form of glucose when food was scarce (Cahill 1971). The four rules (Cahill 1971) are described here:

1. *Glucose level has to maintain within a very narrow limit, neither low nor high*: This is done by reduction of insulin and simultaneously increase in glucagon secretion during fasting. It results reduction of peripheral

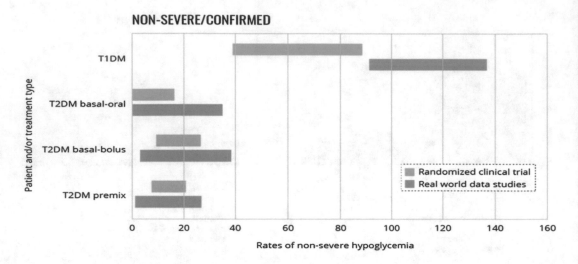

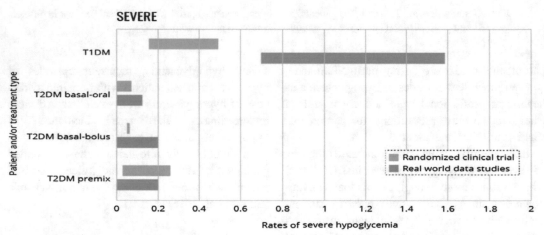

Fig. 2 Non-severe and severe hypoglycaemia rates of data of real-world and RCT in T1D and different insulin regimen of T2D (Elliott et al. 2016). (The figure reproduced here with permission)

glucose utilisation and increase hepatic glycogenolysis and gluconeogenesis. This process is opposite in postprandial state.

2. *The 'Fight or Flight' response, body have to maintain an energy source which can be tapped quickly in an acute shortage*: During overnight fast, glycogenolysis provides 60–80% and gluconeogenesis 20–40% to maintain adequate glucose concentration in brain. In prolonged fasts, glycogen storage depletes and gluconeogenesis is only source of glucose. So glycogen is the emergency fuel source.

3. *'Waste not, want not', maintain a storage of fat and protein when it is plenty*: This is done during fed state excess food is stored as glycogen, fat and protein.

4. *Utilise every trick in the book to maintain good protein reserves*: This is done by reduced insulin and glucagon ratio during fasting state which creates a catabolic state. The resultant effect on fat metabolism is greater than protein which relatively preserves muscle. This adaptation makes hunter-gatherer to have sufficient muscle power to pursue his next move for food. Moreover, at the same time his brain function is optimally maintained to plan that.

Glucose Sensing Neurons Insulin and glucagon are the key hormones that control glucose homeostasis. A number of physiological mechanisms ensure that plasma glucose level is maintained within narrow normal range. Normal arterial fasting plasma glucose value 3.9–6.1 mmol/L (70–110 mg/dL) with a mean 5 mmol/L (90 mg/dL). The lowest value goes to >3 mmol/L (>54 mg/dL) after exercise or a moderate fasting for 60 h (Cryer 1993). CNS acts as a great integrator of counterregulatory responses by central and peripheral glucose sensing neurons. Central neurons are widely distributed in ventromedial hypothalamus and hind brain. The distribution mostly in areas involved in the regulation of neuroendocrine function, nutrient metabolism and energy homeostasis. The peripheral glucose sensing neurons are located in oral cavity, gut, the hepatic portal-mesenteric vein and carotid body. These are to coordinate with the neuroendocrine, autonomic and behavioural reactions to hypoglycaemia. These neurons act through vagus nerve and spinal cord to hind brain to further higher centres (Watts and Donovan 2010) (Fig. 3).

A falling glucose is detected by central and peripheral glucose sensing neurons. These contain GE (glucose-excited), GI (glucose-inhibited) neurons and an astrocytic support structure with glucose sensing neurons. These neurons signal to activate efferent pathways to initiate counterregulatory mechanism to hypoglycaemia.

Glucose Uptake The obligate oxidative fuel of brain is only glucose and it utilises 25% of whole-body glucose though it weigh only 1/40th of body weight. Glucose levels in brain ranges between 0.8 and 1.5 mmol/L which is significantly lower than systemic circulation. Glucose transport across the blood-brain barrier occurs through facilitative diffusion. So, it needs a constant normal concentration of glucose in the arteries. Brain can neither synthesize glucose nor utilise non-glucose fuels. It can only store few minutes supply as glycogen, so brain needs continuous supply of glucose. But the delivery of exogenous glucose from dietary carbohydrates is intermittent. So it is needed normal arterial plasma glucose concentration to facilitate blood-brain glucose transport for smooth function of brain. Normally glucose balance is maintained tightly so that hypoglycaemia and hyperglycaemia are prevented. This is done by dynamic minute to minute endogenous glucose production by

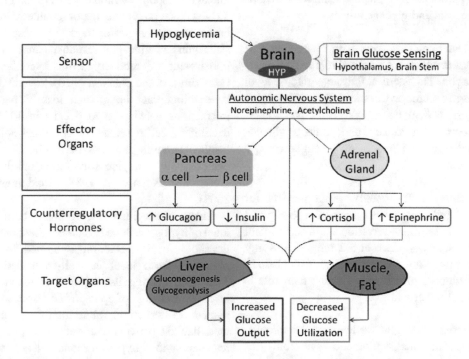

Fig. 3 Integrated circuit of physiological response to hypoglycaemia. (Reproduced with permission (Reno et al. 2013)

gluconeogenesis and glycogenolysis. These processes are mainly carried by liver and partly by kidney. Glucose utilisation other than brain is by skeletal muscles and adipose tissues with the primary action of insulin and glucagon as key regulator (McCrimmon 2008; Cryer 2016).

First Defence When plasma glucose level falls at 4.4–4.7 mmol/L (80–85 mg/dL), then insulin secretion is dramatically decreased. The rate of glucose production mainly from liver and partly from kidneys are increased. The peripheral glucose utilisation is significantly decreased. These three actions are called as '*First defence*' against hypoglycaemia.

Hypoglycaemia counterregulation is shown in Table 1: (Cryer 2016; Levin et al. 2004; McCrimmon and Sherwin 2010; Abi-Saab et al. 2002; Ishihara et al. 2003; Boden 2004; Fisher et al. 2005).

Second Defence This starts when glucose value falls further to 3.6–3.9 mmol/L (65–70 mg/dL), the secretion of glucagon begins. This is the primary glucose counterregulatory hormone. This prompts production of glucose from liver by gluconeogenesis and glycogenolysis.

Third Defence This starts with the secretion of epinephrine which has a critical role in deficiency of glucagon. This starts at the same glucose level of glucagon secretion threshold. It stimulates gluconeogenesis & glycogenolysis from liver, gluconeogenesis from kidney and inhibits glucose clearance by alpha- & beta-adrenergic stimulation.

Forth Defence If hypoglycaemia persists for long and falls around 3.7 mmol/L (66 mg/dL) then growth hormone secretion starts and cortisol hormone secretion starts at 3.2 mmol/L (58 mg/dL). This two limit glucose utilisation and help in hepatic glucose production. Though their role is not critical rather additive.

Fifth Defence At glucose level of 3.2 mmol/L (58 mg/dL), the neurogenic symptoms of hypoglycaemia start to appear and prompts behavioural defence to get food ingestion. This is done by combined action of catecholamine and acetylcholine. If food ingestion happens then the person recovers.

Reduced cognition If food ingestion does not happen then at <2.8 mmol/L (<50 mg/dL) neuroglycopenic symptoms appear and cognition deteriorates with functional brain failure including seizure or coma (Cryer 2016).

4.2 Hypoglycaemia Counterregulation in T1D

Increased risk of hypoglycaemia in T1D results from defects in three major counterregulatory mechanisms described above. Firstly, insulin deficiency in T1D and depending on exogenous insulin for life. This makes hypoglycaemia is more likely to develop due to unregulated and sustained hyperinsulinaemia. This exogenous insulin is not responsive to falling glucose. Secondly, hypoglycaemia fails to stimulate adequate glucagon response within 5 years of diagnosis of T1D. Glucagon is the major counterregulatory hormone responsible for hepatic glucose synthesis. Failure of glucagon response causes marked impairment of recovery from hypoglycaemia (McCrimmon and Sherwin 2010). Several studies have found that the gradual loss of paracrine interactions between α to β cells in T1D. The declining β cell mass and function makes loss of insulin secretion and loss of insulin flux. This dampens the paracrine stimulation of glucagon secretion (Fukuda et al. 1988; Mundinger et al. 2003).

Lastly, the epinephrine response to a given level of hypoglycaemia is blunted. The glycaemic threshold for its secretion is shifted to lower plasma glucose level in well-controlled T1D. Therefore, making its response at much lower level. This change of epinephrine response is not related to structural abnormality in adrenal medullae. As patients with defective epinephrine response to hypoglycaemia has normal

Table 1 Counterregulatory responses in normal person, T1D and T2D

Response	Glucose value	Hormones – normal action	In T1D	In T2D
First defence insulin	4.4–4.7 mmol/L (80–85 mg/dL)	↓↓↓**Insulin**, so inhibitory action on liver is gone resulting: ↑↑↑Glycogenolysis in liver produces ↑glucose and in skeletal muscles produces lactate. ↑↑Gluconeogenesis in liver, ↑in kidney. ↓↓Glycolysis in liver, ↓↓Glucose uptake in skeletal muscle and adipose tissue.	No endogenous insulin but as on exogenous insulin, so its inhibitory action on liver, kidney, and enhancing peripheral glucose uptake remains unopposed.	In insulin-deficient case on exogenous insulin situation like T1D but on sulphonylurea or glinides cases glucose-independent insulin release makes insulin level uninhibited.
Second defence glucagon	3.6–3.9 mmol/L (65–70 mg/dL)	↑↑↑**Glucagon**, which causes: ↑↑↑Glycogenolysis in liver is main role, ↑↑Gluconeogenesis if hypo is prolonged, ↓↓Glycolysis and ↓↓Glycogenesis in liver. It has no significant role in periphery.	Paracrine interactions between α and β cells are gradually lost due to β-cell mass decline so Glucagon response is gradually down to nil.	In insulin-deficient case Glucagon response is near absence but cases on sulphonylurea or glinide induced hypoglycaemia, have almost intact response.
Third defence epinephrine	3.6–3.9 mmol/L (65–70 mg/dL)	↑**Epinephrine**, which causes: ↑Glycogenolysis, ↓Glycolysis, ↑Gluconeogenesis, ↓Glycogenesis in liver resulting ↑glucose. It also increases delivery of substrates for gluconeogenesis from periphery, ↓glucose utilisation and ↓insulin secretion.	Epinephrine response to a given level of glucose is blunted and the threshold for its secretion is lowered in well-controlled cases.	Epinephrine response almost remains intact.
Fourth defence growth hormone cortisol	3.7 mmol/L (66 mg/dL): GH 3.2 mmol/L (58 mg/dL): C	**Growth hormone (GH)** and **Cortisol(C)** work after prolonged hypo for more than one hour. They work to change metabolic processes over longer periods by ↑↑Lipolysis, ↑↑ Ketogenesis and stimulating ↑↑ Gluconeogenesis. Cortisol makes ↑Epinephrine as well.	>10 year duration, 25% T1D patients might have reduced GH and cortisol response. In rare cases, Addison's disease or hypopituitarism may coexist and cause profound hypoglycaemia. Rarely, ante-partum infarction, Sheehan syndrome or Houssay syndrome can cause severe hypo due to panhypopituitarism.	Though GH and cortisol non-responses in T2D are not so much investigated but some case reports showed that in a very elderly, female with less duration of T2D there are insufficient response of GH and cortisol which makes severe hypo. Normally GH and cortisol responses in T2D remain intact.
Fifth defence: behavioural defence norepinephrine/epinephrine acetylcholine	3.2 mmol/L (58 mg/dL)	Sympathoadrenal discharge triggered by hypoglycaemia starting neurogenic symptoms: **Catecholamine (NE):** palpitations, tremor and	The unopposed hyperinsulinaemia, blunted glucagon, attenuated epinephrine response, long duration, asymptomatic nocturnal	In insulin-deficient advanced T2D the responses are more or less similar to T1D but in SU or glinide group

(continued)

Table 1 (continued)

Response	Glucose value	Hormones – normal action	In T1D	In T2D
		anxiety/arousal. **Acetylcholine:** sweating, **hunger** and paresthesias. Hypothalamo-pituitary-adrenal mechanisms also causes **hunger**. This intense hunger drives a person to get food and then recovers from low glucose.	hypo and recent antecedent hypo make hypo threshold lower and precipitates into recurrent hypo. This leads to impaired awareness of hypo and eventually ends up as HAAF. So, food craving comes much later.	the behavioural responses remain intact.
↓↓ Cognition	<2.8 mmol/L (50 mg/dL)	The **failsafe** counterregulatory functions at all five levels maintain brain glucose supply uncompromised so neuroglycopenic symptoms cannot occur in normal state.	Counterregulatory response failure to hypo at all five defences which initiates neuroglycopenic symptoms: Disruption in fine motor coordination, mental speed, concentration and some memory functions occur if outside assistance not done.	In insulin-deficient advanced T2D the responses are more or less similar to T1D. In SU or glinide group hypo rarely goes to neuroglycopenic level.

Hypoglycaemia counterregulation is shown in this table: (Cryer 2016; Levin et al. 2004; McCrimmon and Sherwin 2010; Abi-Saab et al. 2002; Ishihara et al. 2003; Boden 2004; Fisher et al. 2005)

epinephrine response to exercise, standing or to a meal. This autonomic response to hypoglycaemia is too impaired in majority of people with T1D with around 10 years of disease duration (McCrimmon 2008).

4.3 Hypoglycaemia Counterregulation in T2D

In T2D some investigators have shown impaired counterregulatory response to hypoglycaemia and some others have not. A study showed that attenuated glucagon, growth hormone and cortisol response but no change in epinephrine response. It showed rather increased norepinephrine response during subcutaneous insulin-induced hypoglycaemia in T2D without autonomic neuropathy (Boli et al. 1984). In contrast, in another study it was found that basal glucagon was higher in T2D than controls. But there was no difference in glucagon and epinephrine responses between the two groups (Boden et al. 1983). Similar studies done by Polonsky et al. and Heller

et al. found normal responses of glucagon, epinephrine, cortisol and growth hormone in T2D (Polonsky et al. 1984; Heller et al. 1987). In another study with T2D, on only oral hypoglycaemics, on only insulin and on controls. The results showed that the insulin-deficient T2D had a near absence of glucagon response to hypoglycaemia. But those on oral hypoglycaemics and the controls had normal response (Segel et al. 2002). So, we can conclude from these studies that the glucagon response may be normal or blunted, while epinephrine, cortisol and growth hormone response remains intact in T2D.

4.4 Impaired Awareness of Hypoglycaemia

In health, hypoglycaemia triggers activation of sympathetic nervous system, which results in signs and symptoms of hypoglycaemia. This hypoglycaemic awareness enables patients to recognize hypoglycaemia, prompting them to take corrective measures (Heller et al. 1987). Aware-ness of hypoglycaemia are categorised into three:

- **Normal:** where the person is always aware of the onset of hypoglycaemia
- **Partial awareness:** where the symptom pro-file is altered to either reduction in intensity or number
- **No awareness:** where no awareness of symptoms at all.

Hypoglycaemia Associated Autonomic Failure (Fig. 4) In attenuated blood glucose counterre-gulation, an impaired autonomic response leads to compromised behavioural defences against hypoglycaemia. In people with IAH, threshold for autonomic symptoms comes closer to or lower than threshold for neuroglycopenic symptoms. Persistent exposure to iatrogenic

hypoglycaemia leads to progression to IAH. As a result, the first symptom of hypoglycaemia may be confusion or loss of consciousness. Addition-ally, recent hypoglycaemia diminishes the glu-cose counterregulation by attenuating sympatho-adrenal response and leads to reduced symptoms or IAH. It affects sympathetic neural response which is then termed as HAAF. It can end up in a vicious cycle of recurrent hypoglycaemia (Reno et al. 2013; Dagogo-Jack et al. 1993). Altered sensing or utilizing alternative fuels or impaired counterregulatory response by brain can be explained.

IAH is a risk factor for severe hypoglycaemia, increasing the risk by sixfold in T1D and 17-fold in T2D (Gold et al. 1994; Schopman et al. 2010). Prevalence of IAH increases with time, affecting 20–25% people with T1D and 10% people with

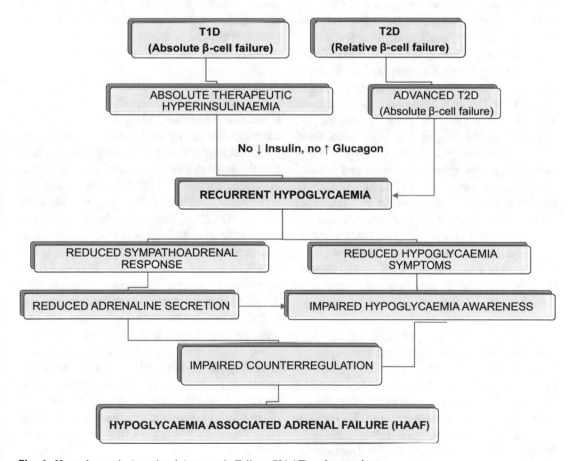

Fig. 4 Hypoglycaemia Associated Autonomic Failure (HAAF) pathogenesis

insulin treated T2D (Pearson and McCrimmon 2018). Strict avoidance of hypoglycaemia can restore hypoglycaemia awareness, primarily by improving sensitivity of β-adrenergic system. Studies have showed hypoglycaemic symptoms were restored after 2 weeks of hypoglycaemia prevention. At 3 months, glycaemic threshold for symptoms and neuroendocrine responses were normalized with some recovery of glucagon response (Fritsche et al. 2001).

4.5 Nocturnal Hypoglycaemia

It is always a major concern to persons with diabetes. Several studies have shown that nocturnal hypoglycaemia occurs commonly in T1D and long standing T2D on insulin during sleep. Asymptomatic episodes may precipitate due to IAH and deficient counterregulation. This may lead to recurrent nocturnal hypoglycaemia and ends up cognitive impairment with the risk of severe hypoglycaemia (Chico et al. 2003; Weber et al. 2007). In the DCCT study half of the severe hypoglycaemia occurred during sleep (Hypoglycemia in the Diabetes Control and Complications Trial 1997). Nighttime sleep is the longest interprandial and longest time in between blood glucose monitoring. It is the time when many mild to moderate episodes occur without records. Moreover, during night, insulin sensitivity increases and sympathoadrenal responses decreases in T1D. This increases the risk of severe episodes if evening exercise done in a tight glycaemic controlled case. In this situation a child may not wake up from hypoglycaemia (Raju et al. 2006). It sometime manifests as disturbed sleep, nightmares and waking up with sweats.

Dead in bed (DIB) syndrome is a very rare condition happens suddenly with unexplained cause in T1D under 40 years in an undisturbed bed. Some evidences to date suggest that the deaths could be caused by ventricular arrhythmias induced by severe hypoglycaemia. This happens particularly in patients with subtle autonomic dysfunction and may cause QTc lengthening.

Table 2 Risk factors for dead in bed syndrome

Previous nocturnal hypoglycaemia	Sleeping alone	Living alone
Intensive glycaemic control	Male gender	T1D
Alcohol ingestion before bedtime	HAAF	Mitral valve prolapse

Epinephrine, a counterregulatory hormone in hypoglycaemia, may contribute to abnormal cardiac repolarization (Tu et al. 2010; Weston 2012).

Table 2 is showing risk factors for DIB syndrome (Weston 2012).

Patient and parental counselling, larger bed time snacks, using CGMS to diagnose nocturnal hypoglycaemia in undiagnosed IAH cases are some measures to avoid DIB syndrome (Weston 2012; Secrest et al. 2011). Some studies have shown that medicines that causes QT prolongation should be avoided. ACE or ARB have beneficial effects on cardiovascular autoregulation and may reduce DIB (Weston 2012).

"Somogyi phenomenon" which was first hypothesised by Michael Somogyi, a Hungarian-American professor. It is about early morning fasting hyperglycaemia preceded by nocturnal hypoglycaemia. He proposed that nocturnal hypoglycaemia provokes counterregulatory response with the secretion of epinephrine, cortisol and growth hormone (Somogyi 1959). However, clinical relevance were challenged and repeated studies have alternate explanations. Fasting hyperglycaemia is due to fall of insulin concentrations during night or excessive hepatic glucose production. Researchers have shown by self-monitoring and CGMS nocturnal hypoglycaemia does not provoke rebound fasting hyperglycaemia (Somogyi 1959; Havlin and Cryer 1987). Rather, it is a physiological process where growth hormone spikes make hepatic glucose production due to waning of insulin which is called **"Dawn phenomenon"**. The inadequate overnight basal insulin replacement precipitates this (Campbell et al. 1985).

5   Hypoglycaemia in Special Groups

5.1   Hypoglycaemia in Pregnancy

Frequent severe hypoglycaemia during first tri-mester of pregnancy with diabetes might have adverse neurocognitive outcome in baby (Pacaud and Dewey 2011). It is crucial to avoid hyperglycaemia induced adverse outcomes. But striving the near-normoglycaemia increases the risk of severe hypoglycaemia (Schwartz and Teramo 1998). Risk of hypoglycaemia is the main obstacle to good glycaemic control during pregnancy in women with diabetes (Rosenn et al. 1995).

In T1D women, most studies showed that severe hypoglycaemia is three to five times more common in early pregnancy but in third trimester the incidences are lower. Up to 45% women with T1D experience severe hypoglycaemia during pregnancy and the first event usually occurs before 20 weeks of gestation (Evers et al. 2002; Nielsen et al. 2009). There are overall higher rates of severe hypoglycaemia in pregnancy. Interest-ingly, the rates of nocturnal hypoglycaemia are similar to non-pregnant state (Mathiesen et al. 2007). A pre-bedtime blood glucose <6 mmol/L (<110 mg/dL) predicts nocturnal hypoglycaemia in first trimester (Akram et al. 2009). In T2D women with pregnancy the incidences are sparse. Because, T2D or GDM cases during their repro-ductive years are having good beta-cell function and counterregulation. But in advance T2D on insulin are having greater risk of severe hypoglycaemia (Akram et al. 2009).

In T1D cases show marked impairment of counterregulation both in glucagon and epineph-rine response (Diamond et al. 1992). This makes T1D more dependent on growth hormone and cortisol (Ringholm et al. 2012). During preg-nancy, increased C-peptide concentrations have been found in long standing T1D with good glycaemic control. Even though the C-peptide was undetectable in early pregnancy (Nielsen et al. 2009). The growth promoting factors during pregnancy might promote rejuvenated beta-cell

function and suppression of immune system (Ilic et al. 2000). Past history of severe hypoglycaemia, long standing diabetes and IAH are major risk factors of hypoglycaemia during pregnancy. Insulin requirement reduces to 60% of prepregnancy level immediately after delivery. It is due to lack of placental hormonal influence. There is risk for severe hypoglycaemia during initial period of breast feeding. As fluctuating glucose levels and titration of insulin dose which settles down gradually. The insulin requirement remains 10% lower than prepregnancy level during breast feeding (Stage et al. 2006; Riviello et al. 2009).

5.2   Exercise and Hypoglycaemia

During exercise cardiovascular and respiratory functions increase to support increased demand of oxygen and fuel requirements of muscles. This is same in people with and without diabetes. In initial stage of exercise, muscle uses glucose from its own glycogen store as primary source. When this source depletes in continuing exercise then glucose comes from hepatic glycogenolysis (Petersen et al. 2004). In prolonged exercise insu-lin secretion reduces significantly. Then glucagon and catecholamine secretion prompt hepatic glu-coneogenesis by lipolysis yielding free fatty acids for gluconeogenesis (Stich et al. 2000; Kreisman 2001). Muscle glucose uptake continues to increase though insulin secretion is reduced. As exercise stimulates translocation of GLUT-4 receptors to cell surface in a similar manner of insulin mediated GLUT-4 translocation (Thorell et al. 1999; Fujii et al. 2000). When exercise is over, glucagon and catecholamine reduces and insulin secretion increases rapidly to maintain euglycaemia. But in T1D and advanced T2D on insulin, the exogenous insulin makes significant derangement in endocrine response. It is due to continuous release of insulin from injected sub-cutaneous insulin depot and not related to endo-crine response. So the insulin: glucagon ratio deranges and hence glucose production from liver hampers. The high insulin during exercise makes non-exercising muscles and other tissues to increase glucose uptake (Marliss et al. 2000). This endocrine derangement makes blood glucose level markedly low partly during and mostly after exercise. Intensive exercise makes profound counterregulatory response whereas moderate intensity does not. So, for diabetes patients a mixture of moderate intensity with few minutes of high intensity exercise is better. Furthermore, titrating insulin dosage with exercise time and pre-exercise snacking are needed to prevent hypoglycaemia (Bussau et al. 2006).

5.3   Hypoglycaemia in Children

Children with diabetes remain at risk of hypoglycaemia due to intensive insulin therapy, metabolic demands of growth and lifestyle. This might risk them detrimental neuropsychological sequel (Allen et al. 2001). Children can tolerate lower levels of glucose as the developing brain can utilise alternate substrates for cerebral metabolism. So, the defining hypoglycaemia is controversial as much lower cut-off value could initiate symptoms. Most guidelines say that glucose level should be kept above 4 mmol/L (72 mg/dL), "4 is the floor", (Jones 2017; Clarke et al. 2009).

International Society for Paediatric and Adolescent Diabetologists (ISPAD) now recommends that hypoglycaemia in childhood to be classified as either moderate or severe (International Diabetes Federation 2011). Studies consistently found that children under 5 years are most vulnerable to hypoglycaemia. It is due to increased insulin sensitivity, irregular and unpredictable eating pattern and difficulty in communicating symptoms. Very young child and toddlers have exaggerated epinephrine response but poor glucagon response. On the other hand, pre-pubertal and pubertal groups have no difference in poor epinephrine response (Amiel et al. 1987; Ross 2005).

Nocturnal hypoglycaemia is common in children mostly due to blunted epinephrine response. Those children who had frequent episodes of hypoglycaemia have some learning ability compromised. In the long term there might be some effects on overall performances. These could be inferior academic achievements and poor performance on measures of intelligence. It could also impact executive function, speed of information processing, cognitive flexibility, psychomotor efficiency and visual perception (Brands et al. 2005). Having diabetes in first 5 years of life bears more clinically significant cognitive deficits (Ryan 2006; Lin et al. 2010). Hypoglycaemic hemiplegia, a rare complication that occurs after recovery from severe episode. It lasts not more than 24 h with any residual abnormality or any permanent disability (Pocecco and Ronfani 2007). Addressing the risk factors and structured education to child & parents are important steps of prevention. Dietary modifications, titrating insulin doses and using CGMS are the additional measures to reduce the episodes of severe hypoglycaemia.

5.4   Hypoglycaemia in Elderly

Due to improvement in early diagnosis, prevention and treatment the elderly population has grown significantly. The number of older people with T2D on insulin also increased significantly due to long duration of T2D. Insistence for good glycaemic control and other comorbidities predispose this elderly group to hypoglycaemia. Apart from usual symptoms, in elderly, neurological symptoms are more common. Blurring of vision, diplopia and incoordination are commonly seen (McAulay et al. 2001). These symptoms are sometime misinterpreted as manifestations of cerebrovascular disease. Most elderly patients and their care givers lack knowledge of hypoglycaemia and its home remedy. Structured education can help to achieve good glycaemic control and reduce the frequency of hypoglycaemia. In studies it was found that there are 50% higher rates of severe hypoglycaemia (Sue Kirkman et al. 2012).

In the INITIATEplus study 4875 insulin-naïve T2D patients were commenced on twice daily biphasic insulin and randomised to get varying levels of telephone counselling. Counselling was

Table 3 Hypoglycaemia and driving

Precautions for diabetes patient who are on insulin	Always have to carry Glucometer and strips even if using RT-CGM (real time continuous glucose monitoring system) or FGM (flash glucose monitoring system). Should keep fast-acting carbohydrate such as Glucotabs or hard candy or sugary juice nearby in vehicle. Personal identification card mentioning that one have diabetes. In long drive, should eat regular meals, rest and avoid alcohol.
Glucose monitoring	Should check glucose <2 h before start and every 2 h after start. If finger prick glucose is <5 mmol/L (< 90 mg/dL) then journey should not start. Should eat a snack and wait for 45 min till glucose reaches $\geq$5 mmol/L ($\geq$90 mg/dL) and then start.
Monitoring device	For **Group 1** (car and motorbike) finger prick, FGM and RT-CGM are allowed. For **Group 2** (Lorry and bus) only finger prick method is allowed.
Hypoglycaemia developing while driving	If symptoms of hypoglycaemia is perceiving or glucose level is <5 mmol/L (90 mg/dL) then have to stop driving safely and park. Engine should be switched off, key should be removed from ignition and have to move from driver's seat. Then have to take a snack / Glucotabs / sugary juice and waiting for 45 min till glucose level returns >5 mmol/L (90 mg/dL). Then only can resume driving. If group 1 is using FGM, RT-CGM and showing low sugar then finger prick should be done to make sure.
Notifying DVLA	Group 1: Have to notify if >1 episode of severe hypoglycaemia in last one year in awake or in sleep or having IHA.
	Group 2: driving must stop if having a single episode of severe hypoglycaemia in awake or in sleep or having IHA.
	Both groups: Have to notify if suffer severe hypoglycaemia during driving, Oneself or medical team think at high risk of hypoglycaemia, comorbidities which might be risk factor or might affect ability to drive.

done from a registered dietitian as no counselling or one sessions or three sessions. Similar improvements in HbA1c were found among all groups but with intensive counselling arm had less hypoglycaemia. Intensive vs no counselling were 4 vs 9 episodes of severe hypoglycaemia/ 100 patient years respectively (Oyer et al. 2011). The increasing prevalence of cognitive impairment in elderly affects the identification of hypoglycaemia. A study with elderly patients aged 75 and above, >20% with diabetes had cognitive impairment. It was evidenced by scoring $\leq$26 in Mini mental State Examination (MMSE). The participants were evaluated on the use of carbohydrate for hypoglycaemia. A 30% made mistakes compared to only 7% with normal cognitive function (Hewitt et al. 2010). In ADVANCE study, the elderly participants with MMSE score < 24 showed severe mental impairment. They had a hazard ratio 2.1 of developing severe hypoglycaemia compared those with MMSE score of 28–30 (de Galan et al. 2009).

5.5 Hypoglycaemia and Driving

Driving is an essential part of daily life which needs complex psychomotor skills. A good visual function, prompt information processing, vigilance, rapid situational judgment and critical thinking are needed to drive safely. Hypoglycaemia rapidly impairs cognitive functions that would interfere driving skills. Most countries have adopted guidelines for diabetes patients for safety of driver and others. Table 3 is showing key points of hypoglycaemia and driving by UK DVLA (Driving and Vehicle Licencing Authority 2019).

6 Hypoglycaemia and Cariovascular Morbidity and Mortality

Potential mechanism by which hypoglycaemia can precipitate cardiovascular events are as follows: (a) inflammation, (b) endothelial

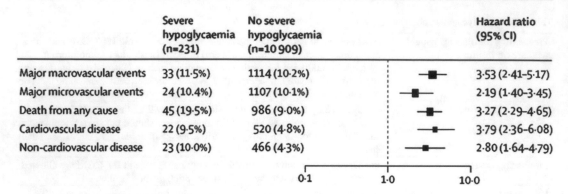

	Severe hypoglycaemia (n=231)	No severe hypoglycaemia (n=10 909)	Hazard ratio (95% CI)
Major macrovascular events	33 (11·5%)	1114 (10·2%)	3·53 (2·41–5·17)
Major microvascular events	24 (10.4%)	1107 (10·1%)	2·19 (1·40–3·45)
Death from any cause	45 (19·5%)	986 (9·0%)	3·27 (2·29–4·65)
Cardiovascular disease	22 (9·5%)	520 (4·8%)	3·79 (2·36–6·08)
Non-cardiovascular disease	23 (10·0%)	466 (4·3%)	2·80 (1·64–4·79)

Fig. 5 Association of severe hypoglycaemia with the risk of adverse clinical outcome or death (Amiel et al. 2019). (The slide is reproduced here with permission)

dysfunction, (c) reduced fibrinolysis, (d) cardiac ischaemia, (e) cardiac autonomic dysfunction and (f) pro-arrhythmic effects (Frier et al. 2011; Hanefeld et al. 2016). Severe hypoglycaemia has consistently shown in studies to be the strongest predictor of morbidity and mortality due to cardiovascular events in T2D (Mellbin et al. 2013; Bedenis et al. 2014; Zoungas et al. 2010).

In T1D, DCCT and EURODIAB studies have not shown increased risk of mortality or fatal cardiovascular disease related to hypoglycaemia (Gruden et al. 2012; The Diabetes Control and Complications Trial Research Group 1993). But a study showed that hazard ratio for all-cause mortality was of 1.98 in T1D. It was for those who experienced at least one episode of severe hypoglycaemia (Khunti et al. 2014b). The ADVANCE study cohorts were analysed for severe hypoglycaemia and adverse events including death. Those patients with episodes of severe hypoglycaemia had hazard ratio for cardiovascular disease were 3.79 (Amiel et al. 2019). It has been debated that confounding factors like pre-existing arrhythmia, renal or liver diseases, malignancy, weight loss and cognitive impairment might have played a role in between hypoglycaemia and mortality (Amiel et al. 2019) (Fig. 5).

Patients with a history of severe hypoglycaemia are at increased risk of fatal cardiovascular events in subsequent months. The landmark studies ACCORD, VADT and ADVANCE randomised 24,000 patients with cardiovascular risks to tight glycaemic control. In ACCORD trial highest numbers of death were found in intensive arm where a threefold increased number of severe hypoglycaemia occurred. One-third of all deaths were attributed to cardiovascular disease (Bonds et al. 2010). In VADT, a recent severe episode of hypoglycaemia was a strong predictor of death at 90 days (Duckworth et al. 2009). A similar pattern also found in ADVANCE trial (Zoungas et al. 2010).

7 Causes of Hypoglycaemia

Hypoglycaemia in T1D and advanced T2D on insulin has many causes which are mostly iatrogenic and other associated factors (Table 4).

8 Risk Factors

There are many established risk factors for hypoglycaemia in diabetes. So healthcare professionals should search for wide ranging factors to find the risk factors.

Table 5, is showing *common risk factors of hypoglycaemia in diabetes.*

9 Clinical Presentation

A person with diabetes on insulin or insulin secretagogues like sulphonylurea or glinides, who behaves oddly, a suspicion of hypoglycaemia should come first. The symptoms

Table 4 Causes of hypoglycaemia in diabetes and in person without diabetes (Hulkower et al. 2014; Cryer 2009a)

Causes of hypoglycaemia in person with diabetes

1. Improper insulin administration: wrong dose, wrong timing, wrong type, and wrong place: pushing over lipohypertrophied skin
2. Altered insulin carbohydrate ratio: unplanned exercise, breastfeeding, gastroparesis, malabsorption
3. Inadequate carbohydrate intake: missed meal or snack but insulin taking on time
4. Increased carbohydrate burn out: prolonged exercise without prior snack or insulin titration
5. Endogenous glycogen reservoir depletion: excessive alcohol without food
6. Increased insulin sensitivity: night time, exercise, weight loss, honeymoon phase in T1D
7. Decreased insulin clearance: renal failure, CCF, hepatic failure, major amputation
8. Old age: shorter life expectancy, long duration of DM, recent hospitalisations
9. Hypoglycaemia defence impairment: defective counterregulation, IAH
10. Diabetes in pregnancy: 1st trimester, lactation

Causes of hypoglycaemia in person without diabetes

1. Drugs: Cibenzoline, Gatifloxacin, Pentamidine, Quinine, Indomethacin...etc.
2. Critical illness: renal failure, hepatic failure, cardiac failure, sepsis, inanition
3. Hormone deficiency: cortisol, growth hormone, glucagon, epinephrine, hypopituitarism
4. Non-islet cell tumour
5. Endogenous hyperinsulinism: insulinoma, nesidioblastosis, post-gastric bypass surgery
6. Insulin autoimmune hypoglycaemia: antibody to insulin, antibody to insulin receptor
7. Intentional/ accidental: surreptitious, malicious, fictitious
8. Infancy and childhood: preterm, infants of DM mother, maternal drugs-sulphonyluea, Rh incompatibility, Beckwith-Wiederman syndrome, exchange transfusions, enzyme defects-glycogen storage disease I, III, VI.

Table 5 Hypoglycaemia risk factors

1. **Intensive therapy**: In DCCT, there were severe hypoglycaemic episodes 0.61 vs 0.19/patient/year in intensive vs conventional group. A threefold increase in intensive arm. HbA1c that approach the upper end on non-diabetic range by intensive therapy showed a quadratic relationship with severe hypoglycaemia (Hypoglycemia in the Diabetes Control and Complications Trial 1997). In SDIS trial there were 1.1 vs 0.4/patient/year in intensive vs conventional group (Reichard and Pihl 1994).
2. **Comorbidities**: Severe hepatic dysfunction, impaired renal function and RRT, sepsis, terminal illness, cognitive dysfunction or dementia (Borzì et al. 2016).
3. **Previous episode**: There is a well-known adage that 'hypoglycaemia begets hypoglycaemia'. Patient who had a severe episode and inadequately treated is vulnerable to have more in coming weeks or months.
4. **Duration of diabetes**: In T1D after 5 years declines its glucagon responses and within further few years its epinephrine response. T2D after 10 years or more follows the same sequences (Bolli et al. 1983; UK Hypoglycaemia Study Group 2007).
5. **Impaired awareness**: When epinephrine responses are attenuated then autonomic symptoms are diminished and neuroglycopenic symptoms predominate (Lin et al. 2019; Geddes et al. 2008; Graveling and Frier 2010).
6. **Negative C-peptide**: Patients who are C-peptide negative are two to fourfold increased risk of severe hypoglycaemia than those who have detectable C-peptide (Hypoglycemia in the Diabetes Control and Complications Trial 1997; Pedersen-Bjergaard et al. 2001).
7. **Lifestyle issues**: Sudden unusual exercise, irregular lifestyle, alcoholism, early pregnancy, breast feeding diabetes mom, inadequate SMBG.
8. **ACE activity**: Though contradicting results were found but in a study higher ACE concentrations are associated with severe hypoglycaemia (Nordfeldt and Samuelsson 2003).
9. **Sleep**: In DCCT 43% episodes of severe hypoglycaemia happened between midnight and 8 AM and 55% occurred when patients were asleep (Hypoglycemia in the Diabetes Control and Complications Trial 1997).
10. **Social and psychological factors**: Low mood, poor emotional coping and low socio-economic status have been found to have link with severe hypoglycaemia (Gonder-Frederick et al. 1997).

Table 6 Evolution of signs and symptoms of hypoglycaemia

1. Flow of thought and perception: Mental function starts to change before patient thinks that s/he is hypoglycaemic. It is like distancing from the world around, changing perception, intensity of sounds may be altered. Time assessment slows, so risk of accidents for pedestrians and drivers. There are short attention span and easy distraction, concentration deteriorates. Interactions become slow and hesitant. As glucose level continues to fall, s/he becomes increasingly confused.
2. Emotional status: Behaves out of character suddenly, irritations and frustration worsened by attempts to help, fast rising rage out of proportion. Looks depressed or tearfulness. A change in personality could be early sign of hypoglycaemia.
3. Refusal of help: Patients might think that they are coping well so usually refuse of help
4. Hunger: Patient feels intense hunger and if possible eats voraciously but at times food may be rejected as well. Patient may feel dry mouth.
5. Hyperactivity and panic: It happens mostly in children and elderly. Catecholamine effects with cerebral irritation may produce panic, terror and a desire to escape. Carers may be perceived as pursuers.
6. Colour of skin changes: A pallor, flushing or blotchy rashes may occur.
7. Sweating: Some patient thinks sweating comes first in hypoglycaemia but it may be a late sign. It is used to awaken a hypoglycaemic patient in alarming tools.
8. Palpitations: Uncomfortable awareness of rapid heartbeat, moderate heart rate increase, there may be increase in systolic blood pressure.
9. Respiratory changes: There may be apnoea, hyperventilation, and shortness of breath. In extreme cases Cheyne-Stokes breathing in comatose patients.
10. Tingling: Paraesthesia occurs fleetingly, often around mouth and lips. It could occur in the median nerve distribution as well.
11. Fine tremor: A fine tremor of the hands but not always noticeable unless it is sought.
12. Slurring of speech: It could well sound like drunk.
13. Headache: A dull throbbing feeling on the temple. Sometime hangover after nocturnal hypoglycaemia.
14. Unsteadiness and incoordination: Patient feels unsteady and stumbles readily, may appear drunk, even bumps into people. A lack of coordination with sweating and tremulous hands may cause spillages and breakages.
15. Weakness: Patient shows limb weakness and sometime generalised weakness.
16. Seizures: There might be tonic-clonic seizures but not a fit unless patient is known epileptic.
17. Weariness, sleep and coma: A compulsion to fall asleep from intense exhaustion and rarely to coma.
18. No symptoms- a loss of warning: A reduction in warning is very common occurring in at least 25% of insulin treated cases and which can rise up to 50% with >25 year duration of diabetes. Lots of patients are unaware of nocturnal hypoglycaemia.

of hypoglycaemia were first recorded by two researchers when insulin was first introduced for the treatment of diabetes (Fletcher and Campbell 1922). After almost a century, now with more structured investigations we find those findings were so true. Hypoglycaemia has impacts on more than one part of our system. The most frequent 6 symptoms are: sweating, trembling, inability to concentrate, weakness, hunger and blurred vision. The response initiates in the cerebral cortex when blood sugar touches the lowest threshold level and then it activates autonomic nervous system. The **autonomic** or **neurogenic** or **sympathoadrenal** symptoms precedes neuroglycopenic symptoms, which are: anxiety or irritability, fine tremor, tachycardia, hunger, cold sweats, paresthesias and headache. The **neuroglycopenic** symptoms are: cognitive impairment, mood and behavioural changes, fatigue and weakness, lightheadedness and dizziness, blurred vision, diplopia, slurred speech, seizures and coma (Yale et al. 2018). As per 'Edinburgh Hypoglycaemia Scale' general feeling of malaise which includes headache and nausea were also included (Deary et al. 1993). Let us elaborate how the signs and symptoms of hypoglycaemia are developed (Table 6) (Hillson 2015).

Children shows marked variability in symptoms but most reported sign is pallor. Children shows behavioural changes along with autonomic, neuroglycopenic symptoms. A list of symptoms in children are shown in Table 7 (McCrimmon et al. 1995; Ross et al. 1998).

Table 7 Symptoms in children

Tearful	Headache	Irritable	Uncoordinated	Naughty
Weak	Aggressive	Trembling	Sleepiness	Nightmares
Sweating	Slurred speech	Blurred vision	Tummy pain	Feeling sick
Hungry	Yawning	Odd behaviour	Warmness	Restless
Daydreaming	Argumentative	Pounding heart	Confused	Tingling lips
Dizziness	Tired	Feeling awful	Disobedience	Difficulty hearing

Elderly patient shows predominantly neurological symptoms along with autonomic and neuroglycopenic symptoms.

10 Diagnosis and Work-Up

As the consequences of hypoglycaemia can be devastating if prolonged and its treatment is readily available, so the treatment must be started immediately as per the level of hypoglycaemia after an initial rapid finger-prick blood glucose reading and a brief assessment. When patient is settled down and recovering then a detailed history and work-up should be done to diagnose the cause and prevent future episodes. Work-up for severe hypoglycaemia in diabetes should include detailed history of medication especially on insulin administration. A comprehensive system review to exclude hepato-renal impairment, any source of infection and assessment of beta-cell mass in advanced T2D are crucial. All possible risk factors should be screened and reviewed as well.

11 Management

11.1 Management in Adult

As soon as the adult patient is diagnosed of having hypoglycaemia, treatment should start immediately. The management approach depends on patient status as follows:

1. Mild to moderate conscious symptomatic adult patient who can swallow
2. Conscious, confused, disoriented, non-cooperative or aggressive adult patient but able to swallow
3. Unconscious, with or without seizures and or aggressive patient
4. 'Nil by mouth' adult patient
5. Adult patient on enteral or parenteral feeding

The Table 8 adapted from JBDS-IP hypoglycaemia in hospital management for adults 4th edition (Walden et al. 2020).

11.2 Management in Children

Severe hypoglycaemia is a life threatening event for children and a frightening incident for family. So management should be prompt and effective. In the management of children, there are two categories as follows:

1. Mild to moderate hypoglycaemia
2. Severe hypoglycaemia

Table 9, showing the management hypoglycaemia in children (Ng et al. 2018).

Table 8 Hypoglycaemia management steps

Management of mild to moderate symptomatic hypoglycaemia in a conscious and able to swallow adult patient (CBG: <4 mmol/L)

SECTION-A: 15–15 rule:

Step-1: 15–20 g of simple carbohydrate: e.g. 4–5 glucose tablets, 1 bottle (60 mL) of Glucose drink, 150–200 mL of pure fruit juice e.g. orange, 3–4 heaped teaspoons of sugar dissolved in water. Chronic renal failure patient on low K diet should not take orange juice. Patient on acarbose should not take sugar dissolved in water. 20 g of simple carbohydrate will raise blood glucose to 2.5–3.6 mmol/L in 15 min.

Step-2: Repeat CBG after 10–15 min, if still <4 mmol/L then repeat step-1 (maximum 3 times).

Step-3: If CBG < 4 mmol/L after 30–40 min or 3 cycles then have to call clinician and consider:

  1 mg of glucagon IM stat. It may be less effective in patients on sulphonylurea or under the influence of alcohol.

  150–200 ml of 10% glucose IV over 15 min bolus

Step-4: Once CBG is above 4 mmol/L and patient is recovered then a long acting carbohydrate of patient's choice, if possible, should be given, e.g. 2 biscuits, one slice of bread/toast and 200-300 ml glass of milk but not alternative milk like soya, almond or coconut. Normal meal can be given if due. In patients given glucagon injection, larger long acting carbohydrate should be given to replenish glycogen stores.

Step-5: Insulin dose should not be omitted if due though a review is ideal.

Step-6: Patients on CSII may not need long acting carbohydrate but should take up to step-3 and adjust the pump setting as per advised hypoglycaemia protocol.

Step-7: If hypoglycaemia is due to a sulphonylurea then have to admit the patient and ensure glucose monitoring for 24–48 h. The usual scenario is an older thin person who may have lost weight recently, is either very careful with diet or whose appetite has gone down. This type of persons have usually low or very low HbA1c as of 5–6.5%. Glucagon has to avoid as it might stimulate residual insulin and worsen hypoglycaemia. After acute treatment, start a 10% IV glucose infusion at 100 ml/h initially and CBG hourly till 12 h. if there is recurrent severe hypoglycaemia despite IV glucose the somatostatin analogue octreotide 50 mcg s.c 12 hourly is needed. This inhibits insulin and glucagon secretion.

Management of conscious, confused, disoriented, non-copetative, or aggressive but able to swallow adults (CBG: <4 mmol/L)

SECTION-B:

Step-1: If patient is cooperative, section A have to follow

Step-2: If patient is not capable or cooperative but able to swallow then 2 tubes of Glucogel 40% be squeezed into mouth between teeth and gum or if this ineffective then glucagon 1 mg IM stat.

Step-3: Step-2 of section-A have to be repeated

Step-4: Step-3, 4, 5, 6 and 7 of section-A have to be repeated

Management of unconscious adult with/ without seizures and/ or aggressive (CBG: <3 mmol/L)

SECTION-C:

Step-1: Have to check ABCDE: Airway-give O_2 , Breathing, Circulation, Disability-GCS and CBG and Exposure-safeguarding and temperature

Step-2: If patient has insulin infusion in situ, have to stop it immediately.

Step-3: Immediate assistance from medical staff urged

Step-4: One of the following three options to be followed as per availability:

  1. If IV access available, give 75-100 ml of 20% glucose over 15 min, (e.g. 300-400 ml/h). A 100 ml preparation of 20% glucose is now available that will deliver the required amount after being run through a standard giving set. If an infusion pump is available use this, but if not readily available the infusion should not be delayed. Repeat capillary blood glucose measurement 10 min later. If it is still less than 4.0 mmol/L, repeat.

  2. If IV access available, give 150-200 ml of 10% glucose (over 15 min, e.g. 600-800 ml/h). If an infusion pump is available use this, but if not readily available the infusion should not be delayed. Care should be taken if larger volume bags are used to ensure that the whole infusion is not inadvertently administered. Repeat capillary blood glucose measurement 10 min later. If it is still less than 4.0 mmol/L, repeat.

  3. If no IV access is available then give 1 mg glucagon IM. Glucagon may be less effective in patients prescribed sulfonylurea therapy and may take up to 15 min to take effect. Glucagon mobilises glycogen from the liver and will be less effective in those who are chronically malnourished (including those who have had a prolonged period of starvation), chronic alcoholics or have severe liver disease. In this situation IV glucose is the preferred option.

Step-5: Step-4, 5, 6, 7 of section A have to be repeated

(continued)

Table 8 (continued)

Step-6: If the patient was on IV insulin, have to continue to check blood glucose every 15 min interval. It should be done till it reaches above 3.5 mmol/L. then have to re-start IV insulin after review of dose regimen to try and prevent hypoglycaemia recurrence. Consider concurrent IV 10% glucose infusion at 100 ml/h and/or stepping down the insulin increments on the variable scale if appropriate.

Management of adults with hypoglycaemia who are 'nil by mouth'

SECTION-D:

Step-1: If the patient has a VRIII (variable rate intravenous insulin infusion), then as per prescribed regimen titration have to be done and expert advice should be sought. Most VRIII should be restarted once blood glucose is above 4.0 mmol/L, although titration in most cases needed.

Step-2: Options 1 or 2 of step-4 of Section-C to be followed

Step-3: Once blood glucose is greater than 4.0 mmol/L and the patient has recovered, consider intravenous infusion of 10% glucose at a rate of 100 ml/h until patient is no longer 'Nil by Mouth' or has been reviewed by a doctor.

Step-4: Same as step-7 of Section-A.

Management of hypoglycaemia in adults requiring enteral / parenteral feeding

SECTION-E:

Risk factors for hypoglycaemia: displaced or blocked tube, changes in feed regimen, enteral feed discontinuation, and TPN or IV glucose discontinuation. Moreover, medicines are given at an inappropriate time to feeding, changes in medicines that cause hyperglycaemia like steroid therapy reduced or stopped, intolerance to feed and vomiting should be reviewed. If patient on enteral tube feed then oral diabetes treatment should be continued via tube but if on TPN then treatment should be administered orally or intravenously as applies.

Step-1: A 15–20 g quick acting carbohydrate of the patient's choice where possible should be given. Such as 2 tubes of 40% glucose gel- it is not for use with fine bore NGT, 1 bottle (60 ml) Glucojuice and 150–200 ml orange juice. These are to give 15–20 g carbohydrate, re-start feed. All treatments should be followed by a 40–50 ml water flush of the feeding tube to prevent tube blockage (Dandeles and Lodolce 2011).

Step-2: Step-2, 3 of Section-1 to be followed.

Step-3: Once blood glucose is above 4.0 mmol/L and the patient has recovered, feeding has to restart. If in bolus feeding, additional bolus feed is needed. Then 10% IV glucose at 100 ml/h have to start and volume should be determined by clinical circumstances.

Step-4: Step-5, 6 and 7 of section-A have to follow.

Table 9 Hypoglycaemia management in children

Treatment of mild to moderate hypoglycaemia in children (CBG <4 mmol/L)

Step-1: If child is co-operative and able to tolerate oral fluids

Give 10–20 g of fast acting oral carbohydrate: 3–4 glucose tablets

200 ml (~1/2 cup) sugary drink not diet, such as cola. Chocolate or milk will not work fast. 9 g of glucose is needed for a 30 kg and 15 g for 50 kg child.

Step-2: If child refuses to drink and uncooperative but is conscious:

Give Glucogel or Dextrogel, a fast acting sugar gel in an easy twist top tube contains 10 g glucose. Squirt tube contents in side of each teeth and gum evenly and massage gently from outside enabling glucose to be swallowed and absorbed quickly.

Act and decide: After 10–15 min recheck CBG:

1. If still <4 mmol/L and able to take fluids then repeat step-1 once

2. If still <4 mmol/L and refuses to take oral but conscious then repeat step-2 once

3. If better and blood glucose >4 mmol/L then follow step-3 below

4. If deteriorated after first run through above ways or not responded after second dose of above then follow step-4 below

Step-3: If feeling better and blood glucose >4 mmol/L

Retest in 20–30 min later to confirm target glucose >4 mmol/L is maintained

Patient on insulin pump needs to adjust basal insulin rates following correction of hypoglycaemia

When hypoglycaemia occurs before meal time then correct first and CBG should be >4 mmol/L and then insulin should be given but should not omit.

Treatment of severe hypoglycaemia in children (CBG <3 mmol/L)

(continued)

Table 9 (continued)

At this stage medical assistance has to be involved
Place in recovery position and check ABCDE
Do not attempt to give any oral fluid or Glucogel
If IV access is present then follow step-5 instead of step-4.

Step-4: Give glucagon injection by IM

Before giving check whether already given at home or not
Give IM or SC in thigh: if age under 12 give 0.5 mg IM, if age 12 or over give 1 mg IM. BNF says:
2–17 years and body weight < 25 kg 0.5 mg IM
2–17 years and body weight > 25 kg 1 mg IM. Glucagon works usually in 5 min in children. After child regains consciousness, child should be placed in recovery position as glucagon causes vomiting as side effect.

Step-5: Give IV 10% glucose: If recovery is not adequate after a dose of glucagon or IV access is readily available and CBG <4 mmol/L, then a 5 mL/kg 10% dextrose IV bolus stat.

Blood glucose monitoring in severe hypoglycaemia:

CBG after 5 min, 15 min and then 30 min should be checked until CBG is stable
Continue to monitor vitals: O_2 saturation, pulse, blood pressure, temperature
Presence or absence of ketones should be recorded
Documentation of management
Diabetes team should be informed during day and if concerns during night
Should not omit scheduled insulin unless instructed to do so.

Step-6: If BG >4 mmol/L and child is able to tolerate oral fluids

Offer clear fluids and if tolerating give complex carbohydrate

Try to identify the cause of hypoglycaemia and discuss with family.

Step-7: If child not improving:

If child has protracted vomiting and unable to tolerate oral fluids then hospital admission and IV glucose infusion must be considered. It is important if child has returned to emergency with further hypoglycaemia during same intercurrent illness.

If child remains unconscious after correction of BG, cerebral oedema, head injury, adrenal insufficiency or drug overdose should be considered.

After successful management, the following key points are to be remembered:

Risk factors or aetiology identification
Taking measures to avoid hypoglycaemia in future with discussing with diabetes team.
If causes are not found in case of severe and recurrent hypoglycaemia then a review with diabetes team is needed.
Next insulin dose should not be omitted or should not start VRIII to stabilise blood glucose. It is safe to omit mealtime bolus dose if patient declines meal and has taken usual basal insulin.

Table 10 Hypoglycaemia prevention strategy

1. Medications should be chosen that have lower risk of hypoglycaemia-if having medicines that do not make hypoglycaemia is cost prohibitive then short acting sulphonylurea should be chosen.
2. Recognising BG alert values which is as per ADA is 3.9 mmol/L should make patient aware and dose adjustment advised.
3. Renal function should be checked annually if normal and more frequently if impaired. Dose adjustment or removal of a well-known harmful drug should be the important step
4. Initiate medication at low dose
5. Warn and educate patient about hypoglycaemic symptoms
6. Patients should be encouraged to report any episodes to parents, caregivers and healthcare professionals.
7. Encourage patients to do frequent SMBG and eat regularly
8. Inj. Glucagon should be prescribed for those who are vulnerable to severe hypoglycaemia
9. Encouraging patients to enrol in structured education like DAFNE in UK.
10. Individualised treatment targets and every visit discussion about hypoglycaemia has great reductions in episodes.
11. SMBG: In case of impaired awareness using real-time CGM with or without sensor augmented CSII, bionic pancreas and other models can be used if possible.

11.3 Management: Recurrent Hypoglycaemia

It is important to find out the reasons behind recurrent hypoglycaemia without compromising the good glycaemic control. The following steps should be taken:

- Any significant comorbidities such as hepatic or renal impairment should be ruled out.
- Malabsorptive problem such as coeliac disease or eating disorder should look for
- Endocrine disorders such as Addison's disease, hypopituitarism or hypothyroidism should be screened
- Insulin administration ways, meal pattern, exercise type, alcohol intake etc. should be carefully explored.
- Structured patient and family education is important to fill the information gap. Düsseldorf education and training and DAFNE are the two examples of structured education (DAFNE Study Group 2002; Sämann et al. 2005).
- Bedtime CBG testing, bedtime snacking and if possible use of CGMS are helpful to reduce recurrent hypoglycaemia.

12   Prevention

To prevent severe hypoglycaemia healthcare professionals should follow three main principles:

1. Individualise treatment goals with appropriate anti-diabetes medicines
2. Incorporate structured education and training for people with diabetes
3. Recognition and being alert for potential risk factors, causes and problems with ongoing diabetes treatment and hypoglycaemia.

To fulfil above principles, in Table 10, the following key points are to be followed (Heller et al. 2015).

13   Conclusion

Hypoglycaemia is a common and fearsome complication of diabetes treatment with significant morbidity and mortality. Fear of hypoglycaemia is a major barrier in achieving glycaemic control. The risk of hypoglycaemia is high in T1D and insulin treated advanced T2D. It is contributed by impaired glucose counterregulatory responses and risk of severe episode increases with impaired hypoglycaemia awareness. Mismatch between insulin availability and requirement can result from physiological conditions like unplanned exercise or diseases like renal and hepatic impairment. Patients should be educated about recognizing symptoms of hypoglycaemia and its management. Milder episodes can be self-treated with fast acting carbohydrate followed by slow acting carbohydrates. Severe episodes require external help and may need institutional management. Prevention of hypoglycaemia should be a part of every diabetes management plan. This is achieved by patient education, selecting appropriate treatment regime and frequent monitoring. Newer technology in the form of CGM and closed loop CSII can help in reducing episodes of hypoglycaemia in selected patients.

References

Abi-Saab WM et al (2002) Striking differences in glucose and lactate levels between brain extracellular fluid and plasma in conscious human subjects: effects of hyperglycemia and hypoglycemia. J Cereb Blood Flow Metab 22(3):271–279

Akram K et al (2009) Prospective and retrospective recording of severe hypoglycaemia, and assessment of hypoglycaemia awareness in insulin-treated type 2 diabetes. Diabetic Med 26(12):1306–1308

Allen C et al (2001) Risk factors for frequent and severe hypoglycemia in type 1 diabetes. Diabetes Care 24 (11):1878–1881

Amiel SA et al (1987) Exaggerated epinephrine responses to hypoglycemia in normal and insulin-dependent diabetic children. J Pediatr 110(6):832–837

Amiel SA et al (2019) Hypoglycaemia, cardiovascular disease, and mortality in diabetes: epidemiology, pathogenesis, and management. Lancet Diabetes Endocrinol 7(5):385–396

Balijepalli C, Druyts E, Siliman G, Joffres M, Thorlund K, Mills EJ (2017) Hypoglycemia: a review of definitions used in clinical trials evaluating antihyperglycemic drugs for diabetes. Clin Epidemiol 9:291–296

Bedenis R et al (2014) Association between severe hypoglycemia, adverse macrovascular events, and inflammation in the edinburgh type 2 diabetes study. Diabetes Care 37(12):3301–3308

Boden G (2004) Gluconeogenesis and glycogenolysis in health and diabetes. J Investig Med 52(6):375–378

Boden G et al (1983) Counterregulatory hormone release and glucose recovery after hypoglycemia in non-insulin-dependent diabetic patients. Diabetes 32 (11):1055–1059

Boli GB et al (1984) Defective glucose counterregulation after subcutaneous insulin in noninsulin-dependent diabetes mellitus. Paradoxical suppression of glucose utilization and lack of compensatory increase in glucose production, roles of insulin resistance, abnormal neuroendocrine responses, and islet paracrine interactions. J Clin Invest 73(6):1532–1541

Bolli G et al (1983) Abnormal glucose counterregulation in insulin-dependent diabetes mellitus. Interaction of anti-insulin antibodies and impaired glucagon and epinephrine secretion. Diabetes 32(2):134–141

Bonds DE et al (2010) The association between symptomatic, severe hypoglycaemia and mortality in type 2 diabetes: retrospective epidemiological analysis of the ACCORD study. BMJ 340(jan08):b4909–b4909

Borzì V et al (2016) Risk factors for hypoglycemia in patients with type 2 diabetes, hospitalized in internal medicine wards: findings from the FADOI-DIAMOND study. Diabetes Res Clin Pract 115:24–30

Brands AMA et al (2005) The effects of type 1 diabetes on cognitive performance: a meta-analysis. Diabetes Care 28(3):726–735

Briscoe VJ (2006) Hypoglycemia in Type 1 and Type 2 diabetes: physiology, pathophysiology, and management. Clin Diabetes 24(3):115–121

Bussau VA et al (2006) The 10-s maximal sprint: a novel approach to counter an exercise-mediated fall in glycemia in individuals with type 1 diabetes. Diabetes Care 29(3):601–606

Cahill GF (1971) Physiology of insulin in man: the Banting memorial lecture 1971. Diabetes 20 (12):785–799. https://doi.org/10.2337/diab.20.12.785

Campbell PJ et al (1985) Pathogenesis of the dawn phenomenon in patients with insulin-dependent diabetes mellitus. N Engl J Med 312(23):1473–1479

Chico A et al (2003) The continuous glucose monitoring system is useful for detecting unrecognized hypoglycemias in patients with type 1 and type 2 diabetes but is not better than frequent capillary glucose measurements for improving metabolic control. Diabetes Care (26, 4):1153–1157

Clarke W et al (2009) Assessment and management of hypoglycemia in children and adolescents with diabetes. Pediatr Diabetes 10:134–145

Cryer PE (1993) Glucose counterregulation: prevention and correction of hypoglycemia in humans. Am J Physiol Endocrinol Metab 264(2):E149–E155

Cryer PE (2016) Hypoglycemia in diabetes: pathophysiology, prevalence, and prevention, 3rd edn. American Diabetic Association

Cryer PE et al (2009a) Evaluation and management of adult hypoglycemic disorders: an endocrine society clinical practice guideline. J Clin Endocrinol Metab 94(3):709–728

Cryer PE, Axelrod L, Grossman AB et al (2009b) Evaluation and management of adult hypoglycemic disorders: an Endocrine Society clinical practice guideline. J Clin Endocrinol Metab 94:709–728

DAFNE Study Group (2002) Training in flexible, intensive insulin management to enable dietary freedom in people with type 1 diabetes: dose adjustment for normal eating (DAFNE) randomised controlled trial. BMJ 325(7367):746–746

Dagogo-Jack SE et al (1993) Hypoglycemia-associated autonomic failure in insulin-dependent diabetes mellitus. Recent antecedent hypoglycemia reduces autonomic responses to, symptoms of, and defense against subsequent hypoglycemia. J Clin Invest 91 (3):819–828

Dandeles LM, Lodolce AE (2011) Efficacy of agents to prevent and treat enteral feeding tube clogs. Ann Pharmacother 45(5):676–680

DCCT Research Group (1993) The effect of intensive treatment of diabetes on the development and progression of long-term complications in insulin-dependent diabetes mellitus. N Engl J Med 329(14):977–986

de Galan BE, Zoungas S, Chalmers J et al (2009) Cognitive function and risks of cardiovascular disease and hypoglycaemia in patients with type 2 diabetes: the action in diabetes and vascular disease: preterax and diamicron modified release controlled evaluation (ADVANCE) trial. Diabetologia 52:2328–2336

Deary IJ et al (1993) Partitioning the symptoms of hypoglycaemia using multi-sample confirmatory factor analysis. Diabetologia 36(8):771–777

Desimone ME, Weinstock RS (2017) Non-diabetic hypoglycemia. www.endotext.org. Accessed 27 Mar 2020

Diamond MP et al (1992) Impairment of counterregulatory hormone responses to hypoglycemia in pregnant women with insulin-dependent diabetes mellitus. Am J Obstet Gynecol 166(1):70–77

Driving and Vehicle Licencing Authority (2019) Drivers who have any form of diabetes treated with any insulin preparation must inform DVLA, pp 1–5

Duckworth W et al (2009) Glucose control and vascular complications in veterans with type 2 diabetes. J Vasc Surg 49(4):1084

Elliott L et al (2016) Hypoglycemia event rates: a comparison between real-world data and randomized controlled trial populations in insulin-treated diabetes. Diabetes Therapy 7(1):45–60

Evers IM et al (2002) Risk indicators predictive for severe hypoglycemia during the first trimester of type 1 diabetic pregnancy. Diabetes Care 25(3):554–559

Fisher SJ et al (2005) Insulin signaling in the central nervous system is critical for the normal sympathoadrenal response to hypoglycemia. Diabetes 54(5):1447–1451

Fletcher AA, Campbell WR (1922) The blood sugar following insulin administration and the symptom complex-hypoglycemia. Journal of Metabolic Research 2:637–649

Frier BM (2014) Hypoglycaemia in diabetes mellitus: epidemiology and clinical implications. Nat Rev Endocrinol 10(12):711–722

Frier BM et al (2011) Hypoglycemia and cardiovascular risks. Diabetes Care 34(Suppl_2):S132–S137

Frier BM et al (2015) Hypoglycaemia in adults with insulin-treated diabetes in the UK: self-reported frequency and effects. Diabetic Med 33(8):1125–1132

Fritsche A et al (2001) Avoidance of hypoglycemia restores hypoglycemia awareness by increasing β-adrenergic sensitivity in type 1 diabetes. Ann Intern Med 134(9_Part_1):729

Fujii N et al (2000) Exercise induces isoform-specific increase in 5′AMP-activated protein kinase activity in human skeletal muscle. Biochem Biophys Res Commun 273(3):1150–1155

Fukuda M et al (1988) Correlation between minimal secretory capacity of pancreatic beta-cells and stability of diabetic control. Diabetes 37(1):81–88

Geddes J et al (2008) Prevalence of impaired awareness of hypoglycaemia in adults with type 1 diabetes. Diabet Med 25(4):501–504

Gold AE et al (1994) Frequency of severe hypoglycemia in patients with type I diabetes with impaired awareness of hypoglycemia. Diabetes Care 17(7):697–703

Gonder-Frederick L et al (1997) The psychosocial impact of severe hypoglycemic episodes on spouses of patients with IDDM. Diabetes Care 20(10):1543–1546

Graveling AJ, Frier BM (2010) Impaired awareness of hypoglycaemia: a review. Diabetes Metab 36:S64–S74

Gruden G et al (2012) Severe hypoglycemia and cardiovascular disease incidence in type 1 diabetes: the EURODIAB prospective complications study. Diabetes Care 35(7):1598–1604

Hamdy O et al (2019) Hypoglycemia. http://emedicine.medscape.com/article/122122-overview#a5. Accessed 26 Mar 2020

Hanefeld M et al (2016) Hypoglycemia and cardiovascular risk: is there a major link? Diabetes Care 39(Suppl 2):S205–S209

Havlin CE, Cryer PE (1987) Nocturnal hypoglycemia does not commonly result in major morning hyperglycemia in patients with diabetes mellitus. Diabetes Care 10(2):141–147

Heller SR (2016) Glucose concentrations of less than 3.0 Mmol/L (54 Mg/DL) should be reported in clinical trials: a joint position statement of the American Diabetes Association and the European Association for the Study of Diabetes: Table 1. Diabetes Care 40(1):155–157

Heller SR et al (1987) Counterregulation in type 2 (non-insulin-dependent) diabetes mellitus. Normal endocrine and glycaemic responses, up to ten years after diagnosis. Diabetologia 30(12):924–929

Heller S et al (2015) Hypoglycaemia, a global cause for concern. Diabetes Res Clin Pract 110(2):229–232

Heller SR et al (2020a) Validation of definitions recently adopted by the American Diabetes Association/European Association for the study of diabetes. Diabetes Care 43(2):398–404

Heller SR et al (2020b) Redefining hypoglycemia in clinical trials: validation of definitions recently adopted by the American Diabetes Association/European Association for the study of diabetes. Diabetes Care 43(2):398–404

Hewitt J et al (2010) Self management and patient understanding of diabetes in the older person. Diabetic Med 28(1):117–122

Hillson R (2015) Diabetes care-a practical manual, 2nd edn. Oxford University Press, Oxford, pp 84–86

Hulkower RD et al (2014) Understanding hypoglycemia in hospitalized patients. Diabetes Manage 4(2):165–176

Hypoglycemia in the Diabetes Control and Complications Trial (1997) The Diabetes Control and Complications Trial Research Group. Diabetes 46(2):271–286

Ilic S, Jovanovic L, Wollitzer AO (2000) Is the paradoxical first trimester drop in insulin requirement due to an increase in C-peptide concentration in pregnant type I diabetic women? Diabetologia 43(10):1329–1330

International Diabetes Federation (2011) Chapter 11: Global IDF/ISPAD guideline for diabetes in childhood and adolescence. International Diabetes Federation, Brussels

Ishihara H et al (2003) Islet beta-cell secretion determines glucagon release from neighbouring alpha-cells. Nat Cell Biol 5(4):330–335

Jabbar A et al (2017) CREED study: hypoglycaemia during Ramadan in individuals with type 2 diabetes mellitus from three continents. Diabetes Res Clin Pract 132:19–26

Jones TW (2017) Defining relevant hypoglycemia measures in children and adolescents with type 1 diabetes. Pediatr Diabetes 19(3):354–355

Khunti K et al (2014a) PO118 self-reported hypoglycemia: a GLOBAL study of 24 countries with 27,585 insulin-treated patients with diabetes: the hat study. Diabetes Res Clin Pract 106:S105–S106

Khunti K et al (2014b) Hypoglycemia and risk of cardiovascular disease and all-cause mortality in insulin-treated people with type 1 and type 2 diabetes: a cohort study. Diabetes Care 38(2):316–322

Kreisman SH (2001) Norepinephrine infusion during moderate-intensity exercise increases glucose production and uptake. J Clin Endocrinol Metab 86(5):2118–2124

Levin BE et al (2004) Neuronal glucosensing: what do we know after 50 years? Diabetes 53(10):2521–2528

Lin A et al (2010) Neuropsychological profiles of young people with type 1 diabetes 12 yr after disease onset. Pediatr Diabetes 11(4):235–243

Lin YK et al (2019) Impaired awareness of hypoglycemia continues to be a risk factor for severe hypoglycemia despite the use of continuous glucose monitoring system in type 1 diabetes. Endocrine Pract 25(6):517–525

Marliss EB et al (2000) Gender differences in glucoregulatory responses to intense exercise. J Appl Physiol 88(2):457–466

Mathiesen ER et al (2007) Maternal glycemic control and hypoglycemia in type 1 diabetic pregnancy: a randomized trial of insulin aspart versus human insulin in 322 pregnant women. Diabetes Care 30(4):771–776

McAulay V et al (2001) Symptoms of Hypoglycaemia in people with diabetes. Diabet Med 18(9):690–705

McCrimmon R (2008) The mechanisms that underlie glucose sensing during hypoglycaemia in diabetes. Diabet Med 25(5):513–522

McCrimmon RJ, Sherwin RS (2010) Hypoglycemia in type 1 diabetes. Diabetes 59(10):2333–2339

McCrimmon RJ et al (1995) Symptoms of hypoglycemia in children with IDDM. Diabetes Care 18(6):858–861

Mellbin LG et al (2013) Does hypoglycaemia increase the risk of cardiovascular events? A report from the ORIGIN trial. Eur Heart J 34(40):3137–3144

Mundinger TO et al (2003) Impaired glucagon response to sympathetic nerve stimulation in the BB diabetic rat: effect of early sympathetic islet neuropathy. Am J Physiol Endocrinol Metabol 285(5):E1047–E1054

Ng S et al (2018) ACDC/ BSPED, Management of hypoglycaemia in children and young people with type-1 diabetes. 2018, 4 May, https://www.bsped.org.uk/clinical-resources/guidelines

Nielsen LR et al (2009) Pregnancy-induced rise in serum C-peptide concentrations in women with type 1 diabetes. Diabetes Care 32(6):1052–1057

Nordfeldt S, Samuelsson U (2003) Serum ACE predicts severe hypoglycemia in children and adolescents with type 1 diabetes. Diabetes Care 26(2):274–278

Östenson CG et al (2013) Self-reported non-severe hypoglycaemic events in Europe. Diabetic Med 31 (1):92–101. https://doi.org/10.1111/dme.12261

Oyer DS et al (2011) Efficacy and tolerability of self-titrated biphasic insulin aspart 70/30 in patients aged >65 years with type 2 diabetes: an exploratory post hoc subanalysis of the INITIATEplus trial. Clin Ther 33 (7):874–883

Pacaud D, Dewey D (2011) Neurocognitive outcome of children exposed to severe hypoglycemiain utero. Diabetes Manage 1(1):129–140

Pearson EM, McCrimmon RJ (2018) Diabetes mellitus. In: Ralston S, Strachan MWJ, Britton R, Penman ID, Hobson RP (eds) Davidson's principles and practice of medicine, 23rd edn. Elsevier, Edinburgh, pp 719–762

Pedersen-Bjergaard U et al (2001) Activity of angiotensin-converting enzyme and risk of severe hypoglycaemia in type 1 diabetes mellitus. Lancet 357 (9264):1248–1253

Petersen KF et al (2004) Regulation of net hepatic glycogenolysis and gluconeogenesis during exercise: impact of type 1 diabetes. J Clin Endocrinol Metab 89 (9):4656–4664

Pocecco M, Ronfani L (2007) Transient focal neurologic deficits associated with hypoglycaemia in children

with insulin-dependent diabetes mellitus. Acta Paediatr 87(5):542–544

Polonsky KS et al (1984) Glucose counterregulation in patients after pancreatectomy. Comparison with other clinical forms of diabetes. Diabetes 33(11):1112–1119

Raju B et al (2006) Nocturnal hypoglycemia in type 1 diabetes: an assessment of preventive bedtime treatments. J Clin Endocrinol Metab 91(6):2087–2092

Ratzki-Leewing A, Harris SB, Mequanint S et al (2018) Real-world crude incidence of hypoglycemia in adults with diabetes: results of the InHypo-DM study, Canada. BMJ Open Diabetes Res Care 6(1):e000503. Published 2018 Apr 24

Reichard P, Pihl M (1994) Mortality and treatment side-effects during long-term intensified conventional insulin treatment in the Stockholm diabetes intervention study. Diabetes 43(2):313–317

Reno CM et al (2013) Defective counterregulation and hypoglycemia unawareness in diabetes. Endocrinol Metab Clin N Am 42(1):15–38

Ringholm L et al (2012) Hypoglycaemia during pregnancy in women with type 1 diabetes. Diabetic Med 29 (5):558–566

Ritz P, Vaurs C, Barigou M, Hanaire H (2016) Hypoglycaemia after gastric bypass: mechanisms and treatment. Diabetes Obes Metab 18:217–223

Riviello C et al (2009) Breastfeeding and the basal insulin requirement in type 1 diabetic women. Endocr Pract 15 (3):187–193

Rosenn B et al (1995) Hypoglycemia: the price of intensive insulin therapy for pregnant women with insulin-dependent diabetes mellitus. Obstet Gynecol 85 (3):417–422

Ross LA (2005) Pubertal stage and hypoglycaemia counterregulation in type 1 diabetes. Arch Dis Child 90(2):190–194

Ross LA et al (1998) Hypoglycaemic symptoms reported by children with type 1 diabetes mellitus and by their parents. Diabet Med 15(10):836–843

Ryan CM (2006) Why is cognitive dysfunction associated with the development of diabetes early in life? The diathesis hypothesis. Pediatr Diabetes 7(5):289–297

Salti I et al (2004) A population-based study of diabetes and its characteristics during the fasting month of Ramadan in 13 countries: results of the epidemiology of diabetes and Ramadan 1422/2001 (EPIDIAR) study. Diabetes Care 27(10):2306–2311

Sämann A et al (2005) Glycaemic control and severe hypoglycaemia following training in flexible, intensive insulin therapy to enable dietary freedom in people with type 1 diabetes: a prospective implementation study. Diabetologia 48(10):1965–1970

Saremi A et al (2016) A link between hypoglycemia and progression of atherosclerosis in the veterans affairs diabetes trial (VADT). Diabetes Care 39(3):448–454

Schopman JE et al (2010) prevalence of impaired awareness of hypoglycaemia and frequency of hypoglycaemia in insulin-treated type 2 diabetes. Diabetes Res Clin Pract 87(1):64–68

Schwartz R, Teramo KA (1998) Pregnancy outcome, diabetes control and complications trial, and intensive glycemic control. Am J Obstet Gynecol 178 (2):416–417

Seaquist ER et al (2011) The impact of frequent and unrecognized hypoglycemia on mortality in the ACCORD study. Diabetes Care 35(2):409–414

Seaquist ER et al (2013) Hypoglycemia and diabetes: a report of a workgroup of the American Diabetes Association and The Endocrine Society. Diabetes Care 36 (5):1384–1395. https://doi.org/10.2337/dc12-2480

Secrest AM, Becker DJ, Kelsey SF, LaPorte RE, Orchard TJ (2011) Characterizing sudden death and dead-in-bed syndrome in type 1 diabetes: analysis from two childhood-onset type 1 diabetes registries. Diabet Med 28:293–300

Segel SA et al (2002) Hypoglycemia-associated autonomic failure in advanced type 2 diabetes. Diabetes 51(3):724–733

Service FJ (1999) Classification of hypoglycemic disorders. Endocrinol Metab Clin N Am 28:501–517

Service GJ, Thompson GB, Service FJ, Andrews JC, Collazo-Clavell ML, Lloyd RV (2005) Hyperinsulinemic hypoglycemia with nesidioblastosis after gastric-bypass surgery. N Engl J Med 353:249–254

Somogyi M (1959) Exacerbation of diabetes by excess insulin action. Am J Med 26(2):169–191

Stage E et al (2006) Long-term breast-feeding in women with type 1 diabetes. Diabetes Care 29(4):771–774

Stich V et al (2000) Adipose tissue lipolysis is increased during a repeated bout of aerobic exercise. J Appl Physiol 88(4):1277–1283

Sue Kirkman M et al (2012) Diabetes in older adults: a consensus report. J Am Geriatr Soc 60(12):2342–2356

The Diabetes Control and Complications Trial Research Group (1993) The effect of intensive treatment of diabetes on the development and progression of long-term complications in insulin-dependent diabetes mellitus. N Engl J Med 329:977–986

The NICE-SUGAR Study Investigators (2012) Hypoglycemia and risk of death in critically ill patients. N Engl J Med 367(12):1108–1118

Thorell A et al (1999) Exercise and insulin cause GLUT-4 translocation in human skeletal muscle. Am J Physiol Endocrinol Metab 277(4):E733–E741

Tu E et al (2010) Post-mortem pathologic and genetic studies in 'dead in bed syndrome' cases in type 1 diabetes mellitus. Hum Pathol 41(3):392–400

Turchin A et al (2009) Hypoglycemia and clinical outcomes in patients with diabetes hospitalized in the general ward: response to Ng et Al. Diabetes Care 32 (12):e152–e152

UK Hypoglycaemia Study Group (2007) Risk of hypoglycaemia in types 1 and 2 diabetes: effects of treatment modalities and their duration. Diabetologia 50:1140–1147

UKPDS Study Group (1998) Intensive blood-glucose control with sulphonylureas or insulin compared with conventional treatment and risk of complications in patients with type 2 diabetes (UKPDS 33). Lancet 352(9131):837–853

Walden E et al (2020) Hospital management of hypoglycaemia in adults with diabetes I ABCD (Diabetes Care) Ltd. Abcd.Care, Jan. 2020, abcd.care/resource/hospital-management-hypoglycaemia-adults-diabetes

Watts AG, Donovan CM (2010) Sweet talk in the brain: glucosensing, neural networks, and hypoglycemic Counterregulation. Front Neuroendocrinol 31 (1):32–43

Weber K et al (2007) High frequency of unrecognized hypoglycaemias in patients with type 2 diabetes is discovered by continuous glucose monitoring. Exp Clin Endocrinol Diabetes 115(08):491–494

Weston PJ (2012) The dead in bed syndrome revisited: a review of the evidence. Diabetes Management 2 (3):233–241

Whipple AO, Frantz VK (1935) Adenoma if islet cells with hyperinsulinism: a review Annals of Surgery, 101:1299–1335

Yale J-F et al (2018) Hypoglycemia, diabetes Canada Clinical Practice Guidelines Expert Committee. Can J Diabetes 42:S104–S108

Zoungas S et al (2010) Severe hypoglycemia and risks of vascular events and death. N Engl J Med 363 (15):1410–1418

Adv Exp Med Biol - Advances in Internal Medicine (2020) 4: 71–84
https://doi.org/10.1007/5584_2020_526
© Springer Nature Switzerland AG 2020
Published online: 24 April 2020

Hypoglycemia, Malnutrition and Body Composition

I. Khanimov, M. Shimonov, J. Wainstein, and Eyal Leibovitz

Abstract

Hypoglycemia is one of the most significant factors to affect prognosis, and is detrimental to patients regardless of diabetes mellitus (DM) status. The classical paradigms dictate that hypoglycemia is a result of overtreatment with glucose lowering agents (iatrogenic hypoglycemia), or, as among patients without DM, this condition is attributed to disease severity.

The original version of this chapter was revised: subtitle was removed. The correction to this chapter is available at https://doi.org/10.1007/5584_2020_559

I. Khanimov
Sackler Faculty of Medicine, Tel Aviv University, Tel Aviv, Israel

M. Shimonov
Sackler Faculty of Medicine, Tel Aviv University, Tel Aviv, Israel

Diabetes Unit, Edith Wolfson Medical Center, Holon, Israel

J. Wainstein
Sackler Faculty of Medicine, Tel Aviv University, Tel Aviv, Israel

Department of Surgery "A", Edith Wolfson Medical Center, Holon, Israel

E. Leibovitz (✉)
Department of Internal Medicine "A", Yoseftal Hospital, Eilat, Israel
e-mail: heartman@matav.net.il

New information shows that hypoglycemia occurs among patients that have a tendency for it. Incident hypoglycemia is very prevalent in the hospital setting, occurring in 1:6 patients with DM and in 1:17 patients without DM (Leibovitz E, Khanimov I, Wainstein J, Boaz M; Diabetes Metab Syndr Clin Res Rev. 13:222–226, 2019).

One of the major factors associated with incidence of hypoglycemia is the nutritional status on hospital admission and during the hospitalization. Assessment of nutritional status using questionnaires and biomarkers might be helpful in determining risk of hypoglycemia. Moreover, administration of oral nutritional supplements was shown to decrease this risk.

It is also well known that a high burden of comorbidities is associated with an increased risk of hypoglycemia. For example, kidney disease, whether acute or chronic, was shown to increase the risk for hypoglycemia, as well as some endocrine disorders.

In this review we elaborate on specific findings that are characteristic of patients at risk for developing hypoglycemia, as well as treatment aimed at preventing its occurrence.

Keywords

Albumin · Body composition · Cholesterol · Diabetes mellitus · Hypoglycemia · Malnutrition · Muscle glucose metabolism · NRS-2002 · Nutrition · ONS

1 Introduction

Hypoglycemia can be defined as serum glucose level equal or under 70 mg/dL ($\leq$ 3.9 mmol/L) (Seaquist et al. 2013a; Cryer 2009), and is usually accompanied by specific clinical presentation as sweating, trembling, warmness, weakness, and drowsiness (Hepburn et al. 1991). Among patients with diabetes mellitus, and in accordance with the International Hypoglycemia Study Group guidelines, a serum glucose under 54 mg/dL ($\leq$ 3.0 mmol/L), is sufficiently low to indicate serious, clinically important hypoglycemia (International Hypoglycaemia Study Group IHS 2017). For decades, it was assumed that serum glucose level is mostly regulated by two hormones- insulin and glucagon. Nowadays, it is well known that glucose homeostasis is regulated by various glucoregulatory hormones which effect multiple target tissues, such as muscle, brain, liver, and adipocyte (Wasserman 2009; Aronoff et al. 2004). In the hospital setting, hypoglycemia has been associated with increased length of hospital stay as well as increased mortality both in-hospital and at 1- follow-up (Leibovitz et al. 2019a).t

2 Classification of Hypoglycemia and Association with Prognosis

According to the report of the workgroup of the American Diabetes Association and The Endocrine Society, hypoglycemia can be classified as follows (Seaquist et al. 2013b):

1. Severe hypoglycemia: Severe hypoglycemia is an event requiring assistance of another person to actively administer carbohydrates, glucagon, or take other corrective actions. Plasma glucose concentrations may not be available during an event, but neurological recovery following the return of plasma glucose to normal is considered sufficient evidence that the event was induced by a low plasma glucose concentration.
2. Documented symptomatic hypoglycemia: Documented symptomatic hypoglycemia is an event during which typical symptoms of hypoglycemia are accompanied by a measured plasma glucose concentration $\leq$ 70 mg/dL ($\leq$ 3.9 mmol/L).
3. Asymptomatic hypoglycemia: Asymptomatic hypoglycemia is an event not accompanied by typical symptoms of hypoglycemia but with a measured plasma glucose concentration $\leq$ 70 mg/dL ($\leq$ 3.9 mmol/L).
4. Probable symptomatic hypoglycemia: Probable symptomatic hypoglycemia is an event during which symptoms typical of hypoglycemia are not accompanied by a plasma glucose determination but that was presumably caused by a plasma glucose concentration $\leq$ 70 mg/dL ($\leq$ 3.9 mmol/L).
5. Pseudo-hypoglycemia∗: Pseudo-hypoglycemia is an event during which the person with diabetes reports any of the typical symptoms of hypoglycemia with a measured plasma glucose concentration > 70 mg/dL (> 3.9 mmol/L) but approaching that level.

∗ According to the previous report of the American Diabetes Association workgroup on hypoglycemia from 2004, this state was defined as **"Relative hypoglycemia"** (Workgroup on Hypoglycemia, American Diabetes Association 2005).

In some articles, the term pseudo-hypoglycemia describes a condition wherein the glucose measurement in fingerstick is significantly lower compared to a measurement according to basic metabolic panel (BMP) (El Khoury et al. 2008; Lee and Abadir 2015). A possible mechanism of pseudo hypoglycemia may be a result of impaired blood flow in the digital microcirculation, leading to local increase in glucose consumption. This phenomenon was described among patients suffering from mixed connective tissue disease, Raynaud's phenomenon and acrocyanosis (El Khoury et al. 2008; Lee and Abadir 2015; Crevel et al. 2009; Rushakoff and Lewis 2001).

It is not clear why some patients experience the classical symptoms of hypoglycemia, while others remain a-symptomatic. This may be associated with the actual degree of hypoglycemia (Hepburn et al. 1991), or co-administration of drugs that inhibit the symptoms associated with hypoglycemia (Casiglia and Tikhonoff 2017), but not necessarily. It is also not clear whether patients with asymptomatic hypoglycemia have a different prognosis than patients with symptoms.

Documented hypoglycemia is associated with poor short and long term prognosis among patients admitted to internal medicine departments regardless of diabetes mellitus (DM) status (Turchin et al. 2009). Among patients with DM, hypoglycemia was repeatedly shown to be associated with a three-fold increase of length of hospital stay and decreased short and long term survival (Turchin et al. 2009; Zapatero et al. 2014; Brodovicz et al. 2013; Boucai et al. 2011). Mortality rates are also associated with the degree of hypoglycemia, and patients with lower glucose have the worst survival rates (Cryer 2012; Yun et al. 2019). Moreover, hypoglycemia was found to be associated with an increased risk for dementia among older adults with DM (Whitmer et al. 2009; Chaytor et al. 2019; Yaffe et al. 2013). Association between morbidity and hypoglycemia among patients without DM was documented as well (Fischer et al. 1986; Mannucci et al. 2006; Shilo et al. 1998). While the prevalence of hypoglycemia is higher among DM patients, the prognosis of patients with hypoglycemia is worse among patients without DM (Leibovitz et al. 2019a). This raises the possibility that the etiology of hypoglycemia, rather than hypoglycemia per-se, is responsible for the poor outcome observed.

3 Spontaneous Hypoglycemia

The most significant determinant of hypoglycemia occurrence in DM patients is glucose lowering medications. Tight glycemic control and increasing number of antidiabetic agents (Boucai et al. 2011; Krinsley and Grover 2007) were shown to cause hypoglycemia. Iatrogenic hypoglycemia is associated with administration of insulin and insulin-secretagogues (sulfonylureas and meglitinides) (Bonaventura et al. 2015). However, as described previously, hypoglycemia can also occur among hospitalized patients

without DM, as well as DM patients not receiving glucose lowering medications. This is termed spontaneous hypoglycemia. Glucose levels of spontaneous hypoglycemia can be very low as well. In non-diabetic patients, rate of spontaneous severe (below 55 mg/dL, 3 mmol/L) hypoglycemia outside the intensive care unit was 36 per 10,000 admissions (95% CI 24–64), and more than 90% of the patients were admitted as emergency cases (Nirantharakumar et al. 2012).

Risk factors for spontaneous hypoglycemia among hospitalized patients include older age, renal insufficiency, infection, shock and need of mechanical ventilation.

Hypoglycemia was more prevalent among patients with sepsis, alcohol dependence, pneumonia, liver disease and cancer. Spontaneous hypoglycemia may also occur among patients with DM, as was found among patients suffering from infection, with or without sepsis (Toda et al. 2014). Postprandial (reactive) hypoglycemia may also be the cause of spontaneous hypoglycemia among patients with DM (Nydick et al. 1964; Conn et al. 1956).

4 Hypoglycemia and Malnutrition

It has been recently shown that, one of the most significant factors to affect the rate of hypoglycemia is malnutrition. Malnutrition is as an imbalance between consumption and expenditure of either energy, protein or any other nutrient that damages body function (Stratton et al. 2003). Several suggestions for classification of malnutrition were suggested over the years, mostly in infants and children in developing countries. In 1955, Gomez at al. (1955) suggested a classification of malnutrition in infancy and childhood based on child's weight compared to that of a child in the 50th percentile of the same age. A later classification by Waterlow at al. (1972) was based on z- scores (SD) which took into account both weight and height of the child. The World Health Organization (WHO) classification is based on Waterlow's system with some modifications (Grover and Ee 2009).

Two major clinical syndromes of protein-energy malnutrition include kwashiorkor (almost normal weight for age, marked generalized edema) and marasmus (depletion of subcutaneous fat stores, muscle wasting, and absence of edema), though a mixed variant is also frequent (Grover and Ee 2009).

Micronutrient deficiencies could be ascertained by lab tests; however, there is no gold standard for diagnosis of protein-energy malnutrition. As a result, several nutrition screening and assessment tools were developed over the years (American Society of Parenteral and Enteral Nutrition 2002; Skipper et al. 2012; Kondrup et al. 2003). All validated tools consist of questions regarding patient's diet, global assessment and a series of body measurements. The most known and widely used tools include the Mini Nutrition Assessment (MNA) (Charlton et al. 2007; Vellas et al. 1999), the Mini Nutritional Assessment-Short Form (MNA-SF) (Kaiser et al. 2009), Malnutrition Universal Screening Tool (MUST) (Cansado et al. 2009; Boléo-Tomé et al. 2011), The Nutrition Risk Screen 2002 (NRS-2002) (Kondrup et al. 2003) and The Subjective Global Assessment (SGA) (Detsky et al. 1987).

All validated tools were tested on specific patient populations, and therefore, are considered most specific in those populations. For instance, the SGA was found to be reliable among patients on hemodialysis and liver-transplant candidates (Hasse et al. 1993; Steiber et al. 2007). The MUST was shown to be effective among patients with cancer and in the elderly, and The Nutrition risk screen 2002 (NRS-2002) was validated among hospitalized individuals (Kondrup et al. 2003). However, it was shown that all screening tools generally perform well (Young et al. 2013), but may vary when studied in specific populations (Ye et al. 2018).

Several pathogeneses were described to cause malnutrition (Jensen et al. 2009). For example, malnutrition can be induced by reduced food intake due to diminished appetite, or as a result of impaired nutrient absorption (Campbell 1999). Other causes include older age, socio-economic status and comorbidities (Saunders and Smith 2010). In 2015, about 795 million people were undernourished worldwide (FAO et al. 2015). Prevalence of malnutrition is high among inpatients as well. In a study including 504 newly hospitalized adult patients, 159 of them (31.5%) were identified as being at high risk for malnutrition according to the NRS-2002 (Rasmussen et al. 2010; Giryes et al. 2012). Other studies showed that as many as 50% of all patients admitted to hospitals in western countries were malnourished or had an increased risk of malnutrition (Edington et al. 2000; Correia and Campos 2003).

In the developing world, it has been demonstrated that patients with severe malnutrition (and hypoglycemia) may have Malnutrition-Related Diabetes Mellitus (MRDM), a rare type of diabetes associated with long term malnutrition (Taksande et al. 2008; Chattopadhyay et al. 1995). Additionally, we showed recently that one of the risk factors for hypoglycemia among patients with and without DM was high risk for malnutrition as measured by the NRS-2002 (Leibovitz et al. 2018a).

The mechanism responsible for hypoglycemia among malnourished patients is still unknown. An association between acute viral hepatitis and hypoglycemia has been previously reported (Felig et al. 1970). It might be possible that among those patients, both impaired gluconeogenesis and depletion of hepatic glycogen stores are responsible for hypoglycemia (Aldridge et al. 2015). Malnutrition due to chronic disease and liver dysfunction may be important predisposing factors in development of hypoglycemia in the elderly (Mori and Ito 1988). Moreover, it has been found that chronically fasted patients failed to increase epinephrine secretion in response to insulin induced hypoglycemia (Drenick et al. 1972). Thus, it could be that malnourished patients are lacking in adrenergic symptoms and, as a result, are more susceptible to hypoglycemia (Mori and Ito 1988). Another evidence to support the association of glycogen storage with hypoglycemia can be found in patients that suffer from Glycogen Storage Disorders (GSDs). These disorders are inborn errors of metabolism with abnormal storage or utilization of glycogen, with a cardinal presenting feature of hypoglycemia

(Burda and Hochuli 2015). In spite of the different pathophysiology of both conditions, they may have the same clinical outcome, i.e., hypoglycemia. Other factors that may increase the likelihood of glycogen depletion include a shift towards anaerobic metabolism.

There are several surrogate markers for malnutrition that can be measured in the blood stream, serum albumin being the most frequently used. Albumin level was shown to be influenced by nutritional status (Rothschild et al. 1973), however, there is still a controversy whether albumin level is a marker of nutrition status (Lee et al. 2015), because it plays a major role in various medical conditions (Fanali et al. 2012). Interestingly, serum albumin is as accurate a predictor of outcome as the APACHE II Score among Intensive Care Unit (ICU) patients (McCluskey et al. 1996). Low albumin level is associated with increased morbidity and mortality in many patient populations (Numeroso et al. 2008; Vincent et al. 2003).

Another marker suggested as an indicator of malnutrition is serum prealbumin (i.e. transthyretin). The major source of this protein is in the liver (Tormey and O'Brien 1993) and it has a half-life of 2 days (Beck and Rosenthal 2002). Moreover, serum level of prealbumin is not altered by hydration status (Mears 1996). A reverse correlation was found between prealbumin level and mortality among hospitalized elderly patients with a decreased nutrient intake (Sullivan et al. 1999) and patients treated with hemodialysis and peritoneal dialysis (Sreedhara et al. 1996). On the other hand, among critically ill patients with inflammation, serum prealbumin level was not a sensitive marker for evaluating the adequacy of nutrition support. It was found that only change in CRP level was able to significantly predict changes in level of prealbumin, indicating that increase in prealbumin was as a result of improvement in inflammation, rather than nutrient intake (Davis et al. 2012).

Additional markers of nutritional status include serum cholesterol. It was shown that among patients with DM, both total and LDL cholesterol levels were lower in undernourished patients compared to well-nourished (Das et al. 1984). This was true regardless of treatment with insulin. In a recently published systematic review and meta-analysis regarding the association between blood biomarkers and risk of malnutrition in older adults, it was found that serum albumin, prealbumin and total cholesterol are useful biochemical indicators of malnutrition, even with the presence of chronic inflammation (Zhang et al. 2017).

Recently, admission levels of serum albumin and cholesterol were predictive of hypoglycemia in patients admitted to general internal medicine units, regardless of DM status (Leibovitz et al. 2018b). Serum albumin and cholesterol were also indicative of the severity of hypoglycemia (Fig. 1), and the two parameters had an additive effect on incident hypoglycemia. The study showed that combination of serum albumin below 3.5 g/dL (527 µmol/L) and serum cholesterol below 130 mg/dL (3.4 mmol/L) were associated with a 2.5-fold increase in the risk of incident hypoglycemia. Interestingly, serum albumin and the NRS-2002 have also additive effects on incident hypoglycemia. Combination of NRS-2002 and serum albumin below 3.5 g/dL (527 µmol/L) are associated with more than three-fold increase in the likelihood of hypoglycemia (submitted for publication). Moreover, more than 70% of patients with hypoglycemia had either positive NRS-2002 and/or low albumin upon admission (Fig. 2). These results indicate that the nutritional status upon admission to the hospital is a significant determinant of the risk for hypoglycemia during hospitalization.

A recent study found that nutritional intervention using oral nutritional supplements (ONS) was significantly associated with reduction of the risk of hypoglycemia (Fig. 3). The study was performed on patients with DM that had a low serum albumin upon hospitalization. The ONS prescribed was Glucerna® (manufactured by Abbott, 330 KCAL, 28 g carbohydrates, 17 g protein, 17 g fat and micronutrients). Results show that one can of ONS was associated with a 60% decrease in the risk (Leibovitz et al. 2019b). Moreover, every day without nutritional support carried approximately a 10% increase in risk for

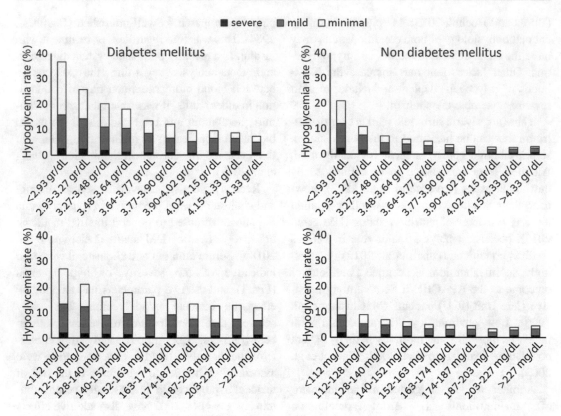

Fig. 1 Association between deciles of admission serum albumin (top panels) and cholesterol (bottom panels) with incident hypoglycemia and its severity among patients with (left panels) and without diabetes mellitus (right panels). Severe hypoglycemia – glucose <55 mg/dL. Mild hypoglycemia – glucose <69 mg/dL and $\geq$ 55 mg/dL. Minimal hypoglycemia – glucose between 69 and 70 mg/dL

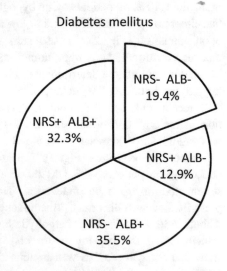

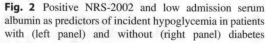

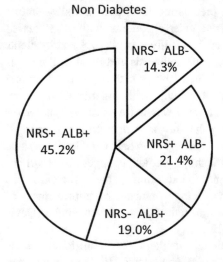

Fig. 2 Positive NRS-2002 and low admission serum albumin as predictors of incident hypoglycemia in patients with (left panel) and without (right panel) diabetes mellitus. NRS – Nutritional risk screen 2002. ALB – Low (below 3.5 g/dL or 527 µmol/L) serum albumin

Fig. 3 Forest plot of regression model showing the parameters associated with incident hypoglycemia. Significant parameters are marked in BOLD

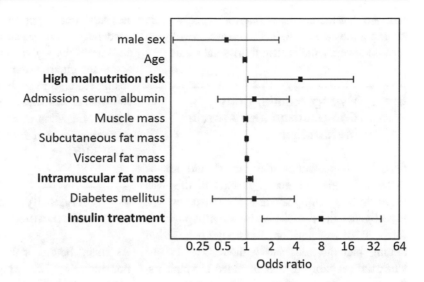

incident hypoglycemia. The conclusions from that study call for the addition of ONS for all patients with DM that are at risk of malnutrition (as measured with serum albumin). Consumption of the ONS should be encouraged to all patients, and it should be dispensed as soon as the patients are hospitalized.

5 Kidney Function

Kidney disease, both acute kidney injury (AKI) and chronic kidney disease (CKD), were found to be strong predictors of hypoglycemia among hospitalized individuals. This was true, again, regardless of DM status.

The presence of AKI during the hospitalization increases the likelihood of hypoglycemia by a factor of 2. It was found also that, among the patients with AKI, every increase of 0.1 mg/dL (8.8 µmol/L) of creatinine during the hospitalization is associated with 20% increase in the likelihood of patients having a hypoglycemic event, and a 4% increase of the risk for multiple events. For patients with chronic kidney disease, every 1 mL/min/1.73 m^2 of eGFR is associated with 1.2% change in the risk of developing incident hypoglycemia during the hospitalization period. This was true for patients with and without diabetes mellitus. Moreover, every 1 mL/min/1.73 m^2

of eGFR is also associated with about 4% chance of having additional events.

The kidney function was also a determinant for hypoglycemia among community dwelling older individuals with DM. It has been previously shown by Hodge et al. (Hodge et al. 2017) that lower eGFR is associated with incident hypoglycemia. The study population consisted of diabetes outpatients over 65 years old, users and nonusers of anti-diabetic agents. It may be that the prolonged clearance of antidiabetic agents or the accumulation of either the parent compound or its metabolites may occur (Snyder and Berns 2004; Biesenbach et al. 2003), leading to their higher bioavailability. However, Hodge et al., like us, also documented the association among DM patients not taking glucose lowering medications. We found a strong correlation between eGFR and admission serum albumin levels. This may indicate that patients with CKD may have a deterioration of their nutritional status, and this in turn may lead to increased risk of hypoglycemia. In addition, we also documented a correlation between eGFR and CRP. It is well known that patients with CKD have higher concentrations of pro-inflammatory cytokines (Maltzman and Berns 2005; Lam 2009; Neirynck et al. 2015), which are associated with higher disease severity. For patients with AKI, no correlation was found to either serum albumin or CRP levels. This may indicate that the increased risk for hypoglycemia among patients with AKI is more associated with disease severity rather than nutritional status.

6 Hypoglycemia, Body Composition and Muscle Metabolism

One of the questions that one should ask is "where the glucose goes". A clue to this was found in body composition analysis studies. We investigated the association between hypoglycemia, risk of malnutrition, and body composition among patients with and without DM. In our study not yet published that included 155 patients, muscle and fat masses were assessed at the level of third lumbar vertebra (L3) using computed tomography (CT). This level provides precise estimates of body composition, including regional abdominal adipose tissue and skeletal muscle (Gomez-Perez et al. 2016).

According to our study, intra-muscular fat was found to be associated with incident hypoglycemia, irrespective of the nutritional status, and insulin treatment. We found that with every increase of 1 cm^2/m^2 of intramuscular fat, there is a ~9% increase in the chances of hypoglycemia (data on file). Moreover, it has been found that higher fat mass in the trunk is a strong determinant of disturbed glucose metabolism (Snijder et al. 2004). This may suggest a possible involvement of muscle energy balance in pathophysiology of hypoglycemia.

The fact that an-aerobic muscle metabolism may play a role in the development of hypoglycemia could also be found in studies of alanine aminotransferase (ALT) levels. This enzyme that is prevalent in many tissues, most notably in the liver, is also abundant in muscle, where it participates in the Cahill cycle (alanine cycle) (Karmen et al. 1955). Low levels of ALT were shown to be an independent marker for poor prognosis among hospitalized (Irina et al. 2018) and non-hospitalized (Vespasiani-Gentilucci et al. 2017) individuals. We also showed that low ALT values were observed in patients with hypoglycemia (Khanimov et al. 2019). This issue is still under investigation. Another entity associated with muscle metabolism is Exercise-induced hypoglycemia. Physical activity increases glucose demand by muscles, and, as a result, leads to hypoglycemia. In order to reduce the risk, It was suggested that reducing doses of insulin, ingestion of carbohydrates and vigilance around glucose monitoring is required (Riddell et al. 2017).

7 Hypoglycemia and Endocrine Disorders

As stated before, it is well known that glucose homeostasis is regulated by various glucoregulatory hormones which effect multiple target tissues (Wasserman 2009; Aronoff et al. 2004). Glucagon is one of the major hormones involved in glucose regulation, and provides primary defense against hypoglycemia (Service 1995). Levels of glucagon are regulated by, among others, insulin and glucose (Cooperberg and Cryer 2010). Glucagon increases blood glucose concentrations by stimulation of hepatic glucose synthesis and mobilization (Quesada et al. 2008). Thus, the glycemic response to glucagon during hypoglycemia may aid in determining the etiology of hypoglycemia (Finegold et al. 1980). As a result of the major involvement of glucagon in glucose homeostasis, a decreased response or deficiency of glucagon play a role in incidence of hypoglycemia (Segel et al. 2002; Grollman et al. 1964; Vidnes and Oyasaeter 1977; Cherrington et al. 1979; Foà et al. 1980).

Another hormone involved in glucose homeostasis is cortisol, and hypoglycemia is one of the findings in patients with adrenal insufficiency (a deficiency in cortisol). Adrenal insufficiency can be a result of an impairment of the adrenal gland (primary adrenal insufficiency), impairment of hypothalamus or pituitary (secondary adrenal insufficiency), or impairment of the hypothalamic-pituitary-adrenal (HPA) axis by exogenous glucocorticoid therapy (tertiary adrenal insufficiency) (Martin-Grace et al. 2020). Hypoglycemia may occur in any of the syndromes described, however, more commonly seen in patients with secondary adrenal insufficiency (Todd et al. 2002; Burke 1985; Stacpoole et al. 1982).

On the other hand, treatment with steroids (glucocorticoids) leads to an increase in both postprandial and fasting glucose in patients with and without diabetes mellitus (Olefsky and Kimmerling 1976; Gurwitz et al. 1994); however, we recently showed that systemic steroid treatment was associated with increased incidence of hypoglycemia during the hospitalization of non-critically ill patients (Khanimov et al. 2020). However, the mechanism of this phenomenon is not yet established.

8 Treatment and Prevention of Hypoglycemia

Initial treatment of hypoglycemia depends on patients' symptoms, and the goal is to raise plasma glucose level to normal values.

For asymptomatic patients with serum glucose level under 70 mg/dL (3.9 mmol/L), it is advised to ingest carbohydrates and repeat glucose measurement more frequently. It is also recommended to avoid actions as exercise or driving, adjust treatment regimen to prevent future drops in glucose levels (Workgroup on Hypoglycemia, American Diabetes Association 2005; Cryer 2009). Early symptoms of hypoglycemia can be treated by consuming of 15 g of a fast-acting carbohydrate. Serum glucose level should be rechecked after 15 min. If glucose level still under 70 mg/dL (3.9 mmol/L), other 15 g should be administered.

As mentioned above, severe hypoglycemia is an event requiring assistance of another person to actively administer carbohydrates, glucagon, or take other corrective actions (Seaquist et al. 2013b). Patients with impaired consciousness should be treated with IV glucose (if IV access is available), or glucagon. Administration of buccal or sublingual preparations or foods is not recommended (Gunning and Garber 1978).

Hypoglycemic events should be prevented, and it is reasonable to estimate risk for hypoglycemia and act accordingly. This fact should

dictate glucose control among diabetes patients, since intensive glucose reduction to HbA1c levels below 6% was shown to be associated with more than two-fold increase in the risk of hypoglycemia (Turnbull et al. 2009). Moreover, it was shown that 20% of American patients are receiving intensive treatment (HbA1c <6%) but do not require this (McCoy et al. 2016). It is therefore recommended to tailor treatment goals according to individual patients, and pursue higher HbA1c levels in patients at risk.

There are several classical risk factors that are associated with hypoglycemia, all of which were described in community dwelling DM patients. The classical risk factors for hypoglycemia among DM patients include a previous documented hypoglycemic event, age and duration of DM, renal impairment and cognitive state of the patient. A previous hypoglycemic event is associated with more than a five-fold increased risk for recurrence of hypoglycemia (Davis et al. 2010), and therefore, history taking should be encouraged to ascertain this information. Additionally, both age and DM disease duration are probably associated with hypoglycemia because of defective hormonal counter-regulatory response to hypoglycemia (Yun and Ko 2015; Cryer 2013).

Renal impairment is also a predisposing factor for hypoglycemia. In the ACCORD trial, data showed that an albumin/creatinine ratio greater than 300 mg/g or a serum creatinine greater than 1.3 mg/dL (115 μmol/L) were strong predictors of hypoglycemia (Miller et al. 2010). According to the ADVANCE trial, for each 1 μmol/L increase in serum creatinine, the risk of severe hypoglycemia event increased by 1% (Zoungas et al. 2010). These results stress the importance of tailoring glucose lowering medications to patients. Strategic modulation of glycemic targets is a treatment option that should be considered in the month following hypoglycemia.

To date, the chief, if not only, way to prevent hypoglycemia is to lower glucose lowering medications in treated DM patients. This however will not affect the incidence of hypoglycemia among hospitalized non-critically ill patients without DM, nor will it affect hypoglycemia among DM patients that do not receive glucose lowering medications. The identification of new surrogate markers for incident hypoglycemia, like admission serum albumin and malnutrition screening may indicate that nutritional intervention may play a role in preventing hypoglycemia. Currently, a multi-center randomized study is underway to examine this possibility.

9   Conclusions and Recommendations

In conclusion, hypoglycemia is associated with poor short and long-term prognosis and increased morbidity and mortality. Hypoglycemia is more common among patients with DM, probably as a result of anti-diabetic agents. However, it can occur among patients without DM as well, and among DM patients that are not on glucose lowering medications. After diagnosis of hypoglycemia is established, a thorough evaluation of patient's nutritional status is recommended. This is true regardless of DM status. We recommend that nutritional assessment includes both laboratory tests as well as a referral to nutrition nurse, dietitian or expert clinician for further evaluation, including nutrition risk screening questionnaire and other nutritional and metabolic variables. A detailed care plan should be established following careful monitoring of outcomes.

Conflict of Interest All authors declare no conflict of interest.

Bibliography

Aldridge DR, Tranah EJ, Shawcross DL (2015) The management of acute hepatic failure. J Clin Exp Hepatol 5 (549):S7–S20

American Society of Parenteral and Enteral Nutrition (2002) Guidelines for the use of parenteral and enteral nutrition in adult and pediatric patients. J Parenter Enter Nutr 26(1_suppl):1SA–138SA

Aronoff SL, Berkowitz K, Shreiner B, Want L (2004) Glucose metabolism and regulation: beyond insulin and glucagon. Diabetes Spectr 17(3):183–190

Beck FK, Rosenthal TC (2002) Prealbumin: a marker for nutritional evaluation. Am Fam Physician 65 (8):1575–1578

Biesenbach G, Raml A, Schmekal B, Eichbauer-Sturm G (2003) Decreased insulin requirement in relation to GFR in nephropathic type 1 and insulin-treated type 2 diabetic patients. Diabet Med 20(8):642–645

Boléo-Tomé C, Chaves M, Monteiro-Grillo I, Camilo M, Ravasco P (2011) Teaching nutrition integration: MUST screening in cancer. Oncologist 16(2):239–245

Bonaventura A, Montecucco F, Dallegri F (2015) Update on strategies limiting iatrogenic hypoglycemia. Endocr Connect 4(3):R37–R45

Boucai L, Southern WN, Zonszein J (2011) Hypoglycemia-associated mortality is not drug-associated but linked to comorbidities. Am J Med 124(11):1028–1035

Brodovicz KG, Mehta V, Zhang Q, Zhao C, Davies MJ, Chen J et al (2013) Association between hypoglycemia and inpatient mortality and length of hospital stay in hospitalized, insulin-treated patients. Curr Med Res Opin 29(2):101–107

Burda P, Hochuli M (2015) Hepatic glycogen storage disorders. Curr Opin Clin Nutr Metab Care 18 (4):415–421

Burke CW (1985) Adrenocortical insufficiency. Clin Endocrinol Metab 14(4):947–976

Campbell IT (1999) Limitations of nutrient intake. The effect of stressors: trauma, sepsis and multiple organ failure. Eur J Clin Nutr 53(Suppl 1):S143–S147

Cansado P, Ravasco P, Camilo M (2009) A longitudinal study of hospital undernutrition in the elderly: comparison of four validated methods. J Nutr Health Aging 13 (2):159–164

Casiglia E, Tikhonoff V (2017) Long-standing problem of β-blocker-elicited hypoglycemia in diabetes mellitus. Hypertens (Dallas, Tex 1979) 70(1):42–43

Charlton KE, Kolbe-Alexander TL, Nel JH (2007) The MNA, but not the DETERMINE, screening tool is a valid indicator of nutritional status in elderly Africans. Nutrition 23(7–8):533–542

Chattopadhyay PS, Gupta SK, Chattopadhyay R, Kundu PK, Chakraborti R (1995) Malnutrition-related diabetes mellitus (MRDM), not diabetes-related malnutrition. A report on genuine MRDM. Diabetes Care 18 (2):276–277

Chaytor NS, Barbosa-Leiker C, Ryan CM, Germine LT, Hirsch IB, Weinstock RS (2019) Clinically significant cognitive impairment in older adults with type 1 diabetes. J Diabetes Complicat 33(1):91–97

Cherrington AD, Liljenquist JE, Shulman GI, Williams PE, Lacy WW (1979) Importance of hypoglycemia-induced glucose production during isolated glucagon deficiency. Am J Phys 236(3):E263–E271

Conn JW, Fajans SS, Selyzer HS (1956) Spontaneous hypoglycemia as an early manifestation of diabetes mellitus. Diabetes 5(6):437–442

Cooperberg BA, Cryer PE (2010) Insulin reciprocally regulates glucagon secretion in humans. Diabetes 59 (11):2936–2940

Correia MITD, Campos ACL (2003) ELAN Cooperative Study. Prevalence of hospital malnutrition in Latin America: the multicenter ELAN study. Nutrition 19 (10):823–825

Crevel E, Ardigo S, Perrenoud L, Vischer UM (2009) Acrocyanosis as a cause of pseudohypoglycemia: letters to the editor. J Am Geriatr Soc 57:1519–1520

Cryer PE (2009) Preventing hypoglycaemia: what is the appropriate glucose alert value? Diabetologia 52 (1):35–37

Cryer PE (2012) Severe hypoglycemia predicts mortality in diabetes. Diabetes Care 35(9):1814–1816

Cryer PE (2013) Mechanisms of hypoglycemia-associated autonomic failure in diabetes. N Engl J Med. (Massachussetts Medical Society) 369:362–372

Das S, Tripathy BB, Samal KC, Panda NC (1984) Plasma lipids and lipoprotein cholesterol in undernourished diabetic subjects and adults with protein energy malnutrition. Diabetes Care 7(6):579–586

Davis TME, Brown SGA, Jacobs IG, Bulsara M, Bruce DG, Davis WA (2010) Determinants of severe hypoglycemia complicating type 2 diabetes: the Fremantle diabetes study. J Clin Endocrinol Metab 95 (5):2240–2247

Davis CJ, Sowa D, Keim KS, Kinnare K, Peterson S (2012) The use of Prealbumin and C-reactive protein for monitoring nutrition support in adult patients receiving enteral nutrition in an urban medical center. J Parenter Enter Nutr 36(2):197–204

Detsky AS, Mclaughlin J, Baker JP, Johnston N, Whittaker S, Mendelson RA et al (1987) What is subjective global assessment of nutritional status? J Parenter Enter Nutr 11(1):8–13

Drenick EJ, Alvarez LC, Tamasi GC, Brickman AS (1972) Resistance to symptomatic insulin reactions after fasting. J Clin Invest 51(10):2757–2762

Edington J, Boorman J, Durrant ER, Perkins A, Giffin CV, James R et al (2000) Prevalence of malnutrition on admission to four hospitals in England. Clin Nutr 19 (3):191–195

El Khoury M, Yousuf F, Martin V, Cohen RM (2008) Pseudohypoglycemia: a cause for unreliable finger-stick glucose measurements. Endocr Pract 14 (3):337–339

Fanali G, Di Masi A, Trezza V, Marino M, Fasano M, Ascenzi P (2012) Human serum albumin: from bench to bedside. Mol Asp Med 33(3):209–290

FAO, IFAD, WFP (2015) The State of Food Insecurity in the World 2015. Meeting the 2015 international hunger targets: taking stock of uneven progress. FAO, Rome

Felig P, Brown WV, Levine RA, Klatskin G (1970) Glucose homeostasis in viral hepatitis. N Engl J Med 283 (26):1436–1440

Finegold DN, Stanley CA, Baker L (1980) Glycemic response to glucagon during fasting hypoglycemia: an aid in the diagnosis of hyperinsulinism. J Pediatr 96 (2):257–259

Fischer KF, Lees JA, Newman JH (1986) Hypoglycemia in hospitalized patients. N Engl J Med 315 (20):1245–1250

Foà PP, Dunbar JC, Klein SP, Levy SH, Malik MA, Campbell BB et al (1980) Reactive hypoglycemia and A-cell ('pancreatic') glucagon deficiency in the adult. JAMA 244(20):2281–2285

Giryes S, Leibovitz E, Matas Z, Fridman S, Gavish D, Shalev B et al (2012) MEasuring nutrition risk in hospitalized patients: MENU, a hospital-based prevalence survey. Isr Med Assoc J 14(7):405–409

Gomez F, Galvan RR, Cravioto J, Frenk S (1955) Malnutrition in infancy and childhood, with special reference to kwashiorkor. Adv Pediatr Infect Dis 7:131–169

Gomez-Perez SL, Haus JM, Sheean P, Patel B, Mar W, Chaudhry V et al (2016) Measuring abdominal circumference and skeletal muscle from a single cross-sectional computed tomography image: a step-by-step guide for clinicians using National Institutes of Health ImageJ. J Parenter Enter Nutr 40(3):308–318

Grollman A, McCaleb WE, White FN (1964) Glucagon deficiency as a cause of hypoglycemia. Metabolism 13 (8):686–690

Grover Z, Ee LC (2009) Protein Energy Malnutrition. Pediatr Clin N Am 56(5):1055–1068

Gunning RR, Garber AJ (1978) Bioactivity of instant glucose: failure of absorption through oral mucosa. JAMA 240(15):1611–1612

Gurwitz JH, Bohn RL, Glynn RJ, Monane M, Mogun H, Avorn J (1994) Glucocorticoids and the risk for initiation of hypoglycemic therapy. Arch Intern Med 154 (1):97–101

Hasse J, Strong S, Gorman MA, Liepa G (1993) Subjective global assessment: alternative nutrition-assessment technique for liver-transplant candidates. Nutrition 9(4):339–343

Hepburn DA, Deary IJ, Frier BM, Patrick AW, Quinn JD, Fisher BM (1991) Symptoms of acute insulin-induced hypoglycemia in humans with and without IDDM. Factor-analysis approach. Diabetes Care 14 (11):949–957

Hodge M, McArthur E, Garg AX, Tangri N, Clemens KK (2017) Hypoglycemia incidence in older adults by estimated GFR. Am J Kidney Dis 70(1):59–68

International Hypoglycaemia Study Group IHS (2017) Glucose concentrations of less than 3.0 mmol/L (54 mg/dL) should be reported in clinical trials: a joint position statement of the American Diabetes Association and the European Association for the Study of Diabetes. Diabetes Care 40(1):155–157

Irina G, Refaela C, Adi B, Avia D, Liron H, Chen A et al (2018) Low blood ALT Activity and High FRAIL questionnaire scores correlate with increased mortality and with each other. A prospective study in the internal medicine department. J Clin Med 7(11):386

Jensen GL, Bistrian B, Roubenoff R, Heimburger DC (2009) Malnutrition syndromes: a conundrum vs continuum. J Parenter Enter Nutr 33(6):710–716

Kaiser MJ, Bauer JM, Ramsch C, Uter W, Guigoz Y, Cederholm T et al (2009) Validation of the mini nutritional assessment short-form (MNA®-SF): a practical tool for identification of nutritional status. J Nutr Heal Aging 13(9):782–788

Karmen A, Wroblewski F, Ladue JS (1955) Transaminase activity in human blood. J Clin Invest 34(1):126–131

Khanimov I, Segal G, Wainstein J, Boaz M, Shimonov M, Leibovitz E (2019) High-intensity statins are associated with increased incidence of hypoglycemia during hospitalization of individuals not critically ill. Am J Med 132:1305–1310

Khanimov I, Boaz M, Shimonov M, Wainstein J, Leibovitz E (2020) Systemic treatment with glucocorticoids is associated with incident hypoglycemia and mortality: a historical prospective analysis. Am J Med (in press)

Kondrup J, Allison SP, Elia M, Vellas B, Plauth M (2003) ESPEN guidelines for nutrition screening 2002. Clin Nutr 22(4):415–421

Krinsley JS, Grover A (2007) Severe hypoglycemia in critically ill patients: risk factors and outcomes∗. Crit Care Med 35(10):2262–2267

Lam CWK (2009) 2. Inflammation, cytokines and chemokines in chronic kidney disease. EJIFCC 20 (1):12–20

Lee KT, Abadir PM (2015) Failure of glucose monitoring in an individual with pseudohypoglycemia. J Am Geriatr Soc (Blackwell Publishing Inc) 63:1706–1708

Lee JL, Oh ES, Lee RW, Finucane TE (2015) Serum albumin and prealbumin in calorically restricted, nondiseased individuals: a systematic review. Am J Med 128(9):1023.e1–1023.e22

Leibovitz E, Adler H, Giryes S, Ditch M, Burg NF, Boaz M (2018a) Malnutrition risk is associated with hypoglycemia among general population admitted to internal medicine units. Results from the MENU study. Eur J Clin Nutr 72(6):888–893

Leibovitz E, Wainstein J, Boaz M (2018b) Association of albumin and cholesterol levels with incidence of hypoglycaemia in people admitted to general internal medicine units. Diabet Med 35:1735–1741

Leibovitz E, Khanimov I, Wainstein J, Boaz M (2019a) Documented hypoglycemia is associated with poor short and long term prognosis among patients admitted to general internal medicine departments. Diabetes Metab Syndr Clin Res Rev 13(1):222–226

Leibovitz E, Moore F, Mintser I, Levi A, Dubinsky R, Boaz M (2019b) Consumption of nutrition supplements is associated with less hypoglycemia during admission-results from the MENU project. Nutrients 11(8):1832

Maltzman JS, Berns JS (2005) Are inflammatory cytokines the "evil humors" that increase morbidity and cardiovascular mortality in chronic kidney disease? Semin Dial 18(5):441–443

Mannucci E, Monami M, Mannucci M, Chiasserini V, Nicoletti P, Gabbani L et al (2006) Incidence and prognostic significance of hypoglycemia in hospitalized non-diabetic elderly patients. Aging Clin Exp Res 18(5):446–451

Martin-Grace J, Dineen R, Sherlock M, Thompson CJ (2020) Adrenal insufficiency: physiology, clinical presentation and diagnostic challenges. Clin Chim Acta 505:78–91

McCluskey A, Thomas AN, Bowles BJ, Kishen R (1996) The prognostic value of serial measurements of serum albumin concentration in patients admitted to an intensive care unit. Anaesthesia 51(8):724–727

McCoy RG, Lipska KJ, Yao X, Ross JS, Montori VM, Shah ND (2016) Intensive treatment and severe hypoglycemia among adults with type 2 diabetes. JAMA Intern Med 176(7):969–978

Mears E (1996) Outcomes of continuous process improvement of a nutritional care program incorporating serum prealbumin measurements. Nutrition 12(7–8):479–484

Miller ME, Bonds DE, Gerstein HC, Seaquist ER, Bergenstal RM, Calles-Escandon J et al (2010) The effects of baseline characteristics, glycaemia treatment approach, and glycated haemoglobin concentration on the risk of severe hypoglycaemia: post hoc epidemiological analysis of the ACCORD study. BMJ 340 (7738):138

Mori S, Ito H (1988) Hypoglycemia in the elderly. Jpn J Med 27(2):160–166

Neirynck N, Glorieux G, Schepers E, Dhondt A, Verbeke F, Vanholder R (2015) Pro-inflammatory cytokines and leukocyte oxidative burst in chronic kidney disease: culprits or innocent bystanders? Nephrol Dial Transplant 30(6):943–951

Nirantharakumar K, Marshall T, Hodson J, Narendran P, Deeks J, Coleman JJ et al (2012) Hypoglycemia in non-diabetic in-patients: clinical or criminal? Sesti G, editor. PLoS One 7(7):e40384

Numeroso F, Barilli AL, Delsignore R (2008) Prevalence and significance of hypoalbuminemia in an internal medicine department. Eur J Intern Med 19(8):587–591

Nydick M, Samols E, Kuzuya T, Williams RH (1964) A difficult diagnostic problem in spontaneous hypoglycemia. Reactive hypoglycemia in mild diabetes mellitus. Ann Intern Med 61:1122–1127

Olefsky JM, Kimmerling G (1976) Effects of glucocorticoids on carbohydrate metabolism. Am J Med Sci 271(2):202–210

Quesada I, Tudurí E, Ripoll C, Nadal A (2008) Physiology of the pancreatic α-cell and glucagon secretion: Role in glucose homeostasis and diabetes. J Endocrinol 199:5–19

Rasmussen HH, Holst M, Kondrup J (2010) Measuring nutritional risk in hospitals. Clin Epidemiol 2:209

Riddell MC, Gallen IW, Smart CE, Taplin CE, Adolfsson P, Lumb AN et al (2017) Exercise management in type 1 diabetes: a consensus statement. Lancet Diabetes Endocrinol 5(5):377–390

Rothschild MA, Oratz M, Schreiber SS (1973) Progress in gastroenterology albumin metabolism. Gastroenterology 64(642):324–337

Rushakoff RJ, Lewis SB (2001) Case of pseudohypoglycemia. Diabetes Care 24:2157–2158

Saunders J, Smith T (2010) Malnutrition: causes and consequences. Clin Med 10(6):624–627

Seaquist ER, Anderson J, Childs B, Cryer P, Dagogo-Jack S, Fish L et al (2013a) Hypoglycemia and diabetes: a report of a workgroup of the American Diabetes Association and the Endocrine Society. J Clin Endocrinol Metab 98(5):1845–1859

Seaquist ER, Anderson J, Childs B, Cryer P, Dagogo-Jack S, Fish L et al (2013b) Hypoglycemia and diabetes: a report of a workgroup of the American Diabetes Association and the Endocrine Society. Diabetes Care 36(5):1384–1395

Segel SA, Paramore DS, Cryer PE (2002) Hypoglycemia-associated autonomic failure in advanced type 2 diabetes. Diabetes 51(3):724–733

Service FJ (1995) Hypoglycemic disorders. N Engl J Med 332(17):1144–1152

Shilo S, Berezovsky S, Friedlander Y, Sonnenblick M (1998) Hypoglycemia in hospitalized nondiabetic older patients. J Am Geriatr Soc 46(8):978–982

Skipper A, Ferguson M, Thompson K, Castellanos VH, Porcari J (2012) Nutrition screening tools: an analysis of the evidence. J Parenter Enter Nutr 36(3):292–298

Snijder MB, Dekker JM, Visser M, Bouter LM, Stehouwer CDA, Yudkin JS et al (2004) Trunk fat and leg fat have independent and opposite associations with fasting and postload glucose levels: the hoorn study. Diabetes Care 27(2):372–377

Snyder RW, Berns JS (2004) Use of insulin and oral hypoglycemic medications in patients with diabetes mellitus and advanced kidney disease. Semin Dial 17(5):365–370

Sreedhara R, Avram MM, Blanco M, Batish R, Avram MM, Mittman N (1996) Prealbumin is the best nutritional predictor of survival in hemodialysis and peritoneal dialysis. Am J Kidney Dis 28(6):937–942

Stacpoole PW, Interlandi JW, Nicholson WE, Rabin D (1982) Isolated ACTH deficiency: a heterogeneous disorder. Critical review and report of four new cases. Medicine (Baltimore) 61(1):13–24

Steiber A, Leon JB, Secker D, McCarthy M, McCann L, Serra M et al (2007) Multicenter study of the validity and reliability of subjective global assessment in the hemodialysis population. J Ren Nutr 17(5):336–342

Stratton RJ, Green CJ, Elia M (eds) (2003) Disease-related malnutrition: an evidence-based approach to treatment. CABI, Wallingford

Sullivan DH, Sun S, Walls RC (1999) Protein-energy undernutrition among elderly hospitalized patients: a prospective study. JAMA 281(21):2013–2019

Taksande A, Taksande B, Kumar, Vilhekar KY (2008) Malnutrition related diabetes mellitus. J Mahatma Gandhi Inst Med Sci 13:19–24

Toda G, Fujishiro M, Yamada T, Shojima N, Sakoda H, Suzuki R et al (2014) Lung abscess without sepsis in a patient with diabetes with refractory episodes of spontaneous hypoglycemia: a case report and review of the literature. J Med Case Rep 8(1):51

Todd GRG, Acerini CL, Ross-Russell R, Zahra S, Warner JT, McCance D (2002) Survey of adrenal crisis associated with inhaled corticosteroids in the United Kingdom. Arch Dis Child 87(6):457–461

Tormey WP, O'Brien PA (1993) Clinical associations of an increased transthyretin band in routine serum and urine protein electrophoresis. Ann Clin Biochem Int J Biochem Lab Med 30(6):550–554

Turchin A, Matheny ME, Shubina M, Scanlon JV, Greenwood B, Pendergrass ML (2009) Hypoglycemia and clinical outcomes in patients with diabetes hospitalized in the general ward. Diabetes Care 32(7):1153–1157

Turnbull FM, Abraira C, Anderson RJ, Byington RP, Chalmers JP, Duckworth WC et al (2009) Intensive glucose control and macrovascular outcomes in type 2 diabetes. Diabetologia 52(11):2288–2298

Vellas B, Guigoz Y, Garry PJ, Nourhashemi F, Bennahum D, Lauque S et al (1999) The mini nutritional assessment (MNA) and its use in grading the nutritional state of elderly patients. Nutrition 15(2):116–122

Vespasiani-Gentilucci U, De Vincentis A, Ferrucci L, Bandinelli S, Antonelli Incalzi R, Picardi A (2017) Low alanine aminotransferase levels in the elderly. J Gerontol Ser A 00(00):1–6

Vidnes J, Oyasaeter S (1977) Glucagon deficiency causing severe neonatal hypoglycemia in a patient with normal insulin secretion. Pediatr Res 11(9 I):943–949

Vincent J-L, Dubois M-J, Navickis RJ, Wilkes MM (2003) Hypoalbuminemia in acute illness: is there a rationale for intervention? A meta-analysis of cohort studies and controlled trials. Ann Surg 237(3):319–334

Wasserman DH (2009) Four grams of glucose. Am J Physiol Endocrinol Metab 296(1):E11–E21

Waterlow JC (1972) Classification and definition of protein-calorie malnutrition. Br Med J 3(5826):566–569

White JV, Guenter P, Jensen G, Malone A, Schofield M (2012) Consensus statement: academy of nutrition and dietetics and American society for parenteral and enteral nutrition: characteristics recommended for the identification and documentation of adult malnutrition (undernutrition). J Parenter Enter Nutr 36(3):275–283

Whitmer RA, Karter AJ, Yaffe K, Quesenberry CP, Selby JV (2009) Hypoglycemic episodes and risk of dementia in older patients with type 2 diabetes mellitus. JAMA 301(15):1565–1572

Workgroup on Hypoglycemia, American Diabetes Association (2005) Defining and reporting hypoglycemia in diabetes: a report from the American Diabetes Association Workgroup on Hypoglycemia. Diabetes Care 28(5):1245–1249

Yaffe K, Falvey CM, Hamilton N, Harris TB, Simonsick EM, Strotmeyer ES et al (2013) Association between hypoglycemia and dementia in a biracial cohort of older adults with diabetes mellitus. JAMA Intern Med 173(14):1300–1306

Ye XJ, Ji YB, Ma BW, Huang DD, Chen WZ, Pan ZY, Shen X, Zhuang CL, Yu Z (2018) Comparison of three common nutritional screening tools with the new European Society for Clinical Nutrition and Metabolism (ESPEN) criteria for malnutrition among patients with geriatric gastrointestinal cancer: a prospective study in China. BMJ Open 8(4):e019750

Young AM, Kidston S, Banks MD, Mudge AM, Isenring EA (2013) Malnutrition screening tools: comparison against two validated nutrition assessment methods in older medical inpatients. Nutrition 29(1):101–106

Yun JS, Ko SH (2015) Avoiding or coping with severe hypoglycemia in patients with type 2 diabetes. Korean J Intern Med 30(1):6–16

Yun J-S, Park Y-M, Han K, Cha S-A, Ahn Y-B, Ko S-H (2019) Severe hypoglycemia and the risk of cardiovascular disease and mortality in type 2 diabetes: a nationwide population-based cohort study. Cardiovasc Diabetol 18(1):103

Zapatero A, Gómez-Huelgas R, González N, Canora J, Asenjo Á, Hinojosa J et al (2014) Frequency of hypoglycemia and its impact on length of stay, mortality, and short-term readmission in patients with diabetes hospitalized in internal medicine wards. Endocr Pract 20(9):870–875

Zhang Z, Pereira S, Luo M, Matheson E (2017) Evaluation of blood biomarkers associated with risk of malnutrition in older adults: a systematic review and meta-analysis. Nutrients 9(8):829

Zoungas S, Patel A, Chalmers J, De Galan BE, Li Q, Billot L et al (2010) Severe hypoglycemia and risks of vascular events and death. N Engl J Med 363 (15):1410–1418

Adv Exp Med Biol - Advances in Internal Medicine (2020) 4: 85–114
https://doi.org/10.1007/5584_2020_545
© Springer Nature Switzerland AG 2020
Published online: 3 Jun 2020

Acute Metabolic Emergencies in Diabetes: DKA, HHS and EDKA

Muhammad Muneer and Ijaz Akbar

Abstract

Emergency admissions due to acute metabolic crisis in patients with diabetes remain some of the most common and challenging conditions. DKA (Diabetic Ketoacidosis), HHS (Hyperglycaemic Hyperosmolar State) and recently focused EDKA (Euglycaemic Diabetic Ketoacidosis) are life-threatening different entities. DKA and HHS have distinctly different pathophysiology but basic management protocols are the same. EDKA is just like DKA but without hyperglycaemia. T1D, particularly children are vulnerable to DKA and T2D, particularly elderly with comorbidities are vulnerable to HHS. But these are not always the rule, these acute conditions are often occur in different age groups with diabetes. It is essential to have a coordinated care from the multidisciplinary team to ensure the timely delivery of right treatment. DKA and HHS, in many instances can present as a mixed entity as well. Mortality rate is higher for HHS than DKA but incidences of DKA are much higher than HHS. The prevalence of HHS in children and young adults are increasing due to exponential growth of obesity and increasing T2D cases in this age group. Following introduction of SGLT2i (Sodium-GLucose co-Transporter-2 inhibitor) for T2D and off-label use in T1D, some incidences of EDKA has been reported. Healthcare professionals should be more vigilant during acute illness in diabetes patients on SGLT2i without hyperglycaemia to rule out EDKA. Middle aged, mildly obese and antibody negative patients who apparently resemble as T2D without any precipitating causes sometime end up with DKA which is classified as KPD (Ketosis-prone diabetes). Many cases can be prevented by following 'Sick day rules'. Better access to medical care, structured diabetes education to patients and caregivers are key measures to prevent acute metabolic crisis.

M. Muneer (✉)
Cardiff University, Cardiff, Heath Park, Cardiff, UK
e-mail: MuneerM1@cardiff.ac.uk;
genome2006@yahoo.ca

I. Akbar
Shukat Khanam Cancer Hospital and Research Centre, Lahore, Pakistan

Keywords

Anion gap · ARDS (Acute Respiratory Distress Syndrome) · Cerebral Oedema · CSII (Continuous Subcutaneous Insulin Infusion) · Fixed Rate Intravenous Insulin Infusion (FRIII) · Hypokalaemia · Osmolality · Variable Rate Intravenous Insulin Infusion (VRIII)

1 Introduction

DKA and HHS are two similar yet in many ways different metabolic emergencies of diabetes are encountered in emergency departments. Hyperglycemia, despite being the common ground for both conditions, is different in magnitude for each emergency, being more severe in HHS. Ketoacidosis is the hallmark of DKA found mostly in T1D due to absolute insulin deficiency. In HHS, ketoacidosis is nominal unless a mixed variety, which is due to residual insulin sufficient to prevent ketosis. It was thought that DKA is specific condition for T1D and HHS for T2D but this does not hold anymore. More and more cases of DKA are being reported in T2D and HHS in T1D. Similarly, the characteristic age distribution of acute hyperglycemic emergencies is not valid anymore. It is also not uncommon to find out mixture of two entities presenting in same patient.

Both of these conditions require immediate hospitalization and therefore have negative impact on the economy of a country. DKA primarily affects T1D and may be the first manifestation in up to 25% of cases (Dabelea et al. 2014; Jefferies et al. 2015; Rewers et al. 2008). More recently, EDKA is being found in T1D and T2D patient on SGLT2i (Peters et al. 2015). So there should be high index of suspicion in an unwell person with diabetes without hyperglycaemia on SGLT2i and EDKA should be ruled out. Up to 42% of DKA hospitalisations are due to readmissions for DKA within 1 year (Edge et al. 2016). It is a matter of solace that DKA mortality rates have fallen significantly in last 20 years from 7.96% to less than 1% (Umpierrez and Korytkowski 2016) Unfortunately, the mortality rates are still higher in patients over 60 years old with comorbidities, in low-income countries and in non-hospitalised patients (Otieno et al. 2006). The recent updated mortality in HHS is around 5–16% globally (Umpierrez and Korytkowski 2016). This high mortality rates necessitates early diagnosis and effective prevention programmes. Cheaper insulin should be readily available globally (Greene and Riggs 2015). The most common cause of mortality is cerebral oedema in children and young adults. On the other hand, the main causes of mortality in adults and elderly with HHS are diverse and many. The major causes are severe hypokalaemia, cardiac dysrhythmia, severe hypoglycaemia, ARDS, pneumonia, ACS (Acute Coronary Syndrome) and sepsis (Wolfsdorf et al. 2018).

Efforts should be directed to decrease hospitalization rate and acute metabolic crisis of diabetes by introducing structured diabetes education and provision of better healthcare to less developed areas. There exist some subtle differences in management protocols of DKA, EDKA and HHS patients. The purpose of this review is to provide the latest insights of epidemiology, pathophysiology, management and prevention of acute metabolic emergencies of diabetes.

2 Classification and Diagnostic Criteria

History, clinical examination, signs & symptoms and biochemical tests are required to aid diagnosis of condition. The classical triad of DKA includes hyperglycemia, ketonaemia and high anoin gap metabolic acidosis. The biochemical criterion set by JBDS, BSPED and ISPAD for diagnosis of DKA (Dhatariya 2014; Wolfsdorf et al. 2018; BSPED 2020) are as follows:

- Ketonaemia (blood level > 3 mmol/l) or ketonuria (2+ on dipstick)
- Hyperglycemia (>11 mmol/l) or known diabetic patient)
- Acidosis (HCO_3^- < 15 mmol/l and/or venous pH <7.3)

Classification of DKA is generally based upon anion gap, HCO_3, pH and cognitive status of patient.

Table 1 Classification of DKA in adults and children (BSPED 2020; Kitabchi et al. 2009; Sheikh-Ali et al. 2008; Kelly 2006).

Table 2 ADA, JBDS and AACE/ACE diagnostic criteria of DKA (Karslioglu French et al. 2019). The table is reproduced with permission.

Table 1 Classification of DKA in adults and children

Variables	Mild	Moderate	Severe
Blood glucose	Adult: >13.9 mmol/L (> 250 mg/dL)		
	Children: >11 mmol/L (> 200 mg/dL)		
Vitals (Pulse, SBP, SpO_2)	P < 100 or > 60 bpm, SBP >100; SpO_2 > 95%	P < 100 or > 60 bpm, BP >100; SpO_2 > 95%	P > 100 or < 60; BP <90; SpO_2 < 92%
Anion gap (mEq/L; mmol/L)	>10	>12	>16
Dehydration	5%	> 5 to 7%	>7 to $\geq$10%
Venous pH[a]	Adult: 7.24 to 7.3	Adult: 7.00 to <7.24	Adult: <7.00
	Children: 7.2 to 7.29	Children: 7.1 to 7.19	Children: <7.1
Serum osmolality mOsm/kg	Variable	Variable	Variable
Mental status	Alert	Alert/ drowsy	Stupor/ coma
Venous HCO_3[b]	Adult: 15 to 18	Adult: 10 to <15	Adult: < 10
(mEq/L, mmol/L)	Children: < 15	Children: < 10	Children: < 5
Serum/capillary BOHB[c] (mmol/L)	$\geq$ 3.8 to <6 in adults; $\geq$ 3 to <6 in children	$\geq$ 3.8 to <6 in adults; $\geq$ 3 to <6 in children	$\geq$ 6 both adult and children
Urine STICKS-AcAc[d]	> 2+ in urine sticks	> 2+ in urine sticks	> 2+ in urine sticks
GCS	14–15	14–15	< 12

[a]Venous pH: just 0.02–0.15 units higher than arterial; [b]venous HCO_3: just 1.88 mmol/L lower than arterial; [c]BOHB: 3 β-hydroxybutyrate found in blood mainly; [d]*AcAc* acetoacetate found in urine mainly

Table 2 Diagnostic criteria of DKA in adults

Criteria	ADA	JBDS	AACE/ACE
Publication year	2009	2013	2016
Plasma glucose	>13.9 mmol/L	>11 mmol/L	NA
	(250 mg/dL)	(>200 mg/dL) or known diabetes	
pH	Mild: 7.25–7.30; moderate: 7.00–7.24; severe: <7.00	Mil & moderate <7.3 Severe: <7.0	<7.3
Bicarbonate, mmol/L or mEq/L	Mild: 15–18; moderate: 10–14.9; severe: <10	<15 but >5 Severe: <5	NA
Anion gap: Na^+- ($Cl^- + HCO_3^-$)	Mild: >10; moderate: >12; severe: >12	Mild & moderate >10 but <16 Severe: >16)	>10
Urine acetoacetate (nitroprusside reaction)	Positive	Positive	Positive
Blood BOHB, mmol/L	NA	Mild & moderate $\geq$3 Severe: >6	$\geq$3.8
Mental status	Mild: alert; moderate: alert or drowsy; severe: stupor or coma	NA	Drowsy, stupor, or coma

The diagnostic criteria of DKA differs in many ways between societies. There is no consensus on all four key parameters such as ketonaemia/ketonuria, HCO_3, pH and glucose values. Table 2 shows the diagnostic criteria formulated by key societies of diabetes.

JBDS criterion to classify severe DKA is slightly different and includes both physical and biochemical parameters.

Table 3 Diagnostic criteria of HHS by ADA and JBDS (Karslioglu French et al. 2019), reproduced with permission.

For DKA the prominent biochemical features are ketonemia and high anion gap acidosis. In HHS the circumstances are different as this condition is characterized by high osmolality and severe dehydration secondary to severe hyperglycemia. However in clinical practice, a mixed

Table 3 Diagnostic criteria of HHS in adults

Criteria	ADA	UK
Publication year	2009	2015
Plasma glucose	> 33.3 mmol/L (600 mg/dL)	≥30 mmol/L (540 mg/dL)
pH	>7.30	>7.30
Bicarbonate	>18 mmol/L	>15 mmol/L
Anion gap: $Na^+-(Cl^- + HCO_3^-)$	NA	NA
Urine acetoacetate (nitroprusside reaction)	Negative or low positive	NA
Blood BOHB	NA	<3
Osmolality, mmol/kg	>320	≥320
Presentation	Stupor or coma	Severe dehydration and feeling unwell

Table 4 Key differences between DKA, HHS AND EDKA

Variables	DKA	HHS	EDKA
Predominant feature	Ketonemia and high anion gap metabolic acidosis	Very high glucose and serum osmolality	Ketonemia and high anion gap metabolic acidosis
Glucose level mmol/L (mg/dL)	High: >13.9 (> 250)	Very high: ≥ 33.3 (≥600)	Normal: <11 (<200)
Ketones mmol/L	High (>3 mmol/L in blood or 2+ in urine)	Normal	High (>3 mmol/L in blood or 2 + in urine)
Serum osmolality mOsm/kg	Raised	> 320	Raised
Predominant diabetes type and comorbidities	T1D, less frequently in T2D and GDM	T2D, less frequently in T1D, T2D in children and in 6q24 genotype TND[a]	T1D, LADA, T2D on SGLT-2i, with pregnancy, glycogen storage disease, alcoholism, very low calorie diet, severe liver diseases etc.
Age	Young patient	Older patient	Mostly young patients but adults might also have
Predominant phenotype	Lean	Obese	Lean to obese
Complications	.7–10% risk of cerebral oedema, iatrogenic hypoglycaemia, hypokalaemia, ARDS and a small risk of arterial or venous thromboembolism	Iatrogenic hypoglycaemia, hypokalaemia, MI, DIC, higher risk of pulmonary, arterial or venous thromboembolism	Same like DKA

[a]*TND* transient neonatal diabetes

picture of DKA and HHS can also be encountered. It is also important to remember that some degree of acidosis in HHS can also be found due to nominal ketogenesis in some cases.

Table 4 Differences between DKA, HHS and EDKA (Rosenstock et al. 2018; Dandona et al. 2018; Mathieu et al. 2018; Blau et al. 2017; Rosenstock and Ferrannini 2015).

Euglycaemic DKA (EDKA/euDKA) was first reported in 1973 (Munro et al. 1973). It should be suspected in any diabetes patients who is on SGLT2i with classic symptoms. Diagnosis can be made with the classical cutoff level of pH, HCO3 and ketone with normoglycaemia (Dhatariya 2016). Many studies have shown evidence that use of SGLT2i in T1D and in LADA cases have higher incidence of EDKA. In advanced T2D there are few case reports published recently (Rosenstock and Ferrannini 2015). Reduced carbohydrate intake, depleted glycogen reserve due to alcoholism, SGLT2i in T1D, reduction of insulin might precipitate EDKA.

Ketosis-Prone Diabetes (KPD) KPD also called 'Flatbush diabetes' is found in certain ethnic minorities like African-Americans, Asians, sub-Saharan Africans and African-Caribbean. The genotype looks like idiopathic T1D but phenotype looks like T2D. Usually a middle aged obese man presents with DKA at diagnosis of new onset diabetes. Initial aggressive insulin therapy settles the acute stage. Subsequently, diet alone or a combination with oral hypoglycaemics can achieve glycaemic without need of insulin (Lebovitz and Banerji 2018). There are four types of classification of KPD such as 'ADA', 'modified ADA', 'BMI based' and 'Aβ' exist.

KPD classification 'Aβ' is based on the presence/ absence of autoantibodies and the presence/ absence of β-cell functional reserve. This is the most used and the most acceptable classification. The four subgroups are:

1. **A + β −** (present autoantibodies but absent β-cell function)
2. **A + β +** (present autoantibodies but present β-cell functional reserve)
3. **A − β −** (absent autoantibodies and absent β-cell function) and
4. **A − β +** (absent autoantibodies but present β-cell functional reserve).

A + β − and A − β − patients are immunologically and genetically distinct from each other but they share clinical characteristics of T1D with very low β-cell function. Whereas, A + β + and A − β + patients are immunologically and genetically distinct from each other but they share clinical characteristics of T2D with preserved β-cell functional reserve. Group 4 has the largest share of KPD with 76% (Balasubramanyam et al. 2006).

3 Epidemiology

DKA is no more considered as a metabolic emergency of only T1D (Bedaso et al. 2019; Jabbar et al. 2004; Takeuchi et al. 2017; Mudly et al. 2007). It is estimated that out of all DKA cases, around 34% occur in T2D (Kitabchi et al. 2009). Up to 25–40% cases of T1D present as DKA at first diagnosis (Duca et al. 2017). Apart from key precipitating factors, socio-economic factors also act as a causative factor of DKA. These are low income, limited access to health care facilities and illiteracy (Dabelea et al. 2014).

Desai et al. made a very large retrospective observational study that included 56.7 million hospitalisations during 2007–2014 with diabetes out of which 0.9% had DKA and HHS. Younger patients show (Fig. 1) highest rate of admissions with lowest mortality and the trend is reverse for older patients (Desai et al. 2019). According to a survey in USA, more than two-thirds of children presenting with HHS have T1D (Ng et al. 2020). These are mostly obese T1D and adolescent T2D.

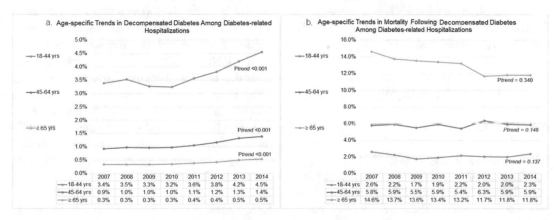

Fig. 1 Age-specific prevalence and morality of DKA and HHS (Desai et al. 2019). The figure is reproduced with permission

Table 5 SGLT2i use and EDKA rates in T1D

Studies	SGLT2i Molecule	Drug ARM	Placebo ARM	Event Rate/1000 PT YR
'EASE' Trial	Empagliflozin 2.5 mg, 10 mg and 25 mg	0.8%, 4.3% and 3.3%	1.2%	59.4, 50.5 and 17.7 events respectively
DEPICT-1 & 2 52-wk & 24-wk study	Dapagliflozin 5 mg and 10 mg	2.6% and 4% for 5 mg; 2.2% and 3.4% for 10 mg	1.9% (21.5 events/ 1 K patient yr)	58.3 and 47.6 events respectively
TANDEM-1 & 2 study	Sotagliflozin 200 mg and 400 mg	2.3%–3.4% for 200 mg and 3%-3.4%	0–0.6% (0–3.8 events/1 K patient year)	30–34 events
'FEARS' DATA	SGLT2i: Canagliflozin, Empagliflozin, Dapagliflozin	CANA-48 cases, DAPA-21 cases, EMPA-4 cases	N/A	Overall: 14-fold increase of DKA out of which 71% EDKA

From Prospective Diabetes registry in Germany comprising 31,330 patients, DKA admission rate was 4.81/100 patient-years (Karges et al. 2015). A multinational data from 49,859 children with T1D across three registries and five nations found higher odds of DKA in females (OR 1.23), in ethnic minorities (OR 1.27) and in those with HbA1c $\geq$ 7.5% (OR 2.54) (Maahs et al. 2015).

Table 5 Studies on SGLT2i in T1D with EDKA incidences (Rosenstock et al. 2018; Mathieu et al. 2018; Blau et al. 2017).

The hospitalization rate of HHS is less compared to DKA. According to an estimate it accounts for only 1% of diabetes related hospitalizations. In contrast to DKA, the mortality rate in HHS is considerably higher. It is 15% for HHS compared to 2–5% for DKA (Umpierrez et al. 2002). The possible explanations for this higher mortality rate for HHS are older age and presence of co-morbid conditions (Kitabchi et al. 2009).

4 Pathophysiology

4.1 DKA

Glucose homeostasis is maintained by the intricate balance of two hormones, insulin and glucagon. There are four axes (Fig. 2) which control glucose homeostasis. These are 'Brain-islet axis', 'Liver-islet axis', 'Gut-islet axis' and 'Adipocytes/ myocytes-islet axis'. These axes interplay with positive and negative feedback to maintain glucose homeostasis (Röder et al. 2016).

In DKA this hormonal balance of body is tilted towards counter-regulatory hormones due to absolute insulin deficiency. Due to this shift in balance, while liver uninhibitedly keeps on producing more glucose, peripheral tissues are not able to utilize glucose from blood in the absence of insulin (Gosmanov and Kitabchi 2000). The liver is able to secrete high amounts of glucose due to presence of two metabolic pathways namely gluconeogenesis and glycogenolysis. The gluconeogenic enzymes fructose 1, 6 bisphosphatase, phosphoenolpyruvate carboxykinase (PEPCK), glucose-6-phosphatase, and pyruvate carboxylase are mainly involved. These are stimulated by an increase in the glucagon to insulin ratio and by an increase in circulating cortisol concentrations (DeFronzo and Ferrannini 1987; Stark et al. 2014). This insulin counterregulatory hormone mismatch activates hormone sensitive lipase activity which leads to increase formation of FFAs from the triglycerides (Fig. 3). This FFAs are then beta oxidised to form acetylcoenzyme A into AcAc (Acetoacetic acid) and BOHB (Beta hydroxybutyrate) in hepatic mitochondria. These are major ketone bodies, resulting ketonemia and acidosis (Dhatariya 2016; Barnett and Barnett 2003).

Ketogenesis Conversion of FFAs into ketones in the hepatic mitochondria needs certain conditions. These are lower insulin to glucagon ratio, reduction

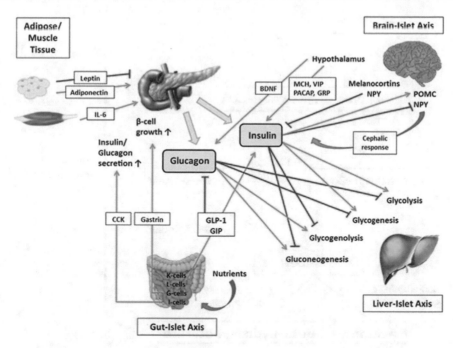

Fig. 2 Interplay of pancreas with brain, liver, gut, adipose and muscle tissue to maintain glucose homeostasis (Röder et al. 2016). The figure is reproduced with permission

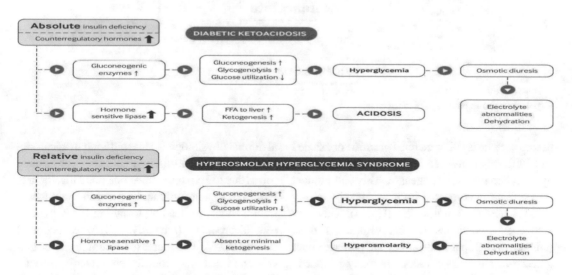

Fig. 3 Pathogenesis of DKA and HHS. Reproduced with permission (Karslioglu French et al. 2019)

in activity of acetyl CoA carboxylase and low levels of malonyl CoA. These eventually trigger transportation of FFAs inside mitochondria by CPT-1 (Carnitine Palmitoyltransferase-1) for conversion to ketones. FFAs in hepatic mitochondria are then broken down into acetyl CoA by beta-oxidation. Two acetyl-CoA molecules are converted into acetoacetyl-CoA by enzyme thiolase. Then this acetoacetyl-CoA is converted into HMG-CoA by HMG-CoA synthase. Then HMG-CoA is converted into acetoacetate by HMG-CoA lyase. Acetoacetate then converted into either acetone through nonenzymatic decarboxylation or into BOHB by beta-hydroxybutyrate dehydrogenase. In extra-hepatic

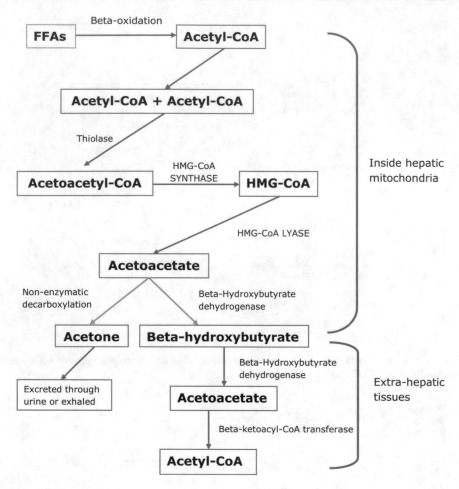

Fig. 4 Metabolic pathways of ketogenesis

tissues acetone is either excreted via urine or exhaled and BOHB is converted into acetoacetate by beta-hydroxybutyrate dehydrogenase. This end product acetoacetate is converted back to Acetyl CoA by beta-ketoacyl-CoA transferase (Fig. 4). This way ketogenesis continues till intervention is done (Dhillon and Gupta 2019). They are preferred over glucose in absence of insulin by many peripheral tissues like brain, and skeletal muscles (Barnett and Barnett 2003; Dhillon and Gupta 2019).

Acidosis As the concentration of Acetoacetic acid and β-hydroxybutyric acid increases in blood, they dissociates completely at physiological pH converting into acetoacetate and β-hydroxybytyrate respectively. This conversion yields hydrogen ion with each molecule which is

at normal physiological state buffered by bicarbonate. In DKA the enormous amount of hydrogen ion forms due to ketogenesis. At one point bicarbonate buffering system fails and hydrogen ion concentration shoots up leading to falling blood pH and low bicarbonate (Dhatariya 2016; Barnett and Barnett 2003). The increased serum levels of glucose and ketones contributes towards osmotic diuresis and hence electrolytes disturbances and dehydration (Karslioglu French et al. 2019).

4.2 HHS

The pathogenesis of HHS differs from DKA significantly. Measurable insulin in T2D is higher than in DKA patients, which is sufficient to

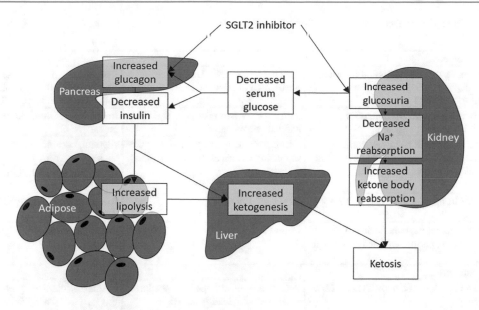

Fig. 5 Mechanism of development of EDKA with SGLT2i (Diaz-Ramos et al. 2019). The slide is reproduced with permission

suppress lipolysis and ketogenesis but inadequate to regulate hepatic glucose production and promote peripheral glucose utilization. Studies have shown the half-maximal concentration of insulin for anti-lipolysis is lower than for glucose utilization by peripheral tissues (Pasquel and Umpierrez 2014; Miles et al. 1983; Umpierrez et al. 1996). The counter-regulatory hormones are high in HHS due to presence of stresses like infection, and myocardial infarction. The reason behind more severe dehydration and hyperglycemia in HHS is that it develops over several days of continued osmotic diuresis. This leads to hypernatremia, especially in older patients with renal impairment. This is worsened by inability to drink adequate water to keep up urinary losses resulting in profound dehydration. Furthermore, this hypovolemia deteriorates glomerular filtration rate as clinically shown by higher creatinine values in HHS and it eventually leads to severe hyperglycaemic state (Umpierrez et al. 2002; Kitabchi et al. 2006).

4.3 EDKA

In T2D on SGLT2i, the lower insulin to glucagon ratio stimulates lipolysis which makes around 20% enhanced lipid oxidation. This happens at the expense of markedly reduced carbohydrate oxidation, which falls by 60%. In the face of lower glucose concentrations, nonoxidative glucose disposal by glycogen synthesis and lactate release also fall by 15% (Rosenstock and Ferrannini 2015). Reduced insulin level causes reduced formation of acetyl-CoA, so inhibition of CPT-I is less. This promotes transport of FFAs into mitochondria and hence ketogenesis (Diaz-Ramos et al. 2019). BOHB levels rises by two folds in fasting and fed state. Plasma lactate level decreases to 20% which reflects reduced carbohydrate utilization. In T1D where absolute insulin deficiency prevails and if carbohydrate availability is drastically reduced then the mild ketosis would lead to ketoacidosis (Rosenstock and Ferrannini 2015) (Fig. 5).

5 Precipitating Factors

The predominant trigger for an acute DKA episode is insulin omission or non-adherence, whereas in HHS infections are most common precipitating factors. In EDKA, there is common association with the use of SGLT2i which ensures good

Table 6 Precipitating factors of DKA, HHS AND EDKA

DKA	HHS	EDKA
Strong factors		
Reduction or repeated omission of insulin dose. Infection: most common are respiratory and UTI	Infections especially UTI and pneumonia in 30–60% cases	Reduction in insulin dose at the context of good glycaemic control in patients on SGLT2i
Non-compliance with insulin, poor control, h/o previous episode	Non adherence to insulin or oral antidiabetic medications	Reduction in carbohydrate intake
Gastroenteritis with persistent vomiting, dehydration. Binge alcohol intake, cocaine or substance abuse	Alcohol abuse, restricted water intake in nursing home residents with diabetes	Alcohol, cocaine or substance abuse
		Failure to stop SGLT2i prior surgery
Acute MI in middle aged T1D or T2D	Acute MI, CVA	Acute MI in middle aged T1D or T2D
Eating disorders, psychiatric disorders, parental abuse, peripubertal and adolescent girls	Depression	Eating disorders, psychiatric disorders
Weak factors		
Pancreatitis, CVA in elderly T1D or T2D, pregnancy	Post cardiac or orthopedic procedure where osmotic load is increased, pregnancy	Chronic alcoholism, pregnancy
Endocrine diseases: Acromegaly, hyperthyroidism, Cushing syndrome	Delay in insulin initiation postoperatively, TPN	Glycogen storage disease
Drugs: corticosteroids, thiazides, Pentamidine, sympathomimetics, second-generation antipsychotics, cocaine, immune checkpoint inhibitors	Drugs: corticosteroids, thiazides, beta-blockers, didanosine, phenytoin, Gatifloxacin, cimetidine, atypical antipsychotics-clozapine, olanzapine	
Remembering 7 'I' can be easy to remember the causes	1. Insulin: Deficiency/insufficiency	5. Infarction: ACS, stroke
	2. Iatrogenic: Steroids, thiazides, atypical antipsychotic drugs	6. Inflammation: Acute pancreatitis, cholecystitis
	3. Infection: Commonest cause	1.7. Intoxication: Alcohol, cocaine
	4. Ischaemia: Gut, foot	

glycaemic control that might lead to reduction in insulin dosage. These triggers higher secretion of counter-regulatory hormones in DKA, HHS and EDKA. Table 6: Precipitating factors in DKA, HHS and EDKA (Pasquel and Umpierrez 2014; Laine 2010; Adeyinka and Kondamudi 2020).

6 Clinical Presentation

DKA and EDKA evolve in hours to days but HHS develops gradually over several days to weeks. Results of a systematic review which included 24,000 children from 31 countries suggested that those who are very young and belong to ethnic minorities are more to present acutely in such manner. Other major risk factors include lean built of body, errors in diagnosis of

T1D, treatment delays, infection as presenting complaint etc. Presence of T1D in family history, on the other hand, makes it unlikely that DKA would be the first presentation (Usher-Smith et al. 2011). Up to 25% patients with new-onset T1D present with DKA (Choleau et al. 2014; Jefferies et al. 2015; Usher-Smith et al. 2011).

Table 7 Features of clinical presentations of DKA, HHS and EDKA (BSPED 2020; Hardern 2003; Trence and Hirsch 2001; Hamdy 2019; Nyenwe et al. 2010; Usher-Smith et al. 2011).

In an otherwise healthy child without any family history of diabetes who presents as with urinary symptoms, it is challenging to diagnose T1D (Rafey et al. 2019). Up to 20% people present with HHS as first presentation of new-onset T2D (Pasquel and Umpierrez 2014). EDKA is difficult to diagnose without detail history and work-up as

Table 7 Clinical presentations of DKA, HHS AND EDKA

Parameter	DKA	HHS	EDKA
History	Short h/o unwellness, hours to days	Unwellness for days to weeks	Moderate length
	h/o failure to comply with insulin therapy	Often a preceding illness like dementia, immobility predisposes	h/o SGLT2i intake and not stopped prior surgery
	h/o mechanical failure of CSII		h/o alcoholism, poor carb intake
Most common early features	Polyuria, polydipsia and polyphagia	Polyuria, polydipsia	Lesser osmotic symptoms but other DKA features are present
	Nausea, vomiting and anorexia	weight loss, weakness, lethargy	
	loss of appetite, diffuse abdominal pain	Seizures which may be resistant to anticonvulsive and phenytoin may worsen HHS, muscle cramps	
	Malaise, generalized weakness, fatigue		
	Rapid weight loss in new-onset T1D	Uncommon: Abdominal pain	
Late features	Dry mucous membrane, poor skin turgor	Same as DKA but dehydration is profound	Same as DKA but dehydration is moderate
	Sunken eyes, hypothermia	Acute focal or global neurologic changes: drowsiness, delirium, focal or generalized seizures, coma, visual changes, hemiparesis, sensory deficits	
	Tachycardia, hypotension,		
	Kussmaul breathing, acetone breath, laboured breath, tachypnea		
	Altered mental status, reduced reflexes		
Features of possible intercurrent infection	Constitutional symptoms: fever, cough, chills, chest pain, dyspnea, arthralgia	Same as DKA and infectious components are most common precipitants in HHS	Same as DKA but very less frequently any infectious components are seen

there is usually very low index of suspicion with normoglycaemia.

7   Laboratory Investigations

The laboratory investigations in DKA, HHS and EDKA are aimed primarily for diagnosis and to determine its severity. Then the next step is to identify the underlying causes, early identification of complication and monitoring of response to therapy. As acute metabolic decompensation is an emergency so just after history, clinical examination and provisional diagnosis management starts. Blood and urine samples are taken to do initial investigations without delay. Work-up for DKA and EDKA are same. Table 8 Initial and subsequent work-ups in the management of DKA

and HHS (Wolfsdorf et al. 2018; Savage et al. 2011; Scott et al. 2015; Khazai and Umpierrez 2020; Gosmanov and Nimatollahi 2020; Rawla et al. 2017).

8   Differential Diagnosis

The acute hyperglycemic emergencies DKA and HHS are at the top of differential list for each other. The only biochemical feature that differentiates DKA from EDKA is serum glucose level: normal in EDKA and raised in DKA. The distinguishing feature of HHS is presence of high serum osmolality due to extremely high serum glucose levels. The hyperglycemia in HHS is generally greater than 30 mmol/l (Scott et al. 2015; Westerberg 2013).

Table 8 Work-up for DKA, HHS and EDKA

Initial tests to order

Tests	DKA results	HHS results	Comments
Plasma glucose	> 13.9 mmol/L (>250 mg/dL)	>33.3 mmol/L (>600 mg/dL)	Elevated except in EDKA it is <11 mmol/L (<200 mg/dL)
Venous blood gas	pH varies from 7.00 to 7.30	Usually >7.30	Venous pH is just 0.03 lower than arterial pH. As arterial sampling is painful and risky so venous sample is now commonly taken
Capillary or serum ketones	BOHB ≥3.8 mmol/L in adults and ≥ 3.0 mmol/L in children	BOHB is negative or low	Out of three ketones, BOHB (beta hydroxybutyrate) is early and abundant ketone that is checked first from serum or point-of-care device. In early DKA, acetoacetate (AcAc) can be measured, its result has a high specificity but low sensitivity. Acetone not done as it is volatile
HbA1$_c$ level	Usually high	Usually high	To evaluate the glycaemic control it is done but in good glycaemic control acute hyperglycaemic crisis may precipitate
Urinalysis	Positive for glucose and ketones. Positive for leukocytes and nitrites if there is an infection	Positive for glucose and but usually not ketones. Positive for leukocytes and nitrites if there is an infection	In mixed presentation of DKA and HHS urine ketone also found in HHS
Serum bicarbonate	From 18 mEq/L or mmol/L to <10 depending the grade	> 15 mEq/L	Bicarbonate is an important test to diagnose and grade
Serum bun	Increased due to dehydration	Markedly increased due to severe dehydration	Pre-renal azotaemia
Serum creatinine	Increased due to dehydration	Markedly increased due to severe dehydration	Once dehydration is corrected the creatinine level returns to normal
Serum sodium	Usually low	Variable, usually low but can be high	Total Na deficit in DKA is 7–10 mEq/kg and in HHS is 5–13 mEq/kg. Hypernatraemia with hyperglycaemia indicates profound dehydration
Serum potassium	Usually elevated	Usually elevated but decreased in severe cases	Total deficit of K in DKA is 3–5 mEq/kg and in HHS 4–6 mEq/kg. K is elevated initially due to extracellular shift caused by insulin deficiency or insufficiency, hypertonicity and acidaemia. Low K on admission is a sign of severe case
Serum chloride	Usually low	Usually low	Cl loss is 3–5 mEq/kg in DKA and 5–15 mEq/kg in HHS
Serum magnesium	Usually low	Usually low	Mg deficit is 1–2 mEq/kg in DKA and 0.5 to 1 mEq/kg in HHS
Serum calcium	Usually low	Usually low	Ca deficit is 1–2 mEq/kg in DKA and 0.5–1 mEq/kg in HHS
Serum phosphate	Normal or elevated	Usually low	1 mmol/L deficit in DKA but initially shows normal or elevated. After insulin therapy it decreases. In HHS 3–7 mmol/kg is lost due to diuresis

(continued)

Table 8 (continued)

Tests	DKA results	HHS results	Comments
Initial tests to order			
CBC	Usually elevated	Usually elevated	Leukocytosis correlates with ketones but >25,000/ microliter indicates infection and indicates further evaluations
LFT	Usually normal	Usually normal	Abnormal results due to fatty liver or congestive heart failure may be found
Serum amylase	Usually elevated	Usually elevated	In majority of DKA cases amylase is elevated but mostly due to nonpancreatic causes. In HHS if elevated then pancreatitis should be ruled out
Serum lipase	Usually normal	Usually normal	In elevated amylase level measuring lipase level is useful to differentiate pancreatitis
Serum osmolality	Variable	High, usually ≥320 mOsm/L	Twice measured Na and K plus Glucose makes the osmolality. Urea usually not counted as it is freely permeable. A linear relationship is there between effective osmolality and mental state in HHS. Neurological deficits begin above 320 and stupor or coma come above 340 mOsm/L
Additional tests to consider			
Chest X-RAY	May have findings of pneumonia	Variable, may be compatible with pneumonia	The commonest infections are pneumonia and UTI
ECG	May show findings of MI or hyperkalaemia or hypokalaemia	May show findings of MI or hyperkalaemia or hypokalaemia	Precipitating CAD and severe electrolyte abnormalities are common in both DKA and HHS
Cardiac biomarkers	In suspected MI should be done	In suspected MI should be done	Cardiac troponins are elevated in suspected MI
Body fluid culture	To rule out sepsis blood, urine or sputum culture are needed	To rule out sepsis blood, urine or sputum culture are needed	Fever, leukocyte count >25,000/ microliter should raise the question of infective focus
Creatinine phosphokinase	In cocaine abuse rhabdomyolysis is common	Less common	↑ In rhabdomyolysis. pH and serum osmolality mildly elevated. Blood glucose and ketone are normal. Myoglobinuria/ hemoglobinuria + in urine
Serum lactate	Normal if concomitant lactic acidosis absent	Normal if concomitant lactic acidosis absent	Elevated in lactic acidosis

Ketoacidosis besides being present in diabetes can also occur during starvation and alcoholism. It is important to rule out other causes of high anion gap acidosis like salicylate poisoning, methanol intoxication and lactic acidosis (Keenan et al. 2007). Since abdominal pain and vomiting episodes are often present in such patients, other etiological causes of acute abdomen like pancreatitis and gastroenteritis should also be considered in differential diagnosis (Keenan et al. 2007).

Due to presence of focal neurological deficit, HHS is very commonly confused with stroke (Umpierrez et al. 2002).

Table 9 Differential diagnosis of acute hyperglycaemic crisis (Westerberg 2013; Rawla et al. 2017; Keenan et al. 2007).

Table 9 Differential diagnosis of DKA, HHS and EDKA

Condition	Differentiating features	Tests to rule out
HHS	HHS patients are usually older and commonly with T2D. Symptoms evolve insidiously, more frequently mental obtundation and shows focal neurological signs. Blood glucose is very high in HHS whereas in EDKA it is normal, other distinguishing features are similar to DKA	Blood glucose >33.3 mmol/L, serum osmolality is >320 mOsm/kg and ketones are normal or mildly elevated. Anion gap is variable but usually <12 mEq/L, pH is >7.30 and bicarbonate is >15 mEq/L
DKA	DKA patients are younger and leaner T1D, usually present with abdominal pain and vomiting	pH < 7.30, HCO3 < 15 mmol/L, anion gap >12 mEq/L and ketones are strongly positive
Lactic acidosis	DKA and HHS like presentation but in pure form of lactic acidosis blood glucose and ketone are normal but lactate is raised. History of diabetes may not be there	In T1D with sepsis, lactic acidosis sometime precipitate. Bicarbonate, pH and anion gap are similar to DKA but lactic acid >5 mmol/L. Blood glucose and ketones are normal
Starvation ketosis	Starvation ketosis mimics partly with DKA. It is the consequence of prolonged inadequate availability of carbohydrate. Which results compensatory lipolysis and ketogenesis to provide fuel substrate for muscle	Blood glucose is normal or low, blood ketone is normal but urine contains huge amount of ketones. Blood pH is normal and anion gap is just mildly elevated
Alcoholic ketoacidosis	It results in chronic alcoholics who skips meals and depends on ethanol as main source of calorie for days to weeks. Ketoacidosis is triggered when alcohol and calorie intake abruptly decreases. Signs of chronic liver disease such as spider naevi, palmer erythema, leukonychia, easy bruising, jaundice and hepatomegaly might be present	There is mild to moderate metabolic acidosis with elevated anion gap. Serum and urine ketones are positive. There might be hypoglycaemia
Salicylate poisoning	History is crucial to differentiate. Salicylate poisoning results an anion gap metabolic acidosis with respiratory alkalosis	Salicylate is positive in blood and urine, blood glucose is normal or low, ketones are negative, osmolality is normal. Interestingly, salicylate makes false-positive and false-negative urinary glucose presence
Paracetamol overdose	History is very crucial to differentiate. Confusion, hyperventilation, tinnitus and signs of pulmonary oedema might be found	A positive result for serum and urine paracetamol could be found but might not be in toxic range. Blood sugar may be normal or low
Toxic substance ingestion	History is crucial to differentiate. Common toxic substances are methanol, ethanol, ethylene glycol and propylene glycol. Paraldehyde ingestion makes strong odour in breath	Serum screening for toxic substances might yield the clue. Calcium oxalate and hippurate crystals in urine indicate ethylene glycol ingestion. These organic toxins can produce anion gap and osmolar gap due to low molecular weight
Stroke	Symptoms develop rapidly, in seconds to minutes. There might be limb and facial weakness	Cranial CT or MRI is diagnostic
Uremic acidosis	High BUN and creatinine but normal glucose. A history also important	Very high serum creatinine and BUN are found

9 Management: General

Successful management of DKA, HHS and EDKA needs 5 major components to rectify as follows (Hamdy 2019):

1. Correction of dehydration
2. Correction of hyperglycaemia and ketoacidosis
3. Correction of electrolyte abnormalities
4. Identification of comorbid and precipitating factors and
5. Frequent monitoring and prevention of complications

Once acute metabolic crisis of diabetes is recognized the patient needs to be hospitalized in emergency or acute medical unit or in HDU (High Dependency Unit) or in ICU depending on grading.

Table 10 Markers of severity that requires HDU/ICU admission

Markers of severity	DKA/EDKA	HHS
Venous pH	pH < 7.1	< 7.1
Blood ketones	> 6 mmol/L	> 1 mmol/L
Serum bicarbonate, anion gap	< 5 mmol/L, > 16 mmol/L	
Potassium	< 3.5 mmol/L or > 6 mmol/L	< 3.5 mmol/L or > 6 mmol/L
Systolic BP, pulse	< 90 mmHg, >100 or < 60 bpm	< 90 mmHg, >100 or < 60 bpm
Urine output	< 0.5 mL/kg/h or evidence of AKI	< 0.5 mL/kg/h or evidence of AKI
Mental status, SpO_2	GCS <12 or abnormal AVPU, <92%	GCS <12 or abnormal AVPU, <92%
Sodium, osmolality		>160 mmol/L, >350 mOsm/kg
Comorbidities	Hypothermia, ACS, CHF or stroke	Hypothermia, ACS, CHF or stroke

Table 11 Typical water and electrolyte deficits in DKA, EDKA and HHS (Umpierrez et al. 2002; Savage et al. 2011; Scott et al. 2015)

Variables	DKA/EDKA (deficit/kg body wt)	HHS (deficit/kg body wt)
Water	100 ml	100–200 ml
Na^+	7–10 mEq	5–13 mEq
K^+	3–5 mEq	5–15 mEq
Cl^-	3–5 mEq	4–6 mEq
$PO4^-$	5–7 mEq	3–7 mEq
Mg^{2+}	1–2 mEq	1–2 mEq
Ca^{2+}	1–2 mEq	1–2 mEq

Table 10 The markers of severity in DKA, HHS and EDKA for HDU/ICU admission (Savage et al. 2011; Scott et al. 2015).

The markers of severity should be assessed and recorded (Table 11).

Management should start with prompt assessment of ABCDE (Airway, Breathing, Circulation, Disability-conscious level and Exposure-clinical examination) at emergency department. Acute metabolic crisis in diabetes leads to profound water and electrolyte loss due to osmotic diuresis by hyperglycaemia. In EDKA due to very nominal rise of glucose, water deficit is not profound like DKA but is significant due to ketoacidosis. Without finding the cause of acute metabolic crisis, the management is not complete. Without preceding a febrile illness or gastroenteritis, DKA in a known diabetes patient is usually due to psychiatric disorders such as eating disorders and failure of appropriately administering insulin (Wolfsdorf et al. 2018). Comparatively more aggressive fluid replacement in HHS is needed than DKA to expand intra and extra vascular volume. The purpose is to restore normal kidney perfusion, to normalize sodium concentration and osmolality. DKA usually resolves in 24 h but in HHS correction of electrolytes and osmolality takes 2–3 days. Usually HHS occurs in elderly with multiple co-morbidities, so recovery largely depends on previous functional level and precipitating factors. In EDKA if SGLT2i is suspected, it should be stopped immediately and should not restart unless another cause for DKA is found and resolved (Evans 2019).

9.1 Management: From Admission to 24–48 Hours

Table 12 Shows the details of management from admission onwards (Wolfsdorf et al. 2018; BSPED 2020; Savage et al. 2011; Scott et al. 2015; Evans 2019) (Fig. 6).

Table 12 Management of DKA, EDKA and HHS IN adults and children

Intervention	DKA, EDKA and HHS 0–60 min: Resuscitate, diagnose and treatment	Monitoring, ongoing lab work-up
ABCDE	Fast assessment to grade patient: Shocked, comatose, moderate or mild cases	**First tests:** CBC, U & E, and venous blood gas: pH, HCO3, CRP, glucose, ECG, CXR, infection screen if indicated by blood and urine culture
	In shocked and comatose patients with vomiting an airway, N/G tube have to insert	
	100% oxygen by face mask	**HOURLY**: Capillary blood glucose, ketones, cardiac monitoring, BP, pulse, respirations, pulse oximetry, fluid input/output chart, neurological observations
	IV cannula have to put and blood and urine sample have to take. Cardiac monitor with pulse oximetry have to attach to assess pulse, BP, T wave etc.	
	Blood and urine sample to send for culture for infection screening	**TARGET:** Reduction of glucose by 3 mmol/L/h, ketones by 0.5 mmol/L/h and increasing HCO3 by 3 mmol/L/h
	Elderly HHS patients are at high risk of pressure sore. Foot assessment should be done and should apply heel protectors in those with neuropathy, PVD or lower limb deformity	
Bedside diagnosis	Capillary blood test, point of care blood ketone test and if not available urine dipsticks for 15 s where a > ++ indicates positive	
	Comatose and shocked patients should move to HDU/ICU immediately after starting IV fluid	
	Use of blood gas machine at bedside can promptly test pH, urea, electrolytes, glucose etc. while first blood sample is sent to laboratory	
Initial fluid replacement	Crystalloid fluid such as normal saline is best for volume expansion rather than colloid fluid. Typical fluid deficit is 100 mL/kg and should be corrected within 24–48 h	
	All children with mild, moderate or severe DKA who are not shocked should receive an initial bolus of 10 mL/kg 0.9% NaCl IV over 60 min stat	
	Shocked children should get bolus of 20 mL/kg 0.9% NaCl IV over 15 min stat	
	The maintenance fluid in children should be calculated from Holliday-Segar formula. It is: 100 mL/kg/day for first 10 kg body weight, then 50 mL/kg/day for next 10 kg and 20 mL/kg/day for each kg above 20 kg. This amount should be divided by 24 to get hourly maintenance amount	
	A 5%, 7% and 10% fluid deficit is assumed for mild, moderate and severe DKA respectively. Initial bolus should be subtracted from deficit and then divided by 48 h and adding this to hourly maintenance fluid volume	

(continued)

Table 12 (continued)

Intervention	DKA, EDKA and HHS 0–60 min: Resuscitate, diagnose and treatment	Monitoring, ongoing lab work-up
	HOURLY RATE = [DEFICIT- INITIAL BOLUS] /48 + MAINTANANCE PER HOUR	
	Alert, not clinically dehydrated, no nausea or vomiting children do not always need IV fluids even their ketone is high. They might tolerate oral rehydration and s.c insulin but they do require continuous monitoring to ensure improvement and ketone is falling	
	Adult DKA, EDKA patients should get 1–1.5 L 0.9% NaCl saline in first hour. In DKA average 6 L fluid loss occurs. Slower administration in young, elderly, pregnant, heart and renal failure cases	
	Adult HHS patients should get 1–1.5 L 0.9% NaCl in first hour provided cardio-renal status allows. In HHS average 7–9 L fluid loss occurs	
Insulin therapy	Insulin should start immediately in DKA and HHS if potassium level is >3.3 mEq/L. A 50 units of soluble insulin (e.g. Actrapid) in 49.5 mL of 0.9% NaCl saline to be mixed to make 1 unit/mL to administer through infusion pump	
	Two types of insulin regimens are used in DKA and HHS. First one is fixed rate IV regular insulin infusion as known as **FRIII** (fixed rate intravenous insulin infusion) at 0.14 units/ kg/ h with no initial bolus. Second one is 0.1 units/kg/h IV bolus followed by FRIII at a rate of 0.1 units/kg/h continuous IV infusion	
	In EDKA insulin infusion with 5–10% dextrose in saline helps to settle ketoacidosis	
	In young children with mild to moderate DKA 0.05 units/ kg/h is sufficient to control and in severe DKA and in adolescent patients 0.1 units/kg/h should start **after 1 h of fluid replacement therapy**	
	In children with HHS insulin need is less. So a dose of 0.025 to 0.05 units/kg/h should start **after 1 h of fluid replacement**	
	Insulin pump should stop when FRIII is started. Long acting basal insulin should continue at the usual dose throughout the treatment, it helps to shorten the length of stay after recovery	
	If blood glucose does not fall by 10% or 3 mmol/L in first hour then a dose of	

(continued)

Table 12 (continued)

Intervention	DKA, EDKA and HHS 0–60 min: Resuscitate, diagnose and treatment	Monitoring, ongoing lab work-up
	0.14 units/kg of regular insulin should be given IV bolus and then to continue FRIII at running dose	
	Once blood glucose falls near 13.9 mmol/L (250 mg/dL), then insulin infusion should be reduced to 0.02–0.05 units/kg/h and a 5% dextrose in saline have to add while maintaining blood glucose 11–17 mmol/L (200–300 mg/dL)	
	Rapid reduction of blood glucose should be avoided to prevent sudden osmolar changes and cerebral oedema	
	Insulin injection by a sliding scale is no longer recommended	
Potassium replacement	In acute metabolic crisis in diabetes potassium loss is around 3–15 mEq/kg. Insulin therapy, correction of acidosis and hyperosmolality drive potassium into cells causing serious hypokalemia. So to prevent complications of hypokalemia like respiratory paralysis and cardiac dysrhythmia insulin therapy should be withheld if K level is <3.3 mEq/L at baseline while fluid therapy is going on	
	If K is >5.5 mmol/L = NO potassium	
	If K is 3.5–5.5 mmol/L = 20–40 mmol/L mixed with 0.9% NaCl saline	
	If K is <3.5 mmol/L = 40 mmol/L over 1–2 h with cardiac monitoring	
	Urine output of >50 mL/h should be there while patient on K therapy. The hydration status should be evaluated clinically regularly. If eGFR is <15 mL/min then consultation with renal team is needed before adding K	
	If K level falls <3.3 mEq/L in any time during therapy, insulin should be withheld and K 40 mmol/L should be added in each liter of infusion fluid	
Vesopressor and anticoagulant therapy	If hypotension persists after initial forced hydration, then a vasopressor agent should be administered. Dopamine or Noradrenaline can be used. Dopamine increases stroke volume and heart rate whereas Noradrenaline increases mean arterial pressure	
	Dopamine 5–20 micrograms /kg/min IV infusion, subject to adjustment as per response	
	Noradrenaline 0.5–3 micrograms/min IV infusion and titration as per response. Can be used maximum 30 micrograms/min	

(continued)

Table 12 (continued)

Intervention	DKA, EDKA and HHS 0–60 min: Resuscitate, diagnose and treatment	Monitoring, ongoing lab work-up
	Diabetes and hyperosmolality make increased risk of venous thromboembolism (VTE). It is similar to patients with acute renal failure, acute sepsis or acute connective tissue disease	
	The risk of VTE is greater in HHS than DKA. Hypernatraemia and increased antidiuretic hormone promote thrombogenesis	
	Patients with HHS who are at risk or suspected with thrombosis or ACS should receive prophylactic low molecular weight heparin (LMWH) during admission. There are increased risk of VTE beyond the discharge, so LMWH should continue for 3 months after discharge (Keenan et al. 2007)	
	1–6 Hour: Assessment and monitoring therapy	
Fluid Replacement Continues, FRIII Continues, K replacement if needed	0.9% NaCl 1 l over 2 h, then	**WORK-UP:** **2 HOURLY** serum K, HCO3, venous blood gas for pH **HOURLY**: Capillary blood glucose, ketones, cardiac monitoring, BP, pulse, respirations, pulse oximetry, fluid input/output chart, neurological observations
	0.9% NaCl 1 l over 2 h, then	
	0.9% NaCl 1 l over 4 h	
	After first hour therapy of 1–1.5 L if signs of severe dehydration such as orthostatic hypotension or supine hypotension, poor skin turgor etc. persists then 1 l per hour have to continue till signs resolved	
	These patients' when symptoms are resolved then continue to receive infusion fluid on the basis of corrected sodium	
	CORRECTED Na$^+$ = MEASURED Na$^+$ + (GLUCOSE in mmol/L- 5.6)/3.5	
	In hyponatraemic patients: 0.9% NaCl at 250–500 mL/h and when blood glucose reaches 11 mmol/L (200 mg/dL), fluid should be changed to 5% dextrose with 0.45% NaCl at 150–250 mL/h	
	In hypernatraemic or eunatraemic patients: 0.45% NaCl at 250–500 mL/h and when blood glucose reaches 11 mmol/L (200 mg/dL), it should be changed to 5% dextrose with 0.45% NaCl at 150–250 mL/h	
	Continue FRIII	
	In young children with mild to moderate DKA 0.05 units/ kg/h is sufficient to control and in severe DKA and in adolescent patients 0.1 units/kg/h should start **after 1 h of fluid replacement therapy**	
	In children with HHS insulin need is less. So a dose of 0.025 to 0.05 units/kg/h should start **after 1 h of fluid replacement**	

(continued)

Table 12 (continued)

Intervention	DKA, EDKA and HHS 0–60 min: Resuscitate, diagnose and treatment	Monitoring, ongoing lab work-up
	Continue basal insulin if taking before K replacement if needed	
	If infection is suspected by and evidenced by CXR, DC >25,000, neutrophil >80% then a broad spectrum injectable antibiotic have to start. Culture report takes time so need not wait for that	
Bicarbonate therapy	At pH >7.0 insulin therapy blocks lipolysis and resolves ketoacidosis without use of HCO3. Use of HCO3 in these cases may cause hypokalemia, decreased tissue oxygen uptake and risk of cerebral oedema	
	Arterial pH 6.9–7.0 = 50 mmol NaHCO3 in 200 mL sterile water with 10 mEq KCl may be administered over an hour till pH >7.0	
	Arterial pH < 6.9 = 100 mL of NaHCO3 in 400 mL sterile water with 20 mEq KCL at the rate of 200 mL/h for 2 h until pH >7.0	
Phosphate, magnesium and calcium therapy	Very rarely used though there are some nominal deficits. But in symptomatic cases these are supplemented	
	Significant malnutrition is associated with such deficits	
	6–24 HR: Improvement & resolution monitoring	
Fluid Replacement Continues, FRIII Continues, K replacement if needed	0.9% NaCl 1 l over 4 h, then	**WORK-UP:** **6 HOURLY** and then **12 HOURLY** serum K, HCO3, venous blood gas for pH **HOURLY:** Capillary blood glucose, ketones, cardiac monitoring, BP, pulse, respirations, pulse oximetry, fluid input/output chart, neurological observations
	0.9% NaCl 1 l over 6 h, then	
	0.9% NaCl 1 l over 6 h.	
	Continue FRIII	
	K replacement if needed	
	Once blood glucose falls near 13.9 mmol/L (250 mg/dL), then insulin infusion should be reduced to 0.02–0.05 units/kg/h and a 5% dextrose in saline have to add while maintaining blood glucose 11–17 mmol/L (200–300 mg/dL)	
	Resolution criteria FOR DKA, EDKA and HHS	
	Criteria for resolution in DKA, EDKA (except glucose)	
	1. Plasma glucose <11 mmol/L (< 200 mg/dL)	
	2. Serum HCO3 is >18 mEq/ L	
	3. Blood ketones <0.6 mmol/L	
	4. Venous pH is >7.3, and	
	5. Anion gap is <10	
	Criteria for resolution of HHS:	
	1. Plasma glucose <14–16.7 mmol/L (250–300 mg/dL)	
	2. Plasma osmolality <315 mOsm/kg	

(continued)

Table 12 (continued)

Intervention	DKA, EDKA and HHS 0–60 min: Resuscitate, diagnose and treatment	Monitoring, ongoing lab work-up
	3. Improvement in haemodynamic and mental status	
	Resolution pitfalls: Urinary ketone clearance takes time even after resolution. As BOHB from blood converts to form AcAc after resolution which is abundant in urine	
	HCO3 alone cannot be relied as resolution of DKA. It is due to high amount of 0.9% NaCl saline infusion causes hypercholeraemic acidosis which lowers HCO3	
	24–48 Hours: resolution & discontinuation of FRIII	
FRIII to VRIII	If DKA/ HHS is resolved: Ketones <0.6 mmol/L but NOT eating & drinking then switch from FRIII to **VRIII** (Variable Rate Intravenous Insulin Infusion)	
	VRIII is based on standard rate such as glucose <4 mmol/L = 0 units/kg/h, 4.1–8 mmol/L = 1 units/kg/h, 8.1–12 mmol/L = 2 units/kg/h and so on	
VRIII to S.C. Insulin	VRIII can be discontinued at mealtime. If earlier taking subcutaneous insulin the same insulin can restart with the diabetes team advice of titration	
	VRIII have to continue 30–60 min after first subcutaneous insulin injection	
For newly diagnosed T1D and T2D: Insulin therapy	Total last 24 h insulin should be added and 30% reduction is done. This value have to divide by 5 and 1/5th is given with each meal as rapid acting insulin and 2/5th can be given as basal analogue insulin which is called BASAL BOLUS REGIMEN	
	The 30% reduced amount from last 24 h total insulin use can be used as TWICE DAILY REGIMEN. The amount have to divide by 3 and 2/3 have to take with breakfast and 1/3 with evening meal within the interval of 12 h	
VRIII TO CSII	To reconnect the insulin pump, normal basal rate have to start and a mealtime bolus have to be given. VRII then have to stop 1 h later	

9.2 Management: Acute Hyperglycaemic Crisis Due to COVID-19

The pandemic COVID-19 infection increases the risk of precipitating atypical DKA, HHS or mixed crisis and stress hyperglycaemia with ketones. The recent guideline from ABCD (Association of British Clinical Diabetologists) named 'COVID: Diabetes' (**CO**ncise ad**V**ice on **I**npatient **D**iabetes) has outlined to manage COVID-19 in hyperglycaemic crisis in diabetes (ABCD

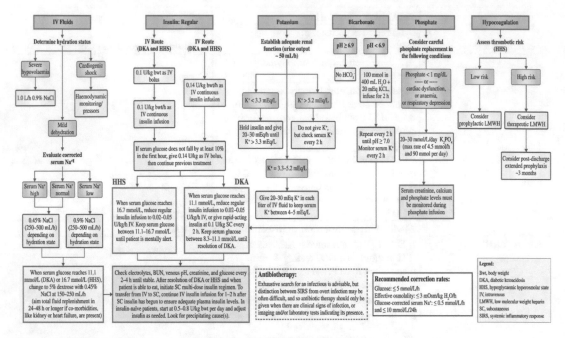

Fig. 6 DKA and HHA management algorithm reproduced with permission (Cardoso et al. 2017)

2020). This guideline is based on UK experience of COVID-19 management and will be updated further when more evidences will be available. COVID-19 infection in known or unknown people with diabetes increases the risk of acute hyperglycaemia with ketones, DKA and HHS. Poorly controlled elderly diabetes patients are more susceptible to COVID-19 and its complications. Because hyperglycaemia can subdue immunity by disrupting the normal function of WBC and other immune cells. Good glycaemic control and following sick day rules are key to reduce risk apart from taking personal protection and social distancing.

Table 13 shows COVID-19 specific management of acute hyperglycaemic crisis paraphrased from the 'COVID: Diabetes' guideline (ABCD 2020).

9.3 Management: Some Controversial Issues

- **0.9% NaCl vs Hartmann's solution**: In a recent RCT (Yung et al. 2017), comparing Hartmann's solution with 0.9% NaCl in 77 children with DKA, it was observed that slightly quicker resolution of acidosis can be achieved with Hartmann's solution in severe DKA. There was however, no difference regarding time required to shift from intravenous to subcutaneous insulin.

- **0.9% NaCl vs Ringer Lactate solution**: A RCT showed no benefit from using Ringer Lactate solution compared with 0.9% NaCl in terms of pH normalization. But Ringer Lactate solution made longer time to reach blood glucose level of 14 mmol/L because lactate converts into glucose (Van Zyl et al. 2011).

- **0.9% NaCl vs Plsma-Lyte 148**: The concern regarding excessive administration of normal saline in DKA is hyperchloremia which can lead to non-anion gap metabolic acidosis. Although self-limiting in nature, this hyperchloremic metabolic acidosis is now believed to have a harmful impact on multiple organs of body like kidneys, myocardium etc. (Eisenhut 2006; Kraut and Kurtz 2014). Plasma-Lyte 148 when compared to normal saline has shown to decrease occurrence of

Table 13 COVID-19 and acute hyperglycaemic crisis in diabetes (ABCD 2020)

Changes seen	Key difference with COVID-19	Action suggested
Risk of early admission	T2D and those on SGLT2i are greater risk	On admission blood glucose checking for everyone
	COVID-19 precipitates DKA or HHS or atypical mixed type	Ketones for all diabetes admission
		Ketones for everybody with admission glucose >12 mmol/L
	Risk of hyperglycaemia with moderate ketones due to stress hyperglycaemia	SGLT2i and Metformin tablets should be immediately stopped on admission
		Safety of ACEi, ARB and NSAID should be reviewed
		10–20% glucose should be used where ketosis persists even usual protocol of DKA is used
Severely sick on admission	Fluid infusion rate may differ in DKA/ HHS and there is evidence of 'lung leak' or myocarditis	After correcting dehydration, rate of fluid infusion should be adjusted in lung leak or myocarditis cases
		Early diabetes specialist team and critical care team involvement needed
Inpatient area	Due to huge demand infusion pumps may not be enough as huge need in ICU	Subcutaneous insulin have to start with basal insulin support to manage hyperglycaemia, DKA or HHS or mixed cases
ICU	Insulin resistance is significantly increased in T2D admitted in ICU	Insulin infusion protocols need amendment. It is seen patients need 20 units/h even
	Higher doses of insulin is required	Patients sometime nursed prone so feeding may be interrupted accidentally with risk of hypoglycaemia

hyperchloremia (Andrew and Patrick 2018; Chua et al. 2012). A systematic review by Gershkovich et al. (Gershkovich et al. 2019) might help aid the decision regarding fluid choice in future.

- **0.9% NaCl vs Ringer Acetate solution:** Though Ringer Acetate is not a popular choice but it has almost similar composition like Ringer Lactate. But its use in hepato-renal emergencies are established (Ergin et al. 2016). Figure 7 shows water shift in hyperglycaemic emergencies with different infusion fluids. The figure is reproduced with permission (Cardoso et al. 2017).
- **Infusion rate**: Regarding infusion rate, rapid administration is feared to increase likelihood of cerebral edema especially in children and young adults. The JBDS guidelines therefore recommend gradual correction of fluid deficit over 48 h unless clear signs of hypovolemic shock are present (Dhatariya 2014).
- **Arterial or venous sample**: the difference between venous and arterial pH is 0.02–0.15 and the difference between arterial and venous HCO3 is 1.88 mmol/L. These neither affect the diagnosis nor the treatment. But getting

arterial sample is risky and painful. So venous sample is widely accepted.
- The target with fluid administration in HHS is to achieve an hourly drop of 3–8 mOsm/kg in osmolality and 5 mmol/L in glucose. Some adjustments in fluid administration rate and solution type are required if these targets are not being met (Scott et al. 2015). These scenarios are mentioned in Table 14 (Scott et al. 2015).

9.4 Management: DKA and EDKA IN Pregnancy

DKA is an emergency during pregnancy and may cause fetal loss which is around 10–25%. The incidence rate is 1–3%. The main causes and precipitating factors are (Savage et al. 2011):

- Starvation: accelerated maternal response ends up in DKA in women with diabetes
- Increased flux of glucose from mother to fetus and placenta: due to increased transporter GLUT-1.

Fig. 7 *ICC* Intracellular compartment, *ISC* Interstitial compartment, *IVC* Intravascular compartment. **Panel A**: Total body water distribution in normal state; **Panel B**: After correction of water deficit with 5% Dextrose water shows suboptimal replenishment of IVC, ISC and excessive rehydration of ICC; **Panel C**: Correction with 0.9% NaCl made exclusive distribution in extracellular compartment resulting excessive hydration of IVC and ISC; **Panel D**: Correction with 0.45% NaCl shows replenishment similar to fluid lost from IVC, ISC and ICC. It is probably the best option; **Panel E**: Correction with 0.225% resulted in suboptimal replenishment of IVC, ISC but excessive hydration of ICC

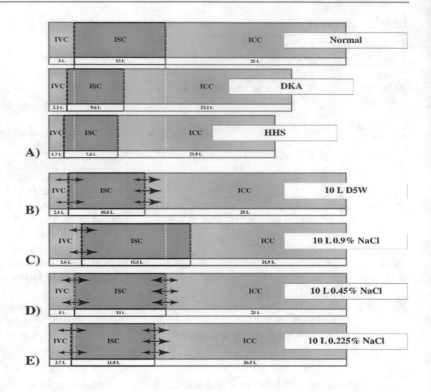

Table 14 Scenarios with serum osmolality and fluid infusion

Scenario	Solution
Plasma osmolality declining at appropriate rate but plasma sodium increasing[a]	Continue 0.9% normal saline
Plasma osmolality declining inappropriately (<3 mOsm/kg/h) or increasing with inadequate fluid balance	Increase rate of 0.9% normal saline
Plasma osmolality increasing with adequate fluid balance	Switch to 0.45% normal saline
Osmolality falling at rate > 8 mOSm/kg/h	Decrease rate of 0.9% normal saline

[a]With fall in serum glucose level, rise in serum sodium level is expected due to shift of water in intracellular space. Drop of blood glucose by 5.5 mmol/L = rise of Na by 2.4 mmol/L (Scott et al. 2015)

- Higher progesterone level: induces respiratory alkalosis which results in metabolic acidosis that reduces buffering capacity.
- Precipitating factors: UTI, hyperemesis gravidarum, new onset T1D, KPD, insulin omission, insulin pump malfunction, glucocorticoid use for inducing fetal lung maturity, use of terbutaline to prevent premature labour.
- DKA and EDKA management is same like non-pregnant cases.

Table 15 Key calculations

Anion gap	Anion gap = Na – (Cl + HCO3); normal is 12 ± 2 mmol/L
	In DKA anion gap is 20–30 mmol/ L.
	An anion gap >35 mmol/L suggests concomitant lactic acidosis
Corrected sodium	Corrected Na = measured Na +2 (Glucose-5.6)/5.6
	Corrected Na is needed to estimate fluid replacement in DKA/HHS when dehydration is mild to moderate
	Web based calculation: https://www.mdcalc.com/sodium-correction-hyperglycemia
Effective osmolality	Serum osmolality = 2Na + glucose + urea
Fluid calculation in children	REQUIREMENT = DEFICIT + MAINTENANCE
	Holliday – Segar formula:
	100 mL/kg/day for first 10 kg
	50 mL/ kg/day for next 10 to 20 kg
	20 mL /kg/day for each kg above 20 kg
	Hourly rate = ({deficit – Initial bolus} /48 h) + maintenance/h

9.5 Management: Key Calculations

Table 15 shows the key calculations needed during management of acute hyperglycaemic crisis (Wolfsdorf et al. 2018).

10 Complications

The probable complications of DKA and HHS are tabulated in Table 16 (Savage et al. 2011; Scott et al. 2015; Khazai and Umpierrez 2020).

Cerebral Edema

Cerebral edema (CE)' is rare and most feared iatrogenic complication of DKA in younger children and in newly diagnosed T1D. It is associated with high mortality and neurodisability & neurocognitive difficulties in survived cohorts. Headache, lethargy, papillary changes and seizure are key manifestation.

Risk of CE found in a study with higher plasma urea, lower arterial pCO_2 and $NaHCO_3$ therapy in DKA (Glaser et al. 2001). Interleukin-1 and 6 (IL-1 and IL 6) are the cytokines that initiate the inflammatory response accompanied by DKA. It is postulated that this IL-1 is linked with the pathogenesis of CE. NLRP3 (nucleotide-binding domain and leucine-rich repeat pyrin 3 domain) is an inflammasome which generates active form of IL-1 in response to hyperglycaemia acts as osmosensors to cause CE in DKA. It contributes to CE and infarction by making tight junctions leaky (Eisenhut 2018). Some studies have found that initial bolus of rehydration fluid and bolus insulin might have a role (Carlotti 2003).

Table 17 shows the diagnosis of cerebral edema in DKA (Wolfsdorf et al. 2018).

The management of cerebral edema is difficult and involves careful administration of fluids with strict blood pressure control and infusion of mannitol or hypertonic saline. Mannitol is administered at dose of 0.5–1 g/kg body weight. The calculated dose is administered over a period of 10–15 min and if necessary repeated after 30 min (Wolfsdorf et al. 2018). If mannitol is not available or if there is no response to mannitol, 3% hypertonic saline can be given at calculated dose of 2.5–5 ml/kg. The time for administration is again 10–15 min (Wolfsdorf et al. 2018).

Regarding mannitol versus hypertonic saline selection, controversies exist but recent data suggests lower mortality rate with mannitol (Wolfsdorf et al. 2018).

Figure 8 Pathogenesis of cerebral edema in DKA. The figure is reproduced with permission (Carlotti 2003).

Table 16 Complications of DKA, EDKA AND HHS

Complications	Cause and remedy	Risk probability
Hypoglycaemia	High dose insulin can cause	In HHS risk probability is more than DKA as insulin sensitivity is more in HHS
	Management protocol should follow throughout and frequent monitoring is needed. 5–10% dextrose saline is needed with FRIII when sugar came down	
	The episode happens for short duration only	
Hypokalemia	Use of excessive high dose of insulin and use of HCO3 can cause it	Risk is high in both DKA and HHS
	Potassium level should be monitored frequently and replacement should be done if inadequate	
Pulmonary or arterial or venous thromboembolism	DKA and HHS patients are at risk of thromboembolism especially in case of central venous catheter use in shock patients	Risk is medium to low. Messenteric vessel thrombosis in extreme rare cases may be found
	Prophylactic LMWH should be given in high risk patients based on clinical evaluation	
Nonanion gap hyperchloremic acidosis	It occurs due to loss of ketoanions through urine which are needed for HCO3 formation	The risk is low
	Moreover, due to high amount of 0.9% NaCl saline infusion, increased amount of chloride reabsorption occurs. Hyperchloremic acidosis resolves during management	
	In DKA in pregnancy this is seen sometime	
Cerebral edema, central pontine myelinolysis	Cerebral edema (CE) incidence is 0.7–10% of children under 5 years of age. It is rare in adults with DKA and in HHS	Avoidance of aggressive hydration and maintaining blood glucose <11 mmol/L can prevent
	Headache, lethargy, papillary changes and seizure are key manifestation	Risk of CE is low if following guidelines properly
	Mortality rate of CE is high and it is around 57–87% of all deaths of DKA (Kitabchi et al. 2009)	
ARDS, DIC	Iatrogenic reduction in colloid osmotic pressure may lead to accumulation of water in lungs, decrease lung compliance and hypoxemia	Risks are very low
	Monitoring blood oxygen saturation, lowering fluid intake and adding colloid fluid can correct ARDS	
	DIC is a rare complication of HHS	
Stroke, AMI	Stroke and MI are rare complication in HHS. Predisposing factors are volume depletion with increased viscosity, increased levels of PAI-1, hyperfibrinogenaemia etc.	Risk is low in cases where the guideline for fluid repletion is followed properly
	Early adequate hydration is helpful	
Coma	Rarely associated in HHS with serum osmolality <330–340 mOsm/kg and in hypernatraemic than hyperglycaemic	Risk is very low
	ICU management is needed	
Foot ulceration	Rarely occurs in DKA in children and young adults buy in elderly could happen in obtunded or uncooperative cases. The heels should be protected and daily foot checks should be done	High risk in elderly cases of HHS
	In HHS patients who are usually elderly with comorbidities it is a high risk especially in those who are obtunded or need to long stay to recover. The heels should be protected and daily foot checks should be done	

Table 17 Diagnosis of cerebral EDEMA

A. Diagnostic criteria
Abnormal verbal or motor response to pain
Decorticate or decerebrate posture
Cranial nerve palsy
Abnormal neurogenic breathing pattern (like grunting, tachypnea, Cheyne-Stoke respiration, apneusis)
B. Major criteria
Altered/fluctuating state of consciousness
Sustained decreasing heart rate (>20 beats per minute) not attributable to any other reason
Age-inappropriate incontinence
C. Minor criteria
Vomiting
Headache
Lethargy
DBP >90 mmHg
Age < 5 years

If one diagnostic criterion or 2 major criteria or 1 major and 2 minor criteria are present, then diagnostic sensitivity for cerebral edema is 92% with false positive rate of only 4%. However signs that occur before start of treatment should not be included

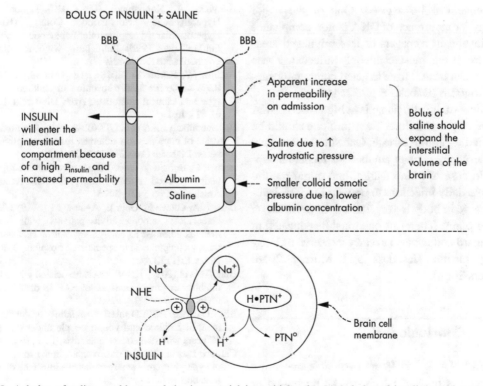

Fig. 8 A bolus of saline could expand the intracranial interstitial volume. A bolus of insulin could expand the intracerebral ICF volume (Scott et al. 2015)

11 Prevention

Management of acute hyperglycemic emergencies is not complete until steps are taken to prevent recurrence of future episodes. Diabetes education is an important component of prevention strategy. The education should be tailored to the individual's requirement. This is only possible after trigger has been identified. Ideally this should be delivered by a specialized diabetes educator (Karslioglu French et al. 2019).

Proper education regarding sick day rules is essential to prevent recurrence. The important components of sick day management include education regarding hydration, glucose and ketones monitoring, continuation of basal insulin and timely contact with health care provider (Karslioglu French et al. 2019). Since the process of ketogenesis occurs in the absence or deficiency of insulin, its recurrence can be avoided. One of the major reasons for recurrence of DKA is non-compliance with insulin in teenagers of less privileged areas who are being most commonly affected. These patients can benefit from targeted community support programs (Dabelea et al. 2014).

Patient on insulin pump is at high risk of DKA in case of pump failure. Therefore one should be educated regarding its care. One should also have an emergency contact number for technical support. In case of pump failure, one might require multiple daily injections to prevent DKA as insulin reserve in body is very limited for a patient on insulin pump. Therefore one must be educated in this regard and advised to carry a reserve of long-acting insulin (Jesudoss and Murray 2016; Rodgers 2008).

12 Conclusion

DKA, EDKA and HHS are avoidable metabolic emergencies both of which can be decreased in incidences with education regarding diabetes management in sick days. The management principles are different for each condition but generally require hospitalization and intravenous fluids with electrolytes. While close monitoring during episode has decreased mortality rate, there are still some controversial areas like fluid choice for rehydration. Due to availability of updated guidelines management is much better now. The structured diabetes education and abiding by sick day rules made significant improvement in reducing the recurrences of acute metabolic crisis of diabetes.

References

ABCD (2020) COncise AdVice on Inpatient Diabetes (COVID: Diabetes): Front Door Guidance National Inpatient Diabetes COVID-19 Response Team COVID-19 Infection in People with or without Previously Recognised Diabetes Increases the Risk of the EMERGENCY States of Hyperglycaemia with Ketones, Diabetic Keto Acidosis (DKA) and Hyperosmolar Hyperglycaemic State (HHS), 9 April 2020

Adeyinka A, Kondamudi NP (2020) Hyperosmolar Hyperglycemic Nonketotic Coma (HHNC), Hyperosmolar hyperglycemic nonketotic syndrome. PubMed, Stat Pearls Publishing. www.ncbi.nlm.nih.gov/books/NBK482142/#

Andrew W, Patrick D (2018) P18 plasma-Lyte 148 vs 0.9% saline for fluid resuscition in children: electrolytic and clinical outcomes. Arch Dis Child 103(2):e1.22–e1.e1

Balasubramanyam A et al (2006) Accuracy and predictive value of classification schemes for ketosis-prone diabetes. Diabetes Care 29(12):2575–2579

Barnett C, Barnett Y (2003) Ketone bodies. In: Encyclopedia of food sciences and nutrition. Academic, Amsterdam, pp 3421–3425

Bedaso A, Oltaye Z, Geja E, Ayalew M (2019) Diabetic ketoacidosis among adult patients with diabetes mellitus admitted to emergency unit of Hawassa university comprehensive specialized hospital. BMC Res Notes 12(1):137

Blau JE et al (2017) Ketoacidosis associated with SGLT2 inhibitor treatment: analysis of FAERS data. Diabetes Metab Res Rev 33(8):e2924

BSPED (2020) BSPED interim guideline for the management of children and young people under the age of 18 years with diabetic ketoacidosis, 1 January 2020

Cardoso L et al (2017) Controversies in the management of hyperglycaemic emergencies in adults with diabetes. Metabolism 68(68):43–54

Carlotti APCP (2003) Importance of timing of risk factors for cerebral oedema during therapy for diabetic ketoacidosis. Arch Dis Child 88(2):170–173

Choleau C, Maitre J, Filipovic Pierucci A, Elie C, Barat P et al (2014) Ketoacidosis at diagnosis of type 1 diabetes in French children and adolescents. Diabetes Metab 40(2):137–142. Elsevier Masson

Chua H-R et al (2012) Plasma-Lyte 148 vs 0.9% saline for fluid resuscitation in diabetic ketoacidosis. J Crit Care 27(2):138–145

Dabelea D, Rewers A, Stafford J, Standiford D, Lawrence J, Saydah S et al (2014) Trends in the prevalence of ketoacidosis at diabetes diagnosis: the SEARCH for diabetes in youth study. Pediatrics 133 (4):e938–e945

Dandona P et al (2018) Efficacy and safety of dapagliflozin in patients with inadequately controlled type 1 diabetes: the DEPICT-1 52-week study. Diabetes Care 41 (12):2552–2559

DeFronzo RA, Ferrannini E (1987) Regulation of hepatic glucose metabolism in humans. Diabetes Metab Rev 3 (2):415–459

Desai R et al (2019) Temporal trends in the prevalence of diabetes decompensation (diabetic ketoacidosis and hyperosmolar hyperglycemic state) among adult patients hospitalized with diabetes mellitus: a Nationwide analysis stratified by age, gender, and race. Cureus 11(4). https://doi.org/10.7759/cureus.4353

Dhatariya K (2014) Diabetic ketoacidosis and hyperosmolar crisis in adults. Medicine 42 (12):723–726

Dhatariya K (2016) Blood ketones: measurement, interpretation, limitations, and utility in the management of diabetic ketoacidosis. Rev Diabet Stud 13(4):217–225

Dhillon KK, Gupta S (2019) Biochemistry, Ketogenesis. [online] Nih.gov. Available at: https://www.ncbi.nlm.nih.gov/books/NBK493179/. Accessed 15 May 2020

Diaz-Ramos A et al (2019) Euglycemic diabetic ketoacidosis associated with sodium-glucose cotransporter-2 inhibitor use: a case report and review of the literature. Int J Emerg Med 12(1):27

Duca LM et al (2017) Diabetic ketoacidosis at diagnosis of type 1 diabetes predicts poor long-term glycemic control. Diabetes Care 40(9):1249–1255

Edge JA et al (2016) Diabetic ketoacidosis in an adolescent and young adult population in the UK in 2014: a national survey comparison of management in paediatric and adult settings. Diabet Med 33 (10):1352–1359

Eisenhut M (2006) Causes and effects of hyperchloremic acidosis. Crit Care 10(3):413

Eisenhut M (2018) In diabetic ketoacidosis brain injury including cerebral oedema and infarction is caused by interleukin-1. Med Hypotheses 121:44–46

Ergin B et al (2016) The role of bicarbonate precursors in balanced fluids during haemorrhagic shock with and without compromised liver function. Br J Anaesth 117 (4):521–528

Evans K (2019) Diabetic ketoacidosis: update on management. Clin Med 19(5):396–398

Gershkovich B et al (2019) Choice of crystalloid fluid in the treatment of hyperglycemic emergencies: a systematic review protocol. Syst Rev 8(1):228

Glaser N, Barnett P, McCaslin I et al (2001) Risk factors for cerebral edema in children with diabetic ketoacidosis. N Engl J Med 344:264–269

Gosmanov AR, Kitabchi AE (2000) Diabetic ketoacidosis [Updated 2018 April 28]. In: Feingold KR, Anawalt B, Boyce A et al (eds) Endotext [Internet]. South Dartmouth (MA): MDText.com, Inc. Available from: https://www.ncbi.nlm.nih.gov/books/NBK279146/

Gosmanov AR, Nimatollahi LR (2020) Diabetic ketoacidosis – symptoms, diagnosis and treatment I BMJ Best Practice. Bestpractice.Bmj.Com. February. https://bestpractice.bmj.com/topics/en-us/162. Accessed 15 Apr 2020

Greene JA, Riggs KR (2015) Why is there no generic insulin? Historical origins of a modern problem. N Engl J Med 372(12):1171–1175

Hamdy O (2019) Diabetic Ketoacidosis (DKA): practice essentials, background, pathophysiology. Medscape. Com. 31 May. https://emedicine.medscape.com/article/118361-overview

Handelsman Y et al (2016) American association of clinical endocrinologists and American college of endocrinology position statement on the association of SGLT-2 inhibitors and diabetic ketoacidosis. Endocr Pract 22 (6):753–762

Hardern RD (2003) Emergency management of diabetic ketoacidosis in adults. Emerg Med J 20(3):210–213

Jabbar A, Farooqui K, Habib A, Islam N, Haque N, Akhter J (2004) Clinical characteristics and outcomes of diabetic ketoacidosis in Pakistani adults with type 2 diabetes mellitus. Diabet Med 21(8):920–923

Jefferies C et al (2015) 15-year incidence of diabetic ketoacidosis at onset of type 1 diabetes in children from a regional setting (Auckland, New Zealand). Sci Rep 5(1):P3

Jesudoss M, Murray R (2016) A practical guide to diabetes mellitus, 7th edn. Jaypee Brothers, New Delhi

Karges B, Rosenbauer J, Holterhus PM et al (2015) Hospital admission for diabetic ketoacidosis or severe hypoglycemia in 31,330 young patients with type 1 diabetes. Eur J Endocrinol 173(3):341–350

Karslioglu French E et al (2019) Diabetic ketoacidosis and hyperosmolar hyperglycemic syndrome: review of acute decompensated diabetes in adult patients. BMJ 365(l1114):l1114

Keenan CR et al (2007) High risk for venous thromboembolism in diabetics with hyperosmolar state: comparison with other acute medical illnesses. J Thromb Haemost 5(6):1185–1190

Kelly A-M (2006) The case for venous rather than arterial blood gases in diabetic ketoacidosis. Emerg Med Australas 18(1):64–67

Khazai N, Umpierrez G (2020) Hyperosmolar hyperglycaemic state – symptoms, diagnosis and treatment I BMJ best practice. Beta-Bestpractice.Bmj.Com. March. https://beta-bestpractice.bmj.com/topics/en-gb/1011. Accessed 15 Apr 2020

Kitabchi AE et al (2006) Hyperglycemic crises in adult patients with diabetes: a consensus statement from the American Diabetes Association. Diabetes Care 29 (12):2739–2748

Kitabchi A, Umpierrez G, Miles J, Fisher J (2009) Hyperglycemic crises in adult patients with diabetes. Diabetes Care 32(7):1335–1343

Kraut JA, Kurtz I (2014) Treatment of acute non-anion gap metabolic acidosis. Clin Kidney J 8(1):93–99

Laine C (2010) Diabetic Ketoacidosis. Ann Intern Med 152(1):ITC1

Lebovitz HE, Banerji MA (2018) Ketosis-prone diabetes (Flatbush diabetes): an emerging worldwide clinically important entity. Curr Diab Rep 18(11):120

Maahs DM, Hermann JM, Holman N et al (2015) Rates of diabetic ketoacidosis: international comparison with 49,859 pediatric patients with type 1 diabetes from England, Wales, the U.S., Austria, and Germany. Diabetes Care 38(10):1876–1882

Mathieu C et al (2018) Efficacy and safety of dapagliflozin in patients with inadequately controlled type 1 diabetes (the DEPICT-2 study): 24-week results from a randomized controlled trial. Diabetes Care 41(9):1938–1946

Miles JM et al (1983) Effects of free fatty acid availability, glucagon excess, and insulin deficiency on ketone body production in postabsorptive man. J Clin Investig 71(6):1554–1561

Mudly S, Rambiritch V, Mayet L (2007) An identification of the risk factors implicated in diabetic ketoacidosis (DKA) in type 1 and type 2 diabetes mellitus. S Afr Fam Pract 49(10):15-15b

Munro JF et al (1973) Euglycaemic diabetic ketoacidosis. BMJ 2(5866):578–580

Ng S, Edge J, Timmis A (2020) Practical management of hyperglycemic hyperosmolar state (HHS) in children [Internet] [cited 6 April 2020]. Available from: http://www.a-c-d-c.org/wp-content/uploads/2012/08/Practical-Management-of-Hyperglycaemic-Hyperosmolar-State-HHS-in-children-2.pdf

Nyenwe EA et al (2010) Acidosis: the prime determinant of depressed sensorium in diabetic ketoacidosis. Diabetes Care 33(8):1837–1183

Otieno CF et al (2006) Diabetic ketoacidosis: risk factors, mechanisms and management strategies in Sub-Saharan Africa: a review. East Afr Med J 82(12). https://doi.org/10.4314/eamj.v82i12.9382

Pasquel FJ, Umpierrez GE (2014) Hyperosmolar hyperglycemic state: a historic review of the clinical presentation, diagnosis, and treatment. Diabetes Care 37(11):3124–3131

Peters AL et al (2015) Euglycemic diabetic ketoacidosis: a potential complication of treatment with sodium–glucose cotransporter 2 inhibition. Diabetes Care 38(9):1687–1693

Rafey MF et al (2019) Prolonged acidosis is a feature of SGLT2i-induced euglycaemic diabetic ketoacidosis. Endocrinol Diabetes Metab Case Rep 1:1–5

Rawla P et al (2017) Euglycemic diabetic ketoacidosis: a diagnostic and therapeutic dilemma. Endocrinol Diabetes Metab Case Rep 2017(1):1–4. www.ncbi.nlm.nih.gov/pmc/articles/PMC5592704/

Rewers A et al (2008) Presence of diabetic ketoacidosis at diagnosis of diabetes mellitus in youth: the search for diabetes in youth study. Pediatrics 121(5):e1258–e1266

Röder PV et al (2016) Pancreatic regulation of glucose homeostasis. Exp Mol Med 48(3):e219–e219

Rodgers J (2008) Using insulin pumps in diabetes: a guide for nurses and other health care professionals. Wiley, Chichester

Rosenstock J, Ferrannini E (2015) Euglycemic diabetic ketoacidosis: a predictable, detectable, and preventable safety concern with SGLT2 inhibitors. Diabetes Care 38(9):1638–1642

Rosenstock J et al (2018) Empagliflozin as adjunctive to insulin therapy in type 1 diabetes: the EASE trials. Diabetes Care 41(12):2560–2569

Savage M et al (2011) Joint British diabetes societies guideline for the management of diabetic ketoacidosis. Diabet Med 28(5):508–515

Scott AR et al (2015) Management of hyperosmolar hyperglycaemic state in adults with diabetes. Diabet Med J Br Diabet Assoc 32(6):714–724

Sheikh-Ali M et al (2008) Can serum –hydroxybutyrate be used to diagnose diabetic ketoacidosis? Diabetes Care 31(4):643–647

Stark R, Guebre-Egziabher F, Zhao X, Feriod C, Dong J, Alves T et al (2014) A role for mitochondrial phosphoenolpyruvate carboxykinase (PEPCK-M) in the regulation of hepatic gluconeogenesis. J Biol Chem 289(11):7257–7263

Takeuchi M, Kawamura T, Sato I, Kawakami K (2017) Population-based incidence of diabetic ketoacidosis in type 2 diabetes: medical claims data analysis in Japan. Pharmacoepidemiol Drug Saf 27(1):123–126

Trence DL, Hirsch IB (2001) Hyperglycemic crises in diabetes mellitus type 2. Endocrinol Metab Clin N Am 30(4):817–831

Umpierrez G, Korytkowski M (2016) Diabetic emergencies – ketoacidosis, hyperglycaemic hyperosmolar state and hypoglycaemia. Nat Rev Endocrinol 12(4):222–232

Umpierrez GE et al (1996) Diabetic ketoacidosis and hyperglycemic hyperosmolar nonketotic syndrome. Am J Med Sci 311(5):225–233

Umpierrez GE et al (2002) Diabetic ketoacidosis and hyperglycemic hyperosmolar syndrome. Diabetes Spectr 15(1):28–36

Usher-Smith JA et al (2011) Factors associated with the presence of diabetic ketoacidosis at diagnosis of diabetes in children and young adults: a systematic review. BMJ 343(1):d4092–d4092

Van Zyl DG et al (2011) Fluid management in diabetic-acidosis – Ringer's lactate versus normal saline: a randomized controlled trial. QJM 105(4):337–343

Westerberg DP (2013) Diabetic ketoacidosis: evaluation and treatment. Am Fam Physician 87(5):337–346

Wolfsdorf JI et al (2018) ISPAD clinical practice consensus guidelines 2018: diabetic ketoacidosis and the hyperglycemic hyperosmolar state. Pediatr Diabetes 19:155–177

Yung M et al (2017) Controlled trial of Hartmann's solution versus 0.9% saline for diabetic ketoacidosis. J Paediatr Child Health 53(1):12–17

Adv Exp Med Biol - Advances in Internal Medicine (2020) 4: 115–128
https://doi.org/10.1007/5584_2020_513
© Springer Nature Switzerland AG 2020
Published online: 7 April 2020

The Role of the Mediterranean Dietary Pattern on Metabolic Control of Patients with Diabetes Mellitus: A Narrative Review

Jéssica Abdo Gonçalves Tosatti, Michelle Teodoro Alves, and Karina Braga Gomes

Abstract

Diabetes mellitus (DM) is a metabolic disorder characterised by hyperglycemia and abnormalities in carbohydrate, fat and protein metabolism. Several studies demonstrated that foods typical of the Mediterranean diet (MedDiet), including vegetables, fruits, oilseeds, extra virgin olive oil and fish, can promote health benefits for individuals at risk of or with type 2 diabetes (T2DM). In this review, we summarised randomised clinical trials, cohort studies, meta-analyses and systematic reviews that evaluated the effects of the MedDiet on metabolic control of T2DM. The data suggest that the MedDiet influences cardiovascular risk factors, including blood pressure, lipid profile, insulin resistance, inflammation and glucose metabolism, in T2DM patients. In conclusion, the MedDiet appears to protect patients from macro- and microangiopathy and should be considering in the management of diabetic patients.

Keywords

Cardiovascular disease · Diabetes mellitus · Diabetic complications · Mediterranean diet

Abbreviations

ADA	American Diabetes Association
CRP	C-reactive protein
CVD	Cardiovascular disease
DKD	Diabetic kidney disease
DM	Diabetes mellitus
EPIC	European Prospective Investigation of Cancer and Nutrition
GLUT-4	Glucose transporter type 4
HbA1c	Glycated hemoglobin
HDL-c	High-density lipoprotein cholesterol
IL-6	Interleukin-6
LDL	Low-density lipoprotein
MedDiet	Mediterranean diet
T1DM	Type 1 diabetes mellitus
T2DM	Type 2 diabetes mellitus
TNF-α	Tumor necrosis factor-alpha

J. A. G. Tosatti, M. T. Alves, and K. B. Gomes (✉)
Clinical and Toxicological Analyzes Department, Faculty of Pharmacy, Federal University of Minas Gerais, Belo Horizonte, Minas Gerais, Brazil
e-mail: karinabgb@ufmg.br

1   Introduction

The term diabetes mellitus (DM) refers to a group of metabolic disorders characterized by the presence of hyperglycemia in the absence of treatment (World Health Organization 2019). The heterogeneous etiopathology of DM includes failure in insulin secretion and/or action and changes in carbohydrate, fat and protein metabolism (Silva et al. 2009). In 2014, it was estimated that 422 million adults lived with DM. Besides, without interventions to halt the increase in diabetes, there will be at least 629 million people living with diabetes by 2045 (World Health Organization 2019). Indeed, it is estimated that a significant percentage of DM cases (30–80%, depending on the country) are undiagnosed (International Diabetes Federation 2017).

DM is commonly associated with, among other factors, changes in dietary patterns, such as increased consumption of ultra-processed foods and sugary drinks, as well as low consumption of fruits, vegetables and fibre, besides an unhealthy lifestyle (World Health Organization 2016). Therefore, the maintenance of controlled plasma glucose levels is essential to prevent or delay the onset of chronic complications associated with DM. The focus in the management of diabetic patients is to obtain an intensified glycemic control, associated with the regulation of serum lipid levels and blood pressure, as well as the maintenance of adequate body weight; these factors are all related to the patient's diet (Silva et al. 2009).

The Mediterranean Diet (MedDiet) is a dietary pattern characterized by high intake of vegetables, fruits, oilseeds, extra virgin olive oil and fish, moderate consumption of wine— depending on the patient's religious beliefs— and rare consumption of red meat, ultra-processed foods, butter and sugary drinks (Itsiopoulos et al. 2011). Recent evidence (Annuzzi et al. 2014; Kastorini et al. 2011; Finicelli et al. 2019) demonstrated the positive relationship between the MedDiet pattern and improvement in T2DM-related parameters, including glycated hemoglobin (HbA1c) (Annuzzi et al. 2014),

inflammatory markers, endothelial function (Kastorini et al. 2011) and risk factors related to cardiovascular diseases (CVD) (Finicelli et al. 2019). Nevertheless, the health benefits of the eating pattern in DM patients and key elements that contribute to those benefits are still being investigated.

2   MedDiet: General Aspects

The MedDiet was created over the last 5000 years in the Fertile Crescent region, where the first established agricultural communities in the Middle East and the Mediterranean basin are supposed to originate in the early ninth millennium BC. The dietetic pattern was influenced by the achievements of many different civilizations, with consolidated dietary rules coming from the three major monotheistic religions and continuous interactions, additions and exchanges within and outside the region. Consequently, the MedDiet is an expression of the different food cultures present in this region; it represents the historical and environmental diversity that defines the Mediterranean (Dernini and Berry 2015).

The MedDiet was well characterized scientifically by Ancel Keys in the 1960s (Trichopoulou 2001). An important finding of this study was that the low saturated fat content typical of the MedDiet could explain the low incidence of CVD in Mediterranean countries by means of lowering blood cholesterol, one of the major known risk factors for CVD at that time (Trichopoulou 2001). Further work demonstrated that the traditional MedDiet had several additional beneficial health effects (Altomare et al. 2014). Consequently, the MedDiet has been widely studied and disseminated as a healthy eating pattern; it is associated with benefits in the prevention and control of chronic diseases such as cancer, CVD and T2DM (Altomare et al. 2014).

The MedDiet was described on the basis of some dietary characteristics common to Mediterranean countries, with olive oil as the main source of fat, low to moderate intake of fish, dairy products, poultry and wine with meals, low con-sumption of red or processed meat and use of herbs and spices as an alternative to refined salt (Trichopoulou et al. 2014). Total lipid intake can be characterized as high (equal to or greater than 40% of total energy consumption) if considering the Greek standard or moderate (about 30% of total energy consumption) considering the Italian dietary standard. In the MedDiet, the ratio of monounsaturated to saturated lipids will be high due to the elevated content of monounsaturated lipids from olive oil; the ratio of polyunsaturated to monounsaturated will also be high due to the elevated consumption of fish and nuts (Trichopoulou et al. 2014).

In 2009 and 2010, based on international sci-entific consensus, a new pyramid was proposed to the MedDiet standard so that is could be adapted to contemporary lifestyles. This new pyramid proposal was developed in a simplified manner to be adapted to different countries, and specific variations are related to the various geographical, socioeconomic, cultural and contemporary Medi-terranean lifestyle contexts. It also considers differences in food portions. Furthermore, the concepts of frugality and regular physical activity were emphasized due to the current major public health challenge: obesity (Bach-Faig et al. 2011). Figure 1 and Table 1 show the adaptation of MedDiet pyramid proposed by Bach-Faig et al. (2011) and a synthesis of the kinds of food, amounts and intake frequency of the MedDiet pattern.

Adherence to the MedDiet pattern is an impor-tant factor, given the beneficial results in the control, prevention and mortality from chronic diseases. In a prospective cohort study that involved approximately 22,000 adults, the findings showed that increased MedDiet adher-ence is associated with a reduction in total mor-tality. This reduction was evident in both coronary disease and cancer deaths (Trichopoulou et al. 2003).

A randomized clinical trial of approximately 7000 patients at high cardiovascular risk evaluated adoption of the MedDiet pattern and primary prevention of cardiovascular events. It demonstrated a relative risk (RR) reduction of 30% in both intervention groups: MedDiet supplemented with nuts (RR = 0.70; 95% confi-dence interval [CI]: 0.53–0.94; p = 0.02) and MedDiet supplemented with extra virgin olive oil (RR = 0.70; 95% CI: 0.53–0.91; p = 0.0015). The results support the benefits of the MedDiet for cardiovascular risk reduction (Estruch et al. 2013).

The MedDiet can also improve glycemic con-trol and cardiovascular risk factors, with an evi-dence rating of B, based on well-conducted cohort, meta-analysis or case-control studies, in a scientific statement published by the American Heart Association and the American Diabetes Association (ADA) (American Diabetes Associa-tion (ADA) 2018). Furthermore, the ADA, in its "Standards of Medical Care in Diabetes – 2019", recommends the MedDiet in the prevention and follow-up of patients with T2DM. Regarding the follow-up of patients with type 1 diabetes mellitus (T1DM), the same document argues that there is not enough evidence to support application of the MedDiet standard in this group (American Dia-betes Association (ADA) 2018).

3   MedDiet and its Effects on DM

3.1   Effects of the MedDiet on Postprandial Glycemia and HbA1c

Postprandial blood glucose concentrations refer to plasma glucose levels after ingestion of a meal. The magnitude and peak time of plasma glucose concentration depend directly on the amount and quality of the consumed carbohydrate and factors involved in the response to this carbo-hydrate (Silva et al. 2009).

HbA1c values close to or below 7% can reduce microvascular and neuropathic complications in patients with DM and possibly macrovascular complications (American Diabetes Association 2009). Postprandial blood glucose is a determin-ing factor in HbA1c values and may account for up to 50% or more of the values of this test

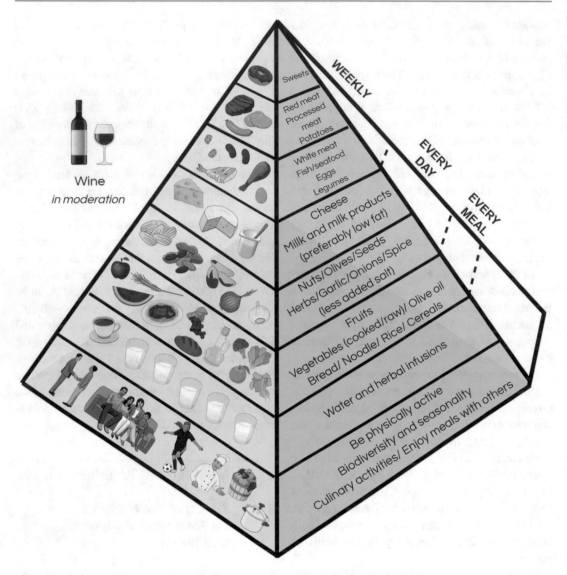

Fig. 1 Mediterranean diet pyramid: a lifestyle for today; based on Bach-Faig et al. (2011)

(Monnier et al. 2003). Evidence from the European Prospective Investigation of Cancer and Nutrition (EPIC-Norfolk) study revealed that a 1% increase in the HbA1c level increases the risk of death due to all causes by 28% (Khaw et al. 2001).

In randomized controlled trials, reduced glycemic-load diets, including the MedDiet, were associated with better control of weight and risk factors for T2DM (Esposito et al. 2015; Roldan et al. 2019). Besides, a systematic review with meta-analysis assessed three long-term

randomized controlled trials in order to evaluate the evidence on the effectiveness of the MedDiet standard in T2DM management. The results showed an estimated global effect of the MedDiet for HbA1c reduction in patients diagnosed with T2DM equal to −0.47% (95% CI: −0.56 to −0.38; p = 0.0001) when compared to usual care or a low-fat diet that aimed the glycemic control (Esposito et al. 2015).

In a descriptive, observational study conducted on 107 patients (45.55% men; age: 61.16 ± 23 years) diagnosed with T2DM, with

Table 1 Types of food, amounts and intake frequency of the Mediterranean dietary pattern (Bach-Faig et al. 2011)

Food group	Serving size	Frequency
Sweets	In small amounts	For special occasions
Red meat	Less than 2 portions	Weekly
Processed meat	Less than 1 portion	
Potatoes	Less than 3 portions	
White meat	2 portions	Weekly
Fish/Seafood	2 or more portions	
Eggs	2–4 portions	
Legumes	2 or more portions	
Cheese/Milk and milk products	2 portions	Daily
Nuts/Olives/Seeds	1–2 portions	Daily
Herbs/Garlic/Onions/Spice	A reasonable consumption	
Fruits	1–2 portions	Per meal
Vegetables	2 or more portions	
Olive oil	One tablespoon per person	
Bread/Noodle/Rice/Cereals	1–2 portions	
Water	1.5–2.0 l	Daily
Herbal Infusions	To complete the water requirements	
Wine	1 glass for women and 2 glasses for men	Daily

poor blood glucose control and a body mass index (BMI) greater than 25 kg/m^2, the authors aimed to analyze the relationship between the level of adherence to the MedDiet (low and high dietary adherence) and the control of cardiovascular risk factors. After the 6-month educational intervention, the HbA1c level in the low dietary adherence group was 8.62%, compared to 6.99% in the high dietary adherence group ($p < 0.03$) (Roldan et al. 2019).

In a randomized clinical trial conducted with 27 adults (47–77 years old) with T2DM that aimed to investigate the impact of a diet modeled on the MedDiet pattern on metabolic control and vascular risk in T2DM, the patients were randomly assigned to consume either the intervention MedDiet *ad libitum* or their usual diet for 12 weeks. Compared with the usual diet, the *ad libitum* MedDiet pattern intervention was associated with a reduction in HbA1c levels from 7.1% (95% CI: 6.5–7.7) to 6.8% (95% CI: 6.3–7.3; $p = 0.012$). The authors concluded that a traditional moderate-fat MedDiet improves glycemic control and diet quality in well-controlled T2DM (Itsiopoulos et al. 2011).

In a cross-sectional analysis with 901 outpatients diagnosed with T2DM who attended a diabetes clinic, the authors aimed to evaluate the association of MedDiet pattern with better glycemic control in T2DM patients. The mean HbA1c and 2-h post-meal glucose concentrations were significantly lower in diabetic patients with high adherence to the MedDiet compared to those with low adherence (mean difference: HbA1c 0.9%, 95% CI: 0.5–1.2%; $p < 0.001$; 2-h glucose 2.2 mmol/l, 95% CI: 0.8–2.9 mmol/l; $p < 0.001$) (Esposito et al. 2009a).

These findings suggest that adherence to the MedDiet could benefit glycemic control, as evidenced by the HbA1c results (Estruch et al. 2006; Martins et al. 2014). However, there was no important association when considering each of the MedDiet components, except for a modest association with whole grains and the ratio of monounsaturated/saturated lipids with lower HbA1c levels (Esposito et al. 2009a). Individually, these MedDiet components present small effects that can only promote overall positive changes when they are integrated, for example, into a meal (Trichopoulou et al. 2003).

3.2 MedDiet, Inflammation and Oxidative Stress

T2DM is characterized by a chronic and subclinical proinflammatory state and oxidative stress, with increased inflammatory markers, such as C-reactive protein (CRP), tumor necrosis factor-alpha (TNF-α) and interleukin-6 (IL-6). These makers are directly related to insulin resistance, a determining factor for T2DM development (Martins et al. 2014). Several studies demonstrated the beneficial effects of diets with a relatively high monounsaturated fatty acid content on DM risk factors. In a prospective study of adiponectin plasma concentrations and dietary data from 987 women with T2DM from the Nurses' Health Study, closer adherence to the MedDiet pattern was associated with higher adiponectin concentrations, which have anti-inflammatory properties (Mantzoros et al. 2006). In fact, the MedDiet significantly increases adiponectin levels and reduces proinflammatory and insulin resistance marker levels (Mattar and Obeid 2009).

Several mechanisms are assumed to explain the metabolic benefits associated with this eating pattern. These benefits include a better lipid profile, rich in monounsaturated fatty acids from regular consumption of extra virgin olive oil and fish; decreased insulin resistance and peripheral inflammation; improved oxidative stress and endothelial function (Estruch et al. 2006). The imbalance between free radical formation and antioxidant factors plays a major role in the onset and development of insulin resistance and pancreatic beta cell dysfunction (Martins et al. 2014). A clinical trial was conducted with 41 healthy individuals submitted to three dietary periods, each lasting 4 weeks, in which the first period comprised a diet rich in saturated fat, the second period a hypolipidaemic diet and the last period a MedDiet. MedDiet consumption reduced the atherogenic index (total cholesterol/high-density lipoprotein cholesterol [HDL-c]) and increased the resistance of low-density lipoproteins (LDL) to oxidation when compared to the low-fat diet period (López-Miranda et al. 2000).

3.3 MedDiet Benefits in the Management of DM-Related Complications

Individualized nutritional orientation, which respects patients' preferences and is directed at the prevention and control of overweight and obesity, is one of the most important strategies to achieve the necessary glycemic control for each patient. One of the main goals, and the last step of nutritional therapy recommended for diabetic patients, is to help them reach and maintain metabolic indexes within ideal values in order to prevent chronic complications, including retinopathy, diabetic kidney disease (DKD) and other microvascular complications (Rahati et al. 2014).

The daily intake of varied fruits and vegetables, indicated by the MedDiet pattern, is one of the diet-related DM control aspects that may attenuate the severity and progression of the complications associated with the disease. These flavonoid-rich foods are related to the maintenance of two main factors: better glycemic control and reduced systemic inflammation (Rahati et al. 2014). The mechanism of action of flavonoids *in vitro* reveals that, in addition to their widely known anti-inflammatory characteristic, these compounds exert a beneficial effect on muscle glucose uptake that is mediated by glucose transporter type 4 (GLUT-4) translocation (Mahoney and Loprinzi 2014).

Evidence shows that inflammatory cytokines play a pro-degenerative role in diabetic complications, including proliferative diabetic retinopathy (Donath and Shoelson 2011). Given the anti-inflammatory properties of flavonoids, it is plausible to suggest that adequate intake of quercetin-, kaempferol- and myricetin-containing foods may be an effective strategy for moderating secondary complications associated with diabetes (Mahoney and Loprinzi 2014).

A 7-year, prospective, population-based and observational multicenter study included 192 Spanish patients diagnosed with T1DM and T2DM. The study aimed to evaluate the adherence to the ADA nutritional recommendations—based on 7-day food diaries—and its relation to targets of metabolic control and onset of diabetic complications. The data did not reveal an association between adherence to each ADA nutritional recommendation and reduction in the onset or progression of diabetic complications. However, a 3.43-to-8.24-fold reduction in the risk of onset of each diabetic complication evaluated (neuropathy, nephropathy, retinopathy, cardiovascular and microvascular complications status) was related with a polyunsaturated fatty acids/saturated fatty acids ratio > 0.4 (Diabetes and Nutrition Study Group of the Spanish Diabetes Association (GSEDNu) 2006).

A randomized clinical trial with approximately 3700 patients diagnosed with T2DM aimed to evaluate the effects of three dietary patterns: MedDiet supplemented with olive oil, MedDiet supplemented with a nut mix and a control diet (following the guidelines based on a low-fat diet, according to ADA) on diabetic complications. After a median 6-year follow-up, there was a significant 43% reduction in the risk of diabetic retinopathy in the group of patients who adhered to the MedDiet standard supplemented with olive oil (hazards ratio [HR] = 0.57; 95% CI: 0.32–0.98). However, there were no significant effects observed in the three dietary patterns in reducing the risk of DKD (Díaz-López et al. 2015).

3.4 MedDiet and CVD Risk in T2DM

The risk for coronary heart disease is increased in diabetic patients (Juutilainen et al. 2005). An increase in the prevalence of insulin resistance syndrome may partly explain the recent plateau or increase in CVD rates after several decades of decline (Rosamond et al. 1998). Since the initial studies in the 1970s, the MedDiet has been recognized as a dietary pattern associated with decreased all-cause mortality and reduced cardiovascular risk factor levels and other health outcomes (Martinez-Gonzalez et al. 2015; Ros et al. 2014). Two controlled trials (Shai et al. 2008; Esposito et al. 2009b) evaluated the effects of the MedDiet on cardiovascular risk factors in T2DM patients and found more marked declines

in traditional cardiovascular risk factors, including systolic blood pressure, triglyceride levels and the ratio of total cholesterol/HDL-c, in those patients allocated to the MedDiet compared to diabetic subjects who received a control diet.

According to the Standards of Medical Care in Diabetes-2019 from the ADA (American Diabetes Association (ADA) 2018), with regards to the nutrition therapy recommendations for the management of adult diabetics, both the macronutrient composition and the combinations of foods or food groups consumed in a diet should be based on individualized assessment of current eating patterns, personal preferences and metabolic goals. A MedDiet pattern is recommended as an effective alternative to a low-fat and low-carbohydrate diet for T2DM patients due to emerging evidence about the beneficial effects of the MedDiet on glycemic control and CVD risk factors (evidence rating B, indicating supportive evidence from well-conducted cohort studies or case-control studies) (American Diabetes Association (ADA) 2018). Indeed, there is robust epidemiological evidence in the general population that indicates greater adherence to a MedDiet is significantly associated with a reduced risk of both overall and cardiovascular mortality (Sofi et al. 2008).

Other data also indicate a beneficial effect of the MedDiet on the diabetes control. Indeed, this diet reduces blood pressure and modifies the components of fibrinolysis (Bowen et al. 2016). Accumulated evidence from long-term (12–48 months) randomised controlled trials showed that the MedDiet improves glycemic control and cardiovascular risk in individuals with established diabetes (Esposito and Giugliano 2014).

Apparently, the benefits observed in the CVD risk under MedDiet adherence involve fish intake and heterogeneity in the effects of omega-3 fatty acids (Bowen et al. 2016). Metabolic studies demonstrated that these factors exert beneficial effects on surrogate measures associated with coronary heart disease, including serum levels of triglycerides and thrombotic factors—markers of endothelial dysfunction—and prevention of cardiac arrhythmias (Matsuzawa and Lerman 2014). In the report of a cohort of T2DM patients who participated in the *PREvención con DIeta MEDiterránea* (PREDIMED) study, a multicenter randomized nutritional intervention trial conducted in a population at high cardiovascular risk, the MedDiet exhibited protective effects on traditional cardiovascular risk factors, including blood pressure, lipid profile and glucose metabolism, and on novel risk factors such as markers of oxidation, inflammation and endothelial dysfunction (Martins et al. 2014).

Current evidence strongly indicates that oxidative modifications of LDL are the key factor in the onset and development of arteriosclerosis. Recent studies in humans and animal models support this hypothesis and show that a diet rich in polyunsaturated fat increases the susceptibility of LDL to oxidation, when compared to a diet rich in monounsaturated fat. The incidence and prevalence of ischaemic heart disease are low in Mediterranean countries, despite a high percentage of their calories (35–40%) coming from fat (Degirolamo and Rudel 2010). This benefit may be due to the high percentage of consumed monounsaturated fat, mainly from olive oil. The Mediterranean experience indicates that the problem of the atherogenic diet depends on the quality rather than the quantity of the ingested fat (Siri-Tarino et al. 2015).

A study was conducted with 41 normolipidemic volunteers submitted to three diets, each lasting 4 weeks: a diet high in saturated fat; a low-fat diet; MedDiet. The results showed that the MedDiet reduced LDL oxidation, in addition to the effect on the already known lipid profile, when compared with the high saturated fat and low-fat diets. In fact, the benefits of the monounsaturated-fat-rich MedDiet highlights its action on plasma lipoproteins, as the risk of CVD is related to the relative tendency of atherogenic lipoproteins, mainly LDL, to suffer oxidative modification (Siri-Tarino et al. 2015). Taken together, these findings suggest that the prevention of atherosclerosis should be based on the type of fat rather than reduction of total fat (López-Mirandaa et al. 2000).

3.5   MedDiet and Obesity Control

The prevalence of obesity has increased worldwide, and it is estimated that by 2030 nearly 40% of world's population will be overweight and one in five people will be obese. Obesity is mostly caused by poor health habits related to what is called the "obesogenic environment", including highly processed food, high consumption of red meat and sugary beverages, low consumption of vegetables and fruits and the reduction or replacement of physical activity (Jurado-Fasoli et al. 2019).

As previously described, obesity-induced oxidative stress and insulin resistance activate a signaling cascade that involves a proinflammatory pathway. In general, adherence to the MedDiet is associated with a reduction in different obesity-associated factors, in addition to a high quality of life. There are several physiological explanations that could elucidate why key components of the MedDiet pattern may protect against weight gain. The MedDiet is rich in plant-based foods that provide a large quantity of dietary fiber that increase satiation through various mechanisms, including prolonged mastication, increased gastric detention and enhanced cholecystokinin release (Schroder 2007). The MedDiet pattern also has a low energy density (Schroder 2007) and a low glycemic load (Willett and Leibel 2002) compared with many other dietary patterns. These characteristics, together with its high-water content, lead to increased satiation and lower calorie intake, and thus help to prevent weight gain (Buckland et al. 2008).

The MedDiet pattern is composed of large amounts of dietary fiber and features a high ratio of monounsaturated to saturated fat. It is recognized as one of the healthiest dietary patterns for the treatment of metabolic syndrome. The MedDiet pattern is beneficial for helping to reduce the associated cardiovascular risk and other health outcomes, and evidence from a previous meta-analysis also indicated that the MedDiet can decrease the risk of diabetes in healthy individuals and improve glycemic control, weight loss and cardiovascular risk factors for T2DM patients (Pan et al. 2019; Huo et al. 2015).

A meta-analysis that aimed to compare the differences among major dietary patterns in improving glycemic control, cardiovascular risk and weight loss for T2DM patients included 10 randomized clinical trials that involved five dietary patterns. Compared to the low-fat diet, the MedDiet pattern showed better effects in reducing body weight (kg) in T2DM patients (RR = −1.18; 95% CI: −1.99–0.37; p = 0.08) and BMI (kg/m^2; RR = −0.63; 95% CI: −1.29–0.02; p = 0.0007) (Pan et al. 2019).

In fact, a systematic review and meta-analysis of randomized controlled trials, which aimed to evaluate the effects of the MedDiet pattern on glycemic control, weight loss and cardiovascular risk factors in T2D patients, included nine studies with 1178 total patients. Compared with control diets (usual diet), the MedDiet pattern led to greater reductions in HbA1c, BMI and body weight (mean differences: HbA1c, −0.30, 95% CI: −0.46 to −0.14, p = 0.001; BMI, −0.29 kg/m^2, 95% CI: −0.46 to −0.12, p = 0.924; body weight, −0.29 kg, 95% CI: −0.55 to −0.04, p = 0.976). Additionally, the MedDiet pattern was associated with a 1.45 mmHg decline (95% CI: −1.97 to −0.94; p = 0.58) in systolic blood pressure and 1.41 mmHg reductions (95% CI: −1.84 to −0.97; p = 0.95) for diastolic blood pressure (Huo et al. 2015).

4   MedDiet and DM: Current Evidence

According to the ADA guideline for T2DM prevention (American Diabetes Association (ADA) 2018), the nutritional instruction to the diabetic patient should be structured in order to include a low-calorie diet that promotes weight loss and stimulate physical activity. These factors are of fundamental importance for those with high risk of developing overweight or obese T2DM. Furthermore, based on intervention testing, dietary patterns should also be useful for patients with pre-diabetes, which include a MedDiet or hypocaloric and hypolipemic diet plan (American Diabetes Association (ADA) 2018).

A European cohort study of prospective population-based research on cancer and nutrition (EPIC) indicated that Mediterranean population groups with good adherence to the MedDiet presented results inversely associated with T2DM risk (odds ratio [OR] = 0.88; 95% CI: 0.77–0.99; p = 0.021). When the MedDiet was combined with a low-glycemic-load dietary profile, the association became stronger (OR = 0.82; 95% CI: 0.71–0.95; p = 0.722). These results suggest that a low-glycemic-load dietary profile combined with a traditional MedDiet can result in 18% increase in T2DM protection (Rossi et al. 2013).

Tripp et al. (Tripp et al. 2019) developed an observational study with 50 healthy overweight and obese subjects with cardiometabolic risk factors and aimed to confirm the safety, tolerability and efficacy of the restricted calorie quantity in the MedDiet. The participants were assigned to a modified MedDiet for 12 weeks; it included protein shakes and targeted supplementation that provided 68–76% of the subject's estimated calorie requirements. The subjects exhibited the following decreases from baseline: 12% in body weight, 18% in body fat and 8.8% in waist circumference. Additionally, inflammation biomarkers, namely oxidized LDL and high-sensitivity C-reactive protein, were reduced by 17% (p < 0.01) and 30% (p < 0.05), respectively, with the modified diet (Tripp et al. 2019).

Another study conducted by Maiorino et al. (2017) assessed the long-term effects of a MedDiet on circulating levels of endothelial progenitor cells (EPCs) and the carotid intima-media thickness (CIMT) in patients with T2DM in a randomized trial with 215 men and women. At the end point, changes in the CIMT were inversely correlated with the changes in EPC and HbA1c levels (mean difference: -0.3; 95% CI: −0.6–0; p = 0.050) and HOMA-IR (mean difference: -0.7; 95% CI: −1.2 to −0.2; p = 0.043) in patients who consumed the MedDiet compared to a low-fat diet (Maiorino et al. 2017).

Another study developed by Tepper et al. (2018) aimed to evaluate whether adherence to

MedDiet was associated with physical function in older T2DM patients. They evaluated 117 patients (age 70.6 ± 6.5 years) at the Center for Successful Aging with Diabetes at Sheba Medical Center. The group with low adherence to MedDiet pattern presented higher HbA1c levels than the group with high adherence to the MedDiet (7.59 ± 1.19 versus 7.35 ± 0.82; p = 0.7003) (Tepper et al. 2018).

Grimaldi et al. (2018) in a non-randomized study that assessed the efficacy and durability of a 3-month intensive dietary intervention, aimed to evaluate body weight and cardiometabolic risk factors after implementing the MedDiet in 116 subjects at high risk for cardiac disease. The intensive intervention consisted of 12 weekly group educational meetings and a free-of-charge supply of meals prepared according to the MedDiet pattern. The conventional intervention (control) consisted of an individual education session along with monthly reinforcements of nutritional messages by the general practitioner. All participants were followed for 9 months. In the subgroup of participants with T2DM (n = 40), there was a significant reduction in HbA1c levels in the group that followed the intensive intervention group (n = 24; from 7.73 to 6.82%; p < 0.0001), while this measure remained essentially unchanged in the control group (n = 16; from 7.73 to 7.40%) (Grimaldi et al. 2018).

5 Other Foods Patterns and Their Effects on T2DM

There are many types of dietary patterns emphasizing the consumption of plant foods besides MedDiet, as Dietary Approaches to Stop Hypertension (DASH), vegetarian and vegan diets (Salas-Salvadó et al. 2019).

The DASH diet was originated in the 1990s and promotes the consumption of vegetables and fruits, lean meat and dairy products, and the inclusion of micronutrients, as vitamins and minerals. It also advocates the reduction of sodium in the diet to about 1500 mg/day and the consumption of minimally processed, stimulating the consumption of fresh food (Kerley 2019). The

adherence to DASH dietary pattern is often related to the nutritional management of individuals with hypertension and its role in the T2DM control has also been studied (Campbell 2017). There is a considerable overlap of components of DASH and the MedDiet, such as vegetables, fruits, nuts, and legumes as beneficial components, considering red and processed meat as a rather detrimental component, although evidence for association between DASH diet and diabetes is limited (Mantzioris and Villani 2019). However, unlike the MedDiet, DASH diet includes whole grains, low-fat dairy, a broad group of fats and oils, and sugary drinks, but does not consider alcohol in your composition (Tangney 2014).

A randomized crossover clinical trial that aimed to determine the effects of the DASH eating pattern on cardiometabolic risks in T2DM patients, observed an inverse association between the intake of DASH diet, when compared with control diet, on fasting blood glucose and HbA1c levels (Azadbakht et al. 2011). In fact, a systematic review with meta-analysis that summarized evidences from prospective studies, which examined association of dietary patterns with T2DM, showed an estimated global effect of DASH diet in T2DM incidence equal to 0.82 when compared the highest with the lowest intake of DASH diet (Jannasch et al. 2017).

The term plant-based diet has been extensively used to refer not only to vegan diets, which do not include any food from animal sources, but also to other diets such as vegetarian, which can include eggs and dairy products, or to semi vegetarian diets, which contain small amounts of meat and fish or other animal products (Salas-Salvadó et al. 2019). Several factors may explain the effect of the vegan diet on the glycemic control, such as reduced fat content, especially saturated fat, which can result in decreased accumulation of intracellular fat, improving insulin sensitivity (Salas-Salvadó et al. 2019). Besides, a high intake of fruits and vegetables and adherence to dietary patterns emphasizing the consumption of plant foods has been associated with reduced body weight and lower abdominal adiposity (Hopping et al. 2010). These benefits are also attributed to a high fiber content and the low glycemic index (Hopping et al. 2010).

The vegetarian dietary pattern may present as an option of easier adhesion in relation to vegan pattern. The Academy of Nutrition and Dietetics of the United States emphasizes that the vegetarian dietary pattern, when properly planned, have positive effects on the treatment and risk of chronic diseases, including T2DM (Melina et al. 2016). In addition, a meta-analysis of six controlled clinical trials observed that the consumption of vegetarian diets was associated with a significant reduction of 0.39% in HbA1c when compared with consumption of other control diets, as omnivorous, conventional diet for diabetics or low fat diets (Yokoyama et al. 2014). Another study that aimed to compare the effects of calorie-restricted vegetarian (experimental) and conventional diabetic diet (controls group) in T2DM patients, observed that the vegetarian diet presented greater capacity to improve insulin sensitivity compared with a conventional diabetic diet over 24 weeks. The metabolic clearance rate of glucose increased more in the experimental group from baseline to 24 weeks (30%) than in the control group (20%) (Kahleova et al. 2011).

Therefore, evidences suggest that healthy patterns, including higher intakes of vegetables and fruits, whole grains, fish, and low-fat dairy, may decrease T2DM risk, and that the unhealthy patterns, including frequent intakes of sugars, processed/red meats, and fried foods, may increase the T2DM risk or harm the treatment. In addition, lifestyle changes are crucial for metabolic control, as well as they may reduce the risk of diabetes complications caused by poor glycemic control (Huang et al. 2019).

6   Conclusion

The MedDiet is related to decreased glycemia levels and reduced cardiovascular risk in patients with T2DM. Studies have also shown favourable changes in glycemic control by reducing HbA1c. Regarding CVD, the MedDiet promotes improvements in the lipid profile, blood pressure, insulin resistance and inflammation. Thus,

adoption of this dietary pattern is recommended for T2DM patients because of its beneficial metabolic control and its role in reducing the risk of macrovascular complication, such as atherosclerosis.

There are controversial data related to the MedDiet improvement on microvascular complications. Notably, the studies indicate a positive relationship between this dietary pattern and the prevention of diabetic retinopathy and DKD. Crucially, almost all studies were performed in T2DM patients or nondiabetic individuals. Thus, there is a lack of data about its effect on T1DM. Consequently, it is necessary studies with a larger sample size and long-term follow-up for analysis of the MedDiet influence on DM complications.

Acknowledgments KBG and MTA are grateful to Conselho Nacional de Desenvolvimento Científico e Tecnológico – CNPq for the research fellowship. JAGT is grateful to Coordenação de Aperfeiçoamento de Pessoal de Nível Superior – CAPES for the research fellowship.

Funding CNPq AND CAPES.

Declarations of Interest The authors declare no conflict of interest.

References

Altomare R, Cacciabaudo F, Damiano G, Palumbo VD, Gioviale MC, Bellavia M, Tomasello G, Lo Monte AI (2014) The Mediterranean diet: a history of health. Iran J Public Health 42(5):449–457

American Diabetes Association (2009) Standards of medical care in diabetes—2009. Diabetes Care 32(Suppl 1):S13–S61

American Diabetes Association (ADA) (2018) Standards of medical care in diabetes—2019. Abridged for primary care providers. Diabetes Care 42(Suppl. 1):S1–S194

Annuzzi G, Rivellese AA, Bozzetto L, Riccardi G (2014) The results of look AHEAD do not row against the implementation of lifestyle changes in patients with type 2 diabetes. Nutr Metab Cardiovasc Dis 24(1):4–9

Azadbakht L, Fard NR, Karimi M, Baghaei MH, Surkan PJ, Rahimi M, Esmaillzadeh A, Willett WC (2011) Effects of the Dietary Approaches to Stop Hypertension (DASH) eating plan on cardiovascular risks among type 2 diabetic patients: a randomized crossover clinical trial. Diabetes Care 34(1):55–57

Bach-Faig A, Berry EM, Lairon D, Reguant J, Trichopoulou A, Dernini S, Medina FX, Battino M, Belahsen R, Miranda G, Serra-Majem L, Mediterranean Diet Foundation Expert Group (2011) Mediterranean diet pyramid today. Science and cultural updates. Public Health Nutr 14(12A):2274–2284

Bowen KJ, Harris WS, Kris-Etherton PM (2016) Omega-3 fatty acids and cardiovascular disease: are there benefits? Curr Treat Options Cardiovasc Med 18 (11):69

Buckland G, Bach A, Serra-Majem L (2008) Obesity and the Mediterranean diet: a systematic review of observational and intervention studies. Obes Rev 9 (6):582–593

Campbell AP (2017) DASH eating plan: an eating pattern for diabetes management. Diabetes Spectr 30(2):76–81

Degirolamo C, Rudel LL (2010) Dietary monounsaturated fatty acids appear not to provide cardioprotection. Curr Atheroscler Rep 12(6):391–396

Dernini S, Berry EM (2015) Mediterranean diet: from a healthy diet to a sustainable dietary pattern. Front Nutr 2:1–15

Diabetes and Nutrition Study Group of the Spanish Diabetes Association (GSEDNu) (2006) Diabetes Nutrition and Complications Trial: adherence to the ADA nutritional recommendations, targets of metabolic control, and onset of diabetes complications. A 7-year, prospective, population-based, observational multicenter study. J Diabetes Complicat 20(6):361–366

Díaz-López A, Babio N, Martínez-González MA, Corella D, Amor AJ, Fitó M, Estruch R, Arós F, Gómez-Gracia E, Fiol M, Lapetra J, Serra-Majem L, Basora J, Basterra-Gortari FJ, Zanon-Moreno V, Muñoz MÁ, Salas-Salvadó J, PREDIMED Study Investigators (2015) Mediterranean diet, retinopathy, nephropathy, and microvascular diabetes complications: a post Hoc analysis of a randomized trial. Diabetes Care 38(11):2134–2141

Donath MY, Shoelson SE (2011) Type 2 diabetes as an inflammatory disease. Nat Rev Immunol 11(2):98–107

Esposito K, Giugliano D (2014) Mediterranean diet and type 2 diabetes. Diabetes Metab Res Rev 30(Suppl 1):34–40

Esposito K, Maiorino MI, Di Palo C, Giugliano D, Campanian Postprandial Hyperglycemia Study Group (2009a) Adherence to a Mediterranean diet and glycaemic control in type 2 diabetes *mellitus*. Diabet Med 26(9):900–907

Esposito K, Maiorino MI, Ciotola M, Di Palo C, Scognamiglio P, Gicchino M, Petrizzo M, Saccomanno F, Beneduce F, Ceriello A, Giugliano D (2009b) Effects of a Mediterranean-style diet on the need for antihyperglycemic drug therapy in patients with newly diagnosed type 2 diabetes. A randomized trial. Ann Intern Med 151:306–314

Esposito K, Maiorino MI, Bellastella G, Chiodini P, Panagiotakos D, Giugliano D (2015) A journey into a Mediterranean diet and type 2 diabetes: a systematic review with meta-analyses. BMJ Open 5(8):e008222

Estruch R, Martínez-González MA, Corella D, Salas-Salvadó J, Ruiz-Gutiérrez V, Covas MI, Fiol M, Gómez-Gracia E, López-Sabater MC, Vinyoles E, Arós F, Conde M, Lahoz C, Lapetra J, Sáez G, Ros E, PREDIMED Study Investigators (2006) Effects of a Mediterranean-style diet on cardiovascular risk factors: a randomized trial. Ann Intern Med 145 (1):1–11

Estruch R, Ros E, Salas-Salvadó J, Covas MI, Corella D, Arós F, Gómez-Gracia E, Ruiz-Gutiérrez V, Fiol M, Lapetra J, Lamuela-Raventos RM, Serra-Majem L, Pintó X, Basora J, Muñoz MA, Sorlí JV, Martínez JA, Martínez-González MA, PREDIMED Study Investigators (2013) Primary prevention of cardiovascular disease with a Mediterranean diet. N Engl J Med 368(14):1279–1290

Finicelli M, Squillaro T, Di Cristo F, Di Salle A, Melone MAB, Galderisi U, Peluso G (2019) Metabolic syndrome, Mediterranean diet, and polyphenols: evidence and perspectives. J Cell Physiol 234(5):5807–5826

Grimaldi M, Ciano O, Manzo M, Rispoli M, Guglielmi M, Limardi A, Calatola P, Lucibello M, Pardo S, Capaldo B, Riccardi G (2018) Intensive dietary intervention promoting the Mediterranean diet in people with high cardiometabolic risk: a non-randomized study. Acta Diabetol 55(3):219–226

Hopping BN, Erber E, Grandinetti A, Verheus M, Kolonel LN, Maskarinec G (2010) Dietary fiber, magnesium, and glycemic load alter risk of type 2 diabetes in a multiethnic cohort in Hawaii. J Nutr 140:68–74

Huang MC, Chang C, Chang WT, Liao YL, Chung HF, Hsu CC, Shin SJ, Lin KD (2019) Blood biomarkers of various dietary patterns correlated with metabolic indicators in Taiwanese type 2 diabetes. Food Nutr Res 63:1–9

Huo R, Du T, Xu Y, Xu W, Chen X, Sun K, Yu X (2015) Effects of Mediterranean-style diet on glycemic control, weight loss and cardiovascular risk factors among type 2 diabetes individuals: a meta-analysis. Eur J Clin Nutr 69(11):1200–1208

International Diabetes Federation (2017) IDF diabetes atlas, 8th edn. International Diabetes Federation, Brussels, pp 1–150. http://www.diabetesatlas.org

Itsiopoulos C, Brazionis L, Kaimakamis M, Cameron M, Best JD, O'Dea K, Rowley K (2011) Can the Mediterranean diet lower HbA1c in type 2 diabetes? Results from a randomized cross-over study. Nutr Metab Cardiovasc Dis 21(9):740–747

Jannasch F, Kröger J, Schulze MB (2017) Dietary patterns and type 2 diabetes: a systematic literature review and meta-analysis of prospective studies. J Nutr 147 (6):1174–1182

Jurado-Fasoli L, De-la-O A, Castillo MJ, Amaro-Gahete FJ (2019) Dietary differences between metabolically healthy overweight-obese and metabolically unhealthy overweight-obese adults. Br J Nutr 23:1–21

Juutilainen A, Lehto S, Ronnemaa T, Pyorala K, Laakso M (2005) Type 2 diabetes as a "coronary heart disease equivalent": an 18-year prospective population-based study in Finnish subjects. Diabetes Care 28:2901–2907

Kahleova H, Matoulek M, Malinska H, Oliyarnik O, Kazdova L, Neskudla T, Skoch A, Hajek M, Hill M, Kahle M, Pelikanova T (2011) Vegetarian diet improves insulin resistance and oxidative stress markers more than conventional diet in subjects with type 2 diabetes. Diabet Med 28(5):549–559

Kastorini CM, Milionis HJ, Esposito K, Giugliano D, Goudevenos JA, Panagiotakos DB (2011) The effect of Mediterranean diet on metabolic syndrome and its components: a meta-analysis of 50 studies and 534,906 individuals. J Am Coll Cardiol 57(11):1299–1313

Kerley CP (2019) Dietary patterns and components to prevent and treat heart failure: a comprehensive review of human studies. Nutr Res Rev 32(1):1–27

Khaw KT, Wareham N, Luben R, Bingham S, Oakes S, Welch A, Day N (2001) Glycated haemoglobin, diabetes, and mortality in men in Norfolk cohort of European prospective investigation of cancer and nutrition (EPIC-Norfolk). BMJ 322(7277):15e8

López-Miranda J, Gómez P, Castro P, Marín C, Paz E, Bravo MD, Blanco J, Jiménez-Perepérez J, Fuentes F, Pérez-Jiménez F (2000) Mediterranean diet improves low density lipoprotein susceptibility to oxidative modifications. Med Clin 115(10):361–365

López-Mirandaa J, Gómez P, Castro P, Marín C, Paz E, Bravo MD, Blanco J, Jiménez-Perepérez J, Fuentesa F, Pérez-Jiménez F (2000) Mediterranean diet improves low density lipoprotein susceptibility to oxidative modifications. Med Clin 115:361–365

Mahoney SE, Loprinzi PD (2014) Influence of flavonoid-rich fruit and vegetable intake on diabetic retinopathy and diabetes-related biomarkers. J Diabetes Complicat 28(6):767–771

Maiorino MI, Bellastella G, Petrizzo M, Gicchino M, Caputo M, Giugliano D, Esposito K (2017) Effect of a Mediterranean diet on endothelial progenitor cells and carotid intima-media thickness in type 2 diabetes: follow-up of a randomized trial. Eur J Prev Cardiol 24(4):399–408

Mantzioris E, Villani A (2019) Translation of a Mediterranean-style diet into the Australian dietary guidelines: a nutritional, ecological and environmental perspective. Nutrients 11(10):2507

Mantzoros CS, Williams CJ, Manson JE, Meigs JB, Hu FB (2006) Adherence to the Mediterranean dietary pattern is positively associated with plasma adiponectin concentrations in diabetic women. Am J Clin Nutr 84(2):328–335

Martinez Gonzalez MA, Salas-Salvado J, Estruch R, Corella D, Fito M, Ros E, PREDIMED Investigators (2015) Benefits of the Mediterranean diet: insights from the PREDIMED study. Prog Cardiovasc Dis 58:50–60

Martins EA, Correia AC, Lemos ET (2014) Functionality of the Mediterranean diet in diabetes type 2. Rev Port Diabetes 9(2):83–91

Matsuzawa Y, Lerman A (2014) Endothelial dysfunction and coronary artery disease: assessment, prognosis, and treatment. Coron Artery Dis 25(8):713–724

Mattar M, Obeid O (2009) Fish oil and the management of hypertriglyceridemia. Nutr Health 20(1):41–49

Melina V, Craig W, Levin S (2016) Position of the academy of nutrition and dietetics: vegetarian diets. J Acad Nutr Diet 116(12):1970–1980

Monnier L, Lapinski H, Colette C (2003) Contributions of fasting and postprandial plasma glucose increments to the overall diurnal hyperglycemia of type 2 diabetic patients: variations with increasing levels of HbA(1c). Diabetes Care 26(3):881–885

Pan B, Wu Y, Yang Q, Ge L, Gao C, Xun Y, Tian J, Ding G (2019) The impact of major dietary patterns on glycemic control, cardiovascular risk factors, and weight loss in patients with type 2 diabetes: a network meta-analysis. J Evid Based Med 12(1):29–39

Rahati S, Shahraki M, Arjomand G, Shahraki T (2014) Food pattern, lifestyle and diabetes *mellitus*. Int J High Risk Behav Addict 3(1):e8725

Roldan CC, Marco MLT, Marco FM, Albero JS, Rios RS, Rodriguez AC, Royo JMP, Lopez PJT (2019) Adhesion to the Mediterranean diet in diabetic patients with poor control. Clin Investig Arterioscler 31(5):210–217

Ros E, Martinez-Gonzalez MA, Estruch R et al (2014) Mediterranean diet and cardiovascular health: teachings of the PREDIMED study. Adv Nutr 5:330S–336S

Rosamond WD, Chambless LE, Folsom AR, Cooper LS, Conwill DE, Clegg L, Wang CH, Heiss G (1998) Trends in the incidence of myocardial infarction and in mortality due to coronary heart disease, 1987 to 1994. N Engl J Med 339(13):861–867

Rossi M, Turati F, Lagiou P, Trichopoulos D, Augustin LS, La Vecchia C, Trichopoulou A (2013) Mediterranean diet and glycaemic load in relation to incidence of type 2 diabetes: results from the Greek cohort of the population-based European Prospective Investigation into Cancer and Nutrition (EPIC). Diabetologia 56(11):2405–2413

Salas-Salvadó J, Becerra-Tomás N, Papandreou C, Bulló M (2019) Dietary patterns emphasizing the consumption of plant foods in the management of type 2 diabetes: a narrative review. Adv Nutr 10:S320–S331

Schroder H (2007) Protective mechanisms of the Mediterranean diet in obesity and type 2 diabetes. J Nutr Biochem 18:149–160

Shai I, Schwarzfuchs D, Henkin Y et al (2008) Weight-loss with a low carbohydrate. Mediterranean, or low-fat diet. N Engl J Med 359:229–241

Silva FM, Steemburgo T, Azevedo MJ, Mello VD (2009) Glycemic index and glycemic load in the prevention and treatment of type 2 diabetes mellitus. Arq Bras Endocrinol Metabol 53(5):560–571

Siri-Tarino PW, Chiu S, Bergeron N, Krauss RM (2015) Saturated fats versus polyunsaturated fats versus carbohydrates for cardiovascular disease prevention and treatment. Annu Rev Nutr 35:517–543

Sofi F, Cesari F, Abbate R, Gensini GF, Casini A (2008) Adherence to Mediterranean diet and health status: meta-analysis. BMJ 337:a1344

Tangney CC (2014) DASH and Mediterranean-type dietary patterns to maintain cognitive health. Curr Nutr Rep 3(1):51–61

Tepper S, Alter Sivashensky A, Rivkah Shahar D, Geva D, Cukierman-Yaffe T (2018) The association between Mediterranean diet and the risk of falls and physical function indices in older type 2 diabetic people varies by age. Nutrients 10(6):767

Trichopoulou A (2001) Mediterranean diet: the past and the present. Nutr Metab Cardiovasc Dis 11 (4 Suppl):1–4

Trichopoulou A, Costacou T, Bamia C, Trichopoulos D (2003) Adherence to a Mediterranean diet and survival in a Greek population. N Engl J Med 348 (26):2599–2608

Trichopoulou A, Martínez-González MA, Tong TY, Forouhi NG, Khandelwal S, Prabhakaran D, Mozaffarian D, de Lorgeril M (2014) Definitions and potential health benefits of the Mediterranean diet: views from experts around the world. BMC Med 24 (12):1–16

Tripp ML, Dahlberg CJ, Eliason S, Lamb JJ, Ou JJ, Gao W, Bhandari J, Graham D, Dudleenamjil E, Babish JG (2019) A low-glycemic, Mediterranean diet and lifestyle modification program with targeted nutraceuticals reduces body weight, improves cardiometabolic variables and longevity biomarkers in overweight subjects: a 13-week observational trial. J Med Food 22(5):479–489

Willett WC, Leibel RL (2002) Dietary fat is not a major determinant of body fat. Am J Med 113(Suppl. 9B):47S–59S

World Health Organization (2016) Global report on diabetes. World Health Organization, Geneva, pp 1–88

World Health Organization (2019) Classification of diabetes mellitus. World Health Organization, Geneva, pp 1–40

Yokoyama Y, Barnard ND, Levin SM, Watanabe M (2014) Vegetarian diets and glycemic control in diabetes: a systematic review and meta-analysis. Cardiovasc Diagn Ther 4(5):373–382

Adv Exp Med Biol - Advances in Internal Medicine (2020) 4: 129–152
https://doi.org/10.1007/5584_2020_514
© Springer Nature Switzerland AG 2020
Published online: 8 April 2020

Glycaemic Control and Vascular Complications in Diabetes Mellitus Type 2

Francesco Maranta, Lorenzo Cianfanelli, and Domenico Cianflone

Abstract

Diabetes mellitus is constantly increasing worldwide. Vascular complications are the most common in the setting of long-standing disease, claiming the greatest burden in terms of morbidity and mortality. Glucotoxicity is involved in vascular damage through different metabolic pathways, such as production of advanced glycation end-products, activation of protein kinase C, polyol pathway activation and production of reactive oxygen species. Vascular complications can be classified according to the calibre of the vessels involved as microvascular (such as diabetic retinopathy, nephropathy and neuropathy) or macrovascular (such as cerebrovascular, coronary and peripheral artery disease). Previous studies showed that the severity of vascular complications depends on duration and degree of hyperglycaemia and, as consequence, early

F. Maranta (✉)
Cardiac Rehabilitation Unit, San Raffaele Scientific Institute, Milan, Italy
e-mail: maranta.francesco@hsr.it

L. Cianfanelli and D. Cianflone
Vita-Salute San Raffaele University and San Raffaele Scientific Institute, Milan, Italy

trials were designed to prove that intensive glucose control could reduce the number of vascular events. Unfortunately, results were not as satisfactory as expected. Trials showed good results in reducing incidence of microvascular complications but coronary heart diseases, strokes and peripheral artery diseases were not affected despite optimal glycemia control. In 2008, after the demonstration that rosiglitazone increases cardiovascular risk, FDA demanded stricter rules for marketing glucose-lowering drugs, marking the beginning of cardiovascular outcome trials, whose function is to demonstrate the cardiovascular safety of anti-diabetic drugs. The introduction of new molecules led to a change in diabetes treatment, as some new glucose-lowering drugs showed not only to be safe but also to ensure cardiovascular benefit to diabetic patients. Empaglifozin, a sodium-glucose cotransporter 2 inhibitor, was the first molecule to show impressing results, followed on by glucagon-like peptide 1 receptor agonists, such as liraglutide. A combination of anti-atherogenic effects and hemodynamic improvements are likely explanations of the observed reduction in cardiovascular events and mortality. These evidences have opened a completely new era in the field of glucose-lowering drugs and of diabetes treatment in particular with respect to vascular complications.

Keywords

Atherosclerosis · Cardiovascular
complications · CVOT · Diabetes · Glycaemic
control · Nephropathy · Neuropathy ·
Peripheral artery disease

1 Introduction

During last years, prevalence of diabetes mellitus (DM) has risen considerably both in developed and developing countries. The high number of patients involved and the significant impact on prognosis determined by its complications make diabetes a key health priority from a global point of view.

It is however important to notice that, thanks to the spread of preventive strategies, the incidence of clinically diagnosed DM has remained stable or even dropped in the majority of populations studied since 2006 (Magliano et al. 2019). On the other hand, the prevalence of DM is constantly increasing and the cases of patients with DM are expected to rise from 9.1% of the population in 2014, to 13.8% in 2030 and 17.9% in 2060 (Lin et al. 2018). The aging of population and the high incidence of obesity and metabolic syndrome are major causes of the increase in DM prevalence. These projections have been done using statistical models mainly applied to the USA population but projections from other high-income countries are similar.

Vascular complications are by far the most common complication in the setting of long-standing diabetes, representing the most important determinant of morbidity and mortality (Tseng 2004). Atherosclerosis in different body districts, both micro and macrovascular, is the reason for reduced life expectancy and poor quality of life owing to its functional consequences.

2 Pathophysiology and Molecular Mechanism at the Basis of Vascular Insult in Diabetes

The origin of vascular complications in type 1 and type 2 DM is certainly multifactorial but persistently elevated glycemia seems to be the key mediator in organ injury through a mechanism called glucotoxicity.

Glucotoxicity refers to the structural and functional damage occurring both in beta pancreatic cells and in target tissues of insulin. It is caused by chronic elevation of glycemia levels leading to disruption of normal cellular mechanisms involved in carbohydrate management and to build-up of toxic metabolic by-products. These alterations represent a double-edged sword as from one side they cause a reduction in insulin secretion from affected beta cells and from the other side they cause reduction in insulin action at peripheral level, inducing the so-called insulin resistance.

Many different metabolic pathways are involved in the development of the vascular insult at the basis of long-term diabetic injuries (Kumar et al. 2010; Rask-Madsen and King 2013). It is important to notice that some pathogenic mechanisms are preferentially active in one organ but generally they are responsible for the development of vascular complications in more than one district.

1. Production of AGEs (Advanced Glycation End-products). AGEs are produced by means of non-enzymatic reaction between di-carbonyl compounds derived from glucose (such as methylglyoxal, glyoxal, and 3-deoxyglucone) and the amino-groups derived from intra and extracellular proteins. Their formation is followed by the interaction with specific receptors, called RAGEs (Receptor of Advanced Glycation End-products) whose activation leads to a chain of metabolic cellular consequences enhancing tissue injury. The RAGEs are found on inflammatory cells and smooth muscle cells of blood vessels. The detrimental effects depend on the release of inflammatory cytokines, the activation of fibroblasts for the deposition of extracellular matrix, the entrapment of certain molecules in the media of arterial vessels (such as LDL particles), the production of reactive oxygen species (ROS) and the increased procoagulant activity of endothelial cells: the final net result is the acceleration of atherosclerosis and a predisposition to atherothrombosis.

2. Activation of Protein Kinase C (PKC) pathway. The numerous glycotic intermediates derived by the constant state of hyperglycaemia can activate PKC signalling pathway. One of the main downstream effects consists in the increased production of Vascular Endothelial Growth Factor (VEGF), responsible for example for the retinal neovascularization which is a typical feature of diabetic retinopathy. Another well studied consequence is the increased release of Tissue Growth Factor (TGF)-beta which is a potent stimulator for fibroblast release of extracellular matrix. The increased deposition of interstitial material is at the base of vascular fibrosis in all body districts, ranging from large-sized blood vessels to nephroangio-sclerosis at glomerular level.

3. Polyol pathway activation. Cellular glucose uptake may be significantly increased because of the high extracellular concentration. The excess of glucose may be shunted to the polyol pathway (also known as sorbitol pathway). Glucose is converted to sorbitol and eventually to fructose in a reaction produced by aldose reductase that uses NADPH as a cofactor. When the polyol pathway is highly active the intracellular storage of NADPH is rapidly depleted with detrimental consequences. NADPH is involved by glutathione-reductase enzyme in a reaction whose aim is to regenerate reduced glutathione (GSH), which is one of the main anti-oxidant molecules at cellular level. When intracellular NADPH level drops, also the level of GSH is reduced because it cannot be regenerated. In this setting the cell loses its primary antioxidant protection from oxidative stress becoming susceptible to multiple injuries. This seems to be the primary mechanism of neuron damage leading to peripheral diabetic neuropathy.

4. Oxidative stress. Oxidative stress at cellular level is multifactorial and depends primarily on the activation of AGE-related intracellular pathways, the depletion of NADPH and GSH storage, the increased production of ROS and free radicals. Oxidative damage may lead to a change in cellular phenotype increasing LDL-R on endothelial cells and promoting the establishment of a procoagulant state in blood vessels.

5. Hexosamine pathway. The presence of elevated glucose concentration inside the cell may lead to the shift of this molecule in unusual pathways such as the hexosamine pathway. The products of this pathway may lead to endoplasmic reticulum stress which can cause altered transcription of molecules involved in accelerated atherosclerosis and insulin resistance.

In the end, hypertension and dyslipidaemia are frequently present in diabetic patients, as well as other cardiovascular (CV) risk factors. Their co-existence not only adds other mechanisms of vascular damage but also enhances the diabetic specific detrimental effects. Endothelial dysfunction is a central and well known final pathophysiological element common to the various factors described above (Avogaro et al. 2011; De Vriese et al. 2000; Hadi and Suwaidi 2007).

3 Classification of Vascular Complications

Atherosclerosis in the context of DM can affect all vascular districts and it is the central pathological mechanism at the basis of vascular complications. Importantly, the risk of developing vascular complication depends on both the severity and the duration of hyperglycaemia, similarly to what happens with LDL exposure (Ference et al. 2018). Patients with long-standing elevation of blood sugar and higher level of glycemia are the ones that will present with earlier, more severe and more diffuse forms of vascular complications.

Vascular complications are conventionally classified as microangiopathies, involving small-sized blood vessels (such as arterioles and capillaries) and macroangiopathies, involving medium and large-sized blood vessels (such as aorta, coronary arteries, lower limb vessels and cerebral vessels) (Kumar et al. 2010). Therefore, according to the size and location of the blood vessels involved authors generally recognize the following:

- microvascular complications: retinopathy, nephropathy, neuropathy;
- macrovascular complications: coronary artery disease, cerebrovascular disease, peripheral artery disease.

3.1 Diabetic Retinopathy

Diabetic retinopathy is probably the most common microvascular complication. It is responsible for as many as 10000 new cases of blindness every year in the USA (Fong et al. 2004). The main mechanism involved in the pathogenesis seems to be the production of AGEs, the increase in local production of VEGF and oxidative stress due to ROS. It is generally classified as non-proliferative (background) retinopathy and proliferative retinopathy. Non-proliferative (background) retinopathy consists of small haemorrhages, referred to as "dot haemorrhages", in the middle layer of the retina, whose margins are characterized by the presence of hard exudates formed by lipid deposition. Microaneurisms are very common together with retinal oedema resulting from fluid extravasation. Proliferative retinopathy is characterized by florid neoangiogenesis with the formation of new blood vessels sprouting in a disorganized manner on the surface of the retina. The new vessels are clearly visible as white areas called "cotton wool spots". Vitreous haemorrhage and retinal detachment are two complications of long-standing retinopathy leading to abrupt or progressive blindness.

3.2 Diabetic Nephropathy

Diabetic nephropathy is one of the leading causes of renal failure and end-stage renal disease

(ESRD) requiring dialysis and its incidence is on the rise due to the high prevalence of diabetes worldwide. It is defined as the presence of overt proteinuria >500 mg in 24 h in the context of DM without other specific causes. It is usually preceded by a long period of microalbuminuria, consisting in albuminuria of 30–300 mg in 24 h. Microalbuminuria signals the presence of an underlying glomerular damage that can be reversed in case of optimal glycaemic and blood pressure control. The onset of overt proteinuria on the other hand, indicates an irreversible damage. Seven percent of type 2 diabetic patients presents with nephropathy at the time of diagnosis. It occurs in up to 12% of patients with type 1 diabetes mellitus by 7 years, and in 25% of patients with type 2 DM by 10 years after the diagnosis (Adler et al. 2003). The origin of diabetic nephropathy stems from a combination of metabolic and hemodynamic alterations contributing to alteration of podocytes function, increasing basement membrane thickening, reduction of filtration rate and reduced tubular function (Cao and Cooper 2011). Pathological changes observable in histological kidney specimen are the presence of thickened glomerular basement membrane, mesangial nodules distorting glomerular architecture (called Kimmelsteil-Wilson nodules), thickening of arteriolar medial wall, capillary microaneurysm formation and progressive extension of interstitial fibrosis. Aggressive treatment consisting in glycaemic control and anti-hypertensive strategies using ACE-inhibitors and angiotensin-receptor blockers can prevent progression toward further damage and delay the need for dialysis.

It is worth to mention a form of diabetes-related nephropathy named non-proteinuric diabetic kidney disease: a variable proportion of patients (around 35–40%) presents with advanced renal impairment (eGFR <60 mL/mq) in the absence of proteinuria or albuminuria (microalbuminuria <300 mg/g) (Robles et al. 2015). This entity is associated with a higher incidence of cardiovascular diseases. However, it is not yet clear the underlying pathological mechanisms and the risk of progression toward end-stage renal disease.

3.3 Diabetic Neuropathy

Diabetic neuropathy has been defined by the American Diabetes Association (ADA) as the presence of symptoms and signs of peripheral nerve dysfunction in the setting of diabetes after the exclusion of other causes. Both vascular and non-vascular abnormalities have been advocated in the establishment of nerve injury. Histological findings show that several pathological changes occur in nerve structure, such as basement membrane thickening and pericyte loss. Notably, there is evidence of reduced density of capillaries, resulting in attenuated perfusion and eventually endoneurial hypoxia. Finally, this contributes to the axonal thickening and loss of neurons seen in most advanced forms of neuropathy (Tavakoli et al. 2008). Polyol accumulation and oxidative stress seem to be the two most important contributors for nerve damage. Many forms of neuropathy can occur, including sensory, motor and autonomic neuropathies. They can be focal or multifocal. It is important to recognize neuropathies as early as possible because of the significant morbidity and mortality they are associated with. Eighty percent of amputations occurs after foot ulceration or injury secondary to peripheral neuropathy and impaired healing due to lower limb perfusion defects (Boulton et al. 2005). Chronic sensorimotor distal symmetric neuropathy is the most common form, presenting with burning tingling sensation at the extremities that is worse at night. Some patients present with hypoesthesia and numbness and they are the ones at higher risk for foot ulceration. Pure sensory neuropathy is rare. Mononeuropathy have sudden onset and can involve every nerve, even though the most common are ulnar, radial and median. Autonomic neuropathy can manifest with gastroparesis, constipation or diarrhoea, erectile dysfunction, bladder dysfunction, orthostatic hypotension; moreover, it affects patient perception of anginal pain, leading to the high incidence of silent ischemia reported in diabetic patients (Maser et al. 2003).

3.4 Coronary Artery Disease

Coronary artery disease has been linked to diabetes mellitus in many studies starting from the Framingham study (Kannel and McGee 1979). DM increases the risk of myocardial infarction more than any other risk factor (except for cigarette smoking) and coronary artery disease is the most common macrovascular complication registered (Anand et al. 2008). From a pathological point of view, DM promotes atheroma formation in coronary arteries and at the same time the constant state of hyperglycaemia promotes a procoagulant state favouring the occurrence of thrombotic events. Notably, from an anatomical point of view, coronary artery lesions tend to be more diffuse and more distal relative to lesions observed in non-diabetic patients (Morgan et al. 2004). Even though the classical lesions concern epicardial vessels, it is important to recognize the role of coronary microvascular dysfunction in diabetes (also known as diabetic coronary microangiopathy) as a large number of diabetic patients with normal epicardial vessels shows reduced coronary flow reserve (Kibel et al. 2017). These patients tend to be symptomatic, have a worse prognosis and tend to progress to overt CAD.

The CV risk in diabetic population is much higher than in normal population and specifically the risk of CV events in many cases is equivalent to the risk of non-diabetic patients who have a history of previous myocardial infarction (Haffner et al. 1998). Therefore, the European Society of Cardiology (ESC), the European Association for the Study of Diabetes (EASD) and ADA consider diabetic patients mainly at high and very-high risk (in particular for this case it is evident how diabetes is considered a sort of coronary artery disease equivalent rather than a simple risk factor) (Buse et al. 2007; Piepoli et al. 2016; Mach et al. 2019). Moreover, the consequences of a myocardial infarction are more pronounced in patients suffering diabetes, whose incidence of cardiovascular death or stroke after a cardiovascular event is higher than in general population (Wallentin et al. 2009). Despite persistently higher rate of coronary artery disease, the rate of mortality has dropped significantly in last two decades (Roger et al. 2012). This improvement is likely the consequence of effective medical treatment and early revascularization strategies. However, the prevalence of DM is increasing over time and patients are living longer: therefore, the overall burden of CAD attributable to DM will rise over time, making strategies to mitigate the risk of CAD in diabetics a fundamental goal for the future (Fox et al. 2007).

3.5   Cerebrovascular Disease

Stroke incidence is elevated in diabetic population claiming a high cost in terms of morbidity and mortality. Patients with type 2 DM have a 150–400% higher risk of stroke relative to non-diabetic population. As for coronary artery disease, DM itself worsen the outcome of stroke as the severity of the cerebrovascular events tend to be higher and the risk of vascular dementia and recurrences are higher as well (Beckman et al. 2002).

3.6   Peripheral Artery Disease

Diabetes is strongly related to peripheral artery disease. The risk of developing PAD is two- to four-fold increased in diabetes mellitus relative to non-diabetic patients. As for other complications the duration and severity of hyperglycaemia influence the extent and severity of PAD. Notably, as observed in coronary arteries, lesions are more diffuse and more distal relative to patients who are not affected by diabetes. Around 20–30% of patients with diabetes have prevalent PAD described as ankle-brachial index (ABI) below <0.9 (Marso and Hiatt 2006). Most patients are asymptomatic and only 20% show symptoms. Importantly, one fourth of patients with PAD demonstrates progression of symptoms over 5 years and a rate of amputation of around 4%. It is important to stress out that diabetes does not only affect large-calibre peripheral vessels, but it

affects distal arterioles as well (the so-called peripheral microangiopathy). This form of distal arteriopathy is thought to be the pathogenetic mechanism at the base of pigmented pretibial patches, necrobiosis lipoidica and erysipelas-like erythema observed in diabetic patients.

4   Efficacy of Glycaemic Control on Coronary Artery Disease

As already mentioned, the elevation of blood sugar strictly correlates with severity of vascular complications. Therefore, a reasonable target to reduce these complications would be the reduction of glycemia. On one hand, many studies have shown how reaching the target of a better glycaemic control can reduce the number of microvascular complications. On the contrary, the evidence regarding the effect of glycaemic control on macrovascular complications has been more controversial.

Coronary artery disease is the most common macrovascular complication in the setting of diabetes, being elevated blood glucose and high glycated haemoglobin (HbA1c) two major well-known risk factors. From this ground, it may seem logical that strategies able to decrease blood glucose could be the mainstem in prevention of cardiovascular events in diabetic population. However, the relationship between glucose-lowering therapies and cardiovascular outcome is not straightforward, as shown by many studies where the reduction of glycaemic parameters did not clearly associate with a reduction of patient cardiovascular events. Since patient outcomes could not be easily predicted by the effect of interventions on surrogate measures, this discrepancy called into question the possible CV benefits of glucose-lowering strategies. Nowadays, a new era of glucose-lowering therapies has been opened by the use of some glucose-lowering drugs (GLDs), such as empaglifozin and liraglutide. These drugs have been demonstrated to impact significantly on CV mortality as shown by the cardiovascular outcome trials. The study of their effects, that extend well beyond the simple reduction of glycemia, could shed light on the "common soil" from where DM and coronary heart disease stem from (Stern 1995).

4.1 Relationship Between Hyperglycaemia and Cardiovascular Outcome

Most studies have established a strong relationship between cardiovascular risk and blood glucose level (measured by means of different parameters, such as fasting glucose, 2-h glucose during oral glucose tolerance test and HbA1c). The relationship reported by the studies is usually linear and continuous with a progressive increase in CV events as glycaemic parameters are increasing.

In the Study of Norfolk, 10232 patients from UK were followed up for 6–8 years showing a linear correlation between the level of HbA1c and CV disease and CV mortality (Khaw et al. 2004). The risk was lower in patient with HbA1c <5% and increased continuously with the elevation of HbA1c: each percentage point of HbA1c over 5% corresponded to a rise in CV relative risk of 20%. Most of the events occurred in patients with moderately elevated HbA1c suggesting that the reduction of HbA1c could be beneficial for CV protection. The Atherosclerosis Risk in Communities (ARIC) study conducted in a US population of adults without prior history of diabetes, showed similar linear trend between CV disease and HbA1c (Selvin et al. 2010). Relative to patients with normal HbA1c values (<5.5%), CV events increased by 23% in those with HbA1c 5.5–6%, by 78% in those with HbA1c 6–6.5% and by 95% in those with HbA1c >6.5%. Similarly, in a diabetic patient population, the United Kingdom Prospective Diabetes Study (UKPDS-35) (Stratton et al. 2000) found that each 1% increase in HbA1c was associated with a 14% relative risk increase for myocardial infarction. On the other side, every 1% decrease in HbA1c was associated with clinically important reductions in the incidence of diabetes-related death (<21%; p-value <0.0001), myocardial

infarction (<14%; p-value <0.0001), microvascular complications (<37%; p-value <0.0001) and peripheral vascular disease (<43%; p-value <0.0001).

The metanalysis of Selvin et al. published in 2004 put together data from all the available observational studies to estimate the association between glycated haemoglobin and cardiovascular events (Selvin et al. 2004). In type 2 diabetes mellitus, there was an increase in relative risk of 18% every 1% point of glycated haemoglobin. This result confirmed the evidences already observed in the single studies.

To be noticed from UKPDS-35 study it appears not to be a lower limit beyond which reductions in HbA1c cease to be of benefit in terms of reduction of CV events and other diabetes-related endpoints. However, the concept "the lower HbA1c the better" cannot be applied in clinical practice because at lower goals of HbA1c the threat of hypoglycemia stands up.

While these initial studies conveyed the message that the lower HbA1c the better for the patient outcome, the UK General Practice Research Database (GPRD) was one of the first study showing that lowering glycemia too much could have harmful consequences (Currie et al. 2010). 27965 patients with type 2 DM whose oral therapy was intensified to oral combinational therapy and 20005 whose oral therapy was intensified adding insulin were followed for 4.5 years monitoring for CV events and mortality. The pattern of risk was U-shaped with an increased number of events at lower and higher HbA1c levels. The same point was confirmed by the results coming from the Kaiser Permanente North Carolina Register (Huang et al. 2011). Data from 71092 patients with diabetes mellitus type 2 were analysed to evaluate the association between HbA1c and CV events and mortality. The authors showed a U-shaped relationship between HbA1c level and mortality with higher risk in those with HbA1c below 6% and over 10%. Again, a third study (Colayco et al. 2011) showed similar results with a U-shaped relationship with increased number of events when HbA1c level was lower than <6% and higher than >8%. All in all, the results of the aforementioned studies added a little more piece to a complex puzzle. In fact, they showed that achievement of low glycemia confers protection from CV events but very low levels of glycemia may result in increased harm, probably due to severe complications of hypoglycaemia.

From these premises glycemia reduction appeared to be the key point to obtain a significant reduction in cardiovascular events, but evidences from later studies were not as satisfactory as expected.

to be linked to the feedback mechanism that is triggered by overactivation of the sympathetic autonomic nervous system. Acute hypoglycaemia stimulates the release of epinephrine which subsequently increases cardiac rate and contractility, induces easier platelet aggregation, worsens vasoconstriction and afterload, heightens cardiac muscle excitability and arrhythmia risk, exacerbates myocardial oxygen consumption leading to ischemia. Moreover, hypoglycaemia may induce hypokalemia as result of potassium shift from extracellular space to intracellular space, leading to worrisome prolongation of QT interval.

4.2 Hypoglycemia and the Possible Explanation of the U-Shaped Mortality Curve

If a strict control of glycemia can have positive prognostic impact on patients, the drawback of excessive glycemia control is an increase in patient mortality. One of the proposed explanations is the higher incidence of hypoglycaemia that is a dreadful complication of intensive glucose-lowering strategies. Hypoglycemia is associated to a higher number of falls, fall-related fractures, cardiovascular events, poor quality of life, dementia and higher number of deaths. The proposed mechanism by which hypoglycaemia could increase mortality seems

4.3 Intensive Versus Conventional Glucose-Lowering Strategies

The first randomized clinical trials were developed to test whether interventions aimed at reducing glycemia were able to reduce micro and macrovascular complications and mortality in population with overt DM. An intensive glucose control group versus a conventional group was usually set to study the effect of the interventions (see Table 1 for summary).

The first landmark trial was the Diabetes Control and Complications Trial (DCCT) published in 1993 showing that intensive glycaemic control reduced the incidence of microvascular

Table 1 Summary of the most important trials comparing intensive versus conventional glycaemic control in type 2 diabetes mellitus

Trials	Population	Follow-up (Years)	Effect on microvascular complications	Effect on macrovascular complications	Effect on mortality
UKPDS (1998)	T2DM N = 3867	11.0	Reduced microvascular endpoints	No difference	No difference
ACCORD (2008)	T2DM N = 10251	3.5	Reduce retinopathy, nephropathy, neuropathy	No difference	Increased
ADVANCE (2008)	T2DM N = 11140	5.5	Reduced nephropathy	No difference	No difference
VADT (2009)	T2DM N = 1791	5.6	Reduced progression of albuminuria	No difference	No difference

Primary endpoints. UKPDS: an aggregate endpoint of any diabetes-related complications. ACCORD: a composite of non-fatal myocardial infarction, non-fatal stroke, and fatal myocardial infarction and stroke. ADVANCE: combined microvascular and macrovascular disease. VADT: time to occurrence of a composite of major cardiovascular events. See text for details

complications in patients affected by type 1 DM (Diabetes Control and Complications Trial Research Group et al. 1993).

Later in 1999, the United Kingdom Prospective Study (UKPDS-33) was targeting patients with type 2 DM (UK Prospective Diabetes Study (UKPDS) Group 1998a). 3867 patients diagnosed with DM type 2 were randomized to receive intensive treatment with sulfonylureas or with insulin versus conventional therapy plus diet alone. Patients were assessed for 10 years follow-up. At the end of follow-up period, the HbA1c in conventional group was 7.9% while in intensive group was 7.0%. The result showed that intensive group had significantly lower incidence (-12%, p-value $= 0.03$) of diabetes-related complications (micro, macrovascular and metabolic) relative to conventional group. However, there was non-significant reduction of diabetes-related death (-10%, p-value $= 0.34$) and non-significant reduction in overall mortality (-6%, p-value $= 0.44$). Moreover, the reduction of diabetes-related complication was dependent on a 25% reduction of microvascular complications (p-value $= 0.0099$) while the overall risk reduction for MI in the two groups was only of borderline significance (p-value $= 0.052$). A significant reduction in macrovascular complications was reached only in a subgroup of obese patients treated with metformin (UK Prospective Diabetes Study (UKPDS) Group 1998b). The conclusion drawn was that an intensive glucose-lowering strategy reduces diabetes-related complications but does not change the overall survival of patients. Furthermore, the reduction of diabetes-related complications was mainly the result of the decrease of microvascular complications in a group of naïve patients, while macrovascular complications were reduced only in a subset of overweight patients. It is interesting to notice that in a 10-year post-trial monitoring study of UKPDS, conducted on patients who survived after the end of the study, a sustained modest effect in reduction of diabetes-related complications in the intensive group control was still present (even if there was no more difference in glycated haemoglobins) (Holman et al. 2008).

Moreover, a significant risk reduction in terms of myocardial infarction (15%, p $= 0.005$) and mortality (27%; p $= 0.002$) emerged in the intensive group, probably suggesting the need for a longer time to evaluate an effect on atherosclerotic outcomes.

One of the lessons learned from this trial is that when a strict glycemia control was initiated early in the history of diabetes and with low CV risk the effect was a longstanding cardiovascular benefit that was not observed for the same degree of glycemia control in older patients with years of uncontrolled diabetes. This protection coming from early diabetes control is thought to come from tissue "metabolic memory". The term metabolic memory refers to the idea that exposure to high levels of glucose for long time is "remembered" by the tissues in terms of damage, because of epigenetic and metabolic long-term effects. Immediate intensive treatment reducing not only the degree of hyperglycemia but also the duration of tissue exposure has protective effects that are maintained for years.

In the Action to Control Cardiovascular Risk in Diabetes (ACCORD) 10251 patients were randomised to intensive control over conventional control (Action to Control Cardiovascular Risk in Diabetes Study Group et al. 2008). The trial was stopped prematurely after 3.5 years because of higher mortality rate in the intensive arm; the rate of serious hypoglycaemia requiring medical treatment was three-fold higher than in conventional group (10.5% Vs 3.5%).

In the Action in Diabetes and Vascular Disease: Preterax and Diamicron Modified Release Controlled Evaluation (ADVANCE) trial 11140 patients were randomised to intensive versus conventional control (ADVANCE Collaborative Group et al. 2008). After 5 years there was a reduction in the primary endpoint (composite of micro and macrovascular complications) but there was no significant effect on MI, suggesting the contribution was mainly from the reduction of the microvascular complications, similarly to the UKPDS trial.

The Veteran Affairs Diabetes Trial (VADT) confirmed the lack of benefit of intensive glucose lowering strategies on major cardiovascular events (Duckworth et al. 2009). 1791 US veterans with type 2 DM were randomly assigned to the intensive and conventional treatment group. There was no significant difference in terms of composite outcomes and cardiovascular endpoints.

In this scenario, it is important to remember the STENO-2 trial published in 2008 because it presents similarities but substantial differences from previous studies (Gæde et al. 2008). The trial enrolled 160 patients presenting with diabetes type 2 and microalbuminuria. Patients were followed-up for a median time of 7.8 years showing net beneficial effect on vascular complications and mortality. The key innovation relative to previous studies was the randomization to multifactorial intensive treatment of risk factors, including stricter glycemia control, blood pressure control, aspirin and statin. Patients were randomized to either intensive or conventional control arms: intensive therapy was associated with a lower risk of CV death (HR 0.43; 95% CI, 0.19–0.94; P = 0.04) and of cardiovascular events (HR 0.41; 95% CI, 0.25–0.67; P < 0.001). Despite positive results, it is not clear if the beneficial effect was simply determined by glycemia control or by the combined reduction of multiple risk factors. It is surely an important trial in defining that combinational control can truly gain CV benefit but it did not clarify if glycemia control alone could impact on CV events.

The net result of these trials is that intensive glucose control failed to improve cardiovascular outcome despite the strong relationship established between glycemia and cardiovascular events. These disappointing results could be explained by several considerations. First of all, glycemia is probably a weaker risk factor for CAD compared to LDL-cholesterol. Cholesterol decrease of 1 mg/dL obtains a relative risk reduction (RRR) of 23% in the incidence of myocardial infarction (Silverman et al. 2016). The expectations of early DM trials were largely based on assumption that glycemia reduction could lead to a similar impact. On a population level based on the data from UKPDS, fewer people should be treated with strict BP control (NNT

23) rather than intensive blood glucose control (NNT 46) to prevent one MI (Vijan and Hayward 2003). In the same way, cholesterol control in primary (NNT 34) and secondary (NNT 13) prevention seems more effective than intensive glucose lowering (Vijan et al. 2004). Secondly, adverse effects secondary to glycemia lowering may counterbalance potential benefits. As shown before, hypoglycaemia is a dreadful complication of intensive control and mortality has been shown to be increased in trials where glycaemic target was set too low. Thirdly, the use of glucose-lowering strategies in advanced diabetes may result useless because the disease and atherosclerosis could be too advanced. Lastly, the effects on macrovascular complications with some drugs may be evident only on the long-term (maybe because of the lack of specific anti-atherogenic effects) and some trials may be too short to adequately observe them.

It should be noted that the aim of these previous studies was to demonstrate the effect of glucose reduction on a certain outcome irrespective of the pharmacological strategy adopted: in most trials combination of various drugs had to be used to control glycemia and there was no particular advantage of one strategy over the other. These early glucose lowering trials were not designed to test the effects on outcomes of a specific drugs but to evaluate the efficacy of a stricter or lenient control of glycemia targets irrespective of the strategy that was used. The lessons learned from these early trials was that most drugs have a significant impact on diabetes mellitus onset and control but scarcely have an effect on cardiovascular events, that are the ones claiming the highest number of deaths in diabetic population.

4.4 Previous Evidences from Early Trials Concerning Specific Drug Classes

When trials started to focus on specific glucose-lowering molecules some of them showed cardiovascular positive effects.

Biguanide drug class is well-represented by metformin, that works reducing hepatic glucose production and promoting peripheral insulin sensitivity, without inducing hypoglycaemia. Numerous observational studies and clinical trials have demonstrated CV benefits in terms of reduction of micro-macrovascular complications and CV-related mortality. In the HOME trial (Hyperinsulinemia: the Outcome of its Metabolic Effect) patients treated with metformin demonstrated a 40% reduction of endpoints (a composite of both micro and macrovascular events) (Kooy et al. 2009). In SPREAD-DIMCAD, metformin is compared with glipizide showing that metformin-treated patients have a 46% reduction of CV events (HR 0.54; 95% CI 0.30–0.90; p-value<0.026) (Hong et al. 2013). Therefore, metformin is one of the earliest drugs to show CV benefit. These evidences support the use of metformin as first-line agents in most patients, as recently confirmed by latest guidelines (Cosentino et al. 2019).

For sulfonylureas evidences are conflicting. Sulfonylureas are the oldest oral agents in the treatment of hyperglycemia. As insulin secretagogues they favour insulin secretion, being effective in reducing glycemia at expenses of a significant risk of hypoglycaemia and weight gain. An early warning concerning safety comes from a study in 1970 (UGDP), in which tolbutamide-treated patients showed increased CV mortality (University Group Diabetes Program 1976). However, later studies which compared different treatment arms containing sulfonylureas, found no difference in the CV events. After 50 years of studies, whether sulfonylureas are associated with adverse events is still debatable.

Intestinal alfa-glucosidase inhibitors act inhibiting carbohydrate breakdown in intestine reducing absorption after meals. In STOP-NIDDM, 1429 participants with glucose intolerance were randomised to receive acarbose or placebo. The acarbose allowed to delay the onset of DM in people with glucose intolerance (Chiasson et al. 2002). After 3 years of follow-up the trial reported 49% RRR in CV events with an incredible 91% reduction of MIs. From this evidence, ACE trial was devised to observe the real impact of acarbose on CV outcome (Holman et al. 2017). The trial enrolled 6522 participants over

176 Centres in China. It did not reduce the risk of major CV events but did reduce the incidence of DM.

Thiazolidinediones lower glucose by activating the nuclear transcription factor peroxisome proliferator-activated receptor gamma (also known as PPR-gamma agonists). The two major drugs, pioglitazone and rosiglitazone, despite the efficacy raised concerns about safety. In PRO-ACTIVE 5238 patients with previous CV disease were treated with pioglitazone or placebo. After a median follow-up of 2.9 years, pioglitazone showed a significant benefit on secondary endpoints (death for all-causes, MI and stroke) that has been reduced by 16% (HR 0.84; 95% CI 0.72–0.98, p-value = 0.027) (Dormandy et al. 2005). However there has been an increased risk of heart failure in the intervention group (16% versus 11.5%). The use of rosiglitazone has been investigated by numerous observational studies however only one trial, the RECORD trial, investigated the action of rosiglitazone on CV endpoints. 4447 diabetic patients with inadequately controlled hyperglycaemia were treated with rosiglitazone showing no difference in the primary end-point (CV hospitalization or CV death) among the groups (Home et al. 2009). Notably, the risk of heart failure was increased approximately two-fold with rosiglitazone and this opened the path for further assessments due to concerns related to safety. In 2007, Nissen et al. published a metanalysis which demonstrated an increased risk of MI and mortality with the use of rosiglitazone (Nissen and Wolski 2007). This paper will be a turning point in the history of anti-diabetic drugs for the consequences generated.

4.5 The Cardiovascular Outcome Trials (CVOTs) in the Era of Novel Glucose-Lowering Drugs

4.5.1 History and Concepts Behind CVOTs in Diabetes

Previous studies were heterogeneous in design and the main outcome was the demonstration of the efficacy in controlling glucose-related parameters. This paradigm however changed in 2007 when Nissen et al. raised great concerns regarding possible unexpected CV risks of anti-hyperglycaemic medications (Nissen and Wolski 2007). The meta-analysis showed a significant 43% increase in MI and 64% increase in CV mortality with the use of rosiglitazone. After rosiglitazone experience, in 2008 Food and Drug Administration (U.S. Food and Drug Administration 2008) mandated that every GLD should be demonstrated not only efficacious in reducing glycaemic level but also safe from a CV point of view. Guidelines for drug acceptance now require randomized double-blinded pla-cebo-control trials to assess drug safety, that is to say non-inferiority relative to placebo. The drugs now need to be tested against placebo on top of background therapy. The primary outcomes are combined in 3-points MACE that are similar for all trials and include CV mortality, non-fatal MI and non-fatal stroke. Some studies use 4-points MACE adding hospitalization for unstable angina as one of the primary endpoints. Secondary outcomes variably include all-cause mortality, hospitalization for heart failure and renal outcome. Median follow-up should be at least of 2 years. FDA specifies that non-inferiority is defined with the upper bound of 95% CI for the risk ratio of CV events being <1.3. An upper 95% CI >1.8 would require further pre-marketing trials for approval. Agents showing CI <1.8 but >1.3 requires a large post-marketing CV outcome trial to define risk (U.S. Food and Drug Administration 2018). As the main task of these randomized studies is to assess the prognostic impact of GLDs on cardio-vascular endpoints they are called CardioVascu-lar Outcome Trials (CVOTs).

It is important to emphasize that the studies were indeed not designed to assess superiority. They were designed to be sure anti-diabetic drugs were not harmful from a cardiovascular point of view. However, something unexpected happened when the results from EMPA-REG trial first came out. The astonishing results nourished the idea that some antidiabetic medications may not only be safe but may even reduce the risk of cardio-vascular diseases in a subset of population at high risk for cardiovascular events. From that moment on a special attention has been focused on these new drugs, especially when different pharmaco-logical classes yielded similar results. Therefore, most recent trials are now powered enough to estimate superiority relative to placebo (upper bound of 95% CI for risk ratio of CV events <1.0) in case non-inferiority (upper bound 95% CI for risk ratio of CV events <1.3) is demonstrated.

At the present time, encouragingly, all the completed trials have reached the non-inferiority standard for the primary cardiovascular end-point relative to placebo, demonstrating reassuring car-diovascular safety. Notably, six trials demonstrated superiority over placebo providing evidence of cardiovascular benefit in addition to safety (EMPA-REG OUTCOME for empaglifozin, CANVAS for canaglifozin, LEADER for liraglutide, SUSTAIN-6 for semaglutide, Harmony Outcomes for albiglutide and REWIND for dulaglutide). These evidences open a completely new era in the field of GLDs.

The following sections describe the main trials and results concerning the most important phar-macological classes (see Table 2 for summary).

4.5.2 Dipeptidyl-Peptidase-4 Inhibitors (DPP-4i)

After demonstration of the increased risk of mor-tality due to hypoglycaemic events in ACCORD trial, dipeptidyl-peptidase-4 inhibitors (DPP-4i) were launched on the market. DDP4 is an enzyme involved in degradation of incretins, like GLP-1, the molecules that favour insulin release. Inhibitors act stopping the action of DPP-4, increasing the systemic level of incretins, mainly GLP-1. Since incretin action results in a glucose-balanced insulin release, the risk of hypoglycaemia with this class of drug is therefore very low.

DPP-4i have minimal adverse effects (most common being nasopharyngitis, headache, and upper respiratory infections). Differently from GLP-1 receptor agonists, they do not slow GI motility and have weight neutral effect, which is still beneficial for most patients with type 2 DM. DPP-4i primarily target the postprandial plasma

Table 2 Summary table of the main cardiovascular outcome trials (CVOTs)

Trials	Active	Patients		Endpoints	Follow-up	Outcome	Superiority	Details
Year	treatment	Features; number			Years	HR (95% CI); p-value	Yes or No	
Dipeptidyl-peptidase-4 inhibitors (DPP-4i)								
SAVOR-TIMI 53 (2013)	Saxagliptin	T2DM + CVD or high CVR	N = 16492	3-MACE	2.1	1.00 (0.89–1.12); p = 0.99	No	Increase HF hospitalization
EXAMINE (2013)	Alogliptin	T2DM + ACS	N = 5380	3-MACE	1.5	0.96 (≤1.16; p = 0.32	No	No increase in HF
TECOS (2015)	Sitagliptin	T2DM + CVD	N = 14671	4-MACE	3.0	0.98 (0.89–1.08); p = 0.65	No	No increase in HF
CARMELINA (2018)	Linagliptin	T2DM + high CVR or renal risk	N = 6979	3-MACE	2.2	1.02 (0.89–1.17); p < 0001	No	No increase in HF
Glucagon-like peptide 1 receptor agonist (GLP-1 RA)								
ELIXA (2015)	Lixisenatide	T2DM + ACS	N = 6068	4-MACE	2.1	1.02 (0.89–1.17); p = 0.81	No	/
LEADER (2016)	Liraglutide	T2DM + CVD or high CVR	N = 9340	3-MACE	3.8	0.87 (0.78–0.97); p = 0.01	Yes	Reduction of CV death
SUSTAIN-6 (2016)	Semaglutide	T2DM + CVD, renal disease or high CVR	N = 2735	3-MACE	1.9	0.74 (0.58 to 0.95); p < 0.001	Yes	Reduction of stroke
EXSCEL (2017)	Exenatide	T2DM +/- CVD	N = 14752	4-MACE	3.2	0.91 (0.83–1.00); p < 0.001	No	/
Harmony outcomes (2018)	Albiglutide	T2DM + CVD	N = 9463	3-MACE	1.6	0.78 (0.68–0.90); p < 0.0001	Yes	Reduction of MI
REWIND (2019)	Dulaglutide	T2DM + CVD or high CVR	N = 9622	3-MACE	5.4	0.88 (0.79–0.99); p = 0.026	Yes	Reduction of stroke
Sodium-glucose linked transporter 2 inhibitors (SGLT-2i)								
EMPA-REG outcome (2015)	Empagliflozin	T2DM + CVD	N = 7020	3-MACE	3.1	0.86 (0.74–0.99); p = 0.0382)	Yes	Reduced HF hospitalization
CANVAS (2017)	Canagliflozin	T2DM + high CVR	N = 10142	3-MACE	2.4	0.86 (0.75–0.97); p = 0.02	Yes	Reduced HF hospitalization
								Increased risk of amputation

(continued)

Table 2 (continued)

Trials	Active	Patients		Follow-up	Outcome	Superiority	Details
Year	treatment	*Features; number*	Endpoints	*Years*	*HR (95% CI); p-value*	*Yes or No*	
DECLARE-TIMI 58 (2018)	Dapaglifozin	T2DM + CVD or high CVR N = 17160	3-MACE	4.2	0.93 (0.84–1.03); p = 0.17	No	Reduced HF hospitalization
Insulin							
ORIGIN (2012)	Glargine	T2DM, IGT + high CVR N = 12537	3-MACE	6.2	1.02 (0.94–1.11); p = 0.63	No	/
DEVOTE (2017)	Degludec	T2DM + CVD, renal disease or high CVR N = 7637	3-MACE	1.9	0.91 (0.78–1.06); p < 0.001	No	/

CVD cardiovascular disease, *CVR* cardiovascular risk, *T2DM* type 2 diabetes mellitus, *ACS* acute coronary syndrome, *IGT* impaired glucose tolerance, *3-MACE* Composite of CV death and nonfatal MI or stroke, *4-MACE* Composite of CV death and nonfatal MI or stroke or hospitalization for UA. See text for details

glucose, and have less impact in reducing HbA1c relative to other drug classes, such as GLP-1 agonists.

Four main drugs are recognized in this class: saxagliptin, alogliptin, sitagliptin and linagliptin.

In the study SAVOR-TIMI 53, 16492 patients were treated with saxagliptin or placebo in addition to conventional therapy (Scirica et al. 2013). Enrolled patients with DM type 2 presented with history of CV diseases (85%) or high CV risk profile. Median follow-up was 2.1 years. No significant differences in outcome were observed for primary endpoints. The safety of the drug was therefore confirmed. However, a significant increase in hospitalization for heart failure was observed, even though the data were not confirmed by further analysis.

Alogliptin was tested against placebo in EXAMINE trial, conducted on 5380 patients with DM type 2 and recent acute coronary syndrome (White et al. 2011). The drug showed CV safety with no increase in risk of MACE.

In TECOS study, 14671 patients with type 2 DM and high CV risk were treated with sitagliptin for a median follow-up period of 3 years (Green et al. 2015). The trial showed no increase in risk for 3-points MACE and for hospitalization for heart failure. Glycaemic control was similar in the two arms.

In CARMELINA, linagliptin was tested on 6979 patients with DM type 2 and high CV risk resulting to be non-inferior relative to usual care (Rosenstock et al. 2019). No increased risk of heart failure was observed.

DPP-4i clearly showed to have cardiovascular safety but no one of them demonstrated cardiovascular benefit. One explanation may be that trials were designed to test for non-inferiority and not adequately powered to evidence superiority. Additionally, it is possible that the increase in incretins generated by DPP-4i acts simply on the reduction of glycemia without a direct CV influence. Notably, DPP-4i should be used cautiously in patients with history of heart failure, due to unclear evidence.

4.5.3 Sodium-Glucose Linked Transporter-2 Inhibitors (SGLT-2i)

Sodium glucose cotransporters-2 are located at the level of the proximal convoluted tubules and are involved in the combined reabsorption of glucose and sodium, being responsible for the 90% of glucose reabsorption of the kidney. Their inhibition leads to significant glycosuria helping in normalization of glycemia. General infections seem to be the most common adverse effect of this class of drug, particularly urinary tract infections due to the induced osmotic diuresis.

Three main drugs are recognized in this class: empaglifozin, canaglifozin, dapaglifozin.

In EMPA-REG OUTCOME trial, 7020 patients with diabetes mellitus type 2 and history of CV disease were treated with empaglifozin versus conventional therapy for a median follow-up of 3.1 years (Zinman et al. 2015). Patients randomized to empaglifozin showed a significant reduction of primary endpoint with a reduction of 14% of risk of 3-points MACE (HR 0.86; p-value = 0.04 for superiority). Moreover, treatment group showed a 38% decrease in CV death (HR 0.62; p-value<0.001), a 32% decrease in all-cause mortality (HR 0.68; p-value<0.001) and a 35% reduction of hospitalization for heart failure (HR 0.65; p-value = 0.002). Following these astonishing results, empaglifozin was the first drugs to demonstrate CV benefit, ensuring a significant reduction of CV events. Interestingly, the reduction of CV mortality was already evident at only 15 weeks from randomization and depended largely on reduction of heart failure numbers, while myocardial infarction incidence was largely unaffected (5.4% in placebo group vs 4.8% in treatment group). These results cannot be explained by the only modest decrease in HbA1c (−0.24% relative to conventional treatment) and suggest beneficial CV effect beyond glucose lowering.

Similar results were presented for canaglifozin. In CANVAS trial, canaglifozin showed a reduction of 14% of primary endpoint (HR 0.86; p-value<0.01 for non-inferiority and p-value<0.02 for superiority) confirming the CV benefit already demonstrated by empaglifozin (Neal et al. 2017). Furthermore, the risk of hospitalization was significantly reduced relative to placebo (HR 0.67). To be noted, there was concern regarding an increased risk of amputation in canaglifozin arm (HR 1.97).

The lastly published SGLT-2i trial is the DECLARE-TIMI 58 (Wiviott et al. 2019). 17160 patients were randomized to dapaglifozin or placebo for a median follow-up of 4.2 years. In primary safety outcome analysis, dapaglifozin met the criteria for non-inferiority. Differently from the previous two drugs, dapaglifozin did not demonstrate superiority with improved CV benefit. However, it did result in reduction of heart failure hospitalization and death (HR 0.83; p-value = 0.005).

Despite dapaglifozin did not result in reduced primary endpoint, empaglifozin and canaglifozin demonstrated strong CV benefit with marked and rapid reduction of CV death and hospitalization for HF. Notably, all three SGLT-2i tested have demonstrated improvements in renal endpoints.

4.5.4 Glucagon-Like Peptide-1 Receptor Agonists (GLP-1 RA)

GLP-1 is an incretin produced by intestinal cells in response to glucose concentration rise. It acts on GLP-1 receptors exposed on the surface of pancreatic cells, favouring insulin release and inhibiting glucagon secretion. Moreover, GLP-1 slows gastric emptying and increases satiety acting on intestinal and gastric receptors. GLP-1 RA mimic the structure of GLP-1 in order to obtain receptor activation and stimulate physiological responses. GLP-1 RA are administered in concentrations that are 6–10 times greater than endogenous levels. This causes significant slowing of GI motility leading to nausea and sometimes vomiting, that are the two most common side effects. Subsequently, weight loss is a frequently observed adverse event under treatment with GLP-1 RA, being beneficial for overweight or obese patients. GLP-1 receptor agonists target fasting plasma glucose as well as postprandial one. This is the reason why a higher HbA1c lowering effect is observed with GLP-1 RA relative to other agents such as DPP-4i.

Six main drugs are recognized in this class: liraglutide, lixisenatide, exenatide, semaglutide, dulaglutide, albiglutide.

In ELIXA trial, 6068 patients were randomized to lixisenatide or conventional therapy, showing no differences in MACE (HR 1.02; p < 0.001 for non-inferiority), confirming its CV safety (Marso et al. 2016a).

LEADER trial was conducted on 9340 patients with DM type 2 and previous CV disease, randomized to receive liraglutide or placebo plus conventional therapy for a median follow-up of 3.8 years. Liraglutide showed a significant reduction of 13% in primary endpoint (HR 0.87; p-value = 0.01 for superiority), with a reduction of 22% of CV death and 13% of all-cause mortality (Marso et al. 2016b). There was a non-significant reduction of MI and stroke. Notably, the survival curves begin to diverge after 12 months, suggesting that liraglutide effect requires more time to become evident from a CV point of view and this may be related to the presence of an anti-atherosclerotic action.

In EXSCEL trial, 10782 patients were randomized to receive exenatide or placebo for a median follow-up of 3.2 years (Mentz et al. 2018). The exenatide treatment demonstrated a reduction of primary endpoint that was significant for CV safety but not for CV benefit (HR 0.91; p-value<0.001 for non-inferiority; p-value = 0.06 for superiority).

SUSTAIN-6 trial was conducted on 3297 patients randomized to receive either semaglutide or conventional therapy plus placebo for a median follow-up of 2.1 years (Marso et al. 2016a). Semaglutide demonstrated 26% reduction in primary endpoint (HR 0.74; p-value = 0.02 for superiority), showing clear CV benefit. This result was mainly driven by the reduction 39% in fatal stroke.

In Harmony Outcomes trial, albiglutide versus placebo was tested in 9463 patients (Hernandez et al. 2018). It reduced by 22% the 3-points

MACE combined endpoint (HR 0.78; p-value < 0.0001 for non-inferiority; p-value = 0.0006 for superiority). This result was mainly driven by a significant reduction of 25% in myocardial infarction.

In REWIND trial, dulaglutide showed once again the great potential of GLP-1 RA (Gerstein et al. 2019). 9901 patients with DM type 2 and high CV risk were randomized to receive dulaglutide or placebo. After a median follow-up of 5.4 years, dulaglutide group showed a significantly lower number of MACE (HR 0.88; p-value = 0.026), mainly dependent on the reduction of the number of non-fatal stroke.

All GLP-1 RA demonstrated CV safety. Additionally, liraglutide, semaglutide, albiglutide and dulaglutide proved their CV benefit reducing CV events.

4.5.5 Insulin

Novel insulin molecules tested in CVOTs are Glargine and Degludec molecules. ORIGIN trial evaluated the use of long-acting basal insulin Glargine against placebo in 12537 patients with pre-diabetes and overt DM type 2 (ORIGIN Trial Investigators et al. 2012). The trial found no significant reduction in two co-primary outcomes, major cardiovascular events and major cardiovascular events plus revascularization and heart failure. DEVOTE trial tested Degludec insulin against Glargin insulin in head-to-head trial showing no differences in outcome (Marso et al. 2017).

4.5.6 Effects on CV Risk Beyond Simple Glycaemic Control

In the end, two classes of drugs (SGLT-2i and GLP-1 RA) demonstrated to provide cardiovascular benefit in diabetic patients, significantly reducing CV events. Their peculiarity is the ability to ensure a reduction of CV events with only a modest decrease in glycated haemoglobin relative to conventional therapy. The understanding of their cardioprotective mechanism is still incomplete and the explanation of their effect is not straightforward as they exert heterogeneous modifications at different levels going beyond simple glycemia control. Here a short and concise overview of the main pathophysiologic mechanisms at the basis of their effect.

SGLT-2 inhibition reduce CV outcome by means of a combination of hemodynamic and metabolic positive effects (Sattar et al. 2016, 2017). SGLT-2i prevent reabsorption of glucose from the proximal convoluted tubule inducing osmotic diuresis due to increased glycosuria and natriuria. Interestingly, reduction of CV events was limited to patients with T2DM and established atherosclerotic cardiovascular disease in secondary prevention, whereas reduction of HF and progression of renal disease occurred even in primary prevention in patients without history of CV disease or HF. Five main mechanisms seem to be involved in CV benefits:

– Modulation of traditional risk factors. SGLT2-2i causes loss of body weight, reduction of HbA1c, reduction in systolic blood pressure and diastolic BP.
– Reduction on LV loading conditions. SGLT2-2i causes a significant reduction on plasma volume due to osmotic diuresis reducing pre-load, LV filling pressure and afterload. The effect is greater relative to diuretics because of selective interstitial volume reduction shown by this class of drugs (Verma and McMurray 2018).
– Reverse cardiac remodelling. The positive effect on cardiac filling pressure may help in reducing LV mass due to reduction of LV wall stress according to Laplace's law (Verma et al. 2019).
– Improvement of cardiac energetics. SGLT2-2i favour the increase of ketone bodies due to a generalized state of starvation derived from glucose depletion. Locally, at heart muscle levels they promote the use of ketone bodies and fatty acid oxidation as main energy source (Garcia-Ropero et al. 2019). Being the metabolic pathway more favourable in terms of ATP production this improves myocardial work performance.

– Inhibition of Na+/H+ exchanger. SGLT2-2i act as ionic exchanger inhibiting the Na+/H+ exchange present on cardiomyocyte surface (Uthman et al. 2018). This causes a drop in sodium and calcium intracellularly and an increase in calcium in sarcoplasmic reticulum leading to improvement in cardiac contractility and mechanics.

Summarizing, the osmotic diuresis favours a decrease in intravascular volume with a drop in blood pressure and peripheral decongestion. These changes significantly reduce the cardiac stressors (both preload and afterload), improving myocardial oxygen supply and decreasing left ventricular stretch that is thought to be an important trigger for arrhythmias and remodelling. Additionally, SGLT2-2i improves myocardial performance thanks to the use of alternative energy sources, such as ketone bodies and fatty acid oxidation, and increasing sarcoplasmic calcium level that favours cardiac contractility. Moreover, renal dysfunction is slowed thanks to the improvement of hemodynamic conditions and the reversal of the maladaptive tubulo-glomerular feedback. Putting all these mechanisms together, they generate the observed positive effect in reducing CV events, HF and renal disease progression.

As already described, GLP-1 receptor agonists stimulate receptors exposed on pancreatic beta cells, gastric and intestinal cells mimicking the action of endogenous GLP-1. Even though the biochemical action is well-known, it is not yet clear how GLP-1 RA may help in reducing CV events. In line with the evidence that the relative benefit over CV mortality appears later after treatment initiation (compared with SGLT-2i), most experts believe that GLP-1 RA action is, at least in part, an anti-atherothrombotic effect, derived from modulation of endothelial function, anti-inflammatory properties and anti-atherosclerotic actions.

Interestingly, despite the common biochemical pathways, DPP-4i and GLP-1 RA did not show the same results. If GLP-1 RA showed positive effect in terms of superiority in the context of coronary artery disease, DPP-4i appeared to be neutral. One possible explanation is that DPP-4i are involved in degradation of incretins and their action can potentiate additional peptides that are shown to have adverse CV effects due to the involvement in inflammation and fibrosis (Packer 2018).

In contrast to SGLT-2i that showed their protective effect reducing the incidence of heart failure, it is important to underline that GLP-1 RA are neutral in this context. One possible explanation is that GLP-1 receptors are localized in sino-atrial node as well. The use of GLP-1 RA cause an increase in heart rate (6–10 bpm for long-acting agents and 3–4 bpm for short-acting agents) (Lorenz et al. 2017). This has been advocated as a possible reason for neutral effect on HF events. However, despite the raised concern, currently there is no evidence of harm derived from this slight increase in heart rate.

Five main mechanisms are involved in CV benefits:

– Modulation of risk factors. GLP-1 RA cause significant reduction in blood pressure, body weight, HbA1c and lipid status. As already discussed this drug class causes significant weight reduction as it slows gastric motion leading to an increased sense of satiety and sometimes to vomiting.
– Modulation of endothelial cells. GLP-1 RA are involved in modulation of endothelial function. This action is obtained thanks the reduction of expression of ICAM-1 and VCAM-1 on endothelial cells, reducing leukocyte translocation (Liu et al. 2009).
– Anti-atherosclerotic and anti-inflammatory action. GLP-1 RA reduce release of pro-inflammatory cytokines reducing local inflammation responsible for plaque formation, expansion and vulnerability (Liu et al. 2009).

– Reduction of pro-thrombotic state. GLP-1 RA reduce coagulation cascade activation, by decreasing PAI-1 release (Liu et al. 2009), and platelet aggregation, reducing expression of platelet surface receptors (Cameron-Vendrig et al. 2016).
– Direct action on heart. GLP-1 could directly protect the heart against ischemic injuries via pro-survival signalling pathways activated by specific kinases, such as PKA, PI3K, p42/44 (Bose et al. 2005).

All things considered, a combination of anti-atherogenic effects and hemodynamic improvements are likely explanations of the reduction of CV events and mortality observed in patients treated with these two classes of drugs.

5 Efficacy of Glycaemic Control on Other Vascular Complications

No prospective trials have been performed to assess whether optimal glycaemic control could reduce the incidence of peripheral artery disease. Similarly, looking to the past studies, no drug has been proven effective in significantly reduce the rate of stroke and coronary artery disease in diabetics, even with intensive glycaemic control. This scenario has changed with the arrival on the market of novel GLDs.

One of the problems reported in the past – and still present nowadays in the context of CVOTs – is the lack of peripheral artery disease among the clinical endpoint under evaluation. No data at hand are present to evaluate the efficacy of novel GLDs in reducing PAD events and progression. To be notice the warning raised for canaglifozin because of the increased risk of limb amputation with the use of this drug.

Stroke is instead well represented by MACEs in all trials. SGLT-2i does not affect the incidence of stroke in any of the trials. On the other hand,

GLP-1 agonists are the only drugs among novel GLDs reducing stroke incidence. In REWIND, dulaglutide treatment arm showed a significantly lower number of total stroke relative to placebo (3.2% vs 4.1%; HR 0.76; 95% CI 0.62–9.94; p-value 0.010) (Gerstein et al. 2019). Similarly, in SUSTAIN-6 trial semaglutide showed a significant reduction in the risk of stroke (1.6% vs 2.7%; HR 0.61; 95% CI 0.38–0.99; p-value 0.04) (Marso et al. 2016a).

As discussed previously, renal function was already improved by intensive glycemia control as patients in treatment arms with stricter control were associated with reduced progression toward CKD and reduction of proteinuria. New GLDs have shown further nephroprotective effect improving renal outcomes. In DPP-4i experience, saxagliptin significantly reduced microalbuminuria (Scirica et al. 2013), but other DPP-4i did not report similar effects on renal function or albuminuria. In LEADER trial, liraglutide group showed a reduction of the composite renal endpoint (new-onset macro-albuminuria, persistent doubling of creatinine, ESRD or death due to renal disease) that was mainly dependent on reduction in macro-albuminuria (HR 0.74, 95% CI 0.60–0.91) (Mann et al. 2017). In the SUSTAIN-6, semaglutide showed similar effect. In ELIXA trial, lixisenatide-treated patients showed lower levels of microalbuminuria compared with placebo (Marso et al. 2016a). All in all, GLP-1 RA demonstrated ability in reduction of albuminuria but clear evidence in reduction of worsening renal function is missing. SGLT-2i are the class of novel GLDs that obtained the best results in terms of improvement of renal function. Canaglifozin showed a 27% reduction in progression of albuminuria and a reduction of the composite renal endpoints (HR 0.53; 95% CI 0.33–0.84), consisting of 40% reduction in eGFR, renal replacement therapy or death from acute kidney injury (Neal et al. 2017). Similarly, empaglifozin demonstrated to decrease the incidence of progression to macro-albuminuria,

Table 3 Summary of recent ESC guidelines 2019 recommendations concerning use of GLDs in diabetic patients according to CV profile

Recommendations	Class of recommendation	Level of evidence
Empagliflozin, canagliflozin, or dapagliflozin are recommended in patients with T2DM and CVD, or at very high/high CV risk, to reduce CV events	I	A
Empagliflozin is recommended in patients with T2DM and CVD to reduce the risk of death	I	B
Liraglutide, semaglutide, or dulaglutide are recommended in patients with T2DM and CVD, or at very high/high CV risk, to reduce CV events	I	A
Liraglutide is recommended in patients with T2DM and CVD, or at very high/high CV risk, to reduce the risk of death	I	B
Metformin should be considered in overweight patients with T2DM without CVD and at moderate CV risk.	IIa	C
Insulin-based glycaemic control should be considered in patients with ACS with significant hyperglycaemia (>10 mmol/L or >180 mg/dL), with the target adapted according to comorbidities	IIa	C
Thiazolidinediones are not recommended in patients with HF	III	A
Saxagliptin is not recommended in patients with T2DM and a high risk of HF	III	B

GLDs glucose lowering drugs, *CV* cardiovascular

doubling of serum creatinine, initiation of renal replacement therapy and death from renal disease (HR 0.61; 95% CI 0.53–0.70) (Wanner et al. 2016). Recently, also patients treated with dapaglifozin showed significant a reduction of renal endpoints, namely 40% decrease in eGFR, ESRD and renal death (HR 0.53; 95% CI 0.43–0.66) (Wiviott et al. 2019).

All these evidences clearly show that new GLDs move diabetes treatment well beyond simple glycemia control.

6 Conclusions and Guidelines Recommendations

After years of disappointing results about the effects of glycaemic control on CV hard endpoints, data from several CVOTs suggest that clear benefits in terms of CV outcomes can be obtained using some of the novel GLDs in patients with already established CVD or in patients at high/very high risk of CV disease. These new evidences have been received and incorporated in recently published 2019 ESC guidelines on diabetes and CV diseases (see Table 3 for reference) (Cosentino et al. 2019). The strongest recommendations concern mainly SGLT-2 inhibitors (empagliflozin, canaglifozin

and dapaglifozin) and GLP-1 RA (liraglutide, semaglutide and dulaglutide). In both cases they are recommended as first line anti-diabetic agents in patients with type 2 DM and with CVD or high CV risk profile to reduce CV events (class I, level A) and to reduce mortality (class I, level B). In case of patients with no history of CV events and moderate-to-low CV risk profile they are recommended on top of metformin whether the HbA1c target is not reached.

Bibliography

Action to Control Cardiovascular Risk in Diabetes Study Group, Gerstein HC, Miller ME et al (2008) Effects of intensive glucose lowering in type 2 diabetes. N Engl J Med 358:2545–2559. https://doi.org/10.1056/NEJMoa0802743

Adler AI, Stevens RJ, Manley SE et al (2003) Development and progression of nephropathy in type 2 diabetes: the United Kingdom Prospective Diabetes Study (UKPDS 64). Kidney Int 63:225–232. https://doi.org/10.1046/j.1523-1755.2003.00712.x

ADVANCE Collaborative Group, Patel A, MacMahon S et al (2008) Intensive blood glucose control and vascular outcomes in patients with type 2 diabetes. N Engl J Med 358:2560–2572. https://doi.org/10.1056/NEJMoa0802987

Anand SS, Islam S, Rosengren A et al (2008) Risk factors for myocardial infarction in women and men: insights from the INTERHEART study. Eur Heart J 29:932–940. https://doi.org/10.1093/eurheartj/ehn018

Avogaro A, Albiero M, Menegazzo L, de Kreutzenberg S, Fadini GP (2011) Endothelial dysfunction in diabetes: the role of reparatory mechanisms. Diabetes Care 34 (Suppl 2):S285–S290. https://doi.org/10.2337/dc11-s239

Beckman JA, Creager MA, Libby P (2002) Diabetes and atherosclerosis. JAMA 287:2570. https://doi.org/10.1001/jama.287.19.2570

Bose AK, Mocanu MM, Carr RD, Brand CL, Yellon DM (2005) Glucagon-like peptide 1 can directly protect the heart against ischemia/reperfusion injury. Diabetes 54:146–151. https://doi.org/10.2337/diabetes.54.1.146

Boulton AJM, Vinik AI, Arezzo JC et al (2005) Diabetic neuropathies: a statement by the American Diabetes Association. Diabetes Care 28:956–962. https://doi.org/10.2337/diacare.28.4.956

Buse JB, Ginsberg HN, Bakris GL et al (2007) Primary prevention of cardiovascular diseases in people with diabetes mellitus: a scientific statement from the American Heart Association and the American Diabetes Association. Diabetes Care 30:162–172. https://doi.org/10.2337/dc07-9917

Cameron-Vendrig A, Reheman A, Siraj MA et al (2016) Glucagon-like peptide 1 receptor activation attenuates platelet aggregation and thrombosis. Diabetes 65:1714–1723. https://doi.org/10.2337/db15-1141

Cao Z, Cooper ME (2011) Pathogenesis of diabetic nephropathy. J Diabetes Investig 2:243–247. https://doi.org/10.1111/j.2040-1124.2011.00131.x

Chiasson J-L, Josse RG, Gomis R et al (2002) Acarbose for prevention of type 2 diabetes mellitus: the STOP-NIDDM randomised trial. Lancet 359:2072–2077. https://doi.org/10.1016/S0140-6736(02)08905-5

Colayco DC, Niu F, McCombs JS, Cheetham TC (2011) A1C and cardiovascular outcomes in type 2 diabetes: a nested case-control study. Diabetes Care 34:77–83. https://doi.org/10.2337/dc10-1318

Cosentino F, Grant PJ, Aboyans V et al (2019) 2019 ESC guidelines on diabetes, pre-diabetes, and cardiovascular diseases developed in collaboration with the EASD. Eur Heart J 41:255–323. https://doi.org/10.1093/eurheartj/ehz486

Currie CJ, Peters JR, Tynan A et al (2010) Survival as a function of HbA1c in people with type 2 diabetes: a retrospective cohort study. Lancet 375:481–489. https://doi.org/10.1016/S0140-6736(09)61969-3

De Vriese AS, Verbeuren TJ, Van de Voorde J, Lameire NH, Vanhoutte PM (2000) Endothelial dysfunction in diabetes. Br J Pharmacol 130:963–974. https://doi.org/10.1038/sj.bjp.0703393

Diabetes Control and Complications Trial Research Group, Nathan DM, Genuth S et al (1993) The effect of intensive treatment of diabetes on the development and progression of long-term complications in insulin-dependent diabetes mellitus. N Engl J Med 329:977–986. https://doi.org/10.1056/NEJM199309303291401

Dormandy JA, Charbonnel B, Eckland DJA et al (2005) Secondary prevention of macrovascular events in patients with type 2 diabetes in the PROactive Study (PROspective pioglitAzone Clinical Trial in macroVascular Events): a randomised controlled trial. Lancet (Lond, Engl) 366:1279–1289. https://doi.org/10.1016/S0140-6736(05)67528-9

Duckworth W, Abraira C, Moritz T et al (2009) Glucose control and vascular complications in veterans with type 2 diabetes. N Engl J Med 360:129–139. https://doi.org/10.1056/NEJMoa0808431

Ference BA, Graham I, Tokgozoglu L, Catapano AL (2018) Impact of lipids on cardiovascular health. J Am Coll Cardiol 72:1141–1156. https://doi.org/10.1016/j.jacc.2018.06.046

Fong DS, Aiello LP, Ferris FL, Klein R (2004) Diabetic retinopathy. Diabetes Care 27:2540–2553. https://doi.org/10.2337/diacare.27.10.2540

Fox CS, Coady S, Sorlie PD et al (2007) Increasing cardiovascular disease burden due to diabetes mellitus: the Framingham Heart Study. Circulation 115:1544–1550. https://doi.org/10.1161/CIRCULATIONAHA.106.658948

Gæde P, Lund-Andersen H, Parving H-H, Pedersen O (2008) Effect of a multifactorial intervention on mortality in type 2 diabetes. N Engl J Med 358:580–591. https://doi.org/10.1056/NEJMoa0706245

Garcia-Ropero A, Santos-Gallego CG, Zafar MU, Badimon JJ (2019) Metabolism of the failing heart and the impact of SGLT2 inhibitors. Expert Opin Drug Metab Toxicol 15:275–285. https://doi.org/10.1080/17425255.2019.1588886

Gerstein HC, Colhoun HM, Dagenais GR et al (2019) Dulaglutide and cardiovascular outcomes in type 2 diabetes (REWIND): a double-blind, randomised placebo-controlled trial. Lancet 394:121–130. https://doi.org/10.1016/S0140-6736(19)31149-3

Green JB, Bethel MA, Armstrong PW et al (2015) Effect of sitagliptin on cardiovascular outcomes in type 2 diabetes. N Engl J Med 373:232–242. https://doi.org/10.1056/NEJMoa1501352

Hadi HAR, Suwaidi JA (2007) Endothelial dysfunction in diabetes mellitus. Vasc Health Risk Manag 3:853–876

Haffner SM, Lehto S, Rönnemaa T, Pyörälä K, Laakso M (1998) Mortality from coronary heart disease in subjects with type 2 diabetes and in nondiabetic subjects with and without prior myocardial infarction. N Engl J Med 339:229–234. https://doi.org/10.1056/NEJM199807233390404

Hernandez AF, Green JB, Janmohamed S et al (2018) Albiglutide and cardiovascular outcomes in patients with type 2 diabetes and cardiovascular disease (Harmony Outcomes): a double-blind, randomised placebo-controlled trial. Lancet 392:1519–1529. https://doi.org/10.1016/S0140-6736(18)32261-X

Holman RR, Paul SK, Bethel MA, Matthews DR, Neil HAW (2008) 10-year follow-up of intensive glucose Control in type 2 diabetes. N Engl J Med 359:1577–1589. https://doi.org/10.1056/NEJMoa0806470

Holman RR, Coleman RL, Chan JCN et al (2017) Effects of acarbose on cardiovascular and diabetes outcomes in patients with coronary heart disease and impaired glucose tolerance (ACE): a randomised, double-blind, placebo-controlled trial. Lancet Diabetes Endocrinol 5:877–886. https://doi.org/10.1016/S2213-8587(17)30309-1

Home PD, Pocock SJ, Beck-Nielsen H et al (2009) Rosiglitazone evaluated for cardiovascular outcomes in oral agent combination therapy for type 2 diabetes (RECORD): a multicentre, randomised, open-label trial. Lancet 373:2125–2135. https://doi.org/10.1016/S0140-6736(09)60953-3

Hong J, Zhang Y, Lai S et al (2013) Effects of metformin versus glipizide on cardiovascular outcomes in patients with type 2 diabetes and coronary artery disease. Diabetes Care 36:1304–1311. https://doi.org/10.2337/dc12-0719

Huang ES, Liu JY, Moffet HH, John PM, Karter AJ (2011) Glycemic Control, complications, and death in older diabetic patients: the diabetes and aging study. Diabetes Care 34:1329–1336. https://doi.org/10.2337/dc10-2377

Kannel WB, McGee DL (1979) Diabetes and cardiovascular disease. The Framingham study. JAMA 241:2035–2038. https://doi.org/10.1001/jama.241.19.2035

Khaw K-T, Wareham N, Bingham S et al (2004) Association of hemoglobin A1c with cardiovascular disease and mortality in adults: the European Prospective Investigation into Cancer in Norfolk. Ann Intern Med 141:413. https://doi.org/10.7326/0003-4819-141-6-200409210-00006

Kibel A, Selthofer-Relatic K, Drenjancevic I et al (2017) Coronary microvascular dysfunction in diabetes mellitus. J Int Med Res 45:1901–1929. https://doi.org/10.1177/0300060516675504

Kooy A, de Jager J, Lehert P et al (2009) Long-term effects of metformin on metabolism and microvascular and macrovascular disease in patients with type 2 diabetes mellitus. Arch Intern Med 169:616. https://doi.org/10.1001/archinternmed.2009.20

Kumar A, Fausto A, Robbins SL, Cotran RS (2010) Pathological basis of disease, 8th edn. Saunders Elsevier, Philadelphia

Lin J, Thompson TJ, Cheng YJ et al (2018) Projection of the future diabetes burden in the United States through 2060. Popul Health Metrics 16:1–9

Liu H, Dear AE, Knudsen LB, Simpson RW (2009) A long-acting glucagon-like peptide-1 analogue attenuates induction of plasminogen activator inhibitor type-1 and vascular adhesion molecules. J Endocrinol 201:59–66. https://doi.org/10.1677/JOE-08-0468

Lorenz M, Lawson F, Owens D et al (2017) Differential effects of glucagon-like peptide-1 receptor agonists on heart rate. Cardiovasc Diabetol 16:6. https://doi.org/10.1186/s12933-016-0490-6

Mach F, Baigent C, Catapano AL et al (2019) 2019 ESC/EAS guidelines for the management of dyslipidaemias: lipid modification to reduce cardiovascular risk. Eur Heart J 41:111–188. https://doi.org/10.1093/eurheartj/ehz455

Magliano DJ, Islam RM, Barr ELM et al (2019) Trends in incidence of total or type 2 diabetes: systematic review. BMJ 366:l5003. https://doi.org/10.1136/bmj.l5003

Mann JFE, Ørsted DD, Brown-Frandsen K et al (2017) Liraglutide and renal outcomes in type 2 diabetes. N Engl J Med 377:839–848. https://doi.org/10.1056/NEJMoa1616011

Marso SP, Hiatt WR (2006) Peripheral arterial disease in patients with diabetes. J Am Coll Cardiol 47:921–929. https://doi.org/10.1016/j.jacc.2005.09.065

Marso SP, Bain SC, Consoli A et al (2016a) Semaglutide and cardiovascular outcomes in patients with type 2 diabetes. N Engl J Med 375:1834–1844. https://doi.org/10.1056/NEJMoa1607141

Marso SP, Daniels GH, Brown-Frandsen K et al (2016b) Liraglutide and cardiovascular outcomes in type 2 diabetes. N Engl J Med 375:311–322. https://doi.org/10.1056/NEJMoa1603827

Marso SP, McGuire DK, Zinman B et al (2017) Efficacy and safety of degludec versus glargine in type 2 diabetes. N Engl J Med 377:723–732. https://doi.org/10.1056/NEJMoa1615692

Maser RE, Mitchell BD, Vinik AI, Freeman R (2003) The association between cardiovascular autonomic neuropathy and mortality in individuals with diabetes: a meta-analysis. Diabetes Care 26:1895–1901. https://doi.org/10.2337/diacare.26.6.1895

Mentz RJ, Bethel MA, Merrill P et al (2018) Effect of once-weekly Exenatide on clinical outcomes according to baseline risk in patients with type 2 diabetes mellitus: insights from the EXSCEL Trial. J Am Heart Assoc 7:e009304. https://doi.org/10.1161/JAHA.118.009304

Morgan KP, Kapur A, Beatt KJ (2004) Anatomy of coronary disease in diabetic patients: an explanation for poorer outcomes after percutaneous coronary intervention and potential target for intervention. Heart 90:732–738. https://doi.org/10.1136/hrt.2003.021014

Neal B, Perkovic V, Mahaffey KW et al (2017) Canaglifozin and cardiovascular and renal events in type 2 diabetes mellitus. N Engl J Med 377:644–657. https://doi.org/10.1056/NEJMoa1611925

Nissen SE, Wolski K (2007) Effect of rosiglitazone on the risk of myocardial infarction and death from cardiovascular causes. N Engl J Med 356:2457–2471. https://doi.org/10.1056/NEJMoa072761

ORIGIN Trial Investigators, Gerstein HC, Bosch J et al (2012) Basal insulin and cardiovascular and other outcomes in dysglycemia. N Engl J Med 367:319–328. https://doi.org/10.1056/NEJMoa1203858

Packer M (2018) Have dipeptidyl peptidase-4 inhibitors ameliorated the vascular complications of type 2 diabetes in large-scale trials? The potential confounding effect of stem-cell chemokines. Cardiovasc Diabetol 17:9. https://doi.org/10.1186/s12933-017-0648-x

Piepoli MF, Hoes AW, Agewall S et al (2016) 2016 European Guidelines on cardiovascular disease prevention in clinical practice. Eur Heart J 37:2315–2381. https://doi.org/10.1093/eurheartj/ehw106

Rask-Madsen C, King GL (2013) Vascular complications of diabetes: mechanisms of injury and protective factors. Cell Metab 17:20–33. https://doi.org/10.1016/j.cmet.2012.11.012

Robles NR, Villa J, Gallego RH (2015) Non-proteinuric diabetic nephropathy. J Clin Med 4:1761–1773. https://doi.org/10.3390/jcm4091761

Roger VL, Go AS, Lloyd-Jones DM et al (2012) Heart disease and stroke statistics—2012 update. Circulation 125:e2–e220. https://doi.org/10.1161/CIR.0b013e31823ac046

Rosenstock J, Perkovic V, Johansen OE et al (2019) Effect of linagliptin vs placebo on major cardiovascular events in adults with type 2 diabetes and high cardiovascular and renal risk. JAMA 321:69–79. https://doi.org/10.1001/jama.2018.18269

Sattar N, McLaren J, Kristensen SL, Preiss D, McMurray JJ (2016) SGLT2 inhibition and cardiovascular events: why did EMPA-REG outcomes surprise and what were the likely mechanisms? Diabetologia 59:1333–1339. https://doi.org/10.1007/s00125-016-3956-x

Sattar N, Petrie MC, Zinman B, Januzzi JL (2017) Novel diabetes drugs and the cardiovascular specialist. J Am Coll Cardiol 69:2646–2656. https://doi.org/10.1016/j.jacc.2017.04.014

Scirica BM, Bhatt DL, Braunwald E et al (2013) Saxagliptin and cardiovascular outcomes in patients with type 2 diabetes mellitus. N Engl J Med 369:1317–1326. https://doi.org/10.1056/NEJMoa1307684

Selvin E, Marinopoulos S, Berkenblit G et al (2004) Meta-analysis: glycosylated hemoglobin and cardiovascular disease in diabetes mellitus. Ann Intern Med 141:421–431. https://doi.org/10.7326/0003-4819-141-6-200409210-00007

Selvin E, Steffes MW, Zhu H et al (2010) Glycated hemoglobin, diabetes, and cardiovascular risk in nondiabetic adults. N Engl J Med 362:800–811. https://doi.org/10.1056/NEJMoa0908359

Silverman MG, Ference BA, Im K et al (2016) Association between lowering LDL-C and cardiovascular risk reduction among different therapeutic interventions. JAMA 316:1289–1297. https://doi.org/10.1001/jama.2016.13985

Stern MP (1995) Diabetes and cardiovascular disease: the "common soil" hypothesis. Diabetes 44:369–374. https://doi.org/10.2337/diab.44.4.369

Stratton IM, Adler AI, Neil HA et al (2000) Association of glycaemia with macrovascular and microvascular complications of type 2 diabetes (UKPDS 35): prospective observational study. BMJ 321:405–412. https://doi.org/10.1136/bmj.321.7258.405

Tavakoli M, Mojaddidi M, Fadavi H, Malik RA (2008) Pathophysiology and treatment of painful diabetic neuropathy. Curr Pain Headache Rep 12:192–197. https://doi.org/10.1007/s11916-008-0034-1

Tseng C-H (2004) Mortality and causes of death in a National Sample of diabetic patients in Taiwan. Diabetes Care 27:1605–1609. https://doi.org/10.2337/diacare.27.7.1605

U.S. Food and Drug Administration (2008) Guidance for industry diabetes mellitus – evaluating cardiovascular risk in new antidiabetic therapies to treat type 2 diabetes. Available at https://www.fda.gov/regulatoryinformation/search-fda-guidance-documents

U.S. Food and Drug Administration (2018) Guidance for industry: diabetes mellitus – evaluating cardiovascular risk in new antidiabetic therapies to treat type 2 diabetes

UK Prospective Diabetes Study (UKPDS) Group (1998a) Intensive blood-glucose control with sulphonylureas or insulin compared with conventional treatment and risk of complications in patients with type 2 diabetes (UKPDS 33). Lancet 352:837–853

UK Prospective Diabetes Study (UKPDS) Group (1998b) Effect of intensive blood-glucose control with metformin on complications in overweight patients with type 2 diabetes (UKPDS 34). Lancet 352:854–865

University Group Diabetes Program (1976) A study of the effects of hypoglycemia agents on vascular complications in patients with adult-onset diabetes. VI. Supplementary report on nonfatal events in patients treated with tolbutamide. Diabetes 25:1129–1153. https://doi.org/10.2337/diab.25.12.1129

Uthman L, Baartscheer A, Schumacher CA et al (2018) Direct cardiac actions of sodium glucose cotransporter 2 inhibitors target pathogenic mechanisms underlying heart failure in diabetic patients. Front Physiol 9:1575. https://doi.org/10.3389/fphys.2018.01575

Verma S, McMurray JJV (2018) SGLT2 inhibitors and mechanisms of cardiovascular benefit: a state-of-the-art review. Diabetologia 61:2108–2117. https://doi.org/10.1007/s00125-018-4670-7

Verma S, Mazer CD, Yan AT et al (2019) Effect of empagliflozin on left ventricular mass in patients with type 2 diabetes mellitus and coronary artery disease. Circulation 140:1693–1702. https://doi.org/10.1161/CIRCULATIONAHA.119.042375

Vijan S, Hayward RA (2003) Treatment of hypertension in type 2 diabetes mellitus: blood pressure goals, choice of agents, and setting priorities in diabetes care. Ann Intern Med 138:593–602. https://doi.org/10.7326/0003-4819-138-7-200304010-00018

Vijan S, Hayward RA, American College of Physicians (2004) Pharmacologic lipid-lowering therapy in type 2 diabetes mellitus: background paper for the American College of Physicians. Ann Intern Med 140:650. https://doi.org/10.7326/0003-4819-140-8-200404200-00013

Wallentin L, Becker RC, Budaj A et al (2009) Ticagrelor versus clopidogrel in patients with acute coronary syndromes. N Engl J Med 361:1045–1057. https://doi.org/10.1056/NEJMoa0904327

Wanner C, Inzucchi SE, Zinman B (2016) Empagliflozin and progression of kidney disease in type 2 diabetes. N Engl J Med 375:1799–1802. https://doi.org/10.1056/NEJMc1611290

White WB, Bakris GL, Bergenstal RM et al (2011) EXamination of CArdiovascular OutcoMes with AlogliptIN versus standard of CarE in patients with type 2 diabetes mellitus and acute coronary syndrome (EXAMINE).

Am Heart J 162:620–626.e1. https://doi.org/10.1016/j.ahj.2011.08.004

Wiviott SD, Raz I, Bonaca MP et al (2019) Dapagliflozin and cardiovascular outcomes in type 2 diabetes. N Engl J Med 380:347–357. https://doi.org/10.1056/NEJMoa1812389

Zinman B, Wanner C, Lachin JM et al (2015) Empagliflozin, cardiovascular outcomes, and mortality in type 2 diabetes. N Engl J Med 373:2117–2128. https://doi.org/10.1056/NEJMoa1504720

Adv Exp Med Biol - Advances in Internal Medicine (2020) 4: 153–169
https://doi.org/10.1007/5584_2020_481
© Springer Nature Switzerland AG 2020
Published online: 5 February 2020

Diabetes Mellitus and Acute Myocardial Infarction: Impact on Short and Long-Term Mortality

Valentina Milazzo, Nicola Cosentino, Stefano Genovese, Jeness Campodonico, Mario Mazza, Monica De Metrio, and Giancarlo Marenzi

Abstract

Diabetes mellitus (DM) is an important risk factor for acute myocardial infarction (AMI) and a frequent co-morbidity in patients hospitalized with AMI, being present in about 30% of cases. Although current treatment of AMI has considerably improved survival in both patients with and without DM, the presence of DM still doubles the case fatality rate during both the acute phase of AMI and at long-term follow-up. This higher mortality risk of DM patients strongly indicates a particular need for better treatment options in these patients and suggests that intensive medical treatment, prolonged surveillance, and stringent control of other risk factors should be carefully pursued and maintained for as long as possible in them.

In this review, we will focus on the close association between DM and in-hospital and long-term mortality in AMI patients. We will also aim at providing current evidence on the mechanisms underlying this association and on emerging therapeutic strategies, which may reduce the traditional mortality gap that still differentiates AMI patients with DM from those without.

Keywords

Acute hyperglycemia · Acute myocardial infarction · Diabetes mellitus · In-hospital mortality · Long-term mortality · Non-ST-elevation myocardial infarction · Percutaneous coronary intervention · Pre-diabetes mellitus · ST-elevation myocardial infarction · Unknown diabetes mellitus

V. Milazzo, N. Cosentino, S. Genovese, J. Campodonico, M. Mazza, M. De Metrio, and G. Marenzi (✉)
Centro Cardiologico Monzino IRCCS, Milan, Italy
e-mail: giancarlo.marenzi@ccfm.it

1   Introduction

Diabetes mellitus (DM), in particular type 2 DM, constitutes one of the largest emerging threats to health in the twenty-first century. It is estimated that by 2030 as many as 360 million people world-wide will be affected (Wild et al. 2004). The cause of death in patients with DM is largely due to coronary artery disease (CAD), along with increased rates of stroke and peripheral vascular disease: the so called macro-vascular complications (Kannel and McGee 1979). Notably, at least two-thirds of deaths in DM patients are due to athero-thrombotic events and their sequelae (Fuller et al. 1983; Geiss et al. 1995).

Compared to individuals without DM, those with DM have a three-fold increased risk of acute myocardial infarction (AMI), which usually occurs 15 years earlier, as compared to their non-DM counterpart (Haffner 2000; Booth et al. 2006). Moreover, AMI may even represent the first clinical manifestation of DM (Norhammar et al. 2002). Indeed, in about 5–10% of AMI patients, the presence of DM, until then unknown, is detected during index hospitalization (Marenzi et al. 2018a). Not only DM is a frequent comorbidity among AMI patients, but it also carries a significantly higher morbidity, mortality, and AMI recurrence risk than non-DM patients (Lee et al. 1995; Abbott et al. 1988; Miettinen et al. 1998). This prognostic gap characterizes DM patients in the acute phase of AMI, as well as during the following years.

In this review, we will focus on the close association between DM and in-hospital and long-term mortality in AMI patients, providing also current evidence on its likely underlying mechanisms and on emerging potential therapeutic strategies.

2   Prevalence of DM Among AMI Patients

Diabetes mellitus has been for a long time a recognized risk factor for AMI. In the 1970s, the Framingham Study showed that DM conferred a two- to four-times greater risk for AMI (Kannel and McGee 1979). The Effect of potentially modifiable risk factors associated with myocardial infarction in 52 countries (INTERHEART) study confirmed this greater risk for AMI on a global scale for DM patients (Yusuf et al. 2004).

Although in the last decades there was an almost 70% reduction in the rates of AMI in patients with DM, compared to a 30% reduction in those without DM (Gregg et al. 2014), the AMI burden in the DM population continues to rise, as a result of the substantial increase in its prevalence. Thus, it is not surprising that the frequency of DM among AMI patients steadily increased from 18% in 1997 to 30% in 2016 (Ovbiagele

et al. 2011; Arnold et al. 2016; Ahmed et al. 2014). The latter figure is even higher when we consider pre-DM, as defined by glycated hemoglobin (HbA$_{1c}$) between 5.7%–6.4% (39–46 mmol/mol), and unknown DM (Marenzi et al. 2018a). Recent reports demonstrated that 25% of AMI patients has pre-DM (Marenzi et al. 2018a), a metabolic profile that is also associated with adverse outcomes in this clinical setting (Bartnik et al. 2004). Moreover, the prevalence of previously unrecognized DM in the AMI population is reported to range between 4% and 22%, depending on the test used for its diagnosis (Marenzi et al. 2018a; Aguilar et al. 2004; Mozaffarian et al. 2007). Taken together, these data clearly demonstrate that abnormal glucose metabolism is a frequent co-morbidity in AMI patients (Fig. 1).

3   In-Hospital Mortality of AMI Patients with DM

Prior to the advent of thrombolytic therapy, studies in AMI patients with DM showed a greater than two-fold in-hospital mortality rate in men, and an even higher rate in women, compared with their non-DM counterpart (Bradley and Bryfogle 1956; Savage et al. 1988; Soler et al. 1974; Radke and Schunkert 2010; Yudkin and Oswald 1988; Jaffe et al. 1984; Granger et al. 1993). In the thrombolytic era, observational, epidemiological, and randomized studies confirmed that in-hospital mortality is two times higher in patients with DM. This was mainly due to their higher rates of early re-infarction and congestive heart failure (Lee et al. 1995; Zuanetti et al. 1993; Barbash et al. 1993; Mueller et al. 1992; Hillis et al. 1990; Murphy et al. 1995; Klein et al. 1993). In particular, in DM patients, acute heart failure or cardiogenic shock accounted for over 80% of in-hospital mortality, while arrhythmias and conduction defects for almost 20% of mortality (Savage et al. 1988). A subgroup analysis of the Global Utilization of Streptokinase and t-PA for Occluded Coronary Arteries (GUSTO)-1 trial demonstrated a significantly higher 30-day mortality in ST-elevation

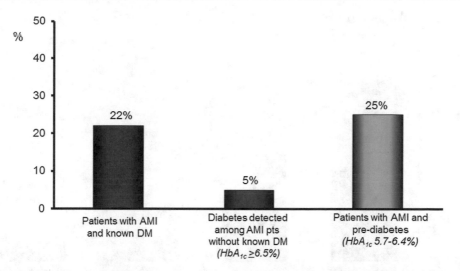

Fig. 1 Percentage of patients with abnormal glucose metabolism among 1553 patients with acute myocardial infarction (AMI; 747 STEMI and 806 NSTEMI) hospitalized between 2010 and 2016 (Ref. Marenzi et al. 2018a). *DM* diabetes mellitus, *HbA₁c* glycated hemoglobin

myocardial infarction (STEMI) patients with DM when compared with those without (10% vs. 6%) (Savage et al. 1988). Similarly, the Global Registry of Acute Coronary Events (GRACE) registry reported an almost twice-higher in-hospital case fatality rate for STEMI patients with DM (Franklin et al. 2004). More recently, among 93,569 AMI patients (in most cases treated with percutaneous coronary intervention [PCI]) included in the National Cardiovascular Data Registry (NCDR) Acute Coronary Treatment and Intervention Outcomes Network-Get with the Guidelines (ACTION Registry-GWTG), the presence of DM was associated with a higher risk of in-hospital mortality, also after adjustment for major confounders (OR 1.17; 95% CI 1.07–1.27) (Rousan et al. 2014). The adverse effect of DM on in-hospital mortality during AMI has been further confirmed in a recent cohort of more than 5,000 STEMI patients undergoing primary PCI (Marenzi et al. 2019). Again, DM was associated with an about two-fold higher in-hospital mortality, as compared to non-DM patients. Notably, in this study, the different mortality rate was not related to differences in the extent of myocardial infarct size, as estimated by enzymatic peak value (Marenzi et al. 2019). Thus, despite evidence for improvement in

outcomes in the general AMI population over the past 30 years, as well as in DM patients, a two-fold higher mortality in DM patients has been consistently reported across decades (Fig. 2).

The reasons for the observed excess early mortality in AMI patients with DM have not been clearly defined, yet. Traditionally, several factors have been held responsible for their higher mortality (Fig. 3): (1) the presence of other cardiovascular risk factors featuring DM subjects (hypertension, dyslipidemia, obesity, and kidney disease) (Kip et al. 1996); (2) a more diffuse and severe coronary atherosclerosis (Kip et al. 1996); (3) the higher incidence of painless infarction, possibly due to cardiac autonomic sensory neuropathy, and atypical symptoms, which may lead to delay in first medical contact and initiation of recommended therapies (Coronado et al. 2004); (4) increased platelet activation and coagulation factors expression, and reduced intrinsic thrombolytic activity, which characterize DM, enhancing the ongoing pro-thrombotic and pro-coagulant state (Odegaard et al. 2016); (5) endothelial dysfunction associated with insulin resistance and metabolic syndrome, which may worsen coronary vasoconstriction (Odegaard et al. 2016); (6) the

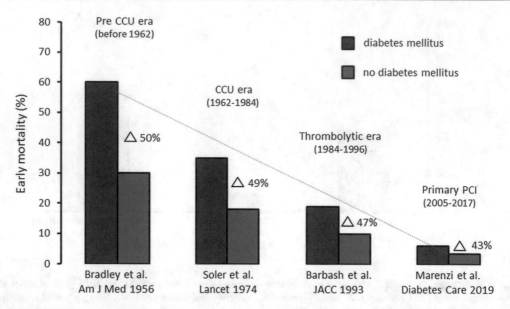

Fig. 2 Difference (Δ) in early mortality rate in acute myocardial infarction patients with and without diabetes mellitus across decades, going from pre coronary care unit (CCU) to primary percutaneous coronary intervention (PCI) era

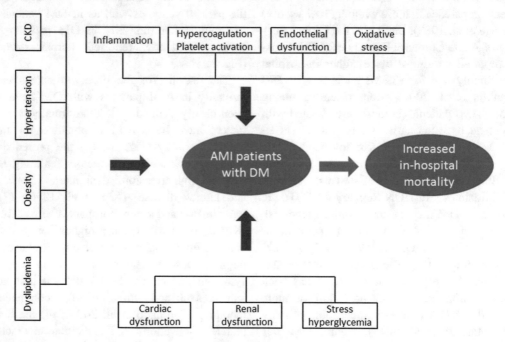

Fig. 3 Possible mechanisms contributing to the increased in-hospital mortality of acute myocardial infarction (AMI) patients with diabetes mellitus (DM). *CKD* chronic kidney disease

sub-clinical chronic inflammatory milieu, a typical feature of DM (Odegaard et al. 2016); (7) the impaired compensatory hyperdynamic response of the remaining non-ischemic myocardium, which has been shown in DM patients with AMI (Woodfield et al. 1996), possibly as a reflection of the diabetic systolic and diastolic cardiomyopathy. Finally, a very recent study, focusing on STEMI patients treated with primary PCI, specifically investigated the impact of cardiac (as evaluated by admission left ventricular ejection fraction) and renal (as assessed by admission estimated glomerular filtration rate) function on in-hospital mortality in DM patients (Marenzi et al. 2019). Interestingly, the higher in-hospital mortality rate of STEMI patients with DM (6.1% vs. 3.5%) was mainly driven by their more frequent cardio-renal dysfunction, as the prognostic power of DM was no longer confirmed after adjustment for cardiac and renal function. These findings cannot be considered fully unexpected, since cardiac and renal functions are the two most important predictors of in-hospital mortality in AMI (Marenzi et al. 2019). Notably, cardiac function evaluated at hospital admission in AMI patients incorporates several clinical information, including pre-existing cardiac dysfunction, extent of the ongoing ischemic process, and the related hemodynamic effects. Similarly, the evaluation of renal function reflects a variable combination of acute (hemodynamic impairment) and chronic (underlying co-morbidities) information. These data do not allow clarifying whether and to what extent the more likely cardio-renal impairment observed in DM patients is due to a pre-existing dysfunction or whether it is the acute consequence of a more severe AMI. Possibly, the evaluation not only of renal function but also of admission microalbuminuria might help to discriminate between these two possibilities (Berton et al. 1997). Indeed, a body of evidence has shown that microalbuminuria is associated with an increased risk of mortality, beyond that yielded by renal function in patients with AMI (Berton et al. 2008) and, particularly, in those with DM (Berton et al. 2004). The mechanisms underlying the adverse prognosis associated with microalbuminuria are not well known. Microalbuminuria may represent an index of generalized vascular damage because it has been correlated with markers of endothelial dysfunction and inflammation that are directly involved in atherogenesis (Naidoo 2002; Stehouwer et al. 1992).

More studies are needed to confirm this intriguing cardio-renal hypothesis and to further investigate the mechanisms underlying this association in AMI patients with DM. This research might pave the way to novel therapeutic strategies, aiming at reducing the mortality gap still existing between DM and non-DM patients.

4 In-Hospital Clinical Relevance of Admission Glycemia in AMI

Acute Hyperglycemia Elevated levels of plasma glucose at hospital admission (acute hyperglycemia) are common among patients with AMI, occurring in up to 50% of all AMI patients, according to the considered glycemic threshold (Wahab et al. 2002; Kosiborod et al. 2008; Ishihara et al. 2005; Marenzi et al. 2010). Although there is currently no uniform definition of hyperglycemia in the setting of AMI, as prior studies used various hyperglycemia cut-off values ranging from 110 mg/dL to 200 mg/dL, a threshold of 200 mg/dL is usually considered based on prior large studies in patients with AMI (Wahab et al. 2002; Kosiborod et al. 2008; Ishihara et al. 2005; Marenzi et al. 2010). Moreover, admission glucose has been identified as a major independent predictor of both in-hospital morbidity and mortality in AMI for DM and non-DM patients (Marenzi et al. 2010; Zeller et al. 2005). Of note, for every 18 mg/dL (1 mmol/L) increase in glucose level above 200 mg/dL, it has been reported a 4% and a 5% increase in hospital mortality risk in patients without and with DM, respectively (Stranders et al. 2004). For the same increase in glucose level, an adjusted increase in mortality risk of 10% has been reported for STEMI patients undergoing primary PCI (Ishihara et al. 2005). Among studies showing that blood glucose levels predict the outcome of patients with AMI, most of them relied on the blood glucose level detected at hospital admission (Capes et al. 2000; Foo et al. 2003), whereas others used fasting blood glucose (Ishihara et al. 2003) or average glucose levels during the admission period (Svensson et al. 2005; Suleiman et al. 2005; Goyal et al. 2006).

Importantly, patients with both elevated admission and next day fasting glucose levels have a three-fold increase in hospital mortality, as compared to those with admission hyperglycemia only (Suleiman et al. 2005).

During AMI, counter regulatory hormones, like catecholamine, growth hormone, glucagon and cortisol, are released in proportion to the degree of cardiovascular stress and may cause hyperglycemia and an elevation of free fatty acids, both of which lead to an increase in hepatic gluconeogenesis and a decrease in insulin-mediated peripheral glucose disposal (Kosiborod 2018). In addition to reflect the ongoing cardiovascular stress associated with AMI, acute hyperglycemia may directly contribute to a poor outcome through several adverse effects. They include suppression of flow-mediated vasodilatation, increased production of oxygen-derived free radicals (Kosiborod 2018; Kawano et al. 1999), and activation of pro-inflammatory factors (Kosiborod 2018). Importantly, the degree of oxidative stress has been shown to correlate most closely with acute, rather than chronic, glucose fluctuations (Monnier et al. 2006). Finally, acute hyperglycemia has been shown to have pro-thrombotic effects (enhanced thrombin formation, platelet activation, and fibrin clot resistance to lysis), which may amplify the risk of thrombotic complications in this clinical setting (Undas et al. 2008; Worthley et al. 2007). From a clinical point of view, it has been reported that acute hyperglycemia in AMI patients is independently associated with lower rate of Thrombolysis In Myocardial Infarction (TIMI) flow grade 3 before primary PCI, with impairment of epicardial coronary flow after primary stent implantation, and with "no-reflow" phenomenon (Iwakura et al. 2003; Niccoli et al. 2009). Moreover, acute hyperglycemia is associated with increased left ventricular dysfunction, larger infarct size, and higher risk of acute heart failure, cardiogenic shock, acute kidney injury, and in-hospital mortality (Ishihara et al. 2005; Marenzi et al. 2010; Zeller et al. 2005; Iwakura et al. 2003; Timmer et al. 2005). Noteworthy, glucose normalization after admission in hyperglicemic patients

hospitalized with AMI seems to be associated with better survival (Kosiborod et al. 2009a).

The close association between admission high glucose levels and poor outcome in AMI has been shown to be particularly robust in patients without DM when compared with those with DM (Umpierrez et al. 2002; Krinsley et al. 2013; Egi et al. 2011). Moreover, this association is detectable at lower glycemic values in non-DM patients. Indeed, in them, mortality risk progressively increases when blood glucose is higher than 120 mg/dL. Conversely, in patients with DM, a blood glucose >200 mg/dL has been associated with a poor outcome (Kosiborod et al. 2005; Deedwania et al. 2008). This emphasizes the role of an acute rise of glucose level, compared to its chronic elevation, in predisposing AMI patients towards a worse prognosis. In fact, in DM patients, elevated glucose levels at hospital admission may not indicate the occurrence of a stress hyperglycemia, but they may reflect a poor chronic glycemic control. Accordingly, it has been recently demonstrated that in AMI patients with DM, the ability of admission glycemia to predict in-hospital mortality, morbidity, infarct size, and acute kidney injury significantly improves when the average chronic glucose level, as estimated by HbA_{1c}, value, is also taken into account (Marenzi et al. 2018a, b). Thus, in DM patients, the evaluation of acute glycemia, considered not in absolute terms but in relation to chronic glycemia, may better reflect "true" stress hyperglycemia and help physicians to accurately discriminate high-risk from low-risk AMI patients. Moreover, it could help to customize treatment for intensive glucose control during the acute phase of AMI. Indeed, the detrimental effects of the glycemic disorder are not limited to stress hyperglycemia but they also include fluctuations of glycemic values, with acute glucose changes in both directions (Kosiborod et al. 2009a). In line with this, it has been shown that acute variability of glucose values, as assessed by measuring the mean amplitude of glycemic excursion with a continuous glucose monitoring system, negatively correlated with the myocardial salvage index (i.e., the proportion of reversibly injured tissue that does not progress to infarction) in AMI (Teraguchi et al. 2012). Moreover, in a very recent study, in which measurement of glycemic variability was evaluated during AMI in DM patients, a glucose variability of >2.70 mmol/L (49 mg/dL) was demonstrated to be the strongest independent predictive factor for mid-term (mean follow-up time was 17 months) major adverse cardiac events (death for cardiac cause, new-onset AMI, acute heart failure) (Gerbaud et al. 2019).

Future studies are needed to investigate whether in AMI patients with DM a strategy based on glucose normalization only in patients with a high acute/chronic glycemic ratio, an accurate index of stress hyperglycemia, and/or with an elevated glycemic variability, may be more beneficial on infarct size and outcomes than an approach centered on the treatment of admission hyperglycemia alone.

Acute Hypoglycemia During hospitalization for AMI, hypoglycemia, usually defined as a blood glucose level < 4.5 mmol/L (<81 mg/dL), may also occur, both spontaneously and as a consequence of glucose lowering therapy (Svensson et al. 2005; Suleiman et al. 2005). Its incidence at hospital admission in non-DM patients is rare (<2%), but it may significantly increase during hospitalization (Kadri et al. 2006). On the other hand, hypoglycemia is more likely in AMI patients with DM in whom the majority of hypoglycemic episodes are asymptomatic (Hay et al. 2003; Chow et al. 2014).

Acute myocardial infarction patients with hypoglycemia appear to have worse outcomes (Kosiborod et al. 2009b; Pinto et al. 2008). Pinto et al. (Pinto et al. 2008) demonstrated that an admission blood glucose <81 mg/dL is associated with a three-fold increased rate of adverse outcomes in patients with STEMI. Clinical data in STEMI patients pooled from the Thrombolysis In Myocardial Infarction (TIMI)-10A/B, Limitation of Myocardial Infarction Following Thrombolysis in Acute Myocardial Infarction (LIMIT-AMI), and Optimal Angioplasty versus Primary Stenting (OPUS)-TIMI-16

studies showed that AMI patients with a blood glucose level <81 mg/dL represent 8% of all patients (Pinto et al. 2005). Admission glycemia was considered in these studies except in the OPUS-TIMI-16 study, where blood glucose levels were drawn within 48 h from admission (Pinto et al. 2005). Interestingly, both hypoglycemia and hyperglycemia were associated with a three-fold increased risk of death at 30 days when compared with euglycemic patients. Thus, a U-shaped relationship between blood glucose levels and adverse outcomes was demonstrated. Of note, death occurred in 4.6% of patients with hypoglycemia, 4.7% of those with a blood glucose level > 200 mg/dL, and 1.0% of those with euglycemia. Recurrent AMI or death occurred in 10.5%, 7.2%, and 4.2%, respectively (Pinto et al. 2005). In a retrospective cohort study using data from patients hospitalized across the United States in 40 hospitals between 2000 and 2005, Kosiborod et al. (Kosiborod et al. 2009b) confirmed that hypoglycemia is associated with increased mortality in patients with AMI. However, this risk was found only in patients who developed hypoglycemia spontaneously. In contrast, iatrogenic hypoglycemia after insulin therapy was not associated with a higher mortality risk (Kosiborod et al. 2009b). In the China Patient-centered Evaluative Assessment of Cardiac Events-Retrospective AMI (China PEACE-Retrospective AMI) study (Zhao et al. 2017), a large, nationally representative sample of patients hospitalized with AMI in China in 2001, 2006, and 2011, it was found that, after adjustment for patient characteristics, hypoglycemia (<70 mg/dL) increased risk of in-hospital death only in DM patients (OR 3.02 [95% CI 1.20–7.63]; P = 0.02) but not in those without DM (OR 1.12 [95% CI 0.60–2.08]; P = 0.73). Mechanisms underlying the detrimental association between hypoglycemia and adverse outcomes in AMI have not been fully clarified, yet. Chow et al. (Chow et al. 2014), by using continuous glucose monitoring systems and Holter monitoring, found that hypoglycemia is associated with ischemic changes and various cardiac arrhythmias, suggesting a possible link

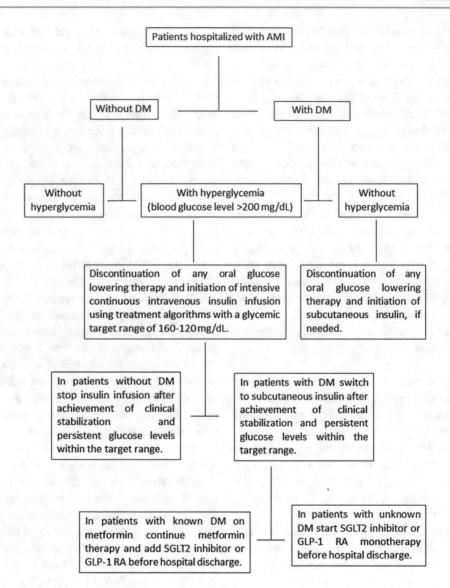

Fig. 4 A suggested algorithm for the management of admission hyperglycemia in patients with acute myocardial infarction (AMI). DM = diabetes mellitus; GLP-1 RA = glucagon-like peptide-1 receptor agonists; *SGLT-2* sodium glucose co-transporter 2

between these events. Some potential mechanisms by which hypoglycemia might lead to myocardial ischemia and arrhythmias have been listed, including increased sympathetic activity, endothelial dysfunction, oxidative stress, prolongation in QT-interval and blood coagulation abnormalities like increased platelet and neutrophil activation (Rana et al. 2014; Desouza et al. 2010; Clark et al. 2014). Despite these controversial aspects, the general consensus is that not only hyperglycemia but also hypoglycemia should be avoided in critically ill patients, including AMI

(Mesotten et al. 2015). A suggested algorithm for the management of admission hyperglycemia in AMI patients is reported in Fig. 4.

5   Long-Term Mortality of AMI Patients with DM

Several studies have shown that patients with AMI are at increased risk of death and recurrent ischemic events after hospital discharge, despite current optimal medical therapy (Fox et al. 2010; Jernberg et al. 2015). The incidence of recurrent ischemic events after AMI is higher in the first year and continues in the subsequent years, in parallel with the number of cardiovascular risk factors (Fox et al. 2010; Jernberg et al. 2015; Mauri et al. 2014; Bhatt et al. 2006; Jernberg et al. 2015). In a large Swedish registry, which included 97,254 patients discharged after AMI, the risk of non-fatal AMI, non-fatal stroke, or cardiovascular death during the first year after the index event was 18% (Jernberg et al. 2015). After the first year, the risk remained relatively high with about 20% of patients experiencing a cardiovascular event during the following 3 years (Jernberg et al. 2015). Similarly, in a recent four-country analysis, death, stroke, or recurrent AMI after the first year occurred in about 30% of patients during the following 3 years (Rapsomaniki et al. 2016). In all these studies, DM patients, again, showed a two- to four-fold higher risk for recurrent AMI and long-term mortality as compared to their non-DM counterpart (Arnold et al. 2016; Rapsomaniki et al. 2016; Nauta et al. 2012; Ahmed et al. 2014). In the Valsartan In Acute Myocardial Infarction (VALIANT) trial, 14,703 AMI patients with left ventricular dysfunction and/or heart failure were enrolled. At 1 year, the risk of death was significantly higher in patients with DM (hazard ratio [HR] 1.42) than in those without DM (Aguilar et al. 2004). In this study, also patients with unknown DM had a higher one-year mortality (HR 1.50). In the Gulf Registry of Acute Coronary Events (Gulf RACE)-2, enrolling 6,362 AMI patients, one-year survival for DM patients was substantially worse than for non-DM, and it was comparable to that of patients with unknown DM (Alfaleh et al. 2014). This higher mortality risk associated with DM persists also at longer follow-up. In the Corpus Christi Heart Project (Orlander et al. 1994), a community-based study, the rate of death at a median follow-up of 44 months after AMI was 37% in DM patients and 23% in those without. Finally, in the recent analysis of patients from the Harmonizing Outcomes with Revascularization and Stents in Acute Myocardial

Infarction (HORIZONS-AMI) trial, DM status, regardless of whether DM was established or newly diagnosed, was associated with a significantly higher risk of late adverse events after STEMI treated with primary PCI (Ertelt et al. 2017). In particular, patients with known DM and those with newly diagnosed DM had higher three-year rates of death compared with non-DM patients (11%, 12%, and 6%, respectively) and major adverse cardiac events (30%, 30%, and 20%, respectively) (Ertelt et al. 2017).

6   Current and Emerging Therapeutic Strategies

The higher mortality risk of DM patients during and after AMI strongly indicates a particular need for better treatment options in these patients and suggests that intensive medical treatment, prolonged surveillance, and stringent control of other risk factors should be carefully pursued and maintained for as long as possible in them (James et al. 2010; Wiviott et al. 2008; Yeh et al. 2016; Bhatt et al. 2016; Cosentino et al. 2019a, b; Giugliano et al. 2018; Sabatine et al. 2017; Ray et al. 2019; Furtado et al. 2019; Bohula et al. 2015; Butler and Vaduganathan 2018).

Intensive glucose management in the critical care setting, including AMI, has led to improved outcomes and intravenous insulin infusion, using a validated protocol to minimize hypoglycemia, is the preferred approach in this setting (Malmberg et al. 1999; Gunst et al. 2019). Indeed, a linear correlation between the degree of hyperglycemia and the risk of death has been demonstrated, even after adjustment for insulin dose suggesting that most of the clinical benefit of intensive insulin therapy is due to the blood glucose control achievement. Glucose-lowering therapies other than insulin have been investigated in preliminary experimental and clinical studies providing intriguing results (Cosentino et al. 2019b; Butler and Vaduganathan 2018). Preclinical studies have repeatedly demonstrated that dipeptidyl peptidase-4 (DPP-4) inhibitors, glucagon-like peptide (GLP-1) receptor agonists, and sodium-glucose cotransporter 2 (SGLT2) inhibitors reduce infarct size in rats (Bayrami et al. 2018; Basalay et al. 2016) and pigs (Timmers et al. 2009) undergoing acute myocardial ischemia and reperfusion. Treatment with sitagliptin, a DPP-4 inhibitor, was associated with a lower risk of 30-days complications than other oral glucose-lowering therapies in patients with DM enrolled in the Acute Coronary Syndrome Israeli Survey (Leibovitz et al. 2013). Moreover, exenatide, a GLP-1 receptor agonist, reduced infarct size and improved left ventricular function in patients with STEMI (Lønborg et al. 2012, 2014; Woo et al. 2013). Similar results were observed for liraglutide (Chen et al. 2015). Conversely, metformin use was not associated with reduction in infarct size in STEMI (Basnet et al. 2015). These results highlight the need for more dedicated studies to determine whether these glucose-lowering agents are as effective or superior to insulin to control blood glucose and to reduce cardiac injury during AMI.

These new classes of glucose-lowering agents are also emerging as potential options to further reduce the cardiovascular mortality in DM patients after AMI (Butler and Vaduganathan 2018). Their positive clinical results seem to be related to mechanisms beyond their glucose lowering effect. In particular, GLP-1 receptor agonists showed anti-inflammatory effects and reduction of atherosclerotic plaques and microalbuminuria, while SGLT2 inhibitors improved cardiac and renal hemodynamic, reducing microalbuminuria and glomerular filtration rate deterioration (Butler and Vaduganathan 2018). Thus, considering that the higher mortality observed in DM patients is in large part explained by their more likely cardio-renal dysfunction (Marenzi et al. 2019), a clinical advantage of these drugs can be hypothesized also during and after AMI and should be the focus of future investigation. Despite these assumptions, controversial results have been provided. Two of the cardiovascular outcome studies have been performed in DM patients with a recent AMI. The Examination of Cardiovascular Outcomes with Alogliptin versus Standard of Care (EXAMINE) trial showed that the addition of the DPP-4 inhibitor alogliptin in DM patients with a recent AMI (between 15 and 90 days before recruitment) did not reduce the rate of major cardiovascular events at a median 18 months follow-up, when compared to placebo (White et al. 2013). However, in this trial, among those receiving metformin plus sulfonylurea therapies, cardiovascular death and all-cause mortality rates were lower in those receiving alogliptin compared with those receiving placebo (White et al. 2018). The Evaluation of LIXisenatide in Acute coronary syndromes (ELIXA) trial demonstrated that the use of lixisenatide (a GLP-1 receptor agonist) in patients with DM and a recent (within the previous 180 days) hospitalization for AMI did not improve cardiovascular outcomes (Pfeffer et al. 2015). The results of the ELIXA trial is different from those observed in cardiovascular outcome trials with other GLP-1 receptor agonists. Indeed, liraglutide, semaglutide, albiglutide and dulaglutide showed superiority to placebo in reducing major adverse cardiovascular events in patients with a previous event. A possible suggested explanation is the pharmacokinetics of lixisenatide, which is a short-acting molecule with shorter half-life than the other drugs of this class (Avogaro et al. 2019). Finally, SGLT2 inhibitors have not been tested in patients with recent AMI, thus far.

The use of more potent dual antiplatelet therapy (James et al. 2010; Wiviott et al. 2008) and its prosecution beyond the first year after AMI (Yeh et al. 2016; Bhatt et al. 2016; Cosentino et al. 2019a) have been shown to be beneficial in terms of preventing cardiovascular events in DM patients. Moreover, a significantly greater mortality benefit derived from intensive lipid lowering therapies has been demonstrated among patients with DM (Giugliano et al. 2018; Sabatine et al. 2017; Ray et al. 2019). As DM is considered a state of chronic low-grade inflammation, part of the clinical benefit associated with lipid lowering therapies may be due to their anti-inflammatory effects. Indeed, in a pre-specified analysis of the Improved Reduction of Outcomes: Vytorin Efficacy International Trial (IMPROVE-IT), patients were significantly more likely to achieve low LDL-cholesterol and high-sensitivity C reactive protein levels with ezetimibe and simvastatin than with simvastatin

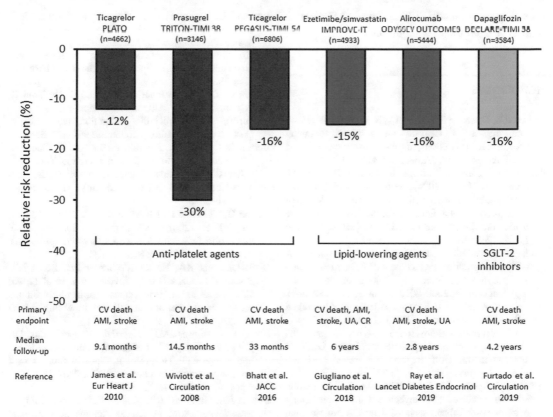

Fig. 5 Cardiovascular (CV) benefit derived from novel antiplatelet, lipid lowering, and glucose lowering treatments observed in diabetic patients with recent or previous acute myocardial infarction (AMI), as evaluated in subgroup analyses of major randomized clinical trials. *CR* coronary revascularization, *SGLT-2* sodium glucose co-transporter 2, *UA* unstable angina requiring hospitalization

monotherapy (Bohula et al. 2015). Notably, patients who reached both targets had improved outcomes also after multivariable adjustment.

Thus, these current pharmacological options should be strongly considered in DM patients after AMI to reduce their ischemic vulnerability and improve their overall long-term prognosis (Fig. 5). All the above-mentioned pharmacological therapies have been endorsed by the current international guidelines (Cosentino et al. 2019b).

The awareness of the risk associated with DM, the improved knowledge of the underlying pathophysiological mechanisms, and the implementation of new therapeutic strategies have markedly reduced mortality during and after AMI in both DM and non-DM patients (Norhammar et al. 2017). However, there still remains a significantly higher mortality in DM patients with AMI so that additional efforts are needed in order to reduce this gap.

7   Conclusions

Patients with DM are at high risk of AMI and, when it occurs, of early death and adverse long-term outcomes. An increased awareness is therefore essential in the face of a rising DM population worldwide. By using evidence-based therapies, physicians can improve both glucose control and cardiovascular outcome in AMI patients with DM. A greater collaboration between cardiologists and diabetologists is crucial to achieving optimal implementation of the current evidence and control of risk factors, with the ultimate goal to reduce the traditional mortality gap that still differentiates AMI patients with DM from those without DM.

References

Abbott RD, Donahue RP, Kannel WB, Wilson PW (1988) The impact of diabetes on survival following myocardial infarction in men vs women. The Framingham Study. JAMA 260:3456–3460

Aguilar D, Solomon SD, Køber L, Rouleau JL, Skali H, McMurray JJ et al (2004) Newly diagnosed and previously known diabetes mellitus and 1-year outcomes of acute myocardial infarction: the VALsartan in acute myocardial iNfarcTion (VALIANT) trial. Circulation 110:1572–1578

Ahmed B, Davis HT, Laskey WK (2014) In-hospital mortality among patients with type 2 diabetes mellitus and acute myocardial infarction: results from the national inpatient sample, 2000–2010. J Am Heart Assoc 26:3

Alfaleh HF, Alhabib KF, Kashour T, Ullah A, Alsheikhali AA, Al Suwaidi J et al (2014) Short-term and long-term adverse cardiovascular events across the glycaemic spectrum in patients with acute coronary syndrome: the Gulf Registry of Acute Coronary Events-2. Coron Artery Dis 25:330–338

Arnold SV, Spertus JA, Jones PG et al (2016) Predicting adverse outcomes after myocardial infarction among patients with diabetes mellitus. Circ Cardiovasc Qual Outcomes 9:372–379

Avogaro A, Bonora E, Consoli A, Del Prato S, Genovese S, Giorgino F (2019) Glucose-lowering therapy and cardiovascular outcomes in patients with type 2 diabetes mellitus and acute coronary syndrome. Diab Vasc Dis Res 16:399–414

Barbash GI, White HD, Modan M, Van de Werf F (1993) Significance of diabetes mellitus in patients with acute myocardial infarction receiving thrombolytic therapy. Investigators of the International Tissue Plasminogen Activator/Streptokinase Mortality Trial. J Am Coll Cardiol 22:707–713

Bartnik M, Malmberg K, Hamsten A, Efendic S, Norhammar A, Silveira A et al (2004) Abnormal glucose tolerance–a common risk factor in patients with acute myocardial infarction in comparison with population-based controls. J Intern Med 256:288–297

Basalay MV, Mastitskaya S, Mrochek A, Ackland GL, Del Arroyo AG, Sanchez J et al (2016) Glucagonlike peptide-1 (GLP-1) mediates cardioprotection by remote ischaemic conditioning. Cardiovasc Res 112:669–676

Basnet S, Kozikowski A, Makaryus AN, Pekmezaris R, Zeltser R, Akerman M et al (2015) Metformin and myocardial injury in patients with diabetes and ST-segment elevation myocardial infarction: a propensity score matched analysis. J Am Heart Assoc 4: e002314

Bayrami G, Karimi P, Agha-Hosseini F, Feyzizadeh S, Badalzadeh R (2018) Effect of ischemic postconditioning on myocardial function and infarct size following reperfusion injury in diabetic rats pretreated with vildagliptin. J Cardiovasc Pharmacol Ther 23:174–183

Berton G, Citro T, Palmieri R, Petucco S, De Toni R, Palatini P (1997) Albumin excretion rate increases during acute myocardial infarction and strongly predicts early mortality. Circulation 96:3338–3345

Berton G, Cordiano R, Palmieri R, De Toni R, Guarnieri GL, Palatini P (2004) Albumin excretion in diabetic patients in the setting of acute myocardial infarction: association with 3-year mortality. Diabetologia 47:1511–1518

Berton G, Cordiano R, Mazzuco S, Katz E, De Toni R, Palatini P (2008) Albumin excretion in acute myocardial infarction: a guide for long-term prognosis. Am Heart J 156:760–768

Bhatt DL, Fox KA, Hacke W, Berger PB, Black HR, Boden WE, CHARISMA Investigators et al (2006) Clopidogrel and aspirin versus aspirin alone for the prevention of atherothrombotic events. N Engl J Med 354:1706–1717

Bhatt DL, Bonaca MP, Bansilal S, Angiolillo DJ, Cohen M, Storey RF et al (2016) Reduction in ischemic events with Ticagrelor in diabetic patients with prior myocardial infarction in PEGASUS-TIMI 54. J Am Coll Cardiol 67:2732–2740

Bohula EA, Giugliano RP, Cannon CP, Zhou J, Murphy SA, White JA et al (2015) Achievement of dual low-density lipoprotein cholesterol and high-sensitivity C-reactive protein targets more frequent with the addition of ezetimibe to simvastatin and associated with better outcomes in IMPROVE-IT. Circulation 132:1224–1233

Booth GL, Kapral MK, Fung K, TU JV. (2006) Relation between age and cardiovascular disease in men and women with diabetes compared with non-diabetic people: a population-based retrospective cohort study. Lancet 368:29–36

Bradley RF, Bryfogle JW (1956) Survival of diabetic patients after myocardial infarction. Am J Med 20:207–216

Butler J, Vaduganathan M (2018) Glucose-lowering therapies in patients with concomitant diabetes mellitus and heart failure: finding the "Sweet Spot". JACC Heart Fail 6:27–29

Capes SE, Hunt D, Malmberg K, Gerstein HC (2000) Stress hyperglycaemia and increased risk of death after myocardial infarction in patients with and without diabetes: a systematic overview. Lancet 355:773–778

Chen WR, Hu SY, Chen YD, Zhang Y, Qian G, Wang J et al (2015) Effects of liraglutide on left ventricular function in patients with ST-segment elevation myocardial infarction undergoing primary percutaneous coronary intervention. Am Heart J 170:845–854

Chow E, Bernjak A, Williams S, Fawdry RA, Hibbert S, Freeman J et al (2014) Risk of cardiac arrhythmias during hypoglycemia in patients with type 2 diabetes and cardiovascular risk. Diabetes 63:1738–1747

Clark AL, Best CJ, Fisher SJ (2014) Even silent hypoglycemia induces cardiacarrhythmias. Diabetes 63:1457–1459

Coronado BE, Pope JH, Griffith JL, Beshansky JR, Selker HP (2004) Clinical features, triage, and outcome of patients presenting to the ED with suspected acute coronary syndromes but without pain: a multicenter study. Am J Emerg Med 22:568–574

Cosentino N, Campodonico J, Faggiano P, De Metrio M, Rubino M, Milazzo V et al (2019a) A new score based on the PEGASUS-TIMI 54 criteria for risk stratification of patients with acute myocardial infarction. Int J Cardiol 278:1–6

Cosentino F, Grant PJ, Aboyans V, Bailey CJ, Ceriello A, Delgado V et al (2019b) 2019 ESC Guidelines on diabetes, pre-diabetes, and cardiovascular diseases developed in collaboration with the EASD. Eur Heart J 31:pii: ehz486. (Epub ahead of print)

Deedwania P, Kosiborod M, Barrett E, Ceriello A, Isley W, Mazzone T et al (2008) American Heart Association Diabetes Committee of the Council on Nutrition, Physical Activity, and Metabolism. Hyperglycemia and acute coronary syndrome: a scientific statement from the American Heart Association Diabetes Committee of the Council on Nutrition, Physical Activity, and Metabolism. Circulation 117:1610–1619

Desouza CV, Bolli GB, Fonseca V (2010) Hypoglycemia, diabetes, and cardiovascular events. Diabetes Care 33:1389–1394

Egi M, Bellomo R, Stachowski E, French CJ, Hart GK, Taori G et al (2011) The interaction of chronic and acute glycemia with mortality in critically ill patients with diabetes. Crit Care Med 39:105–111

Ertelt K, Brener SJ, Mehran R, Ben-Yehuda O, McAndrew T, Stone GW (2017) Comparison of outcomes and prognosis of patients with versus without newly diagnosed diabetes mellitus after primary percutaneous coronary intervention for ST-elevation myocardial infarction (the HORIZONS-AMI Study). Am J Cardiol 119:1917–1923

Foo K, Cooper J, Deaner A, Knight C, Suliman A, Ranjadayalan K et al (2003) A single serum glucose measurement predicts adverse outcomes across the whole range of acute coronary syndromes. Heart 89:512–516

Fox KA, Carruthers KF, Dunbar DR, Graham C, Manning JR, De Raedt H et al (2010) Underestimated and underrecognized: the late consequences of acute coronary syndrome (GRACE UK Belgian Study). Eur Heart J 31:2755–2764

Franklin K, Goldberg RJ, Spencer F, Klein W, Budaj A, Brieger D, GRACE investigators et al (2004) Implications of diabetes in patients with acute coronary syndromes. The Global Registry of Acute Coronary Events. Arch Intern Med 164:1457–1463

Fuller JH, Shipley MJ, Rose G, Jarrett RJ, Keen H (1983) Mortality from coronary heart disease and stroke in relation to degree of glycaemia: the Whitehall study. BMJ 287:867–701

Furtado RHM, Bonaca MP, Raz I, Zelniker TA, Mosenzon O, Cahn A et al (2019) Dapagliflozin and cardiovascular outcomes in patients with type 2 diabetes mellitus and previous myocardial infarction. Circulation 139:2516–2527

Geiss LS, Herman WH, Smith PJ (1995) Mortality in non-insulin dependent diabetes. In: Harris M (ed) Diabetes in America, 2nd edn. National Institute of Diabetes and Digestive and Kidney diseases, Bethesda, pp 233–255

Gerbaud E, Darier R, Montaudon M, Beauvieux MC, Coffin-Boutreux C, Coste P et al (2019) Glycemic variability is a powerful independent predictive factor of midterm major adverse cardiac events in patients with diabetes with acute coronary syndrome. Diabetes Care 42:674–681

Giugliano RP, Cannon CP, Blazing MA, Nicolau JC, Corbalán R, Špinar J, IMPROVE-IT (Improved Reduction of Outcomes: Vytorin Efficacy International Trial) Investigators et al (2018) Benefit of adding Ezetimibe to Statin Therapy on Cardiovascular outcomes and safety in patients with versus without diabetes mellitus: results from IMPROVE-IT (Improved reduction of outcomes: Vytorin Efficacy International Trial). Circulation 137:1571–1582

Goyal A, Mahaffey KW, Garg J, Nicolau JC, Hochman JS, Weaver WD et al (2006) Prognostic significance of the change in glucose level in the first 24 h after acute myocardial infarction: results from the CARDINAL study. Eur Heart J 27:1289–1297

Granger CB, Califf RM, Young S, Candela R, Samaha J, Worley S et al (1993) Thrombolysis and Angioplasty in Myocardial Infarction (TAMI) Study Group. Outcome of patients with diabetes mellitus and acute myocardial infarction treated with thrombolytic agents. J Am Coll Cardiol 21:920–925

Gregg EW, Li Y, Wang J, Burrows NR, Ali MK, Rolka D et al (2014) Changes in diabetes-related complications in the United States, 1990–2010. N Engl J Med 370:1514–1523

Gunst J, De Bruyn A, Van den Berghe G (2019) Glucose control in the ICU. Curr Opin Anaesthesiol 32:156–162

Haffner SM (2000) Coronary heart disease in patients with diabetes. N Engl J Med 342:1040–1042

Hay LC, Wilmshurst EG, Fulcher G (2003) Unrecognized hypo- and hyperglycemia in well-controlled patients with type 2 diabetes mellitus: the results of continuous glucose monitoring. Diabetes Technol Ther 5:19–26

Hillis LD, Forman S, Braunwald E, TIMI Phase II Co-investigators (1990) Risk stratification before thrombolytic therapy in patients with acute myocardial infarction. J Am Coll Cardiol 16:313–315

Ishihara M, Inoue I, Kawagoe T, Shimatani Y, Kurisu S, Nishioka K et al (2003) Impact of acute hyperglycemia on left ventricular function after reperfusion therapy in patients with a first anterior wall acute myocardial infarction. Am Heart J 146:674–678

Ishihara M, Kojima S, Sakamoto T, Asada Y, Tei C, Kimura K, Japanese Acute Coronary Syndrome Study Investigators et al (2005) Acute hyperglycemia is associated with adverse outcome after acute myocardial infarction in the coronary intervention era. Am Heart J 150:814–820

Iwakura K, Ito H, Ikushima M et al (2003) Association between hyperglycemia and the no-reflow phenomenon in patients with acute myocardial infarction. J Am Coll Cardiol 41:1–7

Jaffe AS, Spadaro JJ, Schechtman K, Roberts R, Geltman EM, Sobel BE (1984) Increased congestive heart failure after myocardial infarction of modest extent in diabetes mellitus patients. Am Heart J 108:31–37

James S, Angiolillo DJ, Cornel JH, Erlinge D, Husted S, Kontny F et al (2010) Ticagrelor vs. clopidogrel in patients with acute coronary syndromes and diabetes: a substudy from the PLATelet inhibition and patient outcomes (PLATO) trial. Eur Heart J 31:3006–3016

Jernberg T, Hasvold P, Henriksson M, Hjelm H, Thuresson M, Janzon M (2015) Cardiovascular risk in post-myocardial infarction patients: nationwide real world data demonstrate the importance of a long-term perspective. Eur Heart J 36:1163–1170

Kadri Z, Danchin N, Vaur L, Cottin Y, Guéret P, Zeller M, USIC 2000 Investigators et al (2006) Major impact of admission glycaemia on 30 day and one year mortality in non-diabetic patients admitted for myocardial infarction: results from the nationwide French USIC 2000 study. Heart 92:910–915

Kannel WB, McGee DL (1979) Diabetes and cardiovascular disease. The Framingham study. JAMA 241:2035–2038

Kawano H, Motoyama T, Hirashima O, Hirai N, Miyao Y, Sakamoto T et al (1999) Hyperglycemia rapidly suppresses flow-mediated endothelium-dependent vasodilation of brachial artery. J Am Coll Cardiol 34:146–154

Kip KE, Faxon DP, Detre KM, Yeh W, Kelsey SF, Currier JW (1996) Coronary angioplasty in diabetic patients. The National Heart, Lung, and Blood Institute Percutaneous Transluminal Coronary Angioplasty Registry. Circulation 94:1818–1825

Klein HH, Hengstenberg C, Peuckert M, Jurgensen R (1993) Comparison of death rates from acute myocardial infarction in a single hospital in two different periods (1977–1978 versus 1988–1989). Am J Cardiol 71:518–523

Kosiborod M (2018) Hyperglycemia in acute coronary syndromes: from mechanisms to prognostic implications. Endocrinol Metab Clin N Am 47:185–202

Kosiborod M, Rathore SS, Inzucchi SE, Masoudi FA, Wang Y, Havranek EP et al (2005) Admission glucose and mortality in elderly patients hospitalized with acute myocardial infarction: implications for patients with and without recognized diabetes. Circulation 111:3078–3086

Kosiborod M, Inzucchi SE, Krumholz HM, Xiao L, Jones PG, Fiske S et al (2008) Glucometrics in patients hospitalized with acute myocardial infarction: defining the optimal outcomes-based measure of risk. Circulation 117:1018–1027

Kosiborod M, Inzucchi SE, Krumholz HM, Masoudi FA, Goyal A, Xiao L et al (2009a) Glucose normalization and outcomes in patients with acute myocardial infarction. Arch Intern Med 169:438–446

Kosiborod M, Inzucchi SE, Goyal A, Krumholz HM, Masoudi FA, Xiao L et al (2009b) Relationship between spontaneous and iatrogenic hypoglycemia and mortality in patients hospitalized with acute myocardial infarction. JAMA 301:1556–1564

Krinsley JS, Egi M, Kiss A, Devendra AM, Schuetz P, Maurer P et al (2013) Diabetes status and the relation of the three domains of glycemic control to mortality in critically ill patients: an international multicenter cohort study. Crit Care 7:R37

Lee KL, Woodlief LH, Topol EJ, Weaver WD, Betriu A, Col J et al (1995) Predictors of 30-day mortality in the era of reperfusion for acute myocardial infarction. Results from an international trial of 41,021 patients. GUSTO-I investigators. Circulation 91:1659–1668

Leibovitz E, Gottlieb S, Goldenberg I, Gevrielov-Yusim N, Matetzky S, Gavish D (2013) Sitagliptin pretreatment in diabetes patients presenting with acute coronary syndrome: results from the Acute Coronary Syndrome Israeli Survey (ACSIS). Cardiovasc Diabetol 12:53

Lønborg J, Vejlstrup N, Kelbæk H, Bøtker HE, Kim WY, Mathiasen AB (2012) Exenatide reduces reperfusion injury in patients with ST-segment elevation myocardial infarction. Eur Heart J 33:1491–1499

Lønborg J, Vejlstrup N, Kelbæk H, Nepper-Christensen L, Jørgensen E, Helqvist S et al (2014) Impact of acute hyperglycemia on myocardial infarct size, area at risk, and salvage in patients with STEMI and the association with exenatide treatment: results from a randomized study. Diabetes 63:2474–2485

Malmberg K, Norhammar A, Wedel H, Rydén L (1999) Glycometabolic state at admission: important risk marker of mortality in conventionally treated patients with diabetes mellitus and acute myocardial infarction: long-term results from the Diabetes and Insulin-Glucose Infusion in Acute Myocardial Infarction (DIGAMI) study. Circulation 99:2626–2632

Marenzi G, De Metrio M, Rubino M, Lauri G, Cavallero A, Assanelli E et al (2010) Acute hyperglycemia and contrast-induced nephropathy in primary percutaneous coronary intervention. Am Heart J 160:1170–1177

Marenzi G, Cosentino N, Milazzo V, De Metrio M, Cosore M, Mosca S et al (2018a) Prognostic value of the acute-to-chronic glycemic ratio at admission in acute myocardial infarction: a prospective study. Diabetes Care 41:847–853

Marenzi G, Cosentino N, Milazzo V, De Metrio M, Rubino M, Campodonico J et al (2018b) Acute kidney injury in diabetic patients with acute myocardial infarction: role of acute and chronic glycemia. J Am Heart Assoc 7(8):pii: e008122

Marenzi G, Cosentino N, Genovese S, Campodonico J, De Metrio M, Rondinelli M et al (2019) Reduced cardiorenal function accounts for most of the in-hospital morbidity and mortality risk among patients with type 2 diabetes undergoing primary percutaneous coronary intervention for ST-segment elevation myocardial infarction. Diabetes Care 42:1305–1311

Mauri L, Kereiakes DJ, Yeh RW, Driscoll-Shempp P, Cutlip DE, Steg PG, DAPT Study Investigators et al (2014) Twelve or 30 months of dual antiplatelet therapy after drug-eluting stents. N Engl J Med 371:2155–2166

Mesotten D, Preiser JC, Kosiborod M (2015) Glucose management in critically ill adults and children. Lancet Diabetes Endocrinol 3:723–733

Miettinen H, Lehto S, Salomaa V, Mähönen M, Niemelä M, Haffner SM et al (1998) Impact of diabetes on mortality after the first myocardial infarction. The FINMONICA Myocardial Infarction Register Study Group. Diabetes Care 21:69–75

Monnier L, Mas E, Ginet C, Michel F, Villon L, Cristol JP et al (2006) Activation of oxidative stress by acute glucose fluctuations compared with sustained chronic hyperglycemia in patients with type 2 diabetes. JAMA 295:1681–1687

Mozaffarian D, Marfisi R, Levantesi G, Silletta MG, Tavazzi L, Tognoni G et al (2007) Incidence of new-onset diabetes and impaired fasting glucose in patients with recent myocardial infarction and the effect of clinical and lifestyle risk factors. Lancet 370:667–675

Mueller HS, Cohen LS, Braunwald E, Forman S, Feit F, Ross A et al (1992) Predictors of early mortality and morbidity after thrombolytic therapy of acute myocardial infarction: analyses of patient subgroups in the Thrombolysis in Myocardial Infarction (TIMI) trial, phase II. Circulation 85:1254–1264

Murphy JF, Kahn MG, Krone RJ (1995) Prethrombotic versus thrombolytic era risk stratification of patients with acute myocardial infarction. Am J Cardiol 76:827–829

Naidoo DP (2002) The link between microalbuminuria, endothelial dysfunction and cardiovascular disease in diabetes. Cardiovasc J S Afr 13:194–199

Nauta ST, Deckers JW, Akkerhuis KM, van Domburg RT (2012) Short- and long-term mortality after myocardial infarction in patients with and without diabetes: changes from 1985 to 2008. Diabetes Care 35:2043–2047

Niccoli G, Burzotta F, Galiuto L, Crea F (2009 Jul 21) Myocardial no-reflow in humans. J Am Coll Cardiol 54:281–292

Norhammar A, Tenerz A, Nilsson G, Hamsten A, Efendíc S, Rydén L et al (2002) Glucose metabolism in patients with acute myocardial infarction and no previous diagnosis of diabetes mellitus: a prospective study. Lancet 359:2140–2144

Norhammar A, Mellbin L, Cosentino F (2017) Diabetes: prevalence, prognosis and management of a potent cardiovascular risk factor. Eur J Prev Cardiol 24:52–60

Odegaard AO, Jacobs DR Jr, Sanchez OA, Goff DC Jr, Reiner AP, Gross MD (2016) Oxidative stress, inflammation, endothelial dysfunction and incidence of type 2 diabetes. Cardiovasc Diabetol 15:51

Orlander PR, Goff DC, Morrissey M, Ramsey DJ, Wear ML, Labarthe DR et al (1994) The relation of diabetes to the severity of acute myocardial infarction and postmyocardial infarction survival in Mexican-Americans and non-Hispanic whites: the Corpus Christi Heart Project. Diabetes 43:897–902

Ovbiagele B, Markovic D, Fonarow GC (2011) Recent US patterns and predictors of prevalent diabetes among acute myocardial infarction patients. Cardiol Res Pract 2011:145615

Pfeffer MA, Claggett B, Diaz R, Dickstein K, Gerstein HC, Køber LV et al (2015) Lixisenatide in patients with type 2 diabetes and acute coronary syndrome. N Engl J Med 373:2247–2257

Pinto DS, Skolnick AH, Kirtane AJ, Murphy SA, Barron HV, Giugliano RP et al (2005) TIMI Study Group. U-shaped relationship of blood glucose with adverse outcomes among patients with ST-segment elevation myocardial infarction. J Am Coll Cardiol 46:178–180

Pinto DS, Kirtane AJ, Pride YB, Murphy SA, Sabatine MS, Cannon CP, CLARITY-TIMI 28 Investigators et al (2008) Association of blood glucose with angiographic and clinical outcomes among patients with ST-segment elevation myocardial infarction (from the CLARITY-TIMI-28 study). Am J Cardiol 101:303–307

Radke PW, Schunkert H (2010) Diabetics with acute coronary syndrome: advances, challenges, and uncertainties. Eur Heart J 31:2971–2973

Rana OA, Byrne CD, Greaves K (2014) Intensive glucose control and hypoglycaemia: a new cardiovascular risk factor? Heart 100:21–27

Rapsomaniki E, Thuresson M, Yang E, Blin P, Hunt P, Chung SC et al (2016) Using big data from health records from four countries to evaluate chronic disease outcomes: a study in 114 364 survivors of myocardial infarction. Eur Heart J Qual Care Clin Outcomes 2:172–183

Ray KK, Colhoun HM, Szarek M, Baccara-Dinet M, Bhatt DL, Bittner VA, ODYSSEY OUTCOMES

Committees and Investigators et al (2019) Effects of alirocumab on cardiovascular and metabolic outcomes after acute coronary syndrome in patients with or without diabetes: a prespecified analysis of the ODYSSEY OUTCOMES randomised controlled trial. Lancet Diabetes Endocrinol 7:618–628

Rousan TA, Pappy RM, Chen AY, Roe MT, Saucedo JF (2014) Impact of diabetes mellitus on clinical characteristics, management, and in-hospital outcomes in patients with acute myocardial infarction (from the NCDR). Am J Cardiol 114:1136–1144

Sabatine MS, Leiter LA, Wiviott SD, Giugliano RP, Deedwania P, De Ferrari GM et al (2017) Cardiovascular safety and efficacy of the PCSK9 inhibitor evolocumab in patients with and without diabetes and the effect of evolocumab on glycaemia and risk of new-onset diabetes: a prespecified analysis of the FOURIER randomised controlled trial. Lancet Diabetes Endocrinol 5:941–950

Savage MP, Krolewski AS, Kenien GG, Lebeis MP, Christlieb AR, Lewis SM (1988) Acute myocardial infarction in diabetes mellitus and significance of congestive heart failure as a prognostic factor. Am J Cardiol 62:665–669

Soler NG, Pentecost BL, Bennett MA, FitzGerald MG, Lamb P, Malins JM (1974) Coronary care for myocardial infarction in diabetics. Lancet 1:475–477

Stehouwer CD, Nauta JJ, Zeldenrust GC, Hackeng WH, Donker AJ, den Ottolander GJ (1992) Urinary albumin excretion, cardiovascular disease, and endothelial dysfunction in non–insulin-dependent diabetes mellitus. Lancet 340:319–323

Stranders I, Diamant M, van Gelder RE, Spruijt HJ, Twisk JW, Heine RJ et al (2004) Admission blood glucose level as risk indicator of death after myocardial infarction in patients with and without diabetes mellitus. Arch Intern Med 164:982–988

Suleiman M, Hammerman H, Boulos M, Kapeliovich MR, Suleiman A, Agmon Y et al (2005) Fasting glucose is an important independent risk factor for 30-day mortality in patients with acute myocardial infarction: a prospective study. Circulation 111:754–760

Svensson AM, McGuire DK, Abrahamsson P, Dellborg M (2005) Association between hyper- and hypoglycaemia and 2 year all-cause mortality risk in diabetic patients with acute coronary events. Eur Heart J 26:1255–1261

Teraguchi I, Imanishi T, Ozaki Y, Tanimoto T, Kitabata H, Ino Y et al (2012) Impact of stress hyperglycemia on myocardial salvage following successfully recanalized primary acute myocardial infarction. Circ J 76:2690–2696

Timmer JR, Ottervanger JP, de Boer MJ, Dambrink JH, Hoorntje JC, Gosselink AT (2005) Hyperglycemia is an important predictor of impaired coronary flow before reperfusion therapy in ST-segment elevation myocardial infarction. J Am Coll Cardiol 45:999–1002

Timmers L, Henriques JP, de Kleijn DP, Devries JH, Kemperman H, Steendijk P et al (2009) Exenatide reduces infarct size and improves cardiac function in a porcine model of ischemia and reperfusion injury. J Am Coll Cardiol 53:501–510

Umpierrez GE, Isaacs SD, Bazargan N, You X, Thaler LM, Kitabchi AE (2002) Hyperglycemia: an independent marker of in-hospital mortality in patients with undiagnosed diabetes. J Clin Endocrinol Metab 87:978–982

Undas A, Wiek I, Stêpien E, Zmudka K, Tracz W (2008) Hyperglycemia is associated with enhanced thrombin formation, platelet activation, and fibrin clot resistance to lysis in patients with acute coronary syndrome. Diabetes Care 31:1590–1595

Wahab NN, Cowden EA, Pearce NJ, Gardner MJ, Merry H, Cox JL, ICONS Investigators (2002) Is blood glucose an independent predictor of mortality in acute myocardial infarction in the thrombolytic era? J Am Coll Cardiol 40:1748–1754

White WB, Cannon CP, Heller SR, Nissen SE, Bergenstal RM, Bakris GL et al (2013) Alogliptin after acute coronary syndrome in patients with type 2 diabetes. N Engl J Med 369:1327–1335

White WB, Heller SR, Cannon CP, Howitt H, Khunti K, Bergenstal RM, EXAMINE Investigators et al (2018) Alogliptin in patients with type 2 diabetes receiving metformin and sulfonylurea therapies in the EXAMINE trial. Am J Med 131:813.e5–819.e5

Wild S, Roglic G, Green A, Sicree R, King H (2004) Global prevalence of diabetes: estimates for the year 2000 and projections for 2030. Diabetes Care 27:1047–1053

Wiviott SD, Braunwald E, Angiolillo DJ, Meisel S, Dalby AJ, Verheugt FW et al (2008) Greater clinical benefit of more intensive oral antiplatelet therapy with prasugrel in patients with diabetes mellitus in the trial to assess improvement in therapeutic outcomes by optimizing platelet inhibition with prasugrel-Thrombolysis in Myocardial Infarction 38. Circulation 118:1626–1636

Woo JS, Kim W, Ha SJ, Kim JB, Kim SJ, Kim WS et al (2013) Cardioprotective effects of exenatide in patients with ST-segment-elevation myocardial infarction undergoing primary percutaneous coronary intervention: results of exenatide myocardial protection in revascularization study. Arterioscler Thromb Vasc Biol 33:2252–2260

Woodfield SL, Lundergan CF, Reiner JS, Greenhouse SW, Thompson MA, Rohrbeck SC et al (1996) Angiographic findings and outcome in diabetic patients treated with thrombolytic therapy for acute myocardial infarction: the GUSTO-I experience. J Am Coll Cardiol 28:1661–1669

Worthley M, Holmes AS, Willoughby SR, Kucia AM, Heresztyn T, Stewart S et al (2007) The deleterious effects of hyperglycemia on platelet function in diabetic patients with acute coronary syndromes. Mediation by superoxide production, resolution with intensive insulin administration. J Am Coll Cardiol 49:304–310

Yeh RW, Secemsky EA, Kereiakes DJ, Normand SL, Gershlick AH, Cohen DJ, DAPT Study Investigators et al (2016) Development and validation of a prediction rule for benefit and harm of dual antiplatelet therapy beyond 1 year after percutaneous coronary intervention. JAMA 315:1735–1749

Yudkin JS, Oswald GA (1988) Determinants of hospital admission and case fatality in diabetic patients with myocardial infarction. Diabetes Care 11:351–358

Yusuf S, Hawken S, Ounpuu S, Dans T, Avezum A, Lanas F, INTERHEART Study Investigators et al (2004) Effect of potentially modifiable risk factors associated with myocardial infarction in 52 countries (the INTERHEART study): case-control study. Lancet 364:937–952

Zeller M, Steg PG, Ravisy J, Laurent Y, Janin-Manificat L, L'Huillier I et al (2005) Observatoire des Infarctus de Côte-d'Or Survey Working Group. Prevalence and impact of metabolic syndrome on hospital outcomes in acute myocardial infarction. Arch Intern Med 165:1192–1198

Zhao S, Murugiah K, Li N, Li X, Xu ZH, Li J et al (2017) Admission glucose and in-hospital mortality after acute myocardial infarction in patients with or without diabetes: a cross-sectional study. Chin Med J 130:767–775

Zuanetti G, Latini R, Maggioni AP, Santoro L, Franzosi MG; for the GISSI-2 Investigators (1993) Influence of diabetes on mortality in acute myocardial infarction: data from the GISSI-2 study. J Am Coll Cardiol 22:1788–1794

Adv Exp Med Biol - Advances in Internal Medicine (2020) 4: 171–192
https://doi.org/10.1007/5584_2020_496
© Springer Nature Switzerland AG 2020
Published online: 20 February 2020

Effects of GLP-1 and Its Analogs on Gastric Physiology in Diabetes Mellitus and Obesity

Daniel B. Maselli and Michael Camilleri

Abstract

The processing of proglucagon in intestinal L cells results in the formation of glucagon, GLP-1, and GLP-2. The GLP-1 molecule becomes active through the effect of proconvertase 1, and it is inactivated by dipeptidyl peptidase IV (DPP-IV), so that the half-life of endogenous GLP-1 is 2–3 min. GLP-1 stimulates insulin secretion from β cells in the islets of Langerhans. Human studies show that infusion of GLP-1 results in slowing of gastric emptying and increased fasting and postprandial gastric volumes. Retardation of gastric emptying reduces postprandial glycemia. Exendin-4 is a peptide agonist of the GLP-1 receptor that promotes insulin secretion. Chemical modifications of exendin-4 and GLP-1 molecules have been accomplished to prolong the half-life of GLP-1 agonists or analogs. This chapter reviews the effects of GLP-1-related drugs used in treatment of diabetes or obesity on gastric motor functions, chiefly gastric emptying. The literature shows that diverse methods have been used to measure effects of the GLP-1-related drugs on gastric emptying, with most studies using the acetaminophen absorption test which essentially measures gastric emptying of liquids during the first hour and capacity to absorb the drug over 4–6 h, expressed as AUC. The most valid measurements by scintigraphy (solids or liquids) and acetaminophen absorption at 30 or 60 min show that GLP-1-related drugs used in diabetes or obesity retard gastric emptying, and this is associated with reduced glycemia and variable effects on food intake and appetite. GLP-1 agonists and analogs are integral to the management of patients with type 2 diabetes mellitus and obesity. The effects on gastric emptying are reduced with long-acting preparations or long-term use of short-acting preparations as a result of tachyphylaxis. The dual agonists targeting GLP-1 and another receptor (GIP) do not retard gastric emptying, based on reports to date. In summary, GLP-1 agonists and analogs are integral to the management of patients with type 2 diabetes mellitus and obesity, and their effects are mediated, at least in part, by retardation of gastric emptying.

D. B. Maselli and M. Camilleri (✉)
Clinical Enteric Neuroscience Translational and Epidemiological Research (C.E.N.T.E.R.), Division of Gastroenterology and Hepatology, Mayo Clinic, Rochester, MN, USA
e-mail: camilleri.michael@mayo.edu

Keywords

Accommodation · Albiglutide · Appetite · Dulaglutide · Emptying · Exenatide · Liraglutide · Lixisenatide · Semaglutide

1 Introduction

Secretions from the gastrointestinal tract include hormones and peptides that provide feedback to control gastric function and to stimulate the secretion of insulin from the β cells of the islets of Langerhans in the pancreas. This feedback regulation is referred to as a system of "brakes." The ileal brake is the most recognized and results from feedback regulation of stomach and jejunal function by ileal products such as peptide YY, neurotensin and oxyntomodulin. However, proximal to the ileum, several products of enteroendocrine cells result in inhibitory effects on gastric motor functions that alter gastric reservoir function and induce antral motility and pyloric contractility, leading to retardation of gastric emptying and thereby reducing the rate of delivery of nutrients and their absorption. The upper gastrointestinal hormones and transmitters include cholecystokinin, glucose-stimulated insulinotropic peptide, glucagon and glucagon-like peptide 1 (GLP-1).

GLP-1 analogs or receptor agonists are established treatments for patients with type 2 diabetes mellitus (T2DM) and obesity. Effects of GLP-1 analogs or receptor agonists on gastric emptying are relevant for at least three reasons: first, because the delay in gastric emptying may reduce postprandial glycemia; second, because delay in gastric emptying may reduce kilocalorie intake, providing beneficial effects in obesity; and third, because delay in gastric emptying may cause symptoms that result in the need to slow the increments in doses of these medications. Over the past two decades, there has been increased understanding of the effects of this class of compounds, including the differentiation between the individual medications, as well as probable differences between the effects of short-acting compared to long-acting formulations of the same chemical entity.

This chapter reviews the effects of GLP-1 and its analogs or agonists on gastric physiology in T2DM and obesity. In addition, given the recent introduction of medications with dual effects on GLP-1 and targets of other hormones, the current state of literature is reviewed for changes in gastric functions in anticipation of further applications of dual agonists.

2 Synthesis, Actions, and Degradation of Glucagon-Like Peptide 1

GLP-1 is cleaved from proglucagon, which is expressed in the gut, pancreas, and brain. The processing of proglucagon in intestinal L cells results in the formation of glucagon (a glucose-regulatory hormone), GLP-1, and GLP-2 (an intestinal growth factor). The GLP-1 molecule becomes an active molecule through the effect of proconvertase 1, and it is inactivated by the cleaving of two amino acids at its N terminal by the enzyme, dipeptidyl peptidase IV (DPP-IV) (Moller 2001). There are two equipotent bioactive forms of GLP-1, GLP-1 (17–36) and GLP-1 (17–37), both of which are rapidly inactivated in the circulation by DPP-IV, rendering GLP-1 half-life a mere 2–3 min (Ritzel et al. 1995).

GLP-1 co-localizes in the distal intestine with oxyntomodulin and PYY. The GLP-1 receptor is expressed in the gut, pancreas, brainstem, hypothalamus, and vagal afferent nerves. Ingested nutrients, especially fats and carbohydrates, stimulate GLP-1 secretion, either indirectly through duodenally activated neurohormonal mechanisms or by direct contact of nutrients within the distal intestine (Cummings and Overduin 2007).

GLP-1 actions include activation of the ileal brake, delay in gastric emptying, increase in glucose-dependent insulin release, decrease in glucagon secretion, and increase in pancreatic β cell growth. Studies employing the specific GLP-1 receptor antagonist, exendin-9-39, show that endogenously released GLP-1 controls fasting plasma glucagon, stimulates insulin, and influences mechanisms controlling gastric emptying in humans (Deane et al. 2010a; Schirra and Göke 2005). The reduced glucagon and increased insulin secretion result in diminished postprandial glucose (Drucker 2003). GLP-1 decreases food intake, possibly via vagal and possibly via direct central pathways, mediated specifically by GLP-1 receptors (Cummings and Overduin 2007). Thus, reduction in spontaneous energy intake was demonstrated using an *ad libitum* meal in healthy, normal weight volunteers (Flint et al. 1998).

3 Structures and Formulations of GLP-1 Agonists and Analogs

GLP-1 receptor agonists can be modified from the active fragment of the human GLP-1 (7–36) or derived from exendin-4, a GLP-1 receptor agonist originally isolated from the venom of the Gila monster. Homology with human GLP-1 varies across all GLP-1 receptor agonists, but all can replicate the effects of the human peptide *in vivo*. For example, the exendin-4 derivative, exenatide, shows just 53% amino acid sequence homology with human GLP-1, but binds to the human GLP-1 receptor with affinity equivalent to that of human GLP-1 (Holst 2019). It is thought that, because of this low homology, exenatide would be associated with the most antibodies among the GLP-1 receptor agonists available for clinical use (Garber 2011). Exenatide weekly formulations are more immunogenic than twice daily formulations (Tibble et al. 2013). Those with higher antibody titers were observed to have high incidence of injection site reactions (Fineman et al. 2012).

Given the short half-life of endogenous GLP-1, multiple strategies were used to prolong the duration of action of GLP-1 receptor agonists and, thereby, reduce the need for frequent injections. These strategies include: first, changes in the amino acid sequence of GLP-1 to increase resistance to inactivation by DPP-IV, such as lixisenatide, which is mostly homologous with but slightly modified from exenatide, permitting it to be administered only once instead of twice daily; second, binding to albumin, either covalently (e.g., albiglutide and semaglutide) or noncovalently (e.g., liraglutide), or binding to immunoglobulin G (e.g., dulaglutide), all of which limit renal elimination (Fig. 1). A poly-microsphere preparation allowed exenatide to be

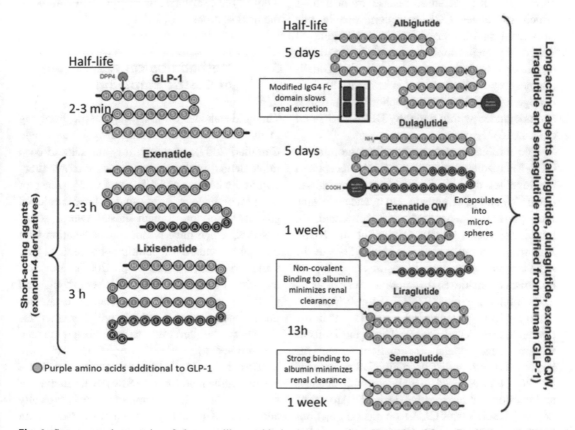

Fig. 1 Structure and properties of glucagon-like peptide 1 receptor agonists. (Reproduced from Gentilella et al. 2019)

continuously released from weekly injections (Gentilella et al. 2019).

4   Effects of GLP-1 on Gastric Functions in Health

Gastrointestinal release of peptides during and after meals has played a critical role in the homeostatic mechanisms regulating caloric intake (Gibbs et al. 1973). GLP-1 has been recognized for a multitude of regulatory functions in humans: secretion of insulin, inhibition of glucagon release, and delay in the emptying of gastric contents into the duodenum. This final feature is referred to as the "ileal brake", a critical inhibitory control mechanism that modulates food consumption and digestive function in health (Read et al. 1984).

GLP-1 secretion is stimulated through a complex cascade of signaling which involves entry of chime into the intestine, release of nesfatin-1 which stimulates CCK secretion, rise in bile salts from gallbladder emptying, and binding of Takeda P protein couple receptor 5 (TGR5) on the basolateral surface of enteric endocrine L cells (Ramesh et al. 2016; Bronden et al. 2018). This endogenous 30-amino acid peptide acts until it is soon degraded within 5 min by DPP-IV (Steinert et al. 2016).

The inhibition of gastrointestinal motility by GLP-1 is mediated through the GLP-1 receptor at the level of the myenteric neurons and downstream signaling of nitregic and cyclic adenosine monophosphate-dependent mechanisms, inhibiting vagal activity (Halim et al. 2018). This results in reduced phasic contractions of the stomach, as well as delay in gastric emptying and diminished gastric acid secretion (Schirra et al. 2002; Imeryüz et al. 1997). GLP-1 also increases fasting and postprandial gastric volumes (Delgado-Aros et al. 2002). These mechanisms require intact vagal innervation; thus, the increased postprandial accommodation induced by GLP-1 is reduced in patients with diabetes and cardiovagal neuropathy (Delgado-Aros et al. 2003). The effects of GLP-1 on gastric functions are confirmed by the reported effects of a GLP-1

receptor antagonist, which increased antral motility and inhibited pyloric tone (Schirra et al. 2006).

The effects of GLP-1 and its analogs are impacted by physiologic principles. First, increases in circulating levels of GLP-1 occur in the fed rather than the fasting state, and they impact the cholinergic mechanisms pertaining to the postprandial upper gastrointestinal motor function, in contrast to the fasting state (Schirra et al. 2009). Second, intragastric calories stimulated far more robust GLP-1 excursion than did intraduodenal infusion (Steinert et al. 2012); however, intraduodenal fat and carbohydrate infusion and absorption stimulate GLP-1 in rats (Lu et al. 2007) and GLP-1 and GIP in humans (Deane et al. 2010a). Third, the effect of GLP-1 to modulate energy intake was augmented substantially by the presence of protein in the stomach (Degen et al. 2006).

These effects of endogenous GLP-1 on gastric motor functions have been co-opted for management of disease states, which will be explored in the next sections.

5   Methodological Assessment of Gastric Emptying

The gold standard for assessing gastric emptying in humans is nuclear scintigraphy (Odunsi and Camilleri 2009); however, this has seldom been implemented in the study of gastric motor function in the context of GLP-1 and GLP-1 receptor agonists or analogs. A few studies have employed gastric emptying breath tests using stable isotopes, which have reasonable correlation with the gold standard of scintigraphy (Szarka and Camilleri 2009; Szarka et al. 2008). The vast majority of studies of gastric emptying and GLP-1 agonists use the paracetamol/acetaminophen absorption test (which will hereafter for simplicity be referred to as acetaminophen absorption test).

This chapter outlines many such studies, but it is worthwhile mentioning a few key limitations of such an assay. First, acetaminophen is typically administered as a liquid such that, even when given with a solid meal, it more closely follows the exponential pattern of gastric emptying of liquids rather than solids, which are governed by a distinct gastric emptying profile and mechanism (that is, initial retention while solids are triturated to a small particle size before emptying at a relatively constant rate). Second, gastric emptying was calculated by acetaminophen absorption at the end of a 4-, 5-, or 6-h appraisal, at which time much of the acetaminophen will have had the opportunity to be absorbed, consequently missing the potential impact of gastric emptying to the plasma acetaminophen profile. Such studies that express acetaminophen area under the curve (AUC) over 4–6 h undervalue the impact on gastric emptying. Some studies have avoided this potential limitation by assessing acetaminophen AUC from 0 to 1 h, which provides a more precise assessment of gastric emptying. Overall, the literature supports the need to avoid the acetaminophen absorption test to estimate gastric emptying due to the limitations mentioned and to use nuclear scintigraphy which is an accurate, safe, and reproducible method of assessing gastric emptying of solids and liquids.

6 Mechanism of Impairment of GI Motor Function by GLP-1 Agonists

The degree of slowing of gastric emptying by GLP-1 analogs or receptor agonists is dependent on the baseline rates of gastric emptying; thus, the induced delay in gastric emptying is more substantial in those with more rapid baseline gastric emptying (Linnebjerg et al. 2008). In contrast, when baseline emptying is already delayed, there is far less of an effect on gastric emptying from these agents (Marathe et al. 2011). The slowing of gastric emptying induced by GLP-1 agonists is dose dependent (Meier 2012). Delay in gastric emptying, as well as effects on appetite, have not been observed with medications that inhibit the enzyme that breaks down GLP-1 (DPP-IV) (Vella et al. 2007, 2008; Stevens et al. 2012). This is likely due to the lower levels of endogenous GLP-1 activity achieved with the

DPP-IV inhibitors compared to the actions of GLP-1 analogs or receptor agonists.

7 Effect of GLP-1 Receptor Agonism in Disease State: Diabetes Mellitus

While much has been published on GLP-1 receptor agonism enhancement in islet cell function (Bunck et al. 2009), the physiologic underpinnings driving improvement in diabetes appear more complex. Indeed, postprandial serum glucose levels were correlated with the degree of slowing of gastric emptying (Linnebjerg et al. 2008; Little et al. 2006; Lorenz et al. 2013; Deane et al. 2010b), and modulation of postprandial serum glucose more closely associated with delay in gastric emptying than it did increase in levels of insulin in the setting of exogenous GLP-1 administration, underscoring the importance of gastric motor functions in homeostatic mechanisms (Little et al. 2006; Willms et al. 1996; Nauck et al. 1997). In addition, when gastric emptying was accelerated by administration of erythromycin, GLP-1 was less able to modulate postprandial serum glucose levels (Meier et al. 2005). This mechanism may also explain why infusion of exogenous GLP-1 improved satiety scores and reduced meal intake in patients with diabetes (Gutzwiller et al. 1999).

The following text and Table 1 summarize key findings of the studies that evaluated gastric emptying in subjects with diabetes mellitus who were exposed to GLP-1 analogs or receptor agonists. The effects of GLP-1 analogs and receptor agonists have ushered in a more "personalized" approach for the management of both type 1 (Marathe et al. 2018) and type 2 diabetes (Holst et al. 2016). Since postprandial glycemic excursions predominate in patients with HbA1c <8% (Monnier et al. 2003), the use of the "short-acting" GLP-1 receptor agonists, exenatide twice daily and lixisenatide alone or in combination with basal insulin, has been proposed as a method to diminish postprandial glycemic excursions, predominantly by slowing gastric emptying (Holst et al. 2016).

Table 1 Summary of clinical trials of the effects of GLP-1 agonists or analogs on gastric emptying and associated features

GLP-1 agonist/ analog	Study design, dose, treatment duration and timing of GE test	Measurement of gastric emptying	Effect on upper GI motor function	References
Dulaglutide (T2DM)	RCT, double-blinded, three-period, cross-over, parallel group, multi-center study of 43 participants with T2DM (F = 65.1%, age 55.3 ± 6.1 years, BMI 30.8 ± 4.0 kg/m², HbA1c 7.6 ± 1.1%). Participants received:	Acetaminophen absorption test (480 mg) with a standardized meal	A significant change in AUC_{0-12h} was observed for the 5 mg dose (decreased by 48%, p < 0.01). The time to peak plasma acetaminophen concentration was delayed by 1.5 h for the 8 mg dose relative to placebo, suggesting delay in GE for the 5 and 8 mg doses of dulaglutide at 5 weeks	Barrington et al. (2011)
	Once weekly placebo or			
	Once weekly Dulaglutide (LY2189265) SC at 0.05 mg (n = 3), 0.3 mg (n = 6), 1 mg (n = 5), 3 mg (n = 3), 5 mg (n = 9), or 8 mg			
	GE test: At baseline and after 5 weeks of study			
Exenatide (obesity)	Double-blinded, placebo-controlled, RCT, single center in 20 participants (F = 65%, age 39.3 ± 3.7 years, and BMI 33.9 ± 1 kg/m²) with obesity and accelerated GE at baseline ($T_{1/2}$ < 90 min). Participants received	Scintigraphy ^{99m}Tc-sulfur colloid solid meal	Delayed GE in exenatide compared to placebo	Acosta et al. (2015a)
			GE at 1 h 12.4% (IQR 8–18.5%) in the exenatide group versus 38.2% (26.6–42.1%) in the placebo group (p < 0.001)	
	Exenatide 5 µg SC twice daily or			
	Placebo		$T_{1/2}$ 187 (141–240) min in the exenatide group versus 86 (73–125) min in the placebo group (p < 0.001)	
	GE test: After 30 days of study			
Exenatide (T2DM and obesity)	Single center, placebo controlled, cross-over study of 12 participants with T2DM and obesity (F = 25%, age 44 ± 2 years, BMI 34 ± 4 kg/m², HbA1c 7.5 ± 1.5%, T2DM duration 6.6 ± 3.5 years). Participants received treatments on 3 separate occasions, 2–4 weeks apart	Acetaminophen absorption with a solid meal over 6 h	After 6 h, a 58% decrease in mean plasma acetaminophen concentration was observed in the exenatide cohort (840 ± 135 µg/ml) compared with the control cohort (1,995 ± 270 µg/ml) (p < 0.001). This suggested a delay in GE in exenatide group	Cervera et al. (2008)
	IV Exenatide (0.05 µg/min 15 min before meal, decreased to 0.025 µg/min 45 min after meal ingestion), or			
	IV saline infusion (control)			
	GE test: after single dose			
Exenatide (T2DM)	RCT, double-blinded, cross over, multi-center study of participants with T2DM (54% female, BMI 33 ± 5 kg/m², HbA1c 8.5 ± 1.2%, T2DM duration 7 ± 5 years). Participants received:	Acetaminophen absorption test (liquid, 1,000 mg) over 4 h, with standardized solid meal	Delay in GE seen with exenatide but not sitagliptin.	DeFronzo et al. (2008)
	Exenatide 5 µg twice daily for 1 week, then 10 ug twice daily for 1 week) or		Acetaminophen AUC_{0-4h} ratio exenatide to sitagliptin: 0.56 ± 0.05 (95% CI 0.46–0.67, p < 0.0001)	

	Sitagliptin 100 mg daily for 2 weeks GE test: at end of each 2 week treatment arm		Delay in GE coincided with a decreased caloric intake	
Exenatide (T2DM and obesity)	Randomized, open-label study of 50 participants with T2DM and obesity. Participants received Exenatide 2.0 mg SC weekly in 26 patients or SC twice daily (F = 45%, age 55 ± 10 years, BMI 35 ± 5 kg/m², HbA1c 8.3 ± 1%, T2DM duration 7 ± 6 years) or Exenatide 5 µg SC twice daily, titrated up to 10 µg twice daily after 28 days in 24 patients (F = 49%, age 55 ± 10 years, BMI 35 ± 5 kg/m², HbA1c 8.3 ± 1%, T2DM duration 6 ± 5 years) GE test: at baseline and week 14	Acetaminophen absorption test (1,000 mg) with standardized solid meal, tested over 5 h	GE at week 14 delayed compared to baseline with exenatide twice daily but not with once weekly formulation, suggesting tachyphylaxis in effect on GE with long-acting GLP-1 receptor agonists	Drucker et al. (2008)
Exenatide (T2DM)	RCT, single-blind placebo-controlled study in 8 participants (F = 37.5%, age 52 ± 8 years, BMI 28.6 ± 5.0 kg/m², HbA1c 7.6 ± 1.6%). Participants received on 4 consecutive days a daily SC dose of: Exenatide daily: 0.02, 0.05, and 0.10 µg/kg or Placebo GE test: during each of 4 consecutive days of injection	Acetaminophen absorption test with 20 mg/kg liquid acetaminophen over 2 h	Dose-dependent slowing of GE compared to placebo. Mean plasma acetaminophen AUC$_{0-2h}$ lower for exenatide 0.05 µg/kg (907 µg*min/mL) versus placebo (1,288 µg*min/mL) (p = 0.0329), as well as for 0.10 µg/kg exenatide (645 µg*min/mL) (p = 0.0011). Mean plasma acetaminophen concentrations of 13.8, 11.4, and 7.7 µg/mL for 0.02, 0.05, and 0.1 µg/mL for 0.02, 0.05, and 0.1 µg/kg exenatide, respectively, compared with 15.6 µg/mL for the placebo group with Cmax lower with exenatide 0.05, and 0.1 µg/kg compared to placebo	Kolterman et al. (2005)
Exenatide (T2DM)	RCT, single-blind, 3 period, cross over of 17 participants (F = 5.9%, age 57 ± 10.1 years, BMI 29.2 ± 3.6 kg/m², HbA1c 8.5 ± 1.1%, T2DM duration 6.7 ± 4.5 years). Participants received twice daily: 5 µg Exenatide, 10 µg Exenatide, or Placebo GE test: After 5 days of intervention	Scintigraphy with solid meal, 450 kCal, ^{99m}Tc-labeled eggs with ^{111}In-labeled water	Delayed GE for both doses of exenatide. T$_{1/2}$ of solids was 60 (90% CI 50–70) min for placebo, 111 (90% CI 94–132) min for exenatide 5 µg, and 169 (90% CI 143–201) min for exenatide 10 µg (both vs placebo p < 0.01)	Linnebjerg et al. (2008)

(continued)

Table 1 (continued)

GLP-1 agonist/ analog	Study design, dose, treatment duration and timing of GE test	Measurement of gastric emptying	Effect on upper GI motor function	References
Exenatide QW (obesity)	RCT of 32 healthy participants randomized to receive: Exenatide QW (2 mg per week subcutaneously) (6 M,10F; age: 59.9 ± 0.9 yr.; BMI: 29.6 ± 0.6 kg/m2) or Placebo (6 M,10F; age: 60.6 ± 1.2 yr.; BMI: 29.5 ± 1.0 kg/m2) for 8 weeks GE test: At baseline and after 8 weeks' treatment	Scintigraphy with solid-liquid meal, 330 kCal, ^{99m}Tc-labeled beef with ^{67}Ga-labeled 10% glucose	Exenatide QW slowed GE of solids (AUC 0-120 min: $P < 0.05$) and liquids (AUC 0-120 min: $P = 0.01$), attenuated glucose absorption (iAUC 0-30 min: $P = 0.001$), postprandial rise in plasma glucose (iAUC 0-30 min: $P = 0.001$),and plasma glucagon at 2 h ($P = 0.001$) with no difference in scores for nausea (which were consistently low) The reduction in plasma glucose at 30 min from baseline to 8 weeks with ExQW was related inversely to $t_{1/2}$ of the glucose drink ($r = -0.55$, $P = 0.03$)	Jones et al. (2020)
Exenatide (T2DM and obesity)	RCT, double-blinded, cross over study of 13 participants with T2DM (f = 38.5%, age 56.4 ± 9.2 years, BMI 31.2 ± 3.6 kg/m2, HbA1c 7.3 ± 0.4%, T2DM duration 3.0 ± 2.6 years); Participants received daily for 9 days in each arm of trial: Liraglutide 6 µg/kg SC or Placebo GE test: Day 8 of each arm	Acetaminophen absorption test over 4 h, with a solid meal	No change in GE from placebo was observed based on acetaminophen AUC_{0-4h}	Degn et al. (2004)
Liraglutide (T2DM)	RCT, double-blinded, parallel, single center study of 100 participants with T2DM who received: Liraglutide (n = 50, F = 40%, age 7 ± 13 years, BMI 30.3 ± 3.5 kg/m2, HbA1c 8.7 ± 0.7%, T2DM duration 20 ± 12 years) or Placebo (n = 50, F = 30%, age 49 ± 12 years, BMI 29.8 ± 3.1 kg/m2, HbA1c 8.7 ± 0.7%, duration 25 ± 12 years) Participants received placebo or liraglutide at 0.6 mg SC per day, increased to 1.2 mg SC per day after 1 week, then increased to 1.8 mg per day after 1 week	Acetaminophen absorption test (1,500 mg) with standardized liquid meal over 4 h	Week 3: Difference in acetaminophen AUC_{0-4h} for liraglutide vs placebo was 2.7 (0.2–5.1) mmol/L great for liraglutide ($p = 0.0332$). Time to peak plasma acetaminophen was 19.9 (0.8–39.0) min faster for placebo ($p = 0.0412$) Week 24: Difference in acetaminophen AUC_{0-4h} for liraglutide vs placebo was 3.1 (0.6–5.5) mmol/L greater for liraglutide ($p = 0.0332$). Time to peak plasma acetaminophen was equal for liraglutide and placebo ($p = 0.8793$)	Dejgaard et al. (2016)

Drug (population)	Study	Measurement	Results	Reference
Liraglutide (T2DM)	RCT, double-blinded, two period, cross-over single center study of 18 participants with T2DM (F = 0%, age 58.6 ± 6.9 years, BMI 29.7 ± 4.2 kg/m², HbA1c 7.8 ± 0.5%) Participants received either placebo or liraglutide 0.6 mg daily for 1 week, then 1.2 mg daily for 1 week, then 1.8 mg daily for 1 week. After a washout period, the alternate intervention was given GE test: At baseline and after 3 weeks and 24 weeks of intervention	Acetaminophen absorption test over 5 h	Compared to placebo, acetaminophen AUC_{0-5h} was 17% lower for 1.2 mg dose ($p < 0.001$) but was not significant for 0.6 mg ($p = 0.287$) or 1.8 mg doses ($p = 0.301$) Acetaminophen AUC_{0-1h} 43% lower than placebo for 1.2 mg ($p < 0.001$) and 30% lower than placebo for 1.8 mg liraglutide ($p = 0.028$)	Flint et al. (2011)
Liraglutide (obesity)	RCT, double-blinded, single center study of subjects with obesity (n = 21, age 37 years [IQR 26–51 years], BMI 34.6 kg/m² [IQR 33.4–38.9 kg/m²]), randomized to receive liraglutide titrated up to 3.0 mg SC daily or placebo. (n = 19, age 42 years [IQR 32–51 years], BMI 37.2 kg/m² [IQR 33.6–41.0 kg/m²]) GE test: At baseline, 5 weeks, and 16 weeks	Nuclear scintigraphy of a solid meal over 4 h	$T_{1/2}$ of solids change from baseline to 5 weeks was more delayed with liraglutide (median 70 min [IQR 32–151]) compared with placebo (median 4 min [IQR −21–18]) ($p < 0.0001$) 16 weeks was delayed with liraglutide (median 30.5 min [IQR −11.0–54.0 min]) compared with placebo (median −1 min [IQR −19–7 min]) (p 0.025)	Halawi et al. (2017)
Liraglutide (T2DM)	RCT, double-blinded, incomplete cross over, two center study of 46 participants (F = 41%, age 53.5 years [range 38–65 years], BMI 32.6 kg/m² [range 27.0–39.9 kg/m²], HbA1c 7.4% [range 6.5–9.2%]): Participants were randomized to two of three arms each given for 4 weeks, followed by a 3 week washout: Liraglutide followed by placebo, Placebo followed by glimepiride, Glimepiride followed by liraglutide. Liraglutide was dosed at 0.6 mg SC daily, escalated by 0.6 mg increments weekly to a maximum dose of 1.8 mg daily, maintained for 2 weeks GE test: At end of 4 weeks treatment	Acetaminophen absorption test (1,000 mg) over 5 h, with a standardized liquid meal	GE delayed in liraglutide compared to control groups: Acetaminophen AUC_{0-1h} liraglutide/placebo ratio 0.62) and glimepiride (acetaminophen AUC_{0-1h} liraglutide/glimepiride ratio 0.67, ($p < 0.001$ for both) Maximum serum acetaminophen concentrations were 20% lower than with liraglutide compared to those with placebo and 15% lower than those with glimepiride ($p \leq 0.006$ for both) Also decreased hunger but not sensations of fullness, satiety, and hunger during test meals	Horowitz et al. (2012)

(continued)

Table 1 (continued)

GLP-1 agonist/ analog	Study design, dose, treatment duration and timing of GE test	Measurement of gastric emptying	Effect on upper GI motor function	References
Liraglutide (T2DM)	Single center observational study of 14 participants with T2DM (F = 28.6%, age 60 ± 13.6 years, BMI 26.9 ± 3.8 kg/m², HbA1c 9.9 ± 2.6%, T2DM duration 10.4 ± 12.1 years)	Transit time of capsule endoscopy (PillCam®)	Overall gastric transit time (n = 14) was 1:11:53 ± 1:03:17 h at baseline and 1:45:46 ± 1:40:46 h after liraglutide (p = 0.16)	Nakatani et al. (2017)
	Participants received SC liraglutide 0.3 mg, titrated up by 0.3 mg weekly to final dose 0.9 mg. GE test at baseline and 1 week after reaching final dose of 0.9 mg		Participants with: No diabetic neuropathy (n = 7) had gastric transit time of 1:01:30 ± 0:52:59 h at baseline and 2:33:29 ± 1:37:24 h after liraglutide (p = 0.03) Diabetic neuropathy (n = 7) had gastric transit time of 1:12:36 ± 1:04:30 h at baseline and 0:48:40 ± 0:32)52 h after liraglutide (p = 0.19)	
Liraglutide (obesity)	RCT, double-blinded, two-period, incomplete cross over, single center trial of 49 participants with obesity (F = 40.8%, age 48.3 ± 13.2 years, BMI 34.2 ± 2.7 kg/m²)	Acetaminophen absorption test with standardized solid meal, over 5 h	No difference in GE, based on acetaminophen AUC_{0-5h}, was observed between either liraglutide dose, as well as for liraglutide and placebo	van Can et al. (2014)
	Participants randomly chosen to 2 treatment periods of 5 weeks each, with 6–8 week wash-out		However, acetaminophen AUC_{0-1h} was 23% less with liraglutide 3.0 mg than placebo (p = 0.007), as well as 13% less than liraglutide 1.8 mg than placebo (p = 0.14), indicative of delayed GE from either liraglutide dose	
	Liraglutide SC 1.8 mg,			
	Liraglutide 3.0 mg, or			
	Placebo			
	GE test: At the end of each 5 week treatment period			
Liraglutide and Lixisenatide (T2DM)	RCT, open-label, multi-center trial in 142 participants with T2DM, divided into three treatment arms given once daily for 8 weeks in combination with insulin glargine:	^{13}C-sodium-octanoic acid containing solid meal with breath test over 4 h	A delay in GE at 8 weeks was observed for all GLP-1 receptor agonists studied. The mean change in $T_{1/2}$ from baseline was as follows:	Meier et al. (2015)
	Lixisenatide 20 µg SC daily (n = 48, F = 31.2%, age 61.6 ± 7.4 years, BMI 30.7 ± 4.3 kg/m², HbA1c 7.8 ± 0.7%, T2DM duration 11.4 [range 2.1–32.4] years)		Lixisenatide: 453.6 ± 58.2 min (p < 0.001)	
	Liraglutide 1.2 mg SC (n = 47, F = 17%, age 61.4 ± 7.9 years, BMI 30.5 ± 4.0 kg/m², HbA1c 7.8 ± 0.8%, duration 10.5 [range 0.2–12.0] years)		Liraglutide 1.2 mg: 175.3 ± 58.5 min (p < 0.05)	
			Liraglutide 1.8 mg 130.5 ± 60.3 min (p < 0.05)	

	Participant details	GE test	Results	Reference
	Liraglutide 1.8 mg (n = 47, F = 29.8%, 62.6 ± 9.4 years, BMI 31.2 ± 4.3 kg/m², HbA1c 7.9 ± 0.8%, duration 12.5 [range 4.0–31.6] years) GE test: At baseline and 8 weeks later			
Lixisenatide (healthy)	RCT, open-label, cross-over, single-center of 20 participants (F = 50%, age 31 ± 7.3 years, BMI 22.8 ± 2.7 kg/m²). Participants received lixisenatide SC at 2.5, 5, 10, or 20 µg with 2–7 day washout period GE test: 60 min after single dose of lixisenatide or placebo	Acetaminophen (1000 mg) absorption over 6 h, liquid meal	For lixisenatide doses of 5 ug or more, AUC$_{0-1h}$ showed delayed GE vs. placebo (p < 0.05) Cumulative acetaminophen absorption reduced at 6 h compared to placebo for all doses of lixisenatide, including the 2.5 µg dose	Becker et al. (2015)
Lixisenatide (T2DM)	RCT, double-blinded, cross-over study of 15 healthy participants (F = 40%, age = 67.2 ± 2.3 years, BMI 24.5 ± 0.8 kg/m²) and 15 subjects with T2DM (F = 40%, age 61.9 ± 2.3 years, BMI 30.3 ± 07 kg/m², HbA1c 6.9 ± 0.2%, T2DM duration 5.3 ± 1.2 years). Participants received 10 µg SC lixisenatide or placebo on two separate days GE test: after single dose of drug	Nuclear scintigraphy over 3 h, with liquid 75 g glucose meal.	Because T$_{1/2}$ could not be determined due to a substantial portion of both cohorts having T$_{1/2}$ > 180 min, GE was instead measured by GE rate (kcal/min) for the first 120 min and showed delayed GE in both cohorts with lixisenatide compared to placebo: Healthy: 1.45 ± 0.10 kcal/min (placebo) vs 0.60 ± 0.14 kcal/min (lixisenatide) (p < 0.001) T2DM: 1.57 ± 0.06 kcal/min (placebo) and 0.75 ± 0.13 kcal/min (lixisenatide) (p < 0.001)	Jones et al. (2019)
Lixisenatide (T2DM)	RCT, double-blinded, parallel-group study of participants with T2DM randomized to Lixisenatide SC initiated at 5 µg and increased in increments of 2.5 µg every fifth day to a maximum of 20 µg daily. (n = 21, F = 47.6%, BMI = 31.4 ± 4.1 kg/m², HbA1c 8.5 ± 1.0%, T2DM duration 6.1 ± 4.0 y) or Placebo (n = 22, F = 50%, age 53.8 ± 6.6 years, HbA1c 8.9 ± 1.1%, T2DM duration 5.7 ± 3.8 years). GE test: Baseline and 28 days later (4 days at maximum dose of 20 µg lixisenatide)	^{13}C-sodium-octanoic acid containing solid meal with breath test over 4 h	Change in T$_{1/2}$ from baseline to day 28 was 211.5 ± 67.6 min (i.e. delayed) for lixisenatide and − 24.1 ± 32.3 min for placebo (p = 0.0031) Post-prandial serum glucose inversely related to degree of delay in GE for lixisenatide 20 µg daily	Lorenz et al. (2013)

(continued)

Table 1 (continued)

GLP-1 agonist/analog	Study design, dose, treatment duration and timing of GE test	Measurement of gastric emptying	Effect on upper GI motor function	References
Semaglutide (obesity)	RCT, double-blinded, two-period, cross over trial (with 5–7 week washout) of 30 participants with obesity (F = 33.3%, age 42 ± 11 years, BMI 33.8 ± 2.5 kg/m²) who received weekly injections of: Placebo or Semaglutide 0.25 mg for 4 weeks, 0.5 mg for 4 weeks, then 1.0 mg for 4 weeks GE test: After 12 weeks of study	Acetaminophen absorption test (1,500 mg) over 5 h, with a standardized solid meal	Delay in GE suggested by 27% lower acetaminophen $AUC_{0\text{-}1h}$ reported in the semaglutide arm No differences in the acetaminophen $AUC_{0\text{-}5h}$ for semaglutide vs placebo	Hjerpsted et al. (2018)

Paracetamol absorption test is written as acetaminophen absorption test. *F* female, *T2DM* type 2 diabetes mellitus, *HbA1c* hemoglobin A1c, *RCT* randomized controlled trial, *GE* gastric emptying, $T_{1/2}$ time to half gastric emptying, *SC* subcutaneous, *AUC* area under the curve, *IQR* interquartile range, *h* hours

Dulaglutide Dulaglutide is a long-acting GLP-1 receptor agonist. One study of dulaglutide at multiple subcutaneous (SC) doses (0.05, 0.3, 1, 3, and 5 mg once weekly) in subjects with T2DM revealed a decrease in acetaminophen AUC_{0-12h} after 5 weeks of treatment (Barrington et al. 2011). Given the protracted assessment of acetaminophen absorption, the clinical significance of this AUC is not clear. It is possible, as has been seen with other long-acting GLP-1 receptor agonists, that the effect on gastric emptying is minimal to non-existent.

Exenatide Several studies have evaluated the effect of exenatide in short- or long-acting formulations on gastric emptying in subjects with T2DM. Delayed gastric emptying by the gold standard assessment—nuclear scintigraphy of a solid meal—was observed after 5 days' of either 5 μg or 10 μg SC exenatide, twice daily, compared to placebo (Linnebjerg et al. 2008). Four studies evaluated exenatide using acetaminophen absorption. These revealed delayed gastric emptying over a 6-h meal after an intravenous infusion of exenatide (roughly equivalent to one-half the peak concentration of a 5 μg SC dose exenatide) compared to placebo (Cervera et al. 2008); after 4 days of infusion of 0.05 μg/kg and 0.10 μg/kg compared to placebo (Kolterman et al. 2005); after 2 weeks of 10 μg SC, twice daily compared to placebo (DeFronzo et al. 2008); and after 14 weeks of exenatide, 10 μg SC, twice daily, compared to placebo (Drucker et al. 2008). Delay in gastric emptying based on a plasma acetaminophen absorption test was not observed with the once-weekly long-acting formulation compared to baseline (Drucker et al. 2008). However, exenatide QW substantially slowed gastric emptying measured scintigraphically and this relates to the reduction in postprandial glucose (Jones et al. 2020).

Liraglutide Assessments of liraglutide's effects on gastric emptying in subjects with T2DM have primarily used acetaminophen absorption. After eight doses of 6 μg/kg SC daily liraglutide, no difference in gastric emptying was observed, based on acetaminophen AUC_{0-4h} (Degn et al. 2004). Using a similar treatment duration of 1 week and acetaminophen AUC_{0-5h}, delayed gastric emptying was observed with 1.2 mg liraglutide daily, but not with 0.6 mg or 1.8 mg daily. However, when acetaminophen AUC_{0-1h} was instead used in the same study, gastric emptying was delayed for both 1.2 mg daily and 1.8 mg daily (Dejgaard et al. 2016). When longer treatment duration was studied, liraglutide, 1.8 mg SC daily, was shown to delay gastric emptying at 3 and 24 weeks, also based on acetaminophen AUC_{0-4h} (Flint et al. 2011). Similarly, after 4 weeks of treatment with liraglutide, gastric emptying was delayed with liraglutide, 1.8 mg SC daily, compared to placebo, based on acetaminophen AUC_{0-1h} (Horowitz et al. 2012).

Finally, one study examined gastric emptying effects of liraglutide in subjects with T2DM using gastric transit time of capsule endoscopy. While overall gastric emptying time was not changed from baseline after liraglutide use, when subjects were stratified by presence or absence of diabetic neuropathy, gastric transit time was significantly increased compared to baseline for those without diabetic neuropathy after liraglutide; whereas, those with diabetic neuropathy saw no significant delay in gastric emptying from liraglutide (Nakatani et al. 2017). While there are challenges in interpreting gastric emptying profiles of a solid meal and a capsule, this finding nevertheless illustrates the role of vagal mechanisms in the delay of gastric emptying induced by GLP-1 agonism.

Overall, the reported differences in effects of exenatide and liraglutide on gastric emptying may be more likely related to differences in methods of measurement rather than biological differences, given relatively minor structural differences between the two molecules, as well as the common mechanism of action of binding to the same G-protein-coupled, 7-transmembrane domain GLP-1 receptor. Details of each study are summarized in Table 1.

Lixisenatide Lixisenatide is a relatively more novel, short-acting, once-daily SC GLP-1 receptor agonist. Despite its short half-life (3 h), it is nonetheless administered once daily, most likely due to its ability to delay gastric emptying (Lorenz et al. 2013; Horowitz et al. 2013). Lixisenatide's effect on gastric emptying was appraised with nuclear scintigraphy using a liquid meal and a scan over 3 h. While this method could not calculate $T_{1/2}$ because $T_{1/2}$ exceeded 3 h in the majority of both healthy and diabetic participants of this study, gastric emptying rate (in kcal/min) for the first 2 h was observed to be delayed in both healthy participants and those with T2DM exposed to lixisenatide compared to placebo (Jones et al. 2019).

Using a ^{13}C-sodium-octanoic acid-containing solid meal with breath test over 4 h, lixisenatide was further observed to delay gastric emptying after 4 weeks at a dose of 20 μg daily, compared to placebo (Lorenz et al. 2013). Lixisenatide, 20 μg daily, after 8 weeks was also shown to delay gastric emptying by the same type of breath test. Because the gastric emptying delay was so profound, particularly compared to the liraglutide doses studied in the same trial, the face values of the gastric emptying times present challenges in interpretation (Meier et al. 2015), especially in view of the documented differences in estimated time to half gastric emptying using different mathematical formulas (Odunsi et al. 2009). Nevertheless, these findings do align with delay in gastric emptying by acetaminophen AUC_{0-1h} seen after a single dose of lixisenatide (5, 10, or 20 μg) compared to placebo in healthy participants (Becker et al. 2015).

8   Effects on Gastric Emptying with GLP-1 Receptor Agonists in Obesity

The secretion of GLP-1 in obesity has been reported to be reduced in some studies (Carr et al. 2010; Verdich et al. 2001; Adam and Westerterp-Plantenga 2005), although the results are inconsistent, as documented in a comprehensive review of the literature (Steinert et al. 2017). As outlined throughout this chapter, GLP-1 receptor agonism has a multitude of effects which may be useful to exploit for management of obesity. Indeed, several trials have investigated its role in weight management (le Roux et al. 2017; Pi-Sunyer et al. 2015). One such trial of 3.0 mg daily liraglutide observed a weight loss of 8.4 ± 7.3 kg compared to 2.8 ± 6.5 kg in the placebo arm (p < 0.001) after 56 weeks of intervention (Pi-Sunyer et al. 2015). It is likely that a contribution to weight loss results from appetite mediation by delay in gastric emptying, which has been observed in both human and animal models (Szayna et al. 2000).

The use of GLP-1 receptor agonism in obesity may be partly related to effects on gastric motor function. While the gastric motor functions in obesity are heterogeneous, a substantial portion of patients has accelerated gastric emptying (Acosta et al. 2015a, b), which may provide an opportunity for "personalized" treatment with a medication that delays gastric emptying. Infusion of exogenous GLP-1 in subjects with obesity, resulted in reduced hunger and calorie intake, and these measures correlated with the degree of gastric emptying delay, measured by the acetaminophen absorption test (Flint et al. 2001; Näslund et al. 1999). Nevertheless, weight loss associated with GLP-1 receptor agonism may be independent of gastric motor changes, and there was similar weight loss with liraglutide and exenatide, despite the differences in gastric emptying (Holst 2013). Weight loss with GLP-1 receptor agonists was not related to adverse gastrointestinal effects (which are largely driven by delays in gastric emptying) in several reports (Nauck et al. 2009; Buse et al. 2004; DeFronzo et al. 2005; Garber et al. 2011; Russell-Jones et al. 2009; Zinman et al. 2007). Another confounder is the fact that there appears to be tachyphylaxis in the retardation of gastric emptying from 5 to 16 weeks of liraglutide treatment, even though, at both times, the gastric emptying delay was significantly correlated with the degree of weight loss (le Roux et al. 2017). Table 1 summarizes key findings in gastric motor functions in the studies that evaluated gastric emptying in subjects with obesity exposed to GLP-1 receptor agonists.

Given the focus of this chapter on gastric effects of GLP-1 and its analogs and receptor agonists, the central mechanisms will not be extensively discussed. GLP-1 receptors are expressed throughout the central nervous system (Vrang and Larsen 2010), particularly the hypothalamus and brainstem, and they play a role in regulation of appetite (Holst 2013), as well as blood glucose (Alvarez et al. 2005), independent of gastrointestinal effects of GLP-1 and its analogs and receptor agonists. Nevertheless, peripheral stimuli have been shown to interact with central GLP-1 mechanisms to induce weight loss in preclinical studies. For example, gastric body or fundus distention activated GLP-1 containing neurons in the nucleus of the solitary tract (NTS) of rats (Vrang et al. 2003) and this decreased food intake, an effect that was reversed with exendin-9-39, a GLP-1 receptor antagonist, administered directly into the fourth ventricle (Hayes et al. 2009). Apart from the GLP-1 effects on appetite, which appear to have a significant central component, there is evidence that GLP-1 inhibits gastric emptying through mechanisms that involve vagal afferents (Imeryüz et al. 1997), as well as inhibition of central parasympathetic outflow (Wettergren et al. 1998).

Exenatide Exenatide's effects on gastric emptying have been examined in obesity (without concomitant type 2 diabetes mellitus) in one study that measured gastric emptying of solids by nuclear scintigraphy. All subjects in that study had accelerated gastric emptying at baseline. Gastric emptying was delayed compared to placebo after 30 days of exenatide, 5 μg SC twice daily, based on both 1-h gastric emptying and time to half gastric emptying (Acosta et al. 2015a).

Liraglutide One study of liraglutide, 3.0 mg SC daily, showed delayed gastric emptying compared to placebo at both 5 and 16 weeks, measured by nuclear scintigraphy of a solid meal (Halawi et al. 2017). Notably, the delay in gastric emptying was less at 16 weeks compared to 5 weeks, consistent with the tachyphylaxis of GLP-1 agonism on gastric emptying described later in this chapter. Using acetaminophen AUC_{0-5h}, a separate study showed no difference

between liraglutide, 1.8 mg or 3.0 mg SC daily, after 5 weeks; however, when acetaminophen AUC_{0-1h} was used, both doses showed delayed gastric emptying compared with placebo (van Can et al. 2014). This provides a fitting example of the shortcomings of the assessment of gastric emptying based on a prolonged acetaminophen absorption test and how this can be potentially mitigated by testing the absorption over the first hour.

Semaglutide Semaglutide is a long-acting GLP-1 receptor agonist. Despite our understanding of the diminished effects on gastric emptying induced by long-acting GLP-1 agonists (see below), one study did observe a significant reduction in acetaminophen AUC_{0-1h} (but not acetaminophen AUC_{0-5h}) with 1.0 mg weekly SC semaglutide compared to placebo after 12 weeks of intervention, suggesting that semaglutide may delay gastric emptying (Hjerpsted et al. 2018).

9 Short- vs. Long-Acting GLP-1 Receptor Agonists and Gastric Emptying and Tachyphylaxis

Nauck and colleagues demonstrated that gastric emptying of liquid meals in healthy participants, assessed by double-sampling dye dilution technique over 4 h, was delayed with administration of exogenous GLP-1; they also observed that deceleration of gastric emptying was subject to tachyphylaxis during ingestion of a second meal (Nauck et al. 2011). This loss of the delay in gastric emptying was associated with a statistically significant increase in postprandial glycemia during the second meal. Given this time frame, it is postulated that this tachyphylaxis phenomenon is driven more by the response of the vagal nerve function rather than by GLP-1 receptor downregulation or desensitization. Umapathysivam and colleagues observed similar tachyphylaxis in the delay of gastric emptying from prolonged or intermittent GLP-1 agonism compared to short-acting GLP-1 agonism (Umapathysivam et al. 2014).

There are multiple examples of tachyphylaxis in prolonged GLP-1 agonism. For instance, delay in gastric emptying was observed with short-acting exenatide, but not with the once-weekly, long-acting formulation, compared to placebo (Drucker et al. 2008). In addition, in a separate study, the delay in gastric emptying compared to baseline from 16 weeks of liraglutide (a short-acting GLP-1 receptor agonist) was less substantial than that of 5 weeks of liraglutide (Halawi et al. 2017).

Thus, delay in gastric emptying appears to be more characteristic of short-acting GLP-1 receptor agonists than long-acting GLP-1 receptor agonists (Madsbad 2016; Uccellatore et al. 2015). Tachyphylactic effects on delayed gastric emptying have not been observed with short-acting GLP-1 receptor agonists (Linnebjerg et al. 2008; Drucker et al. 2008; Flint et al. 2011), and may explain the decreased burden of upper gastrointestinal symptoms such as nausea and vomiting observed with long-acting formulations of GLP-1 agonists (Trujillo and Nuffer 2014). On the other hand, it is likely that long-acting formulations, such as albiglutide, dulaglutide, and exenatide long-acting release, improve glycemic control through restoration of balance between insulin and glucagon, rather than robust delays in gastric emptying (Meier 2012).

10   Effects of GLP-1 Agonist on Pharmacokinetics and Pharmacodynamics of Other Medications

An important consideration when prescribing medications in patients with diabetes and/or obesity is the potential of polypharmacy pharmacokinetics. Given the delays in gastric emptying observed with GLP-1 receptor agonists, there are hypothesized effects on other commonly prescribed medications for this demographic. For example, exenatide has been observed to have variable effects on several medications. In healthy volunteers, exenatide did not change steady concentration of digoxin, but it did cause a 17% decrease in mean plasma digoxin and a delay in

time to reach steady state (Kothare et al. 2005). Similarly, exenatide was associated with decreased mean lovastatin plasma concentration AUC and time to maximum plasma concentration, although this did not affect 30-week changes in lipid profile (Kothare et al. 2007). Pharmacodynamics of warfarin in healthy volunteers (Soon et al. 2006) or lisinopril in subjects with mild-to-moderate hypertension (Linnebjerg et al. 2009) were not substantially affected by exenatide. Semaglutide was not observed to derange AUC plasma concentrations for lisinopril, warfarin, and digoxin, although the AUC was increased by 32% for metformin, and this may be of limited clinical concern, given the wide therapeutic index of metformin (Bækdal et al. 2019).

11   Variations in GLP-1 Receptor and Responses to GLP-1 Agonists

The minor A allele of GLP-1R (rs6923761) is associated with greater delay in time to half gastric emptying in response to liraglutide and exenatide. These studies provide data to plan pharmacogenetics testing of the hypothesis that GLP-1R influences weight loss in response to GLP-1R agonists (Chedid et al. 2018). The significance of target receptor genetic variation requires further study with other GLP-1 agonists.

12   Oral Semaglutide

Given that most commercially available GLP-1 receptor agonists have been studied in intravenous or subcutaneous injection formulations, oral semaglutide warrants specific mention. The adverse effects from semaglutide are similar to those of other GLP-1 receptor agonists, namely, nausea and vomiting, and these mirror the safety profile of once-weekly injectable semaglutide (Davies et al. 2017). The gastrointestinal side effects appear most consistently with the 14 mg dose, suggestive of a dose-limiting gastrointestinal side effect profile. Overall, there was a small increase in discontinuations compared to other active drug treatment arms in clinical trials, including liraglutide (le Roux et al. 2017). This underscores the importance of understanding the gastrointestinal related adverse effects from GLP-1 receptor agonists, as well as the potential for therapeutic choice when GLP-1 agents are being considered for treatment of obesity. Given these side effects, it is recommended that dose escalation of oral semaglutide be carried out over 4 weeks or longer (Davies et al. 2017).

13 Effects of Combined GLP-1 and Other Hormone Agonism on Gastric Motor Functions

The evolving landscape of pharmacologic therapy has now incorporated combination therapies with efficacy of GLP-1 receptors and receptors of other hormones, including glucose dependent insulinotropic polypeptide (GIP).

GIP is released from intestinal K cells and, like GLP-1, the release of GIP is triggered by ingestion of nutrients and its activity is modulated by degradation by DPP-IV (Diakogiannaki et al. 2012; Vilsbøll et al. 2006). Dual infusions of GIP and GLP-1 receptor antagonists in healthy participants showed that the combination infusion not only caused poor postprandial glycemic control (compared with placebo and either infusion alone), but also that the combination antagonists accelerated gastric emptying, although perhaps not notably more than the GLP-1 receptor infusion alone (Gasbjerg et al. 2019). In a randomized, cross-over study of overweight or obese subjects, co-infusion of GIP and GLP-1 did not enhance the energy intake or appetite modulating effects of GLP-1 monotherapy, suggesting that GIP likely has little role in altering gastric motor functions, particularly gastric emptying (Gasbjerg et al. 2019). This finding is supported by preliminary data in patients with T2DM, subjected to gastric emptying of a standardized liquid meal (Mathiesen et al. 2019), as well as a phase 1 study of tirzepatide (a novel combination GIP and GLP-1 receptor agonist) using the acetaminophen absorption test (Urva et al. 2019). While many studies of these novel combination agents cite upper gastrointestinal symptoms of nausea and vomiting as relatively frequent and dose-dependent adverse effects from these medications (Coskun et al. 2018; Schmitt et al. 2017; Frias et al. 2018), it is unlikely that these result from a synergistic effect of GLP-1 and GIP on delay in gastric emptying.

There is also a GLP-1 and glucagon receptor dual agonist which results in clinically meaningful reductions in blood glucose, appetite and body weight in obese or overweight individuals with type 2 diabetes mellitus, as well as increase in treatment-emergent gastrointestinal disorders (Ambery et al. 2018). Another GLP-1 and glucagon dual agonist is cotagutide, which enhances insulin release and delays gastric emptying (Parker et al. 2019).

14 Conclusion

GLP-1 agonists and analogs are integral to the management of patients with type 2 diabetes mellitus and obesity. Overall, it appears that their effects are mediated at least in part by retardation of gastric emptying, although the effects on gastric emptying are reduced with long-acting preparations or long-term use of short-acting preparations as a result of tachyphylaxis.

Acknowledgements The authors thank Mrs. Cindy Stanislav for excellent secretarial assistance.

References

Acosta A, Camilleri M, Burton D, O'Neill J, Eckert D, Carlson P et al (2015a) Exenatide in obesity with accelerated gastric emptying: a randomized, pharmacodynamics study. Physiol Rep 3(11):e12610

Acosta A, Camilleri M, Shin A, Vazquez-Roque MI, Iturrino J, Burton D et al (2015b) Quantitative gastrointestinal and psychological traits associated with obesity and response to weight-loss therapy. Gastroenterology 148(3):537–546.e4

Adam TCM, Westerterp-Plantenga MS (2005) Glucagon-like peptide-1 release and satiety after a nutrient challenge in normal-weight and obese subjects. Br J Nutr 93(6):845–851

Alvarez E, Martínez MD, Roncero I, Chowen JA, García-Cuartero B, Gispert JD et al (2005) The expression of

GLP-1 receptor mRNA and protein allows the effect of GLP-1 on glucose metabolism in the human hypothalamus and brainstem. J Neurochem 92(4):798–806

Ambery P, Parker VE, Stumvoll M, Posch MG, Heise T, Plum-Moerschel L et al (2018) MEDI0382, a GLP-1 and glucagon receptor dual agonist, in obese or overweight patients with type 2 diabetes: a randomised, controlled, double-blind, ascending dose and phase 2a study. Lancet 391(10140):2607–2618

Bækdal TA, Borregaard J, Hansen CW, Thomsen M, Anderson TW (2019) Effect of Oral Semaglutide on the pharmacokinetics of lisinopril, warfarin, digoxin, and metformin in healthy subjects. Clin Pharmacokinet 58(9):1193–1203

Barrington P, Chien JY, Showalter HDH, Schneck K, Cui S, Tibaldi F et al (2011) A 5-week study of the pharmacokinetics and pharmacodynamics of LY2189265, a novel, long-acting glucagon-like peptide-1 analogue, in patients with type 2 diabetes. Diabetes Obes Metab 13(5):426–433

Becker RHA, Stechl J, Steinstraesser A, Golor G, Pellissier F (2015) Lixisenatide reduces postprandial hyperglycaemia via gastrostatic and insulinotropic effects. Diabetes Metab Res Rev 31(6):610–618

Bronden A, Alber A, Rohde U, Gasbjerg LS, Rehfeld JF, Holst JJ et al (2018) The bile acid-sequestering resin sevelamer eliminates the acute GLP-1 stimulatory effect of endogenously released bile acids in patients with type 2 diabetes. Diabetes Obes Metab 20 (2):362–369

Bunck MC, Diamant M, Cornér A, Eliasson B, Malloy JL, Shaginian RM et al (2009) One-year treatment with exenatide improves beta-cell function, compared with insulin glargine, in metformin-treated type 2 diabetic patients: a randomized, controlled trial. Diabetes Care 32(5):762–768

Buse JB, Henry RR, Han J, Kim DD, Fineman MS, Baron AD et al (2004) Effects of exenatide (exendin-4) on glycemic control over 30 weeks in sulfonylurea-treated patients with type 2 diabetes. Diabetes Care 27 (11):2628–2635

Carr RD, Larsen MO, Jelic K, Lindgren O, Vikman J, Holst JJ et al (2010) Secretion and dipeptidyl peptidase-4-mediated metabolism of incretin hormones after a mixed meal or glucose ingestion in obese compared to lean, nondiabetic men. J Clin Endocrinol Metab 95(2):872–878

Cervera A, Wajcberg E, Sriwijitkamol A, Fernandez M, Zuo P, Triplitt C et al (2008) Mechanism of action of exenatide to reduce postprandial hyperglycemia in type 2 diabetes. Am J Physiol Endocrinol Metab 294(5): E846–E852

Chedid V, Vijayvargiya P, Carlson P, Van Malderen K, Acosta A, Zinsmeister A et al (2018) Allelic variant in the glucagon-like peptide 1 receptor gene associated with greater effect of liraglutide and exenatide on gastric emptying: a pilot pharmacogenetics study. Neurogastroenterol Motil 30(7):e13313-e

Coskun T, Sloop KW, Loghin C, Alsina-Fernandez J, Urva S, Bokvist KB et al (2018) LY3298176, a novel dual GIP and GLP-1 receptor agonist for the treatment of type 2 diabetes mellitus: from discovery to clinical proof of concept. Mol Metab 18:3–14

Cummings DE, Overduin J (2007) Gastrointestinal regulation of food intake. J Clin Invest 117(1):13–23

Davies M, Pieber TR, Hartoft-Nielsen M-L, Hansen OKH, Jabbour S, Rosenstock J (2017) Effect of oral semaglutide compared with placebo and subcutaneous semaglutide on glycemic control in patients with type 2 diabetes: a randomized clinical trial. JAMA 318 (15):1460–1470

Deane AM, Nguyen NQ, Stevens JE, Fraser RJL, Holloway RH, Besanko LK et al (2010a) Endogenous glucagon-like peptide-1 slows gastric emptying in healthy subjects, attenuating postprandial glycemia. J Clin Endocrinol Metab 95(1):215–221

Deane AM, Chapman MJ, Fraser RJL, Summers MJ, Zaknic AV, Storey JP et al (2010b) Effects of exogenous glucagon-like peptide-1 on gastric emptying and glucose absorption in the critically ill: relationship to glycemia. Crit Care Med 38(5):1261–1269

DeFronzo RA, Ratner RE, Han J, Kim DD, Fineman MS, Baron AD (2005) Effects of exenatide (exendin-4) on glycemic control and weight over 30 weeks in metformin-treated patients with type 2 diabetes. Diabetes Care 28(5):1092–1100

DeFronzo RA, Okerson T, Viswanathan P, Guan X, Holcombe JH, MacConell L (2008) Effects of exenatide versus sitagliptin on postprandial glucose, insulin and glucagon secretion, gastric emptying, and caloric intake: a randomized, cross-over study. Curr Med Res Opin 24(10):2943–2952

Degen L, Oesch S, Matzinger D, Drewe J, Knupp M, Zimmerli F et al (2006) Effects of a preload on reduction of food intake by GLP-1 in healthy subjects. Digestion 74(2):78–84

Degn KB, Juhl CB, Sturis J, Jakobsen G, Brock B, Chandramouli V et al (2004) One week's treatment with the long-acting glucagon-like peptide 1 derivative liraglutide (NN2211) markedly improves 24-h glycemia and alpha- and beta-cell function and reduces endogenous glucose release in patients with type 2 diabetes. Diabetes 53(5):1187–1194

Dejgaard TF, Frandsen CS, Hansen TS, Almdal T, Urhammer S, Pedersen-Bjergaard U et al (2016) Efficacy and safety of liraglutide for overweight adult patients with type 1 diabetes and insufficient glycaemic control (Lira-1): a randomised, double-blind, placebo-controlled trial. Lancet Diabetes Endocrinol 4 (3):221–232

Delgado-Aros S, Kim D-Y, Burton DD, Thomforde GM, Stephens D, Brinkmann BH et al (2002) Effect of GLP-1 on gastric volume, emptying, maximum volume ingested, and postprandial symptoms in humans. Am J Physiol Gastrointest Liver Physiol 282(3):G424–GG31

Delgado-Aros S, Vella A, Camilleri M, Low PA, Burton DD, Thomforde GM et al (2003) Effects of glucagon-like peptide-1 and feeding on gastric volumes in diabetes mellitus with cardio-vagal dysfunction. Neurogastroenterol Motil 15(4):435–443

Diakogiannaki E, Gribble FM, Reimann F (2012) Nutrient detection by incretin hormone secreting cells. Physiol Behav 106(3):387–393

Drucker DJ (2003) Glucagon-like peptides: regulators of cell proliferation, differentiation, and apoptosis. Mol Endocrinol 17(2):161–171

Drucker DJ, Buse JB, Taylor K, Kendall DM, Trautmann M, Zhuang D et al (2008) Exenatide once weekly versus twice daily for the treatment of type 2 diabetes: a randomised, open-label, non-inferiority study. Lancet 372(9645):1240–1250

Fineman MS, Mace KF, Diamant M, Darsow T, Cirincione BB, Booker Porter TK et al (2012) Clinical relevance of anti-exenatide antibodies: safety, efficacy and cross-reactivity with long-term treatment. Diabetes Obes Metab 14(6):546–554

Flint A, Raben A, Astrup A, Holst JJ (1998) Glucagon-like peptide 1 promotes satiety and suppresses energy intake in humans. J Clin Invest 101(3):515–520

Flint A, Raben A, Ersbøll AK, Holst JJ, Astrup A (2001) The effect of physiological levels of glucagon-like peptide-1 on appetite, gastric emptying, energy and substrate metabolism in obesity. Int J Obes Relat Metab Disord 25(6):781–792

Flint A, Kapitza C, Hindsberger C, Zdravkovic M (2011) The once-daily human glucagon-like peptide-1 (GLP-1) analog liraglutide improves postprandial glucose levels in type 2 diabetes patients. Adv Ther 28 (3):213–226

Frias JP, Nauck MA, Van J, Kutner ME, Cui X, Benson C et al (2018) Efficacy and safety of LY3298176, a novel dual GIP and GLP-1 receptor agonist, in patients with type 2 diabetes: a randomised, placebo-controlled and active comparator-controlled phase 2 trial. Lancet 392 (10160):2180–2193

Garber AJ (2011) Long-acting glucagon-like peptide 1 receptor agonists: a review of their efficacy and tolerability. Diabetes Care 34(Suppl 2):S279–SS84

Garber A, Henry RR, Ratner R, Hale P, Chang CT, Bode B et al (2011) Liraglutide, a once-daily human glucagon-like peptide 1 analogue, provides sustained improvements in glycaemic control and weight for 2 years as monotherapy compared with glimepiride in patients with type 2 diabetes. Diabetes Obes Metab 13 (4):348–356

Gasbjerg LS, Helsted MM, Hartmann B, Jensen MH, Gabe MBN, Sparre-Ulrich AH et al (2019) Separate and combined glucometabolic effects of endogenous glucose-dependent insulinotropic polypeptide and glucagon-like peptide 1 in healthy individuals. Diabetes 68(5):906–917

Gentilella R, Pechtner V, Corcos A, Consoli A (2019) Glucagon-like peptide-1 receptor agonists in type 2 diabetes treatment: are they all the same? Diabetes Metab Res Rev 35(1):e3070-e

Gibbs J, Young RC, Smith GP (1973) Cholecystokinin decreases food intake in rats. J Comp Physiol Psychol 84(3):488–495

Gutzwiller JP, Drewe J, Göke B, Schmidt H, Rohrer B, Lareida J et al (1999) Glucagon-like peptide-1 promotes satiety and reduces food intake in patients with diabetes mellitus type 2. Am J Phys 276(5): R1541–R15R4

Halawi H, Khemani D, Eckert D, O'Neill J, Kadouh H, Grothe K et al (2017) Effects of liraglutide on weight, satiation, and gastric functions in obesity: a randomised, placebo-controlled pilot trial. Lancet Gastroenterol Hepatol 2(12):890–899

Halim MA, Degerblad M, Sundbom M, Karlbom U, Holst JJ, Webb D-L et al (2018) Glucagon-like peptide-1 inhibits prandial gastrointestinal motility through myenteric neuronal mechanisms in humans. J Clin Endocrinol Metab 103(2):575–585

Hayes MR, Bradley L, Grill HJ (2009) Endogenous hindbrain glucagon-like peptide-1 receptor activation contributes to the control of food intake by mediating gastric satiation signaling. Endocrinology 150 (6):2654–2659

Hjerpsted JB, Flint A, Brooks A, Axelsen MB, Kvist T, Blundell J (2018) Semaglutide improves postprandial glucose and lipid metabolism, and delays first-hour gastric emptying in subjects with obesity. Diabetes Obes Metab 20(3):610–619

Holst JJ (2013) Incretin hormones and the satiation signal. Int J Obes 37(9):1161–1168

Holst JJ (2019) From the incretin concept and the discovery of GLP-1 to today's diabetes therapy. Front Endocrinol (Lausanne) 10:260

Holst JJ, Gribble F, Horowitz M, Rayner CK (2016) Roles of the gut in glucose homeostasis. Diabetes Care 39 (6):884–892

Horowitz M, Flint A, Jones KL, Hindsberger C, Rasmussen MF, Kapitza C et al (2012) Effect of the once-daily human GLP-1 analogue liraglutide on appetite, energy intake, energy expenditure and gastric emptying in type 2 diabetes. Diabetes Res Clin Pract 97(2):258–266

Horowitz M, Rayner CK, Jones KL (2013) Mechanisms and clinical efficacy of lixisenatide for the management of type 2 diabetes. Adv Ther 30(2):81–101

Imeryüz N, Yeğen BC, Bozkurt A, Coşkun T, Villanueva-Peñacarrillo ML, Ulusoy NB (1997) Glucagon-like peptide-1 inhibits gastric emptying via vagal afferent-mediated central mechanisms. Am J Phys 273(4): G920–G9G7

Jones KL, Rigda RS, Buttfield MDM, Hatzinikolas S, Pham HT, Marathe CS et al (2019) Effects of lixisenatide on postprandial blood pressure, gastric emptying and glycaemia in healthy people and people with type 2 diabetes. Diabetes Obes Metab 21 (5):1158–1167

Jones KL, Huynh LQ, Hatzinikolas S, Rigda RS, Phillips LK, Pham HT et al (2020) Exenatide once weekly slows gastric emptying of solids and liquids in healthy, overweight, subjects under steady-state concentrations. Diabetes Obes Metab. https://doi.org/10.1111/dom.13956

Kolterman OG, Kim DD, Shen L, Ruggles JA, Nielsen LL, Fineman MS et al (2005) Pharmacokinetics, pharmacodynamics, and safety of exenatide in patients with type 2 diabetes mellitus. Am J Health Syst Pharm 62 (2):173–181

Kothare PA, Soon DKW, Linnebjerg H, Park S, Chan C, Yeo A et al (2005) Effect of exenatide on the steady-state pharmacokinetics of digoxin. J Clin Pharmacol 45 (9):1032–1037

Kothare PA, Linnebjerg H, Skrivanek Z, Reddy S, Mace K, Pena A et al (2007) Exenatide effects on statin pharmacokinetics and lipid response. Int J Clin Pharmacol Ther 45(2):114–120

le Roux CW, Astrup A, Fujioka K, Greenway F, Lau DCW, Van Gaal L et al (2017) 3 years of liraglutide versus placebo for type 2 diabetes risk reduction and weight management in individuals with prediabetes: a randomised, double-blind trial. Lancet 389 (10077):1399–1409

Linnebjerg H, Park S, Kothare PA, Trautmann ME, Mace K, Fineman M et al (2008) Effect of exenatide on gastric emptying and relationship to postprandial glycemia in type 2 diabetes. Regul Pept 151 (1–3):123–129

Linnebjerg H, Kothare P, Park S, Mace K, Mitchell M (2009) The effect of exenatide on lisinopril pharmacodynamics and pharmacokinetics in patients with hypertension. Int J Clin Pharmacol Ther 47(11):651–658

Little TJ, Pilichiewicz AN, Russo A, Phillips L, Jones KL, Nauck MA et al (2006) Effects of intravenous glucagon-like peptide-1 on gastric emptying and intragastric distribution in healthy subjects: relationships with postprandial glycemic and insulinemic responses. J Clin Endocrinol Metab 91 (5):1916–1923

Lorenz M, Pfeiffer C, Steinstrasser A, Becker RH, Rutten H, Ruus P et al (2013) Effects of lixisenatide once daily on gastric emptying in type 2 diabetes--relationship to postprandial glycemia. Regul Pept 185:1–8

Lu WJ, Yang Q, Sun W, Woods SC, D'Alessio D, Tso P (2007) The regulation of the lymphatic secretion of glucagon-like peptide-1 (GLP-1) by intestinal absorption of fat and carbohydrate. Am J Physiol Gastrointest Liver Physiol 293(5):G963–GG71

Madsbad S (2016) Review of head-to-head comparisons of glucagon-like peptide-1 receptor agonists. Diabetes Obes Metab 18(4):317–332

Marathe CS, Rayner CK, Jones KL, Horowitz M (2011) Effects of GLP-1 and incretin-based therapies on gastrointestinal motor function. Exp Diabetes Res 2011:279530

Marathe CS, Rayner CK, Wu T, Jones KL, Horowitz M (2018) Gastric emptying and the personalized

management of type 1 diabetes. J Clin Endocrinol Metab 103(9):3503–3506

Mathiesen DS, Bagger JI, Bergmann NC, Lund A, Christensen MB, Vilsbøll T et al (2019) The effects of dual GLP-1/GIP receptor agonism on glucagon secretion-a review. Int J Mol Sci 20(17):4092

Meier JJ (2012) GLP-1 receptor agonists for individualized treatment of type 2 diabetes mellitus. Nat Rev Endocrinol 8(12):728–742

Meier JJ, Kemmeries G, Holst JJ, Nauck MA (2005) Erythromycin antagonizes the deceleration of gastric emptying by glucagon-like peptide 1 and unmasks its insulinotropic effect in healthy subjects. Diabetes 54 (7):2212–2218

Meier JJ, Rosenstock J, Hincelin-Mery A, Roy-Duval C, Delfolie A, Coester HV et al (2015) Contrasting effects of lixisenatide and liraglutide on postprandial glycemic control, gastric emptying, and safety parameters in patients with type 2 diabetes on Optimized insulin glargine with or without metformin: a randomized, open-label trial. Diabetes Care 38(7):1263–1273

Moller DE (2001) New drug targets for type 2 diabetes and the metabolic syndrome. Nature 414(6865):821–827

Monnier L, Lapinski H, Colette C (2003) Contributions of fasting and postprandial plasma glucose increments to the overall diurnal hyperglycemia of type 2 diabetic patients: variations with increasing levels of HbA(1c). Diabetes Care 26(3):881–885

Nakatani Y, Maeda M, Matsumura M, Shimizu R, Banba N, Aso Y et al (2017) Effect of GLP-1 receptor agonist on gastrointestinal tract motility and residue rates as evaluated by capsule endoscopy. Diabetes Metab 43(5):430–437

Näslund E, Barkeling B, King N, Gutniak M, Blundell JE, Holst JJ et al (1999) Energy intake and appetite are suppressed by glucagon-like peptide-1 (GLP-1) in obese men. Int J Obes Relat Metab Disord 23 (3):304–311

Nauck MA, Niedereichholz U, Ettler R, Holst JJ, Orskov C, Ritzel R et al (1997) Glucagon-like peptide 1 inhibition of gastric emptying outweighs its insulinotropic effects in healthy humans. Am J Phys 273(5):E981–E9E8

Nauck M, Frid A, Hermansen K, Shah NS, Tankova T, Mitha IH et al (2009) Efficacy and safety comparison of liraglutide, glimepiride, and placebo, all in combination with metformin, in type 2 diabetes: the LEAD (liraglutide effect and action in diabetes)-2 study. Diabetes Care 32(1):84–90

Nauck MA, Kemmeries G, Holst JJ, Meier JJ (2011) Rapid tachyphylaxis of the glucagon-like peptide 1-induced deceleration of gastric emptying in humans. Diabetes 60(5):1561–1565

Odunsi ST, Camilleri M (2009) Selected interventions in nuclear medicine: gastrointestinal motor functions. Semin Nucl Med 39(3):186–194

Odunsi ST, Camilleri M, Szarka LA, Zinsmeister AR (2009) Optimizing analysis of stable isotope breath tests to estimate gastric emptying of solids. Neurogastroenterol Motil 21(7):706–e38

Parker VER, Robertson D, Wang T, Hornigold DC, Petrone M, Cooper AT et al (2019) Efficacy, safety, and mechanistic insights of cotadutide a dual receptor glucagon-like peptide-1 and glucagon agonist. J Clin Endocrinol Metab. https://doi.org/10.1210/clinem/dgz047

Pi-Sunyer X, Astrup A, Fujioka K, Greenway F, Halpern A, Krempf M et al (2015) A randomized, controlled trial of 3.0 mg of Liraglutide in weight management. N Engl J Med 373(1):11–22

Ramesh N, Mortazavi S, Unniappan S (2016) Nesfatin-1 stimulates cholecystokinin and suppresses peptide YY expression and secretion in mice. Biochem Biophys Res Commun 472(1):201–208

Read NW, McFarlane A, Kinsman RI, Bates TE, Blackhall NW, Farrar GB et al (1984) Effect of infusion of nutrient solutions into the ileum on gastrointestinal transit and plasma levels of neurotensin and enteroglucagon. Gastroenterology 86(2):274–280

Ritzel R, Orskov C, Holst JJ, Nauck MA (1995) Pharmacokinetic, insulinotropic, and glucagonostatic properties of GLP-1 [7-36 amide] after subcutaneous injection in healthy volunteers. Dose-response-relationships. Diabetologia 38(6):720–725

Russell-Jones D, Vaag A, Schmitz O, Sethi BK, Lalic N, Antic S et al (2009) Liraglutide vs insulin glargine and placebo in combination with metformin and sulfonyl-urea therapy in type 2 diabetes mellitus (LEAD-5 met +SU): a randomised controlled trial. Diabetologia 52 (10):2046–2055

Schirra J, Göke B (2005) The physiological role of GLP-1 in human: incretin, ileal brake or more? Regul Pept 128 (2):109–115

Schirra J, Wank U, Arnold R, Göke B, Katschinski M (2002) Effects of glucagon-like peptide-1(7-36)amide on motility and sensation of the proximal stomach in humans. Gut 50(3):341–348

Schirra J, Nicolaus M, Roggel R, Katschinski M, Storr M, Woerle HJ et al (2006) Endogenous glucagon-like peptide 1 controls endocrine pancreatic secretion and antro-pyloro-duodenal motility in humans. Gut 55 (2):243–251

Schirra J, Nicolaus M, Woerle HJ, Struckmeier C, Katschinski M, Göke B (2009) GLP-1 regulates gastroduodenal motility involving cholinergic pathways. Neurogastroenterol Motil 21(6):609–e22

Schmitt C, Portron A, Jadidi S, Sarkar N, DiMarchi R (2017) Pharmacodynamics, pharmacokinetics and safety of multiple ascending doses of the novel dual glucose-dependent insulinotropic polypeptide/glucagon-like peptide-1 agonist RG7697 in people with type 2 diabetes mellitus. Diabetes Obes Metab 19 (10):1436–1445

Soon D, Kothare PA, Linnebjerg H, Park S, Yuen E, Mace KF et al (2006) Effect of exenatide on the pharmacokinetics and pharmacodynamics of warfarin in healthy Asian men. J Clin Pharmacol 46(10):1179–1187

Steinert RE, Meyer-Gerspach AC, Beglinger C (2012) The role of the stomach in the control of appetite and the secretion of satiation peptides. Am J Physiol Endocrinol Metab 302(6):E666–E673

Steinert RE, Beglinger C, Langhans W (2016) Intestinal GLP-1 and satiation: from man to rodents and back. Int J Obes 40(2):198–205

Steinert RE, Feinle-Bisset C, Asarian L, Horowitz M, Beglinger C, Geary N (2017) Ghrelin, CCK, GLP-1, and PYY(3-36): secretory controls and physiological roles in eating and Glycemia in health, obesity, and after RYGB. Physiol Rev 97(1):411–463

Stevens JE, Horowitz M, Deacon CF, Nauck M, Rayner CK, Jones KL (2012) The effects of sitagliptin on gastric emptying in healthy humans – a randomised, controlled study. Aliment Pharmacol Ther 36 (4):379–390

Szarka LA, Camilleri M (2009) Methods for measurement of gastric motility. Am J Physiol Gastrointest Liver Physiol 296(3):G461–G475

Szarka LA, Camilleri M, Vella A, Burton D, Baxter K, Simonson J et al (2008) A stable isotope breath test with a standard meal for abnormal gastric emptying of solids in the clinic and in research. Clin Gastroenterol Hepatol 6(6):635–643.e1

Szayna M, Doyle ME, Betkey JA, Holloway HW, Spencer RG, Greig NH et al (2000) Exendin-4 decelerates food intake, weight gain, and fat deposition in Zucker rats. Endocrinology 141(6):1936–1941

Tibble CA, Cavaiola TS, Henry RR (2013) Longer acting GLP-1 receptor agonists and the potential for improved cardiovascular outcomes: a review of current literature. Expert Rev Endocrinol Metab 8(3):247–259

Trujillo JM, Nuffer W (2014) Albiglutide: a new GLP-1 receptor agonist for the treatment of type 2 diabetes. Ann Pharmacother 48(11):1494–1501

Uccellatore A, Genovese S, Dicembrini I, Mannucci E, Ceriello A (2015) Comparison review of short-acting and long-acting glucagon-like peptide-1 receptor agonists. Diabetes Ther 6(3):239–256

Umapathysivam MM, Lee MY, Jones KL, Annink CE, Cousins CE, Trahair LG et al (2014) Comparative effects of prolonged and intermittent stimulation of the glucagon-like peptide 1 receptor on gastric emptying and glycemia. Diabetes 63(2):785–790

Urva S, Nauck MA, Coskun T, Cui X, Haupt A, Benson C et al (2019) 58-OR: the novel dual GIP and GLP-1 receptor agonist tirzepatide transiently delays gastric emptying similarly to a selective long-acting GLP-1 receptor agonist. Diabetes 68(Supplement 1):58-OR

van Can J, Sloth B, Jensen CB, Flint A, Blaak EE, Saris WHM (2014) Effects of the once-daily GLP-1 analog liraglutide on gastric emptying, glycemic parameters, appetite and energy metabolism in obese, non-diabetic adults. Int J Obes 38(6):784–793

Vella A, Bock G, Giesler PD, Burton DB, Serra DB, Saylan ML et al (2007) Effects of dipeptidyl peptidase-4 inhibition on gastrointestinal function, meal appearance, and glucose metabolism in type 2 diabetes. Diabetes 56(5):1475–1480

Vella A, Bock G, Giesler PD, Burton DB, Serra DB, Saylan ML et al (2008) The effect of dipeptidyl peptidase-4 inhibition on gastric volume, satiation and enteroendocrine secretion in type 2 diabetes: a double-blind, placebo-controlled crossover study. Clin Endocrinol 69(5):737–744

Verdich C, Toubro S, Buemann B, Lysgård Madsen J, Juul Holst J, Astrup A (2001) The role of postprandial releases of insulin and incretin hormones in meal-induced satiety--effect of obesity and weight reduction. Int J Obes Relat Metab Disord 25(8):1206–1214

Vilsbøll T, Agersø H, Lauritsen T, Deacon CF, Aaboe K, Madsbad S et al (2006) The elimination rates of intact GIP as well as its primary metabolite, GIP 3-42, are similar in type 2 diabetic patients and healthy subjects. Regul Pept 137(3):168–172

Vrang N, Larsen PJ (2010) Preproglucagon derived peptides GLP-1, GLP-2 and oxyntomodulin in the CNS: role of peripherally secreted and centrally produced peptides. Prog Neurobiol 92(3):442–462

Vrang N, Phifer CB, Corkern MM, Berthoud H-R (2003) Gastric distension induces c-Fos in medullary GLP-1/2-containing neurons. Am J Physiol Regul Integr Comp Physiol 285(2):R470–R4R8

Wettergren A, Wøjdemann M, Holst JJ (1998) Glucagon-like peptide-1 inhibits gastropancreatic function by inhibiting central parasympathetic outflow. Am J Phys 275(5):G984–GG92

Willms B, Werner J, Holst JJ, Orskov C, Creutzfeldt W, Nauck MA (1996) Gastric emptying, glucose responses, and insulin secretion after a liquid test meal: effects of exogenous glucagon-like peptide-1 (GLP-1)-(7-36) amide in type 2 (noninsulin-dependent) diabetic patients. J Clin Endocrinol Metab 81(1):327–332

Zinman B, Hoogwerf BJ, Durán García S, Milton DR, Giaconia JM, Kim DD et al (2007) The effect of adding exenatide to a thiazolidinedione in suboptimally controlled type 2 diabetes: a randomized trial. Ann Intern Med 146(7):477–485

Adv Exp Med Biol - Advances in Internal Medicine (2020) 4: 193–212
https://doi.org/10.1007/5584_2020_494
© Springer Nature Switzerland AG 2020
Published online: 8 February 2020

GLP-1 Receptor Agonists and SGLT2 Inhibitors for the Treatment of Type 2 Diabetes: New Insights and Opportunities for Cardiovascular Protection

Laura Bertoccini and Marco Giorgio Baroni

Abstract

The risk of cardiovascular disease (CVD) (myocardial infarction, stroke, peripheral vascular disease) is twice in type 2 diabetes (T2D) patients compared to non-diabetic subjects. Furthermore, cardiovascular disease (CV) is the leading cause of death in patients with T2D.

In the last years several clinical intervention studies with new anti-hyperglycaemic drugs have been published, and they have shown a positive effect on the reduction of mortality and cardiovascular risk in T2D patients. In particular, these studies evaluated sodium/glucose-2 cotransporter inhibitors (SGLT2i) and Glucagon-like peptide-1 receptor agonists (GLP-1RA).

In secondary prevention, it was clearly demonstrated that SGLT2i and GLP-1RA drugs reduce CV events and mortality, and new guidelines consider now these drugs as first choice (after metformin) in the treatment of T2D; there are also some signs that they may be effective also in primary prevention of CVD. However, the mechanisms involved in cardiovascular protection are not yet fully understood, but they appear to be both "glycaemic" and "extra-glycaemic".

In this review, we will examine the fundamental results of the clinical trials on SGLT2i and GLP-1RA, their clinical relevance in term of treatment of T2D, and we will discuss the mechanisms that may explain how these drugs exert their cardiovascular protective effects.

Keywords

Cardiovascular disease (CVD) · CVD outcome trials · Heart failure · Ketogenesis · MACE · Primary prevention · Real-world trials · Sodium/Hydrogen Exchanger (NHE) · Tubuloglomerular feedback

L. Bertoccini
Department of Experimental Medicine, Sapienza University of Rome, Rome, Italy

M. G. Baroni (✉)
Department of Experimental Medicine, Sapienza University of Rome, Rome, Italy

IRCCS Neuromed-Pozzilli (IS), Pozzilli, Italy
e-mail: marco.baroni@uniroma1.it

1 Cardiovascular Risk in Type 2 Diabetes

People with type 2 diabetes (T2D) have a risk of cardiovascular disease (CVD) (myocardial infarction, stroke, peripheral vascular disease) that is two or more times higher than non-diabetic subjects, and cardiovascular disease (CV) is the leading cause of death in patients with T2D (Morrish et al. 2001).

Several trials (Action to Control Cardiovascular Risk in Diabetes Study Group et al. 2008; ADVANCE et al. 2008; Duckworth et al. 2009; UK Prospective Diabetes Study (UKPDS) Group 1998) have shown that the lowering of HbA1c in patients with T2D has only a modest (Action to Control Cardiovascular Risk in Diabetes Study Group et al. 2008; ADVANCE et al. 2008) or no effect (Duckworth et al. 2009; UK Prospective Diabetes Study (UKPDS) Group 1998) on the reduction of cardiovascular risk. In contrast, the correction of traditional CVD risk factors, such as blood pressure and cholesterol levels, reduces the risk of CVD and mortality in patients with T2D (Gaede et al. 2008; Holman et al. 2008).

Indeed, hyperglycaemia has a low impact as risk factor for CVD (UK Prospective Diabetes Study (UKPDS) Group 1998; Holman et al. 2008) and large intervention trials (Action to Control Cardiovascular Risk in Diabetes Study Group et al. 2008; ADVANCE et al. 2008; Duckworth et al. 2009) aimed at the intensive treatment of blood glucose have failed to significantly reduce (and some times even worsened) CV risk and mortality (Fig. 1a), especially in secondary prevention studies.

Furthermore, in primary prevention studies, such as the UKPDS (UK Prospective Diabetes Study (UKPDS) Group 1998) and the Veterans Affairs Diabetes Trial (VADT) (Hayward et al. 2015), a CV benefit associated with improved

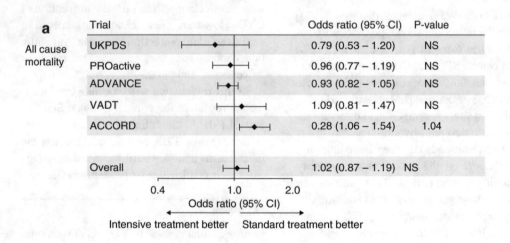

Fig. 1 Intensive vs. Standard Treatment on Blood Glucose Reduction. (**a**) Odds ratios on all cause mortality. (**b**) Effects on MACE (Mortality, non-fatal MI, non-fatal stroke) of Pioglitazone, DPP4-inhibitors and Insulins

glycaemic control was observed after more than 10 years, the so called "legacy effect".

Also, in the last decade, trials assessing the safety of DPP-4 inhibitors and new generation insulins have reported a lack of superiority for major cardiovascular events compared with placebo (White et al. 2013; Green et al. 2015; Scirica et al. 2013; Rosenstock et al. 2019) as well as an increase in the risk of hospitalization for heart failure (Fig. 1b).

Since 2008 The US Food and Drug Administration (FDA) requires that all new glucose-lowering agents must undergo post-marketing endpoint trials with the aim of verifying cardiovascular safety and mortality (Guidance for Industry on Diabetes Mellitus 2008), following the negative experience observed in the study of rosiglitazone (Nissen 2010). All new drugs must therefore be studied in populations at high cardiovascular risk, to demonstrate the safety of the drug.

Amongst all the trials that were started after FDA's recommendations, only the Proactive study with pioglitazone (Dormandy et al. 2005) showed a positive effect, although only on a secondary end-point (Fig. 1b), which was a 3-point MACE (Major Cardiovascular Events) composed by all-cause mortality, myocardial infarction, or stroke. This trial involved more than 5000 patients with T2D who had evidence of macrovascular disease. The primary end-point was not significantly different between patients receiving pioglitazone and those who received placebo, mostly because it included too many outcomes (death from any cause, non-fatal myocardial infarction, including silent myocardial infarction, stroke, acute coronary syndrome, leg amputation, coronary revascularisation, or revascularisation of the leg). Thus, if the primary end-point was designed as in recent trials, in particularly not including peripheral arterial disease and leg amputation, the PROACTIVE trial would have been considered significant, demonstrating a protective effect of pioglitazone on CVD risk.

On the other side, a multifactorial intervention with the aim of reducing CV risk factors has been shown to be effective in reducing CV events and mortality in diabetes (Gaede et al. 2003). The probable explanation is that insulin-resistance is a shared determinant of many of the CV risk factors, such as hypertension, dyslipidaemia and abdominal adiposity. Therefore, anti-diabetic drugs that have an effect only (or mostly) on reducing blood glucose, such as sulfonylureas, insulin and DPP4-i, without effects on the other risk factors, are not able to reduce CV risk. For this reason it is possible to explain the partial results of pioglitazone (Dormandy et al. 2005), that, by acting on insulin resistance and its components (such as hypertension and dyslipidaemia, in addition to glycaemic control), was able to reduce the number of events in treated patients.

In the last years, several clinical trials with new classes of anti- hyperglycaemic drugs have been published, showing a positive and significant effect on mortality and cardiovascular risk in diabetes that surprised the scientific community.

These studies focused in particular on the effects on CV risk of GLP-1 receptor agonists (GLP-1-RA) and sodium-glucose co-transporter-2 inhibitors (SGLT2i) drugs.

2 SGLT2 Inhibitors

2.1 Mechanisms of Action of SGLT-2 Inhibitors

SGLT2 is a sodium-glucose cotransporter that is highly expressed in the first part of the proximal renal tubule (S1 segment). SGLT2 receptors are responsible for approximately 90% of renal glucose reabsorption in the renal tubules (Perreault 2017), whereas residual glucose is reabsorbed by SGLT1 receptors situated in the more distal part of the proximal renal tubule (S3 segment). The result is the absence of glycosuria and the recovery of all the glucose filtered by the glomerulus (Fig. 2).

In T2D patients, SGLT2 reabsorption capacity is amplified compared to healthy subjects, and this amplification, associated with an increased concentration of systemic glucose, results in the upholding of hyperglycaemia and glucotoxicity,

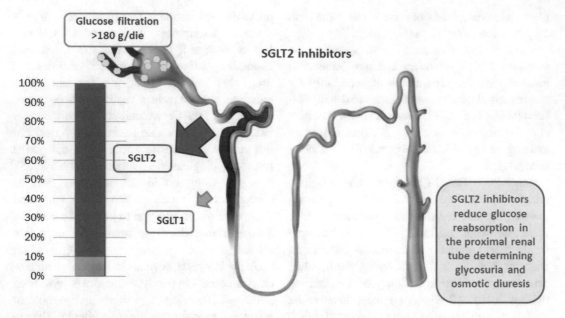

Fig. 2 Actions of sodium/glucose transporters

leading to beta-cell dysfunction (Rahmoune et al. 2005; Vallon 2015).

Based on these mechanisms it was hypothesised that, by inhibiting SGLT2 co-transporters, it is possible to "force" glucose into the urine, thus reducing plasma glucose concentrations. Indeed, SGLT2 inhibitors decrease the maximum glucose reabsorption rate and lower the threshold for glycosuria, two actions which cause increased glucose excretion according to rising plasma glucose concentration (DeFronzo et al. 2012; Ferrannini and Solini 2012). Importantly, SGLT2 inhibitors exert their glucose-lowering effects by increasing urinary glucose excretion in an insulin-independent manner, leading to reduced glycaemia and body weight without risk of hypoglycaemia.

2.2   Cardiovascular Outcome Trials

Several major trials exploring the CV benefits of SGLT2 inhibitors, with an emphasis on HF outcome, have been conducted in recent years and are discussed below (Table 1).

The first published study is The Empagliflozin Cardiovascular Outcome Event Trial in Type 2 Diabetes Mellitus Patients Removing Excess Glucose (EMPA-REG OUTCOME) (Zinman et al. 2015). This trial enrolled 7020 participants with T2D and established CVD, who were then randomly assigned to empagliflozin 10 or 25 mg or placebo. During a median follow-up of 3.1 years, empagliflozin demonstrated significant risk reduction in 3-point MACE (cardiovascular death, nonfatal myocardial infarction [MI] and nonfatal stroke) (HR 0.86, 95% CI 0.74–0.99, p = 0.04) compared to placebo. The primary outcome was largely driven by a 38% reduction in cardiovascular death (HR 0.62, 95% CI 0.49–0.77, p < 0.001) and there was a non-significant decrease in nonfatal MI (HR 0.87, 95% CI 0.70–1.09, p = 0.22). The occurrence of nonfatal stroke was higher although the difference was not significant (HR 1.24, 95% CI 0.92–1.67, p = 0.16). There was also a reduction of 38% in the incidence of hospitalization for heart failure and of 32% in death from all causes in the empagliflozin arm. Furthermore, the effect of SGLT2i was observed already in the first 3 months of treatment, suggesting a mechanism not exclusively linked to glycaemic reduction. Indeed, Hba1c showed a difference of only −0.5% between the two arms, a difference that

Table 1 Principal results of cardiovascular outcome trials with SGLT2 inhibitors

Trial	Drug	N. pazients	MACE hazard ratio (95% CI)	CV death hazard Ratio (95% CI)	All cause mortality hazard ratio (95% CI)
EMPA-REG	empaglifozin	7.020	0.86 (0.74-0.99) ↓	0.62 (0.49-0.77) ↓	0.68 (0.57-0.82) ↓
CANVAS	canaglifozin	10.141	0.86 (0.75-0.97) ↓	0.87 (0.72-1.06) ↓	0.87 (0.74-1.01) ↓
CREDENCE	canaglifozin	4.401	0.80 (0.67-0.95) ↓	0.78 (0.61-1.00) ↓	0.83 (0.68-1.02) ↓
DECLARE-TIMI 58	dapaglifozin	17.160	0.93 (0.84-1.03) →	0.83 (0.73-0.95) ↓	0.93 (0.82-1.04) →
DAPA-HF	dapaglifozin	4.744	0.74 (0.65-0.85) ↓	0.82 (0.69-0.98) ↓	0.83 (0.71-0.97) ↓

Horizontal arrows indicate neutral effects; downward arrows indicate positive (protective) effects on specific end-points

was observed only after 8–12 months from the beginning of treatment, and insufficient to demonstrate causality (Table 1).

The beneficial outcome for cardiovascular risk observed in the EMPA-REG OUTCOME trial was confirmed in the Canagliflozin Cardiovascular Assessment Study (CANVAS) Program (Neal et al. 2017). The CANVAS Program enrolled 10.142 participants with T2D and high cardiovascular risk. A significant cardiovascular protection, with a significant reduction in the relative risk of the primary cardiovascular endpoint by 14% (HR 0.86, 95% CI 0.75–0.97, p = 0.02 for superiority) and a significant 33% reduction in the secondary endpoint of hospitalization for heart failure (HR 0.67, 95% CI 0.52–0.87) was observed in the CANVAS Program, whereas neither the risk of MI nor that of stroke was significantly reduced. Canaglifozin did not demonstrate a significant reduction in cardiovascular and all-cause mortality, in contrast to empagliflozin in the EMPA-REG OUTCOME trial (Table 1).

Both in EMPAREG and in CANVAS trials, the effect on nephroprotection was also significant, assessed as a reduced progression of albuminuria or maintenance of eGFR. The renal outcome with Canagliflozin in patients with impaired kidney function was confirmed in the recently reported Canagliflozin and Renal Endpoints in Diabetes with Established Nephropathy Clinical Evaluation (CREDENCE) trial (Perkovic et al. 2019). In this trial 4401 T2D patients with chronic kidney disease (CKD), defined as an eGFR ranging from 30 to 90 ml/min/ 1.73 m^2 and macroalbuminuria (urine albumin [mg] to creatinine [g] ratio; UACR > 300–5000) have been enrolled. All 4401

participants who underwent randomization were required to have background use of an angiotensin-converting enzyme inhibitor or angiotensin-receptor blocker (ARB) for at least 4 weeks. Canagliflozin significantly reduced the relative risk of the primary composite outcome comprising end-stage renal disease (ESRD), doubling of serum creatinine, or death from renal or CVD by 30%, compared with the placebo group (HR 0.70, 95% CI 0.59–0.82, p = 0.00001) (Perkovic et al. 2019) (Table 1).

Most recently the results of the DECLARE-TIMI 58 (Dapagliflozin Effect on Cardiovascular Events–Thrombolysis in Myocardial Infarction 58) trial were published. The DECLARE trial enrolled 17.160 participants, and included a cohort of subjects in primary prevention (59.4% of the participants) (Wiviott et al. 2019). In DECLARE, dapagliflozin did not show a significant risk reduction of MACE or cardiovascular death. However, consistent with the results of the previous two trials with empagliflozin and canagliflozin mentioned above, hospitalization for heart failure was significantly reduced (HR 0.73, 95% CI 0.61–0.88), and improvement of renal outcomes was also observed in the DECLARE-TIMI 58 trial (HR 0.53, 95% CI 0.43–0.66) (Ferrannini and Solini 2012). The neutral effects on MACE observed in the DECLARE trial may reflect the high proportion of patients without established CVD at baseline, indicating that the impact of an SGLT2 inhibitor on primary prevention might be marginal. In contrast, an apparent beneficial effect on the reduction of cardiovascular risk in established CVD was suggested by a recent meta-analysis (Zelniker et al. 2019) (Table 1).

Very recently, the DAPA-HF (Dapagliflozin and Prevention of Adverse Outcomes in Heart Failure) trial showed that dapagliflozin was superior to placebo in preventing cardiovascular death and heart failure even in subjects without diabetes (McMurray et al. 2019). In this trial 4744 patients with New York Heart Association (NYHA) class II, III, or IV heart failure and an ejection fraction of 40% or less were randomly assigned to receive either dapagliflozin (at a dose of 10 mg once daily) or placebo, in addition to recommended therapy. The primary outcome was a composite of worsening heart failure (hospitalization or an urgent visit resulting in intravenous therapy for heart failure) or cardiovascular death. Among patients with heart failure and a reduced ejection fraction, the risk of worsening heart failure or death from cardiovascular causes was lower among those who received dapagliflozin than among those who received placebo (HR 0.74, 95% CI 0.65–0.85;P < 0.001) (McMurray et al. 2019). It is worth highlighting that the magnitude of benefit was similar regardless of the presence or absence of diabetes (only 41% of the patients had T2D), providing support for prior suggestions that such treatment has beneficial actions other than glucose lowering (Packer et al. 2017; Verma and McMurray 2018) (Table 1).

The results of these randomized clinical trials have also been confirmed by "real world" data that demonstrated the positive effects of SGLT2 on CV risk reduction in diabetes.

The Comparative Effectiveness of Cardiovascular Outcomes in New Users of Sodium- Glucose Cotransporter-2 Inhibitors (CVD-REAL 2) study has involved 235.064 patients across six countries of USA and Europe, without previous CV events or high CV risk. The protective effect of SGLT2 inhibitors was confirmed in this cohort. Treatment with SGLT2 inhibitors (Dapaglifozin, Empaglifozin and Canaglifozin) was associated with significant risk reduction in hospitalization for heart failure and all-cause death (HR 0.64, 95% CI 0.50–0.82, p = 0.001 and HR 0.51 95% CI 0.37–0.70, p < 0.001, respectively), with a directionally similar trend regardless of the existence of prior CVD (Kosiborod et al. 2018).

The EASEL study, another "real world" study that compared SGLT2 inhibitors with standard therapies in patients with T2D, added its results to these data, showing a 43% reduction in heart failure and mortality. In this study all three SGLT2 inhibitors (Dapagliflozin, Empagliflozin and Canagliflozin) have been used, suggesting that the observed effects are likely to be class-specific, and not exclusive to the individual SGLT2 inhibitor (Udell et al. 2018).

All these studies consistently demonstrated that SGLT2 inhibitors modify the cardiovascular risk (especially hospitalization for heart failure and all-cause death) in patients with T2D, and possibly also in non-diabetic subjects; this protection is certain in secondary prevention, but perhaps, at least in the "real world" data, even in primary prevention. The open question therefore concerns the mechanisms involved in cardiovascular protection, given that the reduction of glycaemia does not seem to be the principal mechanism of SGLT2i to obtain CV risk reduction.

2.3 Cardiovascular Protection Mechanisms of SGLT2 Inhibitors

There are several hypothesized mechanisms that can lead to the reduction of cardiovascular risk in subjects treated with SGLT2 inhibitors. The supposed mechanisms may be metabolic and/or cardio-hemodynamic (Table 1).

2.3.1 Metabolic Effects of SGLT2i

Among the metabolic non-glycaemic effects of SGLT2i (Table 2), the effects on lipolysis are particularly noteworthy. Indeed, SGLT2 inhibitors promote a shift to fatty substrate utilization in response to the decreased plasma glucose caused by glycosuria, leading to enhanced fatty oxidation, lipolysis and ketogenesis.

Lower plasma glucose subsequently stimulates glucagon secretion, and suppression of plasma insulin level may partially contribute to this mechanism (Ferrannini et al. 2016a), while the direct effect of SGLT2 inhibitors on alpha

Table 2 Mechanisms of cardiovascular protection of SGLT-2 inhibitors

Metabolic effects	Cardiac effects
Adipose tissue: increase of fatty acid (FFA) mobilization and reduction of adipocyte inflammation	Improvement of ventricular preload (secondary to natriuresis and osmotic diuresis) and post-load through blood pressure reduction
Weight loss (about 3-5 kg)	
Pancreas: increase of glucagon secretion which leads to increased lipolysisand FFA uptake in the liver	Improvement of cardiac metabolism
Liver: increase of glycogenolysis, gluconeogenesis, FFA uptake and ketogenesis	Inhibition of sodium-hydrogen exchange in the myocardium
Heart: Substrate exchange with increased FFA and ketones use ("superfuel") and reduction of glucose use. Epicardial fat decrease	Reduction of cardiac fibrosis and necrosis

cells is not yet clear (Bonner et al. 2015; Kuhre et al. 2019). The presence of the SGLT2 co-transporters has been shown on alpha cells, whose inhibition in vitro leads to an increase of glucagon release. The increased glucagon, whatever the mechanism, causes lipolysis and augmented use of ketone bodies. The rise in lipolysis leads also to an increase in energy expenditure and consumption of adipose reserves, which is associated with weight loss (on average 3–5 kg). Furthermore, the increase in hepatic ketogenesis promotes the consumption of fatty acids, with positive effects on hepatic steatosis and inflammation (Table 2).

With regards to heart metabolism, Ferranini et al. hypothesized that under conditions of mild but persistent hyperketonemia, such as during treatment with SGLT2 inhibitors, β-hydroxybutyrate is freely taken up by the heart and oxidized in preference to fatty acids. This substrate selection improves the transduction of oxygen consumption into work efficiency at the mitochondrial level (β-hydroxybutyrate as a sort of "superfuel") (Ferrannini et al. 2016b), given the fact that b-hydroxybutyrate oxidation compares favorably with the oxidation of glucose and pyruvate, and the oxygen cost of the energy output is 27% decreased.

In T2D patient, throughout a variety of cardiac injuries (ischemic, myopathic and reperfusion) (Chouchani et al. 2014) there is a common effect that converges on insufficient mitochondrial energy output and contractile failure as the basic mechanism. Thus, in these conditions of cardiac injuries, whole-body and myocardial insulin-mediated glucose utilization are impaired, and a larger than normal proportion of energy is derived from the oxidation of fatty substrates. FFAs require 8% more oxygen than glucose to produce the same number of calories (Lopaschuk et al. 2010). The increased β-hydroxybutyrate induced by SGLT2i can replace glucose and FFAs as a fuel, increasing external cardiac work at the same time as reducing oxygen consumption, thereby improving cardiac efficiency by 24%. In this context, β-hydroxybutyrate can be viewed as a constitutive mitochondrial helper (Kolwicz Jr et al. 2016).

2.3.2 Cardio-Hemodynamic Effects of SGLT2i

The cardio-hemodynamic effects of SGLT2 inhibitors (Table 2) can also explain the positive effects on cardiovascular risk, in particular on heart failure. Indeed, the results of clinical trials showed that many of the positive effects were determined by the reduction of heart failure (−38%), a frequent complication of T2D (Einarson et al. 2018).

The cardio-hemodynamic direct effects of SGLT2 inhibitors (Table 2) include improvement in ventricular preload (secondary to natriuresis and osmotic diuresis) and afterload through pressure reduction, improvement of cardiac metabolism, inhibition of sodium-hydrogen exchange, reduction of cardiac fibrosis and modification of adipokines and cytokines levels and epicardial fat accumulation.

All these mechanisms are particularly attractive. For example, the activation of the tubuloglomerular feedback, mediated by the increase in sodium to the macula dense due to the inhibition of sodium/glucose co-transport, followed by the reduction of the activation of the renin-angiotensin system, seems to determine a contraction of volume and consequently a reduction of cardiac edema (Hallow et al. 2018). In particular, it appears that SGLT2 inhibitors have a more pronounced effect on interstitial fluid volume (Hallow et al. 2018), that is reduced more than intravascular volume. A differential effect in regulating interstitial fluid may be particularly important in patients with heart failure. The ability to selectively reduce interstitial fluid may be a unique feature of SGLT2 inhibitors vs other diuretics and they may more effectively relieve signs and symptoms of interstitial congestion, providing some relief of elevated cardiac filling pressures, without the deleterious effects of excessive blood volume depletion, including neurohumoral activation, that occurs in response to traditional diuretics.

The reduction of blood sodium concentrations and volume has positive effects on blood pressure, which shows a mean reduction of 4 mmHg in all clinical trials (Briasoulis et al. 2018). All these effects appear very early, and might help to explain the rapidity of the effect on CV risk observed in clinical trials (Vallon 2015; DeFronzo et al. 2012).

Regarding the modulation of sodium/hydrogen transport in cardiomyocytes, it is known that diabetes-associated heart failure is characterized by an increase in myocardial expression of NHEs (Na+/H+ exchangers). This increase causes high sodium and calcium concentrations in the myocardium, which may contribute to cardiac dysfunction. Recent data demonstrated an affinity between SGLT2 inhibitors and the sodium-binding site of NHE-1, with a consequent block in sodium transport (Uthman et al. 2018) as a mechanism of cardioprotection. As a proof of the potential positive effects of NHE inhibition, in the past it was demonstrated that Cariporide (Mentzer Jr et al. 2008), a selective inhibitor of NHE-1, reduced the risk of myocardial infarction in patients undergoing cardiac bypass, although an increase of deaths by cerebral bleeding associated with Cariporide caused the withdrawal of the drug.

The effects of SGLT2 inhibitors on the adipocytokine profile can be related to the effects on lipolysis discussed above. A reduction in epicardial fat (a known CV risk factor) was observed in animal models treated with SGLT2i (Sato et al. 2018), and an increase in adiponectin and a reduction in TNF-alpha were observed in patients treated with SGLT2i, with an improvement in insulin sensitivity and reduction of inflammation (Garvey et al. 2018).

In conclusion, SGLT2 inhibitors have cardiovascular and renal effects that are independent from glucose reduction, and may explain the reduction in cardiovascular risk observed in patients with T2D treated with SGLT2i. The cardio-protective benefits of SGLT2 inhibitors were observed very early in the cardiovascular outcome trials, and also the reduction in the risk of hospitalizations for HF occurred early after initiation of therapy (Fig. 3). These effects are consistent across all SGLT2 inhibitors, suggesting a class effect.

The observations of putative non-glycaemic effects of SGLT2i have led to a series of targeted trials to address more specific outcomes in HF, such as mortality benefit, improvement in left ventricular remodelling, diastolic function, right ventricular function, WHO functional class and 6-min walk test, even in patients without T2D. Among these, the DAPA-HF trial (McMurray et al. 2019) is the first to demonstrate a significant effect in patients with HF, independently form the presence of diabetes. Furthermore, mechanistic studies on non-alcoholic hepatic steatosis (NAFLD), hypertension and peripheral vascular pathologies are in progress to clarify all aspects concerning SGLT2 inhibitors.

Fig. 3 Cardiovascular protection by SGLT2 inhibitors

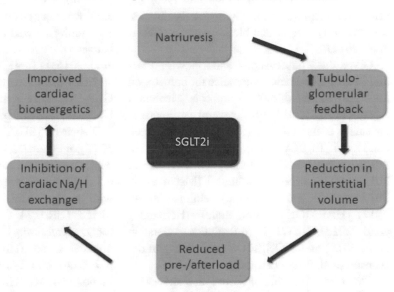

3 GLP-1 Receptor Agonists (GLP-1RA)

3.1 Mechanisms of Action of GLP-1RA

Recent clinical intervention studies have also demonstrated a positive effect of GLP-1 receptor agonists on mortality and cardiovascular risk in diabetes.

Glucagon-like peptide-1 is synthesized in intestinal L-cells, situated in the distal ileum, where it is secreted in response to nutrient intake (Anandhakrishnan and Korbonits 2016), and it is quickly inactivated within 2–3 min by the enzyme dipeptydil peptidase-4 (DPP-4) (Meier 2012). Following administration of a GLP-1RA, postprandial insulin secretion is stimulated and glucagon secretion is reduced in a glucose-dependent manner, thereby delaying gastric emptying, which in turn induces satiety with consequent reduction of postprandial hyperglycaemia. Furthermore, GLP-1RA suppresses directly and indirectly endogenous glucose production and increases sensitivity to insulin due to weight loss.

In T2D patients, a reduction of GLP-1 production in response to food intake has been determined, and several clinical trials with GLP-1RA have demonstrated the efficacy in glycaemic control, usually associated with a significant weight loss.

3.2 Cardiovascular Outcome Trials

The currently available formulations are administered by subcutaneous injection and are classified in short-acting (lixisenatide, exenatide short-acting), intermediate-acting (liraglutide) and with a long duration of action (exenatide long-acting, dulaglutide, semaglutide and albiglutide) agonists. Very recently, an oral formulation of semaglutide has entered commercialization. They are extensively used in the treatment of T2D as they lower blood sugar with weight reduction, and present a very low risk of hypoglycaemia (Nauck 2016). Furthermore, treatment with GLP-1RA drugs has favorable effects on well-known cardiovascular risk factors, such as body weight and blood pressure (Nauck et al. 2017).

Given these observations, and following FDA and EMA requirements (Guidance for Industry on Diabetes Mellitus 2008; European Medicines Agency (EMA) 2012, 2016) several cardiovascular safety trials with GLP-1RA drugs have been conducted: ELIXA (lixisenatide) (Pfeffer et al. 2015), LEADER (liraglutide) (Marso et al. 2016a), SUSTAIN (semaglutide) (Marso et al. 2016b), EXSCEL (exenatide) (Holman et al. 2017) HARMONY (albiglutide) (Hernandez et al. 2018), REWIND (dulaglutide) (Gerstein et al. 2019) and PIONEER 6 (oral semaglutide) (Husain et al. 2019) (Table 3).

The only trial that involved a short-acting GLP1-RA, the Evaluation of Lixisenatide in Acute Coronary Syndrome (ELIXA) trial, enrolled 6068 patients who had experienced acute coronary syndrome within the preceding 180 days, and demonstrated full safety of lixisenatide but failed to obtain a significant reduction in a 4-point MACE composite outcome (CV death, nonfatal MI, nonfatal stroke, or hospitalization for unstable angina) (HR 1.02, 95% CI 0.89–1.17, p = n.s.), any component of MACE, or hospitalization for heart failure over a median follow-up period of 2.1 years (Pfeffer et al. 2015) (Table 3).

The LEADER trial, in which 9340 patients, including 72.4% with established atherosclerotic CVD (ASCVD) were randomly assigned to the intermediate–acting GLP-1RA liraglutide or placebo, demonstrated a 13% reduction in the 3-point MACE composite outcome (HR 0.87, 95% CI 0.78–0.97, P < 0.001 for non-inferiority; P = 0.01 for superiority) with liraglutide versus placebo during a median follow-up period of 3.8 years. Each component of the primary composite outcome showed a directionally similar trend toward a reduction in 3-point MACE, in which cardiovascular death reached statistical significance (HR 0.78, 95% CI 0.66–0.93, p = 0.007). Patients with liraglutide showed a significant reduction in

all-cause mortality (HR 0.85, 95% CI 0.74–0.97, p = 0.02), predominantly driven by a reduction in cardiovascular death. The prespecified analysis at 36 months showed a mean difference in HbA1c between the two treatment groups of 0.4% (Marso et al. 2016a) (Table 3).

Next, trials involving the study of long-acting GLP-1RA were published. In the Semaglutide Unabated Sustainability in Treatment of Type 2 Diabetes 6 (SUSTAIN-6) trial, which enrolled 3297 patients with T2D and previous cardiovascular disease, the administration of semaglutide 0.5 or 1.0 mg per week or placebo was associated with a significant 26% reduction in 3-point MACE (HR 0.74, 95% CI 0.58–0.95, p < 0.001 for non-inferiority), a significant reduction in nonfatal stroke (HR 0.61, 95% CI 0.38–0.99, p = 0.04), and a directionally concordant result in nonfatal MI (HR 0.74, 95% CI 0.51–1.08, p = 0.12). This trial was powered as a non-inferiority study to exclude a preapproval safety margin of 1.8 set by the Food and Drug Administration. The median follow-up period was 2.1 years and the difference of HbA1c compared to placebo group was −0.7% for the group treated with semaglutide 0.5 mg and − 1.0% for semaglutide 1 mg group (Marso et al. 2016b) (Table 3). The Exenatide Study of Cardiovascular Event Lowering (EXSCEL) trial, involving 14,752 subjects (of whom 10,782 [73.1%] had previous cardiovascular disease), showed that in patients treated with once-weekly exenatide the 3-point MACE was lower compared with placebo, although not significant (HR 0.91 95% CI 0.83–1.00). Once-weekly exenatide was also associated with 14% reduction of risk of all cause mortality (HR 0.86, 95% CI 0.77–0.97) compared with placebo, accompanied by a directionally consistent trend in cardiovascular mortality (HR 0.88, 95% CI 0.76–1.02). The median follow-up period was 3.2 years, and the difference in HbA1c between the two treatment groups was 0.53% (Holman et al. 2017) (Table 3).

The Albiglutide and cardiovascular outcomes in patients with T2D and cardiovascular disease (HARMONY OUTCOMES) trial enrolled 9463 patients that were evaluated for a median duration of 1,5 years. Participants with T2D and previous

Table 3 Principal results of cardiovascular outcome trials with GLP1-RAs

Trial	Drug	N. pazients	MACE hazard ratio (95% CI)	CV death hazard ratio (95% CI)	All cause mortality hazard ratio (95% CI)
ELIXA	lixisenctide	6.068	1.02 (0.89–1.17) →	0.98 (0.78–1.22) →	0.94 (0.78–1.13) →
LEADER	liraglut de	9.340	0.87 (0.78–0.97) ↓	0.78 (0.66–0.93) ↓	0.85 (0.74–0.97) ↓
SUSTAIN	semaglutide	3.297	0.74 (0.58–0.95) ↓	0.98 (0.65–1.48) →	1.05 (0.74–1.50) →
EXSCEL	exenatide	14.752	0.91 (0.83–1.00) →	0.88 (0.76–1.02) →	0.86 (0.77–0.97) ↓
HARMONY	albiglutide	9.463	0.78 (0.68–0.90) →	0.93 (0.73–1.19) →	0.95 (0.79–1.16) →
REWIND	dulaglutide	9.901	0.88 (0.79–0.99) ↓	0.91 (0.78–1.06) →	0.90 (0.80–1.01) →
PIONEER	semaglutide (oral)	3.183	0.79 (0.57–1.11) ↓	0.49 (0.27–0.92) ↓	0.51 (0.31–0.84) ↓

Horizontal arrows indicate neutral effects; downward arrows indicate positive (protective) effects on specific end-points

cardiovascular disease were enrolled and randomly assigned to 2 groups: 4731 patients were assigned to receive albiglutide and 4732 patients to placebo. It was demonstrated a significant 22% risk reduction in major adverse cardiovascular events. The primary composite outcome occurred in 338 (7%) of 4731 patients in the albiglutide group and in 428 (9%) of 4732 patients in the placebo group (HR 0.78, 95% CI 0.68–0.90), which indicated that albiglutide was superior to placebo (p < 0.0001 for non-inferiority; p = 0.0006 for superiority) (Hernandez et al. 2018) (Table 3).

In the Researching Cardiovascular Events with a Weekly Incretin in Diabetes (REWIND) trial 9901 participants were enrolled and randomly assigned to receive dulaglutide (n = 4949) or placebo (n = 4952). During a median follow-up of 5.4 years the primary composite outcome occurred in 594 (12%) participants in the dulaglutide group and in 663 (13.4%) participants in the placebo group (HR 0.88, 95% CI 0.79–0.99; p = 0.026). All-cause mortality did not differ between groups (HR 0.90, 95% CI 0.80–1.01; p = 0.067). At follow-up, participants assigned to treatment with dulaglutide showed lower levels of HbA1c, body weight, BMI, blood pressure and total cholesterol and LDL (Gerstein et al. 2019) (Table 3).

Finally, the Peptide Innovation for Early Diabetes Treatment (PIONEER 6) trial, which employed an oral formulation of semaglutide and involved 3183 patients (of whom 84.7% had previous cardiovascular disease or chronic kidney disease), showed that major adverse cardiovascular events occurred in 61 of 1591 patients (3.8%) in the oral semaglutide group and 76 of 1592 (4.8%) in the placebo group (HR 0.79, 95% CI 0.57–1.11; p < 0.001 for non-inferiority). Death from any cause occurred in 23 of 1591 patients (1.4%) in the oral semaglutide group and 45 of 1592 (2.8%) in the placebo group (HR 0.51, 95% CI 0.31–0.84) (Husain et al. 2019) (Table 3).

It has also been observed, in all these trials, that treatment with a GLP-1 RA induces a significant weight loss (Vilsbøll et al. 2012) and a significant reduction in blood pressure (Marso et al. 2016b). Furthermore, data from the studies

of cardiovascular outcomes LEADER (liraglutide) and SUSTAIN (semaglutide) showed that GLP-1RA are useful in preventing the onset and progression of diabetic nephropathy, although remains unclear if this effect is mediated primarily by an improvement of glycaemic control (Marso et al. 2016a, b).

An 18-month "real-life" study also showed that treatment with liraglutide in addition to metformin causes an improvement in several cardiometabolic risk factors (Rizzo et al. 2016). Indeed, the study showed a significant reduction in carotid medial-intimal thickness values in patients with metabolic syndrome treated with liraglutide. As carotid medial-intimal thickness is a surrogate marker of early and subclinical atherosclerosis, its reduction after treatment with liraglutide appears consistent with the emerging cardiovascular effects of this drug (Rizzo et al. 2016).

Overall, these data indicate that GLP-1RA drugs have protective cardiovascular effects and positive effects on other cardiovascular risk factors, such as body weight and blood pressure.

Despite clearly demonstrating cardiovascular protection from these studies, the mechanisms by which GLP-1RA determine a reduction in MACE remain to be fully elucidated (Drucker 2018).

3.3 Cardiovascular Protection Mechanisms of GLP1-RA

Experimental studies in both animal and human models have shown that endogenous GLP-1 has positive effects on many cardiovascular risk parameters, including endothelial function in T2D patients with coronary heart disease and blood pressure levels (Ceravolo et al. 2003).

These effects have also been confirmed with GLP-1 receptor agonists. Indeed GLP-1RA exert potentially favorable effects on cardiovascular outcomes not only through glycaemic control, body weight reduction, and improvement of blood pressure and lipid profiles, but also on cardiovascular parameters as cardiac function and cardiac ischemia and on inflammatory markers, resulting in the prevention or delay of

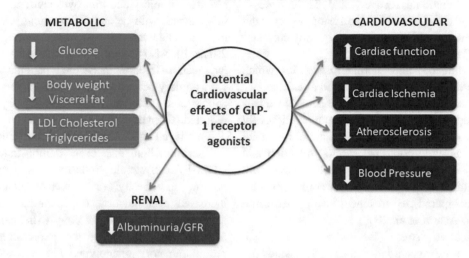

Fig. 4 The potential mechanisms of cardiovascular benefits of Glucagon-like peptide-1 receptor agonists

the atherosclerotic process and on renal function (Bruen et al. 2017) (Fig. 4). However, the exact mechanisms are not yet fully elucidated.

Hypertension is a known cardiovascular risk factor in T2D patients and it has been observed that treatment with GLP-1RA is associated with a significant lowering of systolic blood pressure (Katout et al. 2014). Significant reductions in systolic blood pressure were observed after only 2 weeks of treatment with liraglutide (Fonseca et al. 2014). Recent meta-analyses have confirmed that most, if not all, GLP1-RA determine a significant reduction of blood pressure of 3–4 mmHg (Vilsbøll et al. 2012). The mechanisms by which GLP-1RA reduce blood pressure are not completely clear. It has been hypothesized a natriuretic effect (Nyström et al. 2004; Gutzwiller et al. 2004). Intravenous infusions of GLP-1 enhance sodium excretion, reduce H+ secretion, and reduce glomerular hyperfiltration in healthy subjects (Gutzwiller et al. 2004). These findings suggest an action at the proximal renal tubule and a potential renoprotective effect.

Previous studies suggested that GLP-1 directly induces natriuresis by inhibiting sodium-hydrogen exchanger isoform-3 (NHE3) in the proximal tubule, which may contribute to reducing albuminuria through amelioration of the tubuloglomerular feedback (Skov et al. 2016; Farah et al. 2016; Tonneijck et al. 2016).

Another potential mediator of GLP-1 effects on arterial vasodilation has been identified as the Atrial Natriuretic Peptide (ANP). It has been shown that the effects of GLP1 on arterial vasodilation are mediated by a substance released by the cardiac atria after stimulation by GLP-1, and not by the GLP-1 itself (Kim et al. 2013). The authors demonstrated that cardiac GLP-1 receptors are situated in the cardiac atria, and that GLP-1 receptors activation promotes the secretion of ANP, followed by a reduction of blood pressure. Based on this theory, a new multi-organ axis regulating blood pressure levels was suggested: the starting point is the intestine that produces GLP-1, which acts on the heart causing the release of ANP, which in turn acts on the arterial vessels inducing vasodilation, and in the kidney determining an increase in the urinary elimination of sodium. The identification of a GLP-1R–ANP gut-heart axis detects a new mechanism of actions of GLP-1 in the heart and cardiovascular system (Kim et al. 2013).

Studies on animal models have also shown that a functional GLP-1 receptor is expressed on endothelium and on cardiac and vascular myocytes, and that GLP-1 administration has cardioprotective effects: it increases glucose utilization, functional recovery and cardiomyocyte viability after ischemia-reperfusion injury, and promotes vasodilation and consequently coronary flow (Ban et al. 2008).

In diabetic and normal mice in which myocardial infarction was induced after coronary occlusion, treatment with liraglutide reduced cardiac rupture and infarct size, and significantly improved cardiac output. Furthermore, treatment with liraglutide conferred cardio-protection and increased survival of diabetic mice with myocardial infarction induced by coronary occlusion compared to treatment with metformin (Noyan-Ashraf et al. 2009).

Oxidative stress is another mechanism involved in myocardial damage. In a study by Laviola and co-workers, the protective effects of GLP-1 on oxidative stress-induced apoptosis were investigated in human cardiac progenitor cells (CPCs). Mesenchymal-type cells were isolated from human heart biopsies, exhibited the marker profile of CPCs, differentiated toward the cardiomyocyte, adipocyte, chondrocyte, and osteocyte lineages under appropriate culture conditions, and expressed functional GLP-1 receptors, therefore representing an important indicator of the heart's ability to repair and renew itself following ischemia-induced damage. Upon increase of reactive oxygen species, cardiac progenitor cells undergo apoptosis with reduction of myocardial regenerative potential. This study demonstrated that activation of GLP-1 receptors prevents oxidative stress-mediated apoptosis and increases the survival in human CPCs by interfering with JNK activation (Laviola et al. 2012). These data may represent another important mechanism for the cardio-protective effects of GLP-1.

In humans, several evidences suggest that GLP-1 is able to exert positive effects on endothelial function *in vivo*: intravenous GLP-1 infusion improves flow-mediated vasodilation of about 50% in T2D patients with coronary heart disease (Nyström et al. 2004). Some studies have shown that exenatide protects against ischemia-reperfusion injury and improves cardiac function in patients with acute ST-segment elevation myocardial infarction (STEMI) (Lønborg et al. 2012a, b). Also an improvement in left ventricular ejection fraction in STEMI patients treated with primary percutaneous coronary intervention and treated with liraglutide for 7 days was reported (Chen et al. 2015).

Finally, an effect on the reduction of epicardial fat accumulation has been demonstrated for GLP1-RA. Systemic inflammation (typically present in obesity or diabetes) determines increased accumulation of epicardial fat and adversely influences its biology (Thalmann and Meier 2007), promoting the expression of a proinflammatory phenotype. The accumulation of epicardial adipose tissue is closely associated with the presence, severity, and progression of coronary artery disease (Thalmann and Meier 2007), atrial arrhythmias, and heart failure with preserved ejection fraction (Packer 2018). GLP1 receptors have been shown in epicardial fat (Dozio et al. 2019), and a −35% significant reduction was shown after 6 months treatment with liraglutide (Iacobellis et al. 2017).

In conclusion, both experimental data from cellular and animal models, and data from studies conducted in humans, show that administration of GLP-1 or its analogues determines several positive endothelial and cardiac effects. The direct actions of GLP-1 on blood vessels, inflammation, natriuresis, blood pressure and on the regulation of plasma lipids may impact the development and/or progression of atherosclerotic plaques. Direct actions of GLP-1 on islets result in elevated insulin and reduced glucagon levels, whereas GLP-1 action in the intestine has been associated with reductions in circulating lipids. All these effects, ultimately, determine a reduction of "classic" cardiovascular risk factors (obesity, hypertension, dyslipidemia), independently from the hypoglycaemic mechanisms of GLP1 (Fig. 5).

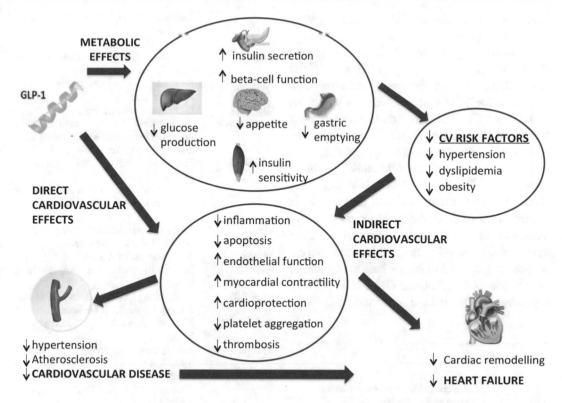

Fig. 5 Metabolic and direct cardiovascular effects of GLP1 for cardiovascular protection

4   Conclusions

A new era in T2D treatment has begun after the demonstration that the anti-hyperglycaemic agents GLP1-RA and SGLT2i consistently reduce the risk of cardiovascular events in patients with T2D (Bethel et al. 2018). Both classes of drugs have also been demonstrated to improve renal outcomes, including a protective effect against albuminuria, are rarely associated with hypoglycemia, have similar reductions of HbA1c, and therefore appear to exert their beneficial cardiovascular effects independently from glucose control through their individual and specific pleiotropic properties.

A series of cardiovascular outcome trials revealed an important benefit on cardiovascular outcome with SGLT2 inhibitors and GLP-1RA, highlighting the advantage of secondary prevention in patients with established CVD. In the trials reported to date, GLP-1RA and SGLT2i reduce atherosclerotic MACE to a similar degree in patients with established atherosclerotic cardiovascular disease, whereas SGLT2i have a more marked effect on preventing hospitalization for heart failure and progression of kidney disease. Their distinct clinical profiles should be considered in the decision-making process when treating patients with T2D.

Thus, recent clinical trials have revealed a novel role of SGLT2 inhibitors and GLP-1 receptor agonists beyond their glucose-lowering effect, and both agents have received great attention, as a paradigm shift from "the lower the glucose, the better" to "how to optimize glycaemic control without hypoglycaemia and overweight, and protecting CVD" has been generated by the new evidences.

With regards to treatment, a consensus report by the ADA/EASD (Davies et al. 2018) recommends for patients with clinical cardiovascular disease an SGLT2 inhibitor or a GLP-1 receptor agonist with proven cardiovascular benefit. For patients with chronic kidney disease or clinical heart failure, an SGLT2 inhibitor with proven benefit is recommended. In patients with atherosclerotic cardiovascular disease (ASCVD), particularly if overweight or obese, GLP-1 receptor agonists are generally recommended (Davies et al. 2018). Very recently, ESC and EASD (Cosentino et al. 2019) have proposed a treatment algorithm for T2D that involves the use of an SGLT2i or GLP1-RA as first-line drugs, even before metformin, in treatment-naïve patients with ASCVD or at very high/high CV risk.

Finally, the "extra-glycaemic" effects observed for SGLT2i and GLP-1RA drugs have encouraged clinical studies aimed, for example, at the treatment of heart failure in patients without diabetes or at the treatment of obesity, NAFLD or hypertension. Results from these mechanistic studies are expected to further help to understand the actions of SGLT2i and GLP1-RA, possibly expanding the clinical use of these drugs.

Glossary

ADA	American Diabetes Association
ANP	atrial natriuretic peptide
ARB	angiotensin-receptor blocker
ASCVD	atherosclerotic cardiovascular disease
CANVAS	Canagliflozin Cardiovascular Assessment Study
CKD	chronic kidney disease
CPCs	cardiac progenitor cells
CREDENCE	Canagliflozin and Renal Endpoints in Diabetes with Established Nephropathy Clinical Evaluation CV cardiovascular
CVD-Real 2	Comparative Effectiveness of Cardiovascular Outcomes in New Users of Sodium- Glucose Cotransporter-2 Inhibitors
CVD	cardiovascular disease
DAPA-HF	Dapagliflozin and Prevention of AdverseOutcomes in Heart Failure
DECLARE-TIMI 58	Dapagliflozin Effect on Cardiovascular Events
DPP4-i	dipeptidyl peptidase-4 inhibitors
EASD	European association for the study of Diabetes
EASEL	Evidence for Cardiovascular Outcomes With Sodium Glucose Cotransporter 2 Inhibitors in the Real World
ELIXA	Evaluation of Lixisenatide in Acute Coronary Syndrome
EMPAREG	The Empagliflozin Cardiovascular Outcome Event Trial in Type 2 Diabetes
	Mellitus Patients–Removing Excess Glucose
ESRD	end-stage renal disease
EXSCEL	Exenatide Study of Cardiovascular Event Lowering Trial
FDA	Food and Drug Administration
FFA	free fatty acid
GLP-1RA	GLP-1 receptor agonists
HARMONY	Albiglutide and cardiovascular outcomes in patients with type 2 diabetes and cardiovascular disease
LEADER	Liraglutide Effect and Action in Diabetes Evaluation of Cardiovascular Outcome Results
MACE	Major Cardiovascular Events
NAFLD	non-alcoholic hepatic steatosis
NYHA	New York Heart Association
PIONEER 6	Peptide Innovation for Early Diabetes Treatment
REWIND	Researching Cardiovascular Events with a Weekly Incretin in Diabetes
SGLT2i	Sodium/glucose-2 cotransporter inhibitors
STEMI	ST-segment elevation myocardial infarction
SUSTAIN	Semaglutide Unabated Sustainability in Treatment of Type 2 Diabetes 6 (SUSTAIN-6)
T2D	type 2 diabetes
UKPDS	United Kingdom Prospective Diabetes Study
VADT	Veterans Affairs Diabetes Trial

References

Action to Control Cardiovascular Risk in Diabetes Study Group, Gerstein HC, Miller ME, Byington RP, Goff DC Jr, Bigger JT, Buse JB et al (2008) Effects of intensive glucose lowering in type 2 diabetes. N Engl J Med 358:2545–2559

ADVANCE, Collaborative Group, Patel A, MacMahon S, Chalmers J, Neal B, Billot L, Woodward M, Marre M et al (2008) Intensive blood glucose control and vascular outcomes in patients with type 2 diabetes. N Engl J Med 358:2560–2572

Anandhakrishnan A, Korbonits M (2016) Glucagon-like peptide 1 in the pathophysiology and pharmacotherapy of clinical obesity. World J Diabetes 7:572–598

Ban K, Noyan-Ashraf MH, Hoefer J, Bolz SS, Drucker DJ, Husain M (2008) Cardioprotective and vasodilatory actions of glucagon-like peptide 1 receptor are mediated through both glucagon-like peptide 1 receptor dependent and independent pathways. Circulation 117(18):2340–2350. Epub 2008 Apr 21. Erratum in: Circulation. 2008 Jul 22;118(4):e81

Bethel MA, Patel RA, Merrill P, Lokhnygina Y, Buse JB, Mentz RJ, EXSCEL Study Group et al (2018) Cardiovascular outcomes with glucagon-like peptide-1 receptor agonists in patients with type 2 diabetes: a meta-analysis. Lancet Diabetes Endocrinol 6:105–113. https://doi.org/10.1016/S2213-8587(17)30412-6

Bonner C, Kerr-Conte J, Gmyr V, Queniat G, Moerman E, Thévenet J et al (2015) Inhibition of the glucose transporter SGLT2 with dapagliflozin in pancreatic alpha cells triggers glucagon secretion. Nat Med 21:512–517

Briasoulis A, Al Dhaybi O, Bakris GL (2018) SGLT2 inhibitors and mechanisms of hypertension. Curr Cardiol Rep 20:1

Bruen R, Curley S, Kajani S, Crean D, O'Reilly ME, Lucitt MB et al (2017) Liraglutide dictates macrophage phenotype in apolipoprotein E null mice during early atherosclerosis. Cardiovasc Diabetol 16(1):143

Ceravolo R, Maio R, Pujia A, Sciacqua A, Ventura G, Costa MC et al (2003) Pulse pressure and endothelial dysfunction in never-treated hypertensive patients. J Am Coll Cardiol 41:1753–1758

Chen WR, Hu SY, Chen YD, Zhang Y, Qian G, Wang J et al (2015) Effects of liraglutide on left ventricular function in patients with ST-segment elevation myocardial infarction undergoing primary percutaneous coronary intervention. Am Heart J 170(5):845–854

Chouchani ET, Pell VR, Gaude E, Aksentijević D, Sundier SY, Robb EL et al (2014) Ischaemic accumulation of succinate controls reperfusion injury through mitochondrial ROS. Nature 515(7527):431–435

Cosentino F, Grant PJ, Aboyans V, Bailey CJ, Ceriello A, Delgado V, ESC Scientific Document Group et al (2019) ESC Guidelines on diabetes, pre-diabetes, and cardiovascular diseases developed in collaboration with the EASD. Eur Heart J 40(39):3215–3217

Davies MJ, D'Alessio DA, Fradkin J, Kernan WN, Mathieu C, Mingrone G et al (2018) Management of Hyperglycemia in type 2 diabetes, 2018. A consensus report by the American Diabetes Association (ADA) and the European Association for the Study of Diabetes (EASD). Diabetes Care 41(12):2669–2701

DeFronzo RA, Davidson JA, Del Prato S (2012) The role of the kidneys in glucose homeostasis: a new path towards normalizing glycaemia. Diabetes Obes Metab 14:5–14

Dormandy JA, Charbonnel B, Eckland DJ, Erdmann E, Massi Benedetti M, Moules IK, The PROactive Investigators (2005) Secondary prevention of macrovascular events in patients with type 2 diabetes in the PROactive study (PROspective pioglitAzone clinical trial in macro vascular events): a randomised controlled trial. Lancet 366:1279–1289

Dozio E, Vianello E, Malavazos AE, Tacchini L, Schmitz G, Iacobellis G, Corsi Romanelli MM (2019)

Epicardial adipose tissue GLP-1 receptor is associated with genes involved in fatty acid oxidation and white-to-brown fat differentiation: a target to modulate cardiovascular risk? Int J Cardiol 292:218–224

Drucker DJ (2018) The ascending GLP-1 road from clinical safety to reduction of cardiovascular complications. Diabetes 67(9):1710–1719

Duckworth W, Abraira C, Moritz T, Reda D, Emanuele N, Reaven PD, VADT Investigators et al (2009) Glucose control and vascular complications in veterans with type 2 diabetes. N Engl J Med 360:129–139

Einarson TR, Acs A, Ludwig C, Panton UH (2018) Prevalence of cardiovascular disease in type 2 diabetes: a systematic literature review of scientific evidence from across the world in 2007–2017. Cardiovasc Diabetol 17(1):83

European Medicines Agency (EMA) (2012) Guideline on clinical investigation of medicinal products in the treatment or prevention of diabetes mellitus

European Medicines Agency (EMA) (2016) Reflection paper on assessment of cardiovascular safety profile of medicinal products

Farah LX, Valentini V, Pessoa TD, Malnic G, McDonough AA, Girardi AC (2016) The physiological role of glucagon-like peptide-1 in the regulation of renal function. Am J Physiol Renal Physiol 310(2): F123–F127

Ferrannini E, Solini A (2012) SGLT2 inhibition in diabetes mellitus: rationale and clinical prospects. Nat Rev Endocrinol 8:495–502

Ferrannini E, Baldi S, Frascerra S, Astiarraga B, Heise T, Bizzotto R et al (2016a) Shift to fatty substrate utilization in response to sodium-glucose cotransporter 2 inhibition in subjects without diabetes and patients with type 2 diabetes. Diabetes 65:1190–1195

Ferrannini E, Mark M, Mayoux E (2016b) CV protection in the EMPA-REG OUTCOME trial: a "thrifty substrate" hypothesis. Diabetes Care 39:1108–1114

Fonseca VA, Devries JH, Henry RR, Donsmark M, Thomsen HF, Plutzky J (2014) Reductions in systolic blood pressure with liraglutide in patients with type 2 diabetes: insights from a patient-level pooled analysis of six randomized clinical trials. J Diabetes Complicat 28:399–405

Gaede P, Vedel P, Larsen N, Jensen GV, Parving HH, Pedersen O (2003) Multifactorial intervention and cardiovascular disease in patients with type 2 diabetes. N Engl J Med 348:383–393

Gaede P, Lund-Andersen H, Parving HH, Pedersen O (2008) Effect of a multifactorial intervention on mortality in type 2 diabetes. N Engl J Med 358(6):580–591

Garvey WT, Van Gaal L, Leiter LA, Vijapurkar U, List J, Cuddihy R et al (2018) Effects of canagliflozin versus glimepiride on adipokines and inflammatory biomarkers in type 2 diabetes. Metabolism 85:32–37

Gerstein HC, Colhoun HM, Dagenais GR, Diaz R, Lakshmanan M, Pais P, REWIND Investigators et al (2019) Dulaglutide and cardiovascular outcomes in type 2 diabetes (REWIND): a double-blind, randomised placebo-controlled trial. Lancet 394 (10193):121–130

Green JB, Bethel MA, Armstrong PW, Buse JB, Engel SS, Garg J, TECOS Study Group et al (2015) Effect of Sitagliptin on cardiovascular outcomes in type 2 diabetes. N Engl J Med 373:232–242

Guidance for Industry on Diabetes Mellitus (2008) Evaluating cardiovascular risk in new antidiabetic therapies to treat type 2 diabetes. U.S. Department of Health and Human Services, Silver Spring, pp 1–5. https://www.fda.gov/regulatory-information/search-fda-guidance-documents/diabetes-mellitus-evaluating-cardiovascular-risk-new-antidiabetic-therapies-treat-type-2-diabetes

Gutzwiller JP, Tschopp S, Bock A, Zehnder CE, Huber AR, Kreyenbuehl M et al (2004) Glucagon-like peptide 1 induces natriuresis in healthy subjects and in insulin-resistant obese men. J Clin Endocrinol Metab 89(6):3055–3061

Hallow KM, Helmlinger G, Greasley PJ, McMurray JJV, Boulton DW (2018) Why do SGLT2 inhibitors reduce heart failure hospitalization? A differential volume regulation hypothesis. Diabetes Obes Metab 20 (3):479–487

Hayward RA, Reaven PD, Wiitala WL, Bahn GD, Reda DJ, Ge L, VADT Investigators et al (2015) Follow-up of glycemic control and cardiovascular outcomes in type 2 diabetes. N Engl J Med 372(23):2197–2206

Hernandez AF, Green JB, Janmohamed S, D'Agostino RB Sr, Granger CB, Jones NP, Harmony Outcomes committees and investigators et al (2018) Albiglutide and cardiovascular outcomes in patients with type 2 diabetes and cardiovasculardisease (Harmony outcomes): a double-blind, randomised placebo-controlled trial. Lancet 392(10157):1519–1529

Holman RR, Paul SK, Bethel MA, Matthews DR, Neil HA (2008) 10-year follow-up of intensive glucose control in type 2 diabetes. N Engl J Med 359(15):1577–1589

Holman RR, Bethel MA, Mentz RJ, Thompson VP, Lokhnygina Y, Buse JB, EXSCEL Study Group et al (2017) Effects of once-weekly exenatide on cardiovascular outcomes in type 2 diabetes. N Engl J Med 377:1228–1239

Husain M, Birkenfeld AL, Donsmark M, Dungan K, Eliaschewitz FG, Franco DR, PIONEER 6 Investigators et al (2019) Oral Semaglutide and cardiovascular outcomes in patients with type 2 diabetes. N Engl J Med 381(9):841–851

Iacobellis G, Mohseni M, Bianco SD, Banga PK (2017) Liraglutide causes large and rapid epicardial fat reduction. Obesity (Silver Spring) 25(2):311–316

Katout M, Zhu H, Rutsky J, Shah P, Brook RD, Zhong J, Rajagopalan S (2014) Effect of GLP-1 mimetics on blood pressure and relationship to weight loss and glycemia lowering: results of a systematic meta-analysis and meta-regression. Am J Hypertens 27 (1):130–139

Kim M, Platt MJ, Shibasaki T, Quaggin SE, Backx PH, Seino S, Simpson JA, Drucker DJ (2013) GLP-1 receptor activation and Epac2 link atrial natriuretic peptide secretion to control of blood pressure. Nat Med 19 (5):567–575

Kolwicz SC Jr, Airhart S, Tian R (2016) Ketones step to the plate: a game changer for metabolic remodelling in heart failure? Circulation 133:689–691

Kosiborod M, Lam CSP, Kohsaka S, Kim DJ, Karasik A, Shaw J, CVD-REAL Investigators and Study Group et al (2018) Cardiovascular events associated with SGLT-2 inhibitors versus other glucose-lowering drugs: the CVD-REAL 2 study. J Am Coll Cardiol 71:2628–2639

Kuhre RE, Ghiasi SM, Adriaenssens AE, Wewer Albrechtsen NJ, Andersen DB, Aivazidis A et al (2019) No direct effect of SGLT2 activity on glucagon secretion. Diabetologia 62:1011–1023

Laviola L, Leonardini A, Melchiorre M, Orlando MR, Peschechera A, Bortone A et al (2012) Glucagon-like peptide-1 counteracts oxidative stress-dependent apoptosis of human cardiac progenitor cells by inhibiting the activation of the c-Jun N-terminal protein kinase signaling pathway. Endocrinology 153(12):5770–5781

Lønborg J, Kelbæk H, Vejlstrup N, Bøtker HE, Kim WY, Holmvang L et al (2012a) Exenatide reduces final infarct size in patients with ST-segment-elevation myocardial infarction and short-duration of ischemia. Circ Cardiovasc Interv 5(2):288–295

Lønborg J, Vejlstrup N, Kelbæk H, Bøtker HE, Kim WY, Mathiasen AB et al (2012b) Exenatide reduces reperfusion injury in patients with ST-segment elevation myocardial infarction. Eur Heart J 33(12):1491–1499

Lopaschuk GD, Ussher JR, Folmes CDL, Jaswal JS, Stanley WC (2010) Myocardial fatty acid metabolism in health and disease. Physiol Rev 90:207–258

Marso SP, Daniels GH, Brown-Frandsen K, Kristensen P, Mann JF, Nauck MA, LEADER Steering Committee; LEADER Trial Investigators et al (2016a) Liraglutide and cardiovascular outcomes in type 2 diabetes. N Engl J Med 375(4):311–322

Marso SP, Bain SC, Consoli A, Eliaschewitz FG, Jódar E, Leiter LA, SUSTAIN-6 Investigators et al (2016b) Semaglutide and cardiovascular outcomes in patients with type 2 diabetes. N Engl J Med 375:1834–1844

McMurray JJV, Solomon SD, Inzucchi SE, Køber L, Kosiborod MN, Martinez FA, DAPA-HF Trial Committees and Investigators et al (2019) Dapagliflozin in patients with heart failure and reduced ejection fraction. N Engl J Med 381(21):1995–2008

Meier JJ (2012) GLP-1 receptor agonists for individualized treatment of type 2 diabetes mellitus. Nat Rev Endocrinol 8:728–742

Mentzer RM Jr, Bartels C, Bolli R, Boyce S, Buckberg GD, Chaitman B, EXPEDITION Study Investigators et al (2008) Sodium-hydrogen exchange inhibition by cariporide to reduce the risk of ischemic cardiac events in patients undergoing coronary artery bypass grafting:

results of the EXPEDITION study. Ann Thorac Surg 85:1261–1270

Morrish NJ, Wang SL, Stevens LK, Fuller JH, Keen H (2001) Mortality and causes of death in the WHO multinational study of vascular disease in diabetes. Diabetologia 44(Suppl. 2):S14–S21

Nauck M (2016) Incretin therapies: highlighting common features and differences in the modes of action of glucagon-like peptide-1 receptor agonists and dipeptidyl peptidase-4 inhibitors. Diabetes Obes Metab 18:203–216

Nauck MA, Meier JJ, Cavender MA, Abd El Aziz M, Drucker DJ (2017) Cardiovascular actions and clinical outcomes with glucagon-like peptide-1 receptor agonists and dipeptidyl peptidase-4 inhibitors. Circulation 136(9):849–870

Neal B, Perkovic V, Mahaffey KW, de Zeeuw D, Fulcher G, Erondu N, CANVAS Program Collaborative Group et al (2017) Canagliflozin and cardiovascular and renal events in type 2 diabetes. N Engl J Med 377(7):644–657

Nissen SE (2010) The rise and fall of rosiglitazone. Eur Heart J 31:773–776

Noyan-Ashraf MH, Momen MA, Ban K, Sadi AM, Zhou YQ, Riazi AM et al (2009) GLP-1R agonist liraglutide activates cytoprotective pathways and improves outcomes after experimental myocardial infarction in mice. Diabetes 58(4):975–983

Nyström T, Gutniak MK, Zhang Q, Zhang F, Holst JJ, Ahrén B, Sjöholm A (2004) Effects of glucagon-like peptide-1 on endothelial function in type 2 diabetes patients with stable coronary artery disease. Am J Physiol Endocrinol Metab 287(6):E1209–E1215. Epub 2004 Sep 7

Packer M (2018) Epicardial adipose tissue may mediate deleterious effects of obesity and inflammation on the myocardium. J Am Coll Cardiol 71(20):2360–2372. https://doi.org/10.1016/j.jacc.2018.03.509. Review

Packer M, Anker SD, Butler J, Filippatos G, Zannad F (2017) Effects of sodium-glucose cotransporter 2 inhibitors for the treatment of patients with heart failure: proposal of a novel mechanism of action. JAMA Cardiol 2:1025–1029

Perkovic V, Jardine MJ, Neal B, Bompoint S, Heerspink HJL, Charytan DM, CREDENCE Trial Investigators et al (2019) Canagliflozin and renal outcomes in type 2 diabetes and nephropathy. N Engl J Med 380:2295–2306

Perreault L (2017) EMPA-REG OUTCOME: the endocrinologist's point of view. Am J Med 130:51–56

Pfeffer MA, Claggett B, Diaz R, Dickstein K, Gerstein HC, Køber LV, ELIXA Investigators et al (2015) Lixisenatide in patients with type 2 diabetes and acute coronary syndrome. N Engl J Med 373 (23):2247–2257

Rahmoune H, Thompson PW, Ward JM, Smith CD, Hong G, Brown J (2005) Glucose transporters in human renal proximal tubular cells isolated from the urine of patients with noninsulin- dependent diabetes. Diabetes 54:3427–3434

Rizzo M, Rizvi AA, Patti AM, Nikolic D, Giglio RV, Castellino G et al (2016) Liraglutide improves metabolic parameters and carotid intima-media thickness in diabetic patients with the metabolic syndrome: an 18-month prospective study. Cardiovasc Diabetol 15 (1):162

Rosenstock J, Perkovic V, Johansen OE, Cooper ME, Kahn SE, Marx N, CARMELINA Investigators et al (2019) Effect of Linagliptin vs placebo on major cardiovascular events in adults with type 2 diabetes and high cardiovascular and renal risk: the CARMELINA randomized clinical trial. JAMA 321(1):69–79

Sato T, Aizawa Y, Yuasa S, Kishi S, Fuse K, Fujita S, Ikeda Y, Kitazawa H, Takahashi M, Sato M, Okabe M (2018) The effect of dapagliflozin treatment on epicardial adipose tissue volume. Cardiovasc Diabetol 17:6

Scirica BM, Bhatt DL, Braunwald E, Steg PG, Davidson J, Hirshberg B, SAVOR-TIMI 53 Steering Committee and Investigators et al (2013) Saxagliptin and cardiovascular outcomes in patients with type 2 diabetes mellitus. N Engl J Med 369(14):1317–1326

Skov J, Pedersen M, Holst JJ, Madsen B, Goetze JP, Rittig S et al (2016) Short-term effects of liraglutide on kidney function and vasoactive hormones in type 2 diabetes: a randomized clinical trial. Diabetes Obes Metab 18(6):581–589

Thalmann S, Meier CA (2007) Local adipose tissue depots as cardiovascular risk factors. Cardiovasc Res 75 (4):690–701. Epub 2007 Mar 14. Review

Tonneijck L, Smits MM, Muskiet MHA, Hoekstra T, Kramer MHH, Danser AHJ, Diamant M, Joles JA, van Raalte DH (2016) Acute renal effects of the GLP-1 receptor agonist exenatide in overweight type 2 diabetes patients: a randomised, double-blind, placebo-controlled trial. Diabetologia 59(7):1412–1421

Udell JA, Yuan Z, Rush T, Sicignano NM, Galitz M, Rosenthal N (2018) Cardiovascular outcomes and risks after initiation of a sodium glucose cotransporter 2 inhibitor: results from the EASEL population-based cohort study (Evidence for cardiovascular outcomes with sodium glucose cotransporter 2 inhibitors in the real world). Circulation 137:1450–1459

UK Prospective Diabetes Study (UKPDS) Group (1998) Intensive blood-glucose control with sulphonylureas or insulin compared with conventional treatment and risk of complications in patients with type 2 diabetes (UKPDS 33). Lancet 352:837–853. Erratum in: Lancet 1999 Aug 14;354:602

Uthman L, Baartscheer A, Bleijlevens B, Schumacher CA, Fiolet JWT, Koeman A et al (2018) Class effects of SGLT2 inhibitors in mouse cardiomyocytes and hearts: inhibition of Na(+)/H(+) exchanger, lowering of cytosolic Na(+) and vasodilation. Diabetologia 61:722–726

Vallon V (2015) The mechanisms and therapeutic potential of SGLT2 inhibitors in diabetes mellitus. Annu Rev Med 66:255–270

Verma S, McMurray JJV (2018) SGLT2 inhibitors and mechanisms of cardiovascular benefit: a state-of-the-art review. Diabetologia 61:2108–2117

Vilsbøll T, Christensen M, Junker AE, Knop FK, Gluud LL (2012) Effects of glucagon-like peptide-1 receptor agonists on weight loss: systematic review and meta-analyses of randomised controlled trials. BMJ 344:d7771

White WB, Cannon CP, Heller SR, Nissen SE, Bergenstal RM, Bakris GL, EXAMINE Investigators et al (2013) Alogliptin after acute coronary syndrome in patients with type 2 diabetes. N Engl J Med 369:1327–1335

Wiviott SD, Raz I, Bonaca MP, Mosenzon O, Kato ET, Cahn A, DECLARE–TIMI 58 Investigators et al (2019) Dapagliflozin and cardiovascular outcomes in type 2 diabetes. N Engl J Med 380:347–357

Zelniker TA, Wiviott SD, Raz I, Im K, Goodrich EL, Bonaca MP et al (2019) SGLT2 inhibitors for primary and secondary prevention of cardiovascular and renal outcomes in type 2 diabetes: a systematic review and meta-analysis of cardiovascular outcome trials. Lancet 393:31–39

Zinman B, Wanner C, Lachin JM, Fitchett D, Bluhmki E, Hantel S, EMPA-REG OUTCOME Investigators et al (2015) Empagliflozin, cardiovascular outcomes, and mortality in type 2 diabetes. N Engl J Med 373 (22):2117–2128

Adv Exp Med Biol - Advances in Internal Medicine (2020) 4: 213–230
https://doi.org/10.1007/5584_2020_479
© Springer Nature Switzerland AG 2020
Published online: 1 February 2020

Glucose Lowering Efficacy and Pleiotropic Effects of Sodium-Glucose Cotransporter 2 Inhibitors

Mohammad Shafi Kuchay, Khalid Jamal Farooqui, Sunil Kumar Mishra, and Ambrish Mithal

Abstract

In type 2 diabetes, the maladaptive upregulation of sodium-glucose cotransporter 2 (SGLT2) protein expression and activity contribute to the maintenance of hyperglycemia. By inhibiting these proteins, SGLT2 inhibitors increase urinary glucose excretion (UGE) that leads to fall in plasma glucose concentrations and improvement in all glycemic parameters. Clinical studies have demonstrated that in patients with type 2 diabetes, SGLT2 inhibitors resulted in sustained reductions in glycated hemoglobin (HbA$_{1C}$), body weight, blood pressure and serum uric acid levels. Interestingly, the cardiovascular (CV) and renal outcome trials revealed the beneficial effects of SGLT2 inhibitors on CV and renal functions. Because the benefits were seen soon after initiation of SGLT2 inhibitors, these observations are explained by effects beyond their glucose lowering capacity. SGLT2 inhibitors also reduce liver fat in patients with nonalcoholic fatty liver disease (NAFLD) and type 2 diabetes. This chapter describes the basic information about SGLT2 inhibitors, current status of SGLT2 inhibitors in the management of type 2 diabetes and their beneficial effects in addition to glycemic control.

Keywords

Canagliflozin · Cardiovascular health · Dapagliflozin · Empagliflozin · Ertugliflozin · Gliflozins · Ipragliflozin · Luseogliflozin · Nonalcoholic fatty liver disease · Remogliflozin · Sodium-glucose cotransporter 2 inhibitors · Tofogliflozin · Type 2 diabetes

M. S. Kuchay (✉), K. J. Farooqui, S. K. Mishra, and A. Mithal
Division of Endocrinology and Diabetes, Medanta The Medicity Hospital, Gurugram, Haryana, India
e-mail: drshafikuchay@gmail.com

Abbreviations

ALT	alanine aminotransferase
AST	aspartate aminotransferase
CANVAS	Canagliflozin Cardiovascular Assessment Study
CK	cytokeratin
CKD	chronic kidney disease
CV	cardiovascular
CVD	cardiovascular disease
CVD-REAL	Comparative Effectiveness of Cardiovascular Outcomes in

	New Users of SGLT-2 Inhibitors
CVOT	cardiovascular outcome trial
DECLARE TIMI 58	Dapagliflozin Effect on Cardiovascular Events-Thrombolysis in Myocardial Infarction 58
DKA	diabetic ketoacidosis
eGFR	estimated glomerular filtration rate
E-LIFT	Effect of Empagliflozin on Liver Fat
EMA	European Medicines Agency
EMPA-REG OUTCOME	Empagliflozin Cardiovascular Outcome Event Trial in Type 2 Diabetes Mellitus Patients-Removing Excess Glucose
FGF 21	fibroblast growth factor 21
FRG	familial renal glucosuria
GGT	gamma glutamyl transpeptidase
GLP-1r	glucagon like peptide-1 receptor
GLUT	glucose transporter
HbA_{1C}	glycated hemoglobin
HGO	hepatic glucose output
MACE	major adverse cardiovascular events
MRI	magnetic resonance imaging
NAFLD	nonalcoholic fatty liver disease
NASH	nonalcoholic steatohepatitis
NHE	sodium hydrogen exchanger
PCT	proximal convoluted tubule
PDFF	proton density fat fraction
SGLT2	sodium-glucose cotransporter 2
UGE	urinary glucose excretion
US FDA	The United States Food and Drug Administration

gliflozins) to treat type 2 diabetes. SGLT2 inhibitors reduce plasma glucose levels by inhibiting reabsorption of glucose in the proximal tubule of kidneys (White 2010; Vick et al. 1973). Among patients with type 2 diabetes, administration of SGLT2 inhibitors result in sustained decreases in HbA_{1C}, body weight, blood pressure and serum uric acid concentrations (van Baar et al. 2018). Gliflozins have beneficial effect on CV and renal functions (Zinman et al. 2015; Perkovic et al. 2019). Recently, SGLT2 inhibitors have been shown to have a salutary effect on nonalcoholic fatty liver disease (NAFLD) (Kuchay et al. 2018). The SGLT2 inhibitors are recommended in the American Diabetes Association (ADA)/European Association for the Study of Diabetes (EASD) Consensus Report as a preferred oral treatment after metformin in patients with type 2 diabetes with established atherosclerotic cardiovascular disease (ASCVD), heart failure (HF), or chronic kidney disease (CKD) and also in patients without ASCVD where there is a compelling need to minimize weight gain and need to minimize risk of hypoglycemia (Davies et al. 2018). The European Society of Cardiology (ESC)/EASD 2019 guidelines recommend SGLT2 inhibitors as first line agents in drug naïve patients with type 2 diabetes and ASCVD or high CV risk (Cosentino et al. 2019). Since 2013, four gliflozin congeners have been approved in the United States, namely Dapagliflozin, Canagliflozin, Empagliflozin and Ertugliflozin; and three in Japan, namely Ipragliflozin, Luseogliflozin and Tofogliflozin. Remogliflozin has been approved in India. This chapter describes the clinical pharmacology of SGLT2 inhibitors, their glucose lowering efficacy and other clinical benefits based on the current evidences from clinical studies.

1   Introduction

Following the finding that phlorizin had SGLT inhibiting properties, the pharmaceutical industry developed selective SGLT2 inhibitors (also called

2   Historical Perspective

Phlorizin was the first compound (isolated in 1835 from the root bark of an apple tree) found to have SGLT inhibiting properties (White 2010). It could not find any clinical use for two reasons.

Firstly, it is a non-selective inhibitor of SGLT1 and SGLT2 proteins. Its clinical use leads to desired glucosuria (SGLT2 inhibition) along with significant gastrointestinal side-effects (SGLT1 inhibition) (Vick et al. 1973). Secondly, it is a highly unstable compound when administered orally, as it is rapidly degraded by glucosidases in the small intestine. However, it served an important role in the development of gliflozin class of medications. Phlorizin is a natural botanical O-aryl glycoside composed of a D-glucose and an aromatic ketone (Vick et al. 1973). Structural modifications made stable compounds of phlorizin possible. This was achieved by conjugating aryl moiety with glucose moiety since C-glucoside derivatives were more stable in the small intestines than O-glucoside derivatives (C-C bond instead of C-O-C bond) (Vick et al. 1973). Currently, there are at least eight gliflozin compounds available for treatment

of type 2 diabetes (Table 1). Earlier evidence for the safety of renal glycosuria on long-term kidney function comes from individuals with familial renal glycosuria (FRG). Individuals with FRG have some or no functional SGLT2 proteins. They present with significant glucosuria. Interestingly, this condition does not cause hypoglycemia or any serious side effects (Santer and Calado 2010).

3 Renal Handling of Glucose and SGLT2 Inhibition

Glucose homeostasis is tightly maintained by regulating glucose production, reabsorption, and utilization in a coordinated manner. In healthy individuals, despite extreme variations in glucose intake, the circulatory glucose concentrations are maintained in a narrow range (Abdul and

Table 1 Pharmacological characteristics of various SGLT2 inhibitors

Generic name	Dose range	Protein binding	t-max (hours)	C-max	$t^{1/2}$ (hours)	SGLT2 selectivity over SGLT1
Dapagliflozin[a,b]	5–10 mg once daily	91%	1–1.5	79.6 ng/mL (5 mg dose) 165.0 ng/mL (10 mg dose)	12.9	1200-fold
Canagliflozin[a,b]	100–300 mg once daily	99%	1–2	1096 ng/mL (100 mg dose) 3480 ng/mL (300 mg dose)	10.6 (100 mg dose) 13.1 (300 mg dose)	250-fold
Empagliflozin[a,b]	10–25 mg once daily	86.2%	1.5	259 nmol/L (10 mg dose) 687 nmol/L (25 mg dose)	12.9	2500-fold
Ertugliflozin[a,b]	5–15 mg once daily	95%	0.5–1.5	268 ng/mL (15 mg dose)	11–17	2000-fold
Ipragliflozin[c]	25–50 mg once daily	96.3%	1.0	975 ng/mL	15–16 (50 mg dose)	360-fold
Luseogliflozin[c]	2.5–5 mg once daily	96%	0.625	119 + 27.0 ng/mL	9.2	1650-fold
Tofogliflozin[c]	20–40 mg once daily	83%	0.75	489 ng/mL	6.8	2900-fold
Remogliflozin	100 mg twice daily	–	1.0	–	2–4	100–1000-fold

[a]USA FDA approved, [b]EMA approved, [c]approved in Japan by Ministry of Health, Labour and Welfare; *t-max* time to achieve maximum plasma concentration, *C-max* maximum serum concentration that the drug achieves in body after the drug has been administered, $t^{1/2}$ biological half life

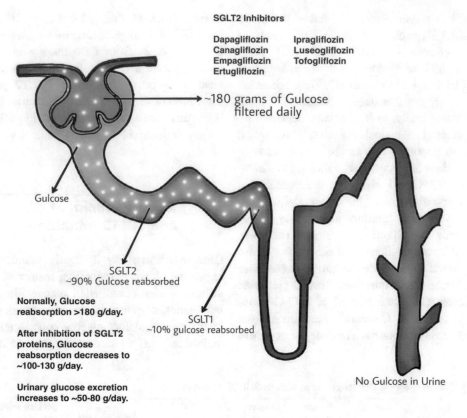

Fig. 1 Nephron showing different segments where SGLT proteins are located

DeFronzo 2013). About 180 g of glucose is filtered daily by the renal glomeruli, and is then reabsorbed from the glomerular filtrate in the proximal convoluted tubule (PCT) and returned to the circulation (Fig. 1). Effectively, no glucose is wasted in the urine of a healthy individual (Abdul and DeFronzo 2013). This efficient renal glucose transport system is carried out by two carrier proteins: the facilitated (passive) glucose transporters (GLUT family), and the active (energy-dependent) sodium-dependent glucose cotransporters (SGLT family). Reabsorption of glucose from the glomerular filtrate is mediated by two SGLT proteins (SGLT1 and SGLT2) (Wright 2001). Reabsorption primarily occurs in the first section (S1 and S2) of the proximal renal tubule at the brush border of cells via the action of SGLT2, which removes the majority (~90%) of the filtered glucose, while the remainder (~10%)

is removed further along the proximal tubule (S3) via the action of SGLT1 (Lee et al. 2007). Difference between the two SGLT proteins are shown in Table 2. Reabsorbed glucose is then released from proximal tubular cells at the basolateral membrane into the circulation via GLUT2 and to a lesser extent by GLUT1.

At normal plasma glucose levels of approximately 100 mg/dL, all of the filtered glucose is reabsorbed. As the plasma glucose concentration rises, the amount of glucose filtered by the kidney increases until a threshold is reached at which the renal glucose transport system is effectively saturated. This "saturation point" is called the transport maximum for glucose, or Tmax glucose. In healthy individuals, the Tm glucose is equivalent to a filtration rate of 260–350 mg/min/1.73m^2. Any further increase in the amount of glucose filtered results in the excess glucose

Table 2 Comparison between SGLT1 and SGLT2

Characteristic	SGLT1	SGLT2
Gene	SLC5A1	SLC5A2
Locus	Chr. 22 q13.1	Chr. 16 p11.2
Distribution in proximal tubule	S3 segment	S1 and S2 segments
Na$^+$: Glucose cotransporter ratio	2:1	1:1
Contribution to glucose reabsorption	10–20%	80–90%
Capacity	Low	High
Affinity	High	Low
Disease state if mutation occurs	Glucose-galactose malabsorption	Familial renal glucosuria

SGLT1 sodium-glucose cotransporter 1, *SGLT2* sodium-glucose cotransporter 2, *SLC5A1* solute carrier family 5 member 1, *SLC5A2* solute carrier family 5 member 2, *Chr.* chromosome

being excreted into the urine. The plasma glucose concentration at which this occurs is approximately 200 mg/dL (Hummel et al. 2011).

Among patients with type 2 diabetes, glucose handling by the kidney is impaired. The kidneys continue to reabsorb glucose even when plasma glucose concentrations are high, with levels that usually exceed the Tmax of healthy individuals. The mean Tmax glucose is increase by up to 20% in patients with diabetes vs healthy individuals. This is because of up-regulation of SGLT2 and GLUT2 expression and activity in diabetes. As a consequence of elevated Tmax glucose and diminished UGE in patients with type 2 diabetes, there is increased reabsorption of urinary glucose and continuous backflow of glucose from the kidneys into the bloodstream, thus potentiating hyperglycaemia and increasing the risk for diabetes-associated complications (Nauck 2014).

After oral delivery, gliflozins are rapidly absorbed and bind to plasma proteins. They remain in the circulation with a half-life of 8–16 h (Table 1). They are filtered at the renal glomeruli and reversibly block SGLT2 proteins in the apical membrane of the S1 and S2 segments of the proximal tubule. The maximal inhibition of renal glucose reabsorption is about 50% because the glucose that spills into the S3 segment of the proximal tubule is reabsorbed by SGLT1. In monotherapy, the SGLT2-inhibiting drugs lower fasting and postprandial plasma glucose levels by excreting glucose into the urine (Kasichayanula et al. 2014). There drugs also lead to some reduction in body weight and

arterial blood pressure due to calorie loss and natriuresis, respectively.

4 Glucose Lowering Efficacy of SGLT2 Inhibitors

At present, four oral agents (i.e., canagliflozin, dapagliflozin, empagliflozin, and ertugliflozin) are approved for the treatment of type 2 diabetes by the U.S. Food and Drug Administration (FDA) and the European Medicines Agency (EMA), either as monotherapy or in combination with other anti-diabetic drug classes. In Japan, three oral agents (i.e., ipragliflozin, luseogliflozin and tofogliflozin) are approved by the Ministry of Health, Labour and Welfare. Recently, remogliflozin has been approved in India.

When used as monotherapy in patients with preserved renal function, the glucose-lowering efficacy of SGLT2 inhibitors at baseline HbA$_{1C}$ of ~7.0–9.0% is on average ~0.8% reduction in HbA$_{1C}$, and ~0.6% when used as add-on therapy (Vasilakou et al. 2013). The efficacy of gliflozins is enhanced at high baseline glycemia as hyperglycemia increases the renal filtered glucose load. Conversely, since the filtered glucose load in patients with impaired renal function is reduced, the glucose-lowering efficacy parallels renal function and gradually declines to HbA$_{1C}$ reductions of 0.3–0.4% in eGFR range 30–59 mL/min/1.73m^2 and no effect <30 mL/min/1.73m^2 (Van Bommel et al. 2017a). SGLT2 inhibitors increase plasma glucagon levels and stimulate hepatic glucose output (HGO), which limits their further glucose-lowering capacity. Considering that their mode of action is independent of insulin action, SGLT2 inhibitors are effective in all individuals with type 2 diabetes and preserved renal function.

5 Pleiotropic Effects of SGLT2 Inhibitors

5.1 Cardiovascular Protection

SGLT2 inhibitors have shown unprecedented CV benefits in large CV outcome trials (CVOTs). The CV benefits of empagliflozin, canagliflozin and dapagliflozin were demonstrated in the EMPA-REG OUTCOME trial (Zinman et al. 2015), the CANVAS Program (Neal et al. 2017) and DECLARE TIMI 58 (Wiviott et al. 2019), respectively (Table 3). Importantly, nearly all included patients were adequately treated with statins and BP-lowering agents, most notably ACE inhibitors or angiotensin II receptor blockers.

In 2015, data from the Empagliflozin Cardiovascular Outcome Event Trial in Type 2 Diabetes Mellitus Patients-Removing Excess Glucose (EMPA-REG OUTCOME) trial changed the paradigm in diabetes management by showing a lower risk of CV death among the patients who received empagliflozin than among the controls (3.7% vs. 5.9%), a 38% reduction in the risk of CV death (Zinman et al. 2015). Following a median observation period of 3.1 years, treatment with empagliflozin versus placebo in addition to standard of care led to a 14% reduction in the risk of 3-point major adverse CV events (MACE, the composite of CV death, nonfatal myocardial infarction, and nonfatal stroke). Empagliflozin also reduced the risk of hospitalization for heart failure by 35% and death from any cause by 32% versus placebo (Table 3). These findings led the U.S. Food and Drug Administration (US FDA) to extend the indication for empagliflozin to include reducing the risk of CV death in patients with type 2 diabetes and established CV disease (U.S. Food and Drug Administration 2019).

In the CANVAS (Canagliflozin Cardiovascular Assessment Study) program (Neal et al. 2017), the 3-point MACE occurred at rates of 26.9 versus 31.5 per 1000 patient–years in the canagliflozin and placebo groups, respectively, giving a significant 14% risk reduction – similar to that seen with empagliflozin (Table 3). The benefit for canagliflozin appeared to be similar for patients with heart failure with reduced or preserved ejection fraction (Neal et al. 2017).

The DECLARE-TIMI 58 (Dapagliflozin Effect on Cardiovascular Events-Thrombolysis in Myocardial Infarction 58) is the largest of the SGLT2 inhibitor CVOTs (Wiviott et al. 2019), and like the CANVAS trial specifically included both a primary and secondary prevention population. The co-primary endpoints were a 3-point MACE and the combination of CV death or hospitalization for HF. Dapagliflozin had a significant effect on the second of these endpoints, with a rate of 4.9% versus 5.8% with placebo, driven entirely by a reduction in HF hospitalization risk (Wiviott et al. 2019). There was no significant effect on the MACE endpoint overall or in patients with multiple CV risk factors. Patients with established CVD had a risk reduction with dapagliflozin which, although nonsignificant, was nevertheless in line with the size of the effect seen in EMPA-REG OUTCOME and CANVAS, at rates of 13.9% versus 15.3%. In the DAPA-HF trial, patients with heart failure (New York Heart Association class II, III or IV) and reduced left ventricular ejection fraction (40% or less) were randomised to dapagliflozin 10 mg per day and placebo. The primary outcome was a composite of worsening of heart failure or cardiovascular death. Over a median of 18.2 months, dapagliflozin significantly reduced the primary outcome versus placebo (16.3% versus 21.2%; hazard ratio, 0.74; 95% confidence interval, 0.65–0.85; P < 0.001). The study also demonstrated that the findings in patients with diabetes were similar to those in patients without diabetes. In other words, the beneficial effect on heart failure were beyond the glucose-lowering effect of dapagliflozin (McMurray et al. 2019).

Further support for CV benefits of SGLT2 inhibitors were noted in two mata-analyses. In the first meta-analysis of all SGLT2 inhibitors which included data of 70,910 patients, there was 16% relative risk reduction in MACE and

Table 3 Cardiovascular outcome trials of SGLT2 inhibitors

CVOTs of SGLT2 inhibitors	Intervention (n)	Population	Follow-up (years)	Primary outcome	CV death	MI	Stroke	HHF	All-cause mortality
EMPA-REG OUTCOME	Empagliflozin vs. placebo (7020)	T2DM patients with established CVD	3.1	3-point MACE 0.86 (0.74–0.99)	0.62 (0.49–0.77)	0.87 (0.70–1.09)	1.18 (0.89–1.56)	0.65 (0.50–0.85)	0.68 (0.57–0.82)
CANVAS Program	Canagliflozin vs. placebo (10,142)	T2DM patients with established CVD or ≥ 2 CV risk factors	2.4	3-point MACE 0.85 (0.75–0.97)	0.96 (0.77–1.18)	0.89 (0.73–1.09)	0.87 (0.69–1.09)	0.67 (0.52–0.87)	0.60 (0.47–0.77)
DECLARE TIMI 58	Dapagliflozin vs. placebo (17,160)	T2DM patients with CVD or ≥ 1 CV risk	4.2	3-point MACE 0.93 (0.84–1.03)	0.98 (0.82–1.17)	0.89 (0.77–1.01)	1.01 (0.84–1.21)	0.73 (0.61–0.88)	0.93 (0.82–1.04)

HHF hospitalization for heart failure, *CVD* cardiovascular disease, *MACE* major adverse cardiovascular events, *MI* myocardial infarction

37% reduction in CV death (Wu et al. 2016). In another meta-analysis of dapagliflozin that analysed data of 9339 patients, there was 23% relative risk reduction in MACE (Sonesson et al. 2016). In a large real-world study, namely CVD-REAL (Comparative Effectiveness of Cardiovascular Outcomes in New Users of SGLT-2 Inhibitors) that analysed data of 309,056 patients, there was 39% reduced risk of heart failure hospitalization and 51% lower risk of all-cause death in patients taking SGLT2 inhibitor (Kosiborod et al. 2017). In another real-world study, called the Health Improvement Network (THIN) trial, that analysed data of 22,124 patients, there was 50% reduced risk of all-cause mortality (Toulis et al. 2017).

Evidence from all the dedicated CVOTs (Zinman et al. 2015; Neal et al. 2017; Wiviott et al. 2019; U.S. Food and Drug Administration 2019), meta-analyses (Wu et al. 2016; Sonesson et al. 2016) and real-world studies (Kosiborod et al. 2017; Toulis et al. 2017) of SGLT2 inhibitors demonstrated similar CV benefits, indicating a class effect of SGLT2 inhibitors.

The mechanism of action of SGLT2 inhibitors in reducing the risk of CV outcomes is currently incompletely understood. However, it is said to be multifactorial. An antiatherosclerotic effect is unlikely given the speed of the observed reduction in CV mortality. There are at least five putative mechanisms underlying CV benefits of SGLT2 inhibitors: (a) improvement in ventricular loading conditions through a reduction in preload (secondary to natriuresis and diuresis) and afterload (reduction in BP and improvement in vascular function). This effect is particularly important in the heart of patients with diabetes that function on a steep Frank-Starling curve. The early hemodynamic benefit would also explain the observed early separation of the Kaplan-Meier curves noted in the EMPA-REG OUTCOME and CANVAS trials (Sattar et al. 2016; Lytvyn et al. 2017); (b) improvement in cardiac metabolism and fuel bioenergetics. In patients with diabetes and/or heart failure, the metabolic flexibility of the heart is impaired. SGLT2 inhibition causes modest increase in the production of ketone body, namely β-hydroxybutyrate. This is

thought to offer an alternative fuel for myocardium in these patients, and is oxidised by the heart in preference to non-esterified fatty acids and glucose. This preferred ketone body utilization of the myocardium is said to improve global cardiac function and also improve its mechanical efficiency (Ferrannini et al. 2016; Lopaschuk and Verma 2016); (c) sodium-hydrogen exchanger (NHE) inhibition. SGLT2 inhibitors are known to work by inhibiting NHE1 in myocardium and NHE3 in the proximal tubule. The end result is the improvement in cardiac and whole-body sodium homeostasis, and reduction in cardiac failure (Packer et al. 2017; Uthman et al. 2018); (d) reduction in cardiac fibrosis. Cardiac fibrosis is a common final pathway through which cardiac failure develops. SGLT2 inhibitors have cardiac antifibrotic effects by suppressing collagen synthesis and inhibiting myofibroblast differentiation (Lee et al. 2017); e, alteration in adipokines, cytokines and epicardial adipose tissue mass. Recently, SGLT2 inhibitors have been shown to reduce proinflammatory adipokine, leptin; and increase the levels of anti-inflammatory adipokine, adiponectin. It is thought that these drugs mediate their beneficial effects, in part, by restoring the balance between pro- and anti-inflammatory adipokines and cytokines (Garvey et al. 2018). Increased epicardial adipose tissue volume has been implicated in the development of heart failure, and SGLT2 inhibitors has been shown to reduce epicardial adipose tissue (Sato et al. 2018). The modulation of aforementioned pathogenetic mechanisms by SGLT2 inhibitors for improving cardiac function needs further confirmation by dedicated pre-clinical and clinical studies.

5.2   Renal Protection

Chronic hyperglycemia causes an increase in proximal tubule filtered glucose load in patients with diabetes. This results in overactivity of SGLT2 and SGLT1 proteins and thereby increased renal tubular reabsorption of glucose and sodium (Thomson et al. 2004; Hannedouche et al. 1990). This increased proximal sodium reabsorption leads to decreased sodium delivery to the macula densa, with consequent reduction in adenosine production. Adenosine is a strong vasoconstrictor, and its reduction causes vasodilation of the afferent arteriole and thus glomerular hyperfiltration (Faulhaber-Walter et al. 2008). Hyperfiltration is an early renal hemodynamic abnormality which is associated with increased intraglomerular pressure (Hostetter et al. 1981). SGLT2 inhibition increases glucose and sodium delivery to the macula densa, which is sensed by the juxtaglomerular apparatus as increased glomerular perfusion. This leads to increased adenosine production and thereby vasoconstriction of the afferent arteriole, which decreases intraglomerular pressure. Although these effects decrease the estimated GFR in the short term, over time that effect stabilizes. The level of angiotensin II in the circulation decreases, as does the level of atrial natriuretic peptide, with a subsequent decrease in inflammation and an increase in intrarenal oxygenation (Škrtić and Cherney 2015; Lovshin and Gilbert 2015). Decreased body weight and sympathetic output, decreased uric acid, and perhaps an increase in glucagon may also contribute. SGLT2 inhibitors also have inhibitory effects on the inflammatory and fibrotic responses of proximal tubular cells to hyperglycemia (Komala et al. 2013). There is an increase in hematocrit during SGLT2 inhibitor treatment, which may partly be mediated by an increase in erythropoiesis. Under diabetic conditions, tubulointerstitial hypoxia causes reductions in erythropoietin production and this tubulointerstitial injury may be attenuated with SGLT2 inhibition, thereby also improving global renal function (Sano et al. 2016). In clinical studies, the changes in eGFR during SGLT2 inhibition are similar in patients with normal renal function and in those with CKD. The time course of changes in renal function is typically characterized by a rapid decline in GFR during the first weeks of treatment, followed by a progressive recovery that is faster and more evident in patients with normal renal function at baseline. In a large study where patients with normal renal function were randomized to canagliflozin or glimepiride on top of metformin for 52 weeks

(Cefalu et al. 2013), the expected pattern of GFR course was observed, with an initial fall followed by stabilization. In patients with mild-to-moderate renal impairment, treatment with dapagliflozin (Kohan et al. 2014), canagliflozin (Yale et al. 2014), and empagliflozin (Barnett et al. 2014) resulted in an initial decrease in eGFR with a trend toward an increase over time. In patients with CKD stage 3, after 1 week of treatment there was a reduction in eGFR (Abdul and DeFronzo 2013; Kohan et al. 2014) with a progressive recovery of eGFR during the following weeks. In a study where patients with eGFR of 30–50 mL/min/1.73 m2 were randomized to canagliflozin or placebo for 26 weeks (Wright 2001; Yale et al. 2014), a similar eGFR course was observed. Finally, in CKD stage 3, patients treated with empagliflozin 25 mg reduced eGFR by ~4 mL/min at 12 weeks, with a slight recovery at 52 weeks. Interestingly, after drug discontinuation eGFR returned to the baseline values, suggesting that the decrease in eGFR during treatment is hemodynamic and not consequent of renal injury (Barnett et al. 2014).

In addition to the effects on eGFR, SGLT2 inhibitors also influence albuminuria. In a post hoc analysis including 116 patients with stage 3 CKD and increased albuminuria (>30 mg/g), dapagliflozin caused placebo-corrected UACR reductions of −57.2% and −43.8% in the dapagliflozin 10 mg and 5 mg groups, respectively, at 104 weeks. Overall, 17.8%, 18.9% and 7.0% of patients improved to normoalbuminuria status in the dapagliflozin 10 mg, 5 mg and placebo groups, respectively (Fioretto et al. 2016). In the secondary analysis of EMPA-REG OUTCOME trial, empagliflozin reduced the risk of progression to macroalbuminuria by 38%, the risk of a doubling of creatinine by 44% and the risk of starting renal-replacement therapy by 55% (Wanner et al. 2016).

In the DELIGHT study, dapagliflozin reduced albuminuria by 21% after 24 weeks of treatment in patients with type 2 diabetes and moderate-to-severe CKD who were receiving ACE inhibitor or angiotensin II receptor blocker (ARB) therapy. The DELIGHT included patients with type 2 diabetes, increased albuminuria (UACR

30–3500 mg/g), and an eGFR of 25–75 mL/min/ $1.73m^2$, who had been receiving stable doses of ACE inhibitor or ARB therapy and glucose-lowering therapy for at least 12 weeks (Pollock et al. 2019).

In the CREDENCE study (Perkovic et al. 2019), canagliflozin reduced risk of kidney failure in patients with type 2 diabetes and CKD. It was a large study (n = 4401) with a median follow-up of 2.62 years. All the patients had an eGFR of 30–90 ml/min/$1.73m^3$ and albuminuria (UACR >300–5000 mg/g) and were treated with renin–angiotensin system blockade. The primary outcome was a composite of end-stage kidney disease (dialysis, transplantation, or a sustained estimated GFR of <15 ml/min/1.73 m²), a doubling of the serum creatinine level, or death from renal or cardiovascular causes. The relative risk of the primary outcome was 30% lower in the canagliflozin group than in the placebo group (P = 0.00001). The relative risk of the renal-specific composite of end-stage kidney disease, a doubling of the creatinine level, or death from renal causes was lower by 34% (P < 0.001), and the relative risk of end-stage kidney disease was lower by 32% (P = 0.002) (Perkovic et al. 2019). The ADA has recently updated its diabetes standards of care to incorporate results from the CREDENCE trial. Among the updates, the patients with type 2 diabetes and diabetic kidney disease, SGLT2 inhibitors should be considered when the eGFR is at or above 30 ml/min/$1.73m^2$, especially with albuminuria above 300 mg/g, to lower renal and CV risk.

The overall evidence from secondary analyses of CVOTs and dedicated renal-outcome trials show that the nephroprotective effect is a class effect of gliflozins and is independent of their glucose-lowering effect.

5.3 Weight Reducing Effect

SGLT2 inhibitors have been shown to induce body weight loss of 1.5–2.5 kg, which starts with a rapid decline of 1–1.5 kg in the first weeks, probably due to osmotic diuresis (Sha et al. 2014). Thereafter, body weight declines more gradually over 20 weeks, which can probably be related to reductions in fat mass, and subsequently reaches a plateau (Ferrannini et al. 2015). The weight reducing property of SGLT2 inhibitors is dose-dependent. In a meta-analysis (Cai et al. 2018), compared with a placebo, dapagliflozin 5 mg and 10 mg led to significant weight reductions (weighted mean difference, WMD, −1.51 kg and − 2.24 kg, respectively). Similarly, canagliflozin 100 mg, 200 mg and 300 mg led to a significant decrease in weight (WMD, −1.82 kg, −1.83 kg and − 2.37 kg, respectively) (Cai et al. 2018). Weight changes from baseline corrected by a placebo with different SGLT2 inhibitors is shown in Table 4.

Reductions in body weight due to SGLT2 inhibitors in the plateau phase may be explained by reduced total body fat mass, visceral adipose tissue, and subcutaneous adipose tissue volume (Bolinder et al. 2012). Interestingly, this 1.5- to 2.5-kg weight loss observed at the plateau phase is less than expected based on the calculated loss of calories excreted in the urine (approximately 240–320 kcal/d), which would equal a weight reduction of ~10–11 kg. Since SGLT2 inhibitors do not alter resting or postprandial energy expenditure (Ferrannini et al. 2014), the discrepancy between expected and observed weight loss implies that caloric intake is probably increased. This compensatory increase in caloric intake has been seen in pre-clinical studies (Devenny et al. 2012).

5.4 Blood Pressure Lowering Effect

SGLT2 inhibitors are known to have beneficial effects on blood pressure (BP). In a study by Ferdinand, et al., empagliflozin consistently reduced office and 24-h ambulatory SBP and DBP, by about −4/−2 mmHg greater than placebo at 12 weeks, and by about −7/−4 mmHg greater than placebo at 24 weeks (Ferdinand et al. 2019). This effect was observed in a black population with type 2 diabetes and hypertension, who were already receiving at least one BP-lowering medication. The SBP lowering effect in the afore-mentioned study was greater than that reported

Table 4 Placebo corrected weight changes from baseline with SGLT2 inhibitors

SGLT2 inhibitor (once daily dose)	WMD from baseline in weight (kg)	95% CI (kg)	BMI <30 kg/m^2 WMD (95% CI) (kg)	BMI >30 kg/m^2 WMD (95% CI) (kg)
Empagliflozin				
10 mg	−1.84	−1.98 to −1.69	−1.89 (−2.29 to −1.48)	−2.03 (−2.34 to −1.72)
25 mg	−1.93	−2.08 to −1.77	−1.93 (−2.04 to −1.82)	−1.94 (−2.34 to −1.55)
Canagliflozin				
100 mg	−1.82	−2.10 to −1.53	−1.74 (−2.14 to −1.34)	−1.85 (−2.26 to −1.44)
300 mg	−2.37	−2.82 to −1.93	−2.01 (−2.80 to −1.21)	−2.47 (−2.97 to −1.96)
Dapagliflozin				
5 mg	−1.51	−1.74 to −1.29	−1.69 (−2.32 to −1.06)	−1.47 (−1.73 to −1.21)
10 mg	−1.79	−2.02 to −1.55	−1.92 (−2.04 to −1.80)	−1.77 (−2.05 to −1.49)
Tofogliflozin				
20 mg	−2.15	−2.82 to −1.48	−2.50 (−2.65 to −2.34)	−1.81 (−1.98 to −1.64)
40 mg	−2.35	−2.87 to −1.83	−2.62 (−2.77 to −2.46)	−2.08 (−2.25 to −1.92)
Ipragliflozin				
50 mg	−1.40	−1.71 to −1.10	−1.35 (−1.66 to −1.04)	−1.58 (−1.77 to −1.38)

WMD weighted mean difference, *CI* confidence interval, *BMI* body mass index defined as body weight in kilograms divided by height in meters squared

with empagliflozin in a predominantly white population with type 2 diabetes and hypertension, in which the changes from baseline to week 12 in mean 24-h ambulatory SBP were − 3.44 mmHg (empagliflozin 10 mg) and − 4.16 mmHg (empagliflozin 25 mg), (both p = 0.001) (Tikkanen et al. 2015). In a meta-analysis of randomised controlled trials, SGLT2 inhibitors significantly reduced 24-h ambulatory SBP and DBP by −3.76 mmHg and − 1.83 mmHg, respectively (Baker et al. 2017), without a potentially harmful increase in heart rate (Vasilakou et al. 2013). Interestingly, the antihypertensive effect of SGLT2 inhibitors has been shown to be independent of eGFR, indicating that other factors then volume depletion also contribute (Baker et al. 2017; Vasilakou et al. 2013). In a study by Kawasoe et al., SBP significantly decreased at 1 month after administration of SGLT2 inhibitors. The decrease in SBP at 1 month was accompanied by an increase in urinary volume and urinary excretion of both glucose and sodium. However, SBP continued to decrease up to 6 months after administration of SGLT2 inhibitors. In contrast with the observations at 1 month, the decrease in SBP at 6 months did not corelated with decreases in UGE. Instead, the 6-month SBP reduction corelated with the increase in urinary sodium excretion, suggesting that the BP-lowering effect of SGLT2 inhibitors at 6 months derives mainly from plasma volume reduction (secondary to a natriuretic effect). These observations suggest that SGLT2 inhibitors may exert both short-term and longer-term effects on BP (Kawasoe et al. 2017). Several mechanisms that may underlie this antihypertensive effect have been suggested: a) plasma volume contraction by osmotic diuresis; b) body weight reduction; c) improvements in vascular stiffness by reductions in body weight, hyperglycemia-associated oxidative stress, and/or endothelial glycocalyx protection from sodium overload; d) reduced sympathetic nervous system activity; and e) lower serum uric acid concentrations (Muskiet et al. 2016).

5.5   Uricosuric Effect

Hyperuricemia is a frequent association of type 2 diabetes, as insulin resistance and hyperinsulinemia reduce renal excretion of uric acid (Richette et al. 2014). Clinical studies indicate that hyperuricemia contribute to the development of renal disease and CVD. SGLT2 inhibitors have been shown to reduce uric acid levels in patients with type 2 diabetes by 0.3–0.9 mg/dl (Bailey et al. 2010; Rosenstock et al. 2012). By increasing glucose levels in the glomerular filtrate, SGLT2 inhibition is proposed to cause glucose transporter 9 (GLUT9) isoform 2, which is located more distally in the proximal tubule, to excrete more uric acid in exchange for glucose reuptake and also, lead to reduced uric acid reabsorption in the collecting duct (Chino et al. 2014). Whether this uricosuric effect of SGLT2 inhibitors is of any clinical relevance is not yet known.

5.6   Changes in Lipid Metabolism

SGLT2 inhibitors modestly alter lipid profiles by reductions in plasma TG (1.0–9.4%) and increases in HDL cholesterol (5.5–9.4%) and LDL cholesterol (1.5–6.3%) (Inzucchi et al. 2015; van Bommel et al. 2017b). This improvement in TG-HDL axis may be related to weight loss and improvement in insulin sensitivity, whereas the increase in LDL could be explained by the switch in energy metabolism from carbohydrate to lipid utilization (Cefalu 2014). Whether this change in TG-HDL axis translates into clinical benefit is not known. The clinical relevance of increase in LDL cholesterol is probably small, as it was accompanied by CV benefit in the EMPA-REG OUTCOME trial.

5.7   Improvement in Nonalcoholic Fatty Liver Disease

The E-LIFT Trial (Effect of Empagliflozin on Liver Fat) was the first randomized controlled trial that demonstrated usefulness of empagliflozin in reducing liver fat (as measured by MRI-proton density fat fraction) in patients with type 2 diabetes and nonalcoholic fatty liver disease (NAFLD) (Kuchay et al. 2018). The trial also demonstrated that serum alanine aminotransferase (ALT) reduced significantly with empagliflozin vs. controls. In the post hoc analysis of EMPA-REG OUTCOME trial (Sattar et al. 2018), changes from baseline ALT and AST were assessed in all treated patients (n = 7020). Patients with type 2 diabetes and established CVD were randomized to receive empagliflozin 10 mg, 25 mg or placebo in addition to standard care. The results were reduction in ALT and AST with empagliflozin vs. placebo, with greater reductions in ALT than AST, in a pattern consistent with reduction in liver fat. This study also demonstrated that reductions in ALT were greatest in the highest tertile of baseline ALT ($p < 0.0001$) (Sattar et al. 2018).

In a recent study from Japan, forty patients with type 2 diabetes and NAFLD were treated with luseogliflozin for 24 weeks. The results were significant reduction in liver fat as measured by MRI hepatic fat fraction (p < 0.001). There were also significant reductions in ALT and AST levels (Sumida et al. 2018). In a post hoc analysis of pooled data, a short-term (12 weeks) as well as long-term (52 weeks) canagliflozin therapy reduced serum ALT levels significantly in patients with baseline serum ALT >30 IU/L (Seko et al. 2018). Dapagliflozin therapy for 8 weeks also reduced liver fat as measured by PDFF (p < 0.01) in patients with type 2 diabetes (Latva-Rasku et al. 2019). In a double-blind, placebo-controlled trial, dapagliflozin therapy for 12 weeks reduced the levels of hepatocyte injury biomarkers, including ALT, AST, GGT, cytokeratin (CK) 18-M30 and CK 18-M65 and plasma fibroblast growth factor 21 (FGF21) (Eriksson et al. 2018).

In a histological study of five patients with nonalcoholic steatohepatitis (NASH) and type 2 diabetes, canagliflozin therapy for 24 weeks showed histological improvement (defined as a decrease in an NAFLD activity score of ≥ 1 point without a worsening of fibrosis stage) in all the patients (Akuta et al. 2017). In another paired biopsy pilot study of nine patients, empagliflozin improved steatosis ($p = 0.025$), ballooning ($p = 0.024$), and fibrosis ($p = 0.008$) compared with historical placebo (Lai et al. 2019).

In conclusion, effect of SGLT2 inhibitors on NAFLD in patients with type 2 diabetes is a class effect and looks promising. A large and long term, paired biopsy study is needed to see the effect of SGLT2 inhibitors on NAFLD-related liver fibrosis.

6 Adverse Effects of SGLT2 Inhibitors

SGLT2 inhibitors are generally well tolerated, the adverse events are clinically manageable and rarely lead to drug discontinuation. However, the important concerns of SGLT2 inhibitors are as following:

6.1 Genital Mycotic Infections

There is a four- to fivefold increased risk of genital mycotic infection, especially in women and uncircumcised men (Van Bommel et al. 2017a). The pooled safety data from 12 randomised clinical trials of dapagliflozin revealed that when administered in daily doses of 2.5, 5 and 10 mg, the incidence of clinically diagnosed urogenital tract infections were 4.1%, 5.7% and 4.8%, respectively, while it was 0.9% with placebo (Johnsson Kristina et al. 2013). The most common infections seen in females (8.5–10.8%) included vulvovaginal mycotic infections, vaginal infections and vulvovaginal pruritus who received dapagliflozin versus 3.4% who received

placebo; while among males, balanitis was more frequent (1.0–1.2%) in the dapagliflozin group compared with the placebo group (0.1%) (Johnsson Kristina et al. 2013). This study also reported that most of the genital infections were mild to moderate in nature and responded to standard antimicrobial therapy. Another study reported that the incidence of genital infections can be significantly reduced by maintaining perineal hygiene and standard antifungal treatment (Yabe et al. 2015). The discontinuation of SGLT2 inhibitor therapy was rarely required (Yabe et al. 2015). Similarly, higher incidence of genital infections were reported in pooled safety data of empagliflozin. The incidence of genital infections in patients on empagliflozin was 3.9% as compared with patients on placebo (0.7%) (Kim et al. 2014). most of the patients had only one episode of genital infection, and one a few of them discontinued empagliflozin therapy due to these adverse events (Kim et al. 2014). Genital infections are probably due to increased glucose load in the urinary tract, which promotes fungal growth.

Fournier's gangrene, a rare condition characterized by necrotizing infection of the external genitalia, perineum and perianal region, is a newly identified safety concern in patients receiving SGLT2 inhibitors. The FDA identified 55 cases of Fournier gangrene in patients receiving SGLT2 inhibitors. All patients required surgical intervention and were severely ill (Bersoff-Matcha et al. 2019). Clinicians prescribing these agents should be aware of this condition and have a high index of suspicion to recognise it in its early stages.

6.2 Orthostatic Hypotension

Since SGLT2 inhibitors cause osmotic diuresis, they have the potential to cause volume depletion and orthostatic hypotension. In clinical trials, however, the incidence of these adverse events has been minimal (<3%) (Ptaszynska et al. 2014).

6.3 Diabetic Ketoacidosis

Rare episodes of diabetic ketoacidosis (DKA), particularly in patients with long-standing type 2 diabetes, have been reported, which prompted the FDA to issue a warning about this potential complication. SGLT2 inhibitors have the propensity to cause DKA due to a reduced availability of carbohydrates caused by gliflozin-induced glycosuria, a shift in substrate utilization from glucose to fat oxidation, and the promotion of hyperglucagonemia, stimulating ketogenesis (Rosenstock and Ferrannini 2015). The increased incidence of DKA was seen with all the approved SGLT2 inhibitors. However, most of the DKA cases were reported in patients with insulin-treated type 2 diabetes. The identified triggering factors for DKA were intercurrent illness, reduced fluid and food intake, reduced insulin doses, and history of alcohol intake (Rosenstock and Ferrannini 2015). In order to mitigate this problem, patients need to be counselled about the risk of euglycemic DKA. In the presence of ketosis or ketonuria, SGLT2 inhibitors should be discontinued, carbohydrate and fluid intake should be maintained to allow continuation of insulin therapy at optimal doses.

6.4 Bone Fractures

In the CANVAS Program, canagliflozin was associated with a higher risk of bone fractures (Neal et al. 2017), which has not been reported with other SGLT2 inhibitors. There are probably several mechanisms for the effect of SGLT2 inhibitors on bone metabolism, including the disordered calcium and phosphate homeostasis. This contributes to increased parathyroid hormone (PTH) levels and reduced 1,25-dihydroxy vitamin D levels, affecting bone metabolism. Following canagliflozin administration, phosphate, FGF23, and PTH levels increase, whereas 1,25-dihydroxyvitamin D level decreases, which might have detrimental effects on bone health (Blau et al. 2018). Theoretically, increased risk of fractures can also be explained by occasional

orthostatic hypotension secondary to SGLT2 inhibitor-associated diuresis and thereby increased risk of falls, especially in elderly patients. However, the effect of SGLT2 inhibitors on fracture risk remains controversial, and evidence indicating the direct effect of SGLT2 inhibitors on fracture risk is lacking. No detrimental effect of SGLT2 inhibitors on bone fracture risk was indicated by a network meta-analysis of 40 RCTs involving 32,343 patients with type 2 diabetes and 466 fracture cases (Azharuddin et al. 2018). Another meta-analysis of 30 studies involving 23,372 patients with type 2 diabetes also did not indicate any increased risk of bone fractures due to SGLT2 inhibitors as compared with placebo (Cheng et al. 2019). Finally, the CREDENCE trial did not indicate any signal of bone fractures associated with canagliflozin treatment (Perkovic et al. 2019).

6.5 Risk of Lower Limb Amputation

The CANVAS trial programme reported a two-fold increase in the occurrence of lower limb amputations (LLAs) in participants receiving canagliflozin (Perkovic et al. 2019). These were predominantly toe or metatarsal amputations and they occurred in individuals without established peripheral vascular disease prior to study commencement. These findings prompted the US FDA to issue a caution on the use of canagliflozin in individuals at risk of amputation (United States Food and Drug Administration 2017). In the EMPA-REG OUTCOME trial, rates of LLAs were not significantly higher in those randomised to empagliflozin (Zinman et al. 2015). Similarly, a pooled analysis of 30 trials did not find a significant association between dapagliflozin and LLAs (Jabbour et al. 2017). However, long term data are lacking, in particular regarding the use of SGLT2 inhibitors amongst people at *high baseline risk* of amputation. On this basis, the European Medicines Agency has adopted a cautious approach, advising against the use of *any* SGLT2 inhibitors in patients at risk of amputation until further data are available (European Medicines Agency 2017). Recently, the CREDENCE trial did not show any signal of LLAs associated with canagliflozin use.

7   Conclusions

SGLT2 inhibitors are antidiabetic agents with good glucose-lowering efficacy. These agents have clinically significant pleiotropic effects especially on CV and renal health. These drugs also cause sustained reductions in body weight, BP and uric acid concentrations. SGLT2 inhibitors seem to reduce liver fat and improve liver inflammation in patients with NAFLD. These agents are usually well tolerated and the adverse effects are easily manageable. The mechanism of action for how SGLT2 inhibitors improve cardiorenal outcomes is incompletely understood, which makes this an important area of investigation.

Acknowledgements The authors would like to acknowledge Ganesh Jevalikar (Medanta The Medicity Hospital) for his valuable inputs for the manuscript.

Funding No funding was received from any source.

Conflict of Interest Mohammad Shafi Kuchay, Khalid Jamal Farooqui, Sunil Kumar Mishra and Ambrish Mithal declare that they have no conflicts of interest.

References

Abdul GM, DeFronzo R (2013) Dapagliflozin for the treatment of type 2 diabetes. Expert Opin Pharmacother 14:1695–1703

Akuta N, Watanabe C, Kawamura Y et al (2017) Effects of a sodium-glucose cotransporter 2 inhibitor in nonalcoholic fatty liver disease complicated by diabetes mellitus: preliminary prospective study based on serial liver biopsies. Hepatol Commun 1.46–52

Azharuddin M, Adil M, Ghosh P, Sharma M (2018) Sodium-glucose cotransporter 2 inhibitors and fracture risk in patients with type 2 diabetes mellitus: a systematic literature review and Bayesian network meta-analysis of randomized controlled trials. Diabetes Res Clin Pract 146:180–190

Bailey CJ, Gross JL, Pieters A, Bastien A, List JF (2010) Effect of dapagliflozin in patients with type 2 diabetes

who have inadequate glycaemic control with metformin: a randomised, double-blind, placebo-controlled trial. Lancet 375:2223–2233

Baker WL, Buckley LF, Kelly MS et al (2017) Effects of sodium-glucose cotransporter 2 inhibitors on 24-hour ambulatory blood pressure: a systematic review and meta-analysis. J Am Heart Assoc 6:e005686

Barnett AH, Mithal A, Manassie J et al (2014) Efficacy and safety of empagliflozin added to existing antidiabetes treatment in patients with type 2 diabetes and chronic kidney disease: a randomised, double-blind, placebo-controlled trial. Lancet Diabetes Endocrinol 2:369–384

Bersoff-Matcha SJ, Chamberlain C, Cao C, Kortepeter C, Chong WH (2019) Fournier gangrene associated with sodium-glucose Cotransporter-2 inhibitors: a review of spontaneous postmarketing cases. Ann Intern Med 170 (11):764–769

Blau JE, Bauman V, Conway EM et al (2018) Canagliflozin triggers the FGF23/1,25-dihydroxyvitamin D/PTH axis in healthy volunteers in a randomized crossover study. JCI Insight 3:99123. https://doi.org/10.1172/jci.insight.99123

Bolinder J, Ljunggren Ö, Kullberg J et al (2012) Effects of dapagliflozin on body weight, total fat mass, and regional adipose tissue distribution in patients with type 2 diabetes mellitus with inadequate glycemic control on metformin. J Clin Endocrinol Metab 97:1020–1031

Cai X, Yang W, Gao X et al (2018) The association between the dosage of SGLT2 inhibitor and weight reduction in type 2 diabetes patients: a meta-analysis. Obesity (Silver Spring) 26:70–80

Cefalu WT (2014) Paradoxical insights into whole body metabolic adaptations following SGLT2 inhibition. J Clin Invest 124:485–487

Cefalu WT, Leiter LA, Yoon KH et al (2013) Efficacy and safety of canagliflozin versus glimepiride in patients with type 2 diabetes inadequately controlled with metformin (CANTATA-SU): 52-week results from a randomised, double-blind, phase 3 non-inferiority trial. Lancet 382:941–950

Cheng L, Li YY, Hu W et al (2019) Risk of bone fracture associated with sodium-glucose cotransporter-2 inhibitor treatment: a meta-analysis of randomized controlled trials. Diabetes Metab 45:436–445. https://doi.org/10.1016/j.diabet.2019.01.010. [Epub ahead of print]

Chino Y, Samukawa Y, Sakai S et al (2014) SGLT2 inhibitor lowers serum uric acid through alteration of uric acid transport activity in renal tubule by increased glycosuria. Biopharm Drug Dispos 35:391–404

Cosentino F, Grant PJ, Aboyans V et al (2019) ESC Guidelines on diabetes, pre-diabetes, and cardiovascular diseases developed in collaboration with the EASD. Eur Heart J pii: ehz486. https://doi.org/10.1093/eurheartj/ehz486. [Epub ahead of print]

Davies MJ, D'Alessio DA, Fradkin J et al (2018) Management of hyperglycaemia in type 2 diabetes, 2018. A consensus report by the American Diabetes

Association (ADA) and the European Association for the Study of diabetes (EASD). Diabetologia 61 (12):2461–2498

Devenny JJ, Godonis HE, Harvey SJ, Rooney S, Cullen MJ, Pelleymounter MA (2012) Weight loss induced by chronic dapagliflozin treatment is attenuated by compensatory hyperphagia in diet-induced obese (DIO) rats. Obesity (Silver Spring) 20:1645–1652

Eriksson JW, Lundkvist P, Jansson PA et al (2018) Effects of dapagliflozin and n-3 carboxylic acids on non-alcoholic fatty liver disease in people with type 2 diabetes: a double-blind randomised placebo-controlled study. Diabetologia 61:1923–1934

European Medicines Agency (2017) SGLT2 inhibitors: information on potential risk of toe amputation to be included in prescribing information. http://www.ema.europa.eu/docs/en_GB/document_library/Referrals_document/SGLT2_inhibitors_Canagliflozin_20/European_Commission_final_decision/WC500227101.pdf. Accessed 12 Nov 2017

Faulhaber-Walter R, Chen L, Oppermann M et al (2008) Lack of A1 adenosine receptors augments diabetic hyperfiltration and glomerular injury. J Am Soc Nephrol 19:722–730

Ferdinand KC, Izzo JL, Lee J et al (2019) Antihyperglycemic and blood pressure effects of empagliflozin in black patients with type 2 diabetes mellitus and hypertension. Circulation 139:2098–2109

Ferrannini E, Muscelli E, Frascerra S et al (2014) Metabolic response to sodium-glucose cotransporter 2 inhibition in type 2 diabetic patients. J Clin Invest 124:499–508

Ferrannini G, Hach T, Crowe S, Sanghvi A, Hall KD, Ferrannini E (2015) Energy balance after sodium–glucose cotransporter 2 inhibition. Diabetes Care 38:1730–1735

Ferrannini E, Mark M, Mayoux E (2016) CV protection in the EMPA-REG OUTCOME trial: a "thrifty substrate" hypothesis. Diabetes Care 39:1108–1114

Fioretto P, Stefansson BV, Johnsson E, Cain VA, Sjöström CD (2016) Dapagliflozin reduces albuminuria over 2 years in patients with type 2 diabetes mellitus and renal impairment. Diabetologia 59:2036–2039

Garvey WT, Van Gaal L, Leiter LA et al (2018) Effects of canagliflozin versus glimepiride on adipokines and inflammatory biomarkers in type 2 diabetes. Metabolism 85:32–37

Hannedouche TP, Delgado AG, Gnionsahe DA, Boitard C, Lacour B, JP G"u (1990) Renal hemodynamics and segmental tubular reabsorption in early type 1 diabetes. Kidney Int 37:1126–1133

Hostetter TH, Olson JL, Rennke HG, Venkatachalam MA, Brenner BM (1981) Hyperfiltration in remnant nephrons: a potentially adverse response to renal ablation. Am J Phys 241:F85–F93

Hummel CS, Lu C, Loo DD et al (2011) Glucose transport by human renal NA+/d-glucose cotransporters SGLT1 and SGLT2. Am J Physiol Cell Physiol 300:C721

Inzucchi SE, Zinman B, Wanner C et al (2015) SGLT-2 inhibitors and cardiovascular risk: proposed pathways and review of ongoing outcome trials. Diab Vasc Dis Res 12:90–100

Jabbour S, Seufert J, Scheen A et al (2017) Dapagliflozin in patients with type 2 diabetes mellitus: a pooled analysis of safety data from phase IIb/III clinical trials. Diabetes Obes Metab 20:620–628

Johnsson Kristina M, Agata P, Bridget S, Jennifer S et al (2013) Vulvovaginitis and balanitis in patients with diabetes treated with dapagliflozin. J Diabetes Complicat 27:479–484

Kasichayanula S, Liu X, LaCreta F, Griffen S, Boulton D (2014) Clinical pharmacokinetics and pharmacodynamics of dapagliflozin, a selective inhibitor of sodium-glucose co-transporter type 2. Clin Pharmacokinet 53:17–27

Kawasoe S, Maruguchi Y, Kajiya S et al (2017) Mechanism of the blood pressure-lowering effect of sodium-glucose cotransporter 2 inhibitors in obese patients with type 2 diabetes. BMC Pharmacol Toxicol 18:23

Kim G, Gerich J, Salsali A, Hach T, Hantel S, Woerle HJ (2014) Empagliflozin (EMPA) increases genital infections but not Urinary Tract Infections (UTIs) in pooled data from four pivotal phase III trials. Diabetologie und Stoffwechsel 9:140

Kohan DE, Fioretto P, Tang W, List JF (2014) Long-term study of patients with type 2 diabetes and moderate renal impairment shows that dapagliflozin reduces weight and blood pressure but does not improve glycemic control. Kidney Int 85:962–971

Komala MG, Panchapakesan U, Pollock C, Mather A (2013) Sodium glucose cotransporter 2 and the diabetic kidney. Curr Opin Nephrol Hypertens 22:113–119

Kosiborod M, Cavender MA, Fu AZ et al (2017) Lower risk of heart failure and death in patients initiated on sodium-glucose Cotransporter-2 inhibitors versus other glucose-lowering drugs: the CVD-REAL study (comparative effectiveness of cardiovascular outcomes in new users of sodium-glucose cotransporter-2 inhibitors). Circulation 136:249–259

Kuchay MS, Krishan S, Mishra SK et al (2018) Effect of Empagliflozin on liver fat in patients with type 2 diabetes and nonalcoholic fatty liver disease: a randomized controlled trial (E-LIFT trial). Diabetes Care 41:1801–1808

Lai LL, Vethakkan SR, Nik Mustapha NR, Mahadeva S, Chan WK (2019) Empagliflozin for the treatment of nonalcoholic Steatohepatitis in patients with type 2 diabetes mellitus. Dig Dis Sci. https://doi.org/10.1007/s10620-019-5477-1. [Epub ahead of print]

Latva-Rasku A, Honka MJ, Kullberg J et al (2019) The SGLT2 inhibitor Dapagliflozin reduces liver fat but does not affect tissue insulin sensitivity: a randomized, double-blind, placebo controlled study with 8-week treatment in type 2 diabetes patients. Diabetes Care 42:931–937. https://doi.org/10.2337/dc18-1569.. [Epub ahead of print]

Lee YJ, Lee YJ, Han HJ (2007) Regulatory mechanisms of Na(+)/glucose cotransporters in renal proximal tubule cells. Kidney Int Suppl 106:S27–S35

Lee TM, Chang NC, Lin SZ (2017) Dapagliflozin, a selective SGLT2 inhibitor, attenuated cardiac fibrosis by regulating the macrophage polarization via STAT3 signaling in infarcted rat hearts. Free Radic Biol Med 104:298–310

Lopaschuk GD, Verma S (2016) Empagliflozin's fuel hypothesis: not so soon. Cell Metab 24:200–202

Lovshin JA, Gilbert RE (2015) Are SGLT2 inhibitors reasonable antihypertensive drugs and renoprotective? Curr Hypertens Rep 17:551

Lytvyn Y, Bjornstad P, Udell JA, Lovshin JA, Cherney DZI (2017) Sodium glucose cotransporter-2 inhibition in heart failure: potential mechanisms, clinical applications, and summary of clinical trials. Circulation 136:1643–1658

McMurray JJV, Solomon SD, Inzucchi SE et al (2019) Dapagliflozin in patients with heart failure and reduced ejection fraction. N Engl J Med 381(21):1995–2008

Muskiet MHA, van Bommel EJM, van Raalte DH (2016) Antihypertensive effects of SGLT2 inhibitors in type 2 diabetes. Lancet Diabetes Endocrinol 4:188–189

Nauck MA (2014) Update on developments with SGLT2 inhibitors in the management of type 2 diabetes. Drug Des Devel Ther 8:1335–1380

Neal B, Perkovic V, Mahaffey KW et al (2017) Canagliflozin and cardiovascular and renal events in type 2 diabetes. N Engl J Med 377:644–657

Packer M, Anker SD, Butler J, Filippatos G, Zannad F (2017) Effects of sodium-glucose cotransporter 2 inhibitors for the treatment of patients with heart failure: proposal of a novel mechanism of action. JAMA Cardiol 2:1025–1029

Perkovic V, Jardine MJ, Neal B et al (2019) Canagliflozin and renal outcomes in type 2 diabetes and nephropathy. N Engl J Med. https://doi.org/10.1056/NEJMoa1811744. [ahead of print]

Pollock C, Stefánsson B, Reyner D et al (2019) Albuminuria-lowering effect of dapagliflozin alone and in combination with saxagliptin and effect of dapagliflozin and saxagliptin on glycaemic control in patients with type 2 diabetes and chronic kidney disease (DELIGHT): a randomised, double-blind, placebo-controlled trial. Lancet Diabetes Endocrinol. https://doi.org/10.1016/S2213-8587(19)30086-5. [ahead of print]

Ptaszynska A, Johnsson KM, Parikh SJ, de Bruin TW, Apanovitch AM, List JF (2014) Safety profile of dapagliflozin for type 2 diabetes: pooled analysis of clinical studies for overall safety and rare events. Drug Saf 37:815–829

Richette P, Perez-Ruiz F, Doherty M et al (2014) Improving cardiovascular and renal outcomes in gout: what should we target? Nat Rev Rheumatol 10:654–661

Rosenstock J, Ferrannini E (2015) Euglycemic diabetic ketoacidosis: a predictable, detectable, and preventable safety concern with SGLT2 inhibitors. Diabetes Care 38(9):1638–1642

Rosenstock J, Aggarwal N, Polidori D et al (2012) Dose-ranging effects of canagliflozin, a sodium-glucose cotransporter 2 inhibitor, as add-on to metformin in subjects with type 2 diabetes. Diabetes Care 35:1232–1238

Sano M, Takei M, Shiraishi Y, Suzuki Y (2016) Increased hematocrit during sodium-glucose cotransporter 2 inhibitor therapy indicates recovery of tubulointerstitial function in diabetic kidneys. J Clin Med Res 8:844–847

Santer R, Calado J (2010) Familial renal glucosuria and SGLT2: from a mendelian trait to a therapeutic target. Clin J Am Soc Nephrol 5:133–141

Sato T, Aizawa Y, Yuasa S et al (2018) The effect of dapagliflozin treatment on epicardial adipose tissue volume. Cardiovasc Diabetol 17:6

Sattar N, McLaren J, Kristensen SL, Preiss D, McMurray JJ (2016) SGLT2 inhibition and cardiovascular events: why did EMPA-REG outcomes surprise and what were the likely mechanisms? Diabetologia 59:1333–1339

Sattar N, Fitchett D, Hantel S, George JT, Zinman B (2018) Empagliflozin is associated with improvements in liver enzymes potentially consistent with reductions in liver fat: results from randomised trials including the EMPA-REG OUTCOME trial. Diabetologia 61:2155–2163. https://doi.org/10.1007/s00125-018-4702-3

Seko Y, Sumida Y, Sasaki K et al (2018) Effects of canagliflozin, an SGLT2 inhibitor, on hepatic function in Japanese patients with type 2 diabetes mellitus: pooled and subgroup analyses of clinical trials. J Gastroenterol 53:140–151

Sha S, Polidori D, Heise T et al (2014) Effect of the sodium glucose co-transporter 2 inhibitor canagliflozin on plasma volume in patients with type 2 diabetes mellitus. Diabetes Obes Metab 16:1087–1095

Škrtić M, Cherney DZ (2015) Sodium-glucose cotransporter-2 inhibition and the potential for renal protection in diabetic nephropathy. Curr Opin Nephrol Hypertens 24:96–103

Sonesson C, Johansson PA, Johnsson E, Gause-Nilsson I (2016) Cardiovascular effects of dapagliflozin in patients with type 2 diabetes and different risk categories: a meta-analysis. Cardiovasc Diabetol 15:37

Sumida Y, Murotani K, Saito M et al (2018) Effect of luseogliflozin on hepatic fat content in type 2 diabetes patients with NAFLD: a prospective, single arm trial. Hepatol Res 49:64–71. https://doi.org/10.1111/hepr.13236

Thomson SC, Vallon V, Blantz RC (2004) Kidney function in early diabetes: the tubular hypothesis of glomerular filtration. Am J Physiol Renal Physiol 286:F8–F15

Tikkanen I, Narko K, Zeller C et al (2015) Empagliflozin reduces blood pressure in patients with type 2 diabetes and hypertension. Diabetes Care 38:420–428

Toulis KA, Willis BH, Marshall T et al (2017) All-cause mortality in patients with diabetes under treatment with dapagliflozin: a population-based, open-cohort study in the health improvement network database. J Clin Endocrinol Metab 102:1719–1725

United States Food and Drug Administration (2017) FDA confirms increased risk of leg and foot amputations with the diabetes medicine canagliflozin. https://www.fda.gov/downloads/Drugs/DrugSafety/UCM558427.pdf. Accessed 12 Nov 2017

U.S. Food and Drug Administration (2019) FDA approves Jardiance to reduce cardiovascular death in adults with type 2 diabetes [Internet]. Available from http://www.fda.gov/NewsEvents/Newsroom/PressAnnouncements/ucm531517.htm. Accessed 10 Mar 2019

Uthman L, Baartscheer A, Bleijlevens B et al (2018) Class effects of SGLT2 inhibitors in mouse cardiomyocytes and hearts: inhibition of Na^+/H^+ exchanger, lowering of cytosolic Na^+ and vasodilation. Diabetologia 61:722–726

van Baar MJB, van Ruiten CC, Muskiet MHA et al (2018) SGLT2 inhibitors in combination therapy: from mechanisms to clinical considerations in type 2 diabetes management. Diabetes Care 41:1543–1556

Van Bommel EJ, Muskiet MH, Tonneijck L et al (2017a) SGLT2 inhibition in the diabetic kidney-from mechanisms to clinical outcome. Clin J Am Soc Nephrol 12:700–710

van Bommel EJ, Muskiet MH, Tonneijck L, Kramer MH, Nieuwdorp M, van Raalte DH (2017b) SGLT2 inhibition in the diabetic kidney-from mechanisms to clinical outcome. Clin J Am Soc Nephrol 12:700–710

Vasilakou D, Karagiannis T, Athanasiadou E et al (2013) Sodium-glucose cotransporter 2 inhibitors for type 2 diabetes: a systematic review and meta-analysis. Ann Intern Med 159:262–274

Vick H, Diedrich D, Baumann K (1973) Reevaluation of renal tubular glucose transport inhibition by phlorizin analogs. Am J Phys 224:552–557

Wanner C, Inzucchi SE, Lachin JM et al (2016) Empagliflozin and progression of kidney disease in type 2 diabetes. N Engl J Med 375:323–334

White J (2010) Apple trees to sodium glucose co-transporter inhibitors: a review of SGLT2 inhibition. Clin Diabetes 28:5–10

Wiviott SD, Raz I, Bonaca MP et al (2019) Dapagliflozin and cardiovascular outcomes in type 2 diabetes. N Engl J Med 380:347–357

Wright EM (2001) Renal Na+-glucose cotransporters. Am J Physiol Renal Physiol 280:F10–F18

Wu JH, Foote C, Blomster J et al (2016) Effects of sodium-glucose cotransporter-2 inhibitors on cardiovascular events, death, and major safety outcomes in adults with type 2 diabetes: a systematic review and meta-analysis. Lancet Diabetes Endocrinol 4:411–419

Yabe D, Nishikino R, Kaneko M, Iwasaki M, Seino Y (2015) Short-term impacts of sodium/glucose co-transporter 2 inhibitors in Japanese clinical practice: considerations for their appropriate use to avoid serious adverse events. Expert Opin Drug Saf 14:795–800

Yale JF, Bakris G, Cariou B et al (2014) Efficacy and safety of canagliflozin over 52 weeks in patients with type 2 diabetes mellitus and chronic kidney disease. Diabetes Obes Metab 16:1016–1027

Zinman B, Wanner C, Lachin JM et al (2015) Empagliflozin, cardiovascular outcomes, and mortality in type 2 diabetes. N Engl J Med 373:2117–2128

Adv Exp Med Biol - Advances in Internal Medicine (2020) 4: 231–255
https://doi.org/10.1007/5584_2020_512
© Springer Nature Switzerland AG 2020
Published online: 21 April 2020

Gestational Diabetes Mellitus Screening and Diagnosis

U. Yasemin Sert ⓘ and A. Seval Ozgu-Erdinc ⓘ

Abstract

An ideal screening test for gestational diabetes should be capable of identifying not only women with the disease but also the women with a high risk of developing gestational diabetes mellitus (GDM). Screening and diagnosis are the main steps leading to the way of management. There is a lack of consensus among healthcare professionals regarding the screening methods worldwide. Different study groups advocate a variety of screening methods with the support of evidence-based comprehensive data. Some of the organizations suggest screening for high risk or all pregnant women, while others prefer to offer definitive testing without screening. Glycemic thresholds are also not standardized to decide GDM among different guidelines. Prevalence rates of GDM vary between populations and with the choice of glucose thresholds for both screening and definitive tests. One-step or two-step methods have been used for GDM diagnosis. However, screening includes selecting patients with historical risk factors, 50 g 1-h glucose challenge test, fasting plasma glucose, random plasma glucose, and hemoglobin A1c with different cutoffs. In this chapter, screening and diagnosis methods of GDM accepted by different study groups will be discussed which will be followed by the evaluation of different glycemic thresholds. Then the advantages and disadvantages of used methods will be explained and the chapter will finish with an evaluation of the current international guidelines.

Keywords

100-g Oral glucose tolerance test · 50-g glucose challenge test · 75-g Oral glucose tolerance test · Diagnosis · Fasting plasma glucose · Gestational diabetes mellitus · Hemoglobin A1c · Postprandial glucose · Random plasma glucose · Screening

1 Introduction

Diabetes mellitus is defined as a metabolic disease presented with the defect of the function of glucose metabolism (American Diabetes A 2013). Pregnancy is naturally associated with pancreatic β-cell hyperplasia, which results in higher fasting and postprandial insulin levels (Butte 2000). While during the first-trimester blood glucose levels decrease with increasing insulin, hormones such as cortisol and estrogen lead to insulin resistance, especially in the second and third trimesters (Carr and Gabbe 1998). In addition to the hormonal changes, placental

U. Y. Sert and A. S. Ozgu-Erdinc (✉)
Ministry of Health-Ankara City Hospital, Universiteler Mahallesi Bilkent Cad, Ankara, Turkey

tumor necrosis factor (TNF)-α, human placental lactogen (HPL), and growth hormone (GH) impair glucose metabolism due to pregnancy-associated insulin resistance (IR) (Barbour et al. 2007; Catalano et al. 1991). The problem can be characterized by the failure of insulin secretion, activity, or both, which results in a diabetogenic state in pregnancy (Mumtaz 2000).

Gestational diabetes mellitus (GDM) is the most common metabolic disease of pregnancy, which is associated with short- and long-term adverse outcomes for the mother and the off-spring. The incidence depends on the population and the diagnostic criteria (2.4–37.7%) (Simmons 2017; Ozgu-Erdinc et al. 2019a), and the prevalence is significantly increasing, mostly due to the obesity epidemic. Uncontrolled hyperglycemia is associated with serious/severe acute and chronic effects. The fetal adverse outcome increases if hyperglycemia which is associated with long term effects on different organ systems occurs during the pregnancy (Fuller and Borgida 2014). Langer et al. demonstrated that every 10 mg/dL increase in fasting plasma glucose (FPG) is associated with a 15% increase in both maternal and fetal adverse outcomes (Langer et al. 2005). The risk of developing Diabetes mellitus (DM) also increases among women with GDM (Herath et al. 2017). As appropriate glycemic control decreases the risk of GDM-related complications, early diagnosis and treatment/management are very essential.

To prevent or decrease the fetal and maternal risks of GDM, prediction, screening, diagnosis, management, and follow up strategies are essential. However, there is currently no consensus on the definition, screening, diagnosis, and management strategies about GDM. Today nearly 30 different guidelines for screening and diagnosing GDM addressed by national and international diabetes organizations, healthcare societies, endocrine groups, and obstetric associations (Agarwal 2018; Committee on Practice B-O 2018; International Association of D, Pregnancy Study Groups Consensus P et al. 2010; Panel NC 2013; Diabetes Canada Clinical Practice Guidelines Expert C

et al. 2018; Haneda et al. 2018; Yang 2012; National Institute for Health and Care Excellence 2015; World Health Organization 2013; Kleinwechter et al. 2014; Negrato et al. 2010; Hod et al. 2018; Mahmood 2018; Seshiah et al. 2009; Blumer et al. 2013; Diabetes Association Of The Republic Of China T 2020; Moyer and Force USPST 2014; Brown 2014; IDF GDM Model of Care 2015; RANZCOG 2017; Nankervis et al. 2014; Guidelines QC 2015; Guidelines H 2016; American Diabetes A 2020; Network SIG 2014).

A disagreement is present even between obstetric and diabetes mellitus organizations of a country (e.g., American Diabetes Association (ADA) and American College of Gynecology and Obstetrics (ACOG)) (Committee on Practice B-O 2018; American Diabetes A 2020). The diversity of recommendations results in different approaches, even within the same hospital, which leads the clinicians to a complicated dilemma. Since this lack of consensus creates major problems in addressing prevalence, complications, the efficacy of treatment, and follow-up of GDM, the need for consensus has been repeatedly expressed by the experts. There is a need for standardization to have global uniformity for diagnosing GDM.

2 History of GDM

The first documented case of hyperglycemia in pregnancy in literature was a 22-year-old multigravida woman with a history of severe fetal macrosomia and stillbirth in 1824. Bennewitz demonstrated that the women were suffering from severe hyperglycemic symptoms in the pregnancy, while all of them were disappearing after the delivery (Hadden 1998). This was the first observation of the existence of the hyperglycemic conditions and their unusual effects on the newborn. Previously, the clinicians had recognized that any degree of hyperglycemia from mild to severe resulted in poor perinatal outcomes (Hoet and Lukens 1954). However, the most crucial step for the definition of "hyperglycemia in pregnancy" was an epidemiologic study performed with a two-step method following an oral glucose tolerance test (OGTT) in 1964 by O'Sullivan and Mahan (1964). New approaches to plasma glucose measurement revealed Carpenter and Coustan criteria based on this study (Carpenter and Coustan 1982). The US National Diabetes Data Group (NDDG) (1979) and the World Health Organization (WHO) (1980) established that the 2-h 75-g OGTT should be the main diagnostic test for glucose intolerance outside pregnancy. WHO issued a recommendation suggesting to use the same criteria of non-pregnant people by using a 75 g oral glucose tolerance test, which was substantially weak for the prediction of perinatal outcomes (Houshmand et al. 2013). In 1980, Freinkel N. demonstrated a study focusing on the effects of fuel metabolism of mother to fetus, and this was an important cornerstone leading The American Diabetes Association (ADA) to define GDM as "glucose intolerance with onset or first recognized during pregnancy" (Metzger and Coustan 1998). This step demonstrated the need for different diagnostic methods and management strategies for pregnant women different from the recommendation of WHO.

(IADPSG) in 2010 (International Association of D, Pregnancy Study Groups Consensus P, et al. 2010). WHO has also announced a recommendation concerning hyperglycemia during pregnancy in 2013 (World Health Organization 2013). WHO classified hyperglycemia during pregnancy as "diabetes mellitus in pregnancy" and "GDM" and the International Federation of Gynecology and Obstetrics (FIGO) endorsed this definition (Chi et al. 2018; Hod et al. 2015). ADA introduced the definition of GDM as the diagnosis of diabetes during the second or third trimester of pregnancy in 2018 (American Diabetes A 2018). Endocrine Society has suggested distinguishing hyperglycemia in terms of "pregnancy-associated" and "antedated to pregnancy" (Blumer et al. 2013). According to the recommendations, diabetes mellitus during pregnancy, whether symptomatic or not, carries a significant risk of adverse perinatal outcomes. Management of diabetes mellitus during pregnancy should be based on the microvascular and chronic effects of hyperglycemia, while GDM management targeted first and foremost perinatal outcomes.

3 Definition

Gestational diabetes mellitus (GDM) was historically defined as any glucose intolerance first recognized in the pregnancy by ADA and NDDG (Metzger and Coustan 1998). This definition also included unrecognized pre-existing diabetes cases, insulin or only diet modification used for treatment, and whether or not condition persists after pregnancy. Studies showed that including such a wide range of glucose abnormalities under one title might cause difficulties in management and postpartum follow-up strategies of the disease (Fuller and Borgida 2014; National Diabetes Data Group 1979). The first determined hyperglycemia during the pregnancy was assessed as "overt diabetes" or "gestational diabetes" by The International Association of Diabetes and Pregnancy Study Groups

4 GDM Screening

The purpose of the screening should be to identify GDM cases and asymptomatic pregnant women who are at risk of GDM progression. Screening is indispensable for the prevention of obstetric outcomes and the future effects of GDM on several organ systems (Committee on Practice Bulletins—Obstetrics 2018).

Controversies on GDM screening among various guidelines mainly stem from the inclusion of different studies on the improvement of perinatal outcomes (Houshmand et al. 2013). These controversies related to the indication for screening (universal or selective-risk-factor based), the timing of screening (early or 24–28 weeks of pregnancy/gestation), how to screen (one or two-step), and criteria for diagnosis (not standardized).

4.1 Indications for Screening

4.1.1 Risk Factor-Based Screening or Universal Screening?

Guidelines that are based on recent reviews have no consensus on performing universal screening or risk-based screening. ADA defines the women at low risk as younger than 25 years, not a member of an ethnic group, body mass index (BMI) <25 kg/m2, no history of previous abnormal glucose tolerance or adverse obstetric outcome, and no known history of glucose metabolism abnormalities in the first degree relatives. ADA recommends not to screen these women since screening provides no additional benefit (American Diabetes A 2020). Risk factors defined for GDM development by different study groups are listed in Table 1 (Committee on Practice B-O 2018; American Diabetes A 2020; Lee et al. 2018; Mirghani Dirar and Doupis 2017). Both ADA and ACOG recommend using these factors to screen GDM (Committee on Practice B-O 2018; American Diabetes A 2020). These factors are quite decisive for increased risk of GDM; however, diagnostic accuracy is limited for one factor alone (Griffin et al. 2000; van Leeuwen et al. 2010). Some authors suggest improving the diagnostic accuracy of risk factors by adding some risk indicators such as FPG and some biochemical markers (Nanda et al. 2011; Savvidou et al. 2010). Risk factors such as BMI have variations between different ethnic groups, and this appears to be a limitation of adopting all the risk factors universally (Shah et al. 2011). There

Table 1 Risk factors for GDM prediction

Previous history of gestational diabetes mellitus
Previously elevated blood glucose levels, pre-diabetes/impaired glucose metabolism
Maternal age ≥25-35-40 years
Family history of diabetes mellitus (first-degree relative with diabetes or a sister with gestational diabetes mellitus)
Body mass index ≥25–30 kg/m^2
Previous macrosomia (baby with birth weight >4500g or >90th centile) or polyhydramnios.
Signs of insulin resistance such as Polycystic ovary syndrome (PCOS) and Acanthosis nigricans
Medications: corticosteroids, antipsychotics
History of congenital anomalies
Pregnancy-induced hypertension (PIH)
History of stillbirth
History of abortion
Multiparity ≥2
History of preterm delivery
History of neonatal death
Current smoking
Current drinking
Excessive weight gain in the index pregnancy
Minority ethnic family origin with a high prevalence of diabetes. (Latino, Native American, Caribbean, Chinese, Asian, Indian subcontinent, Aboriginal, Torres Strait Islander, Pacific Islander, Maori, Middle Eastern, non-white African)
History of PAOD (Peripheral Arterial Occlusive Disease), CVD (cerebral vascular disease).
Hypertension (>140/90 mmHg or on therapy for hypertension)
HDL cholesterol level <35 mg/dL (0.90 mmol/L) and/or a triglyceride level >250 mg/dL (2.82 mmol/L)
A sister with hyperglycemia in pregnancy
Physical inactivity
Short stature
Multifetal pregnancy
Vitamin D deficiency
Maternal history of low birth weight
HbA1c ≥5.7%

are several risk factors with several cutoffs in the literature. Some of the studies endorse population-based risk factor assessment and significant effect of life-style modifications (physical activity, diet, alcohol, and smoking) on developing GDM (Caliskan et al. 2004; Gobl et al. 2012; Zhang et al. 2014).

The main concern about the risk-based selective screening regarding historical or clinical factors is missing the majority of GDM cases (Lavin 1985). In one study, GDM was found in 1.45% of women with risk-based screening, while it increased to 2.7% in the same population with universal screening, which demonstrates that risk-based screening has missed half of the GDM cases (Griffin et al. 2000). For this reason, most of the experts prefer universal screening.

4.2 When to the Screen?

4.2.1 Pre-conceptual Prediction of GDM

In a recent study, it is postulated that GDM can be predicted before conception by using the first degree relatives with type 2 diabetes mellitus, FPG, fasting insulin, androstenedione, and sex hormone-binding globulin (SHBG) levels (Bidhendi Yarandi et al. 2019). Although these factors can predict GDM, it is not suitable to screen all the women with these factors before conception.

4.2.2 First-Trimester Screening

Prediction of GDM during the first trimester is unclear and also controversial because different guidelines suggest different screening methods and follow-up strategies by using various cutoff values (Agarwal 2016). In the first trimester, FPG physiologically tends to decline (Mills et al. 1998). Regardless of this anticipated decline, it is still possible to miss the risky/high-risk patients by using the cut-off values of the third trimester of pregnancy (<92 mg/dL) (McIntyre et al. 2016; Ozgu-Erdinc et al. 2015).

The study of Bhattacharya demonstrated that fasting plasma glucose evaluated during the first trimester of pregnancy cannot be used as an efficient diagnostic test for the prediction of later development of GDM (Bhattacharya 2004). On the contrary, one study from Israel that included 6129 pregnant women demonstrated a significant association between FPG and developing GDM (the group with FPG: 100–105 mg/dL and < 75 mg/dL have GDM prevalence of 11.7% and 1% respectively) (Riskin-Mashiah et al. 2009). Although this study reveals that GDM prevalence is low among pregnant women with FPG < 75 mg/dL, GDM diagnosis would be missed nearly 9% of pregnant women (Riskin-Mashiah et al. 2009). Most of the studies have shown that pre-conception FPG could be used to predict GDM and in the first trimester with different thresholds (Ozgu-Erdinc et al. 2019b; Li et al. 2019). FPG, high sensitive C-reactive protein (hs-CRP), SHBG, tissue plasminogen activator (tPA), TNF-α, leptin, adiponectin, placental protein 13, endoglin, and lipids were identified as the predictors of GDM when combined with risk factors in the first trimester of pregnancy in the literature (Savvidou et al. 2010; Ozgu-Erdinc et al. 2015; Yeral et al. 2014; Kansu-Celik et al. 2019a; Huhn et al. 2018). Many biomarkers have been identified for the prediction of GDM during the first trimester (Table 2) (Kansu-Celik et al. 2019a; Huhn et al. 2018; Kansu-Celik et al. 2019b; Yang et al. 2017; Ravnsborg et al. 2016; Nevalainen et al. 2016; Powe 2017; Donovan et al. 2018; Tu et al. 2017; Yoffe et al. 2019; Wang et al. 2018; Zhu et al. 2019; Mertoglu et al. 2019; Arslan et al. 2019; Gan et al. 2019). It is possible to identify 75% of pregnancies in which GDM will develop, by a combination of maternal factors, biomarkers, and obstetric history (Plasencia et al. 2011).

Although pregnancy is associated with hyperinsulinemia and IR, the level of maternal hormones such as cortisol and estrogen, reach the peak effect between 26–33 weeks of gestation (Carr and Gabbe 1998; Kuhl and Holst 1976). Women who are not able to normalize the glucose level with this increment of insulin develop hyperglycemia (Plows et al. 2018). This condition constitutes the basis for screening between 24–28 weeks of pregnancy. Performing the test too early can lead to missing the cases due to the increase of IR from the first to the third trimester,

Table 2 List of tested biomarkers for early GDM screening

Glycemic related markers:	Glucose
	Hemoglobin A1c
	Fructosamine
	C-peptide
	1.5 Anhydroglucitol
	Glycosylated fibronectin
	Osteocalcin
	Fetuin-A
Insulin resistance markers	Fasting insulin and related indices
	Sex-hormone binding globulin
Growth factors and their binding proteins:	Insulin-like growth factor-1 and 2,
	Insulin-like growth factor-binding protein-1,2,3 and 5
Lipid related markers:	High-density lipoprotein
	Low-density lipoprotein
	Triglycerides
	Plasma phospholipid fatty acids
	Apolipoprotein E, M, D, L1, AIV
Pregnancy-related/Placenta derived markers:	Placental growth factor
	Soluble FMS-like tyrosine kinase
	Pregnancy-associated plasma protein-A
	Placental exosomes
	Follistatin-like 3
	Soluble endoglin
	Human placental protein 13
	Placental lactogen
Adipocytokines:	Adiponectin
	Leptin
	Visfatin
	Resistin
	Retinol binding protein-4
	Chemerin
	Vaspin
	Adipocyte fatty acid-binding protein (Fatty Acid-Binding Protein 4)
	Neutrophil gelatinase-associated lipocalin
	Lipocalin-2
Hormones:	Placental lactogen
	Free β-Human chorionic gonadotropin
	Inhibin A, Activin A
	Testosterone
	Dehydroepiandrosterone sulfate
	Thyroid-stimulating hormone
	Thyroxin
	Natriuretic peptides (Atrial, brain)
Nucleic acids:	Micro RNAs
	Genetic variants
	Omics based biomarkers
Inflammation related markers:	High sensitive C-reactive protein
	Tumor necrosis factor-α
	Interleukin-6

(continued)

Table 2 (continued)

Others:	Soluble (pro)renin
	Alanine aminotransferase
	Ferritin
	Hemoglobin
	Serum iron concentration
	Soluble human leukocyte antigen
	Coagulation factor IX
	Fibrinogen alpha-chains
	Plasminogen activator (inhibitor)
	Afamin
	α2-macroglobulin
	Haptoglobin β chain
	Clustering α chain
	Serum Amino Acids (Arginine, Glycine)
	Acylcarnitines (3-hydroxy-isovalerylcarnitine)

while late screening may result in losing the chance of implementing preventive measures. The clinicians need to be vigilant to identify pregnant women who will develop diabetes mellitus before the first trimester. All the women should be assessed at the initial visit for overt diabetes and should be screened if the suspicion of GDM arises due to the risk factors. GDM screening should be repeated during 24–28 weeks of gestation if the pregnant women have risk factors and GDM were not addressed during the first prenatal visit evaluation (International Association of D, Pregnancy Study Groups Consensus P et al. 2010; Gupta et al. 2015).

IADPSG, which is based on Hyperglycemia and Adverse Pregnancy Outcome (HAPO) study, recommends the first-trimester screening with a 75-g OGTT for GDM screening and to make the diagnosis of GDM if FPG $\geq$ 92 mg/dL (International Association of D, Pregnancy Study Groups Consensus P et al. 2010). However, ADA suggests evaluating high-risk women at the initial visit with FPG, hemoglobin A1c (HbA1c), or random plasma glucose (RPG) (American Diabetes A 2020). If FPG ranges from 92 mg/dL to 126 mg/dL, it must be accepted as pre-diabetes (impaired glucose tolerance), and the patients should be encouraged to have life-style changes to prevent GDM or type-2 DM due to the lack of evidence-based studies (American Diabetes A 2020). A recent study

suggested gestational-age specific OGTT thresholds for early detection of GDM (Jokelainen et al. 2020) and as stated in ADA GDM screening and diagnostic criteria used in the either one or two-step approach were not obtained from the data of the first trimester, therefore the diagnosis of GDM in this period by these tests is not evidence based (McIntyre et al. 2016) and research is needed (American Diabetes A 2020).

First-trimester universal screening for GDM is not internationally recommended currently since randomized controlled trials evaluating the potential benefits and harms of early detection and subsequent treatment are lacking (Committee on Practice Bulletins—Obstetrics 2018). Recommendations of guidelines for trimester-based prediction of GDM are listed in Table 3.

4.2.3 24–28 Weeks Screening

The diabetogenic effect of pregnancy appears more pronounced at 24–28 weeks of pregnancy. Most of the guidelines recommend universal screening at 24–28 weeks of gestation with or without the first-trimester screening. (Table 4) (Yang 2012; Nankervis et al. 2014; Kuo and Li 2019; Li-zhen et al. 2019). International Diabetes Federation (IDF) and Diabetes in Pregnancy Study group in India (DIPSI) recommend repeat screening at 32–34 weeks of gestation (IDF GDM Model of Care 2015; Seshiah et al. 2006).

Table 3 First trimester/early gestational diabetes screening according to the guidelines

Guideline	Target population	Target diagnose	Screening method	Threshold mg/dL (mmol/L)	Diagnostic method	Diagnostic threshold mg/dL (mmol/L)
NICE (RCOG) 2015	Risk-based	GDM			75-g OGTT	**FPG** ≥100 (≥5.6)
						2-h ≥140 (≥7.8)
WHO 2013/ EBCOG-EAPM-FIGO 2018	Universal	Overt DM GDM			75-g OGTT at any time of the gestation	**Overt DM**
						FPG ≥126 (≥7)
						2-h ≥200 (≥11.1)
						RPG ≥200 (11.1)
						GDM
						[a]**FPG** ≥92 (≥5.1)
						1-h ≥180 (≥10)
						2-h ≥153 (≥8.5)
ADIPS 2014 RANZCOG 2017	Risk-based	Overt DM GDM	Moderate risk factors **RPG, FPG, OGTT** (if indicated)	Not clear	75-g OGTT	[a]**FPG** ≥92 (≥5.1)
						1-h ≥180 (≥10)
			One high risk or two moderate risks: **75-g OGTT**			**2-h** ≥153 (≥8.5)
GDA/GAGO 2014	Risk-based	Overt DM	RPG	**FPG** ≥92 (≥5.1) (repeat FPG)	75-g OGTT	[a]**FPG** ≥92 (≥5.1)
			FPG			**1-h** ≥180 (≥10)
				RPG: 140–199 (7.8–11.05) (OGTT)		**2-h** ≥**153** (≥**8.5**)
				RPG ≥200 (11.1) (Overt DM)		
IADPSG 2010/IDF/ FIGO 2015	Risk-based	Overt DM	FPG		FPG	**Overt DM**
	Universal (IDF, FIGO)	GDM	RPG		RPG	**FPG** ≥126 (≥7)
			HbA1c		HbA1c	**HbA1c** ≥6.5%
			75g OGTT (high-risk ethnic groups) (FIGO)			**RPG** ≥200 (11.1)
						GDM
						FPG 92–126 (5.1–7)
						FPG<92 (5.1) 75-g OGTT (24–28 GW)

(continued)

Table 3 (continued)

Guideline	Target population	Target diagnose	Screening method	Threshold mg/dL (mmol/L)	Diagnostic method	Diagnostic threshold mg/dL (mmol/L)
JDS 2016	Risk-based	Overt DM	FPG		FPG	**FPG** ≥126 (≥7)
			RPG		RPG	**HbA1c** ≥6.5%
			75 g OGTT		75-g OGTT	**RPG** ≥200 (11.1)
			HbA1c		HbA1c	**75-g OGTT 2-h** ≥200 (11.1)
DIPSI 2009	Universal	Overt DM	–	–	75-g OGTT in the first trimester (Irrespective of fasting state)	**Overt DM** 2-h ≥200 (11.1)
BSD 2010	Universal	Overt DM	FPG	**IGT:**	75-g OGTT	[a]**FPG** ≥ 92 (≥5.1)
				FPG ≥85 (4.7)		**1-h** ≥180 (≥10)
				Overt DM		**2-h** ≥153 (≥8.5)
				FPG ≥126 (Simmons 2017)		If negative, repeat at 24–28 GW
Queensland 2015	Risk-based	GDM			HbA1c	**GDM**
		Diabetes in pregnancy			75-g OGTT	**HbA1c** 5.9–6.5%
						FPG 92–124 (5.1–6.9)
						1-h>180 (≥10)
						2-h 153–200 (8.5–11)
						DIP
						FPG ≥126 (Simmons 2017)
						2-h ≥200 (11.1)
						RPG ≥200 (11.1)
						HbA1c ≥6.5%

(continued)

Table 3 (continued)

Guideline	Target population	Target diagnose	Screening method	Threshold mg/dL (mmol/L)	Diagnostic method	Diagnostic threshold mg/dL (mmol/L)
MOH of China 2012	Risk-based	GDM		–	75-g OGTT	[a]**FPG** ≥92 (≥5.1)
						1-h ≥180 (≥10)
						2-h ≥153 (≥8.5)
						If negative, repeat at 24–28 GW
DAROC 2018	Universal	Overt DM (type1,2 or other)		–	FPG	Overt DM (type1,2 or other)
					HbA1c	**FPG** mg/dL (mmol/L)
						≥126 (≥7.0)
						HbA1c ≥6.5%
HKCOG 2016	Risk-based	Overt DM (type1,2 or other)		–	75-g OGTT	**GDM**
					FPG	**FPG** ≥92 (≥5.1)
		GDM			HbA1c	**Overt DM**
						FPG ≥126 (7.0)
						75-g OGTT 2-h ≥200 (11.1)
						RPG ≥200 (11.1)
						HbA1c ≥6.5%
CDA 2018	Risk-based	Overt DM (type1,2 or other)	FPG	NA	–	–
			HbA1c			
			50-g GCT			
			75-g OGTT			
SIGN 2017	Risk-based	Overt DM (type1,2 or other)	FPG	**Overt DM**	FPG	**Overt DM**
			2-h PPG	**FPG** ≥126 (≥7.0)	HbA1c	**FPG** ≥126 (≥7.0)
			HbA1c	**HbA1c** ≥6.5%	PPG	**HbA1c** ≥6.5%
		IGT		**2-h PPG** ≥200 (11.1)		**2-h PPG** ≥200 (11.1)
				IGT		
				FPG:92–124 (5.1–6.9)		

(continued)

Table 3   (continued)

Guideline	Target population	Target diagnose	Screening method	Threshold mg/dL (mmol/L)	Diagnostic method	Diagnostic threshold mg/dL (mmol/L)
				2-h PPG: 140–200 (7.8–11)		
				HbA1c: 6–6.4%		
				[a]**IGT** should be evaluated with 75-g OGTT at 24–28 GW		
ACOG 2018/ NIH 2015	Risk-based	Overt DM (type1,2 or other)	75-g OGTT	NA	75-g OGTT	[a]**FPG** ≥92 (≥5.1)
			50-g GCT		or	**1-h** ≥180 (≥10)
			HbA1c (not alone)		100-g OGTT	**2-h** ≥**153** (≥**8.5**)
		GDM				**or**
						[b]**FPG** ≥95 (≥5.3)
						1-h ≥180 (≥10)
						2-h ≥155 (≥8.6)
						3-h ≥140 (≥7.8)
NZSSD 2014	Universal	Overt DM (type 1,2 or other)	HbA1c	**Overt DM** **HbA1c** ≥ 6.7%	**Overt DM** **HbA1c** ≥6.7%	
				GDM	**GDM**	
		GDM		If HbA1c: 5.9–6.6%, apply 2-h 75-g OGTT at 24–28 GW		
Endocrine Society of USA 2013	Universal	Overt DM (type 1,2 or other)	FPG	–	FPG	**Overt DM**
			RPG		RPG	**FPG** ≥126 (≥7.0)
			-HbA1c		HbA1c	**RPG** ≥200 (≥11.1)
		GDM			2-h 75-g OGTT before 13 GW (8–14 h fasting)	**HbA1c** ≥6.5%
						GDM
						FPG: 92–125 (5.1–6.9)
						RPG: NA
						HbA1c: NA

(continued)

Table 3 (continued)

Guideline	Target population	Target diagnose	Screening method	Threshold mg/dL (mmol/L)	Diagnostic method	Diagnostic threshold mg/dL (mmol/L)
ADA 2020	Risk-based	Overt DM	FPG	–	FPG	**FPG** ≥126 (7.0)
			HbA1c		HbA1c	**75-g OGTT**
			75-g OGTT, 2-h value		75-g OGTT, 2-h value	**2-h** ≥200 (11.1)
						RPG ≥200 (11.1)
						HbA1c ≥6.5%

ACOG The American College of Obstetricians and Gynecologists, *ADA* American Diabetes Organization, *ADIPS* Australian Diabetes in Pregnancy Society, *BSD* Brazilian Society of Diabetes, *CDA* Canadian Diabetes Association, *DAROC* The diabetes association of republic of China, *DIPSI* Diabetes in Pregnancy Study group in India, *EAPM* European Association of Perinatal Medicine, *EASD* European Association for the Study of Diabetes, *EBCOG* the European Board and College of Obstetrics and Gynecology, *DGGT* Decreased gestational glucose tolerance, *DIP* Diabetes in pregnancy, *FIGO* International Federation of Gynecology and Obstetrics, *FPG* Fasting plasma glucose, *GAGO* German Association of Gynecologists and obstetricians, *GCT* Glucose challenge test, *GDA* German diabetes Association, *GDM* Gestational diabetes mellitus,, *GW* Gestational week, *HbA1c* Hemoglobin A1c, *HKCOG* Hong Kong College of Obstetricians and Gynecologists, *IADPSG* The International Association of the Diabetes and Pregnancy study groups, *IDF* International Diabetes Federation, *JDS* Japan Diabetes Society, *MOH* The Ministry of Health, *NA* Not applicable, *NICE* National Institute for Health and Care Excellence, *NIH* National Institutes of Health, *NZSSD* New Zealand Society for the Study of Diabetes, *OGTT* Oral glucose tolerance test, *RANZCOG* The Royal Australian and New Zealand College of Obstetricians and Gynecologists, *RCOG* Royal College of Obstetricians and Gynecologists, *RPG* Random plasma glucose, *SIGN* Scottish Intercollegiate Guidelines Network, *UPSTF* The U.S. Preventive Services Task Force, *USA* The United States of America, *WHO* World Health Organization
[a]A diagnosis requires that one or more thresholds be met or exceeded
[b]A diagnosis requires that two or more thresholds be met or exceeded

4.2.4 Screening and Diagnostic Criteria

There is no consensus on screening and diagnosis of GDM in the world. The lack of uniformity, as a consequence of different guidelines, leads clinicians to evaluate the goals and weaknesses and strength of the methods, especially before implementing them to their clinical practice. These methods might be risk-based or universal and either with a one-step or two-step procedure. Since each method acknowledges different studies performed in various populations, the method of choice should be decided based on the population characteristics, the prevalence of diabetes mellitus and obesity, access to healthcare services, and patients' adherence to follow-up.

4.2.5 Universal Testing During the First Trimester

Deteriorated blood glucose levels due to overt diabetes cause congenital fetal defects and maternal worsening on developing retinopathy and nephropathy. The majority of the guidelines recommend screening overt diabetes during the first prenatal visit, particularly for high-risk patients due to the metabolic disturbance on glucose metabolism during the pregnancy (Panel NC 2013; Blumer et al. 2013; Committee on Practice Bulletins—Obstetrics 2018; American Diabetes A 2019; WHO 2014).

Overt diabetes detected in early pregnancy is diagnosed with the same cutoff plasma levels of diabetes of non-pregnant people. The thresholds are FPG ≥ 126 mg/dL (7.0 mmol/L); RPG ≥ 200 mg/dL (11.1 mmol/L); or HbA1c ≥ 6.5% (47 mmol/mol) (International Association of D, Pregnancy Study Groups Consensus P et al. 2010). WHO recommended that HbA1c has no diagnostic accuracy for diabetes in pregnancy (World Health Organization 2013). Screening and treatment before 24 weeks of

Table 4 International guidelines for the diagnosis and screening of GDM between 24–28 gestational weeks

Guideline	Target population	Screening method	Threshold mG/dL (mmol/L)	Diagnosing method	Diagnostic threshold mg/dL (mmol/L)
NICE (RCOG) 2015	Universal			75-g OGTT	**FPG** ≥100 (≥5.6)
					2-h ≥140 (≥7.8)
WHO 2013/ EBCOG-EAPM-FIGO2018	Universal Risk-based	75-g OGTT at any time of the gestation		75-g OGTT at any time of the gestation	[a]**FPG** ≥92 (≥5.1)
					1-h ≥180 (≥10)
					2-h ≥153 (≥8.5)
ADIPS 2014 RANZCOG 2017	Universal			75-g OGTT at 24–28 weeks gestation	[a]**FPG** ≥ 92 (≥5.1)
					1-h ≥180 (≥10)
					2-h ≥153 (≥8.5)
GDA/GAGO 2014	Universal	50-g GCT non-fasting (24–28 GW)	50-g GCT ≥135 (7.5)	50 + 75-g OGTT	[a]**FPG** ≥92 (≥5.1)
				or	**1-h** ≥180 (≥10)
				50-g GCT	**2-h** ≥153 (≥8.5)
					or
					50-g GCT ≥201 (≥11.1)
IADPSG 2010/ IDF/FIGO 2015	Universal			75-g OGTT at 24–28 weeks gestation	[a]**FPG** ≥92 (≥5.1)
					1-h ≥180 (≥10)
					2-h ≥153 (≥8.5)
JDS 2016	Risk-based	FPG		FPG	**Overt DM**
		RPG		RPG	**FPG** ≥126 (≥7)
		75-g OGTT		75-g OGTT	**HbA1c** ≥6.5%
		HbA1c		HbA1c	**RPG** ≥200 (11.1)
					75-g OGTT 2-h ≥200 (11.1)
					[a]**GDM**
					FPG ≥92 (≥5.1)
					1-h ≥180 (≥10)
					2-h ≥153 (≥8.5)

(continued)

Table 4 (continued)

Guideline	Target population	Screening method	Threshold mG/dL (mmol/L)	Diagnosing method	Diagnostic threshold mg/dL (mmol/L)
DIPSI 2009	Universal	75-g OGTT (24–28 GW) (irrespective of fasting state)	**GDM** 2-h ≥140 (7.7)	75-g OGTT (24–28 GW) (irrespective of fasting state	**GDM** 2-h ≥140 (7.7)
			Overt DM ≥200 (11.1)		**Overt DM** ≥200 (11.1)
			DGGT ≥120 (6.6)		**DGGT** ≥120 (6.6)
BSD 2010	Universal	FPG	**IGT:FPG** mg/dL (mmol/L) ≥ 85 (4.7)	75-g OGTT (24–28 GW)	[a]**FPG** mg/dL (mmol/L) ≥92 (≥5.1)
			Overt DM:FPG mg/dL (mmol/L) ≥ 126 (Simmons 2017)		**-1-h Value,** mg/dL (mmol/L) ≥180 (≥10)
					-2-h Value, mg/dL (mmol/L) ≥153 (≥8.5)
QUEENSLAND 2015	Universal	75-g OGTT (24–28 GW)	–	75-g OGTT at 24–28 weeks gestation	[a]**FPG** ≥92 (≥5.1)
					1-h ≥180 (≥10)
					2-h ≥153 (≥8.5)
MOH OF CHINA 2012	Universal	75-g OGTT (24–28 GW)	–	75-g OGTT (24–28 GW)	[a]**FPG** ≥92 (≥5.1)
					1-h ≥180 (≥10)
					2-h ≥153 (≥8.5)
DAROC 2018	Universal	50-g GCT non-fasting (24–28 GW)	50-g GCT mg/dL (mmol/L) 130/140 (7.2/7.8)	50 + 100-g OGTT	[b]**FPG** ≥95 (≥5.3)
					1-h ≥180 (≥10)
				or	**2-h** ≥155 (≥8.6)
					3-h ≥140 (≥7.8)
				75-g OGTT	**or**
					[a]**FPG** ≥92 (≥5.1)
					1-h ≥180 (≥10)
					2-h ≥153 (≥8.5)
KCOG 2016	Universal	–	–	75-g OGTT	[a]**FPG** ≥92 (≥5.1)
					1-h ≥180 (≥10)
					2-h ≥153 (≥8.5)

(continued)

Table 4 (continued)

Guideline	Target population	Screening method	Threshold mG/dL (mmol/L)	Diagnosing method	Diagnostic threshold mg/dL (mmol/L)
UPSTF 2014	Universal	50-g GCT non-fasting (24–28 GW)	50-g GCT mg/dL (mmol/L) 130/135/140 (7.2/7.5/7.8)	50 + 100-g OGTT	[b]FPG ≥95 (≥5.3)
				or	1-h ≥180 (≥10)
					2-h ≥155 (≥8.6)
					3-h ≥140 (≥7.8)
				50 + 75-g OGTT	[a]FPG ≥92 (≥5.1)
					1-h ≥180 (≥10)
					2-h ≥153 (≥8.5)
CDA 2018	Universal	50-g GCT	50-gGCT mg/dL (mmol/L) ≥140 (≥7.8)	75-g OGTT	[a]FPG ≥95 (≥5.3)
		or			1-h ≥190 (≥10.6)
		75-g OGTT (24–28 GW)			2-h ≥162 (≥9)
IGN 2017	Risk-based	FPG (low risk) 75-g OGTT (high risk)	–	FPG	[a]FPG ≥92 (≥5.1)
				75-g OGTT	1-h ≥180 (≥10)
					2-h ≥153 (≥8.5)
ACOG 2018/ NIH 2015	Universal	50-g GCT	50-g GCT 130/135/140 (7.2/7.5/7.8)	100-g OGTT at 24–28 GW	[b]FPG ≥ 95 (≥5.3)
					1-h ≥180 (≥10)
					2-h ≥155 (≥8.6)
					3-h ≥140 (≥7.8)
NZSSD 2014	Universal	If HbA1c: 5.9–6.6% (at first trimester), apply 2-h 75-g OGTT (24–28 GW)	50-g GCT mg/dL (mmol/L) 140–200 (7.8–11)	2-h 75-g OGTT (24–28 GW)	FPG ≥99 (≥5.5)
					2-h ≥162 (≥9)
				If HbA1c: 5.9–6.6% (at first trimester)	or
					50-g GCT ≥200 (≥11.1)
		Others		or	
		50-g GCT		50-g GCT 140–200 (7.8–11) (24–28 GW)	

(continued)

Table 4 (continued)

Guideline	Target population	Screening method	Threshold mg/dL (mmol/L)	Diagnosing method	Diagnostic threshold mg/dL (mmol/L)
Endocrine Society of USA 2013	Universal	–	–	2-h 75-g OGTT (24–28 GW)	**Overt DM** (type1,2 or other):
				(8–14 h fasting)	**FPG** $\geq$126 ($\geq$7.0
					1-h NA
					2-h $\geq$200 ($\geq$11.1)
					[a]**GDM**
					FPG 92–125 (5.1–6.9)
					1-h $\geq$180 ($\geq$10.0)
					2-h 153–199 (8.5–11.0)
ADA 2020	Universal			75-g OGTT	[a]**FPG** $\geq$ 92 ($\geq$5.1)
				or	**1-h** $\geq$180 ($\geq$10)
				50 + 100-g OGTT	**2-h** $\geq$153 ($\geq$8.5)
					[b]**FPG** $\geq$ 95 ($\geq$5.3)
					1-h $\geq$180 ($\geq$10)
					2-h $\geq$155 ($\geq$8.6)
					3-h $\geq$140 ($\geq$7.8)

ACOG The American College of Obstetricians and Gynecologists, *ADA* American Diabetes Organization, *ADIPS* Australian Diabetes in Pregnancy Society, *BSD* Brazilian Society of Diabetes, *CDA* Canadian Diabetes Association, *DAROC* The diabetes association of republic of China, *DIPSI* Diabetes in Pregnancy Study group in India, *EAPM* European Association of Perinatal Medicine, *EASD* European Association for the Study of Diabetes, *EBCOG* the European Board and College of Obstetrics and Gynecology, *DGGT* Decreased gestational glucose tolerance, *FIGO* International Federation of Gynecology and Obstetrics, *FPG* Fasting plasma glucose, *GAGO* German Association of Gynecologists and obstetricians, *GCT* Glucose challenge test, *GDA* German diabetes Association, *GDM* Gestational diabetes mellitus, *GW* Gestational week, *HbA1c* Hemoglobin A1c, *HKCOG* Hong Kong College of Obstetricians and Gynecologists, *IADPSG* The International Association of the Diabetes and Pregnancy study groups, *IDF* International Diabetes Federation, *JDS* Japan Diabetes Society, *MOH* The Ministry of Health, *NA* Not applicable, *NICE* National Institute for Health and Care Excellence, *NIH* National Institutes of Health, *NZSSD* New Zealand Society for the Study of Diabetes, *OGTT* Oral glucose tolerance test, *RANZCOG* The Royal Australian and New Zealand College of Obstetricians and Gynecologists, *RCOG* Royal College of Obstetricians and Gynecologists, *RPG* Random plasma glucose, *SIGN* Scottish Intercollegiate Guidelines Network, *UPSTF* The U.S. Preventive Services Task Force, *USA* The United States of America, *WHO* World Health Organization

[a]A diagnosis requires that one or more thresholds be met or exceeded

[b]A diagnosis requires that two or more thresholds be met or exceeded

gestation have no additional benefit for GDM according to National Institutes of Health (NIH), United States (US) Preventive Services Task Force, Committee of Practice Bulletins, and recommend performing risk factor-based screening in the first trimester (Panel NC 2013; Donovan et al. 2013).

IADPSG recommends diagnosing women with FPG $\geq$ 92 mg/dL (5.1 mmol/L) at any time during pregnancy as GDM (International Association of D, Pregnancy Study Groups Consensus P et al. 2010). This recommendation is debatable due to the fact that the same cutoff value is used to diagnose GDM in later pregnancy with a 75-g OGTT. Pregnancy is associated with changes in glucose levels, and the same cutoffs with late pregnancy may result in diagnosing GDM more often than it is actually present (Carr and Gabbe 1998). A study from China examined 17,186 pregnant women in the first prenatal visit and 24–28 weeks of gestation and recommended to use FPG between 6.10 and 7.00 mmol/L (110–126 mg/dL) for the diagnosis of GDM during the first trimester, which is higher than IADPSG recommendation (Zhu et al. 2013).

IADPSG recommendation on overt diabetes is endorsed by most of the clinicians; however, universal screening needs to be performed at 24–28 weeks of gestation according to the available evidence (Donovan et al. 2013; Long and Cundy 2013). The first prenatal visit should be essential for the detection of overt diabetes and risky patients (Blumer et al. 2013).

4.2.6 100-g OGTT and Two-Step Screening

O'Sullivan and Mahan introduced the diagnostic criteria of GDM, which was based on blood glucose testing before and every 3 h after 100-g oral glucose intake in 1964 (Hoet and Lukens 1954). In one study included 752 pregnant women, two abnormal values were accepted as significant for GDM diagnosis (Houshmand et al. 2013).

In 1979, NDDG suggested converting the whole-body glucose cutoffs to plasma values (approximately 14% higher) (National Diabetes Data Group 1979). The new glucose measurement model by using glucose oxidase and hexokinase combined with new plasma values and Carpenter Coustan criteria emerged (Carpenter and Coustan 1982).

In the two-step approach, which is endorsed by ACOG and NIH, a 50-g glucose challenge test (GCT) irrespective of last meal is followed by a 100-g, fasting 3-h OGTT if required according to GCT results. The screening thresholds for 50-g GCT vary in different study groups and generally either 7.8 mmol/L (140 mg/dL) or 7.2 mmol/L (130 mg/dL) on the 50-g, 1-h oral GCT is acceptable to assume that screening was positive (Benhalima et al. 2018). GDM diagnosis is made if two or more plasma glucose levels equals or exceeds with 100-g OGTT performed while FPG $\geq$ 95 mg/dL (5.5 mmol/L), 1-h plasma glucose$\geq$180 mg/dL (10 mmol/L), 2-h plasma glucose$\geq$155 mg/dL (8.6 mmol/L) and 3-h plasma glucose$\geq$140 mg/dL (7.8 mmol/L) (Rani and Begum 2016). The patients who had positive GCT results but did not meet the thresholds for GDM diagnosis were identified as impaired glucose tolerance (IGT), and this condition is thought to indicate deteriorated glucose regulatory capacity, which might progress to DM in the future (Retnakaran et al. 2008).

4.2.7 75-g OGTT and One Step Screening

WHO recommended using 75-g OGTT for the diagnosis of GDM with the same thresholds of non-pregnant people in their first recommendation (Gupta et al. 2015). This approach was criticized for ignoring the physiological changes in glucose metabolism peculiar to pregnancy, which resulted in new recommendations in 1999 and lastly in 2013. WHO decreased the cutoff value of FPG from 7.8 mmol/L (140 mg/dL) to 5.1 mmol/L (92 mg/dL) for pregnant women with new updates (Chi et al. 2018). These criteria, which were changed by WHO, primarily considered the effects of the metabolic changes during pregnancy on prenatal outcomes (Houshmand et al. 2013). The current IADPSG criteria for the diagnosis of GDM were devised according to the HAPO study, which focused on adverse perinatal outcomes (McIntyre et al. 2015; Group HSCR et al. 2008). HAPO study became a cornerstone for threshold values that are able to predict adverse perinatal outcomes such as large for gestational age (LGA), primary cesarean section, neonatal hypoglycemia, and birth trauma. One study showed a linear association between glycemic values and adverse pregnancy outcomes (Group HSCR et al. 2008). Therefore, IADPSG criteria have been endorsed by several groups including, WHO, ADA, and The Endocrine Society of the USA (Blumer et al. 2013; American Diabetes A 2020; WHO 2014). IADPSG recommends a FPG level of 5.1 mmol/L (92 mg/dL), a 1-h level of 10.0 mmol/L (180 mg/dL) or 2-h value of 8.5 mmol/L (153 mg/dL) for GDM diagnosis after 75-g oral glucose load after overnight fasting of 8–14 h (WHO 2014).

4.2.8 Advantages and Disadvantages of One-Step and Two-Step Methods

Two-Step Method Arguments opposing the two-step method dominate among the study groups. The main concerns about implementing the method are the lack of existence of studies to decide the cutoffs, how to incorporate GCT with OGTT, and why two abnormal values needed (McIntyre et al. 2015; Akgol et al. 2019).

The two-step screening implies an inevitable delay in the diagnosis of GDM (Moses and Cheung 2009). In a systemic review, 50-g GCT was compared with OGTT (75-g or 100-g) to estimate the accuracy of GCT for GDM diagnosis (van Leeuwen et al. 2012). The sensitivity and specificity to predict GDM were 0.74 and 0.85, respectively. The study concluded that the two-step screening implemented with GCT followed by OGTT misses 26% of actual GDM cases, which will result in a delay in initiating the treatment (van Leeuwen et al. 2012). Another concern about the two-step method is that several studies showed that many pregnant women did not proceed with the diagnostic test after the GCT application. The missing rate is 10% in Toronto while it increases to 23% in New Zealand (Sermer et al. 1995; Sievenpiper et al. 2012).

The fasting time also affects the results of GCT. The metabolic capacity of pregnant women is considered to be better when tested in the afternoon. This also suggests that the time of OGTT might cause GDM over-, or delayed- diagnosis. Hence, it would affect fetal and maternal outcomes (Goldberg et al. 2012). According to the study of Hancerliogullari et al. fasting duration of >6.5 h resulted in 2.7 times more unnecessary 100-g OGTT, which is also another disadvantage of the method (Hancerliogullari et al. 2018).

In contrast, the GCT method is advantageous in (1) fever false-positive rates, (2) avoiding OGTT in more than 75% of women (Brown and Wyckoff 2017) (Table 5).

Table 5 Advantages and disadvantages of one-step and two-step screenings

	Two-step	One-step
Method	A 50-g GCT followed by a 100-g, 3-h OGTT. Those who screen positive are followed up by an oral 100-g glucose tolerance test	75-or 100-g OGTT is done in all patients, without the preliminary step by GCT
Advantages	Fewer false positives	Simple to follow
	Avoids OGTT in more than 75% of the women	Easily diagnosed
		Less missing follow up
Disadvantages	Higher missed diagnosis	Poor reproducibility
	Delay in initiating treatment even in those who test positive	All women need to come in a fasting state
	It requires patients to make two visits for testing	
	GCT results differ according to the time applied	

Abbreviations: *GDM* gestational diabetes mellitus, *GCT* glucose challenge test, *OGTT* oral glucose tolerance test

One-Step Method Not only the prevalence of GDM will increase from 10.6% to 35.5% with the one-step approach by using the IADPSG criterion but also pregnancy outcome and cost-effectiveness will be improved (Duran et al. 2014). The debate is about whether the increase allows identifying missed cases or results in over-diagnosis and over-treatment of healthy pregnancies (McIntyre et al. 2014, 2015; Long 2011).

Weak association of the primary adverse outcomes with the glycemic levels is the main obstacle for adopting the IADPSG criteria universally (Akgol and Budak 2019; Ayhan et al. 2016). Although the association with secondary outcomes such as shoulder dystocia, premature delivery, and preeclampsia is significant, these complications are not frequent (Long and Cundy 2013; Group HSCR et al. 2008).

Australian Carbohydrate Intolerance Study in Pregnant Women (ACHOIS) demonstrated and other studies showed that diagnosis of GDM results in increased cesarean section rate, earlier delivery, increased interventions, and more frequent neonatal intensive care unit management/need (Long and Cundy 2013; Ryan 2011; Crowther et al. 2005). Lower diagnostic levels will cause lower glucose targets achieved with diet or glucose-lowering therapies, which can lead to hypoglycemic conditions and fetal growth abnormalities (Crowther et al. 2005).

Arguments favoring IADPSG criteria accused the experts criticizing the one-step method and NIH report for ignoring the increased prevalence of pre-diabetes and undiagnosed type 2 diabetes of childbearing aged young women (Panel NC 2013; McIntyre 2013). The National Health and Nutrition Examination Survey (NHANES) is a program of studies designed to assess the health and nutritional status of adults and children in the US. This program showed that about 30% of women in the reproductive period have IGT (McIntyre 2013; Shin et al. 2015). These alterations of glucose metabolism will undoubtedly reflect GDM prevalence. It is clear that lowering the thresholds is the only way to reduce the number of missed cases (McIntyre 2013).

4.2.9 Role of FPG and Postprandial Plasma Glucose (PPG) for Screening GDM

In the literature, the ideal biomarker to predict GDM was studied by several investigators. Especially diagnostic accuracy of using FPG and PPG was evaluated in different studies to be used instead of the glucose load (Bhattacharya 2004; Powe 2017; Senanayake et al. 2006; Kansu-Celik et al. 2019c). In the first trimester, FPG physiologically tends to decline (Ozgu-Erdinc et al. 2019b; Yeral et al. 2014). Regardless of this anticipated decline, it is possible to miss out the risky/high-risk patients by using the cut-off values of the third trimester of pregnancy (<92 mg/dL) (McIntyre et al. 2016).

First-trimester FPG values between 79 mg/dL and 92 mg/dL have sensitivity and specificity ranging between 55% and 88% to predict GDM in the rest of the pregnancy according to the previous studies (Riskin-Mashiah et al. 2009; Yeral et al. 2014; Kansu-Celik et al. 2019a; Sacks et al. 2003; Reichelt et al. 1998). The study of Zhu et al. demonstrated that FPG between 110 mg/dL and 126 mg/dL should be considered and treated as GDM to improve pregnancy outcomes; However, nutritional and exercise advice will be enough when the FPG level is between 92 mg/dL and 110 mg/dL (Zhu et al. 2013). ADA also recommends nutrition and exercise advice for the women whose FPG is between 92 mg/dL and 110 mg/dL without diagnosing as GDM (American Diabetes A 2020). HAPO study also recommended that FPG >95 mg/dL is correlated with fetal macrosomia at 24–28 weeks of gestation (Group HSCR et al. 2008). Increasing FPG is not only associated with increasing GDM risk but also predicts the need for treatment (Alunni et al. 2015). Although Senanayake et al. demonstrated that postprandial glucose level could be used to predict GDM with lower sensitivity than FPG (Senanayake et al. 2006), most of the studies do not recommend the use of the 2-h PPG level instead of GCT (Agarwal 2018; Bhattacharya 2004). Huddleson et al. advocated that the PPG level is not necessary if FPG is within the normal range (Huddleston et al. 1993).

4.2.10 Role of HbA1c for Screening GDM

The cutoff values for the healthy population are well defined for HbA1c, while it needs to be evaluated for pregnant women (Rafat and Ahmad 2012). Although HAPO study demonstrated that the association between glucose level and the adverse perinatal outcome is more significant than with HbA1c, several studies claimed that 40–60% of pregnant women could be avoided from OGTT by using different cutoffs of HbA1c (Rajput et al. 2012; Lowe et al. 2012; Yerebasmaz et al. 2014). There is no consensus on change in the level of HbA1c during the pregnancy; some studies showed an increase, decrease, and no change in HbA1c level during trimesters (Rafat and Ahmad 2012; Davies and Welborn 1980; Pollak et al. 1979; McFarland et al. 1981). HbA1c should not be considered as an acceptable alternative for screening and diagnosing GDM.

5 Conclusion and Recommendations

The main reason for the diagnostic dilemma between different guidelines is a large number of procedures and glucose thresholds suggested for the diagnosis of glucose metabolism disorders in the pregnancy (Table 4). HAPO study and concomitant IADPSG criteria became a cornerstone for the screening and diagnosis of GDM (Group HSCR et al. 2008). The Endocrine Society of the USA, Australasian Diabetes in Pregnancy Society (ADIPS) and WHO endorsed the criteria of IADPSG, while ACOG and NIH recommended a two-step approach (Committee on Practice B-O 2018; Panel NC 2013; World Health Organization 2013; Blumer et al. 2013; Nankervis et al. 2014). ADA recommended using either the one-step or two-step approach for the diagnosis of GDM (Table 4) (American Diabetes A 2020). GDM is currently diagnosed between 24 and 28 weeks of pregnancy, while the combination of maternal risk factors and

biochemical markers can support early detection of high-risk patients. Table 4 demonstrates the different approaches to GDM screening and diagnosis among different guidelines and countries.

5.1 Overt Diabetes

IADPSG recommends screening for overt diabetes at first prenatal visit or universal screening of high-risk women and leaves the decision to the clinicians. WHO endorses the recommendations of IADPSG (WHO 2014). The Endocrine Society recommends screening using the same criteria used for non-pregnant people (Blumer et al. 2013). ADA, ACOG, NIH, and ADIPS suggested screening before 24 weeks of gestation if only risk factors were determined (Panel NC 2013; Nankervis et al. 2014; American Diabetes A 2020; Committee on Practice Bulletins—Obstetrics 2018).

5.2 One-Step Versus Two-Step Testing

ACOG and NIH still advocated the use of two-step testing for the screening and diagnosis of GDM (Panel NC 2013; Committee on Practice Bulletins—Obstetrics 2018). WHO, ADIPS, and Endocrine Society support the criteria of IADPSG (Blumer et al. 2013; WHO 2014). The current recommendations of ADA point out the need for further researches to build a uniformity among the health professionals and suggest leaving the decision about one-step or two-step to the clinicians (American Diabetes A 2020). IADPSG strongly recommended not using HbA1c for GDM screening, and all other guidelines have endorsed this. HbA1c can be used to rule out overt diabetes in early pregnancy (International Association of D, Pregnancy Study Groups Consensus P et al. 2010).

The optimal screening method is still controversial. Women with high plasma glucose levels are detected more sensitively with a 75-g OGTT and correlated with adverse pregnancy outcomes. However, the increasing rate of GDM diagnosis and lack of studies demonstrating that treatment of women labeled as GDM improved prenatal outcomes are the major obstacles in the way of creating a universal approach. Further researches are needed to be able to generalize the diagnostic methods.

Despite numerous research, multiple international conferences, and major trials, GDM remains a complex and contentious obstetrical and public health problem that deserves to be carefully discussed and studied. The lack of consensus and a single acceptable, evidence-based guideline confuses the health care providers. A simple, easy to follow, and validated recommendation is essential for GDM screening and diagnosis.

References

Agarwal MM (2016) Gestational diabetes mellitus: screening with fasting plasma glucose. World J Diabetes 7 (14):279–289

Agarwal MM (2018) Consensus in gestational diabetes MELLITUS: looking for the Holy Grail. J Clin Med 7(6):123

Akgol S, Budak MS (2019) Obstetric and neonatal outcomes of pregnancies with mild gestational hyperglycemia diagnosed at gestational diabetes mellitus screening. Gynecol Obstet Reprod Med 25(3):138–141

Akgol S, Obut M, Baglı İ, Kahveci B, Budak MS (2019) An evaluation of the effect of a one or two-step gestational diabetes mellitus screening program on obstetric and neonatal outcomes in pregnancies. Gynecol Obstet Reprod Med 25(2):62–66

Alunni ML, Roeder HA, Moore TR, Ramos GA (2015) First trimester gestational diabetes screening – change in incidence and pharmacotherapy need. Diabetes Res Clin Pract 109(1):135–140

American Diabetes A (2013) Diagnosis and classification of diabetes mellitus. Diabetes Care 36(Suppl 1):S67–S74

American Diabetes A (2018) 13. Management of diabetes in pregnancy: standards of medical care in diabetes-2018. Diabetes Care 41(Suppl 1):S137–SS43

American Diabetes A (2019) 14. Management of diabetes in pregnancy: standards of medical care in diabetes-2019. Diabetes Care 42(Suppl 1):S165–SS72

American Diabetes A (2020) 14. Management of diabetes in pregnancy: standards of medical care in diabetes-2020. Diabetes Care 43(Suppl 1):S183–SS92

Arslan E, Gorkem U, Togrul C (2019) Is there an association between kisspeptin levels and gestational diabetes mellitus? Gynecol Obstet Reprod Med 25(1):1–5

Ayhan S, Altınkaya SÖ, Güngör T, Özcan U (2016) Prognosis of pregnancies with different degrees of glucose intolerance. Gynecol Obstet Reprod Med 19(2):76–81

Barbour LA, McCurdy CE, Hernandez TL, Kirwan JP, Catalano PM, Friedman JE (2007) Cellular mechanisms for insulin resistance in normal pregnancy and gestational diabetes. Diabetes Care 30(Suppl 2): S112–S119

Benhalima K, Van Crombrugge P, Moyson C, Verhaeghe J, Vandeginste S, Verlaenen H et al (2018) The sensitivity and specificity of the glucose challenge test in a universal two-step screening strategy for gestational diabetes mellitus using the 2013 World Health Organization Criteria. Diabetes Care 41 (7):e111–e1e2

Bhattacharya SM (2004) Fasting or two-hour postprandial plasma glucose levels in early months of pregnancy as screening tools for gestational diabetes mellitus developing in later months of pregnancy. J Obstet Gynaecol Res 30(4):333–336

Bidhendi Yarandi R, Behboudi-Gandevani S, Amiri M, Ramezani TF (2019) Metformin therapy before conception versus throughout the pregnancy and risk of gestational diabetes mellitus in women with polycystic ovary syndrome: a systemic review, meta-analysis and meta-regression. Diabetol Metab Syndr 11:58

Blumer I, Hadar E, Hadden DR, Jovanovic L, Mestman JH, Murad MH et al (2013) Diabetes and pregnancy: an endocrine society clinical practice guideline. J Clin Endocrinol Metab 98(11):4227–4249

Brown J (2014) Screening, diagnosis and management of gestational Diabetes in New Zealand: a clinical practice guideline. Ministry of Health, Wellington. http://www.health.govt.nz/publication/screening-diagnosis-and-management-gestational-0

Brown FM, Wyckoff J (2017) Application of one-step IADPSG versus two-step diagnostic criteria for gestational diabetes in the real world: impact on health services, clinical care, and outcomes. Curr Diab Rep 17(10):85

Butte NF (2000) Carbohydrate and lipid metabolism in pregnancy: normal compared with gestational diabetes mellitus. Am J Clin Nutr 71(5 Suppl):1256S–1261S

Caliskan E, Kayikcioglu F, Ozturk N, Koc S, Haberal A (2004) A population-based risk factor scoring will decrease unnecessary testing for the diagnosis of gestational diabetes mellitus. Acta Obstet Gynecol Scand 83(6):524–530

Carpenter MW, Coustan DR (1982) Criteria for screening tests for gestational diabetes. Am J Obstet Gynecol 144 (7):768–773

Carr DB, Gabbe S (1998) Gestational diabetes: detection, management, and implications. Clin Diabetes 16 (1):4–12

Catalano PM, Tyzbir ED, Roman NM, Amini SB, Sims EA (1991) Longitudinal changes in insulin release and insulin resistance in nonobese pregnant women. Am J Obstet Gynecol 165(6 Pt 1):1667–1672

Chi C, Loy SL, Chan SY, Choong C, Cai S, Soh SE et al (2018) Impact of adopting the 2013 World Health Organization criteria for diagnosis of gestational diabetes in a multi-ethnic Asian cohort: a prospective study. BMC Pregnancy Childbirth 18(1):69

Committee on Practice B-O (2018) ACOG practice bulletin no. 190: gestational diabetes mellitus. Obstet Gynecol 131(2):e49–e64

Committee on Practice Bulletins—Obstetrics (2018) ACOG practice bulletin no. 190 summary: gestational diabetes mellitus. Obstet Gynecol 131(2):406–408

Crowther CA, Hiller JE, Moss JR, McPhee AJ, Jeffries WS, Robinson JS et al (2005) Effect of treatment of gestational diabetes mellitus on pregnancy outcomes. N Engl J Med 352(24):2477–2486

Davies D, Welborn T (1980) Glycosylated haemoglobin in pregnancy. Aust N Z J Obstet Gynaecol 20(3): 147–150

Diabetes Association Of The Republic Of China T (2020) Executive summary of the DAROC clinical practice guidelines for diabetes care- 2018. J Formos Med Assoc 119(2):577–586

Diabetes Canada Clinical Practice Guidelines Expert C, Feig DS, Berger H, Donovan L, Godbout A, Kader T et al (2018) Diabetes and pregnancy. Can J Diabetes 42 (Suppl 1):S255–SS82

Donovan L, Hartling L, Muise M, Guthrie A, Vandermeer B, Dryden DM (2013) Screening tests for gestational diabetes: a systematic review for the U.S. Preventive Services Task Force. Ann Intern Med 159(2):115–122

Donovan BM, Nidey NL, Jasper EA, Robinson JG, Bao W, Saftlas AF et al (2018) First trimester prenatal screening biomarkers and gestational diabetes mellitus: a systematic review and meta-analysis. PLoS One 13 (7):e0201319

Duran A, Saenz S, Torrejon MJ, Bordiu E, Del Valle L, Galindo M et al (2014) Introduction of IADPSG criteria for the screening and diagnosis of gestational diabetes mellitus results in improved pregnancy outcomes at a lower cost in a large cohort of pregnant women: the St. Carlos Gestational Diabetes Study. Diabetes Care 37(9):2442–2450

Fuller KP, Borgida AF (2014) Gestational diabetes mellitus screening using the one-step versus two-step method in a high-risk practice. Clin Diabetes 32 (4):148–150

Gan WZ, Ramachandran V, Lim CSY, Koh RY (2019) Omics-based biomarkers in the diagnosis of diabetes. J Basic Clin Physiol Pharmacol. https://doi.org/10. 1515/jbcpp-2019-0120

Gobl CS, Bozkurt L, Rivic P, Schernthaner G, Weitgasser R, Pacini G et al (2012) A two-step screening algorithm including fasting plasma glucose measurement and a risk estimation model is an accurate strategy for detecting gestational diabetes mellitus. Diabetologia 55(12):3173–3181

Goldberg RJ, Ye C, Sermer M, Connelly PW, Hanley AJ, Zinman B et al (2012) Circadian variation in the response to the glucose challenge test in pregnancy: implications for screening for gestational diabetes mellitus. Diabetes Care 35(7):1578–1584

Griffin ME, Coffey M, Johnson H, Scanlon P, Foley M, Stronge J et al (2000) Universal vs. risk factor-based screening for gestational diabetes mellitus: detection rates, gestation at diagnosis and outcome. Diabet Med 17(1):26–32

Group HSCR, Metzger BE, Lowe LP, Dyer AR, Trimble ER, Chaovarindr U et al (2008) Hyperglycemia and adverse pregnancy outcomes. N Engl J Med 358 (19):1991–2002

Guidelines H (2016) Guidelines for the management of gestational diabetes mellitus. The Hong Kong College of Obstetricians and Gynaecologists

Guidelines QC (2015) Gestational diabetes mellitus. Queensland Health, Queensland

Gupta Y, Kalra B, Baruah MP, Singla R, Kalra S (2015) Updated guidelines on screening for gestational diabetes. Int J Women's Health 7:539–550

Hadden DR (1998) A historical perspective on gestational diabetes. Diabetes Care 21(Suppl 2):B3–B4

Hancerliogullari N, Celik HK, Karakaya BK, Tokmak A, Tasci Y, Erkaya S et al (2018) Effect of prolonged fasting duration on 50 gram oral glucose challenge test in the diagnosis of gestational diabetes mellitus. Horm Metab Res 50(9):671–674

Haneda M, Noda M, Origasa H, Noto H, Yabe D, Fujita Y et al (2018) Japanese clinical practice guideline for diabetes 2016. J Diabetes Investig 9(3):657–697

Herath H, Herath R, Wickremasinghe R (2017) Gestational diabetes mellitus and risk of type 2 diabetes 10 years after the index pregnancy in Sri Lankan Women-A community based retrospective cohort study. PLoS One 12(6):e0179647

Hod M, Kapur A, Sacks DA, Hadar E, Agarwal M, Di Renzo GC et al (2015) The International Federation of Gynecology and Obstetrics (FIGO) initiative on gestational diabetes mellitus: a pragmatic guide for diagnosis, management, and care. Int J Gynaecol Obstet 131 (Suppl 3):S173–S211

Hod M, Pretty M, Mahmood T, Figo E, Ebcog (2018) Joint position statement on universal screening for GDM in Europe by FIGO, EBCOG and EAPM. Eur J Obstet Gynecol Reprod Biol 228:329–330

Hoet JP, Lukens FD (1954) Carbohydrate metabolism during pregnancy. Diabetes 3(1):1–12

Houshmand A, Jensen DM, Mathiesen ER, Damm P (2013) Evolution of diagnostic criteria for gestational diabetes mellitus. Acta Obstet Gynecol Scand 92 (7):739–745

Huddleston JF, Cramer MK, Vroon DH (1993) A rationale for omitting two-hour postprandial glucose determinations in gestational diabetes. Am J Obstet Gynecol 169(2 Pt 1):257–262; discussion 62-4

Huhn EA, Rossi SW, Hoesli I, Gobl CS (2018) Controversies in screening and diagnostic criteria for gestational diabetes in early and late pregnancy. Front Endocrinol (Lausanne) 9:696

IDF GDM Model of Care (2015) Implementation protocol. Guidelines for healthcare professionals. International Diabetes Federation, pp 7–9

International Association of D, Pregnancy Study Groups Consensus P, Metzger BE, Gabbe SG, Persson B, Buchanan TA et al (2010) International association of diabetes and pregnancy study groups recommendations on the diagnosis and classification of hyperglycemia in pregnancy. Diabetes Care 33 (3):676–682

Jokelainen M, Stach-Lempinen B, Rono K, Nenonen A, Kautiainen H, Teramo K et al (2020) Oral glucose tolerance test results in early pregnancy: a Finnish population-based cohort study. Diabetes Res Clin Pract 162:108077

Kansu-Celik H, Ozgu-Erdinc AS, Kisa B, Findik RB, Yilmaz C, Tasci Y (2019a) Prediction of gestational diabetes mellitus in the first trimester: comparison of maternal fetuin-A, N-terminal proatrial natriuretic peptide, high-sensitivity C-reactive protein, and fasting glucose levels. Arch Endocrinol Metab 63 (2):121–127

Kansu-Celik H, Ozgu-Erdinc AS, Kisa B, Eldem S, Hancerliogullari N, Engin-Ustun Y (2019b) Maternal serum glycosylated hemoglobin and fasting plasma glucose predicts gestational diabetes at the first trimester in Turkish women with a low-risk pregnancy and its relationship with fetal birth weight; a retrospective cohort study. J Matern Fetal Neonatal Med 1–211:1–8

Kansu-Celik H, Ozgu-Erdinc AS, Kisa-Karakaya B, Tasci Y, Erkaya S (2019c) Fasting and post-prandial plasma glucose screening for gestational diabetes mellitus. East Mediterr Health J 25(4):282–289

Kleinwechter H, Schafer-Graf U, Buhrer C, Hoesli I, Kainer F, Kautzky-Willer A et al (2014) Gestational Diabetes Mellitus (GDM) diagnosis, therapy and follow-up care: practice guideline of the German Diabetes Association (DDG) and the German Association for Gynaecology and Obstetrics (DGGG). Exp Clin Endocrinol Diabetes 122(7):395–405

Kuhl C, Holst JJ (1976) Plasma glucagon and the insulin: glucagon ratio in gestational diabetes. Diabetes 25 (1):16–23

Kuo C-H, Li H-Y (2019) Diagnostic strategies for gestational diabetes mellitus: review of current evidence. Curr Diab Rep 19(12):155

Langer O, Yogev Y, Most O, Xenakis EM (2005) Gestational diabetes: the consequences of not treating. Am J Obstet Gynecol 192(4):989–997

Lavin JP Jr (1985) Screening of high-risk and general populations for gestational diabetes. Clin Appl Cost Anal Diabetes 34(Suppl 2):24–27

Lee KW, Ching SM, Ramachandran V, Yee A, Hoo FK, Chia YC et al (2018) Prevalence and risk factors of gestational diabetes mellitus in Asia: a systematic review and meta-analysis. BMC Pregnancy Childbirth 18(1):494

Li P, Lin S, Li L, Cui J, Zhou S, Fan J (2019) First-trimester fasting plasma glucose as a predictor of gestational diabetes mellitus and the association with adverse pregnancy outcomes. Pak J Med Sci 35 (1):95–100

Li-zhen L, Yun X, Xiao-Dong Z, Shu-bin H, Zi-lian W, Sandra DA et al (2019) Evaluation of guidelines on the screening and diagnosis of gestational diabetes mellitus: systematic review. BMJ Open 9(5):e023014

Long H (2011) Diagnosing gestational diabetes: can expert opinions replace scientific evidence? Diabetologia 54 (9):2211–2213

Long H, Cundy T (2013) Establishing consensus in the diagnosis of gestational diabetes following HAPO: where do we stand? Curr Diab Rep 13(1):43–50

Lowe LP, Metzger BE, Dyer AR, Lowe J, McCance DR, Lappin TR et al (2012) Hyperglycemia and Adverse Pregnancy Outcome (HAPO) Study: associations of maternal A1C and glucose with pregnancy outcomes. Diabetes Care 35(3):574–580

Mahmood T (2018) Paris consensus on gestational diabetes mellitus screening 2018. Eur J Obstet Gynecol Reprod Biol 227:75–76

McFarland KF, Catalano EW, Keil JE, McFarland DE (1981) Glycosylated hemoglobin in diabetic and non-diabetic pregnancies. South Med J 74(4):410–412

McIntyre HD (2013) Diagnosing gestational diabetes mellitus: rationed or rationally related to risk? Diabetes Care 36(10):2879–2880

McIntyre HD, Metzger BE, Coustan DR, Dyer AR, Hadden DR, Hod M et al (2014) Counterpoint: establishing consensus in the diagnosis of GDM following the HAPO study. Curr Diab Rep 14(6):497

McIntyre HD, Colagiuri S, Roglic G, Hod M (2015) Diagnosis of GDM: a suggested consensus. Best Pract Res Clin Obstet Gynaecol 29(2):194–205

McIntyre HD, Sacks DA, Barbour LA, Feig DS, Catalano PM, Damm P et al (2016) Issues with the diagnosis and classification of hyperglycemia in early pregnancy. Diabetes Care 39(1):53–54

Mertoglu C, Gunay M, Gungor M, Kulhan M, Kulhan NG (2019) A study of inflammatory markers in gestational diabetes mellitus. Gynecol Obstet Reprod Med 25 (1):7–11

Metzger BE, Coustan DR (1998) Summary and recommendations of the Fourth International Workshop-Conference on Gestational Diabetes Mellitus. The Organizing Committee. Diabetes Care 21(Suppl 2):B161–B167

Mills JL, Jovanovic L, Knopp R, Aarons J, Conley M, Park E et al (1998) Physiological reduction in fasting plasma glucose concentration in the first trimester of normal pregnancy: the diabetes in early pregnancy study. Metab Clin Exp 47(9):1140–1144

Mirghani Dirar A, Doupis J (2017) Gestational diabetes from A to Z. World J Diabetes 8(12):489–511

Moses RG, Cheung NW (2009) Point: universal screening for gestational diabetes mellitus. Diabetes Care 32 (7):1349–1351

Moyer VA, Force USPST (2014) Screening for gestational diabetes mellitus: U.S. Preventive Services Task Force recommendation statement. Ann Intern Med 160 (6):414–420

Mumtaz M (2000) Gestational diabetes mellitus. Malays J Med Sci 7(1):4–9

Nanda S, Savvidou M, Syngelaki A, Akolekar R, Nicolaides KH (2011) Prediction of gestational diabetes mellitus by maternal factors and biomarkers at 11 to 13 weeks. Prenat Diagn 31(2):135–141

Nankervis A MH, Moses R, Ross GP, Callaway L, Porter C, Jeffries W, Boorman C, De Vries B, McElduff A for the Australasian Diabetes in Pregnancy Society (2014) ADIPS consensus guidelines for the testing and diagnosis of gestational diabetes mellitus in Australia 2014. Available from: http://adips.org/downloads/ADIPSConsensusGuidelinesGDM-03.05.13VersionACCEPTEDFINAL.pdf

National Diabetes Data Group (1979) Classification and diagnosis of diabetes mellitus and other categories of glucose intolerance. Diabetes 28(12):1039–1057

National Institute for Health and Care Excellence (2015) Diabetes in pregnancy: management of diabetes and its complications from preconception to the postnatal period. National Institute for Health and Care Excellence: Clinical Guidelines, London

Negrato CA, Montenegro RM Jr, Mattar R, Zajdenverg L, Francisco RP, Pereira BG et al (2010) Dysglycemias in pregnancy: from diagnosis to treatment. Brazilian consensus statement. Diabetol Metab Syndr 2:27

Network SIG (2014) Management of diabetes: a national clinical guideline. Scottish Intercollegiate Guidelines Network, Edinburgh

Nevalainen J, Sairanen M, Appelblom H, Gissler M, Timonen S, Ryynanen M (2016) First-trimester maternal serum amino acids and acylcarnitines are significant predictors of gestational diabetes. Rev Diabet Stud 13(4):236–245

O'Sullivan JB, Mahan CM (1964) Criteria for the Oral glucose tolerance test in pregnancy. Diabetes 13:278–285

Ozgu-Erdinc AS, Yilmaz S, Yeral MI, Seckin KD, Erkaya S, Danisman AN (2015) Prediction of gestational diabetes mellitus in the first trimester: comparison of C-reactive protein, fasting plasma glucose, insulin and insulin sensitivity indices. J Matern Fetal Neonatal Med 28(16):1957–1962

Ozgu-Erdinc AS, Sert UY, Buyuk GN, Engin-Ustun Y (2019a) Prevalence of gestational diabetes mellitus and results of the screening tests at a tertiary referral center: a cross-sectional study. Diabetes Metab Syndr 13(1):74–77

Ozgu-Erdinc AS, Sert UY, Kansu-Celik H, Moraloglu Tekin O, Engin-Ustun Y (2019b) Prediction of gestational diabetes mellitus in the first trimester by fasting plasma glucose which cutoff is better? Arch Physiol Biochem 28(16):1–5

Panel NC (2013) National Institutes of Health consensus development conference statement: diagnosing gestational diabetes mellitus, March 4–6, 2013. Obstet Gynecol 122(2 Pt 1):358–369

Plasencia W, Garcia R, Pereira S, Akolekar R, Nicolaides KH (2011) Criteria for screening and diagnosis of gestational diabetes mellitus in the first trimester of pregnancy. Fetal Diagn Ther 30(2):108–115

Plows JF, Stanley JL, Baker PN, Reynolds CM, Vickers MH (2018) The pathophysiology of gestational diabetes mellitus. Int J Mol Sci 19(11):3342

Pollak A, Widness JA, Schwartz R (1979) 'Minor Hemoglobins': an alternative approach for evaluating glucose control in pregnancy. Neonatology 36 (3–4):185–192

Powe CE (2017) Early pregnancy biochemical predictors of gestational diabetes mellitus. Curr Diab Rep 17 (2):12

Rafat D, Ahmad J (2012) HbA1c in pregnancy. Diabetes Metab Syndr 6(1):59–64

Rajput R, Yogesh Y, Rajput M, Nanda S (2012) Utility of HbA1c for diagnosis of gestational diabetes mellitus. Diabetes Res Clin Pract 98(1):104–107

Rani PR, Begum J (2016) Screening and diagnosis of gestational diabetes mellitus, where do we stand. J Clin Diagn Res 10(4):QE01–QE04

RANZCOG (2017) Diagnosis of Gestational Diabetes Mellitus (GDM) and diabetes mellitus in pregnancy. In: RANZCOG (ed). Available from: https://ranzcog.edu.au/RANZCOG_SITE

Ravnsborg T, Andersen LL, Trabjerg ND, Rasmussen LM, Jensen DM, Overgaard M (2016) First-trimester multimarker prediction of gestational diabetes mellitus using targeted mass spectrometry. Diabetologia 59 (5):970–979

Reichelt AJ, Spichler ER, Branchtein L, Nucci LB, Franco LJ, Schmidt MI (1998) Fasting plasma glucose is a useful test for the detection of gestational diabetes. Brazilian Study of Gestational Diabetes (EBDG) Working Group. Diabetes Care 21(8):1246–1249

Retnakaran R, Qi Y, Sermer M, Connelly PW, Hanley AJ, Zinman B (2008) Glucose intolerance in pregnancy and future risk of pre-diabetes or diabetes. Diabetes Care 31(10):2026–2031

Riskin-Mashiah S, Younes G, Damti A, Auslender R (2009) First-trimester fasting hyperglycemia and adverse pregnancy outcomes. Diabetes Care 32 (9):1639–1643

Ryan EA (2011) Diagnosing gestational diabetes. Diabetologia 54(3):480–486

Sacks DA, Chen W, Wolde-Tsadik G, Buchanan TA (2003) Fasting plasma glucose test at the first prenatal visit as a screen for gestational diabetes. Obstet Gynecol 101(6):1197–1203

Savvidou M, Nelson SM, Makgoba M, Messow CM, Sattar N, Nicolaides K (2010) First-trimester prediction of gestational diabetes mellitus: examining the potential of combining maternal characteristics and laboratory measures. Diabetes 59(12):3017–3022

Senanayake H, Seneviratne S, Ariyaratne H, Wijeratne S (2006) Screening for gestational diabetes mellitus in

Southern Asian women. J Obstet Gynaecol Res 32 (3):286–291

Sermer M, Naylor CD, Gare DJ, Kenshole AB, Ritchie JW, Farine D et al (1995) Impact of increasing carbohydrate intolerance on maternal-fetal outcomes in 3637 women without gestational diabetes. The Toronto Tri-Hospital Gestational Diabetes Project. Am J Obstet Gynecol 173(1):146–156

Seshiah V, Das AK, Balaji V, Joshi SR, Parikh MN, Gupta S et al (2006) Gestational diabetes mellitus – guidelines. J Assoc Physicians India 54:622–628

Seshiah V, Sahay BK, Das AK, Shah S, Banerjee S, Rao PV et al (2009) Gestational diabetes mellitus – Indian guidelines. J Indian Med Assoc 107(11):799–802, 4–6

Shah A, Stotland NE, Cheng YW, Ramos GA, Caughey AB (2011) The association between body mass index and gestational diabetes mellitus varies by race/ethnicity. Am J Perinatol 28(7):515–520

Shin D, Lee KW, Song WO (2015) Dietary patterns during pregnancy are associated with risk of gestational diabetes mellitus. Nutrients 7(11):9369–9382

Sievenpiper JL, McDonald SD, Grey V, Don-Wauchope AC (2012) Missed follow-up opportunities using a two-step screening approach for gestational diabetes. Diabetes Res Clin Pract 96(2):e43–e46

Simmons D (2017) Epidemiology of diabetes in pregnancy. A Practical Manual of Diabetes in pregnancy, (second edition, chapter 1), pp 1–16

Tu WJ, Guo M, Shi XD, Cai Y, Liu Q, Fu CW (2017) First-trimester serum fatty acid-binding protein 4 and subsequent gestational diabetes mellitus. Obstet Gynecol 130(5):1011–1016

van Leeuwen M, Opmeer BC, Zweers EJ, van Ballegooie E, ter Brugge HG, de Valk HW et al (2010) Estimating the risk of gestational diabetes mellitus: a clinical prediction model based on patient characteristics and medical history. BJOG 117 (1):69–75

van Leeuwen M, Louwerse MD, Opmeer BC, Limpens J, Serlie MJ, Reitsma JB et al (2012) Glucose challenge test for detecting gestational diabetes mellitus: a systematic review. BJOG 119(4):393–401

Wang C, Lin L, Su R, Zhu W, Wei Y, Yan J et al (2018) Hemoglobin levels during the first trimester of pregnancy are associated with the risk of gestational diabetes mellitus, pre-eclampsia and preterm birth in Chinese women: a retrospective study. BMC Pregnancy Childbirth 18(1):263

WHO (1980) WHO expert committee on diabetes mellitus: second report. World Health Organ Tech Rep Ser 646:1–80

WHO (2014) Diagnostic criteria and classification of hyperglycaemia first detected in pregnancy: a World Health Organization Guideline. Diabetes Res Clin Pract 103(3):341–363

World Health Organization (2013) Diagnostic criteria and classification of hyperglycaemia first detected in pregnancy. WHO guidelines approved by the guidelines review committee. World Health Organization, Geneva

Yang HX (2012) Diagnostic criteria for gestational diabetes mellitus (WS 331-2011). Chin Med J 125 (7):1212–1213

Yang X, Quan X, Lan Y, Ye J, Wei Q, Yin X et al (2017) Serum chemerin level during the first trimester of pregnancy and the risk of gestational diabetes mellitus. Gynecol Endocrinol 33(10):770–773

Yeral MI, Ozgu-Erdinc AS, Uygur D, Seckin KD, Karsli MF, Danisman AN (2014) Prediction of gestational diabetes mellitus in the first trimester, comparison of fasting plasma glucose, two-step and one-step methods: a prospective randomized controlled trial. Endocrine 46(3):512–518

Yerebasmaz N, Aldemir O, Asıltürk Ş, Esinler D, Karahanoğlu E, Kandemir Ö et al (2014) Is HbA1c predictive for screening and diagnosis of gestational diabetes mellitus? Gynecol Obstet Reprod Med 20 (2):88–91

Yoffe L, Polsky A, Gilam A, Raff C, Mecacci F, Ognibene A et al (2019) Early diagnosis of gestational diabetes mellitus using circulating microRNAs. Eur J Endocrinol/European Federation of Endocrine Societies 181(5):565–577

Zhang C, Tobias DK, Chavarro JE, Bao W, Wang D, Ley SH et al (2014) Adherence to healthy lifestyle and risk of gestational diabetes mellitus: prospective cohort study. BMJ 349:g5450

Zhu WW, Yang HX, Wei YM, Yan J, Wang ZL, Li XL et al (2013) Evaluation of the value of fasting plasma glucose in the first prenatal visit to diagnose gestational diabetes mellitus in China. Diabetes Care 36 (3):586–590

Zhu B, Liang C, Xia X, Huang K, Yan S, Hao J et al (2019) Iron-related factors in early pregnancy and subsequent risk of gestational diabetes mellitus: the Ma'anshan Birth Cohort (MABC) Study. Biol Trace Elem Res 191(1):45–53

Adv Exp Med Biol - Advances in Internal Medicine (2020) 4: 257–272
https://doi.org/10.1007/5584_2020_552
© Springer Nature Switzerland AG 2020
Published online: 17 June 2020

Management of Gestational Diabetes Mellitus

Z. Asli Oskovi-Kaplan and A. Seval Ozgu-Erdinc ⓘ

Abstract

Once a woman is diagnosed with gestational diabetes mellitus (GDM), two strategies are considered for management; life-style modifications and pharmacological therapy. The management of GDM aims to maintain a normoglycemic state and to prevent excessive weight gain in order to reduce maternal and fetal complications. Lifestyle modifications include nutritional therapy and exercise. Calorie restriction with a low glycemic index diet is recommended to avoid postprandial hyperglycemia and to reduce insulin resistance. Blood glucose levels, HbA1c levels, and ketonuria are monitored to analyze the efficacy of conservative management. Pharmacological treatment is initiated if conservative strategies fail to provide expected glucose levels during follow-ups.

Insulin has been the first choice for the treatment of diabetes during pregnancy. Recently, metformin has been used more commonly in diabetic pregnant women in cases when insulin cannot be prescribed, after its safety has been proven. However, a high percentage of women, which may be up to 46% may require additional insulin to maintain expected blood glucose levels. The evidence on the long-term safety of other oral anti-diabetics has been lacking yet.

Z. A. Oskovi-Kaplan and A. S. Ozgu-Erdinc (✉)
Ministry of Health-Ankara City Hospital, Ankara, Turkey
e-mail: aslioskovi@gmail.com; sevalerdinc@gmail.com

Women with diet-controlled GDM can wait for spontaneous labor expectantly in case there are no obstetric indications for birth. However, in women with GDM under insulin therapy or with poor glycemic control, elective induction at term is recommended by authorities.

The women who have GDM during pregnancy should be counseled about their increased risks of impaired glucose tolerance, type 2 diabetes mellitus, hypertensive disorders, cardiovascular diseases, and metabolic syndrome.

Keywords

Delivery · Diet · Exercise · Gestational diabetes mellitus · Glucose monitoring · Insulin · Labor · Metformin · Nutritional therapy · Postpartum counseling · Pregnancy

Gestational diabetes is one of the most common problems of pregnancy. Most of the women are not screened for diabetes mellitus before pregnancy. The routine screening for gestational diabetes mellitus (GDM) is performed between 24th and 28th weeks of pregnancy; therefore, not only the women with GDM but also some of the women with pregestational diabetes mellitus are diagnosed after the second trimester (Committee on Practice B-O 2018; Akgol et al. 2019). Once a woman is diagnosed with diabetes mellitus, nutritional counseling is given and in case of failure to achieve normal glucose levels, medication is prescribed.

GDM is classified into two groups according to its treatment (American Diabetes Association 2020). The terminology, Class A1 GDM is used for diet-controlled GDM which is successful in 70–85% of women. If the patient needs insulin therapy, then it is classified as Class A2 GDM.

There are complications related to GDM which cause maternal and fetal morbidities. Some of them are listed as;

– Large for gestational age infant and macrosomia LGA and macrosomia substantially increase the risk of operative vaginal delivery, cesarean delivery, neonatal hypoglycemia, birth trauma, shoulder dystocia and its associated complications nerve injury, fractures and neonatal asphyxia (Kc et al. 2015; Hancerliogullari et al. 2019).

– Neonatal morbidity and mortality Hypoglycemia, hyperbilirubinemia, hypocalcemia, hypomagnesemia, polycythemia, respiratory distress, cardiomyopathy, and stillbirth may increase in women with poor glycemic control (Billionnet et al. 2017; Farrar et al. 2016; Ayhan et al. 2016).

– Polyhydramnios Fetal hyperglycemia causes fetal polyuria which results in polyhydramnios. Polyhydramnios may increase perinatal morbidity and mortality whether or not it is related to GDM (Kollmann et al. 2014).

– Preeclampsia and gestational hypertension (Weissgerber and Mudd 2015).

– Maternal increased risk of diabetes and cardiovascular diseases later in life (Committee on Practice B-O 2018).

– Long-term increased risk of disorders of glucose metabolism, obesity, increased adiposity and hypertension in fetuses born to GDM mothers (Bianco and Josefson 2019).

Treatment of mild GDM is associated with reduced infant morbidity and improved quality of life for the mothers (Tieu et al. 2014). When GDM is diagnosed and managed properly, the benefits would be decreased newborn complications (perinatal death, birth trauma, shoulder dystocia, fracture/nerve palsy); reduced rates in LGA infants, fetal macrosomia and decreased risk of preeclampsia (Crowther et al. 2005).

Once a patient is diagnosed with GDM, non-pharmacologic treatments including lifestyle modifications, dietary counseling and exercise are recommended and the glucose levels are monitored for follow-ups. Even the life-style modifications provided a reduction of LGA fetuses, macrosomia, neonatal fat mass, postpartum depression and helped to achieve postpartum weight goals (Brown et al. 2017a).

1 Nutritional Therapy

Nutritional therapy is the initial and the main step in the management of GDM (Diabetes Canada Clinical Practice Guidelines Expert C et al. 2018). Maintenance of euglycemic levels is the first aim of dietary counseling (Committee on Practice B-O 2018). The important point in nutrition therapy is the prevention of ketosis while restricting diet (Diabetes Canada Clinical Practice Guidelines Expert C et al. 2018). The gestational weight gain should be controlled in pregnancy follow-ups and a balanced weight gain should be achieved until birth (Committee on Practice B-O 2018). The diet also should contribute to appropriate fetal growth. A closer follow-up than low-risk pregnant population should be planned for the prevention of maternal and fetal complications associated with GDM (Committee on Practice B-O 2018).

1.1 Ideal Diet

While planning the diet; self-glucose monitoring, weight gain, maternal factors such as; work, appetite and exercise program should be considered. Usually, three main meals (small/moderate size) and two to four snacks are recommended. There is limited on different dietary compositions but in general practice, low-glycemic index diet with complex carbohydrates are preferred to simple carbohydrates in order to reduce postprandial hyperglycemia and insulin resistance (Committee on Practice B-O 2018; Han et al. 2017; Tieu et al. 2017). The Dietary Reference Intakes recommends to all pregnant women to consume 175 g of carbohydrate, 71 g of protein and 28 g of fiber (American Diabetes Association 2020).

Carbohydrate intake should be limited to 40% of total calories (Moreno-Castilla et al. 2013). High rates of carbohydrates should be avoided to prevent excessive weight gain and postprandial hyperglycemia. Ketonuria should be avoided while carbohydrates are restricted. Low glycemic index diet with low consumption of flour-based products and potatoes was significantly associated with less insulin need but not with pregnancy outcomes as fetal macrosomia, mode of delivery and maternal macrosomia (Viana et al. 2014). **Proteins** should form 20% of the total calories. **Fat** intake should take part in 40% of calories and saturated fats should be <7% of total calories.

1.2 Ideal Weight Gain According to Pre-pregnancy BMI (Institute of Medicine Recommendations) (Rasmussen and Yaktine 2009; National Academy of Sciences, Institute of Medicine, Food and Nutrition Board, Subcommittee on Nutritional Status and Weight Gain During Pregnancy 1990)

– Underweight women (BMI <18.5) 12.5–18 kg
– Normal weight women (BMI 18.5–24.9) 11.5–16 kg
– Overweight women (BMI 25–29.9) 6.8–11.3 kg
  Obese women (BMI >30) 5–9.1 kg

For twins

– Underweight or normal weight women (BMI 18.5–24.9) 17–25 kg
– Overweight women (BMI 25–29.9) 14–23 kg

– Obese women (BMI >30) 11–19 kg

For triplets, approximately 22.5–27 kg is recommended.

The studies showed that excessive weight gain of women with GDM was associated with increased risk of fetal macrosomia, preterm birth and cesarean delivery (Durnwald 2015; Gou et al. 2019). Also, GDM and maternal obesity were found as risk factors for fetal macrosomia and preeclampsia (Vieira et al. 2018). Weight loss in pregnancy is not recommended except for the severely obese women. Calorie restriction shall be considered for obese women by reducing calories by 30% below the Dietary Reference Intakes for pregnant women (Moreno-Castilla et al. 2013; American Diabetes Association 2004). Pregnant women who don't have excessive weight gain, as well as who had restrictions of continuing excessive weight gain have a lower risk for LGA infants and less insulin requirement (Barnes et al. 2020).

1.3 Recommended Calories

The recommended calorie intake for pregnant women according to the American Diabetes Association are (American Diabetes Association 2004);

– Underweight women 35–40 kcal/kg/day
– Normal weight women 30 kcal/kg/day
– Overweight women 22–25 kcal/kg/day
– Severely obese women 12–14 kcal/kg/day

The target of the diet is to provide 1800–2,500 kcal intake per day. The calories should be increased step by step in the second and third trimesters. Over-restriction of calories in obese women should be avoided to prevent ketosis.

Potential benefits of probiotic supplementation to women with GDM is a current topic. Probiotic supplementation to pregnant women with GDM reduces the risk of neonatal hyperbilirubinemia by 74% and improves glycemic control, lipid profile and blood markers for inflammation and oxidative stress (Zhang et al. 2019).

2 Exercise

Exercise improves glycemic control in women by the following mechanisms; the increase of muscle tissue with exercise reduces insulin resistance, fasting, and postprandial glucose levels and may reduce the need for insulin therapy (Yu et al. 2018; Brown et al. 2017b). Even simple exercise such as 10–15 min of walking after each meal helps improvement in glycemic control.

Moderate-intensity exercise for 30 min, 5 days a week or minimum 150 min per week is recommended to pregnant women with GDM unless they have a medical or obstetrical contraindication for exercise ACOG and ADA (Committee on Practice B-O 2018; American Diabetes Association 2020). Aerobic exercise, resistance strength training, hydrotherapy exercise are safe for pregnant women and help improving blood glucose (Padayachee and Coombes 2015).

3 Monitorization of Laboratory Parameters

3.1 Glucose Monitoring

When a pregnant woman is diagnosed with GDM, glucose measurement is recommended four times a day in the beginning as one fasting glucose and three postprandial measurements after one or two hours of meals (American Diabetes Association 2020). Among women with well-controlled gestational diabetes, monitoring blood glucose levels every other day had similar neonatal outcomes and birth weight while it did not have adverse outcomes and increased rates of macrosomia when compared with four times daily glucose monitoring, and had better compliance (Mendez-Figueroa et al. 2017).

In pregnant women who are under insulin therapy, there are controversies on the timing of postprandial glucose monitoring. One-hour postprandial glucose monitoring resulted in better glycemic control, lower incidence of large for gestational age infants and lower rates of cesarean delivery with indications of cephalopelvic disproportion (de Veciana et al. 1995). However, contradicting studies comparing one and two-hour postprandial glucose monitoring was published which reported similar perinatal outcomes and fetal macrosomia (Ozgu-Erdinc et al. 2016).

Continuous glucose monitoring should not be routinely recommended and should be preferred in cases who experience severe hypoglycemia episodes, to monitor blood glucose variability and in cases with unstable blood glucose levels (National Collaborating Centre for Women's and Children's Health (Great Britain) 2015). Continuous glucose monitoring was associated with more women who needed insulin therapy, less preeclampsia, less primary cesarean delivery rates, fewer large for gestational age infants; and may advantages of detecting higher post-prandial glucose levels and nocturnal hypoglycemic events (Chen et al. 2003; Yu et al. 2014).

3.2 Glucose Target

The target levels recommended by ADA and ACOG are essentially determined for pregnant women with pre-existing diabetes (American Diabetes Association 2020). An optimal level has not been determined, in which the disadvantages and benefits of insulin therapy get balanced. Eventually, the persistence of high blood glucose levels, despite the nutritional counseling and a restricted diet will require insulin therapy. To sum up, if one-third of measured glucose levels are above the target fasting or postprandial levels within a week, insulin therapy is initiated or the dose is increased.

The recommended glucose target levels by ADA and ACOG are

- Fasting blood glucose <95 mg/dL (5.3 mmol/L)
- Postprandial one-hour glucose <140 mg/dL (7.8 mmol/L)

- Postprandial two-hour glucose <120 mg/dL (6 7 mmol/l)

The risk of fetal macrosomia and large for gestational age increases with the increased fasting plasma glucose levels during pregnancy. The Hyperglycemia and Adverse Pregnancy Outcomes study showed that risk of fetal macrosomia increased five times in women with fasting plasma glucose level of 100–105 mg/dL (5.6–5.8 mmol/L) when compared with women with levels of 75 mg/dL (4.2 mmol/L) (Group HSCR 2002).

3.3 Glycated Hemoglobin (HbA1c)

HbA1c values are lower in pregnant women than non-pregnant women (Hughes et al. 2014). There is not a clear recommendation for HbA1c levels on how often it should be measured (Soumya et al. 2015). The HbA1c levels of ≥5.9% (≥41 mmol/mol) may also be predictive for adverse pregnancy outcomes (Hughes et al. 2014). If hypoglycemia occurs with the target HbA1c levels of <6%, the target level may be extended to <7% (National Collaborating Centre for Women's and Children's Health (Great Britain) 2015).

3.4 Ketonuria

Routine monitorization of ketonuria is not recommended for pregnant women and Maternal ketonemia may affect the behavioral and intellectual development of the fetus and may be associated with lower IQ in the child, however, increased risk for adverse effects of ketonuria for fetal development is not clearly proved (Ozgu-Erdinc et al. 2016). Monitorization of ketonuria may be recommended in conditions of blood glucose is above 180 mg/dL (11.1 mmol/L), serious infection, trauma of other stress factors for metabolism especially if symptoms related with ketoacidosis such as nausea, vomiting or abdominal pain; despite diabetic ketoacidosis is seen extremely rare in gestational diabetic ketoacidosis

(Committee on Practice B-O 2018; American Diabetes Association 2020).

4 Medical Therapy

If the blood glucose levels exceed the target levels recommended, two options of medical therapy come into a choice insulin or oral antihyperglycemic agents (metformin and glyburide). Approximately 20% of women with GDM would require medical therapy.

Insulin does not cross the placenta and it is preferred for treatment of GDM by ACOG and ADA (American Diabetes Association 2020). Oral antihyperglycemic agents (metformin and glyburide) are not approved by the FDA for the treatment of GDM. ACOG recommends metformin for women who refuses insulin therapy, who are unable to afford insulin or would fail in treatment compliance (Committee on Practice B-O 2018).

The National Institute for Health and Care Excellence and International Federation of Gynecology and Obstetrics recommends oral antihyperglycemic agents for first-line treatment especially in women with low fasting blood glucose levels (National Collaborating Centre for Women's and Children's Health (Great Britain) 2015; Hod et al. 2015).

4.1 Insulin

Insulin treatment is recommended as the first choice in conditions

- -Pre-gestational diabetes (<20th week)
- After 30th week in case of nutritional therapy fails
- FPG 110 > mg/dL (6.1 mmol/L)
- One-hour postprandial glucose >140 mg/dL (7.8 mmol/L)
- Pregnancy weight gain >12 kg (26.5 pounds); by the International Federation of Gynecology and Obstetrics (Hod et al. 2015).

It was reported that fetuses with fetal abdominal circumference >75th percentile in 28–32 weeks would benefit from insulin treatment to avoid fetal macrosomia (Bonomo et al. 2004; Rossi et al. 2000; Buchanan et al. 1994). Treatment of mild GDM improves neonatal outcomes (mean birth weight, neonatal fat mass, the rise of large for gestational age infants, fetal macrosomia and cesarean section rates) despite the prenatal death rates do not change (Landon et al. 2009). It has been reported that similar outcomes are present when mild gestational diabetes mellitus (fasting glucose <95 mg/dl and two elevated glucose values after glucose challenge test) was diagnosed and treated in the earlier weeks (24–26, 27, 28 or 29 weeks) when compared to ≥30 weeks of gestation (Palatnik et al. 2015).

Hospitalization is not essential for initiating insulin therapy. The patient should be educated about the insulin injection technique of insulin and self-monitorization of blood glucose. Blood glucose should be monitored 4–6 times daily, preferably FPG, pre-lunch and pre-dinner glucose levels; and one or two-hour postprandial blood glucose after each meal.

Theoretically, insulin-pumps should provide better glycemic control by real-time glucose measurements in pregnant women should improve obstetric and neonatal outcomes. However, the studies on insulin-pumps in pregnant women are limited and yet, they are not proved to be more effective than conventional methods (Combs 2012). A meta-analysis of six small randomized controlled trials reported neither advantages nor disadvantages in terms of pregnancy outcomes and glycemic control and the higher number of ketoacidotic episodes in continuous insulin injection group did not reach statistical significance (Mukhopadhyay et al. 2007).

The dosing of insulin is a patient-dependent decision, regarding the timing of their hyperglycemic episodes and the gestational age. Four-times-daily insulin injection provides better glycemic control than two-times daily doses, although it does not, changes the macrosomia rates (Nachum et al. 1999). Administration of rapid-acting and short-acting insulin in 21 proportion is comparable with insulin release in third-trimester normal pregnant women (Lewis et al. 1976). Twin-pregnancies may need approximately the twice dose of singletons (Akiba et al. 2019). An initial dose of insulin of 0.7–1 unit/kg is usually enough, however, if the glucose levels are too high, a dose up to 2 units/kg can be used to achieve appropriate glycemic levels (Committee on Practice B-O 2018). Insulin is given in multiple injections before meals with a combination of short-acting and intermediate/long-acting insulin preparations. The dosage of insulin is individualized adjusted appraised by glucose values during the day. Half of the required insulin dose is administered as intermediate-acting insulin such as NPH in two equal doses which are given before breakfast and bedtime. For the treatment of fasting hyperglycemia, intermediate-acting insulin (NPH) should be administered before bedtime with a starting dose of 0.2 U/kg. The proper adjustment of insulin dosage is essential to avoid hypoglycemia. Dizziness, shaking, sweating, tachycardia, numbness in tongue and lips may be signs of hypoglycemia (Lv et al. 2015).

A systematic review and meta-analysis of four studies of 1214 women which compared insulin therapy and oral anti-diabetic drugs (metformin and glibenclamide) revealed insulin treatment was related with an increased risk for hypertensive disorders of pregnancy when compared with oral anti-diabetic drugs while there was no clear evidence for the increased risk of pre-eclampsia (Brown et al. 2017c).

Insulin lispro and insulin aspart are short-acting insulin analogs that are safe and mostly preferred during pregnancy because of minimal transfer through the placenta and no evidence for teratogenicity. They can be administered just before the meal instead of waiting for 10–15 min as it is for regular insulin (American Diabetes Association 2020). They are associated with better postprandial results and lower risk of delayed post-prandial hypoglycemia when compared with the less immunogenic preparation, human regular insulin (Nicholson et al. 2008).

NPH insulin is the most commonly used intermediate/long acting insulin, therewithal insulin glargine and insulin detemir are also used for long-acting choice (Vellanki and Umpierrez 2016). Insulin aspart, detemir, and glargine are not related to increased pregnancy-related complications; lispro was related to decreased neonatal jaundice and severe maternal hypoglycemia while fetal macrosomia rates slightly increased (Lv et al. 2015).

4.2 Oral Antihyperglycemic Agents

There are two main oral antihyperglycemic agents and the current evidence does not clearly support the benefit of one drug over another. Both agents cross the placenta however, neither difference in main perinatal outcomes nor increased macrosomia rates were reported (Brown et al. 2017d). However, some benefits such as lower mean birth weight, less gestational weight gain and less composite neonatal death or serious morbidity were reported for metformin. In addition, the frequencies of treatment failure were also similar between drugs (Brown et al. 2017d). Therefore the decision for the first-choice drug was left for the clinician's preference, experience and guideline recommendations.

4.2.1 Metformin

Metformin crosses the placenta and reaches approximately to maternal levels in fetal circulation. The evidence of metformin on long-term fetal prognosis is lacking however, similar fetal outcomes by the age of 2 were reported (Wouldes et al. 2016). No teratogenic effects were reported for metformin in animal studies and in limited human studies (Denno and Sadler 1994; Gilbert et al. 2006) When the risk of adverse neonatal outcomes, teratogenicity and major birth defects are analyzed in pregnant women under metformin therapy for several indications, the increased risk was related to underlying pre-gestational diabetes rather than drug use (Panchaud et al. 2018). In one study comparing metformin and insulin with a long-term follow-up, It was reported that total and abdominal fat percent, and metabolic measures were found similar between groups while the offspring exposed to metformin were larger by the means of BMI, waist circumferences and triceps skin-folds at the age of 9 years (Rowan et al. 2018). This study meant, the medication of GDM during pregnancy may have an impact on the metabolic states of children in further lives. In addition, up to 46% of women who take metformin may also require insulin for glycemic control (Rowan et al. 2008; Spaulonci et al. 2013). Recent studies reported some outcomes with metformin which was superior to insulin treatment. Risk of planned elective c-section, LGA infant and neonatal hypoglycemia were found lower in women treated with metformin when compared with insulin in a population-based retrospective cohort study (Landi et al. 2019). Furthermore, metformin treatment has improved patient satisfaction and clinic efficiency (Kumar et al. 2019). The Society for Maternal-Fetal Medicine states since metformin has function on suppression of mitochondrial respiration, growth inhibition and gluconeogenic responses, which may affect fetal and childhood development (Barbour et al. 2018). There is no clear consensus on the superiority of metformin to insulin and because of the mentioned disadvantages of metformin during pregnancy, the woman should be counseled before initiating treatment.

Metformin should be started as 500 mg per day at night for one week and increased to twice daily doses (Committee on Practice B-O 2018). The most common side effect is gastrointestinal complaints such as pain and diarrhea; therefore, gradually increased the dose and taking the pills with meals are recommended.

4.2.2 Glyburide

Glyburide may increase the rates of macrosomia and is correlated with a higher risk of neonatal hypoglycemia when compared with insulin (Song et al. 2017). In literature, higher rates of pre-eclampsia, hyperbilirubinemia, and stillbirth are also reported for glyburide treatment when compared with insulin (Langer et al. 2005). Glyburide also crosses the placenta like metformin, and long-term data on safety in pregnant women is lacking however, the amount that crosses the placenta is minimal (Harper et al. 2016; Elliott et al. 1991, 1994). Women who use glyburide may need the addition of insulin for glycemic control (Anjalakshi et al. 2007). Glyburide is not a first-choice drug in pregnant women with GDM because of the aforementioned evidence, since a superiority to insulin or metformin has not been showed.

4.3 Management of Hypoglycemia

In pregnant women with GDM, blood glucose of <60 mg/dl is defined as hypoglycemia. In cases with hypoglycemia, 15–20 g glucose intake with any carbohydrate source that contains glucose is recommended (SEMT Diabetes Mellitus Working Group 2019). A response for the treatment should be observed in 1–20 min and the blood glucose should be measured again after 1 h. A snack with 15–20 g carbohydrates may be required before the next meal time if it is more than 30 min away to prevent the recurrence of hypoglycemia.

5 Follow-Up for Pregnancy

Pregnant women who have GDM require a multidisciplinary and closer follow-up for having an increased risk for maternal and fetal complications. Technological applications seem to have a promising potential for the follow-up of pregnant women in the future. A recent study proposed a telehomecare program for management of gestational diabetes, which was found cost-effective for reducing the need of medical visits and direct costs without compromising the pregnancy outcomes, despite increasing the bur-

den on nursing time (Lemelin et al. 2020). Artificial-Intelligence-Augmented telemedicine has been proposed for the follow-up of women with GDM, which provided access for guidelines, patients' data and health records and activity sensors and this system showed better compliance for blood glucose monitoring and high acceptance (Rigla et al. 2018).

5.1 Antepartum Fetal Assessment

Women with A1 GDM, who is under nutritional therapy and have regulated blood glucose levels; do not have a demonstrated increased risk for stillbirth. Based on the available data, there is not a consensus on when to start the antepartum fetal assessment (Committee on Practice B-O 2018). For women with medically treated or poorly controlled GDM (A2 GDM), unless there are other risk factors, antepartum fetal assessment is recommended to start at 32 weeks and the frequency of antenatal follow-ups should be determined according to patient characteristics (Committee on Practice B-O 2018). Patients should be evaluated with a non-stress test (NST) and amniotic fluid assessment in each visit. Doppler measurements do not take part in routine antenatal follow up. Umbilical artery Doppler had lower sensitivity and specificity in order to predict poor neonatal outcomes when compared with NST (Niromanesh et al. 2017). However, a small sample-sized study reported, evaluation of middle cerebral artery Doppler and the umbilical-to-cerebral ratio at >37 weeks could be useful to decide which pregnancies would favor from labor induction to provide better neonatal outcomes (Familiari et al. 2018).

6 Delivery

Many studies were issued in the literature comparing elective induction on 38 weeks or expectant management until 41st week in women with GDM either under nutritional therapy or insulin treatment (Alberico et al. 2017). Cesarean rates, macrosomic and LGA fetuses, birth complications such as shoulder dystocia, neonatal complications such as hyperbilirubinemia, hypoglycemia and perinatal mortality rates were analyzed and in the light of these results, ideal timing for delivery was determined by experts for each condition of GDM.

- **A1 GDM** According to ACOG, unless there are other indications for delivery, women with diet-controlled GDM should not be delivered before the 39th week of gestation and expectant management should be continued until 40 6/7 weeks (Committee on Practice B-O 2018).
- **A2 GDM** For women with well glycemic control under medication, delivery is recommended between 39 0/7 and 39 6/7 weeks (Committee on Practice B-O 2018). The most challenging situation is the decision of timing of poorly controlled GDM. Considering the risks of both stillbirth and prematurity, the recommended timing for delivery was between 37 0/7 weeks and 38 6/7 weeks (Committee on Practice B-O 2018).

6.1 Birth Method

The delivery of diabetic mothers may have increased risks than non-diabetic mothers when the fat mass and fat distributions of the infants are considered. Maternal diabetes is associated with greater infant adiposity, higher BMI, larger waist circumference, higher subcutaneous abdominal fat and increased subcapsular to triceps skinfold thickness ratio (Logan et al. 2017; Crume et al. 2011). Higher results of 50-g glucose challenge test and second hour 75-g glucose tolerance tests may be also predictive for LGA infants or excessive fetal weight (Beksac et al. 2018; Fadiloglu et al. 2019). The birth method should be individualized for cases depending on the gesta-

tional week, maternal characteristics and birth weight as well as the progress of labor. The risks and benefits of performing cesarean delivery to LGA infants of mothers with GDM are controversial. A routine cesarean delivery is not recommended for women with GDM as well as there is not a cesarean limit for a definite estimated fetal weight. Although the increased risk of shoulder dystocia in women with GDM is known, the studies estimated that 433 cesarean deliveries would prevent 1 permanent brachial plexus palsy in infants with a 4.500 g estimated fetal weight and even approximately twice cesarean deliveries are needed for an estimated fetal weight of 4.000 g (Rouse et al. 1996).

ACOG recommends considering cesarean delivery in infants of women with GDM, in case the estimated fetal weight is above 4.500 g (Committee on Practice B-O 2018).

When hospitalized for labor, women with a diet-controlled GDM would not require insulin in normal conditions, therefore, they would not need a periodical glucose assessment.

In patients who are under medication (A2 GDM), during labor and delivery, glucose levels of the patient should be assessed. Although rare, neonatal hypoglycemia can occur because of intrapartum maternal hyperglycemia; which may lead fetal increased insulin secretion from the already hyperplastic pancreatic cells of the fetus due to chronic exposure to maternal hyperglycemia during pregnancy (Dude et al. 2018). Recent studies support higher thresholds of maternal glucose levels and assessment of glucose every 4 h instead of lower thresholds and hourly assessment (Denno and Sadler 1994). A blood glucose level of 72–126 mg/dL (4.0–7.0 mmol/L) during labor is recommended by the Endocrine Society (Gilbert et al. 2006).

7 Postpartum Management

Gestational diabetes in pregnancy means an increased risk of diabetes in further life, therefore the patients should be informed about their potential risks, should be invited for screening, and lifestyle modifications should be recommended. A customized postpartum program for women with GDM with behavioral coaching, diet recommendations, physical activity, and low glycemic index education may be feasible for postpartum management of GDM (Panchaud et al. 2018). Type 2 diabetes will develop in 35–60% of women diagnosed with GDM later in life (Rowan et al. 2018). Despite postpartum hyperglycemia is rare in the early postpartum period, a fasting or random capillary glucose level should be measured before discharge and confirmed by laboratory measurements if the results are elevated (Rowan et al. 2018). In postpartum 4-12th weeks, a 75-g oral glucose testing is recommended to all women with GDM (Committee on Practice B-O 2018; American Diabetes Association 2020; Rowan et al. 2008). Fasting plasma glucose test is an easy-applicable and cheap test; however, it has low sensitivity to detect all spectrum of impaired glucose tolerance, therefore a measurement of only fasting plasma glucose is not enough. The oral glucose testing should be repeated after postpartum 1 year and then in every 1–3 years (Committee on Practice B-O 2018; Rowan et al. 2018). The frequency of screening should be determined according to the patient's risk factors (obesity, family history of diabetes, the need for medical therapy during pregnancy) for diabetes (Rowan et al. 2008).

Progestin-only pills, progestin implants, and non-hormonal/hormonal IUD are effective methods for contraception in postpartum women. IUD's can be placed immediately in postpartum women. The literature regarding the further metabolic effects of contraception in women with GDM is limited (Spaulonci et al. 2013). The use of progestin-only pills and injectable depo-medroxyprogesterone acetate in breastfeeding women was found associated with an increased risk of diabetes mellitus when compared with combined oral contraceptives (Rowan et al. 2018). While counseling the women for contraception, the decision should be made depending on the patient's expectations and additional medical comorbidities.

Breastfeeding was reported to slightly improve the results of the glucose tolerance tests performed in postpartum 6–9th week (Rowan et al. 2018). There is a limited number of studies that report breastfeeding reduces the risk of development of type 2 diabetes up to 19 years after delivery (Landi et al. 2019).

The increased risk of neonatal morbidity and mortality and birth complications due to increased risk of fetal macrosomia is preventable by screening and treating pregnant women with GDM. The risk of impaired glucose tolerance, type 2 diabetes, metabolic syndrome, and cardiovascular diseases are also increased in women with GDM; so healthy life-style modifications should be counseled for patients (Committee on Practice B-O 2018). GDM continues to gain more importance each day with the increasing incidence of obesity.

8 Recommendations of Major Societies and Latest Guidelines

The most of the major guidelines mainly mention that pregnant women with GDM should be counseled for diet and exercise and medical therapy should be initiated if life style-modifications fail to achieve normoglycemic levels (Committee on Practice B-O 2018; American Diabetes Association 2020; Hod et al. 2015; Brown et al. 2017c). Tables 1 and 2 summarize the recommendations of guidelines for pharmacological therapy, and the timing and the mode of birth.

Table 1 Recommendations of major guidelines for pharmacological therapy of GDM

Guideline	Pharmacological therapy
ACOG, 2018 (Committee on Practice B-O 2018)	Insulin should be considered as the first-line treatment; however, in women who would be unable to use insulin properly or afford treatment, metformin is an alternative choice while glyburide is not recommended as a first-choice drug
ADA, 2020 (American Diabetes Association 2020)	It does not support metformin for treatment of GDM and also recommends discontinuing metformin initiated for women with polycystic ovary syndrome by the end of the first trimester. Oral anti-diabetics are not recommended by ADA because of the suspects on long-term safety
SEMT, 2019 (SEMT Diabetes Mellitus Working Group 2019)	In cases when glycemic control is not achieved by diet and exercise within 2 weeks, insulin must be initiated.
SMFM, 2018 (SMFM Statement 2018)	Metformin may be a safe, first-line alternative to insulin women with GDM, preferable to glyburide while almost one-half of women would require additional insulin.
DDG, 2018 (Schafer-Graf et al. 2018)	If diet and exercise fail to achieve target glucose levels, insulin should be considered first. Metformin should be considered in women with GDM and with suspicion of severe insulin resistance after the patient is informed about the off-label use. A daily dose of 2 g should not be exceeded.
CDA, 2018 (Schafer-Graf et al. 2018)	If target blood glucose levels are not achieved within 2 weeks with diet, insulin therapy should be initiated as first-line therapy. Metformin is an alternative to insulin in women who decline insulin or not adherent to insulin therapy. Glyburide may be used in women who are nonadherent to insulin therapy, cannot tolerate metformin or when metformin is inadequate for glycemic control.
JDS, 2016 (Haneda et al. 2018)	In women with GDM who fail to achieve glycemic control, insulin therapy should be initiated.
NICE, 2015 (National Collaborating Centre for Women's and Children's Health (Great Britain) 2015)	If target glucose levels are not achieved with diet and exercise within 1–2 weeks, metformin should be offered to patients if metformin is contraindicated, unacceptable or inadequate to achieve target glucose levels, insulin should be prescribed to patients
FIGO, 2015 (Hod et al. 2015)	Insulin, metformin or glyburide can be used as a first-line treatment in second and third trimesters. Metformin is a better choice than glyburide. Insulin should be the first choice in conditions Diagnosis <20 weeks of gestation, need of pharmacologic therapy >30 weeks of gestation, fasting plasma glucose >110 mg/dl, 1-h postprandial glucose >140 mg/dl, pregnancy weight gain >12 kg
Queensland, 2015 (Queensland Health 2015)	Insulin is safe in pregnancy. Metformin is effective in lowering blood glucose and safe for pregnant women and fetuses.
API, 2014 (Seshiah et al. 2014)	If nutritional therapy fails, insulin is recommended for pharmacologic therapy. Oral anti-diabetic drugs are not recommended in pregnant women.
NZGG, 2014 (New Zealand Ministry of Health 2014)	The women who have poor glycemic control should be offered insulin and/or oral antidiabetic drugs. For the decision of the drug, clinical assessment, patient's preferences and the ability to adhering to the medication should be discussed.

(continued)

Table 1 (continued)

Guideline	Pharmacological therapy
Endocrine Society, 2013 (Blumer et al. 2013)	Glyburide (glibenclamide) should be considered in women with GDM who do not achieve target glucose levels after 1 week of diet and exercise. Insulin should be used as first-line treatment in women with a diagnosis of diabetes before 25 weeks of gestation and women with a fasting plasma glucose >110 mg/dl (6.1 mmol/l).
SIGN, 2013 (Scottish Intercollegiate Guidelines Network 2013)	Metformin or glibenclamide may be considered as first-line treatment in women with GDM.

Abbreviations: *ACOG* American College of Obstetricians and Gynecologists, *ADA* American diabetes association, *SEMT* The Society of Endocrinology and Metabolism of Turkey, *SMFM* Society of Maternal-Fetal Medicine, *DDG* German Diabetes Association, *CDA* Canadian Diabetes Association, *JDS* Japanese Diabetes Society, *NICE* National Institute for Health and Care Excellence, *FIGO* The International Federation of Gynecology and Obstetrics, *API* The Association of Physicians of India, *NZGG* New Zealand Guideline Group, *SIGN* Scottish Intercollegiate Guidelines Network

Table 2 Recommendations of major guidelines for timing and method of delivery in women with GDM

Guideline	Timing and method of delivery
ACOG, 2018 (Committee on Practice B-O 2018)	The pregnant women with diet-controlled GDM should not be delivered before 39 weeks in case there are no obstetric indications and should be followed-up until 40 6/7 weeks of gestation. In women with well-controlled GDM under medical treatment, delivery should be performed at 39 0/7 to 39 6/7 weeks of gestation. If the estimated fetal weight is ≥4,500 g, cesarean delivery should be considered.
CDA, 2018 (Diabetes Canada Clinical Practice Guidelines Expert C et al. 2018)	Elective induction of labor in 38–40 weeks should be considered to reduce the risks of stillbirth and cesarean section. The timing of induction should be discussed based on the patient's glycemic control and potential risks.
NICE, 2015 (National Collaborating Centre for Women's and Children's Health (Great Britain) 2015)	Women with GDM should be delivered before 40 6/7 weeks. The method of birth should be decided for women with GDM considering individual factors.
FIGO, 2015 (Hod et al. 2015)	Elective induction should be considered in GDM mothers with an estimated fetal weight of 3,800–400 g or large for gestational age fetuses between 38–39 weeks of gestation and elective cesarean delivery for fetuses with an estimated fetal weight over 4,000 g. in the cases <3,800 g or appropriate for gestational age fetuses and in the absence of poor diabetic control, vascular diseases and previous stillbirth, pregnancy can be continued to 40–41 weeks.
Queensland, 2015 (Queensland Health 2015)	Women with well-managed GDM with diet, without fetal macrosomia and other complications, can wait for spontaneous labor. Induction of labor should be considered in 38–39 weeks for women with macrosomic fetuses. Women who have pharmacologic therapy and with normal glucose levels would not need birth before 39 weeks. Vaginal birth is appropriate for fetuses <4,000 g and cesarean section should be performed for fetuses >4,500 g. when estimated fetal weight is between 4,000–4,500 g, the birth method should be decided depending on the individual factors.
DDG, 2018 (Schafer-Graf et al. 2018)	Induction <39 0/7 weeks of pregnancy increases the neonatal risks and should be avoided. Induction of labor before 40 weeks in women with insulin-dependent GDM should be offered to reduce fetal morbidity. Induction of labor of GDM women with fetuses with an estimated fetal weight > 95th percentile should be discussed for benefits and risks for early gestational age at birth. Cesarean section should be recommended to fetuses with estimated fetal weight ≥ 4,500 g and the birth method should be decided individually for patients with an estimated fetal weight of 4,000–4,500 g.
SIGN, 2013 (Scottish Intercollegiate Guidelines Network 2013)	Women with GDM and require medication should be assessed after 38 weeks and should be delivered before 40 weeks.

References

Akgol S, Obut M, Bağlı İ, Kahveci B, Budak MS (2019) An evaluation of the effect of a one or two-step gestational diabetes mellitus screening program on obstetric and neonatal outcomes in pregnancies. Gynecol Obstet Reprod Med. 25(2):62–66

Akiba Y, Miyakoshi K, Ikenoue S, Saisho Y, Kasuga Y, Ochiai D et al (2019) Glycemic and metabolic features in gestational diabetes singleton versus twin pregnancies. Endocr J 66(7):647–651

Alberico S, Erenbourg A, Hod M, Yogev Y, Hadar E, Neri F et al (2017) Immediate delivery or expectant management in gestational diabetes at term the GINEXMAL randomised controlled trial. BJOG 124 (4):669–677

American Diabetes Association (2004) Gestational diabetes mellitus. Diabetes Care 27(Suppl 1):S88–S90

American Diabetes Association (2020) 14. Management of diabetes in pregnancy standards of medical care in diabetes-2020. Diabetes Care 43(Suppl 1):S183–SS92

Anjalakshi C, Balaji V, Balaji MS, Seshiah V (2007) A prospective study comparing insulin and glibenclamide in gestational diabetes mellitus in Asian Indian women. Diabetes Res Clin Pract 76(3):474–475

Ayhan S, Altınkaya SÖ, Güngör T, Özcan U (2016) Prognosis of pregnancies with different degrees of glucose intolerance. Gynecol Obstet Reprod Med 19(2):76–81

Barbour LA, Scifres C, Valent AM, Friedman JE, Buchanan TA, Coustan D et al (2018) A cautionary response to SMFM statement pharmacological treatment of gestational diabetes. Am J Obstet Gynecol 219 (4):367 e1–e7

Barnes RA, Wong T, Ross GP, Griffiths MM, Smart CE, Collins CE et al (2020) Excessive weight gain before and during gestational diabetes mellitus management what is the impact? Diabetes Care 43:74–81

Beksac MS, Tanacan A, Hakli DA, Ozyuncu O (2018) Use of the 50-g glucose challenge test to predict excess delivery weight. Int J Gynaecol Obstet 142(1):61–65

Bianco ME, Josefson JL (2019) Hyperglycemia during pregnancy and long-term offspring outcomes. Curr Diab Rep 19(12):143

Billionnet C, Mitanchez D, Weill A, Nizard J, Alla F, Hartemann A et al (2017) Gestational diabetes and adverse perinatal outcomes from 716,152 births in France in 2012. Diabetologia 60(4):636–644

Blumer I, Hadar E, Hadden DR, Jovanovic L, Mestman JH, Murad MH et al (2013) Diabetes and pregnancy an endocrine society clinical practice guideline. J Clin Endocrinol Metab 98(11):4227–4249

Bonomo M, Cetin I, Pisoni MP, Faden D, Mion E, Taricco E et al (2004) Flexible treatment of gestational diabetes modulated on ultrasound evaluation of intrauterine growth a controlled randomized clinical trial. Diabetes Metab 30(3):237–244

Brown J, Alwan NA, West J, Brown S, McKinlay CJ, Farrar D et al (2017a) Lifestyle interventions for the treatment of women with gestational diabetes. Cochrane Database Syst Rev 5:CD011970

Brown J, Ceysens G, Boulvain M (2017b) Exercise for pregnant women with gestational diabetes for improving maternal and fetal outcomes. Cochrane Database Syst Rev 6:CD012202

Brown J, Grzeskowiak L, Williamson K, Downie MR, Crowther CA (2017c) Insulin for the treatment of women with gestational diabetes. Cochrane Database Syst Rev 11:CD012037

Brown J, Martis R, Hughes B, Rowan J, Crowther CA (2017d) Oral anti-diabetic pharmacological therapies for the treatment of women with gestational diabetes. Cochrane Database Syst Rev 1:CD011967

Buchanan TA, Kjos SL, Montoro MN, Wu PY, Madrilejo NG, Gonzalez M et al (1994) Use of fetal ultrasound to select metabolic therapy for pregnancies complicated by mild gestational diabetes. Diabetes Care 17 (4):275–283

Chen R, Yogev Y, Ben-Haroush A, Jovanovic L, Hod M, Phillip M (2003) Continuous glucose monitoring for the evaluation and improved control of gestational diabetes mellitus. J Maternal-Fetal Neonatal Med 14 (4):256–260

Combs CA (2012) Continuous glucose monitoring and insulin pump therapy for diabetes in pregnancy. J Maternal-Fetal Neonatal Med 25(10):2025–2027

Committee on Practice B-O (2018) ACOG practice bulletin no. 190 gestational diabetes mellitus. Obstet Gynecol 131(2):e49–e64

Crowther CA, Hiller JE, Moss JR, McPhee AJ, Jeffries WS, Robinson JS et al (2005) Effect of treatment of gestational diabetes mellitus on pregnancy outcomes. N Engl J Med 352(24):2477–2486

Crume TL, Ogden L, West NA, Vehik KS, Scherzinger A, Daniels S et al (2011) Association of exposure to diabetes in utero with adiposity and fat distribution in a multiethnic population of youth the Exploring Perinatal Outcomes among Children (EPOCH) study. Diabetologia 54(1):87–92

de Veciana M, Major CA, Morgan MA, Asrat T, Toohey JS, Lien JM et al (1995) Postprandial versus preprandial blood glucose monitoring in women with gestational diabetes mellitus requiring insulin therapy. N Engl J Med 333(19):1237–1241

Denno KM, Sadler TW (1994) Effects of the biguanide class of oral hypoglycemic agents on mouse embryogenesis. Teratology 49(4):260–266

Diabetes Canada Clinical Practice Guidelines Expert C, Feig DS, Berger H, Donovan L, Godbout A, Kader T et al (2018) Diabetes and pregnancy. Can J Diab 42 (Suppl 1):S255–SS82

Dude A, Niznik CM, Szmuilowicz ED, Peaceman AM, Yee LM (2018) Management of diabetes in the intrapartum and postpartum patient. Am J Perinatol 35(11):1119–1126

Durnwald C (2015) Gestational diabetes linking epidemiology, excessive gestational weight gain, adverse pregnancy outcomes, and future metabolic syndrome. Semin Perinatol 39(4):254–258

Elliott BD, Langer O, Schenker S, Johnson RF (1991) Insignificant transfer of glyburide occurs across the human placenta. Am J Obstet Gynecol 165(4 Pt 1):807–812

Elliott BD, Schenker S, Langer O, Johnson R, Prihoda T (1994) Comparative placental transport of oral hypoglycemic agents in humans a model of human placental drug transfer. Am J Obstet Gynecol 171(3):653–660

Fadiloglu E, Tanacan A, Unal C, Aydin Hakli D, Beksac MS (2019) Clinical importance of the 75-g glucose tolerance test (GTT) in the prediction of large for gestational age (LGA) fetuses in non-diabetic pregnancies. J Perinat Med 47(5):534–538

Familiari A, Neri C, Vassallo C, Di Marco G, Garofalo S, Martino C et al (2018) Fetal Doppler parameters at term in pregnancies affected by gestational diabetes role in the prediction of perinatal outcomes. Ultraschall Med

Farrar D, Simmonds M, Bryant M, Sheldon TA, Tuffnell D, Golder S et al (2016) Hyperglycaemia and risk of adverse perinatal outcomes systematic review and meta-analysis. BMJ 354:i4694

Gilbert C, Valois M, Koren G (2006) Pregnancy outcome after first-trimester exposure to metformin a meta-analysis. Fertil Steril 86(3):658–663

Gou BH, Guan HM, Bi YX, Ding BJ (2019) Gestational diabetes weight gain during pregnancy and its relationship to pregnancy outcomes. Chin Med J 132 (2):154–160

Group HSCR (2002) The hyperglycemia and adverse pregnancy outcome (HAPO) study. Int J Gynaecol Obstet 78(1):69–77

Han S, Middleton P, Shepherd E, Van Ryswyk E, Crowther CA (2017) Different types of dietary advice for women with gestational diabetes mellitus. Cochrane Database Syst Rev 2:CD009275

Hancerliogullari N, Kansu-Celik H, Asli Oskovi Kaplan Z, Oksuzoglu A, Ozgu-Erdinc AS, Engin-Ustun Y (2019) Correlation of maternal neck/waist circumferences and fetal macrosomia in low-risk Turkish pregnant women, a preliminary study. Fetal Pediatr Pathol:1–8

Haneda M, Noda M, Origasa H, Noto H, Yabe D, Fujita Y et al (2018) Japanese clinical practice guideline for diabetes 2016. Diabetol Int 9(1):1–45

Harper LM, Mele L, Landon MB, Carpenter MW, Ramin SM, Reddy UM et al (2016) Carpenter-Coustan compared with National Diabetes Data Group criteria for diagnosing gestational diabetes. Obstet Gynecol 127 (5):893–898

Hod M, Kapur A, Sacks DA, Hadar E, Agarwal M, Di Renzo GC et al (2015) The International Federation of Gynecology and Obstetrics (FIGO) initiative on gestational diabetes mellitus a pragmatic guide for diagnosis, management, and care. Int J Gynaecol Obstet 131 (Suppl 3):S173–S211

Hughes RC, Moore MP, Gullam JE, Mohamed K, Rowan J (2014) An early pregnancy HbA1c >/=5.9%

(41 mmol/mol) is optimal for detecting diabetes and identifies women at increased risk of adverse pregnancy outcomes. Diabetes Care 37(11):2953–2959

Kc K, Shakya S, Zhang H (2015) Gestational diabetes mellitus and macrosomia a literature review. Ann Nutr Metab 66(Suppl 2):14–20

Kollmann M, Voetsch J, Koidl C, Schest E, Haeusler M, Lang U et al (2014) Etiology and perinatal outcome of polyhydramnios. Ultraschall Med 35(4):350–356

Kumar R, Lowe J, Thompson-Hutchison F, Steinberg D, Shah B, Lipscombe L et al (2019) Implementation and evaluation of the "metformin first" protocol for management of gestational diabetes. Can J Diabetes 43:554–559

Landi SN, Radke S, Boggess K, Engel SM, Sturmer T, Howe AS et al (2019) Comparative effectiveness of metformin versus insulin for gestational diabetes in New Zealand. Pharmacoepidemiol Drug Safety 28:1609–1619

Landon MB, Spong CY, Thom E, Carpenter MW, Ramin SM, Casey B et al (2009) A multicenter, randomized trial of treatment for mild gestational diabetes. N Engl J Med 361(14):1339–1348

Langer O, Yogev Y, Xenakis EM, Rosenn B (2005) Insulin and glyburide therapy dosage, severity level of gestational diabetes, and pregnancy outcome. Am J Obstet Gynecol 192(1):134–139

Lemelin A, Pare G, Bernard S, Godbout A (2020) Demonstrated cost-effectiveness of a telehomecare program for gestational diabetes mellitus management. Diabetes Technol Ther 22(3):195–202

Lewis SB, Wallin JD, Kuzuya H, Murray WK, Coustan DR, Daane TA et al (1976) Circadian variation of serum glucose, C-peptide immunoreactivity and free insulin normal and insulin-treated diabetic pregnant subjects. Diabetologia 12(4):343–350

Logan KM, Gale C, Hyde MJ, Santhakumaran S, Modi N (2017) Diabetes in pregnancy and infant adiposity systematic review and meta-analysis. Arch Dis Child Fetal Neonatal Ed 102(1):F65–F72

Lv S, Wang J, Xu Y (2015) Safety of insulin analogs during pregnancy a meta-analysis. Arch Gynecol Obstet 292(4):749–756

Mendez-Figueroa H, Schuster M, Maggio L, Pedroza C, Chauhan SP, Paglia MJ (2017) Gestational diabetes mellitus and frequency of blood glucose monitoring a randomized controlled trial. Obstet Gynecol 130 (1):163–170

Moreno-Castilla C, Hernandez M, Bergua M, Alvarez MC, Arce MA, Rodriguez K et al (2013) Low-carbohydrate diet for the treatment of gestational diabetes mellitus a randomized controlled trial. Diabetes Care 36(8):2233–2238

Mukhopadhyay A, Farrell T, Fraser RB, Ola B (2007) Continuous subcutaneous insulin infusion vs intensive conventional insulin therapy in pregnant diabetic women a systematic review and metaanalysis of randomized, controlled trials. Am J Obstet Gynecol 197(5):447–456

Nachum Z, Ben-Shlomo I, Weiner E, Shalev E (1999) Twice daily versus four times daily insulin dose regimens for diabetes in pregnancy randomised controlled trial. BMJ 319(7219):1223–1227

National Academy of Sciences, Institute of Medicine, Food and Nutrition Board, Subcommittee on Nutritional Status and Weight Gain During Pregnancy (1990) Nutrition during pregnancy part I, weight gain part II, nutrient supplements. National Academy Press, Washington, DC

National Collaborating Centre for Women's and Children's Health (Great Britain) (2015) Diabetes in pregnancy management of diabetes and its complications from preconception to the postnatal period. National Institute for Health and Care Excellence Clinical Guidelines, London

New Zealand Ministry of Health (2014) Screening, diagnosis and management of gestational diabetes in New Zealand a clinical practice guideline. Ministry of Health, Wellington

Nicholson WK, Wilson LM, Witkop CT, Baptiste-Roberts K, Bennett WL, Bolen S et al (2008) Therapeutic management, delivery, and postpartum risk assessment and screening in gestational diabetes. Evid Rep Technol Assess (Full Rep) (162):1–96

Niromanesh S, Shirazi M, Eftekhariyazdi M, Mortazavi F (2017) Comparison of umbilical artery Doppler and non-stress test in assessment of fetal well-being in gestational diabetes mellitus a prospective cohort study. Electron Physician 9(12):6087–6093

Ozgu-Erdinc AS, Iskender C, Uygur D, Oksuzoglu A, Seckin KD, Yeral MI et al (2016) One-hour versus two-hour postprandial blood glucose measurement in women with gestational diabetes mellitus which is more predictive? Endocrine 52(3):561–570

Padayachee C, Coombes JS (2015) Exercise guidelines for gestational diabetes mellitus. World J Diabetes 6 (8):1033–1044

Palatnik A, Mele L, Landon MB, Reddy UM, Ramin SM, Carpenter MW et al (2015) Timing of treatment initiation for mild gestational diabetes mellitus and perinatal outcomes. Am J Obstet Gynecol 213(4):560 e1–8

Panchaud A, Rousson V, Vial T, Bernard N, Baud D, Amar E et al (2018) Pregnancy outcomes in women on metformin for diabetes or other indications among those seeking teratology information services. Br J Clin Pharmacol 84(3):568–578

Queensland Health (2015) Queensland clinical guideline gestational diabetes mellitus

Rasmussen KM, Yaktine AL (eds) (2009) Weight gain during pregnancy reexamining the guidelines. The National Academies Collection Reports funded by National Institutes of Health, Washington, DC

Rigla M, Martinez-Sarriegui I, Garcia-Saez G, Pons B, Hernando ME (2018) Gestational diabetes management using smart mobile telemedicine. J Diabetes Sci Technol 12(2):260–264

Rossi G, Somigliana E, Moschetta M, Bottani B, Barbieri M, Vignali M (2000) Adequate timing of fetal ultrasound to guide metabolic therapy in mild gestational diabetes mellitus. Results from a randomized study. Acta Obstet Gynecol Scand 79 (8):649–654

Rouse DJ, Owen J, Goldenberg RL, Cliver SP (1996) The effectiveness and costs of elective cesarean delivery for fetal macrosomia diagnosed by ultrasound. JAMA 276 (18):1480–1486

Rowan JA, Hague WM, Gao W, Battin MR, Moore MP, Mi GTI (2008) Metformin versus insulin for the treatment of gestational diabetes. N Engl J Med 358 (19):2003–2015

Rowan JA, Rush EC, Plank LD, Lu J, Obolonkin V, Coat S et al (2018) Metformin in gestational diabetes the offspring follow-up (MiG TOFU) body composition and metabolic outcomes at 7-9 years of age. BMJ Open Diabetes Res Care 6(1):e000456

Schafer-Graf UM, Gembruch U, Kainer F, Groten T, Hummel S, Hosli I et al (2018) Gestational diabetes mellitus (GDM) – diagnosis, treatment and follow-up. Guideline of the DDG and DGGG (S3 level, AWMF registry number 057/008, February 2018). Geburtshilfe Frauenheilkd 78(12):1219–1231

Scottish Intercollegiate Guidelines Network (2013) Management of diabetes a national clinical guideline. Scottish Intercollegiate Guidelines Network, Edinburgh

SEMT Diabetes Mellitus Working Group (2019) Clinical practice guidelines for diagnosis, treatment and follow-up of diabetes mellitus and its complications – 2019. The Society of Endocrinology and Metabolism of Turkey (SEMT), Ankara

Seshiah V, Banerjee S, Balaji V, Muruganathan A, Das AK, Diabetes Consensus Group (2014) Consensus evidence-based guidelines for management of gestational diabetes mellitus in India. J Assoc Physicians India 62(7 Suppl):55–62

SMFM Statement (2018) Pharmacological treatment of gestational diabetes. Am J Obstet Gynecol 218(5): B2–B4

Song R, Chen L, Chen Y, Si X, Liu Y, Liu Y et al (2017) Comparison of glyburide and insulin in the management of gestational diabetes a meta-analysis. PLoS One 12(8):e0182488

Soumya S, Rohilla M, Chopra S, Dutta S, Bhansali A, Parthan G et al (2015) HbA1c a useful screening test for gestational diabetes mellitus. Diabetes Technol Ther 17(12):899–904

Spaulonci CP, Bernardes LS, Trindade TC, Zugaib M, Francisco RP (2013) Randomized trial of metformin vs insulin in the management of gestational diabetes. Am J Obstet Gynecol 209(1):34 e1–7

Tieu J, McPhee AJ, Crowther CA, Middleton P (2014) Screening and subsequent management for gestational diabetes for improving maternal and infant health. Cochrane Database Syst Rev 2:CD007222

Tieu J, Shepherd E, Middleton P, Crowther CA (2017) Dietary advice interventions in pregnancy for preventing gestational diabetes mellitus. Cochrane Database Syst Rev 1:CD006674

Vellanki P, Umpierrez G (2016) Detemir is non-inferior to NPH insulin in women with pregestational type 2 diabetes and gestational diabetes mellitus. Evid Based Med 21(3):104–105

Viana LV, Gross JL, Azevedo MJ (2014) Dietary intervention in patients with gestational diabetes mellitus a systematic review and meta-analysis of randomized clinical trials on maternal and newborn outcomes. Diabetes Care 37(12):3345–3355

Vieira MC, Begum S, Seed PT, Badran D, Briley AL, Gill C et al (2018) Gestational diabetes modifies the association between PlGF in early pregnancy and preeclampsia in women with obesity. Pregnancy Hypertens 13:267–272

Weissgerber TL, Mudd LM (2015) Preeclampsia and diabetes. Curr Diab Rep 15(3):9

Wouldes TA, Battin M, Coat S, Rush EC, Hague WM, Rowan JA (2016) Neurodevelopmental outcome at 2 years in offspring of women randomised to metformin or insulin treatment for gestational diabetes. Arch Dis Child Fetal Neonatal Ed 101(6):F488–FF93

Yu F, Lv L, Liang Z, Wang Y, Wen J, Lin X et al (2014) Continuous glucose monitoring effects on maternal glycemic control and pregnancy outcomes in patients with gestational diabetes mellitus a prospective cohort study. J Clin Endocrinol Metab 99(12):4674–4682

Yu Y, Xie R, Shen C, Shu L (2018) Effect of exercise during pregnancy to prevent gestational diabetes mellitus a systematic review and meta-analysis. J Maternal-Fetal Neonatal Med 31(12):1632–1637

Zhang J, Ma S, Wu S, Guo C, Long S, Tan H (2019) Effects of probiotic supplement in pregnant women with gestational diabetes mellitus a systematic review and meta-analysis of randomized controlled trials. J Diabetes Res 2019:5364730

Adv Exp Med Biol - Advances in Internal Medicine (2020) 4: 273–297
https://doi.org/10.1007/5584_2020_480
© Springer Nature Switzerland AG 2020
Published online: 4 February 2020

From Entero-Endocrine Cell Biology to Surgical Interventional Therapies for Type 2 Diabetes

Marta Guimarães, Sofia S. Pereira, and Mariana P. Monteiro

Abstract

The physiological roles of the enteroendocrine system in relation to energy and glucose homeostasis regulation have been extensively studied in the past few decades. Considerable advances were made that enabled to disclose the potential use of gastro-intestinal (GI) hormones to target obesity and type 2 diabetes (T2D). The recognition of the clinical relevance of these discoveries has led the pharmaceutical industry to design several hormone analogues to either to mitigate physiological defects or target pharmacologically T2D.

Amongst several advances, a major breakthrough in the field was the unexpected observation that enteroendocrine system modulation to T2D target could be achieved by surgically induced anatomical rearrangement of the GI tract. These findings resulted from the widespread use of bariatric surgery procedures for obesity treatment, which despite initially devised to induce weight loss by limiting the systemic availably of nutrients, are now well recognized to influence GI hormone dynamics in a manner that is highly dependent on the type of anatomical rearrangement produced.

This chapter will focus on enteroendocrine system related mechanisms leading to improved glycemic control in T2D after bariatric surgery interventions.

M. Guimarães
Endocrine, Cardiovascular & Metabolic Research, Unit for Multidisciplinary Research in Biomedicine (UMIB), University of Porto, Porto, Portugal

Department of Anatomy, Institute of Biomedical Sciences Abel Salazar (ICBAS), University of Porto, Porto, Portugal

Department of General Surgery, Centro Hospitalar de Entre o Douro e Vouga, Santa Maria da Feira, Portugal

S. S. Pereira
Endocrine, Cardiovascular & Metabolic Research, Unit for Multidisciplinary Research in Biomedicine (UMIB), University of Porto, Porto, Portugal

Department of Anatomy, Institute of Biomedical Sciences Abel Salazar (ICBAS), University of Porto, Porto, Portugal

Instituto de Investigação e Inovação em Saúde (I3S), Universidade do Porto, Porto, Portugal

Institute of Molecular Pathology and Immunology of the University of Porto (IPATIMUP), Porto, Portugal

M. P. Monteiro (✉)
Endocrine, Cardiovascular & Metabolic Research, Unit for Multidisciplinary Research in Biomedicine (UMIB), University of Porto, Porto, Portugal

Department of Anatomy, Institute of Biomedical Sciences Abel Salazar (ICBAS), University of Porto, Porto, Portugal
e-mail: mpmonteiro@icbas.up.pt

Keywords

Bariatric surgery · Endoscopic interventions · Enteroendocrine system · Gastrointestinal

hormones · GIP · GLP-1 · Obesity ·
Oxyntomodulin · Peptide YY · Type 2 diabetes

Abbreviations

ADA	American Diabetes Association
BMI	Body mass index
BPD	Biliopancreatic diversion
BPD-DS	Biliopancreatic diversion with duodenal switch
CCK	Cholecystokinin
DJBL	Duodenal-Jejunal Bypass Liner
DMR	Duodenal mucosal resurfacing
DSS-II	Second Diabetes Surgery Summit
EEC	Enteroendocrine cells
ESG	Endoscopic Sleeve Gastroplasty
EWL	Excess weight-loss
Gcg	Preproglucagon gene
GcGR	Glucagon receptor
GERD	Gastroesophageal reflux disease
GI	Gastrointestinal
GIP	Glucose-dependent insulinotropic polypeptide
GJB	Gastrojejunal bypass
GLP-1	Glucagon-like peptide-1
GLP-1R	Glucagon-like peptide-1 receptor
HbA1c	Hemoglobin A1c
IFSO	International Federation for the Surgery of Obesity
LSG	Laparoscopic sleeve gastrectomy
MNU	Neuromedin U
NAFLD	Nonalcoholic fatty liver disease
NASH	Nonalcoholic steatohepatitis
OHS	Obesity-hypoventilation syndrome
OSA	Obstructive sleep apnea
OXM	Oxyntomodulin
PYY	Peptide YY
RCTs	Randomized clinical trials
RYGB	Roux-en-Y Gastric Bypass
SADI-S	Single Anastomosis Duodeno-ileal Bypass with Sleeve Gastrectomy
T2D	Type 2 Diabetes

1 Anatomy and Physiology of the Enteroendocrine System

The gastrointestinal (GI) tract has been known to synthetize and release over twenty different hormones among several other bioactive molecules, which rendered it to be considered 'the largest endocrine organ in the human body' (Ahlman and Nilsson 2001). Although not exclusively, GI hormones are predominantly produced by a discrete endocrine-specialized cells population so called enteroendocrine cells (EEC). The EEC are found as scattered individual cells throughout the mucosa along the entire GI tract and comprise approximately 1% of the overall epithelial cell population (Fig. 1) (Rehfeld 2004; Buffa et al. 1978; Sternini et al. 2008).

EEC cell density is highest in the proximal small intestine and decreases throughout the gut until the distal colon. EEC cell density rises again in the rectum, a location where these cells can be found adjacent to each other in clusters, unlike what is observed in the rest of the gut (Cristina et al. 1978; Sjolund et al. 1983; Gunawardene et al. 2011).

The ECC can be categorized as "open-type" or "closed-type" according to the morphology and location within the GI mucosa. Open-type ECC exhibit prominent microvilli extending to the surface of the GI mucosa that enables the cells to react with the luminal contents by releasing secretory products, which by a variety of mechanisms activate local and distant target tissues and neuronal pathways. By contrast, the "closed-type" EEC are embedded in the GI mucosa with no part of the cell surface exposed to the luminal contents and are mainly regulated by neural and humoral mechanisms acting through the basolateral cell membrane (Sternini et al. 2008; Latorre et al. 2016; Hofer et al. 1999).

The distribution of EEC subpopulations responsible for secreting different hormones exhibits a characteristic pattern within the GI tract with ghrelin, somatostatin and gastrin being

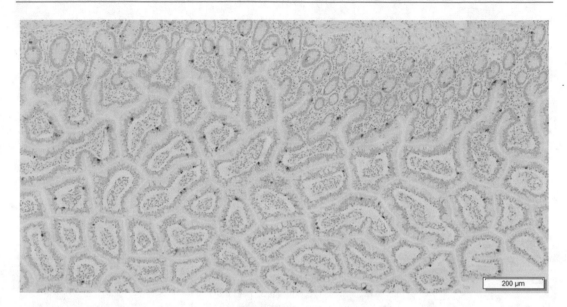

Fig. 1 Human small intestine staining for chromogranin A, used as a molecular marker specifically present in cells that store peptide hormones and monoamides, as occurs in enteroendocrine cells within the gastrointestinal tract (40x)

mainly expressed in the stomach; cholecystokinin (CCK) in duodenum and jejunum; glucose-dependent insulinotropic polypeptide (GIP) in the proximal small intestine; and glucagon-like peptide-1 (GLP1), oxyntomodulin (OXM) and peptide YY (PYY) in the distal small intestine and colon. Classically, EEC were classified according to the predominant hormone secreted. However, more recent evidence has demonstrated that each individual EEC usually expresses a variety of GI hormones. In turn, GI hormone secretion is not only determined by EEC distribution throughout the GI tract, but also modulated by the patterns of food intake, meal composition and nutrient absorption (Sjolund et al. 1983; Stengel and Tache 2009; Lamberts et al. 1991; Itoh 1997; Egerod et al. 2012; Svendsen et al. 2015; Posovszky and Wabitsch 2015; Monteiro and Batterham 2017; Grunddal et al. 2016) (Table 1). Besides that, the same EEC type may have different secretory profiles in different locations. For example, the EEC cells that produce GLP-1 and PYY (L-cells) although mainly present in the distal small intestine, are also present in the proximal gut. However, Svendesen *et al*, found that in the proximal small intestine, there are more GLP-1-positive cells than PYY- positive cells, suggesting that contrarily to the distal L-cells, not all the proximal L-cells produce PYY (Svendsen et al. 2015).

Besides that, the expression and the secretion of the GI hormones do not necessarily correlate. *In vivo* studies found that, although GLP-1-staining cells predominate in the distal small intestine, the same luminal and vascular stimuli of the proximal and distal small intestine induce similar GLP-1 secretion levels, suggesting that the secretory activity and responsiveness to the stimuli may be different in the proximal and distal gut (Svendsen et al. 2015).

GI hormones can act on peripheral and central organs that are reached though the blood stream or through modulation of the electrical activity of the vagal nerve afferent fibers (Browning and Travagli 2014; Ye and Liddle 2017). A considerable number of GI hormones are well known players in the regulation of energy and glucose homeostasis. In the past few decades, GI hormones have been extensively studied in what concerns the physiological roles in glucose homeostasis regulation and potential pharmacological use in the context of obesity and type 2 diabetes (T2D) treatments.

Table 1 Summary of human enteroendocrine cells subsets, predominant GI tract location, hormones secreted, stimuli and physiological function

Cell type	Predominant location	Hormones	Main stimuli	Main functions
P/ D1- like cells	Gastric fundus	Ghrelin Nesfatin-1	Fasting periods	Hunger drive, food intake stimulation, glucose homeostasis, growth hormone release
G cells	Pyloric antrum and duodenum	Gastrin	Gastrin releasing peptides; amino acids, stomach expansion	Stimulation of acid secretion, production of pepsinogen
D cells	Gastric body, pyloric antrum and small intestine	Somatostatin	Gastrin, CCK and H^+	Inhibition of gastrin and acid secretion
I cells	Proximal small intestine	CCK	Long-chain fatty acid, peptides and amino acids	Gallbladder contraction stimulation, gastric emptying and food intake inhibition
S cells	Duodenum	Secretin	Secretin releasing peptide; low duodenal pH and digested proteins	Stimulation of pancreatic secretion and gastric emptying reduction
K cells	Proximal small intestine	GIP	Carbohydrates, long chain fatty acids and some amino acids	Insulin release, inhibition of gastric emptying and gastric acid secretion, food intake reduction
L cells	Along the entire small intestine with distal small intestine and large intestine predominance	GLP-1, GLP-2, PYY, OXM	Fatty acids, carbohydrates, amino acids	Nutrient sensing, gastric and intestinal motility inhibition, stimulation of insulin release, inhibition of glucagon release, appetite suppression and promotion of energy expenditure
M cells	Proximal small intestine	Motilin	Lipids, gastric distension, bile acids and low pH	Enhancement of gut motility
N cells	Distal small intestine	Neurotensin	Fatty acids	Inhibition of gastric secretion and gastric emptying, stimulation of pancreatic and intestinal secretion

CCK cholecystokinin, *GIP* glucose-dependent insulinotropic polypeptide, *GLP-1* glucagon-like peptide-1, *GLP-2* - glucagon-like peptide-2, *NT* neurotensin, *OXM* oxyntomodulin, *PYY* peptide YY

2 Experimental Data Supporting Enteroendocrine System Modulation through Interventional Therapies for Type 2 Diabetes

The unexpected observation that T2D could undergo clinical remission as a consequence of GI tract anatomical rearrangements induced by bariatric surgery interventions has led to the provocative hypothesis that T2D could be an intestinal disease (Rubino 2008).

In fact, bariatric surgery interventions in individuals with obesity can be responsible not only for rapid and sustained weight loss, but also for considerable improvements of several comorbidities and most particularly of T2D. Indeed, bariatric surgery is currently the most effective therapy for patients with obesity and concurrent T2D. However, the rates of T2D improvement and remission can be widely variable dependent not only on the patient clinical characteristics but also on the type of bariatric procedure performed, as described in further detail later in this chapter. Most importantly, these observations have changed the focus of interest of the scientific community working on addressing T2D disease mechanisms from the pancreas towards the gut.

Several lines of research enabled to dissect the mechanisms underlying the improved glycemic control induced by bariatric surgery, however despite deeply investigated these are still far from being completely disclosed. Nevertheless, the available data has made clear that the metabolic effects derived from bariatric surgery cannot be appointed nor justified by any single phenomenon but require the combination of multiple physiological mechanisms. Of notice is the relevance of caloric restriction for the anti-diabetic effects of bariatric surgery interventions, and most particularly in the early post-operative period. However, to estimate the relative contribution of caloric restriction, weight loss and GI hormone profiles for T2D improvement after bariatric surgery can be challenging. In addition, the relative contribution of each physiological mechanism can vary depending on the anatomical modification produced by different surgical procedures (Pérez-Pevida et al. 2019; Stefater et al. 2012).

There is a large amount of evidence that support the hypothesis that the modification of GI hormones secretion profile by EEC induced by bariatric surgery, plays a significant role in the anti-diabetic effects observed in human subjects after these procedures (Madsbad et al. 2014). Although the authors acknowledge the importance of several weight loss dependent- and independent-mechanisms contributing for glucose homeostasis improvement after bariatric surgery, those are summarized in Fig. 2 but will not be addressed in further detail.

Since the first reports that GI tract anatomical rearrangements could modulate glycemic control (Barnes 1947), different levels of evidence have emerged supporting the effects of GI surgical interventions in improving or even inducing complete clinical disease remission in patients with T2D (Kodama et al. 2018; Khorgami et al. 2019; Vetter et al. 2012; Wang et al. 2015). Despite the fact bariatric surgery procedures were technically devised to induce weight loss through a negative energy balance by limiting the systemic availably

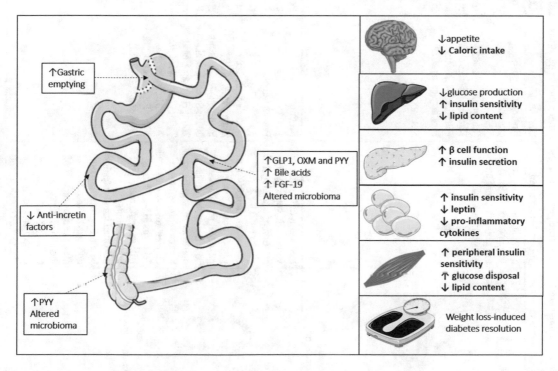

Fig. 2 Mechanisms of weight loss and T2D remission after gastric bypass

Table 2 Summary of GI tract rearrangement and gut hormone profile modification induce by different bariatric surgery procedures

Surgical technique	Surgical description	Ghrelin	CCK	GIP	GLP-1	PYY	OXM
Vertical Sleeve Gastrectomy	Surgical removal of a large portion (~80%) of the stomach along the greater curvature	Fasting and post-prandial: ↓ (Langer et al. 2005; Karamanakos et al. 2008; Peterli et al. 2009; McCarty et al. 2019)	Post-prandial: ↑ (Peterli et al. 2012; Mans et al. 2015; Lee et al. 2011)	Fasting and post-prandial = (McCarty et al. 2019; Wallenius et al. 2018)	Fasting = Post-prandial: ↑ (McCarty et al. 2019; Valderas et al. 2011; Sista et al. 2017)	Fasting ↓ = ↑ Post-prandial: ↑ (Karamanakos et al. 2008; Peterli et al. 2009; McCarty et al. 2019; Valderas et al. 2010)	–
Roux-en-Y gastric bypass	Reduction of the stomach volumetric capacity by creation of a small gastric pouch (~30 mL) Direct anastomosis of the gastric pouch to the proximal small intestine with the exclusion of nutrients passage through the remaining stomach, duodenum, and proximal Jejunum (65–220 cm) Bile and pancreatic juices drainage into the proximal small intestine by a biliopancreatic limb, only mixed with food after the anastomosis of the alimentary (120–150 cm) and biliopancreatic limbs to create the common limb	Fasting: ↓ = ↑ Post-prandial: ↓ = (Peterli et al. 2009; Zhang et al. 2019; Pournaras and le Roux 2010; Stoeckli et al. 2004; Holdstock et al. 2003; Geloneze et al. 2003; Morinigo et al. 2004; Pardina et al. 2009; Jacobsen et al. 2012)	Post-prandial: ↑ = (Peterli et al. 2012; Jacobsen et al. 2012; Kellum et al. 1990)	Fasting = Post-prandial: ↑ = ↓ (Laferrère et al. 2007; Korner et al. 2007; Laferrère et al. 2008; Jirapinyo et al. 2018a)	Fasting = Post-prandial: ↑ (Laferrère et al. 2007; Korner et al. 2007; Korner et al. 2009; Falken et al. 2009; Jorgensen et al. 2012)	Fasting = Post-prandial: ↑ (Peterli et al. 2009; Valderas et al. 2010; Korner et al. 2009; Olivan et al. 2009)	Fasting = Post-prandial: ↑ (Falken et al. 2011; Laferrère et al. 2010)

Biliopancreatic diversion Reduction of the stomach volumetric capacity by creation of a gastric pouch (200–500 ml) Direct anastomosis of the distal part of the small intestine to the gastric pouch Intestinal rearrangement that results in an alimentary limb of ~200-cm, a common limb of ~75–100 cm, and a biliopancreatic limb of unknown length corresponding to the remainder of the small intestine	–	Fasting: ↓ = ↑ Post-prandial: ↑ (Adami et al. 2003; Valera Mora et al. 2007; Garcia de la Torre et al. 2008)	–	Fasting and post-prandial: ↓ (Mingrone et al. 2009; Salinari et al. 2009; Gissey et al. 2017)	Fasting and post-prandial: ↑ (Salinari et al. 2009; Gissey et al. 2017; Guidone et al. 2006; Valverde et al. 2005)	Fasting and post-prandial: ↑ (Garcia-Fuentes et al. 2008; Tsoli et al. 2013)	–
Biliopancreatic diversion with duodenal switch Surgical removal of a large portion (~80%) of the stomach along the greater curvature The duodenum is divided 3 cm distal to the pylorus and anastomosed to the distal 300 cm of ileum Intestinal rearrangement that results in an alimentary limb of ~200-cm, a common limb of ~100 cm, and a biliopancreatic limb of unknown length corresponding to the remainder of the small intestine	–	Fasting: ↓ = Post-prandial: ↑ (Kotidis et al. 2006a, b; Plourde et al. 2014; Michaud et al. 2017)	–	Fasting = Post-prandial: ↓ (Plourde et al. 2014; Michaud et al. 2017)	Fasting = ↑ Post-prandial: ↑ (Plourde et al. 2014; Michaud et al. 2017)	Fasting ↑ = Post-prandial: ↑ = (Plourde et al. 2014; Michaud et al. 2017; Hedberg et al. 2011)	–

= No significant changes in the majority of studies, ↑ Significantly increased in the majority of studies, ↓ Significantly decreased in the majority of the studies, – No studies for this parameter, *CCK* Cholecystokinin, *GLP-1* Glucagon Like Peptide-1, *GIP* Glucose-dependent insulinotropic polypeptide, *OXM* Oxyntomodulin, *PYY* Peptide YY

of nutrients, these interventions are now well recognized to influence GI hormone dynamics in a manner that is highly dependent on the type of surgical procedure performed. In this context, GI hormones known to play a relevant role in the endocrine axis that regulate energy homeostasis, including ghrelin, CCK, incretins (GLP1 and GIP), OXM and PYY, have been more extensively studied This information has been summarized in Table 2.

2.1   Ghrelin

Ghrelin is a GI hormone predominantly produced by gastric fundus that stimulates appetite, increases short-term food intake and reduces fat metabolism (Wren et al. 2001; Ariyasu et al. 2001; Pinkney 2014). Bariatric surgery is reported to influence ghrelin levels in an inconsistent manner that is highly dependent on the type of surgical procedure and timing after the intervention (Pournaras and le Roux 2010; Tymitz et al. 2011).

After laparoscopic sleeve gastrectomy (LSG) results are somehow consistent, with most studies reporting an anticipated reduction in fasting ghrelin levels, since this procedure removes a large portion of the stomach wall responsible for hormone secretion (Langer et al. 2005; Karamanakos et al. 2008; Peterli et al. 2009; McCarty et al. 2019). In contrast, after Roux-en-Y Gastric Bypass (RYGB), different studies have shown either a decrease, an increase or no change in fasting and postprandial ghrelin levels (Peterli et al. 2009; Zhang et al. 2019; Pournaras and le Roux 2010; Stoeckli et al. 2004; Holdstock et al. 2003; Geloneze et al. 2003; Morinigo et al. 2004; Pardina et al. 2009; Jacobsen et al. 2012). Technical differences in the procedure regarding the preservation of the vagus nerve and gastric pouch configuration (vertical vs horizontal pouch) have been appointed as possible explanations to justify the heterogeneity of ghrelin results (Fruhbeck et al. 2004). In a prospective randomized clinical trial, comparing RYGB and LSG, both procedures reduced fasting and post-prandial ghrelin levels, but the magnitude of this effect

was significantly higher after LSG. So, the authors suggested that the resection of the gastric fundus has a more powerful impact on ghrelin levels when compared to bypassing part of the stomach (Peterli et al. 2009; Pournaras and le Roux 2010). On the contrary, after other procedures that limit the stomach volumetric capacity, such as gastric banding and gastric plication, ghrelin levels were observed to increase (Stoeckli et al. 2004; Cummings et al. 2002).

To understand whether ghrelin plays a role in mediating LSG effects on body weight and glucose control, ghrelin-deficient mice and wild-type mice were submitted to vertical sleeve gastrectomy. This procedure lead to similar results in both mice strains suggesting that ghrelin reduction is not imperative for the metabolic improvements observed after this type of bariatric surgery procedure (Chambers et al. 2013).

2.2   Cholecystokinin

CCK is a GI hormone that is mainly secreted in the duodenum in response to the presence of fatty acids and amino acids, which acts as an anorexigenic signal to the central nervous system (Liddle 1995). CCK levels were intuitively hypothesized to decrease after bariatric surgery procedures that exclude the duodenum from the alimentary tract, such as RYGB. However, through mechanisms not yet completely understood, CCK levels were reported to increase after RYGB (Peterli et al. 2012; Jacobsen et al. 2012). One of the mechanisms appointed as a possible explanation for the observed rise in CCK levels, was an increase of CCK producing cells density after RYGB. Indeed, increased proliferation of CCK producing cells was found in the Roux and common limbs of rats submitted to RYGB (Mumphrey et al. 2013). Besides that, after RYGB human intestinal biopsies presented higher numbers of CCK producing cells when compared to biopsies collected during surgery (Rhee et al. 2015). CCK levels were reported to be even higher after surgeries that maintain the duodenum in the alimentary transit, like LSG (Peterli et al. 2012; Mans et al. 2015; Lee et al. 2011).

In vivo studies found that CCK increases pancreatic β-cell proliferation (Kuntz et al. 2004; Chen et al. 2007; Lavine et al. 2010). However, RYGB was able to induce similar metabolic outcomes in rats lacking CCK-1 receptors and controls. So, the positive metabolic effects of RYGB appear to be independent of CCK (Hajnal et al. 2010).

2.3 Incretins

Incretins are hormones responsible for a phenomenon known as the incretin effect, which consists of a greater insulin secretion being elicited after oral glucose administration when compared to an isoglycemic intravenous glucose load. GIP and GLP-1 are the GI hormones so far identified to be implicated in mediating the incretin effect and are thus called incretin hormones.

GIP and GLP-1 secreting cells are differentially distributed throughout the small intestine (Jorsal et al. 2018; Guedes et al. 2015; Palha et al. 2018). GIP is secreted by K cells predominantly located in the proximal small intestine, duodenum and proximal jejunum, while GLP-1 is produced by the L cells mostly found in the distal small intestine (Jorsal et al. 2018; Guedes et al. 2015; Polak et al. 1973). The incretin effect was shown to be significantly reduced in patients with T2D (Nauck et al. 1986; Vilsboll and Holst 2004). Studies aimed to investigate GLP-1 secretion in individuals with and without T2D have shown that the insulinotropic effect of GLP-1 is substantially preserved in individuals in T2D (Nauck et al. 2011; Calanna et al. 2013). Besides that, GLP-1 secreting cell distribution seems to be similar in subjects with and without T2D (Jorsal et al. 2018; Palha et al. 2018). In contrast, a reduced GIP insulinotropic effect was found in patients with T2D, resulting in a possible compensatory increase in *GIP* expression in the gut (Jorsal et al. 2018; Xu et al. 2007; Vilsboll et al. 2003).

In individuals with T2D, an incretin effect recovery towards the magnitude observed in non-diabetic counterparts was demonstrated to occur as early as 1 month after the RYGB surgery and maintained thereafter. This phenomenon seems to be independent of weight loss since the incretin effect was not reestablished in weight-matched individuals after similar weight loss achieved through caloric restriction (Laferrere et al. 2007; Laferrère et al. 2008).

The effect of bariatric surgery on GIP levels appears to be relatively modest and controversial, as different studies have found either an increase, decrease or no change after the interventions (Laferrere et al. 2007; Korner et al. 2007; Mingrone et al. 2009; Salinari et al. 2009; Kim et al. 2014). These findings were relatively unsurprising, since GIP is known to be predominantly secreted in the proximal small intestine portion that is bypassed after RYGB (Jorsal et al. 2018; Guedes et al. 2015). So, the number of GIP secreting cells available for stimulation and GIP release is lower. Nevertheless, GIP cell density seams to increase after RYGB in proportion to the compensatory villous hypertrophy (Rhee et al. 2015). Thus, contradictory results found in GIP levels across different studies could be attributed to technical variants of the RYGB surgery procedures using different limb lengths for intestinal reconstruction (Laferrère 2016). Moreover, the improved GIP insulinotropic effect after bariatric surgery is less likely to be due to modified GIP secretion but more likely secondary to decreased glucotoxicity (Hojberg et al. 2009; Aaboe et al. 2010).

In contrast to GIP, the effect of GI tract surgical manipulations on circulating GLP-1 levels are very consistent. The density of GLP-1 immunoreactive cells in the proximal intestinal limbs increase after surgical manipulation, while fasting and post-prandial GLP-1 levels are also higher after bariatric surgery (RYGB, LSG; biliopancreatic diversion and duodenal switch) (Lee et al. 2011; Laferrere et al. 2007; Guidone et al. 2006; Valverde et al. 2005; Plourde et al. 2014; Kim et al. 2014; Romero et al. 2012). Only a few days after RYGB surgery, GLP-1 levels were observed to increase over 10 times (Falken et al. 2011; Jorgensen et al. 2012).

The putative reasons for the enhanced GLP-1 secretion observed after bariatric surgery could be related to an increased ileal exposure to undigested nutrients, where L cells prevail, thus contributing to the reversion of hyperglycemia, even in the absence of any gastric restriction. This so called "hindgut hypothesis" was proposed by Cummings et al., in 2004 (Cummings et al. 2004).

In a complex study, patients were submitted to a mixed meal tolerance test in two consecutive days before surgery, 1 week and 3 months after RYGB. In the first day patients were infused with a saline solution and in the other day a GLP-1 receptor antagonist (Exendin 9–39) was administrated. The study showed that RYGB resulted in the predicted enhancement of GLP-1 responses: increased beta-cell sensitivity to glucose, increased insulin release and decreased fasting and postprandial glucose levels. However, these effects were completely lost after administration of a GLP-1 antagonist. Moreover, glucose and insulin levels were identical to those observed without the antagonists before the surgery (Jorgensen et al. 2013). In another study the effects of per oral versus gastroduodenal feeding on glucose metabolism were explored in a patient previously submitted to RYGB that presented a gastro-jejunostomy leak. In this study, a standard liquid meal was given through a gastric tube inserted into the bypassed gastric remnant on the first day and the same meal was given per orally on the second day. This study found that per oral compared to gastroduodenal feeding resulted in higher GLP-1 levels, improved glucose metabolism and β-cell function (Dirksen et al. 2010). These studies provided a powerful evidence that GLP-1 plays a major role in insulin response and T2D improvement after bariatric surgery, independently of weight loss and caloric restriction.

2.4 Peptide YY and Oxyntomodulin

PYY and OXM are both GI hormones secreted by the same L-cells responsible for GLP-1 secretion, which predominate in the distal small gut and colon (Jorsal et al. 2018). According to the "hindgut hypothesis", a greater L-cell stimulation should enhance PYY and OXM secretion, which would then contribute to control appetite and improve glycemic regulation. Indeed, postoperative OXM and PYY levels seem to match those observed for GLP-1 levels (Jacobsen et al. 2012; Falken et al. 2011). In fact, postprandial PYY levels seem to increase regardless the bariatric technique (Korner et al. 2009; Tsoli et al. 2013; Michaud et al. 2017). The role of PYY in mediating the effects of bariatric surgery have been mainly attributed to appetite and food intake regulation (le Roux et al. 2005). However a powerful effect on improving islet secretory function was also attributed to PYY, by a recent study that used isolated donor human pancreatic islets and blood samples from patients with T2D before and after bariatric surgery (Guida et al. 2019).

Contrarily to GLP-1 and PYY, the mechanisms of OXM action are not so well characterized. OXM and GLP-1 are both derived from the preproglucagon gene (*Gcg*) and result from post-translational modifications during proglucagon processing (Pocai 2012; Drucker 1998). OXM is a full agonist of the GLP-1 receptor (GLP-1R) and the glucagon receptor (GcGR), but with lower affinity relative to native hormones (Baldissera et al. 1988; Kerr et al. 2010). GcGR activation by glucagon has beneficial metabolic effects that includes the modulation of lipid metabolism through activation of lipolysis and inhibition of lipid synthesis, energy intake regulation, brown fat thermogenesis stimulation and inhibition of gastric motility (Kuroshima and Yahata 1979; Billington et al. 1991; Amatuzio et al. 1962; Mochiki et al. 1998). Nevertheless, exogenous OXM administration was demonstrated to increase energy expenditure while reducing energy intake, body weight and increase insulin secretion (Wynne et al. 2005, 2006; Pocai et al. 2009). These observations triggered the development of GLP-1R/GCGR co-agonists (Brandt et al. 2018; Sánchez-Garrido et al. 2017). Pre-clinical and clinical studies demonstrated the efficacy of GLP-1R/GCGR agonists in the improvement of postprandial glucose control by reducing glucose absorption rate and increasing insulin sensitivity and β-cell function (Pocai et al. 2009; Brandt et al. 2018; Goebel et al. 2018; Zhou et al. 2017).

Furthermore, obese patients with prediabetes/T2D infused with a continuous subcutaneous combination of GLP-1, OXM, and PYY, replicating the postprandial gut hormone levels after RYGB, exhibited superior glucose tolerance and reduced glycemic variability, compared with RYGB and a very low-calorie diet (Behary et al. 2019). These satisfactory results lead to a growing interest in the development of molecules that combine the effects of multiple key metabolic hormones into a single entity of superior and sustained action relative to the native hormones. Indeed, the hormone receptors co-agonists are now emerging as a new pharmacological class for T2D treatment (Brandt et al. 2018).

2.5 Anti-incretin Factors

Based on the evidence that the most potent anti-diabetic effects are observed for bariatric surgery procedures that exclude the proximal small intestine from the passage of the nutrients [RYGB and Biliopancreatic Diversion (BPD)], Rubino et al., proposed the "foregut hypothesis" an attempt to provide an explanation for the phenomenon (Rubino and Gagner 2002). This hypothesis postulates that by excluding the proximal intestine from the alimentary transit, the secretion of unknown "anti-incretin" factors produced in the proximal small intestine, acting as a counterregulatory mechanism to the incretin-mediated insulin secretion to prevent postprandial hypoglycemia, would be reduced (Rubino and Gagner 2002; Rubino et al. 2006). These putative "anti-incretin" factors would contribute to reduce the incretin effect observed in patients with T2D.

Neuromedin U (NMU), an hormone expressed in the brain and GI has been appointed as an anti-incretin factor since it was able to reduce glucose-stimulated insulin secretion from isolated perfused rat pancreas and human pancreatic islets (Brighton et al. 2004; Alfa et al. 2015; Kaczmarek et al. 2006). However, other studies found contradictory results. Peripheral administration of NMU in mice led to the secretion of the incretin hormone GLP-1, which does not support the role of NMU as an anti-incretin (Peier et al. 2011).

Besides that, an *in vivo* study found that NMU was unable to affect the plasmatic levels of glucose, insulin or glucagon. The same study reported that rat and human pancreatic islets do not express NMU receptors (Kuhre et al. 2019).

Several experimental data come in support of the "foregut hypothesis" for the metabolic improvements observed after bariatric interventions that divert the alimentary transit from the proximal small intestine. In a study, conditioned medium of duodenum/jejunum obtained from small intestine explants prevenient from insulin resistance mice (db/db mice) and insulin resistant and insulin sensitive human subjects' was analyzed. Jejunal proteins either from insulin resistant mice and human subjects were demonstrated to be are capable of impairing muscle insulin signaling, thus inducing insulin resistance (Salinari et al. 2013). In addition, an experimental procedure devised to examine the role of excluding the proximal small intestine in the absence of gastric restriction, named gastrojejunal bypass (GJB) was applied to Goto-Kakizaki nonobese T2D rats and was demonstrated to improve glycemic control without inducing weight loss (Rubino and Marescaux 2004). Moreover, the Duodenal-Jejunal Bypass Liner (DJBL) consisting in the endoscopic placement of an impermeable sleeve into the proximal intestine, somehow able to mimic the surgical duodenal and jejunal exclusion induced by RYGB and BPD surgeries, lead to significant improvements in glycemic indexes of patients with obesity and T2D (Jirapinyo et al. 2018b). Besides that, DJBL induced superior weight loss and T2D improvement when compared with diet alone (Koehestanie et al. 2014).

Duodenal mucosal resurfacing (DMR), a novel minimally invasive upper endoscopic procedure, that involves circumferential hydrothermal ablation of the duodenal mucosa also brought solid evidences that support the "foregut hypothesis" (van Baar et al. 2018). Recent clinical trials showed that DMR induces a substantial clinical improvement in glycemic control and insulin resistance up to 12 months in patients with T2D. The length of the ablated segment was positively correlated with the improvement in glycemic control (Rajagopalan et al. 2016). DMR efficacy is unlikely to be related with the malabsorption or the substantial weight loss often observed after the bariatric surgery and so, it supports a role of the duodenum as playing a major part in the development of insulin resistance and T2D (Rajagopalan et al. 2016; van Baar et al. 2019).

In sum, both the "foregut" and "hindgut" hypotheses provide reasonable explanations and have generated supporting evidence for the well-known anti-diabetic effects induced by widely used bariatric surgery procedures, such as RYGB, BPD and BPD with duodenal switch (BPD-DS). However, the relative contribution of each physiological mechanism modified by the different anatomical rearrangements still needs clarification.

2.6   Bile Acids and Gut Microbiota

Bile acids and gut microbiota were both demonstrated to play an active role in several metabolic pathways involved in glucose and energy homeostasis (Ahmad and Haeusler 2019; Shapiro et al. 2018; Gérard and Vidal 2019; Heiss and Olofsson 2018; Taoka et al. 2016). In addition, considerable changes on bile acid and gut microbiota with a potential positive metabolic impact were found to occur after bariatric surgery (Ulker and Yildiran 2019; Ejtahed et al. 2018). However, it is still unclear if whether these are a causal or consequence of diabetes improvement or remission after bariatric surgery.

Bariatric surgery was shown to interfere with bile acid enterohepatic circulation, leading to a consequent increase in circulating bile acid levels (Liu et al. 2018; Wang et al. 2019; Pérez-Pevida et al. 2019; Madsbad et al. 2014; Patti et al. 2009). In turn, bile acids are able to induce fibroblast growth factor 19 (FGF19) secretion, which acts through a G protein–coupled bile acid receptor (TGR5) to stimulate GLP-1 and PYY secretion (Liu et al. 2018; Sachdev et al. 2016). Thus, this bile acids mediated pathway has been appointed as an additional contributor mechanism for the metabolic benefits observed bariatric surgery.

There are several levels of evidence supporting a role of gut microbiota for the metabolic improvement observed after surgery. In fact, insulin sensitivity was significantly increased in subjects with metabolic syndrome that underwent fecal transplantation to receive gut microbiota from normal weight donors (Vrieze et al. 2012). Furthermore, gut microbiota transplantation from subjects' previously submitted to RYGB operation to non-operated ones resulted in significant weight loss although the effects on glucose metabolism were not assessed (Tremaroli et al. 2015; Liou et al. 2013).

3   Clinical Evidence on the Use of Gastro-Intestinal Interventions to Treat Type 2 Diabetes

3.1   Surgical Interventions

The links between T2D and GI surgery date back to the early 30's, when the first cases of dramatic glycemic improvements after GI operations performed for peptic ulcers and cancer in patients with diabetes were reported (Friedman et al. 1955). However, at those times these anecdotal cases were unvalued and were not focus of further attention until the advent of bariatric surgery in the 1950s, when diabetes remission following GI surgery was increasingly reported (Pories et al. 1995). But it not until the 2000s that experimental evidence was raised to support of the influence of GI anatomy on glucose homeostasis regulation, which also provided a mechanistic rationale for the use of GI surgery for the primary goal of treating T2D (Rubino and Marescaux 2004).

Following bariatric surgery interventions most patients experience a rapid weight loss and continue to do so until 18–24 months after the procedure. Patients may lose 60% of excess body weight in the first six months, 77% of excess weight as early as 12 months after surgery and maintain a 50–60% excess weight loss up to 10–14 years after surgery (Wittgrove and Clark 2000). Notwithstanding the overall excess weight-loss (EWL) in patients with T2D tending to be more modest than the observed for subjects without T2D (Carbonell et al. 2008). A systematic review and meta-analysis on the durability of weight loss at and beyond 10 years published in 2019 by O'Brien *et al*, showed that the most impressive outcome was observed for the BPD techniques or its BPD-DS variant with a pooled effect size of 71% EWL followed by RYGB with 60% EWL (O'Brien et al. 2019).

In addition to the body weight reduction effects, RYGB and BPD, are also the most effective bariatric surgical treatments for T2D when compared to other procedures, which are followed by normalization of plasma glucose, insulin, and Hemoglobin A1c (HbA1c) in 80–100% of morbidly obese patients (Kodama et al. 2018; Schauer et al. 2003). In a considerable proportion of patients with T2D, the resumption of euglycemia and normal insulin levels occurs within days after surgery, long before any significant weight loss takes place (Rubino and Gagner 2002). In fact, the Greenville gastric bypass study reported that bariatric surgical interventions restored and maintained normal levels of glucose, insulin, and HbA1c in 91% of the patients for as long as 14 years (Pories et al. 1995).

Since then, several systematic reviews addressing the impact of bariatric surgery on T2D were performed, with disease remission rates varying according to the surgical procedure and duration of follow-up. Historically relevant are those performed by Buchwald et al., in 2004 and 2009, encompassing different surgical techniques that reported a T2D remission rate of 76.8% and 76.2% of insulin free patients, respectively (Buchwald and Williams 2004; Buchwald and Oien 2009). More recent analyses by Goh et al. (2017) and Kodama et al. (2018) report T2D remission rates ranging from 15% to 100% (Kodama et al. 2018; Goh et al. 2017). In 2018, Kodama et al. published a pooled analysis of the results of 25 eligible randomized controlled trials (RCTs) covering non-surgical treatments and surgical procedures performed in Western and Asian populations, with follow-up times ranging from 3 to 5 years. This study showed that patients with obesity and T2D submitted to bariatric surgery when compared to conventional medical interventions achieved higher disease remission rates (Kodama et al. 2018). In contrast, Koliaki et al. the applying a similar model in a meta-analysis that include only 12 RCTs conducted in Western populations only, concluded that bariatric surgery when compared to non-surgical treatments was associated with a greater weight loss, cardiovascular risk factors reduction, glycemic improvement and T2D remission rates from 24% to 95% within follow-up times ranging from 6 months to 3 years (Koliaki et al. 2017). The anti-diabetic effects of bariatric surgery were also found to be long-lasting, since at 5–18 years of postoperative follow-up time T2D remission rates ranging from 24% to 62% were still observed (Koliaki et al. 2017).

One should be aware that across different trials, depending on the criteria used to define T2D improvement and remission, the metabolic outcomes of bariatric surgery can be widely variable. In 2009, as an attempt to overcome this limitation, Sue Kirkman et al. proposed the use of universal criteria to define T2D remission after bariatric interventions (Buse et al. 2009).

The most frequent bariatric surgery procedures performed worldwide, namely LSG, RYGB and BPD or BPD-DS are associated with short- and medium-term T2D remission rates, widely variable across the different studies and type of procedures (Table 3) (Schauer et al. 2016; Koliaki et al. 2017).

In addition to the long lasting improvement in glycemic control observed after bariatric surgery, diabetes-related morbidity and mortality also experience a significant decline (Phillips and Shikora 2018). In post-bariatric patient populations, 30–40% reductions in myocardial infarction and stroke, 42% reduction of cancer in women, and 30–40% reduction in overall mortality were demonstrated (English and Williams 2018; Adams et al. 2007). Moreover, post-surgical weight loss improves overall quality of life along with obesity-related comorbidities and mortality rate (Faria et al. 2017).

Table 3 Diabetes remission rates following bariatric surgery

Author and year	Surgical procedure	Number of patients with T2D	Mean T2D Duration (years)	Follow-up after surgery (years)	T2D remission rate (%)	T2D Remission criteria
Marceau et al. (2007)	BPD-DS	377	NA	7.3	92	T2D medications withdrawal
Hayes et al. (2011)	RYGB	127	4.5	1	84	FG <6.0 mmol/L and HbA1c < 6.0% with no diabetes medications
Blackstone et al. (2012)	RYGB	505	>3	1.2	43.2	FG <100 mg/dL and Hba1c ≤5.7% with no diabetes medications
Mingrone et al. (2012)	RYGB	20	6.0	2	75	Hba1c ≤6.5% with no diabetes medications
	BPD	20	6.0		95	
Schauer et al. (2012)	LSG	50	8.5	1	27	Hba1c ≤6% with no diabetes medications
	RYGB	50	8.2		42	
Ikramuddin et al. (2013)	RYGB	60	8.9	1	44	HbA1c ≤ 6.0%
Liang et al. (2013)	RYGB	31	7.39	1	90	NA
Schauer et al. (2014)	LSG	49	8.3	3	20	Hba1c ≤6% with no diabetes medications
	RYGB	48			35	
Golomb et al. (2015)	LSG	82	NA	1	50.7	FG <100 mg/dL and Hba1c ≤6% with no diabetes medications
				3	38.2	
				5	20	
Ikramuddin et al. (2015)	RYGB	60	8.9	5	20	HbA1c ≤ 6.0%
Mingrone et al. (2015)	RYGB	19	>5	5	42	Hba1c ≤6.5% with no diabetes medications
	BPD	19			68	
Park and Kim (2016)	RYGB	134	4.6	1	46.1	HbA1c level of <6.0% with no diabetes medications
Torres et al. (2017)	RYGB	97	4	3	55.2	FG <100 mg/dL and Hba1c ≤6% with no diabetes medications
	BPD-DS	77	6		70.4	
	SADI-S	97	8		75.8	
Baltasar et al. (2018)	BPD-DS	115	NA	22	98	HbA1c ≤ 6.0%

BPD Biliopancreatic diversion, *BPD-DS* Biliopancreatic diversion with duodenal switch, *FG* Fasting Glucose, *HbA1c* Hemoglobin A1c, *LSG* Laparoscopic sleeve gastrectomy, *NA* Not available, *RYGB* Roux-en-Y Gastric Bypass, *SADI-S* Single Anastomosis Duodeno-ileal Bypass with Sleeve Gastrectomy, *T2D* Type 2 Diabetes

In patients with obesity and T2D, comprehensive reduction of cardiovascular disease risk factors, including blood pressure and lipid profile are vital to prevent microvascular and macrovascular complications. Bariatric surgery also contributes for the resolution or improvement of cardiovascular risk factors other than hyperglycemia, once hypertension and hyperlipidemia is resolved or improved in 78.5 and 61.7% of the patients, respectively (Noria and Grantcharov 2013). In fact, patients with obesity and T2D undergoing bariatric surgical treatment, sustained weight loss along with cardiovascular risk factors control, experience a remarkable 92% decrease in diabetes-related death (Adams et al. 2007).

In a recent meta-analysis, bariatric surgery was also demonstrated to be superior to conventional medical treatment in reducing overall incidence of microvascular complications in patients with T2D, namely the incidence of nephropathy and retinopathy were lower, while no significant difference was observed regarding peripheral neuropathy (Billeter et al. 2018).

Bariatric surgery, through weight loss-dependent and weight loss-independent mechanisms, also induces a significant improvement in NAFLD, including non-invasive parameters and liver histology (Yeo et al. 2019; von Schönfels et al. 2018).

Early intervention with tight glycemic control in patients diagnosed with T2D, is not only paramount for the overall prognosis of the disease, but also influences the propensity of disease remission after surgical interventions. A recent study examined the legacy effect of early glycemic control on diabetic complications and death. The study hypothesized that a glycemic legacy begins as early as the 1st year after diagnosis and depends on the level of glycemic exposure. In patients with newly diagnosed T2D and at least 10 years survival after diagnosis, first year glycemic control was strongly associated with future risk for diabetic complications and mortality, even after adjusting for glycemic control in the following years (Laiteerapong et al. 2019). This study suggested that failure to achieve an HbA1c < 6.5% (<48 mmol/mol) within the 1st year of diabetes diagnosis was enough to establish an irremediable long-term future risk of microvascular and macrovascular complications. In addition, failure to achieve an HbA1c < 7.0% (< 53 mmol/mol) within the 1st year after diabetes diagnosis may lead to an irreversible increased risk of mortality (Laiteerapong et al. 2019). Although most of the available evidence derives from patients with T2D and body mass index (BMI) > 35 kg/m^2, there is a growing amount of data suggesting that individuals with T2D and a preoperative BMI of 30–35 kg/m^2 are likely experience similar benefits (Cummings and Rubino 2018).

Even though long-term T2D remission is observed after bariatric surgery interventions, disease recurrence is recognized to occur 5 years after surgery in 16–43% of patients. Still, relapsed T2D often presents as more "benign", since longer periods of tight glycemic control significantly improve the prognosis of the disease with decreased risk of incident microvascular complications (Maleckas et al. 2015). This data provides further support for a legacy effect of bariatric surgery, where even a transient period of surgically induced T2D remission is associated with lower long-term risks.

3.2   Endoscopic Interventions

More recently, several endoscopic devices and procedures were developed as potential alternatives or adjuvants to bariatric surgery.

Endoscopic Sleeve Gastroplasty (ESG), DMR, Transpyloric Shuttle, EndoBarrier and SatiSphere, Incision less magnetic anastomosis system are some of the multiple endoscopic techniques available, although most are considered experimental (Jirapinyo and Thompson 2017).

Amongst the endoscopic interventions so far described, DMR has gathered a great deal of enthusiasm and high expectations. DMR is a novel, minimally invasive upper endoscopic procedure devised for the primary treatment of patients with T2D. The hypothetical rationale that led to conceive this procedure is based on the "foregut hypothesis" that states that "anti-incretin" factors secreted in the proximal small gut are responsible for blunting the incretin effect observed in patients with T2D. Thus, the procedure aims to deplete the proximal small gut mucosa from enteroencrine cells responsible for secreting putative "anti-incretin" factors, in order to promote the resurfacing of the mucosa with a new and potentially healthier cell population. The mucosal ablation is achieved by using a catheter alongside the endoscope, the duodenal mucosa is first lifted and then hydrothermally ablated. These cycles are repeated until at least 10 cm of the postpapillary duodenum is treated in a single endoscopic session. It has recently been shown that patients with T2D achieve significant improvements of glucose profiles after a single DMR procedure. It is thought that there is an effect of DMR on hepatic glucose production, possibly by an insulin-sensitizing mechanism, which is in line with observations in RYGB surgery. Still, the effect of DMR on HbA1c is less dramatic than observed after bariatric surgery interventions, which are currently the treatment intervention that has demonstrated to be highly effective for long-term body weight reduction and glycemic control in patients with obesity and T2D. Nonetheless, it is interesting and fascinating that an endoscopic procedure involving solely the duodenum can elicit such glycemic improvement (Rajagopalan et al. 2016).

In sum, some endoscopic therapies hold promise to be used as complementary strategies to bariatric surgery or even to be considered as a primary intervention for patients who are unwilling to accept the potential complications associated with surgery or who may not be suitable for surgical intervention. Nevertheless, these interventions are still on an experimental phase and robust long-term data is needed to provide evidence on the log-term effectiveness for T2D treatment.

4 Current Clinical Recommendations and Guidelines

Given the evidence on the efficacy, safety and cost-effectiveness of bariatric surgery on T2D treatment, the second Diabetes Surgery Summit (DSS-II) endorsed by International Federation For the Surgery of Obesity (IFSO) consensus conference, placed bariatric surgery squarely within the overall diabetes treatment algorithm, recommending consideration of this approach for patients with inadequately controlled T2D and a BMI as low as 30 kg/m^2 or 27.5 kg/m^2 for Asian individuals (Rubino et al. 2016). These new guidelines have been formally ratified by 53 leading diabetes and surgery societies worldwide.

Recognizing the need to inform diabetes care providers about the benefits and limitations of bariatric surgery, the DSS-II was convened in collaboration with six leading international diabetes organizations: the American Diabetes Association (ADA), International Diabetes Federation, Chinese Diabetes Society, Diabetes India, European Association for the Study of Diabetes, and Diabetes UK. DSS-II guidelines were incorporated into the ADA Standards of Diabetes Care in 2017 (American Diabetes Association 2017).

The DSS-II concluded that a substantial body of evidence had been accumulated, including numerous, albeit mostly short- and mid-term RCTs, demonstrating that bariatric surgery can achieve excellent glycemic control and reduce cardiovascular risk factors. Although additional studies are needed to further demonstrate long-term benefits, there is now sufficient clinical and mechanistic evidence to support inclusion of bariatric surgery among anti-diabetes interventions for people with T2D and obesity (Cummings and Cohen 2016; Maggard-Gibbons et al. 2013; American Diabetes A 2019). Complementary criteria to the criterion used to select candidates for bariatric surgery, traditionally relying on BMI only, needs to be identified in order to achieve a better algorithm to select the most adequate treatment for patients with T2D. Bariatric surgery procedures when performed with the primary aim of targeting T2D and metabolic disorders are termed as metabolic surgery (Cummings and Cohen 2016).

Metabolic surgery should be performed only by an experienced surgeon working as part of a well-organized and engaged multidisciplinary team including surgeon, endocrinologist, nutritionist, behavioral health specialist, and nurse. People presenting for bariatric surgery should receive a comprehensive readiness and mental health assessment and should be evaluated on the need for ongoing mental health services to help them adjust to medical and psychosocial changes after surgery. Long-term lifestyle support and routine monitoring of micronutrient and nutritional status must be provided to patients after bariatric surgery, according to guidelines for postoperative management by national and international professional societies (Busetto et al. 2017).

According DSS-II, metabolic surgery indications are:

- Metabolic surgery should be a *recommended* option to treat T2D in appropriate surgical candidates with class III obesity (BMI $\geq$40 kg/m^2), regardless of the level of glycemic control or complexity of glucose-lowering regimens, as well as in patients with class II obesity (BMI 35.0–39.9 kg/m²) with inadequately controlled hyperglycemia despite lifestyle and optimal medical therapy.

- Metabolic surgery should also be *considered* to be an option to treat T2D in patients with class I obesity (BMI 30.0–34.9 kg/m²) and inadequate glycemic control despite optimal medical treatment by either oral or injectable medications (including insulin).
- All BMI thresholds should be reconsidered depending on the ancestry of the patient. For patients of Asian ancestry, the BMI values depicted above should be reduced by 2.5 kg/m².

Contraindications include diagnosis of Type 1 Diabetes (unless surgery is indicated for other reasons, such as severe obesity); current drug or alcohol abuse; uncontrolled psychiatric illness; lack of understanding of the risks/benefits of the procedures, expected outcomes, or alternatives; and lack of commitment to nutritional supplementation and long-term follow-up required after surgery.

In addition to these guidance's, the American Society for Metabolic and Bariatric Surgery (ASMBS) practical guidelines, recommend as indications for bariatric surgery:

- Patients with a BMI ≥ 40 kg/m2 without coexisting medical problems and for whom bariatric surgery would not be associated with excessive risk;
- Patients with a BMI ≥ 35 kg/m² and 1 or more severe obesity-related co-morbidity, including T2D, hypertension, hyperlipidemia, obstructive sleep apnea (OSA), obesity-hypoventilation syndrome (OHS), Pickwickian syndrome (a combination of OSA and OHS), nonalcoholic fatty liver disease (NAFLD) or nonalcoholic steatohepatitis (NASH), pseudotumor cerebri, gastroesophageal reflux disease (GERD), asthma, venous stasis disease, severe urinary incontinence, debilitating arthritis, or considerably impaired quality of life, may also be offered a bariatric procedure. Patients with BMI of 30–34.9 kg/m² with diabetes or metabolic syndrome may also be offered a bariatric procedure although current evidence is limited by the number of subjects studied and lack of long-term data demonstrating net benefit.
- There is insufficient evidence for recommending a bariatric surgical procedure specifically for glycemic control alone, lipid lowering alone, or cardiovascular disease risk reduction alone, independent of BMI criteria.

Ultimately, physicians are responsible for providing updated evidence based information as a contribution to the process leading to patient-centered decision among the available treatments options for each given clinical condition. According to Purnell et al., if the number of gastric bypass operations performed in diabetic patients in USA was one million per year, the currently estimated 0.5% risk of perioperative death for all patients undergoing gastric bypass would mean that nearly 5000 patients would be expected to die of surgical related complications. On the other hand, using survey data from 2005 and estimating a per-year mortality rate of 3 per 1000 patients with diabetes, would suggest that approximately 15,600 deaths would occur over 5 years in a cohort of one million medically managed patients with diabetes. These types of competing timelines and risks should be part of risk-benefit discussions with patients (Purnell and Flum 2009).

5 Concluding Remarks

In summary, major advances have been made since the identification of the first gastro-intestinal hormone and discovery of the enteroendocrine system. Amongst the different hundreds of bioactive molecules produced along the GI tract, some have been recognized to play a paramount role in glucose homeostasis regulation. This achievement has led the pharmaceutical industry to design gut hormone analogues that mimic the physiological actions known to be impaired in patients with T2D. These established or emerging drug classes are now considered either a valuable or promising pharmacological tool in diabetes treatment. However, the major breakthrough has been the demonstration that enteroendocrine system modulation can be achieved through surgically induced anatomical rearrangements of the GI tract. These groundbreaking findings have now challenged the paradigms of disease mechanisms, treatment algorithms and beliefs about the natural history of T2D.

References

Aaboe K, Knop FK, Vilsboll T, Deacon CF, Holst JJ, Madsbad S et al (2010) Twelve weeks treatment with the DPP-4 inhibitor, sitagliptin, prevents degradation of peptide YY and improves glucose and non-glucose induced insulin secretion in patients with type 2 diabetes mellitus. Diabetes Obes Metab 12(4):323–333

Adami GF, Cordera R, Marinari G, Lamerini G, Andraghetti G, Scopinaro N (2003) Plasma ghrelin concentratin in the short-term following biliopancreatic diversion. Obes Surg 13(6):889–892

Adams TD, Gress RE, Smith SC, Halverson RC, Simper SC, Rosamond WD et al (2007) Long-term mortality after gastric bypass surgery. N Engl J Med 357 (8):753–761

Ahlman H, Nilsson O (2001) The gut as the largest endocrine organ in the body. Ann Oncol 12(suppl_2):S63–SS8

Ahmad TR, Haeusler RA (2019) Bile acids in glucose metabolism and insulin signalling — mechanisms and research needs. Nat Rev Endocrinol 15(12):701–712

Alfa RW, Park S, Skelly KR, Poffenberger G, Jain N, Gu X et al (2015) Suppression of insulin production and secretion by a decretin hormone. Cell Metab 21 (2):323–334

Amatuzio DS, Grande F, Wada S (1962) Effect of glucagon on the serum lipids in essential hyperlipemia and in hypercholesterolemia. Metab Clin Exp 11:1240–1249

American Diabetes A (2019) Standards of medical Care in Diabetes-2019 abridged for primary care providers. Clin Diabetes 37(1):11–34

American Diabetes Association (2017) Obesity management for the treatment of Type 2 diabetes. Diabetes Care 40(Suppl 1):S57–s63

Ariyasu H, Takaya K, Tagami T, Ogawa Y, Hosoda K, Akamizu T et al (2001) Stomach is a major source of circulating ghrelin, and feeding state determines plasma ghrelin-like immunoreactivity levels in humans. J Clin Endocrinol Metab 86(10):4753–4758

Baldissera FG, Holst JJ, Knuhtsen S, Hilsted L, Nielsen OV (1988) Oxyntomodulin (glicentin-(33-69)): pharmacokinetics, binding to liver cell membranes, effects on isolated perfused pig pancreas, and secretion from isolated perfused lower small intestine of pigs. Regul Pept 21(1–2):151–166

Baltasar A, Serra C, Pérez N (2018) Long-term experience with duodenal switch in the community hospital. Ann Obes Relat Dis 1(1):1002

Barnes C (1947) Hypoglycaemia following partial gastrectomy. Report of three cases. Lancet 253:536–539

Behary P, Tharakan G, Alexiadou K, Johnson N, Wewer Albrechtsen NJ, Kenkre J et al (2019) Combined GLP-1, oxyntomodulin, and peptide YY improves body weight and glycemia in obesity and Prediabetes/Type 2 diabetes: a randomized, single-blinded, placebo-controlled study. Diabetes Care 42(8):1446–1453

Billeter AT, Scheurlen KM, Probst P, Eichel S, Nickel F, Kopf S et al (2018) Meta-analysis of metabolic surgery versus medical treatment for microvascular complications in patients with type 2 diabetes mellitus. Br J Surg 105(3):168–181

Billington CJ, Briggs JE, Link JG, Levine AS (1991) Glucagon in physiological concentrations stimulates brown fat thermogenesis in vivo. Am J Phys 261(2 Pt 2):R501–R507

Blackstone R, Bunt JC, Cortés MC, Sugerman HJ (2012) Type 2 diabetes after gastric bypass: remission in five models using HbA1c, fasting blood glucose, and medication status. Surg Obes Relat Dis 8(5):548–555

Brandt SJ, Götz A, Tschöp MH, Müller TD (2018) Gut hormone polyagonists for the treatment of type 2 diabetes. Peptides 100:190–201

Brighton PJ, Szekeres PG, Willars GB (2004) Neuromedin U and its receptors: structure, function, and physiological roles. Pharmacol Rev 56(2):231–248

Browning KN, Travagli RA (2014) Central nervous system control of gastrointestinal motility and secretion and modulation of gastrointestinal functions. Compr Physiol 4(4):1339–1368

Buchwald H, Oien DM (2009) Metabolic/bariatric surgery Worldwide 2008. Obes Surg 19(12):1605–1611

Buchwald H, Williams SE (2004) Bariatric surgery worldwide 2003. Obes Surg 14(9):1157–1164

Buffa R, Capella C, Fontana P, Usellini L, Solcia E (1978) Types of endocrine cells in the human colon and rectum. Cell Tissue Res 192(2):227–240

Buse JB, Caprio S, Cefalu WT, Ceriello A, Del Prato S, Inzucchi SE et al (2009) How do we define cure of diabetes? Diabetes Care 32(11):2133–2135

Busetto L, Dicker D, Azran C, Batterham RL, Farpour-Lambert N, Fried M et al (2017) Practical recommendations of the Obesity Management Task Force of the European Association for the Study of Obesity for the Post-Bariatric Surgery Medical Management. Obes Facts 10(6):597–632

Calanna S, Christensen M, Holst JJ, Laferrere B, Gluud LL, Vilsboll T et al (2013) Secretion of glucagon-like peptide-1 in patients with type 2 diabetes mellitus: systematic review and meta-analyses of clinical studies. Diabetologia 56(5):965–972

Carbonell AM, Wolfe LG, Meador JG, Sugerman HJ, Kellum JM, Maher JW (2008) Does diabetes affect weight loss after gastric bypass? Surg Obes Relat Dis 4(3):441–444

Chambers AP, Kirchner H, Wilson-Perez HE, Willency JA, Hale JE, Gaylinn DD et al (2013) The effects of vertical sleeve gastrectomy in rodents are ghrelin independent. Gastroenterology 144(1):50–52.e5

Chen S, Turner S, Tsang E, Stark J, Turner H, Mahsut A et al (2007) Measurement of pancreatic islet cell proliferation by heavy water labeling. Am J Physiol Endocrinol Metab 293(5):E1459–E1464

Cristina ML, Lehy T, Zeitoun P, Dufougeray F (1978) Fine structural classification and comparative distribution of endocrine cells in normal human large intestine. Gastroenterology 75(1):20–28

Cummings DE, Cohen RV (2016) Bariatric/metabolic surgery to treat type 2 diabetes in patients with a BMI <35 kg/m2. Diabetes Care 39(6):924–933

Cummings DE, Rubino F (2018) Metabolic surgery for the treatment of type 2 diabetes in obese individuals. Diabetologia 61(2):257–264

Cummings DE, Weigle DS, Frayo RS, Breen PA, Ma MK, Dellinger EP et al (2002) Plasma ghrelin levels after diet-induced weight loss or gastric bypass surgery. N Engl J Med 346(21):1623–1630

Cummings DE, Overduin J, Foster-Schubert KE (2004) Gastric bypass for obesity: mechanisms of weight loss and diabetes resolution. J Clin Endocrinol Metab 89(6):2608–2615

Dirksen C, Hansen DL, Madsbad S, Hvolris LE, Naver LS, Holst JJ et al (2010) Postprandial diabetic glucose tolerance is normalized by gastric bypass feeding as opposed to gastric feeding and is associated with exaggerated GLP-1 secretion: a case report. Diabetes Care 33(2):375–377

Drucker DJ (1998) Glucagon-Like Peptides. Diabetes 47(2):159–169

Egerod KL, Engelstoft MS, Grunddal KV, Nohr MK, Secher A, Sakata I et al (2012) A major lineage of enteroendocrine cells coexpress CCK, secretin, GIP, GLP-1, PYY, and neurotensin but not somatostatin. Endocrinology 153(12):5782–5795

Ejtahed H-S, Angoorani P, Hasani-Ranjbar S, Siadat S-D, Ghasemi N, Larijani B et al (2018) Adaptation of human gut microbiota to bariatric surgeries in morbidly obese patients: a systematic review. Microb Pathog 116:13–21

English WJ, Williams DB (2018) Metabolic and bariatric surgery: an effective treatment option for obesity and cardiovascular disease. Prog Cardiovasc Dis 61(2):253–269

Falken Y, Hellstrom PM, Holst JJ, Naslund E (2011) Changes in glucose homeostasis after Roux-en-Y gastric bypass surgery for obesity at day three, two months, and one year after surgery: role of gut peptides. J Clin Endocrinol Metab 96(7):2227–2235

Faria GFR, Santos JMN, Simonson DC (2017) Quality of life after gastric sleeve and gastric bypass for morbid obesity. Porto Biomed J 2(2):40–46

Friedman MN, Sancetta AJ, Magovern GJ (1955) The amelioration of diabetes mellitus following subtotal gastrectomy. Surg Gynecol Obstet 100(2):201–204

Fruhbeck G, Diez-Caballero A, Gil MJ, Montero I, Gomez-Ambrosi J, Salvador J et al (2004) The decrease in plasma ghrelin concentrations following bariatric surgery depends on the functional integrity of the fundus. Obes Surg 14(5):606–612

Garcia de la Torre N, Rubio MA, Bordiu E, Cabrerizo L, Aparicio E, Hernandez C et al (2008) Effects of weight loss after bariatric surgery for morbid obesity on vascular endothelial growth factor-A, adipocytokines, and insulin. J Clin Endocrinol Metab 93(11):4276–4281

Garcia-Fuentes E, Garrido-Sanchez L, Garcia-Almeida JM, Garcia-Arnes J, Gallego-Perales JL, Rivas-Marin J et al (2008) Different effect of laparoscopic Roux-en-Y gastric bypass and open biliopancreatic diversion of Scopinaro on serum PYY and ghrelin levels. Obes Surg 18(11):1424–1429

Geloneze B, Tambascia MA, Pilla VF, Geloneze SR, Repetto EM, Pareja JC (2003) Ghrelin: a gut-brain hormone: effect of gastric bypass surgery. Obes Surg 13(1):17–22

Gérard C, Vidal H (2019) Impact of Gut microbiota on host glycemic control. Front Endocrinol 10:29

Gissey LC, Mariolo JRC, Castagneto M, Mingrone G, Casella G (2017) The simultaneous increase of insulin sensitivity and secretion can explain the raised incidence of hypoglycemia after gastric bypass. J Am Coll Surg 225(4):S19–S20

Goebel B, Schiavon M, Visentin R, Riz M, Man CD, Cobelli C et al (2018) Effects of the novel dual GLP-1R/GCGR agonist SAR425899 on postprandial glucose metabolism in overweight/obese subjects with Type 2 diabetes. Diabetes 67(Supplement 1): 72-OR

Goh YM, Toumi Z, Date RS (2017) Surgical cure for type 2 diabetes by foregut or hindgut operations: a myth or reality? A systematic review. Surg Endosc 31(1):25–37

Golomb I, Ben David M, Glass A, Kolitz T, Keidar A (2015) Long-term metabolic effects of laparoscopic sleeve gastrectomy. JAMA Surg 150(11):1051–1057

Grunddal KV, Ratner CF, Svendsen B, Sommer F, Engelstoft MS, Madsen AN et al (2016) Neurotensin is coexpressed, coreleased, and acts together with GLP-1 and PYY in enteroendocrine control of metabolism. Endocrinology 157:176–194

Guedes TP, Martins S, Costa M, Pereira SS, Morais T, Santos A et al (2015) Detailed characterization of incretin cell distribution along the human small intestine. Surg Obes Relat Dis 11(6):1323–1331

Guida C, Stephen SD, Watson M, Dempster N, Larraufie P, Marjot T et al (2019) PYY plays a key role in the resolution of diabetes following bariatric surgery in humans. EBioMedicine 40:67–76

Guidone C, Manco M, Valera-Mora E, Iaconelli A, Gniuli D, Mari A et al (2006) Mechanisms of recovery from type 2 diabetes after malabsorptive bariatric surgery. Diabetes 55(7):2025–2031

Gunawardene AR, Corfe BM, Staton CA (2011) Classification and functions of enteroendocrine cells of the lower gastrointestinal tract. Int J Exp Pathol 92 (4):219–231

Hajnal A, Kovacs P, Ahmed T, Meirelles K, Lynch CJ, Cooney RN (2010) Gastric bypass surgery alters behavioral and neural taste functions for sweet taste in obese rats. Am J Physiol Gastrointest Liver Physiol 299(4):G967–G979

Hayes MT, Hunt LA, Foo J, Tychinskaya Y, Stubbs RS (2011) A model for predicting the resolution of type 2 diabetes in severely obese subjects following Roux-en Y gastric bypass surgery. Obes Surg 21(7):910–916

Hedberg J, Hedenstrom H, Karlsson FA, Eden-Engstrom-B, Sundbom M (2011) Gastric emptying and postprandial PYY response after biliopancreatic diversion with duodenal switch. Obes Surg 21(5):609–615

Heiss CN, Olofsson LE (2018) Gut microbiota-dependent modulation of energy metabolism. J Innate Immun 10 (3):163–171

Hofer D, Asan E, Drenckhahn D (1999) Chemosensory perception in the gut. News Physiol Sci 14:18–23

Hojberg PV, Vilsboll T, Rabol R, Knop FK, Bache M, Krarup T et al (2009) Four weeks of near-normalisation of blood glucose improves the insulin response to glucagon-like peptide-1 and glucose-dependent insulinotropic polypeptide in patients with type 2 diabetes. Diabetologia 52(2):199–207

Holdstock C, Engstrom BE, Ohrvall M, Lind L, Sundbom M, Karlsson FA (2003) Ghrelin and adipose tissue regulatory peptides: effect of gastric bypass surgery in obese humans. J Clin Endocrinol Metab 88 (7):3177–3183

Ikramuddin S, Korner J, Lee WJ, Connett JE, Inabnet WB, Billington CJ et al (2013) Roux-en-Y gastric bypass vs intensive medical management for the control of type 2 diabetes, hypertension, and hyperlipidemia: the diabetes surgery study randomized clinical trial. JAMA 309(21):2240–2249

Ikramuddin S, Billington CJ, Lee WJ, Bantle JP, Thomas AJ, Connett JE et al (2015) Roux-en-Y gastric bypass for diabetes (the Diabetes Surgery Study): 2-year outcomes of a 5-year, randomised, controlled trial. Lancet Diabetes Endocrinol 3(6):413–422

Itoh Z (1997) Motilin and clinical application. Peptides 18 (4):593–608

Jacobsen SH, Olesen SC, Dirksen C, Jorgensen NB, Bojsen-Moller KN, Kielgast U et al (2012) Changes in gastrointestinal hormone responses, insulin sensitivity, and beta-cell function within 2 weeks after gastric bypass in non-diabetic subjects. Obes Surg 22 (7):1084–1096

Jirapinyo P, Thompson CC (2017) Endoscopic bariatric and metabolic therapies: surgical analogues and mechanisms of action. Clin Gastroenterol Hepatol 15 (5):619–630

Jirapinyo P, Jin DX, Qazi T, Mishra N, Thompson CC (2018a) A meta-analysis of GLP-1 after Roux-En-Y gastric bypass: impact of surgical technique and measurement strategy. Obes Surg 28(3):615–626

Jirapinyo P, Haas AV, Thompson CC (2018b) Effect of the duodenal-Jejunal bypass liner on glycemic control in patients with type 2 diabetes with obesity: a meta-analysis with secondary analysis on weight loss and hormonal changes. Diabetes Care 41(5):1106–1115

Jorgensen NB, Jacobsen SH, Dirksen C, Bojsen-Moller KN, Naver L, Hvolris L et al (2012) Acute and long-term effects of Roux-en-Y gastric bypass on glucose metabolism in subjects with Type 2 diabetes and normal glucose tolerance. Am J Physiol Endocrinol Metab 303(1):E122–E131

Jorgensen NB, Dirksen C, Bojsen-Moller KN, Jacobsen SH, Worm D, Hansen DL et al (2013) Exaggerated glucagon-like peptide 1 response is important for improved beta-cell function and glucose tolerance after Roux-en-Y gastric bypass in patients with type 2 diabetes. Diabetes 62(9):3044–3052

Jorsal T, Rhee NA, Pedersen J, Wahlgren CD, Mortensen B, Jepsen SL et al (2018) Enteroendocrine K and L cells in healthy and type 2 diabetic individuals. Diabetologia 61(2):284–294

Kaczmarek P, Malendowicz LK, Pruszynska-Oszmalek E, Wojciechowicz T, Szczepankiewicz D, Szkudelski T et al (2006) Neuromedin U receptor 1 expression in the rat endocrine pancreas and evidence suggesting neuromedin U suppressive effect on insulin secretion from isolated rat pancreatic islets. Int J Mol Med 18 (5):951–955

Karamanakos SN, Vagenas K, Kalfarentzos F, Alexandrides TK (2008) Weight loss, appetite suppression, and changes in fasting and postprandial ghrelin and peptide-YY levels after Roux-en-Y gastric bypass and sleeve gastrectomy: a prospective, double blind study. Ann Surg 247(3):401–407

Kellum JM, Kuemmerle JF, O'Dorisio TM, Rayford P, Martin D, Engle K et al (1990) Gastrointestinal hormone responses to meals before and after gastric bypass and vertical banded gastroplasty. Ann Surg 211(6):763–770; discussion 70-1

Kerr BD, Flatt PR, Gault VA (2010) (D-Ser2)Oxm [mPEG-PAL]: a novel chemically modified analogue of oxyntomodulin with antihyperglycaemic, insulinotropic and anorexigenic actions. Biochem Pharmacol 80(11):1727–1735

Khorgami Z, Shoar S, Saber AA, Howard CA, Danaei G, Sclabas GM (2019) Outcomes of bariatric surgery versus medical management for Type 2 diabetes mellitus: a meta-analysis of randomized controlled trials. Obes Surg 29(3):964–974

Kim MJ, Park HK, Byun DW, Suh KI, Hur KY (2014) Incretin levels 1 month after laparoscopic single anastomosis gastric bypass surgery in non-morbid obese type 2 diabetes patients. Asian J Surg 37(3):130–137

Kodama S, Fujihara K, Horikawa C, Harada M, Ishiguro H, Kaneko M et al (2018) Network meta-analysis of the relative efficacy of bariatric surgeries for diabetes remission. Obes Rev 19(12):1621–1629

Koehestanie P, de Jonge C, Berends FJ, Janssen IM, Bouvy ND, Greve JW (2014) The effect of the endoscopic duodenal-jejunal bypass liner on obesity and type 2 diabetes mellitus, a multicenter randomized controlled trial. Ann Surg 260(6):984–992

Koliaki C, Liatis S, le Roux CW, Kokkinos A (2017) The role of bariatric surgery to treat diabetes: current challenges and perspectives. BMC Endocr Disord 17 (1).30

Korner J, Bessler M, Inabnet W, Taveras C, Holst JJ (2007) Exaggerated glucagon-like peptide-1 and blunted glucose-dependent insulinotropic peptide secretion are associated with Roux-en-Y gastric bypass but not adjustable gastric banding. Surg Obes Relat Dis 3(6):597–601

Korner J, Inabnet W, Febres G, Conwell IM, McMahon DJ, Salas R et al (2009) Prospective study of gut hormone and metabolic changes after adjustable gastric banding and Roux-en-Y gastric bypass. Int J Obes (2005) 33(7):786–795

Kotidis EV, Koliakos GG, Baltzopoulos VG, Ioannidis KN, Yovos JG, Papavramidis ST (2006a) Serum ghrelin, leptin and adiponectin levels before and after weight loss: comparison of three methods of treatment--a prospective study. Obes Surg 16 (11):1425–1432

Kotidis EV, Koliakos G, Papavramidis TS, Papavramidis ST (2006b) The effect of biliopancreatic diversion with pylorus-preserving sleeve gastrectomy and duodenal switch on fasting serum ghrelin, leptin and adiponectin levels: is there a hormonal contribution to the weight-reducing effect of this procedure? Obes Surg 16 (5):554–559

Kuhre RE, Christiansen CB, Ghiasi SM, Gabe MBN, Skat-Rørdam PA, Modvig IM et al (2019) Neuromedin U does not act as a decretin in rats. Cell Metab 29 (3):719–726.e5

Kuntz E, Pinget M, Damge P (2004) Cholecystokinin octapeptide: a potential growth factor for pancreatic beta cells in diabetic rats. JOP 5(6):464–475

Kuroshima A, Yahata T (1979) Thermogenic responses of brown adipocytes to noradrenaline and glucagon in heat-acclimated and cold-acclimated rats. Jpn J Physiol 29(6):683–690

Laferrère B (2016) Bariatric surgery and obesity: influence on the incretins. Int J Obes Suppl 6(Suppl 1):S32–SS6

Laferrere B, Heshka S, Wang K, Khan Y, McGinty J, Teixeira J et al (2007) Incretin levels and effect are markedly enhanced 1 month after Roux-en-Y gastric bypass surgery in obese patients with type 2 diabetes. Diabetes Care 30(7):1709–1716

Laferrère B, Teixeira J, McGinty J, Tran H, Egger JR, Colarusso A et al (2008) Effect of weight loss by gastric bypass surgery versus hypocaloric diet on glucose and incretin levels in patients with type 2 diabetes. J Clin Endocrinol Metab 93(7):2479–2485

Laferrère B, Swerdlow N, Bawa B, Arias S, Bose M, Olivan B et al (2010) Rise of oxyntomodulin in response to oral glucose after gastric bypass surgery in patients with type 2 diabetes. J Clin Endocrinol Metab 95(8):4072–4076

Laiteerapong N, Ham SA, Gao Y, Moffet HH, Liu JY, Huang ES et al (2019) The legacy effect in type 2 diabetes: impact of early Glycemic control on future complications (The Diabetes & Aging Study). Diabetes Care 42(3):416–426

Lamberts R, Stumps D, Plumpe L, Creutzfeldt W (1991) Somatostatin cells in rat antral mucosa: qualitative and quantitative ultrastructural analyses in different states of gastric acid secretion. Histochemistry 95 (4):373–382

Langer FB, Reza Hoda MA, Bohdjalian A, Felberbauer FX, Zacherl J, Wenzl E et al (2005) Sleeve gastrectomy and gastric banding: effects on plasma ghrelin levels. Obes Surg 15(7):1024–1029

Latorre R, Sternini C, De Giorgio R, Greenwood-Van Meerveld B (2016) Enteroendocrine cells: a review of their role in brain-gut communication. Neurogastroenterol Motil 28(5):620–630

Lavine JA, Raess PW, Stapleton DS, Rabaglia ME, Suhonen JI, Schueler KL et al (2010) Cholecystokinin is up-regulated in obese mouse islets and expands β-cell mass by increasing β-cell survival. Endocrinology 151(8):3577–3588

le Roux CW, Bloom SR, Peptide YY (2005) Appetite and food intake. Proc Nutr Soc 64(2):213–216

Lee WJ, Chen CY, Chong K, Lee YC, Chen SC, Lee SD (2011) Changes in postprandial gut hormones after metabolic surgery: a comparison of gastric bypass and sleeve gastrectomy. Surg Obes Relat Dis 7 (6):683–690

Liang Z, Wu Q, Chen B, Yu P, Zhao H, Ouyang X (2013) Effect of laparoscopic Roux-en-Y gastric bypass surgery on type 2 diabetes mellitus with hypertension: a randomized controlled trial. Diabetes Res Clin Pract 101(1):50–56

Liddle RA (1995) Regulation of cholecystokinin secretion by intraluminal releasing factors. Am J Phys 269(3 Pt 1):G319–G327

Liou AP, Paziuk M, Luevano J-M Jr, Machineni S, Turnbaugh PJ, Kaplan LM (2013) Conserved shifts in the gut microbiota due to gastric bypass reduce host weight and adiposity. Sci Transl Med 5 (178):178ra41–178ra41

Liu H, Hu C, Zhang X, Jia W (2018) Role of gut microbiota, bile acids and their cross-talk in the effects of bariatric surgery on obesity and type 2 diabetes. J Diabetes Invest 9(1):13–20

Madsbad S, Dirksen C, Holst JJ (2014) Mechanisms of changes in glucose metabolism and bodyweight after bariatric surgery. Lancet Diabetes Endocrinol 2 (2):152–164

Maggard-Gibbons M, Maglione M, Livhits M, Ewing B, Maher AR, Hu J et al (2013) Bariatric surgery for weight loss and glycemic control in nonmorbidly obese adults with diabetes: a systematic review. JAMA 309(21):2250–2261

Maleckas A, Venclauskas L, Wallenius V, Lonroth H, Fandriks L (2015) Surgery in the treatment of type 2 diabetes mellitus. Scand J Surg 104(1):40–47

Mans E, Serra-Prat M, Palomera E, Sunol X, Clave P (2015) Sleeve gastrectomy effects on hunger, satiation, and gastrointestinal hormone and motility responses after a liquid meal test. Am J Clin Nutr 102 (3):540–547

Marceau P, Biron S, Hould F-S, Lebel S, Marceau S, Lescelleur O et al (2007) Duodenal switch: long-term results. Obes Surg 17(11):1421–1430

McCarty TR, Jirapinyo P, Thompson CC (2019, Oct 4) Effect of sleeve gastrectomy on ghrelin, GLP-1, PYY, and GIP gut hormones: a systematic review and meta-analysis. Ann Surg. https://doi.org/10.1097/SLA.0000000000003614. [Epub ahead of print] PubMed PMID: 31592891

Michaud A, Grenier-Larouche T, Caron-Dorval D, Marceau S, Biertho L, Simard S et al (2017) Biliopancreatic diversion with duodenal switch leads to better postprandial glucose level and beta cell function than sleeve gastrectomy in individuals with type 2 diabetes very early after surgery. Metab Clin Exp 74:10–21

Mingrone G, Nolfe G, Castagneto Gissey G, Iaconelli A, Leccesi L, Guidone C et al (2009) Circadian rhythms of GIP and GLP1 in glucose-tolerant and in type 2 diabetic patients after biliopancreatic diversion. Diabetologia 52(5):873

Mingrone G, Panunzi S, De Gaetano A, Guidone C, Iaconelli A, Leccesi L et al (2012) Bariatric surgery versus conventional medical therapy for type 2 diabetes. N Engl J Med 366(17):1577–1585

Mingrone G, Panunzi S, De Gaetano A, Guidone C, Iaconelli A, Nanni G et al (2015) Bariatric-metabolic surgery versus conventional medical treatment in obese patients with type 2 diabetes: 5 year follow-up of an open-label, single-centre, randomised controlled trial. Lancet 386(9997):964–973

Mochiki E, Suzuki H, Takenoshita S, Nagamachi Y, Kuwano H, Mizumoto A et al (1998) Mechanism of inhibitory effect of glucagon on gastrointestinal motility and cause of side effects of glucagon. J Gastroenterol 33(6):835–841

Monteiro MP, Batterham RL (2017) The importance of the gastrointestinal tract in controlling food intake and regulating energy balance. Gastroenterology 152(7):1707–1717.e2

Morinigo R, Casamitjana R, Moize V, Lacy AM, Delgado S, Gomis R et al (2004) Short-term effects of gastric bypass surgery on circulating ghrelin levels. Obes Res 12(7):1108–1116

Mumphrey MB, Patterson LM, Zheng H, Berthoud HR (2013) Roux-en-Y gastric bypass surgery increases number but not density of CCK-, GLP-1-, 5-HT-, and neurotensin-expressing enteroendocrine cells in rats. Neurogastroenterol Motil 25(1):e70–e79

Nauck M, Stockmann F, Ebert R, Creutzfeldt W (1986) Reduced incretin effect in type 2 (non-insulin-dependent) diabetes. Diabetologia 29(1):46–52

Nauck MA, Vardarli I, Deacon CF, Holst JJ, Meier JJ (2011) Secretion of glucagon-like peptide-1 (GLP-1) in type 2 diabetes: what is up, what is down? Diabetologia 54(1):10–18

Noria SF, Grantcharov T (2013) Biological effects of bariatric surgery on obesity-related comorbidities. Can J Surg 56(1):47–57

O'Brien PE, Hindle A, Brennan L, Skinner S, Burton P, Smith A et al (2019) Long-term outcomes after bariatric surgery: a systematic review and meta-analysis of weight loss at 10 or more years for all bariatric procedures and a single-Centre review of 20-year outcomes after adjustable gastric banding. Obes Surg 29(1):3–14

Olivan B, Teixeira J, Bose M, Bawa B, Chang T, Summe H et al (2009) Effect of weight loss by diet or gastric bypass surgery on peptide YY3-36 levels. Ann Surg 249(6):948–953

Palha AM, Pereira SS, Costa MM, Morais T, Maia AF, Guimaraes M et al (2018) Differential GIP/GLP-1 intestinal cell distribution in diabetics' yields distinctive rearrangements depending on Roux-en-Y biliopancreatic limb length. J Cell Biochem 119(9):7506–7514

Pardina E, Lopez-Tejero MD, Llamas R, Catalan R, Galard R, Allende H et al (2009) Ghrelin and apolipoprotein AIV levels show opposite trends to leptin levels during weight loss in morbidly obese patients. Obes Surg 19(10):1414–1423

Park JY, Kim YJ (2016) Prediction of diabetes remission in morbidly obese patients after Roux-en-Y gastric bypass. Obes Surg 26(4):749–756

Patti M-E, Houten SM, Bianco AC, Bernier R, Larsen PR, Holst JJ et al (2009) Serum bile acids are higher in humans with prior gastric bypass: potential contribution to improved glucose and lipid metabolism. Obesity (Silver Spring) 17(9):1671–1677

Peier AM, Desai K, Hubert J, Du X, Yang L, Qian Y et al (2011) Effects of peripherally administered Neuromedin U on energy and glucose homeostasis. Endocrinology 152(7):2644–2654

Pérez-Pevida B, Escalada J, Miras AD, Frühbeck G (2019) Mechanisms underlying Type 2 diabetes remission after metabolic surgery. Front Endocrinol 10:641

Peterli R, Wolnerhanssen B, Peters T, Devaux N, Kern B, Christoffel-Courtin C et al (2009) Improvement in glucose metabolism after bariatric surgery: comparison of laparoscopic Roux-en-Y gastric bypass and laparoscopic sleeve gastrectomy: a prospective randomized trial. Ann Surg 250(2):234–241

Peterli R, Steinert RE, Woelnerhanssen B, Peters T, Christoffel-Courtin C, Gass M et al (2012) Metabolic and hormonal changes after laparoscopic Roux-en-Y gastric bypass and sleeve gastrectomy: a randomized, prospective trial. Obes Surg 22(5):740–748

Phillips BT, Shikora SA (2018) The history of metabolic and bariatric surgery: development of standards for patient safety and efficacy. Metab Clin Exp 79:97–107

Pinkney J (2014) The role of ghrelin in metabolic regulation. Curr Opin Clin Nutr Metab Care 17(6):497–502

Plourde C-É, Grenier-Larouche T, Caron-Dorval D, Biron S, Marceau S, Lebel S et al (2014) Biliopancreatic diversion with duodenal switch improves insulin sensitivity and secretion through caloric restriction. Obesity 22(8):1838–1846

Pocai A (2012) Unraveling oxyntomodulin, GLP1's enigmatic brother. J Endocrinol 215(3):335–346

Pocai A, Carrington PE, Adams JR, Wright M, Eiermann G, Zhu L et al (2009) Glucagon-like peptide 1/glucagon receptor dual agonism reverses obesity in mice. Diabetes 58(10):2258–2266

Polak JM, Bloom SR, Kuzio M, Brown JC, Pearse AG (1973) Cellular localization of gastric inhibitory polypeptide in the duodenum and jejunum. Gut 14 (4):284–288

Pories WJ, Swanson MS, MacDonald KG, Long SB, Morris PG, Brown BM et al (1995) Who would have thought it? An operation proves to be the most effective therapy for adult-onset diabetes mellitus. Ann Surg 222(3):339–350; discussion 50–2

Posovszky C, Wabitsch M (2015) Regulation of appetite, satiation, and body weight by enteroendocrine cells. Part 1: characteristics of enteroendocrine cells and their capability of weight regulation. Horm Res Paediatr 83 (1):1–10

Pournaras DJ, le Roux CW (2010) Ghrelin and metabolic surgery. Int J Pept 2010:217267

Purnell JQ, Flum DR (2009) Bariatric surgery and diabetes: who should be offered the option of remission? JAMA 301(15):1593–1595

Rajagopalan H, Cherrington AD, Thompson CC, Kaplan LM, Rubino F, Mingrone G et al (2016) Endoscopic duodenal mucosal resurfacing for the treatment of type 2 diabetes: 6-month interim analysis from the first-in-human proof-of-concept study. Diabetes Care 39 (12):2254–2261

Rehfeld JF (2004) A centenary of gastrointestinal endocrinology. Horm Metab Res 36(11/12):735–741

Rhee NA, Wahlgren CD, Pedersen J, Mortensen B, Langholz E, Wandall EP et al (2015) Effect of Roux-en-Y gastric bypass on the distribution and hormone expression of small-intestinal enteroendocrine cells in obese patients with type 2 diabetes. Diabetologia 58 (10):2254–2258

Romero F, Nicolau J, Flores L, Casamitjana R, Ibarzabal A, Lacy A et al (2012) Comparable early changes in gastrointestinal hormones after sleeve gastrectomy and Roux-En-Y gastric bypass surgery for morbidly obese type 2 diabetic subjects. Surg Endosc 26(8):2231–2239

Rubino F (2008) Is type 2 diabetes an operable intestinal disease? A provocative yet reasonable hypothesis. Diabetes Care 31(Suppl 2):S290–S296

Rubino F, Gagner M (2002) Potential of surgery for curing type 2 diabetes mellitus. Ann Surg 236(5):554–559

Rubino F, Marescaux J (2004) Effect of duodenal-jejunal exclusion in a non-obese animal model of type 2 diabetes: a new perspective for an old disease. Ann Surg 239 (1):1–11

Rubino F, Forgione A, Cummings DE, Vix M, Gnuli D, Mingrone G et al (2006) The mechanism of diabetes control after gastrointestinal bypass surgery reveals a role of the proximal small intestine in the pathophysiology of type 2 diabetes. Ann Surg 244(5):741–749

Rubino F, Nathan DM, Eckel RH, Schauer PR, Alberti KG, Zimmet PZ et al (2016) Metabolic surgery in the treatment algorithm for type 2 diabetes: a joint statement by international diabetes organizations. Surg Obes Relat Dis 12(6):1144–1162

Sachdev S, Wang Q, Billington C, Connett J, Ahmed L, Inabnet W et al (2016) FGF 19 and bile acids increase following Roux-en-Y gastric bypass but not after medical Management in Patients with type 2 diabetes. Obes Surg 26(5):957–965

Salinari S, Bertuzzi A, Asnaghi S, Guidone C, Manco M, Mingrone G (2009) First-phase insulin secretion restoration and differential response to glucose load depending on the route of administration in type 2 diabetic subjects after bariatric surgery. Diabetes Care 32 (3):375–380

Salinari S, Debard C, Bertuzzi A, Durand C, Zimmet P, Vidal H et al (2013) Jejunal proteins secreted by db/db mice or insulin-resistant humans impair the insulin signaling and determine insulin resistance. PLoS One 8(2):e56258-e

Sánchez-Garrido MA, Brandt SJ, Clemmensen C, Müller TD, DiMarchi RD, Tschöp MH (2017) GLP-1/glucagon receptor co-agonism for treatment of obesity. Diabetologia 60(10):1851–1861

Schauer PR, Burguera B, Ikramuddin S, Cottam D, Gourash W, Hamad G et al (2003) Effect of laparoscopic Roux-en Y gastric bypass on type 2 diabetes mellitus. Ann Surg 238(4):467–484; discussion 84–5

Schauer PR, Kashyap SR, Wolski K, Brethauer SA, Kirwan JP, Pothier CE et al (2012) Bariatric surgery versus intensive medical therapy in obese patients with diabetes. N Engl J Med 366(17):1567–1576

Schauer PR, Bhatt DL, Kirwan JP, Wolski K, Brethauer SA, Navaneethan SD et al (2014) Bariatric surgery versus intensive medical therapy for diabetes — 3-year outcomes. N Engl J Med 370(21):2002–2013

Schauer PR, Mingrone G, Ikramuddin S, Wolfe B (2016) Clinical outcomes of metabolic surgery: efficacy of Glycemic control, weight loss, and remission of diabetes. Diabetes Care 39(6):902–911

Shapiro H, Kolodziejczyk AA, Halstuch D, Elinav E (2018) Bile acids in glucose metabolism in health and disease. J Exp Med 215(2):383–396

Sista F, Abruzzese V, Clementi M, Carandina S, Cecilia M, Amicucci G (2017) The effect of sleeve gastrectomy on GLP-1 secretion and gastric emptying: a prospective study. Surg Obes Relat Dis 13(1):7–14

Sjolund K, Sanden G, Hakanson R, Sundler F (1983) Endocrine cells in human intestine: an immunocytochemical study. Gastroenterology 85(5):1120–1130

Stefater MA, Wilson-Pérez HE, Chambers AP, Sandoval DA, Seeley RJ (2012) All bariatric surgeries are not created equal: insights from mechanistic comparisons. Endocr Rev 33(4):595–622

Stengel A, Tache Y (2009) Regulation of food intake: the gastric X/A-like endocrine cell in the spotlight. Curr Gastroenterol Rep 11(6):448–454

Sternini C, Anselmi L, Rozengurt E (2008) Enteroendocrine cells: a site of 'taste' in gastrointestinal chemosensing. Curr Opin Endocrinol Diabetes Obes 15(1):73–78

Stoeckli R, Chanda R, Langer I, Keller U (2004) Changes of body weight and plasma ghrelin levels after gastric banding and gastric bypass. Obes Res 12(2):346–350

Svendsen B, Pedersen J, Albrechtsen NJ, Hartmann B, Torang S, Rehfeld JF et al (2015) An analysis of cosecretion and coexpression of gut hormones from male rat proximal and distal small intestine. Endocrinology 156(3):847–857

Taoka H, Yokoyama Y, Morimoto K, Kitamura N, Tanigaki T, Takashina Y et al (2016) Role of bile acids in the regulation of the metabolic pathways. World J Diabetes 7(13):260–270

Torres A, Rubio MA, Ramos-Leví AM, Sánchez-Pernaute A (2017) Cardiovascular risk factors after single anastomosis Duodeno-Ileal bypass with sleeve gastrectomy (SADI-S): a new effective therapeutic approach? Curr Atheroscler Rep 19(12):58

Tremaroli V, Karlsson F, Werling M, Ståhlman M, Kovatcheva-Datchary P, Olbers T et al (2015) Roux-en-Y gastric bypass and vertical banded Gastroplasty induce Long-term changes on the human gut microbiome contributing to fat mass regulation. Cell Metab 22(2):228–238

Tsoli M, Chronaiou A, Kehagias I, Kalfarentzos F, Alexandrides TK (2013) Hormone changes and diabetes resolution after biliopancreatic diversion and laparoscopic sleeve gastrectomy: a comparative prospective study. Surg Obes Relat Dis 9(5):667–677

Tymitz K, Engel A, McDonough S, Hendy MP, Kerlakian G (2011) Changes in ghrelin levels following bariatric surgery: review of the literature. Obes Surg 21(1):125–130

Ulker İ, Yildiran H (2019) The effects of bariatric surgery on gut microbiota in patients with obesity: a review of the literature. Biosci Microbiota Food Health 38(1):3–9

Valderas JP, Irribarra V, Boza C, de la Cruz R, Liberona Y, Acosta AM et al (2010) Medical and surgical treatments for obesity have opposite effects on peptide YY and appetite: a prospective study controlled for weight loss. J Clin Endocrinol Metab 95(3):1069–1075

Valderas JP, Irribarra V, Rubio L, Boza C, Escalona M, Liberona Y et al (2011) Effects of sleeve gastrectomy and medical treatment for obesity on glucagon-like peptide 1 levels and glucose homeostasis in non-diabetic subjects. Obes Surg 21(7):902–909

Valera Mora ME, Manco M, Capristo E, Guidone C, Iaconelli A, Gniuli D et al (2007) Growth hormone and ghrelin secretion in severely obese women before and after bariatric surgery. Obesity (Silver Spring) 15(8):2012–2018

Valverde I, Puente J, Martin-Duce A, Molina L, Lozano O, Sancho V et al (2005) Changes in glucagon-like peptide-1 (GLP-1) secretion after biliopancreatic diversion or vertical banded gastroplasty in obese subjects. Obes Surg 15(3):387–397

van Baar ACG, Nieuwdorp M, Holleman F, Soeters MR, Groen AK, Bergman JJGHM (2018) The duodenum harbors a broad untapped therapeutic potential. Gastroenterology 154(4):773–777

van Baar ACG, Holleman F, Crenier L, Haidry R, Magee C, Hopkins D et al (2019) Endoscopic duodenal mucosal resurfacing for the treatment of type 2 diabetes mellitus: one year results from the first international, open-label, prospective, multicentre study. Gut. https://doi.org/10.1136/gutjnl-2019-318349

Vetter ML, Ritter S, Wadden TA, Sarwer DB (2012) Comparison of bariatric surgical procedures for diabetes remission: efficacy and mechanisms. Diabetes Spectr 25(4):200–210

Vilsboll T, Holst JJ (2004) Incretins, insulin secretion and Type 2 diabetes mellitus. Diabetologia 47(3):357–366

Vilsboll T, Knop FK, Krarup T, Johansen A, Madsbad S, Larsen S et al (2003) The pathophysiology of diabetes involves a defective amplification of the late-phase insulin response to glucose by glucose-dependent insulinotropic polypeptide-regardless of etiology and phenotype. J Clin Endocrinol Metab 88(10):4897–4903

von Schönfels W, Beckmann JH, Ahrens M, Hendricks A, Röcken C, Szymczak S et al (2018) Histologic improvement of NAFLD in patients with obesity after bariatric surgery based on standardized NAS (NAFLD activity score). Surg Obes Relat Dis 14(10):1607–1616

Vrieze A, Van Nood E, Holleman F, Salojärvi J, Kootte RS, Bartelsman JFWM et al (2012) Transfer of intestinal microbiota from lean donors increases insulin sensitivity in individuals with metabolic syndrome. Gastroenterology 143(4):913–916.e7

Wallenius V, Dirinck E, Fändriks L, Maleckas A, le Roux CW, Thorell A (2018) Glycemic control after sleeve gastrectomy and Roux-En-Y gastric bypass in obese subjects with type 2 diabetes mellitus. Obes Surg 28(6):1461–1472

Wang G-F, Yan Y-X, Xu N, Yin D, Hui Y, Zhang J-P et al (2015) Predictive factors of type 2 diabetes mellitus remission following bariatric surgery: a meta-analysis. Obes Surg 25(2):199–208

Wang W, Cheng Z, Wang Y, Dai Y, Zhang X, Hu S (2019) Role of bile acids in bariatric surgery. Front Physiol 10:374

Wittgrove AC, Clark GW (2000) Laparoscopic gastric bypass, Roux-en-Y- 500 patients: technique and results, with 3-60 month follow-up. Obes Surg 10(3):233–239

Wren AM, Seal LJ, Cohen MA, Brynes AE, Frost GS, Murphy KG et al (2001) Ghrelin enhances appetite and increases food intake in humans. J Clin Endocrinol Metab 86(12):5992

Wynne K, Park AJ, Small CJ, Patterson M, Ellis SM, Murphy KG et al (2005) Subcutaneous oxyntomodulin reduces body weight in overweight and obese subjects:

a double-blind, randomized, controlled trial. Diabetes 54(8):2390–2395

Wynne K, Park AJ, Small CJ, Meeran K, Ghatei MA, Frost GS et al (2006) Oxyntomodulin increases energy expenditure in addition to decreasing energy intake in overweight and obese humans: a randomised controlled trial. Int J Obes (2005) 30(12):1729–1736

Xu G, Kaneto H, Laybutt DR, Duvivier-Kali VF, Trivedi N, Suzuma K et al (2007) Downregulation of GLP-1 and GIP receptor expression by hyperglycemia: possible contribution to impaired incretin effects in diabetes. Diabetes 56(6):1551–1558

Ye L, Liddle RA (2017) Gastrointestinal hormones and the gut connectome. Curr Opin Endocrinol Diabetes Obes 24(1):9–14

Yeo SC, Ong WM, Cheng KSA, Tan CH (2019) Weight loss after bariatric surgery predicts an improvement in the non-alcoholic fatty liver disease (NAFLD) fibrosis score. Obes Surg 29(4):1295–1300

Zhang Y, Ji G, Li G, Hu Y, Liu L, Jin Q et al (2019) Ghrelin reductions following bariatric surgery were associated with decreased resting state activity in the hippocampus. Int J Obes 43(4):842–851

Zhou J, Cai X, Huang X, Dai Y, Sun L, Zhang B et al (2017) A novel glucagon-like peptide-1/glucagon receptor dual agonist exhibits weight-lowering and diabetes-protective effects. Eur J Med Chem 138:1158–1169

Adv Exp Med Biol - Advances in Internal Medicine (2020) 4: 299–320
https://doi.org/10.1007/5584_2020_487
© Springer Nature Switzerland AG 2020
Published online: 19 February 2020

Laparoscopic Vertical Sleeve Gastrectomy as a Treatment Option for Adults with Diabetes Mellitus

Timothy R. Koch [ID] and Timothy R. Shope [ID]

Abstract

Obesity is a major factor in the worldwide rise in the prevalence of type 2 diabetes mellitus. The obesity "epidemic" will require novel, effective interventions to permit both the prevention and treatment of diabetes caused by obesity. Laparoscopic vertical sleeve gastrectomy is a newer bariatric surgical procedure with a lower risk of complications (compared to Roux-en-Y gastric bypass surgery). Based in part on restriction of daily caloric intake, sleeve gastrectomy has a major role in inducing significant weight loss and weight loss is maintained for at least 10 years. Prior studies have supported the utility of the vertical sleeve gastrectomy for the treatment and management of subgroups of individuals with diabetes mellitus. There are reports of 11% to 76.9% of obese individuals discontinuing use of diabetic medications in studies lasting up to 8 years after vertical sleeve gastrectomy. Major ongoing issues include the preoperative determination of the suitability of diabetic patients to undergo this bariatric surgical procedure. Understanding how this surgical procedure is performed and the resulting anatomy is important when vertical sleeve gastrectomy is being considered as a treatment option for diabetes. In the postoperative periods, specific macronutrient goals and micronutrient supplements are important for successful and safer clinical results. An understanding of immediate- and long term- potential complications is important for reducing the potential risks of vertical sleeve gastrectomy. This includes the recognition and treatment of postoperative nutritional deficiencies and disorders. Vertical sleeve gastrectomy is a component of a long term, organized program directed at treating diabetes related to obesity. This approach may result in improved patient outcomes when vertical sleeve gastrectomy is performed to treat type 2 diabetes in obese individuals.

Keywords

Bariatric surgery · Diabetes and sleeve gastrectomy · Diabetes mellitus · Obesity · Treatment of type 2 diabetes · Type 2 diabetes · Vertical sleeve gastrectomy

T. R. Koch (✉) and T. R. Shope
Center for Advanced Laparoscopic General & Bariatric Surgery, MedStar Washington Hospital Center and Georgetown University School of Medicine, Washington, DC, USA
e-mail: timothy.r.koch@medstar.net

1   Introduction

The obesity "epidemic" is a growing global medical crisis. Much of the focus in studies of individuals with medically-complicated obesity has been based on Class 2 obesity (Body Mass Index or BMI of 35.0–39.9 kg/m^2) and Class

Fig. 1 Temporal trends in obesity and diabetes mellitus. The worldwide estimated numbers of individuals with obesity and diabetes mellitus from the time period 1975–1980 are shown in the left side of the figure, while the rises in the worldwide estimates by 2014 are shown in the right side of the figure

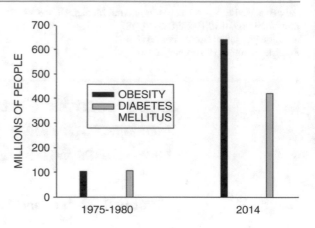

3 obesity (BMI of $>$/$=$ 40.0 kg/m^2). Trends in obesity from 1975 to 2014 have been estimated in 186 countries (NCD Risk Factor Collaboration 2016a). Over these four decades, the worldwide prevalence of Class 2 obesity rose to 2.3% of men and 5.0% of women, while the worldwide prevalence of Class 3 obesity rose to 0.64% of men and 1.6% of women. Based on these trends, it appears that by 2025, Class 2 obesity may surpass 6% of all men and 9% of all women (NCD Risk Factor Collaboration 2016a). A related international study of 195 countries reported that the prevalence of obesity doubled since 1980 in more than 70 countries (Global Burden of Disease 2015 Obesity Collaborators 2017). These studies support the concern that efforts to slow the worldwide rise in obesity have to date been ineffective.

There are multiple reported mechanisms that link obesity to the development of insulin resistance and type 2 diabetes mellitus (Kahn et al. 2006). In parallel to the worldwide rise in obesity (see Fig. 1), global trends in diabetes from 1980 to 2014 have been estimated (NCD Risk Factor Collaboration 2016b). The global prevalence of diabetes increased from 4.3% of men in 1980 to 9.0% of men in 2014, while in women the global prevalence of diabetes increased from 5.0% in 1980 to 7.9% in 2014. There were an estimated 108 million adults (worldwide) with diabetes in 1980, with an increase to 422 million diabetic adults in 2014 (NCD Risk Factor Collaboration 2016b). The number of adults with diabetes mellitus has increased faster in low-income and middle-income countries, placing the highest burden on those countries having the fewest resources for the management of this serious obesity-related medical disorder.

For this review, the terms "diabetes and sleeve gastrectomy" were examined in PubMed from September 1, 1998 up to September 1, 2019 by one of the authors (TRK). Among 1286 abstracts, only those abstracts from English language articles were examined for applicability to this review. Full articles were examined for those abstracts that applied to this review.

2   Medical Management of Type 2 Diabetes Related to Obesity

During management of type 2 diabetes mellitus related to obesity, there is often a hopeful waiting period to see whether there is reversal of an individual's weight gain without an intervention. Unfortunately, in a 9 year cohort study of 76,704 obese men and 99,791 obese women from the United Kingdom (Fildes et al. 2015), for men with BMI 30.0–34.9 kg/m^2 only 1 in 210 attained normal weight each year, and with a BMI of 40.0–44.9 kg/m^2 only 1 in 1290 attained normal weight each year. For women with BMI 30.0–34.9 kg/m^2 only 1 in 124 attained normal weight each year, and with a BMI of 40.0–44.9 kg/m^2 only 1 in 677 attained normal weight each year.

The generally recommended management of type 2 diabetes in obese patients then involves instruction on diet and activity programs, which provide a minimal median reduction (0.5%) in hemoglobin A1c levels (6). Published trials and a systematic review suggest that: (i) among the participants in these programs only about 10% of individuals will lose 10% or more of their total body weight; (ii) mean weight loss with these programs are in the range of 8–9 kg; and (iii) Maintenance therapy, which is designed to prevent weight regain, is often ineffective (Lemacks et al. 2013; Vilar-Gomez et al. 2015; Svetkey et al. 2008). Recent studies from North America and Asia have supported these findings. In a 5 year outcome study of intensive medical therapy for diabetes in the United States, mean total body weight loss (compared to baseline) was only 5% and the mean reduction in hemoglobin A1c level (compared to baseline) was only 0.3% (Schauer et al. 2017). In a second 5 year study from South Korea, conventional treatment achieved a mean of only 2.8% of total body weight loss with no decrease in the prevalence of diabetes (Park et al. 2019).

The next step-up therapy involves consideration of adding pharmacologic therapy. This therapy generally involves the use of nine therapies approved for weight loss by the United States Food and Drug Administration. Four sympathomimetic drugs (phentermine, benzphetamine, diethylpropion, and phendimetrazine) increase adrenergic stimulus in short term medical therapy, while five medical therapies involve long-term pharmacologic treatment (phentermine with topiramate, naltrexone with bupropion, orlistat, liraglutide, and lorcaserin). With pharmacologic treatment, mean weight losses range from 2.6 to 8.8 kg in individuals whose BMI range from 23 kg/m^2 to 39.9 kg/m^2 (Koch et al. 2017).

Because of the minimal responses that result from dietary and activity programs for the treatment of individuals with type 2 diabetes mellitus (Rubino et al. 2016), there was been long standing interest in the development of noninvasive treatments. This interest initiated the use of intragastric balloons to induce weight loss in obese individuals. The United States Food and Drug Administration approved the use of the Garren-Edwards bubble in 1985, and more recently approved use of the Orbera intragastric balloon (Apollo Endosurgery, Inc., Austin, TX USA), the ReShape intragastric balloon (which was purchased on December 17, 2018 by Apollo Endosurgery, Inc., Austin, TX USA), and the Obalon intragastric balloon system (Obalon Therapeutics Inc., San Diego, CA USA) (Koch et al. 2017; Rashti et al. 2014). The 30 year experience using intragastric balloons has been reviewed by multiple groups who have reported incomplete results. Popov and associates have suggested that the small number of subjects and short-term length of follow up limits our understanding of the use of intragastric balloons for improving metabolic risk factors (Popov et al. 2017) while Brethauer and associates suggested that in individuals with type 2 diabetes, additional studies are needed with regards to the role of intragastric balloons (Brethauer et al. 2016).

3 The Role of Bariatric Surgery in Diabetes Mellitus

3.1 Prevention of Diabetes Mellitus in Obese Individuals

Several studies have examined the potential utility of vertical sleeve gastrectomy for preventing the development of diabetes mellitus in individuals with obesity. A national study, using data extracted from the French National Health Service database, has reviewed the results from over 328,000 morbidly obese individuals (Bailly et al. 2019). Over a 7 year period from 2009 to 2016, 9.7% of the population had a diagnosis of type 2 diabetes but a significantly lower percentage of individuals who underwent vertical sleeve gastrectomy (0.9%) had type 2 diabetes compared to controls. In a related study from the United States (Gutierrez-Blanco et al. 2019), vertical sleeve gastrectomy was shown at 12 months of follow up to decrease the risk of developing type 2 diabetes mellitus in a population that predominantly included middle aged females with mean (SD) BMI of 43.1 (6.9) kg/m^2.

3.2 Bariatric Procedures for Treatment of Obesity and Diabetes Mellitus

International diabetes organizations in a joint statement have suggested consideration of bariatric surgery in those individuals with type 2 diabetes mellitus who have a BMI $\geq$40 kg/m^2, a BMI 35–39.9 kg/m^2 with inadequate control of hyperglycemia during optimal medical therapy, or a BMI 30–34.9 kg/m^2 with inadequate control of hyperglycemia during the use of oral or injectable medications (Rubino et al. 2016). Individuals with diabetes mellitus are eligible for bariatric surgery in the United States if they fulfill the National Institutes of Health criterion, which involves the presence of a BMI of $\geq$35 kg/m^2. The American Society for Metabolic and Bariatric Surgery has stated that vertical sleeve gastrectomy is an acceptable option as a primary bariatric procedure (Ali et al. 2017). Their position paper summarizes evidence in individuals with type 2 diabetes that vertical sleeve gastrectomy is superior to intensive medical therapy in glycemic control, weight loss, reduction of medication use for type 2 diabetes, lowering of both blood lipids and blood pressure, stability of renal function, and quality of life scores (Ali et al. 2017).

Contraindications to the performance of vertical sleeve gastrectomy are not well defined. Many insurance companies in the United States do require evaluation by a psychologist or psychiatrist prior to proceeding to vertical sleeve gastrectomy. However, studies of individuals with important psychiatric diagnoses, including depression, anxiety, bipolar disorder, and schizophrenia, have reported similar results when compared to individuals without psychiatric illness (Fuchs et al. 2016; Archid et al. 2019). In a survey of 863 bariatric surgeons who have performed over 500,000 bariatric procedures, 12% of these experienced surgeons stated that there were no absolute contraindications to undergoing vertical sleeve gastrectomy (Adil et al. 2019). Medical conditions that were suggested by these bariatric surgeons to be absolute contraindications to vertical sleeve gastrectomy included Barrett's esophagus, hiatal hernia, gastroesophageal reflux disease, high BMI, diabetes mellitus (suggested by 3.2% of 863 bariatric surgeons), and cirrhosis of the liver (Adil et al. 2019).

With regards to gastroesophageal reflux disease, contradictory postoperative studies have suggested either increased or decreased lower esophageal sphincter pressure following vertical sleeve gastrectomy (Petersen et al. 2012; Valezi et al. 2017). A recent study suggested that preoperative esophageal testing may identity individuals in whom vertical sleeve gastrectomy should be discouraged due to an individual's high risk of postoperative gastroesophageal reflux (Kavanagh et al. 2019). These authors however did not report any postoperative results to support their suggestion (Kavanagh et al. 2019). Prior to vertical sleeve gastrectomy, the potential contraindications to bariatric surgery, the potential risks of this bariatric procedure, and ongoing outcomes obtained following this bariatric procedure are discussed with individuals who are appropriate candidates for bariatric surgery.

The bariatric surgical procedures commonly utilized worldwide include the adjustable gastric band, vertical sleeve gastrectomy, and Roux-en-Y gastric bypass (Angrisani et al. 2017). Long term results upon examining outcomes after placement of an adjustable gastric band are inferior to results obtained in individuals with morbid obesity who have undergone vertical sleeve gastrectomy or Roux-en-Y gastric bypass (Koch et al. 2018). Over time, vertical sleeve gastrectomy has become a more widely utilized bariatric procedure in part due to reduced risk compared to Roux-en-Y gastric bypass. A systematic review of 15 randomized trials has reported a mean complication rate of 12.1% for individuals who underwent vertical sleeve gastrectomy compared to a mean complication rate of 20.9% for individuals who underwent Roux-en-Y gastric bypass surgery (Trastulli et al. 2013). In addition, nutritional complications are reported to be less frequent in individuals who have undergone vertical sleeve gastrectomy (Stroh et al. 2017; Ferraz et al. 2018).

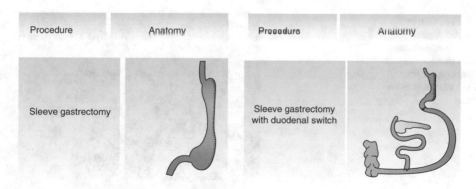

Fig. 2 Bariatric surgical procedures that include the vertical sleeve gastrectomy. The left cartoon depicts vertical sleeve gastrectomy. This procedure limits the volume of consumed food. The right cartoon depicts vertical sleeve gastrectomy with a duodenal switch. In addition to limiting the volume of food consumed, pyloric function has been preserved by surgically connecting the ileum to the duodenal bulb (immediately distal to the pylorus) resulting in partial diversion of biliopancreatic secretions. (Reproduced with the permission of Nature Publishing Group from Bal BS, et al. Nature Rev. Endocrinol 2012; 8: 544–556)

In 2014, 579,000 worldwide bariatric surgical procedures most frequently involved vertical sleeve gastrectomy (45.9%) (Angrisani et al. 2017). In the United States, performance of vertical sleeve gastrectomy rose from 9.3% of bariatric surgical procedures in 2010 to 58.2% of bariatric surgical procedures in 2014 (Khorgami et al. 2017). Diabetes mellitus was present in approximately 22.8% of individuals undergoing vertical sleeve gastrectomy in the United States in 2014 (Khorgami et al. 2017).

The vertical sleeve gastrectomy or gastric sleeve resection can be completed in a single step operation (see Fig. 2). This surgical procedure can be less time consuming compared to Roux en Y gastric bypass, but it is irreversible. The attachments of the greater curvature of the stomach are divided from the antrum to the angle of His to mobilize the body and fundus. Next, a bougie esophageal dilator is passed per orally to the level of the pylorus. The formation of a gastric sleeve then involves the use of multiple stapler firings to surgically excise 60–80% of the stomach using the bougie as a guide to preserve continuity of the gastrointestinal tract. This resection produces a narrow, tubular stomach based on the lesser curve of the stomach (see Fig. 3).

Vertical sleeve gastrectomy should be considered in obese individuals with type 2 diabetes mellitus who have continued significant hyperglycemia or elevation of blood level of hemoglobin A1C despite the use of a diet and activity program, or in individuals being treated with oral medications to treat hyperglycemia. Preoperative endoscopic or radiological imaging of the stomach is important to exclude evidence for diabetic gastropathy, a relative contraindication to performing vertical sleeve gastrectomy. Due to its potential role in release of proximal small intestinal incretins, Roux-en-Y gastric bypass may be a better surgical procedure for treatment of obese individuals with type 2 diabetes who require less than 100 units of insulin daily and may benefit from increased release of pancreatic insulin. Obese individuals requiring more than 100 units of insulin daily more likely have insulin insensitivity. In individuals with insulin insensitivity, it is unclear whether Roux-en-Y gastric bypass may hold a physiological advantage over vertical sleeve gastrectomy. Further work should better define the role for vertical sleeve gastrectomy in those individuals with insulin insensitivity.

Combining restriction with malabsorption by bypassing a portion of the foregut and midgut provides measurable changes in comorbid conditions such as diabetes mellitus that are not proportional to the level of weight loss that is

Fig. 3 Post-operative
gastric sleeve. Upper
endoscopy demonstrates
the narrow, tubular stomach
fundus and corpus after
undergoing vertical sleeve
gastrectomy

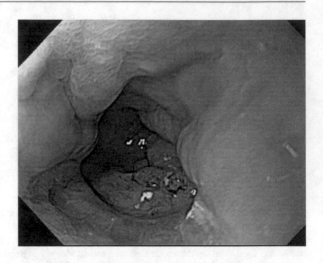

achieved. The duodenal switch takes advantage of this approach. In this combined procedure, vertical sleeve gastrectomy is performed with a small intestinal bypass (see Fig. 2). Vertical sleeve gastrectomy is performed to limit the volume of food consumed. The duodenal switch preserves pyloric function and creates a small intestinal bypass by surgically connecting the ileum to the duodenal bulb (immediately distal to the pylorus). Duodenal switch thus results in substantial diversion of biliopancreatic secretions, which is a potential origin for small intestinal malabsorption as the basis for weight loss. In up to 4 years of follow up, this modification has demonstrated favorable results in the treatment of metabolic diseases (Zaveri et al. 2018).

3.3 Mechanisms of Weight Loss After Vertical Sleeve Gastrectomy

Weight loss after vertical sleeve gastrectomy will result from a restriction in the size of meal portions. However, additional mechanisms of weight loss after vertical sleeve gastrectomy are complex and may include upper gut bacterial overgrowth, a common intestinal disorder in individuals with diabetes mellitus, as well as glucose malabsorption (Sharbaugh et al. 2017). Since plasma levels of the orexigenic hormone,

acyl ghrelin, are suppressed by jejunal administration of glucose (Tamboli et al. 2016) and reduced blood levels of ghrelin have been reported after vertical sleeve gastrectomy (Svane et al. 2019), glucose malabsorption may be an important mechanism for understanding appetite suppression after vertical sleeve gastrectomy. Additional mechanisms of weight loss after vertical sleeve gastrectomy are important because some individuals after bariatric surgery have inadequate long-term weight loss (Pucci and Batterham 2019).

3.4 Nutritional Requirements After Vertical Sleeve Gastrectomy

The immediate and long term nutritional focus after vertical sleeve gastrectomy is based upon the adequate intake of both macronutrients (chiefly protein) and micronutrients.

Protein is commonly the major macronutrient associated with malnutrition after bariatric surgery. Protein deficiency (insufficient protein intake) can result in serious complications. Protein deficiency can commonly induce alopecia, a disorder in bariatric patients which may be compounded by deficiency of zinc or biotin. Similar to kwashiorkor, long-term manifestations of protein deficiency can less commonly include muscle mass wasting and anasarca, emaciation, depigmented hair, and the biochemical findings of hypoalbuminemia and anemia (Lewandowski et al. 2007; Ramirez Prada et al. 2011). To maintain lean body mass during weight loss, postoperative guidelines recommend an average of 60–120 g of daily protein intake (Heber et al. 2010). This can be a difficult daily dietary goal. To improve daily protein intake, after vertical sleeve gastrectomy, patients may first require evaluation and treatment of dysphagia (Nath et al. 2016), followed by the addition of liquid protein supplements if solid food dysphagia persists, and then the ongoing assistance of an experienced registered dietician.

Micronutrients are defined as dietary factors ingested in microgram or milligram quantities that are required to maintain normal physiological metabolic processes and biochemical pathways. Guidelines from The Endocrine Society have supported the postoperative intake of one to two chewable daily multivitamin tablets containing mineral supplements in bariatric patients. Postoperative bariatric patients should chronically maintain intake of 1200 mg per day of elemental calcium (Heber et al. 2010). After repletion of vitamin D deficiency, more recent guidelines have suggested supplementation with at least 3000 units/day of Vitamin D (Parrott et al. 2017).

3.5 Short Term Complications After Vertical Sleeve Gastrectomy

Two immediate major postoperative concerns after performance of vertical sleeve gastrectomy include development of a leak of the gastric staple line or development of swallowing disorders. It has been reported that individuals with elevated preoperative hemoglobin A1C levels have a higher risk of postoperative complications after vertical sleeve gastrectomy, supporting the importance of preoperative optimization of diabetes care (Guetta et al. 2019).

Potential risk factors for the development of a postoperative leak of the gastric sleeve include: ischemia, intraluminal pressure, tobacco use, corticosteroid use, immunosuppressive pharmacotherapy, super morbid obesity (BMI

>49.9 kg/m^2), use of nonsteroidal anti-inflammatory medication, diabetes mellitus, malnutrition, Crohns disease, and revision of a prior weight loss procedure to vertical sleeve gastrectomy (Warner and Sasse 2017). Reduction in the risk of development of a gastric sleeve leak may be obtained by altering the surgical procedure (Warner and Sasse 2017; Loo et al. 2019). One potential change is to use a larger bougie esophageal dilator at the time of gastric stapling which may reduce the risk of developing a gastric sleeve leak (possibly by reducing the pressure on the staple line or by retaining more distensible fundus), but may also reduce the postoperative weight loss that is obtained. A gastric sleeve leak in an unstable patient (the presence of hypotension, fever, and tachycardia) require consideration of surgical reoperation, while stable patients in most instances can be managed with an intervention at therapeutic upper endoscopy (Abou Rached et al. 2014). Upper endoscopic therapeutic procedures that have been described in patients with gastric sleeve leaks include the use of different types of endoclips for closure, placement of a covered endoprosthesis or stent for temporary exclusion, or the application of "glues" to the site of the leak (Abou Rached et al. 2014).

A second common complication is the postoperative development of dysphagia associated with vomiting (Nath et al. 2016). Risk factors for the onset of dysphagia that have been identified in individuals after vertical sleeve gastrectomy include: type 2 diabetes mellitus, gastroesophageal reflux, thiamine deficiency, hypothyroidism, use of nonsteroidal anti-inflammatory medication, and use of opioid narcotics (Nath et al. 2016). Individuals have an increased risk of gastroesophageal reflux after vertical sleeve gastrectomy. Individuals who have narrowing of the gastric sleeve as an origin for dysphagia commonly note symptomatic improvement after dilation of the gastric sleeve by using through-the-endoscope balloon distension (Nath et al. 2016). A small number of patients with refractory dysphagia have undergone dilation using larger diameter pneumatic balloons or have undergone temporary placement of a covered endoprosthesis or stent across a gastric stricture.

3.6   Long Term Complications After Vertical Sleeve Gastrectomy

Long term complications of vertical sleeve gastrectomy include psychosocial disorders and micronutrient deficiencies.

Psychosocial Disorders In a retrospective study on a Danish nationwide register-based cohort of 22,451 patients followed after bariatric surgery, there were increased risks identified for self-harm, the occurrence of mental disorders, and the subsequent use of psychiatric services (Kovacs et al. 2017). In addition, increased alcohol use and increased admissions to substance abuse treatment facilities were described in a systematic review of postoperative bariatric patients (Spadola et al. 2015). Postoperative bariatric care must include ongoing screening for these psychosocial disorders.

Overview of Micronutrients in Morbidly Obese Individuals and After Vertical Sleeve Gastrectomy A major focus in long term complications after vertical sleeve gastrectomy involves prevention of micronutrient deficiencies. Table 1 summarizes the classes of and constituents of micronutrients.

Fat Soluble Vitamins The fat soluble vitamins include vitamin A, vitamin D, vitamin E, and vitamin K. Most fat soluble vitamins have longer lasting body stores.

Vitamin A Retinols, β carotenes, are carotenoids are components of vitamin A. Vitamin A deficiency can cause nyctalopia or poor night vision, decreased vision, dry hair, or pruritus (Bal et al. 2012).

High doses of vitamin A supplements can lead to vomiting, headache, diplopia, hepatitis, bone abnormalities, or alopecia.

Vitamin D Vitamin D improves calcium absorption from small intestine, permitting adequate calcium and phosphate levels that are needed for homeostasis of bone. Individuals who develop

Table 1 Classes of and constituents of micronutrients

Fat soluble vitamins:	Vitamin A
	Vitamin D
	Vitamin E
	Vitamin K
Water soluble vitamins:	Vitamin B1
	Vitamin B2
	Vitamin B3
	Vitamin B5
	Vitamin B6
	Vitamin B7
	Vitamin B9
	Vitamin C
Trace elements:	Copper
	Zinc
	Selenium
	Chromium
	Manganese
Major minerals:	Iron
	Calcium

osteoporosis and osteomalacia as the result of vitamin D deficiency may present with large joint pain or bony pain (Bal et al. 2012). In individuals with vitamin D deficiency, there is reduced calcium absorption and secondary hypocalcemia which can lead to increased secretion of parathyroid hormone.

Vitamin E Tocopherols and tocotrienols are components of vitamin E. Lipid peroxidation can be reduced by vitamin E. Individuals with vitamin E deficiency may present with muscle weakness, visual symptoms, anemia, ataxia, or dysarthria (Bal et al. 2012).

Vitamin K A group of vitamin K compounds are involved in the formation of prothrombin and other blood clotting factors. Vitamin K deficiency may lead to hemorrhage or bleeding (Bal et al. 2012). There is a small whole-body pool of vitamin K, but intestinal flora provides vitamin K via biosynthesis.

Water Soluble Vitamins Minor body stores exist for most water-soluble vitamins. The body stores 14–16 days of vitamin B1. By contrast, vitamin B12 stores may last for 3–5 years.

Table 2 Signs and symptoms related to thiamine deficiency

Beriberi subtype	Common symptoms/findings
Neurologic (Dry beriberi)	Numbness; Muscle Weakness or Cramps; Convulsions; Pain of Lower > Upper Extremities; Increased Tendon Reflexes
High Output	Tachycardia; Respiratory Distress; Lower Extremity Edema;
Cardiovascular (Wet beriberi)	Dyspnea on Exertion; Right Ventricule Dilated; L-Lactic Acidosis
Neuropsychiatric	Nystagmus; Ataxia; Ophthalmoplegia; Hallucinations; Aggressive Behavior; Confusion
Gastrointestinal	Dysphagia; Nausea; Vomiting; Constipation; Megajejunum; Megacolon

Vitamin B1 Vitamin B1 or thiamine is involved in both in the pentose phosphate pathway as well as in early stages of the tricarboxylic acid cycle. Individuals with thiamine deficiency may present with cardiac, neurologic, neuropsychiatric, or gastrointestinal symptoms (Bal et al. 2012). Symptoms of beriberi due to thiamine deficiency are caused by a multiorgan disorder within the body (see Table 2). A clinical diagnosis of thiamine deficiency should be considered in individuals with two positive findings within three defined categories (lower leg edema; loss of balance, muscle deficit, or paresthesias; or dyspnea at rest or exertional).

Vitamin B2 Vitamin B2 or riboflavin is located within flavin adenine dinucleotide, flavocoenzymes, and flavin mononucleotide. Potential symptoms of Vitamin B2 deficiency may include anemia, stomatitis, dermatitis, and pharyngitis (Bal et al. 2012). Vitamin B2 functions in the production of glutathione peroxidase and is present in a number of metabolic pathways.

Vitamin B3 Nicotinic acid and nicotinamide are chemical forms of Vitamin B3 or niacin. Potential symptoms of niacin deficiency (or pellagra) include dermatologic symptoms (scaly dermatitis), neurologic symptoms (headaches, ataxia, hallucinations, delusions, depression, or anxiety), or gastrointestinal symptoms (diarrhea or malabsorption) (Bal et al. 2012).

Vitamin B5 Vitamin B5 or pantothenic acid is involved in coenzyme A functioning. Deficiency of Vitamin B5 may lead to paresthesias, a gait disorder, infections, depression, and hypotension (Bal et al. 2012).

Vitamin B6 The active form of Vitamin B6 is pyridoxal phosphate. Vitamin B6 deficiency can result in confusion, peripheral neuropathy, or dermatitis (Bal et al. 2012). Vitamin B6 serves as a coenzyme in multiple biochemical reactions. Over supplementation of vitamin B6 resulting in hypervitaminosis can induce symptoms of neuropathy.

Vitamin B7 Vitamin B7 or biotin is a coenzyme in five carboxylases. Biotin deficiency may lead to loss of taste, ataxia, hair loss, seizures, and dermatitis (Bal et al. 2012).

Vitamin B9 Vitamin B9 or folic acid is involved in the synthesis of the tetrahydrofolate and the biosynthesis of thymidine nucleotides, methionine, and purine nucleotides. Folic acid deficiency can present with anorexia, weight loss, weakness, or macrocytic anemia (Bal et al. 2012).

Vitamin B12 Vitamin B12 or cobalamin is involved in the biosynthesis of methionine and succinyl-coenzyme A. Signs and symptoms of vitamin B12 deficiency can occur several years after vertical sleeve gastrectomy and may include macrocytic anemia, neuropsychiatric symptoms (including depression), ataxia, and peripheral neuropathy (which is potentially irreversible) (Bal et al. 2012).

Vitamin C Vitamin C or ascorbic acid is a cofactor for iron-dependent dioxygenases and copper-dependent mono-oxygenases. Symptoms of vitamin C deficiency or "scurvy" may include petechia, myalgias, and malaise which can potentially progress to soft tissue or gum disease (Bal et al. 2012). Vitamin C is involved in wound healing and collagen biosynthesis.

Trace Elements Trace elements are cofactors in enzymes and proteins. The transition metals copper, zinc, chromium, and manganese may function as acceptors or donors of electrons.

Copper Copper is present in multiple proteins including cytosolic copper/zinc super oxide dismutase and cytochrome oxidase. In susceptible individuals, low serum copper can induce anemia, pancytopenia, and neutropenia (Bal et al. 2012). The potential role of copper in a myeloneuropathy-like disorder that mimics vitamin B12 deficiency and in visual loss are important issues that are under investigation.

Zinc Zinc is present in cytosolic copper/zinc superoxide dismutase. Zinc deficiency can present with alopecia, hypoalbuminaemia, nail dystrophy, or the skin eruption termed acrodermititis enteropathica (Bal et al. 2012).

Selenium Selenium is present within glutathione peroxidase. Selenium deficiency can lead to cardiomyopathy, which has been termed "Keshan disease" (Bal et al. 2012).

Chromium The initial role of chromium in humans was based on symptoms reported in patients receiving total parenteral nutrition. These potential symptoms include weight loss, peripheral neuropathy, and abnormal intravenous glucose tolerance test (Bal et al. 2012).

Manganese Manganese is a cofactor in the inducible mitochondrial superoxide dismutase. Symptoms of manganese deficiency could include inadequate collagen deposition during wound healing and skeletal deformation (Bal et al. 2012).

Major Minerals Iron and calcium are major minerals that are important in human nutrition after vertical sleeve gastrectomy.

Iron Iron deficiency commonly results in the development of microcytic anemia (Bal et al. 2012).

Calcium Calcium (due to the transfer of calcium ions across cell membranes) is diffusely involved in normal cellular physiology. Calcium deficiency may lead to back pain, aching of the limbs, bony pain, and muscle irritability. Long standing deficiency of calcium can cause osteoporosis, which increases the risk of fractures (Bal et al. 2012). Bones and teeth are major stores of calcium in the body.

3.7   Preoperative Micronutrients

There has been continued interest in examining the notion that repletion of micronutrient deficiencies prior to bariatric surgery may prevent development of or persistence of micronutrient deficiencies postoperatively. This idea has been promoted following the identification of common micronutrient deficiencies in obese adults (Kimmons et al. 2006) and in obese patients seeking bariatric surgery (Flancbaum et al. 2006; de Luis et al. 2013; Carrodeguas et al. 2005; Nath et al. 2017). The major preoperative micronutrient deficiencies have included vitamin B1, vitamin B12, folic acid, vitamin A, vitamin D, and iron, so testing for these micronutrients has been suggested prior to vertical sleeve gastrectomy by the American Society for Metabolic and Bariatric Surgery (Parrott et al. 2017).

Within the National Health and Nutrition Examination Survey III from the United States (Kimmons et al. 2006), among 3831 obese adults, women had low blood levels of vitamin E, alpha- & beta-carotene, vitamin D, folic acid, vitamin C, and selenium, while men has low blood levels of alpha- & beta-carotene, folic acid, vitamin C, and selenium. Other retrospective studies have examined the prevalence of preoperative biochemical deficiencies of micronutrients. Flancbaum and associates reported that among 379 preoperative patients, there were biochemical deficiencies of vitamin B1 (29%), iron (44%), and 25-hydroxy vitamin D (68%) (Flancbaum et al. 2006). In a second preoperative study of 115 obese women from Spain, biochemical deficiencies of iron (5.2%), vitamin D (71.3%), vitamin B12 (9.5%), folic acid (25.2%), copper (67.8%), and zinc (73.9%) were reported by de Luis and associates (de Luis et al. 2013). In a preoperative study of 437 individuals seeking bariatric surgery at the Cleveland Clinic, Florida (USA), the prevalence of biochemical deficiency of vitamin B1 was 15.5% (Carrodeguas et al. 2005). In a retrospective study of 400 individuals seeking bariatric surgery, our group has reported evidence for clinical thiamine deficiency in 16.5% of these individuals (Nath et al. 2017). These studies support the preoperative evaluation of and treatment of micronutrient deficiencies.

3.8 Potential Mechanisms of Micronutrient Deficiencies After Vertical Sleeve Gastrectomy

The potential origins for micronutrient deficiencies in individuals who have undergone vertical sleeve gastrectomy are complex (see Table 3). These potential mechanisms include: altered intestinal absorption or storage of micronutrients, deficiency of bile acid or reduced intraluminal bile acids, inadequate oral intake, utilization of micronutrients during rapid weight loss or during high doses of simple sugars, small intestinal bacterial overgrowth (bacteria can produce thiaminases or deconjugate bile salts), alcohol intake, dietary herbal supplements (such as

Hypericum perforatum **or St. John's wort),** and medications (diuretics increase urinary excretion of B vitamins, while specific medications such as metronidazole interfere with vitamin B1) (Lovette et al. 2012).

Among specific micronutrients, vitamin B12 deficiency can result from inadequate gastric acid production. Hydrochloric acid can deconjugate Pteryl groups from dietary ingested cyanocobalamin. Gastric production of hydrochloric acid and the glycoprotein intrinsic factor require functioning parietal cells. In addition, small intestinal bacterial overgrowth, which is common in individuals who have diabetes and gut neuropathy, may promote bacterial utilization or consumption of vitamin B12 or may interfere with the formation of the cyanocobalamin complex with intrinsic factor (Bal et al. 2010).

Iron absorption has been examined after vertical sleeve gastrectomy (Ruz et al. 2012). In a study of young women, the absorption of Heme-iron was 28% preoperatively but was 8% at 1 year postoperatively, while the absorption of Nonheme-iron was 12% at baseline but was 6% in 1 year postoperatively. Potential mechanisms of decreased absorption included reduced gastric acid secretion and small intestinal bacterial overgrowth. Identification of iron deficiency in a patient after vertical sleeve gastrectomy requires consideration of whether there is another potential cause of blood loss (such as menorrhagia or chronic gastrointestinal blood loss) or iron malabsorption (such as Celiac disease).

Surveillance for calcium malabsorption and vitamin D deficiency should be considered after vertical sleeve gastrectomy. Calcium malabsorption can be induced by steatorrhea due to interaction of calcium in the diet with triglycerides within the small intestine (Bal et al. 2012; Bal et al. 2010). The presence of calcium malabsorption is supported by finding insufficient excretion of calcium (women: < 35 mg/24 h; men: < 55 mg/ 24 h) in a 24 h urinary collection.

Table 3 Potential mechanisms of micronutrient deficiencies

Altered intestinal absorption or storage of micronutrients
Deficiency of bile acid or reduced intraluminal bile acids
Inadequate Oral intake or utilization of micronutrients
Small intestinal bacterial overgrowth
Alcohol intake
Dietary herbal supplements (*Hypericum perforatum* **or St. John's wort**)
Medications (diuretics or the antibiotic metronidazole)

Table 4 Signs and symptoms related to potential micronutrient deficiencies

Signs and symptoms	Micronutrient deficiency
Visual Symptoms	Vitamin A; Vitamin E; Thiamine; Copper
Neurologic Symptoms	Thiamine; Niacin; Vitamin B6; Vitamin B12; Vitamin E; Copper
Edema	Thiamine; Niacin; Selenium
Skin Disorders	Vitamin A; Vitamins B2; Niacin; Vitamin B6; Zinc
Refractory Vitamin D Deficiency	Parathyroid Hormone
Bleeding Disorder	Vitamin K (Prothrombin Time)
Anemia	Vitamin A; Vitamin B12; Folic Acid; Vitamin E; Ferritin; Copper; Zinc

3.9 Postoperative Screening for Micronutrient Deficiencies

Further work is required to support the notion that biochemical blood testing to look for micronutrient deficiencies prevents the development of clinical symptoms and syndromes caused by micronutrient deficiencies. For at risk individuals, it is therefore important to investigate the potential causes of symptoms. As summarized in Table 4, if an individual presents with specific signs and symptoms after vertical sleeve gastrectomy, blood testing may be useful to identify a micronutrient deficiency. In an individual with visual symptoms, blood testing should include vitamin A, vitamin E, vitamin B1, and copper. In an individual with neurological symptoms, blood testing should include vitamin B1, niacin, vitamin B6, vitamin B12, vitamin E, and copper. In an individual with lower leg edema, blood testing should include vitamin B1, niacin, and selenium. In an individual with skin disorders, blood testing should include vitamin A, vitamin B2, niacin, vitamin B6, and zinc. In an individual with refractory vitamin D deficiency, blood testing should include parathyroid hormone and

alkaline phosphatase. In an individual with a bleeding disorder, blood testing should include vitamin K, a prothrombin time, and a complete blood count. In an individual with anemia, blood testing should include vitamin A, vitamin B12, folic acid, vitamin E, ferritin, zinc, and copper.

Intermittent blood tests should be considered after vertical sleeve gastrectomy. The suggested frequency of testing by the American Society for Metabolic and Bariatric Surgery is at 3 months, 6 months, and 12 months after bariatric surgery, and then to measure micronutrients once a year (Parrott et al. 2017). These postoperative micronutrient blood tests could include vitamin B1, vitamin B12, folic acid, iron status, and vitamin D (Parrott et al. 2017). Blood testing for vitamin A is recommended within the first year after bariatric surgery (Parrott et al. 2017).

3.10 Micronutrients Deficiencies After Vertical Sleeve Gastrectomy

Biochemical nutritional deficiencies have been evaluated in patients after vertical sleeve gastrectomy. Early (</= 1 year) biochemical deficiencies

Fig. 4 Short term micronutrient deficiencies. From multiple studies (Damms-Machado et al. 2012; Aarts et al. 2011; Toh et al. 2009), the range of percentages of individuals who </= 1 year after vertical sleeve gastrectomy were found to have deficiencies of fat soluble vitamin D, water soluble vitamins (vitamins B1, B6, B9, and B12), and mineral (iron)

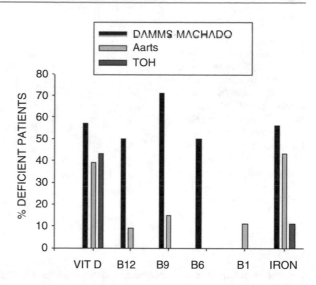

of vitamin D, vitamin B12, folic acid, vitamin B6, vitamin B1, and iron have been reported by multiple groups (see Fig. 4) (Damms-Machado et al. 2012; Aarts et al. 2011; Toh et al. 2009). These early micronutrient deficiencies after vertical sleeve gastrectomy have been confirmed by recent studies. In a study from China, at 1 year after vertical sleeve gastrectomy, despite the use of routine multivitamin supplements, significant biochemical deficiencies of vitamin D, vitamin B1, vitamin B6, folic acid, vitamin C, and iron status were identified (Guan et al. 2018). In a study from the United States, at 1 year after vertical sleeve gastrectomy, biochemical deficiencies of vitamin D (5.2%), vitamin A (9.4%), and vitamin B1 (10.5%) were reported (Johnson et al. 2019).

One year after vertical sleeve gastrectomy, significant bone loss and bone remodeling can develop (Nogués et al. 2010). Nogués and associates measured bone mineral density and bone remodeling markers in 15 women following vertical sleeve gastrectomy and found significant loss of bone mass and marked bone remodeling (Nogués et al. 2010). These findings support ongoing aggressive prevention and treatment of vitamin D deficiency by providing sufficient

vitamin D supplement and daily oral calcium both preoperatively and after vertical sleeve gastrectomy.

In another study from the United States, biochemical thiamine deficiency was identified in 25.7% of individuals within 1 year after vertical sleeve gastrectomy (Tang et al. 2018). These findings are consistent with the multiple case reports describing the development of Wernicke's encephalopathy (which presents with a combination of ophthalmoplegia/nystagmus; ataxia; and/or encephalopathy/confusion) after vertical sleeve gastrectomy. Indeed, both periodic measurement of postoperative thiamine status (for at least 6 months) and providing parenteral thiamine supplementation have been suggested by the European Federation of Neurological Societies (Galvin et al. 2010).

Late (2–3 years) biochemical deficiencies of iron, zinc, vitamin D, vitamin B12, folic acid, and vitamin B1 have been reported by multiple groups (see Fig. 5) (Pech et al. 2012; Saif et al. 2012; Gehrer et al. 2010; Moizé et al. 2013). These late biochemical deficiencies are less likely due to ongoing weight loss. It is presently unclear whether micronutrient deficiencies after vertical sleeve gastrectomy could be related to poor

Fig. 5 Long term micronutrient deficiencies. From multiple studies (Pech et al. 2012; Saif et al. 2012; Gehrer et al. 2010; Moizé et al. 2013), the range of percentages of individuals who 2–3 years after vertical sleeve gastrectomy were found to have deficiencies of fat soluble vitamin D, water soluble vitamins (vitamins B1, B9, and B12), and minerals (iron and zinc)

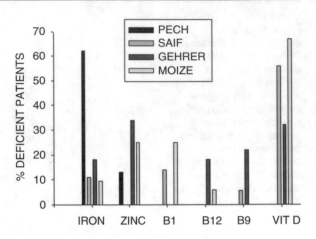

patient compliance, inadequate postoperative micronutrient supplementation, postoperative hypochlorhydria (as the result of resection of stomach), or small intestinal malabsorption (likely related to small intestinal bacterial overgrowth) (Sharbaugh et al. 2017).

3.11 Repletion of Micronutrient Deficiencies

Table 5 summarizes the treatment of micronutrient deficiencies, adapted from nutritional guidelines (Parrott et al. 2017; Bal et al. 2012).

3.11.1 Fat Soluble Vitamins

Vitamin A Treatment of vitamin A deficiency involves oral vitamin A supplementation taken with a meal (10,000–25,000 IU daily depending upon the severity of symptoms) (Bal et al. 2012). Blood levels of vitamin A should be followed due to the risks of hypervitaminosis A. Vitamin A toxicity has not been reported with use of oral β carotene (pre-vitamin A analogue) (Bal et al. 2012).

Vitamin D Vitamin D deficiency should initially be treated with either Vitamin D3 3000–6000 IU daily by mouth or oral Vitamin D2 or ergocalciferol 50,000 IU once weekly taken with a meal (Bal et al. 2012). To confirm repletion,

determination of 25-hydroxy Vitamin D should be repeated 2–3 months later. Some clinician-investigators have suggested then switching patients to Vitamin D3 (cholecalciferol) 2000–4000 IU taken with meals daily in preoperative individuals. After repletion of vitamin D deficiency in postoperative patients, more recent guidelines have suggested supplementation with at least 3000 units/day of Vitamin D (Parrott et al. 2017). In individuals with continued low 25-hydroxy Vitamin D levels, identification of an elevated parathyroid hormone level will support the need for aggressive supplementation with calcium and vitamin D.

Vitamin E Treatment of Vitamin E deficiency involves oral Vitamin E supplementation taken with a meal (100–400 IU daily depending upon the severity of symptoms) (Bal et al. 2012). Individuals can generally tolerate oral doses of 400–1000 mg of daily Vitamin E (1 IU is 0.67 mg of Vitamin E).

3.11.2 Water Soluble Vitamins

Vitamin B1 Mild to moderate symptoms of thiamine deficiency can be treated with 100 mg of oral thiamine 2–3 times daily or with 100–200 mg of intramuscular thiamine once monthly (given by deep injection into the gluteus muscle). When symptoms of acute psychosis or Wernicke encephalopathy (which can present either with a

Table 5 Treatment of micronutrient deficiencies after vertical sleeve gastrectomy[a]

Micronutrient	Suggested repletion therapy[b]
Vitamin A	PO Vit A 10,000–25,000 IU/D
Vitamin D	PO Vit D2 50,000 IU/week or Vit D3 3000–6000 IU/D
Vitamin E	PO Vit E 100–400 IU/D
Vitamin B1	PO Vit B1 100 mg 2–3 times/D or IM 100–200 mg/month
Severe Signs/ Symptoms	IV 200 mg 3 times/D 3–5 days then 200 mg/D 3–5 days or IM Vit B1 200 mg/D for 6–10 days
Vitamin B3	Niacin PO 100–500 mg 3 times/D
Vitamin B6	PO 30 mg/D
Vitamin B9	Folic Acid PO 1–2 mg/D
Vitamin B12	PO Vit B12 1000 mcg/D
Severe Sign/ Symptoms	IM Vit B12 1 mg/D up to 30 days; 1 mg weekly up to 1 Yr
Vitamin C	PO Vit C or Ascorbic Acid 200–500 mg/D
Zinc	PO Zinc Sulfate 220 mg or Zinc Gluconate 50 mg QOD
Copper	PO Copper Gluconate 2–6 mg/D
Selenium	PO Sodium Selenite 100 mcg/D
Magnesium	PO Mg Citrate 175–250 mg/D; Mg Oxide 250–400 mg/D
Iron	PO Iron Sulfate or Iron Gluconate 325 mg 1–3 times/D
Calcium	PO Calcium Carbonate 1.5 g 2–3 times/D

[a]Adapted using Nutritional Guideline from Stroh et al. (2017) and in Tamboli et al. (2016)
[b]*IU* International Units, *PO* Oral Route, *IV* Intravenous Route, *IM* Intramuscular Route, *Vit* Vitamin, *D* Day, *QOD* Every Other Day, *Yr* Year

triad that consists of encephalopathy/confusion, ataxia, and ophthalmoplegia/nystagmus or with blindness) are recognized, these situations are medical emergencies. Hospitalization with supportive care must be considered and the patient should immediately receive high dose parenteral thiamine (200 mg three times daily for 3–5 days by intravenous infusion given over 3–4 h followed by 200 mg daily by intravenous infusion for 3–5 days; or 200 mg daily by deep intramuscular injection into the gluteus muscle for 6–10 days) as summarized in Table 5 (Parrott et al. 2017; Bal et al. 2012).

Vitamin B3 The treatment of vitamin B3 or niacin deficiency involves oral niacin (not the slow-release formulation) 100–500 mg taken three times a day (Parrott et al. 2017). Flushing can be a common adverse effect of this treatment.

Vitamin B9 The treatment of vitamin B9 or folic acid deficiency involves oral folic acid 1–2 mg daily (Bal et al. 2012). Due to an increased

number of pregnancies that have occurred after bariatric surgery, the American College of Obstetrics and Gynecology recommends that women who have undergone bariatric surgery receive counseling prior to conception as well as prenatal treatment of any Vitamin B12, calcium, iron, or folic acid deficiency. Since neural tube defects can occur, it is reasonable to provide 1 mg of oral folic acid daily to women of child bearing age after they undergo bariatric surgery.

Vitamin B12 Mild signs and symptoms of vitamin B12 deficiency can be treated with oral vitamin B12 1000 mcg by mouth daily or intramuscular vitamin B12 1000 mcg once monthly (Bal et al. 2012). Treatment of potential neurological symptoms of subacute combined degeneration should involve parenteral vitamin B12 (1000 mcg daily by intramuscular injection for up to 30 days followed by 1000 mcg intramuscularly once weekly for up to 1 year) as summarized in Table 5.

Vitamin C Treatment of Vitamin C deficiency involves oral Vitamin C or ascorbic acid 200–500 mg daily (Bal et al. 2012).

3.11.3 Minerals

Iron Iron deficiency can be treated with ferrous sulfate 325 mg or ferrous gluconate 325 mg by mouth taken before a meal up to three times daily (Bal et al. 2012). In patients who have an inadequate response to oral iron supplements, parenteral iron is occasionally considered. Large doses of unnecessary iron supplements can produce acquired iron overload and so patients need to be monitored by appropriate blood testing while receiving oral iron supplements.

3.11.4 Trace Elements
When providing trace element supplements to be taken orally, there is a narrow safety window between the risks of deficiency versus the risks of toxicity.

Copper Treatment of hypocupremia involves copper gluconate 2–6 mg daily taken orally before a meal (Bal et al. 2012). Since copper is also present in most multivitamins containing minerals, serum or plasma copper levels should be rechecked frequently until repletion has been confirmed. The daily copper dose can then be reduced. If it is unclear whether the patient has copper deficiency, a 24 h urine collection to confirm decreased copper excretion can be an important confirmatory test. With our collaborators, we have reported a patient who was found to have hypercupremia following vertical sleeve gastrectomy (Koch et al. 2019). The origin of postoperative hypercupremia in this patient appeared to be related to the hepatic production of ceruloplasmin (Koch et al. 2019).

Zinc Low serum or plasma zinc levels are treated with oral zinc sulfate 220 mg or zinc gluconate (50 mg elemental zinc) taken orally before a meal every other day (Bal et al. 2012).

Selenium Low serum or plasma levels of selenium are treated with oral sodium selenite 100 mcg daily (Bal et al. 2012).

4 Outcomes After Vertical Sleeve Gastrectomy in Morbidly Obese Individuals

4.1 Long Term Mortality in Morbidly Obese Individuals After Vertical Sleeve Gastrectomy

Due to its more recent introduction, mortality studies after vertical sleeve gastrectomy have shorter follow up periods compared to Roux-en-Y gastric bypass surgery. However, similar to gastric bypass, survival advantages have been reported after vertical sleeve gastrectomy. A national study of middle aged men and women from Israel followed patients for 2.2–4.1 years after vertical sleeve gastrectomy (Reges et al. 2018). There was a statistically significant decline in death rate after vertical sleeve gastrectomy (2.4/1000 person-years) compared to the matched control group (4.2/1000 person-years) (Reges et al. 2018). In a multicenter study of chiefly middle aged women from France, patients after vertical sleeve gastrectomy were followed for a mean of 6.8 years (Thereaux et al. 2019). Mortality was statistically significantly lower in patients after vertical sleeve gastrectomy (2.32/1000 person-years) compared to the matched control group (5.85/1000 person-years) (Thereaux et al. 2019). It is quite reassuring that the reported mortality rates after vertical sleeve gastrectomy were very similar in these two studies.

4.2 Long Term Weight Loss in Morbidly Obese Individuals After Vertical Sleeve Gastrectomy

Reports of weight loss after vertical sleeve gastrectomy are summarized in Table 6. In

Table 6 Long term weight loss after vertical sleeve gastrectomy

Type of study[a]	Reference	Follow up	Result[b]
SCS	Noel et al. (2017)	8 years	%EWL: 67
SCS	Kowalewski et al. (2018)	8 years	%EWL: 51.1
MCS	Chang et al. (2018)	10 years	%EWL: 70.5
MCS	Aminian et al. (2016)	5 years	%TWL: 16.8
SCS	Misra et al. (2019)	5 years	%TWL: 26.0
SCS	Pucci et al. (2018)	2 years	%TWL: 20.6

[a]*SCS* Single-Center Study, *MCS* Multi-Center Study
[b]%EWL: Mean Percentage Excess Weight Loss; %TWL: Mean Percentage Total Weight Loss

examination of Table 6, significant weight loss has been reported from both single center studies and multicenter studies. Maintenance of weight loss after vertical sleeve gastrectomy has been demonstrated in studies lasting up to 10 years. There is variability in these findings after vertical sleeve gastrectomy. Variability could result from reports published from single center studies versus multicenter studies. In addition, there is not a uniform length of postoperative follow up (Noel et al. 2017; Kowalewski et al. 2018; Chang et al. 2018; Aminian et al. 2016; Misra et al. 2019; Pucci et al. 2018).

Of concern, there is a report that after vertical sleeve gastrectomy, individuals who are 60 years-old or older have lower excess weight loss at 12 months after surgery and have lower remission rate for type 2 diabetes mellitus compared to younger patients (Faucher et al. 2019).

4.3 Long Term Control of Diabetes Mellitus After Vertical Sleeve Gastrectomy

There is also variability in the results that describe control of type 2 diabetes after vertical sleeve gastrectomy. It is difficult from the published reports to fully delineate the factors that are important for achieving remission in individuals with obesity and type 2 diabetes. These studies include obese individuals in whom type 2 diabetes is treated with a diet and activity program, oral medications, or insulin. In addition, there are different ranges of preoperative body mass indices.

Table 7 summarizes remission (as defined by discontinuation of medications for treatment) of type 2 diabetes in obese individuals who have undergone vertical sleeve gastrectomy (Schauer et al. 2017; Kowalewski et al. 2018; Chang et al. 2018; Aminian et al. 2016; Misra et al. 2019; Pucci et al. 2018; Zetu et al. 2018; Abbatini et al. 2013). By contrast, for the treatment of individuals with type 1 diabetes mellitus, vertical sleeve gastrectomy does not appear to be an effective therapy (Al Sabah et al. 2017).

As seen in Table 7, maintenance of remission of type 2 diabetes can last up to 8 years postoperatively. Individuals who lose more weight postoperatively are more likely to obtain remission of type 2 diabetes. Upon examination of Tables 6 and 7, in those references with lower levels of weight loss (Kowalewski et al. 2018; Aminian et al. 2016) the percentage of individuals achieving remission of type 2 diabetes were 11% and 43%, respectively. However, in those references with higher levels of weight loss (Chang et al. 2018; Misra et al. 2019), the percentage of individuals achieving remission of type 2 diabetes were 37% and 66.6%, respectively. This is an important issue because postoperative bariatric patients have decreased incidence of macrovascular and microvascular complications, compared to a control group (Sjöström et al. 2014).

Preliminary work has suggested that scoring systems (DiaBetter and DiaRem) may be useful in predicting which obese individuals with type 2 diabetes may achieve remission after vertical sleeve gastrectomy (Pucci et al. 2018). The DiaRem scoring system includes hemoglobin A1C, age, and diabetes medications, while the

Table 7 Long term control of diabetes mellitus after vertical sleeve gastrectomy

Type of study[a]	Reference	Follow up	Result[b]
SCS	Kowalewski et al. (2018)	8 years	NoRMRxDM: 43.4%
SCS	Chang et al. (2018)	8 years	NoRMRxDM: 37%
MCS	Aminian et al. (2016)	6 years	NoRMRxDM: 11%
SCS	Misra et al. (2019)	5 years	NoRMRxDM: 66.6%
SCS	Pucci et al. (2018)	2 years	NoRMRxDM: 68.6%
NPBCS	Thereaux et al. (2018)	6 years	NoRMRxDM: 41%
MCS	Zetu et al. (2018)	5 years	NoRMRxDM: 63.6%
MCS	Schauer et al. (2017)	5 years	NoRMRxDM: 25%
SCS	Abbatini et al. (2013)	5 years	NoRMRxDM: 76.9%

[a]*SCS* Single-Center Study, *MCS* Multi-Center Study, *NPBCS* Nationwide population-based cohort study
[b]NoRMRxDM: No Requirement for Medical Therapy for Diabetes Mellitus

DiaBetter scoring system includes hemoglobin A1C, age, diabetes medications as well as diabetes duration. This notion certainly deserves further investigation.

There is a report suggesting that liraglutide treatment may be of benefit in obese individuals who have persistent or recurrent type 2 diabetes after undergoing vertical sleeve gastrectomy (Miras et al. 2019). A long-term randomized trial would be required to determine whether this proposed medical therapy would provide an outcome similar to available surgical options, which include proceeding to a duodenal switch as a second stage operation or revision of the sleeve gastrectomy to Roux-en-Y gastric bypass.

5 Conclusions

Obesity is a major factor in the worldwide rise in the prevalence of type 2 diabetes mellitus. Laparoscopic vertical sleeve gastrectomy is a newer bariatric surgical procedure with a lower risk of complications. In obese individuals with type 2 diabetes, significant weight loss occurs after vertical sleeve gastrectomy and weight loss is maintained for at least 10 years. There are reports of 11–76.9% of obese individuals with type 2 diabetes discontinuing their use of diabetic medications in studies lasting up to 8 years after vertical sleeve gastrectomy. Major ongoing issues include the preoperative determination of the suitability of diabetic patients to undergo this bariatric surgical procedure. Postoperatively, specific macronutrient goals and micronutrient supplements are important for successful and safer clinical results. Vertical sleeve gastrectomy is a component of a long term, organized program directed at treating diabetes related to obesity. Further work is needed to determine the best options for management of obese individuals who do not achieve remission of type 2 diabetes after vertical sleeve gastrectomy.

Conflicts of Interest Statement TR Koch and TR Shope have no conflict of interest to report regarding this manuscript.

References

Aarts EO, Janssen IM, Berends FJ (2011) The gastric sleeve: losing weight as fast as micronutrients? Obes Surg 21(2):207–211

Abbatini F, Capoccia D, Casella G, Soricelli E, Leonetti F, Basso N (2013) Long-term remission of type 2 diabetes in morbidly obese patients after sleeve gastrectomy. Surg Obes Relat Dis 9(4):498–502

Abou Rached A, Basile M, El Masri H (2014) Gastric leaks post sleeve gastrectomy: review of its prevention and management. World J Gastroenterol 20 (38):13904–13910

Adil MT, Aminian A, Bhasker AG, Rajan R, Corcelles R, Zerrweck C et al (2019) Perioperative practices concerning sleeve gastrectomy – a survey of 863 surgeons with a cumulative experience of 520,230 procedures. Obes Surg. https://doi.org/10.1007/s11695-019-04195-7

Al Sabah S, Al Haddad E, Muzaffar TH, Almulla A (2017) Laparoscopic sleeve gastrectomy for the management of Type 1 diabetes mellitus. Obes Surg 27 (12):3187–3193

Ali M, El Chaar M, Ghiassi S, Rogers AM (2017) American Society for Metabolic and Bariatric Surgery Clinical Issues Committee. American Society for Metabolic and Bariatric Surgery updated position statement on sleeve gastrectomy as a bariatric procedure. Surg Obes Relat Dis 13(10):1652–1657

Aminian A, Brethauer SA, Andalib A, Punchai S, Mackey J, Rodriguez J et al (2016) Can sleeve gastrectomy "cure" diabetes? Long-term metabolic effects of sleeve gastrectomy in patients with type 2 diabetes. Ann Surg 264(4):674–681

Angrisani L, Santonicola A, Iovino P, Vitiello A, Zundel N, Buchwald H et al (2017) Bariatric surgery and endoluminal procedures: IFSO worldwide survey 2014. Obes Surg 27(9):2279–2289

Archid R, Archid N, Meile T, Hoffmann J, Hilbert J, Wulff D et al (2019) Patients with schizophrenia do not demonstrate worse outcome after sleeve gastrectomy: a short-term cohort study. Obes Surg 29(2):506–510

Bailly L, Schiavo L, Sebastianelli L, Fabre R, Morisot A, Pradier C et al (2019) Preventive effect of bariatric surgery on type 2 diabetes onset in morbidly obese inpatients: a national French survey between 2008 and 2016 on 328,509 morbidly obese patients. Surg Obes Relat Dis 15(3):478–487

Bal BS, Koch TR, Finelli FC, Sarr MG (2010) Management of medical and surgical disorders after divided Roux-en-Y gastric bypass surgery. Nat Rev Gastroenterol Hepatol 7(6):320–334

Bal BS, Finelli FC, Shope TR, Koch TR (2012) Nutritional deficiencies after bariatric surgery. Nat Rev Endocrinol 8:544–556

Brethauer SA, Chang J, Galvao Neto M, Greve JW (2016) Gastrointestinal devices for the treatment of type 2 diabetes. Surg Obes Relat Dis 12(6):1256–1261

Carrodeguas L, Kaidar-Person O, Szomstein S, Antozzi P, Rosenthal R (2005) Preoperative thiamine deficiency in obese population undergoing laparoscopic bariatric surgery. Surg Obes Relat Dis 1(6):517–522

Chang DM, Lee WJ, Chen JC, Ser KH, Tsai PL, Lee YC (2018) Thirteen-year experience of laparoscopic sleeve gastrectomy: surgical risk, weight loss, and revision procedures. Obes Surg 28(10):2991–2997

Damms-Machado A, Friedrich A, Kramer KM, Stingel K, Meile T, Küper MA et al (2012) Pre- and postoperative nutritional deficiencies in obese patients undergoing laparoscopic sleeve gastrectomy. Obes Surg 22(6):881–889

de Luis DA, Pacheco D, Izaola O, Terroba MC, Cuellar L, Cabezas G (2013) Micronutrient status in morbidly obese women before bariatric surgery Surg Obes Relat Dis 9(2):323–327

Faucher P, Aron-Wisnewsky J, Ciangura C, Genser L, Torcivia A, Bouillot JL et al (2019) Changes in body composition, comorbidities, and nutritional status associated with lower weight loss after bariatric surgery in older subjects. Obes Surg. https://doi.org/10.1007/s11695-019-04037-6

Ferraz ÁAB, Carvalho MRC, Siqueira LT, Santa-Cruz F, Campos JM (2018) Micronutrient deficiencies following bariatric surgery: a comparative analysis between sleeve gastrectomy and Roux-en-Y gastric bypass. Rev Col Bras Cir 45(6):e2016

Fildes A, Charlton J, Rudisill C, Littlejohns P, Prevost AT, Gulliford MC (2015) Probability of an obese person attaining normal body weight: cohort study using electronic health records. Am J Public Health 105(9):e54–e59

Flancbaum L, Belsley S, Drake V, Colarusso T, Tayler E (2006) Preoperative nutritional status of patients undergoing Roux-en-Y gastric bypass for morbid obesity. J Gastrointest Surg 10(7):1033–1037

Fuchs HF, Laughter V, Harnsberger CR, Broderick RC, Berducci M, DuCoin C et al (2016) Patients with psychiatric comorbidity can safely undergo bariatric surgery with equivalent success. Surg Endosc 30(1):251–258

Galvin R, Bråthen G, Ivashynka A, Hillbom M, Tanasescu R, Leone MA, EFNS (2010) EFNS guidelines for diagnosis, therapy and prevention of Wernicke encephalopathy. Eur J Neurol 17(12):1408–1418

Gehrer S, Kern B, Peters T, Christoffel-Courtin C, Peterli R (2010) Fewer nutrient deficiencies after laparoscopic sleeve gastrectomy (LSG) than after laparoscopic Roux-Y-gastric bypass (LRYGB)-a prospective study. Obes Surg 20(4):447–453

Global Burden of Disease 2015 Obesity Collaborators et al (2017) Health effects of overweight and obesity in 195 countries over 25 years. N Engl J Med 377(1):13–27

Guan B, Yang J, Chen Y, Yang W, Wang C (2018) Nutritional deficiencies in Chinese patients undergoing gastric bypass and sleeve gastrectomy: prevalence and predictors. Obes Surg 28(9):2727–2736

Guetta O, Vakhrushev A, Dukhno O, Ovnat A, Sebbag G (2019) New results on the safety of laparoscopic sleeve gastrectomy bariatric procedure for type 2 diabetes patients. World J Diabetes 10(2):78–86

Gutierrez-Blanco D, Romero Funes D, Castillo M, Lo Menzo E, Szomstein S, Rosenthal RJ (2019) Bariatric surgery reduces the risk of developing type 2 diabetes in severe obese subjects undergoing sleeve gastrectomy. Surg Obes Relat Dis 15(2):168–172

Heber D, Greenway FL, Kaplan LM, Livingston E, Salvador J, Still C (2010) Endocrine and nutritional management of the post-bariatric surgery patient: an Endocrine Society Clinical Practice Guideline. Jin Endocrinol Metab 95(11):4823–4843

Johnson LM, Ikramuddin S, Leslie DB, Slusarek B, Killeen AA (2019) Analysis of vitamin levels and deficiencies in bariatric surgery patients: a single-institutional analysis. Surg Obes Relat Dis 15(7):1146–1152

Kahn SE, Hull RL, Utzschneider KM (2006) Mechanisms linking obesity to insulin resistance and type 2 diabetes. Nature 444(7121):840–846

Kavanagh R, Smith J, Bashir U, Jones D, Avgenakis E, Nau P (2019) Optimizing bariatric surgery outcomes: a novel preoperative protocol in a bariatric population with gastroesophageal reflux disease. Surg Endosc. https://doi.org/10.1007/s00464-019-06934-4

Khorgami Z, Shoar S, Andalib A, Aminian A, Brethauer SA, Schauer PR (2017) Trends in utilization of bariatric surgery, 2010-2014: sleeve gastrectomy dominates. Surg Obes Relat Dis 13(5):774–778

Kimmons JE, Blanck HM, Tohill BC, Zhang J, Khan LK (2006) Associations between body mass index and the prevalence of low micronutrient levels among US adults. Med Gen Med 8(4):59

Koch TR, Shope TR, Gostout CJ (2017) Organization of future training in bariatric gastroenterology. World J Gastroenterol 23(35):6371–6378

Koch TR, Shope TR, Camilleri M (2018) Current and future impact of clinical gastrointestinal research on patient care in diabetes mellitus. World J Diabetes 9 (11):180–189

Koch TR, Zubowicz EA, Gross JB Jr (2019) Prolonged hypercupremia after laparoscopic vertical sleeve gastrectomy successfully treated with oral zinc. Case Rep GI Med. https://doi.org/10.1155/2019/8175376

Kovacs Z, Valentin JB, Nielsen RE (2017) Risk of psychiatric disorders, self-harm behaviour and service use associated with bariatric surgery. Acta Psychiatr Scand 135(2):149–158

Kowalewski PK, Olszewski R, Walędziak MS, Janik MR, Kwiatkowski A, Gałązka-Świderek N, Cichoń K, Brągoszewski J, Paśnik K (2018) Long-term outcomes of laparoscopic sleeve gastrectomy-a single-center, retrospective study. Obes Surg 28(1):130–134

Lemacks J, Wells BA, Ilich JZ, Ralston PA (2013) Interventions for improving nutrition and physical activity behaviors in adult African American populations: a systematic review, January 2000 through December 2011. Prev Chronic Dis 10:E99

Lewandowski H, Breen TL, Huang EY (2007) Kwashiorkor and an acrodermatitis enteropathica-like eruption after a distal gastric bypass surgical procedure. Endocr Pract 13(3):277–282

Loo GH, Rajan R, Nik Mahmood NRK (2019) Staple-line leak post primary sleeve gastrectomy. A two patient case series and literature review. Ann Med Surg (Lond) 44:72–76

Lovette AS, Shope TR, Koch TR (2012) Origins for micronutrient deficiencies. In: Huang C-K (ed) Bariatric surgery. In Tech d.o.o, Rijeka, pp 229–254

Miras AD, Pérez-Pevida B, Aldhwayan M, Kamocka A, McGlone ER, Al-Najim W et al (2019) Adjunctive liraglutide treatment in patients with persistent or recurrent type 2 diabetes after metabolic surgery (GRAVITAS): a randomised, double-blind, placebo-controlled trial. Lancet Diabetes Endocrinol 7 (7):549–559

Misra S, Bhattacharya S, Saravana Kumar S, Nandhini BD, Saminathan SC, Praveen RP (2019) Long-term outcomes of laparoscopic sleeve gastrectomy from the Indian subcontinent. Obes Surg. https://doi.org/10.1007/s11695-019-04103-z

Moizé V, Andreu A, Flores L, Torres F, Ibarzabal A, Delgado S et al (2013) Long-term dietary intake and nutritional deficiencies following sleeve gastrectomy or Roux-En-Y gastric bypass in a mediterranean population. J Acad Nutr Diet 113(3):400–410

Nath A, Yewale S, Tran T, Brebbia JS, Shope TR, Koch TR (2016) Dysphagia after vertical sleeve gastrectomy: evaluation of risks factors and assessment of endoscopic intervention. World J Gastroenterol 22 (47):10371–10379

Nath A, Tran T, Shope TR, Koch TR (2017) Prevalence of clinical thiamine deficiency in individuals with medically complicated obesity. Nutr Res 37(1):29–36

NCD Risk Factor Collaboration (2016a) Trends in adult body-mass index in 200 countries from 1975 to 2014: a pooled analysis of 1698 population-based measurement studies with 19.2 million participants. Lancet 387(10026):1377–1396

NCD Risk Factor Collaboration (2016b) Worldwide trends in diabetes since 1980: a pooled analysis of 751 population-based studies with 4.4 million participants. Lancet 387(10027):1513–1530

Noel P, Nedelcu M, Eddbali I, Manos T, Gagner M (2017) What are the long-term results 8 years after sleeve gastrectomy? Surg Obes Relat Dis 13(7):1110–1115

Nogués X et al (2010) Bone mass loss after sleeve gastrectomy: a prospective comparative study with gastric bypass. Cir Esp 88:103–109

Park JY, Heo Y, Kim YJ, Park JM, Kim SM, Park DJ et al (2019) Long-term effect of bariatric surgery versus conventional therapy in obese Korean patients: a multicenter retrospective cohort study. Ann Surg Treat Res 96(6):283–289

Parrott J, Frank L, Rabena R, Craggs-Dino L, Isom KA, Greiman L (2017) American Society for metabolic and bariatric surgery integrated health nutritional guidelines for the surgical weight loss patient 2016 update: micronutrients. Surg Obes Relat Dis 13 (5):727–741

Pech N, Meyer F, Lippert H, Manger T, Stroh C (2012) Complications, reoperations, and nutrient deficiencies two years after sleeve gastrectomy. J Obesity 2012. https://doi.org/10.1155/2012/828737

Petersen WV, Meile T, Küper MA, Zdichavsky M, Königsrainer A, Schneider JH (2012) Functional importance of laparoscopic sleeve gastrectomy for the lower esophageal sphincter in patients with morbid obesity. Obes Surg 22(3):360–366

Popov VB, Ou A, Schulman AR, Thompson CC (2017) The impact of intragastric balloons on obesity-related co-morbidities: a systematic review and meta-analysis. Am J Gastroenterol 112(3):429–439

Pucci A, Batterham RL (2019) Mechanisms underlying the weight loss effects of RYGB and SG: similar, yet different. J Endocrinol Investig 42(2):117–128

Pucci A, Tymoszuk U, Cheung WH, Makaronidis JM, Scholes S, Tharakan G et al (2018) Type 2 diabetes remission 2 years post Roux-en-Y gastric bypass and sleeve gastrectomy: the role of the weight loss and comparison of DiaRem and DiaBetter scores. Diabet Med 35(3):360–367

Ramirez Prada D, Delgado G, Hidalgo Patino CA, Perez-Navero J, Gil Campos M (2011) Using of WHO guidelines for the management of severe malnutrition to cases of marasmus and kwashiorkor in a Colombia children's hospital. Nutr Hosp 26(5):977–983

Rashti F, Gupta E, Ebrahimi S, Shope TR, Koch TR, Gostout CJ (2014) Development of minimally invasive techniques for management of medically-complicated obesity. World J Gastroenterol 20(37):13424–13445

Reges O, Greenland P, Dicker D, Leibowitz M, Hoshen M, Gofer I, Rasmussen-Torvik LJ, Balicer RD (2018) Association of bariatric surgery using laparoscopic banding, Roux-en-Y gastric bypass, or laparoscopic sleeve gastrectomy vs usual care obesity management with all-cause mortality. JAMA 319 (3):279–290

Rubino F, Nathan DM, Eckel RH, Schauer PR, Alberti KG, Zimmet PZ, Del Prato S, Ji L, Sadikot SM, Herman WH, Amiel SA, Kaplan LM, Taroncher-Oldenburg G, Cummings DE (2016) Delegates of the 2nd Diabetes Surgery Summit. Metabolic surgery in the treatment algorithm for type 2 diabetes: a joint statement by international diabetes organizations. Surg Obes Relat Dis 12(6):1144–1162

Ruz M, Carrasco F, Rojas P, Codoceo J, Inostroza J, Basfi-Fer K et al (2012) Heme- and nonheme-iron absorption and iron status 12 mo after sleeve gastrectomy and Roux-en-Y gastric bypass in morbidly obese women. Nutr 96(4):810–817

Saif T, Strain GW, Dakin G, Gagner M, Costa R, Pomp A (2012) Evaluation of nutrient status after laparoscopic sleeve gastrectomy 1, 3, and 5 years after surgery. Surg Obes Relat Dis 8(5):542–547

Schauer PR, Bhatt DL, Kirwan JP, Wolski K, Aminian A, Brethauer SA et al (2017) STAMPEDE Investigators. Bariatric surgery versus intensive medical therapy for diabetes – 5-year outcomes. N Engl J Med 376 (7):641–651

Sharbaugh ME, Shope TR, Koch TR (2017) Upper gut bacterial overgrowth is a potential mechanism for glucose malabsorption after vertical sleeve gastrectomy. New Insights Obes Gent Beyond 1.030–035. https://doi.org/10.29328/journal.hodms.1001006

Sjöström L, Peltonen M, Jacobson P, Ahlin S, Andersson-Assarsson J, Anveden Å et al (2014) Association of bariatric surgery with long-term remission of type 2 diabetes and with microvascular and macrovascular complications. JAMA 311(22):2297–2304

Spadola CE, Wagner EF, Dillon FR, Trepka MJ, De La Cruz-Munoz N, Messiah SE (2015) Alcohol and drug use among postoperative bariatric patients: a systematic review of the emerging research and its implications. Alcohol Clin Exp Res 39(9): 1582–1601

Stroh C, Manger T, Benedix F (2017) Metabolic surgery and nutritional deficiencies. Minerva Chir 72 (5):432–441

Svane MS, Bojsen-Møller KN, Martinussen C, Dirksen C, Madsen JL, Reitelseder S, Holm L, Rehfeld JF, Kristiansen VB, van Hall G, Holst JJ, Madsbad S (2019) Postprandial nutrient handling and gastrointestinal hormone secretion after Roux-en-Y gastric bypass vs sleeve gastrectomy. Gastroenterology 156 (6):1627–1641

Svetkey LP, Stevens VJ, Brantley PJ et al (2008) Comparison of strategies for sustaining weight loss: the weight loss maintenance randomized controlled trial. JAMA 299:1139–1148

Tamboli RA, Sidani RM, Garcia AE, Antoun J, Isbell JM, Albaugh VL, Abumrad NN (2016) Jejunal administration of glucose enhances acyl ghrelin suppression in obese humans. Am J Physiol Endocrinol Metab 311 (1):E252–E259

Tang L, Alsulaim HA, Canner JK, Prokopowicz GP, Steele KE (2018) Prevalence and predictors of postoperative thiamine deficiency after vertical sleeve gastrectomy. Surg Obes Relat Dis 14(7):943–950

Thereaux J, Lesuffleur T, Czernichow S, Basdevant A, Msika S, Nocca D et al (2018) Association between bariatric surgery and rates of continuation, discontinuation, or initiation of antidiabetes treatment 6 years later. JAMA Surg 153(6):526–533

Thereaux J, Lesuffleur T, Czernichow S, Basdevant A, Msika S, Nocca D et al (2019) Long-term adverse events after sleeve gastrectomy or gastric bypass: a 7-year nationwide, observational, population-based, cohort study. Lancet Diabetes Endocrinol. pii: S2213–8587(19)30191–3. https://doi.org/10.1016/S2213-8587(19)30191-3

Toh SY, Zarshenas N, Jorgensen J (2009) Prevalence of nutrient deficiencies in bariatric patients. Nutrition 25 (11–12):1150–1156

Trastulli S, Desiderio J, Guarino S, Cirocchi R, Scalercio V, Noya G, Parisi A (2013) Laparoscopic sleeve gastrectomy compared with other bariatric surgical procedures: a systematic review of randomized trials. Surg Obes Relat Dis 9(5):816–829

Valezi AC, Herbella FA, Mali-Junior J, Menezes MA, Liberatti M, Sato RO (2017) Preoperative manometry for the selection of obese people candidate to sleeve gastrectomy. Arq Bras Cir Dig 30(3):222–224

Vilar-Gomez E, Martinez-Perez Y, Calzadilla-Bertot L et al (2015) Weight loss through lifestyle modification significantly reduces features of nonalcoholic steatohepatitis. Gastroenterology 149:367–378

Warner DL, Sasse KC (2017) Technical details of laparoscopic sleeve gastrectomy leading to lowered leak rate: discussion of 1070 consecutive cases. Minim Invasive Surg 2017:4367059

Zaveri H, Surve A, Cottam D, Cottam A, Medlin W, Richards C, Belnap L, Cottam S, Horsley B (2018) Mid-term 4-year outcomes with single anastomosis duodenal-ileal bypass with sleeve gastrectomy surgery at a single US center. Obes Surg 28(10):3062–3072

Zetu C, Popa SG, Popa A, Munteanu R, Mota M (2018) Long-term improvement of glucose homeostasis and body composition in patients undergoing laparoscopic sleeve gastrectomy. Acta Endocrinol (Buchar) 14 (4):477–482

Adv Exp Med Biol - Advances in Internal Medicine (2020) 4: 321–330
https://doi.org/10.1007/5584_2020_511
© Springer Nature Switzerland AG 2020
Published online: 22 March 2020

Surgical Treatment of Type 2 Diabetes Mellitus in Youth

Anna Zenno (iD) and Evan P. Nadler (iD)

Abstract

Bariatric surgery is currently the most effective weight loss treatment of severe obesity and its associated comorbidities and is being increasingly used to treat children and adolescents with severe obesity, including those with Type 2 Diabetes (T2D). This review focuses on the conventional management of T2D in children and adolescents, comparison of various types of bariatric surgeries, effect of bariatric surgery on gastrointestinal physiology and metabolism, current literature on the use of bariatric surgery to treat youth with severe obesity and T2D, and the potential complications of bariatric surgery in this population.

Keywords

Bariatric surgery · Type 2 diabetes · Pediatric obesity · Vertical sleeve gastrectomy · Roux-en-Y gastric bypass · Adjustable gastric banding

A. Zenno (✉)
Division of Endocrinology, Children's National Health System, Washington, DC, USA
e-mail: azenno@childrensnational.org

E. P. Nadler
Division of Pediatric Surgery, Children's National Health System, Washington, DC, USA

The George Washington University School of Medicine & Health Sciences, Washington, DC, USA

1 Introduction

The incidence of type 2 diabetes (T2D) in youth (children and adolescents) has increased rapidly over the past 20 years and is expected to continue to rise in parallel with the global obesity epidemic as T2D is often a complication of obesity with over 85% of youth with T2D being overweight or obese at diagnosis (Pulgaron and Delamater 2014). In the United States, there are up to 5000 new cases of T2D in youth per year with economically disadvantaged racial/ethnic groups (African Americans, Hispanics, and Native American Indians) being most affected (Pettitt et al. 2014) and 1 of 5 adolescents (aged 12–18 years) have prediabetes (Andes et al. 2019). The data with respect to obesity are even more alarming: 35.1% of youth age 2–19 years are overweight (defined as BMI $\geq$ 85th percentile for age and sex); 19% have class I obesity (BMI $\geq$95th percentile); 6% have class II obesity (BMI $\geq$120% of the 95th percentile or BMI $\geq$35); and 2% have class III obesity (BMI $\geq$140% of the 95th percentile or BMI $\geq$40) per the National Health and Nutrition Examination Survey (NHANES), a nationally representative data source of over 3300 individuals (Skinner et al. 2018). Youth aged 12–19 are most affected by the disease, with a prevalence of overweight and obesity of 45% (Skinner et al. 2018). With no end in sight to reversing the obesity epidemic, it is likely that the number of new cases of T2D will continue to grow over the next several years. Thus, strategies to prevent and treat T2D in youth are of vital importance.

The diagnostic laboratory criteria for diabetes mellitus of any type in adults and youth is based on the presence of one or more of the following four determinants: fasting plasma glucose ≥126 mg/dL, random venous plasma glucose ≥200 mg/dL with classic symptoms of hyperglycemia such as polyuria and polydipsia, plasma glucose ≥200 mg/dL after 2-h oral glucose tolerance test, or glycated hemoglobin (A1C) ≥6.5% (American Diabetes A 2018a). Although there can be overlap of the clinical manifestations for Type 1 diabetes (T1D) and T2D, the diseases have different pathophysiology: T1D is characterized by autoimmune beta-cell failure leading to insulin deficiency whereas T2D is characterized by insulin resistance and gradual non-autoimmune beta-cell failure. Compared to T2D in adults, T2D in children is more virulent. There is a faster progressive decline in beta-cell function, more severe insulin resistance, and an accelerated development of diabetes complications in affected youth, such as microalbuminuria and retinopathy (American Diabetes A 2018b; Weiss et al. 2005; Gungor and Arslanian 2004). Moreover age of T2D onset is inversely associated with T2D complication risk and mortality, which further outlines why prevention and early intervention are crucial for children with T2D (Al-Saeed et al. 2016).

2   Current Management of T2D in Youth

Initial management of T2D consists of age-appropriate and culturally sensitive lifestyle modifications that focus on increasing daily physical activity and healthy eating. Lifestyle changes alone may reverse diabetes in a minority of patients, but the majority achieve only mild to moderate weight loss with reduction of BMI by −1 to −2 kg/m2, and unfortunately have difficulty implementing and sustaining their lifestyle

changes (Reinehr 2013). Consequently many patients with T2D are treated with pharmacotherapy in addition to behavioral and dietary modification, with variable benefit.

Currently, the only FDA-approved drugs for youth-onset T2D are metformin, insulin, and liraglutide, a glucagon-like peptide-1 analogue (GLP-1A) (Tamborlane et al. 2019). Metformin is the first line drug in asymptomatic children with A1C of <8.5% and normal renal function, and should be gradually increased from 500 mg to 2 g daily to maximize the clinical response (American Diabetes A 2018b). Metformin's mechanisms of action include: reduction of hepatic glucose production, opposition of glucagon action, and increase in insulin-mediated glucose uptake in peripheral tissues (Pernicova and Korbonits 2014). Its use is often limited by patient tolerance due to common side effects such as nausea, vomiting, diarrhea, and dyspepsia, which are frequently dose-limiting and may lead to the addition of liraglutide and/or insulin to the treatment regimen if A1C values are not well-controlled on Metformin alone. Liraglutide activates the GLP-1 receptor and increases insulin release from pancreatic beta cells in the presence of elevated glucose concentrations. Newer agents such as sodium-glucose cotransporter-2 (SGLT2) inhibitors and inhibitors of dipeptidyl peptidase 4 (DPP4 inhibitors) have also been used off-label in some youth in conjunction with metformin, but these drugs have not yet been FDA-approved for use in patients under the age of 18 (Tamborlane et al. 2019; Laffel et al. 2018; Tamborlane et al. 2018). SGLT2 inhibitors block the glucose reabsorption transporter in the proximal tubule of the kidney leading to increased glucosuria and decreased blood glucose. DPP4 inhibitors prevent the breakdown of endogenous GLP-1 and glucose-dependent insulinotropic polypeptide (GIP) leading to increased glucose-mediated insulin secretion, suppressed glucagon secretion, and decreased gastric emptying.

According to the Treatment Options for Type 2 Diabetes in Adolescents and Youth (TODAY) study, metformin alone provided durable glycemic control (A1C less than or equal to 8%

for 6 months) in only about half of the subjects enrolled (Group TS et al. 2012). Thus, given the challenges of managing diabetes with lifestyle changes and/or medications, an increasing number of youth with severe obesity and T2D are being treated with bariatric surgery with promising results.

3 Metabolic and Bariatric Surgery

The first bariatric surgical procedure was performed by Mason and Ito in an adult patient in 1967 and bariatric surgery has been performed in small cohorts of adolescents since the 1970s (Mason and Ito 1967). With the rapid rise in obesity, there has been an exponential increase in adolescent surgical cases since the early 2000s with 126 bariatric surgeries performed in 2012 and 220 cases performed in 2016 among patients aged greater than 16 years (Kyler et al. 2019).

In general, bariatric surgery mechanically restricts caloric intake by modifying the anatomy of the stomach and in some variants impairs nutrient absorption via alterations of the gastrointestinal tract. The three most commonly performed bariatric surgeries in adolescents are roux-en-Y gastric bypass (RYGB), vertical sleeve gastrectomy (VSG), and adjustable gastric banding (AGB) and they achieve weight loss by either restriction (AGB or VSG), or combined restriction and malabsorption (RYGB). Restrictive procedures limit intake by either creating a small gastric reservoir (VSG, RYGB) or provide a narrow outlet to delay emptying (AGB). Malabsorptive procedures (RYGB) bypass various portions of the small intestine where nutrient and calorie absorption takes place, and generally have a greater degree of weight loss when compared to restrictive procedures. Currently, the vast majority of bariatric procedures worldwide are performed laparoscopically because minimally invasive surgery is associated with shorter length of stay, lower overall complications, and decreased costs compared to open procedures. This approach is also used for adolescents and children.

National Surgical Quality Improvement Quality Program (NSQIP) data from 2005–2015 showed that the most commonly performed procedures were laparoscopic RYGB (43.4%), laparoscopic SG (32.5%) and laparoscopic AGB (18.7%) among 2,625 adolescents ages 18–21 with obesity (average BMI 48) in the United States (Arafat et al. 2019). However the AGB is not approved for patients under the age of 18, so younger adolescents are more often treated with VSG (U.S. Food and Drug Administration 2001).

4 Different Types of Bariatric Surgery

RYGB entails creating a small gastric pouch from the upper stomach that attaches to the esophagus at one end and a section of the small intestine at the other end (Fig. 1). It has been the procedure that has been most studied, and is currently the gold standard bariatric surgery in adults in the United States (Elder and Wolfe 2007). However, it is now the second most common surgical procedure performed for weight loss in adults, mostly due to its surgical complication risk (Angrisani et al. 2015). Postoperative procedure-related complications such as anastomotic leak, strictures, bleeding, bowel obstructions, and wound infections are common with similar prevalence in adults and adolescents (Inge et al. 2014) and frequently lead to hospital readmissions. The procedure is also associated with nutritional complications and dumping syndrome long-term, and therefore, RYGB is often not the preferred primary surgical procedure for adolescents.

Vertical sleeve gastrectomy involves removal of most of the greater curvature of the stomach (about 80–90%) and has been shown to have lower morbidity and mortality compared to RYGB (Fig. 1). Given its relative safety and simplicity, it is being increasingly used to treat adolescents with obesity. In 2010, 11.4% of cases were SG, 47.4% were laparoscopic RYGB and 30.9% were AGB and in 2015, 66.6% were SG, 28.6% were laparoscopic RYGB and 2.6% were AGB among adolescents aged 18–21 who

Fig. 1 Roux-en-Y gastric bypass and sleeve gastrectomy. (From Slomski 2017)

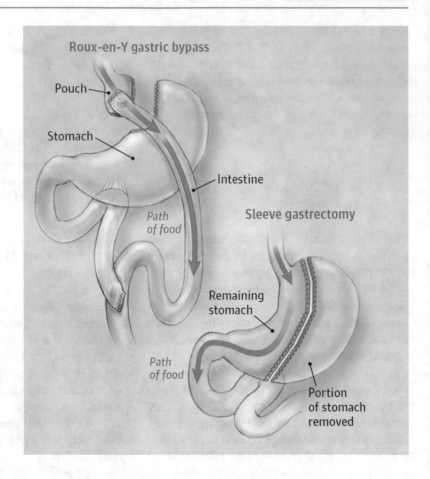

underwent bariatric surgery in the United States (Kyler et al. 2019).

AGB involves placing an adjustable silicone band around the stomach below the gastroesophageal junction to create a small pouch with a narrow outlet and is the least invasive of the purely restrictive bariatric surgery procedures with a lower complication rate than other surgical options. However as stated above, it is only FDA approved for patients age 18 and older which limits its use in younger adolescents.

5   Effect of Bariatric Surgery on GI Physiology and Metabolism

Bariatric surgery effectively treats obesity and improves obesity associated comorbidities such as T2D. Improved glucose homeostasis is evident

within days to weeks following bariatric surgery with acute calorie restriction, which suggests there is a weight loss-independent mechanism for VSG and RYGB. Proposed theories to explain this phenomenon that are under active investigation include: improved beta cell function due to reduced glucotoxicity, reduced islet inflammation, alteration of GI hormonal responses, and gut microbiota (Chen et al. 2019). The intravenous glucose tolerance test provides a way to assess beta cell function by measuring the insulin secretion response (also known as the disposition index) to a standardized dose of intravenous glucose regardless of the anatomical differences between surgical procedures (Hofso et al. 2019). At 1 year after VSG and RYGB, the disposition index increased by 6–8 times with no significant difference between groups in 109 patients supporting the theory that there is improved beta cell function following bariatric surgery; however, the mechanism of action is still not well understood (Hofso et al. 2019).

Both RYGB and VSG increase postprandial levels of distal GI hormones like GLP1 and peptide-YY via rapid nutrient delivery down the GI tract where enteroendocrine L-cells are located (Hutch and Sandoval 2017). It has been suggested that the rise in GLP-1 and peptide-YY subsequently leads to increased satiety, improved postprandial insulin secretion, and decreased hepatic glucose production, which may contribute to remission of T2D. In VSG, ghrelin levels also decrease with complete removal of the fundus of the stomach which is thought to suppress appetite (Chen et al. 2019). It is unclear whether RYGB or VSG leads to higher T2D remission with some studies suggesting no difference between the two (Hutch and Sandoval 2017) whereas others showing that RYGB is more effective (Schauer et al. 2017).

6 The Impact of Bariatric Surgery on Obesity and Type 2 Diabetes

The Teen-Longitudinal Assessment of Bariatric surgery (Teen-LABS) study, the largest prospective multi-center observational study to date of bariatric surgery in pediatric patients ($\leq$ 19 years of age, mean BMI of 53) showed that teenagers achieve similar degrees of weight loss, diabetes remission, and improvement of cardio-metabolic risk factors such as dyslipidemia, elevated blood pressure, renal dysfunction, and prediabetes for at least 3 years after surgery compared to adults (Inge et al. 2016). Among their 242 participants who underwent any surgical intervention, 29 (13%) had T2D at baseline with a median HgbA1C of 6.3%, fasting glucose level of 110 mg/dL, and high fasting endogenous plasma insulin level of 43 IU/mL consistent with glucose intolerance. Of these 29 patients, 6 underwent vertical sleeve gastrectomy (VSG) and 23 underwent gastric bypass (RYGB). At 3 years following surgery, 19 of 20 participants with available data were in remission of type 2 diabetes with a median HgbA1C of 5.3%,

fasting glucose level of 88 mg/dL, and fasting endogenous plasma insulin level of 12 IU/mL. Furthermore, remission of prediabetes occurred in 76% of participants who had the condition at baseline. The 5-year follow up data was recently published and compared only those adolescents who underwent RYGB with a similar cohort of adult patients 5 years after they underwent the same procedure (Inge et al. 2019). These data demonstrated that after 5 years, glycemic control was considerably better and T2D remission was significantly higher (86% vs. 53%) in the adolescent group when compared to the adult group. These data strongly suggest that early surgical intervention for patients with T2D is appropriate regardless of the patient's age.

To date, there have been no randomized trials to compare the effectiveness and safety of surgery to those of conventional T2D treatment options in adolescents. However, comparison of data from adolescents with T2D from the Teen-LABS study to a matched cohort from the TODAY study revealed that surgical treatment of severely obese adolescents with T2D was associated with clinically significant weight reduction, remission of diabetes, and improvement in cardiovascular risk factors despite starting with a higher BMI (Inge et al. 2018) while those on medical treatment experienced modest weight gain, progression of T2D, and no improvement in cardiovascular risk factors in 2 years of follow-up. Randomized prospective studies have however been carried out in adults to compare the effectiveness of bariatric surgery to medications to treat T2D with data supporting the superiority of surgery. Five-year outcome data from the Surgical Treatment and Medications Potentially Eradicate Diabetes Efficiently (STAMPEDE) trial demonstrated that patients with T2D and obesity with a mean BMI of 37 +/− 3.5 who underwent bariatric surgery with RYGB or VSG were significantly more likely to achieve and maintain good glycemic control (HgbA1C < or = to 6.0%) with or without medications, than were those who received intensive medical therapy alone. Furthermore, the surgically treated patients had superior glycemic control throughout the 5-year post-op period while having decreased diabetes medication use (Schauer et al. 2017). Thus, it is likely true that surgery provides better glycemic control in patients with severe obesity and T2D regardless of their age based on the results in adults and the findings from Teen-LABS. However, data from a prospective randomized controlled study in a strictly adolescent cohort is needed to confirm this hypothesis.

7   Criteria for Use of Bariatric Surgery in Youth

In 2018, the American Society for Metabolic and Bariatric Surgery (ASMBS) Pediatric Committee released an update of their evidence-based guidelines on pediatric metabolic surgery with the following recommended indications for MBS in youth between the ages of 10 and 19: BMI >35 kg/m2 and major comorbidities of obesity such as T2D, sleep apnea, or hepatic steatosis with fibrosis; or BMI > 40 kg/mg2 with mild comorbidities such as hypertension or dyslipidemia, extreme obesity, or those who fail to improve comorbidities even after a formal trial of lifestyle modifications, with or without pharmacotherapy (Pratt et al. 2018). In general, previous guidelines have suggested that pediatric patients should attain Tanner 4 to 5 pubertal development and be at final or near-final adult height prior to undergoing bariatric surgery, but there is no evidence to suggest that a patient's linear growth is adversely affected by MBS. Thus, the ASMBS Pediatric Committee recommends that Tanner staging, bone age or height no longer be used to determine eligibility for adolescent MBS. In addition, they suggest that cognitive disabilities, a history of mental illness, or eating disorders that are treated should no longer serve as contraindications given the potential benefit of surgery.

Other organizations such as the American Academy of Pediatrics (AAP) and American Diabetes Association (ADA) also acknowledge the benefit of metabolic surgery in their clinical practice guidelines. The AAP recommends that pediatricians identify patients with class 2 and class 3 obesity who meet criteria for metabolic

and bariatric surgery and provide timely referrals to comprehensive, multidisciplinary, pediatric-focused metabolic and bariatric surgery programs (Armstrong et al. 2019). The ADA recommends that metabolic surgery be considered for treatment of adolescents with type 2 diabetes who are markedly obese (BMI >35 kg/m2) and who have uncontrolled glycemia and/or serious comorbidities despite lifestyle and pharmacologic intervention (American Diabetes A 2019).

8   Potential Complications of Bariatric Surgery

While increasing data support the use of bariatric surgery to treat youth with severe obesity, all surgery is associated with possible post-operative complications that patients and their providers must consider in order to be best informed of the risks associated with the procedures. However, as the adolescent bariatric field has moved away from RYGB and toward VSG, the risks are less likely to be procedurally related for pediatric patients, at least in the short-term.

8.1   Need for Repeat Surgery

In the Teen-LABS study, 13% of patients had repeat abdominal surgeries for a total of 47 procedures with 24% of procedures occurring in the first postoperative year and 55% occurring in the second postoperative year (Inge et al. 2019). These repeat operations were performed due to various conditions such as gastric outlet/bowel obstruction, gastrointestinal leak, wound infection, gastrointestinal bleeding, abdominal pain, and gastroesophageal reflux. The majority of the reoperations resulted from complications of their original RYGB, whereas GERD was more often seen in patients after VSG. Similarly, the Swedish Adolescent Morbid Obesity Study which included only patients undergoing RYGB procedures showed a 15% reoperation rate due to a variety of reasons related to their procedure such as internal hernias and post-operative adhesions, and the rest of the operations were from symptomatic cholelithiasis which can result from rapid weight loss from any of the bariatric surgery procedures (Olbers et al. 2017). One study noted that among 309 patients who underwent laparoscopic SG (LSG), 18 patients (6%) underwent cholecystectomy between 4 weeks and 29 months after LSG (Tashiro et al. 2019).

In adults, late weight regain has been found to occur in up to 20% of patients, especially with extremely elevated BMI of >50 kg/m2 prior to RYGB due to maladaptive eating patterns, gradual enlargement of the gastric pouch, dilatation of the surgical anastomosis, and development of a gastrogastric fistula (Kalarchian et al. 2002; Spaulding 2003; Thompson et al. 2006). Patients with gastrogastric fistulae may require endoscopic procedures to reduce the pouch size, tighten the stoma, or close the fistula if feasible. While revision or reoperation may be needed after any bariatric surgery, the rate of revision is most common following laparoscopic adjustable gastric banding and estimated to be 26% versus 4.9% for RYGB and 9.8% for VSG (Altieri et al. 2018). The need for revisional bariatric surgery for weight regain in pediatric patients remains unknown given absence of long-term data following bariatric surgery in youth. However since VSG has become the mainstay of most adolescent bariatric surgical programs, and one adult series shows at least a 20% revision rate at 7 years post VSG (Clapp et al. 2018), adolescent patients must be followed to ensure that those who may be regaining weight are recognized early and additional support and/or interventions are offered in order to prevent the need for additional surgical procedures.

8.2 Vitamin and Mineral Deficiencies

Bariatric surgery, especially RYGB is associated with multiple nutritional deficiencies and can occur in more than 30% of patients after 5 years from surgery (Lupoli et al. 2017). For example, patients can develop anemia secondary to iron and vitamin B12 deficiency likely due to a combination of factors including restricted nutritional intake, malabsorption, decreased gastric acid, decreased intrinsic factor, and food intolerance with dumping. Other deficiencies that are seen post-surgery include: thiamin, vitamin D, vitamin A, folic acid, iron, copper, and zinc deficiency (Desai et al. 2016). Therefore, pre-op screening, lifelong nutritional monitoring, and start of supplementation are recommended to prevent vitamin/mineral insufficiency or deficiency. For example, standard post-RYGB and VSG supplementation in some adolescent MBS centers includes a multivitamin, calcium, vitamin D, elemental iron, and vitamin B12, and nearly all centers recommend a multivitamin at a minimum (Nogueira and Hrovat 2014).

8.3 Bone Disease

In adults, dramatic and rapid weight loss following RYGB has been associated with loss of bone mass and increased bone fracture risk likely due to decreased vitamin D and calcium intake, decreased calcium absorption, and bone marrow fat changes (Gagnon and Schafer 2018). Similarly, adolescents in the first year after RYGB were found to have Vitamin D deficiency and inadequate calcium levels (Xanthakos 2009). When compared to VSG, RYGB is associated with greater femoral bone loss in adults but similar declines in lumbar spine BMD (Bredella et al. 2017). Whether VSG results in a similar degree of bone mass loss and vitamin D deficiency in adults is unknown. Correction of nutritional deficiencies pre-operatively may be useful in preventing early post-operative micronutrient deficiencies, but serum values still need to be monitored after surgery. Thus studies are needed in adolescents who undergo MBS to determine whether MBS has an adverse acute and long-term impact on their bone health and other nutritional indices, especially if the procedure was a RYGB.

8.4 Psychological Function

The majority of adolescents after RYGB were found to have improvements in mental health and self-esteem mainly during the first post-op year (Jarvholm et al. 2016). However, despite improvements in mental health a year following RGYB, one study showed that 14% had suicidal ideation and 20% had poor mental health so long-term mental health monitoring is recommended (Jarvholm et al. 2016). In the Swedish Adolescent Morbid Obesity Study, 7% had psychological sequelae such as suicidal attempt, suicidal ideation, depression, and anxiety in patients with pre-surgical psychiatric histories (Olbers et al. 2017). Consequently, adolescents with ongoing substance abuse or untreated psychiatric illness are generally not considered candidates for bariatric surgery but rather need those issues addressed first. However, the presence or number of psychiatric illnesses did not predict outcome in our cohort and once patients are properly managed from a mental health perspective, they can undergo bariatric surgery and expect weight loss success (Mackey et al. 2018). Furthermore, there are studies that suggest bariatric surgery may improve neurocognitive functioning within a year following MBS. Improvements in executive function and reward anticipation help guide healthy decision-making about food and physical activity and may consequently help sustain weight loss (Thiara et al. 2017; Pearce et al. 2017). These improvements have been found after VSG as well as RYGB (Thiara et al. 2017; Pearce et al. 2017). Thus, while mental health issues should be identified and treated prior to MBS, the brain may actually function better after MBS in patients who suffer from obesity.

9 Conclusion

The incidence of T2D among youth in the United States is increasing at an alarming rate as a consequence of the obesity epidemic. Lifestyle modifications and medications such as metformin and insulin are the first line of therapy, but may not be sufficient to treat youth with severe obesity and T2D. Studies thus far suggest that bariatric surgery in youth is safe and effective in decreasing weight and improving the comorbidities of obesity including T2D. Questions that still need

to be explored include: whether early surgical intervention can prevent T2D and metabolic disease in youth with severe obesity and pre-diabetes, the duration of diabetes remission following MBS, the potential role of weight loss medications in youth post-MBS to sustain weight loss and remission of comorbidities including T2D, the heterogeneity of obesity pathophysiology and the variability of patient response to MBS. Further understanding of these areas will hopefully enable us to predict which patients would benefit the most from MBS, including those on the insulin-resistance to T2D spectrum.

Author Contributions All authors equally contributed to this paper with literature review and analysis, drafting and critical revision and editing, and final approval of the final version.

Conflict-of-Interest Statement No potential conflicts of interest. No financial support.

References

Al-Saeed AH, Constantino MI, Molyneaux L, D'Souza M, Limacher-Gisler F, Luo C et al (2016) An inverse relationship between age of type 2 diabetes onset and complication risk and mortality: the impact of youth-onset type 2 diabetes. Diabetes Care 39(5):823–829

Altieri MS, Yang J, Nie L, Blackstone R, Spaniolas K, Pryor A (2018) Rate of revisions or conversion after bariatric surgery over 10 years in the state of New York. Surg Obes Relat Dis 14(4):500–507

American Diabetes A (2018a) 2. Classification and diagnosis of diabetes: standards of medical care in diabetes-2018. Diabetes Care 41(Suppl 1):S13–S27

American Diabetes A (2018b) 12. Children and adolescents: standards of medical care in diabetes-2018. Diabetes Care 41(Suppl 1):S126–SS36

American Diabetes A (2019) 13. Children and adolescents: standards of medical care in diabetes-2019. Diabetes Care 42(Suppl 1):S148–S164

Andes LJ, Cheng YJ, Rolka DB, Gregg EW, Imperatore G (2019) Prevalence of prediabetes among Adolescents and Young Adults in the United States, 2005–2016. JAMA Pediatr 174:e194498

Angrisani L, Santonicola A, Iovino P, Formisano G, Buchwald H, Scopinaro N (2015) Bariatric surgery worldwide 2013. Obes Surg 25(10):1822–1832

Arafat M, Norain A, Burjonrappa S (2019) Characterizing bariatric surgery utilization and complication rates in the adolescent population. J Pediatr Surg 54 (2):288–292

Armstrong SC, Bolling CF, Michalsky MP, Reichard KW et al (2019) Pediatric metabolic and bariatric surgery: evidence, barriers, and best practices. Pediatrics 144 (6):1–6

Bredella MA, Greenblatt LB, Eajazi A, Torriani M, Yu EW (2017) Effects of Roux-en-Y gastric bypass and sleeve gastrectomy on bone mineral density and marrow adipose tissue. Bone 95:85–90

Chen X, Zhang J, Zhou Z (2019) Targeting islets: metabolic surgery is more than a bariatric surgery. Obes Surg 29(9):3001–3009

Clapp B, Wynn M, Martyn C, Foster C, O'Dell M, Tyroch A (2018) Long term (7 or more years) outcomes of the sleeve gastrectomy: a meta-analysis. Surg Obes Relat Dis 14(6):741–747

Desai NK, Wulkan ML, Inge TH (2016) Update on adolescent bariatric surgery. Endocrinol Metab Clin N Am 45(3):667–676

Elder KA, Wolfe BM (2007) Bariatric surgery: a review of procedures and outcomes. Gastroenterology 132 (6):2253–2271

Gagnon C, Schafer AL (2018) Bone health after bariatric surgery. JBMR Plus 2(3):121–133

Group TS, Zeitler P, Hirst K, Pyle L, Linder B, Copeland K et al (2012) A clinical trial to maintain glycemic control in youth with type 2 diabetes. N Engl J Med 366(24):2247–2256

Gungor N, Arslanian S (2004) Progressive beta cell failure in type 2 diabetes mellitus of youth. J Pediatr 144 (5):656–659

Hofso D, Fatima F, Borgeraas H, Birkeland KI, Gulseth HL, Hertel JK et al (2019) Gastric bypass versus sleeve gastrectomy in patients with type 2 diabetes (Oseberg): a single-centre, triple-blind, randomised controlled trial. Lancet Diabetes Endocrinol 7(12):912–924

Hutch CR, Sandoval D (2017) The role of GLP-1 in the metabolic success of bariatric surgery. Endocrinology 158(12):4139–4151

Inge TH, Zeller MH, Jenkins TM, Helmrath M, Brandt ML, Michalsky MP et al (2014) Perioperative outcomes of adolescents undergoing bariatric surgery: the teen-longitudinal assessment of bariatric surgery (Teen-LABS) study. JAMA Pediatr 168(1):47–53

Inge TH, Courcoulas AP, Jenkins TM, Michalsky MP, Helmrath MA, Brandt ML et al (2016) Weight loss and health status 3 years after bariatric surgery in adolescents. N Engl J Med 374(2):113–123

Inge TH, Laffel LM, Jenkins TM, Marcus MD, Leibel NI, Brandt ML et al (2018) Comparison of surgical and medical therapy for type 2 diabetes in severely obese adolescents. JAMA Pediatr 172(5):452–460

Inge TH, Courcoulas AP, Jenkins TM, Michalsky MP, Brandt ML, Xanthakos SA et al (2019) Five-year outcomes of gastric bypass in adolescents as compared with adults. N Engl J Med 380(22):2136–2145

Jarvholm K, Karlsson J, Olbers T, Peltonen M, Marcus C, Dahlgren J et al (2016) Characteristics of adolescents with poor mental health after bariatric surgery. Surg Obes Relat Dis 12(4):882–890

Kalarchian MA, Marcus MD, Wilson GT, Labouvie EW, Brolin RE, LaMarca LD (2002) Binge eating among gastric bypass patients at long-term follow-up. Obes Surg 12(2):270–275

Kyler KE, Bettenhausen JL, Hall M, Fraser JD, Sweeney B (2019) Trends in volume and utilization outcomes in adolescent metabolic and bariatric surgery at children's hospitals. J Adolesc Health 65(3):331–336

Laffel LMB, Tamborlane WV, Yver A, Simons G, Wu J, Nock V et al (2018) Pharmacokinetic and pharmacodynamic profile of the sodium-glucose co-transporter-2 inhibitor empagliflozin in young people with type 2 diabetes: a randomized trial. Diabet Med 35 (8):1096–1104

Lupoli R, Lembo E, Saldalamacchia G, Avola CK, Angrisani L, Capaldo B (2017) Bariatric surgery and long-term nutritional issues. World J Diabetes 8 (11):464–474

Mackey ER, Wang J, Harrington C, Nadler EP (2018) Psychiatric diagnoses and weight loss among adolescents receiving sleeve gastrectomy. Pediatrics 142(1):e20173432

Mason EE, Ito C (1967) Gastric bypass in obesity. Surg Clin North Am 47(6):1345–1351

Nogueira I, Hrovat K (2014) Adolescent bariatric surgery: review on nutrition considerations. Nutr Clin Pract 29 (6):740–746

Olbers T, Beamish AJ, Gronowitz E, Flodmark CE, Dahlgren J, Bruze G et al (2017) Laparoscopic Roux-en-Y gastric bypass in adolescents with severe obesity (AMOS): a prospective, 5-year, Swedish nationwide study. Lancet Diabetes Endocrinol 5(3):174–183

Pearce AL, Mackey E, Cherry JBC, Olson A, You X, Magge SN et al (2017) Effect of adolescent bariatric surgery on the brain and cognition: a pilot study. Obesity (Silver Spring) 25(11):1852–1860

Pernicova I, Korbonits M (2014) Metformin – mode of action and clinical implications for diabetes and cancer. Nat Rev Endocrinol 10(3):143–156

Pettitt DJ, Talton J, Dabelea D, Divers J, Imperatore G, Lawrence JM et al (2014) Prevalence of diabetes in U.S. Youth in 2009: the SEARCH for diabetes in youth study. Diabetes Care 37(2):402–408

Pratt JSA, Browne A, Browne NT, Bruzoni M, Cohen M, Desai A et al (2018) ASMBS pediatric metabolic and bariatric surgery guidelines, 2018. Surg Obes Relat Dis 14(7):882–901

Pulgaron ER, Delamater AM (2014) Obesity and type 2 diabetes in children: epidemiology and treatment. Curr Diab Rep 14(8):508

Reinehr T (2013) Lifestyle intervention in childhood obesity: changes and challenges. Nat Rev Endocrinol 9 (10):607–614

Schauer PR, Bhatt DL, Kirwan JP, Wolski K, Aminian A, Brethauer SA et al (2017) Bariatric surgery versus intensive medical therapy for diabetes – 5-year outcomes. N Engl J Med 376(7):641–651

Skinner AC, Ravanbakht SN, Skelton JA, Perrin EM, Armstrong SC (2018) Prevalence of obesity and severe obesity in US children, 1999–2016. Pediatrics 141(3): e20173459

Slomski A (2017) Bariatric surgery has durable effects in controlling diabetes. JAMA 317(16):1615

Spaulding L (2003) Treatment of dilated gastroje-junostomy with sclerotherapy. Obes Surg 13 (2):254–257

Tamborlane WV, Laffel LM, Weill J, Gordat M, Neubacher D, Retlich S et al (2018) Randomized, double-blind, placebo-controlled dose-finding study of the dipeptidyl peptidase-4 inhibitor linagliptin in pediatric patients with type 2 diabetes. Pediatr Diabetes 19(4):640–648

Tamborlane WV, Barrientos-Perez M, Fainberg U, Frimer-Larsen H, Hafez M, Hale PM et al (2019) Liraglutide in children and adolescents with type 2 diabetes. N Engl J Med 381(7):637–646

Tashiro J, Thenappan AA, Nadler EP (2019) Pattern of biliary disease following laparoscopic sleeve gastrectomy in adolescents. Obesity (Silver Spring) 27 (11):1750–1753

Thiara G, Cigliobianco M, Muravsky A, Paoli RA, Mansur R, Hawa R et al (2017) Evidence for neurocognitive improvement after bariatric surgery: a systematic review. Psychosomatics 58(3):217–227

Thompson CC, Slattery J, Bundga ME, Lautz DB (2006) Peroral endoscopic reduction of dilated gastrojejunal anastomosis after Roux-en-Y gastric bypass: a possible new option for patients with weight regain. Surg Endosc 20(11):1744–1748

U.S. Food and Drug Administration (2001) The lap-band adjustable gastric banding system summary of safety and effectiveness data. PMA no: P000008. FDA, United States, 25 p

Weiss R, Taksali SE, Tamborlane WV, Burgert TS, Savoye M, Caprio S (2005) Predictors of changes in glucose tolerance status in obese youth. Diabetes Care 28(4):902–909

Xanthakos SA (2009) Nutritional deficiencies in obesity and after bariatric surgery. Pediatr Clin N Am 56 (5):1105–1121

Adv Exp Med Biol - Advances in Internal Medicine (2020) 4: 331–355
https://doi.org/10.1007/5584_2020_482
© Springer Nature Switzerland AG 2020
Published online: 8 February 2020

Insulin Recommender Systems for T1DM: A Review

Joaquim Massana, Ferran Torrent-Fontbona, and Beatriz López

Abstract

In type 1 diabetes mellitus (T1DM) pancreas beta-cells do not segregate insulin. This hormone is necessary to convert glucose into energy. Thus, people with diabetes are required to maintain blood glucose (BG) levels within a safe range using external control solutions. Insulin recommender systems (IRS's) provide the precise amount of insulin to the patient when needed, reducing the effects of the disease. The goal of this paper is to review and summarize all current proposals of IRS's and, with this purpose, 70 papers have been analysed. The analysis of the works was performed taking the following aspects into account: (i) technology of the recommendation process, (ii) control procedures, (iii) complementary processes, (iv) hardware, testing and assessment, (v) pricing and (vi) results. Those are our main conclusions after the review: There is a lack of published research works providing real experimentation together with simulation processes. Information about the IRS's features is also lacking in a remarkable percentage of the publications. Due to the variability in how experiments are performed and results are presented, research work comparisons become difficult. In summary, this topic requires standards to be able to perform comparison analysis of published papers and therefore, progress adequately.

Keywords

Bolus calculator · Diabetes · Insulin recommender system · Review · T1DM

J. Massana (✉), F. Torrent-Fontbona, and B. López
Polytechnical School, Universitat de Girona, Girona, Catalonia, Spain
e-mail: joaquim.massana@udg.edu; ferran.torrent@udg.edu; beatriz.lopez@udg.edu

Acronyms

1-NN	One nearest neighbour
AG	Age
AI	Artificial intelligence
AL	Alcohol
ANN	Artificial neural network
ATSCH	Actions taken to self-correct for hypoglycaemia
AUCk	Area under curve postprandial
AUCr	Area under curve reference
BB	Black box
BC	Bolus calculator
BDICH	Bolus doses of insulin used to correct for hyperglycaemia
BDIFC	Bolus doses of insulin with food consumption
BE	Beverage
BG	Blood glucose
BGL	Blood glucose low
BGLD	Blood glucose low discount
BI	Basal insulin

CBR	Case-based reasoning	K1	Filter gain
CGM	Continuous glucose monitoring	KF	Kalman filtering
CH	Carbohydrates	KNN	K-nearest neighbours
Cho	Carbohydrates on board	LBGI	Low blood glucose index
CSII	Continuous subcutaneous insulin infusion	LBGT	Low blood glucose target
		LIISPB	Location of insulin infusion set on patient's body
CT	Control trend		
CVGA	Control Variability Grid Analysis	MBG	Mean blood glucose
DFO	Distance from onset	MC	Metabolic control
DI	Digestive illness	MDI	Multiple dose insulin injection
DT	Diet	MF	Meal fat
DM	Diabetes mellitus	MP	Meal proteins
ED	Exercise duration	MT	Meal times
EK	Expert knowledge	NEP	Nephropathies
EL	Exercise level	NOI	Number of injections
ET	Exercise type	NPHI	Isophane insulin
ETS	Equivalent Teaspoons Sugar	NUP	Neuropathies
Ex	Exercise	OCD	Other chronic diseases
FE	Fever	P1 and P2	Gradient descent method parameters
Gmin	Postprandial minimum glucose	PB	Previous bolus
GN	Gender	PDA	Personal Digital Assistant
GRC	Glucose rate of change	PhiM	Postprandial glucose response performance measured
GT	Glucose target		
HAP	Happiness	PhiO	Postprandial glucose response performance optimum
HbA1c	Glycated haemoglobin		
HBGT	High blood glucose target	PI	Premixed insulin
HC	Hormone cycle	PP	Pump problems
HD	Hormonal disorders	PR	Premixed ratio
HE	Height	PS	Pubertal stage
HEA	Number of hyperglycaemic episodes	R2R	Run-to-run
HOA	Number of hypoglycaemic episodes	RI	Regular insulin
HoS	Hours of sleep	RL	Reinforcement learning
HS	Health status	RT	Requirement trend HOD Hour of day
IA	Insulin antibodies		
ICR	Insulin-to-carbohydrates ratio	RTP	Retinopathies
ICRO	Old insulin-to-carbohydrates ratio	SE	Sex
ID	Insulin dosages	SMBG	Self-monitoring of blood glucose
IISVPC	Intra-day insulin sensitivity variability profile class	SS	Safety system
		ST	Stress
IOB	Insulin on board	T1DM	Type 1 diabetes mellitus
IPBIR	Insulin pump basal infusion rate	T2DM	Type 2 diabetes mellitus
IR	Insulin recordings.	TAT	Time above target
IRS	Insulin recommender system	TBT	Time below target
ISF	Insulin sensitivity factor	TBWB	Type of bolus wave for each bolus
ISP	Insulin stacking percentage	TCIIS	Time of change of insulin infusion set
JB	Job	TD	Time data
K	Gain of adaptation	Temp	Temperature

TIR	Tiredness
TTT	Time in target
TM	Type of meal
ToD	Type of day
TOI	Type of insulin
VS	Virtual subjects
W	Weight
WE	Weigh excess
WS	Work schedule
Ym	Deterministic function

1   Introduction

Diabetes mellitus (DM), popularly known as diabetes, is a serious, long-term condition that occurs when the body cannot produce any or enough insulin, or cannot effectively use the insulin it produces (Atlas 2019). Insulin is an hormone produced in the pancreas whose main role is to regulate the glucose levels in the blood, allowing the body's cells to absorb the glucose from the bloodstream. The lack of insulin causes BG concentrations increases beyond safe levels (hyperglycaemia), causing damage to blood vessels, organs, and the whole body (Alberti and Zimmet 1998).

Currently, there is no cure for diabetes, but it can be successfully managed allowing people to lead a healthy life. Nowadays, there are 463 million people with this disease, with an estimated increase of 51% in 2045, reaching 700 million people. Four million people aged 20–79 dies every year as a consequence of diabetes (Atlas 2019).

There are two main types of diabetes: T1DM and type 2 diabetes mellitus (T2DM). T1DM is caused by an autoimmune reaction in which the body's immune system attacks the insulin-producing beta-cells of the pancreas, resulting in a very little or no production of insulin (Atlas 2019). People with T1DM need to take insulin doses every day in order to stay alive. T1DM is usually diagnosed during early ages (Atkinson et al. 2014). T2DM is caused by the inability to the body's cells to respond fully to insulin, what is called insulin resistance. T2DM is usually diagnosed in middle and old ages. The prevalence of T2DM is higher than T1DM (Olokoba et al. 2012).

The main goal of people suffering diabetes is to maintain BG levels within a healthy range, which requires is a tight glycaemic control. Avoiding hyperglycaemia, subjects are able to prevent long term micro-vascular (nephropathy, retinopathy and neuropathy) and macro-vascular health issues (coronary heart disease, peripheral vascular disease and stroke). In the other hand, avoiding hypoglycaemia (low concentration of glucose on the body due to too much injected insulin, eaten too little food, or too much exercise) prevents light-headedness, dizziness or death (Atkinson and Eisenbarth 2001).

The management of diabetes involves insulin injection for T1DM, and some T2DM. There are different types of insulin (Diabetes AA. Insulin basics 2019): rapid-acting insulin, regular or short-acting insulin, intermediate-acting insulin, and long-acting insulin. The differences among them rely on three factors: how long it takes for the insulin to start lowering the blood sugar (onset), when it is at its maximum strength (peak time), and how long it continues to work (duration). There is a continuous investigation on developing new insulin types, and therefore this list is not exhaustive. For example, inhaled insulin has been recently listed by the American Diabetes Association as a possible alternative. Basal-bolus insulin therapy, also called intensive or flexible insulin therapy, is an option for diabetes management that combines different types of long, that is basal, and short, that is bolus, acting insulin (Railton 2019).

Bolus insulin is prescribed to take after meals and in moments of extremely high blood sugar. Insulin dosages (IDs) depend on BG, carbohydrates obtained from meals, insulin-to-carbohydrates ratio (ICR) and insulin sensitivity factor (ISF), among other issues (Torrent-Fontbona and López 2018), and determining the right dosage is a daunting task.

To support diabetic people in such a complex process, IRSs have been developed with the aim to provide accurate and timely bolus dose recommendations, increasing the control over the disease.

Chronologically, the first systems were only plastic cards with formulas, tables or mnemonic techniques. Then, electronic gadgets such as personal digital assistants (PDAs) were used, but nowadays, they are being substituted by smartphones. In the same way, traditional syringes, pens and fingersticks are being substituted by continuous glucose monitoring systems (CGMs) and pumps. In addition, new wearables open the opportunity to gather information about each person in order to improve recommendations. Some systems used simple mathematical equations in order to calculate the bolus, but in recent years, artificial intelligence (AI) is being introduced to respond to the need of having personalized recommendations.

The main goal of the present publication is to provide a complete, useful, exhaustive and updated guide to IRSs. The aim is to introduce the reader to the state-of-art of IRSs, not only being comprehensive in terms of publications, but especially systematic. Publications were analysed extracting the same information for each one and trying to find trends and good practices.

2 Insulin Recommender System Basics

In Fig. 1, and with the aim of introducing the IRS basics, the block diagram of a generic insulin recommender system is shown.

As a general solution, blocks on the left introduce information to the IRS. This data is used for the purpose of calculating the dose of insulin. Information about the user and the meal is usually introduced in the system using the smartphone. In addition, some contextual information, such as exercise intensity, can be collected using activity bands or similar devices. Finally, glucose data, such as BG, can be collected using CGM or self-monitoring of blood glucose (SMBG) systems.

When the IRS has calculated the bolus insulin, its recommendation is sent to the safety system (SS). This system can modify the dose, if necessary, in order to stay within safe limits. In the end, the instruction is sent to the infusion system, that is multiple dose insulin injection (MDI) or continuous subcutaneous insulin infusion (CSII) systems, and the control measures, i.e. postprandial mean glucose (Gmin), can be collected.

3 Methodology

In the following sections, the selection process of the works and criteria used for comparison purposes, are explained.

3.1 Selection of the Works Surveyed

The main used keywords in order to search for publications were: IRS's and bolus calculators (BC's).

In order to write the present paper, 70 works have been analysed. From these 70 works, 50 have been used and 20 have been discarded due to several reasons such as quality, scope and lack of data.

From these 50 analysed papers, 10 of them are focused in the effectiveness of using BC's and do not contain a particular IRS. These papers have been chosen in order to create Table 1, in which the importance of using BC's is pointed out. As seen in Table 1, there is a high amount variability among the proposals, but all of them conclude that the utilization of IRS's improve the users' degree of satisfaction and their quality of life; making the patient's life easier.

The last 40 papers are shown in Tables 2, 3, 4 and 5, in Sect. 3, and are the core of the present publication.

3.2 Comparison Criteria

The IRS's of the selected papers, shown in Tables 2, 3, 4 and 5, have been compared according to several criteria, organized in six blocks: technology of the recommendation process, control procedures, complementary processes, hardware, testing and assessment, pricing and results.

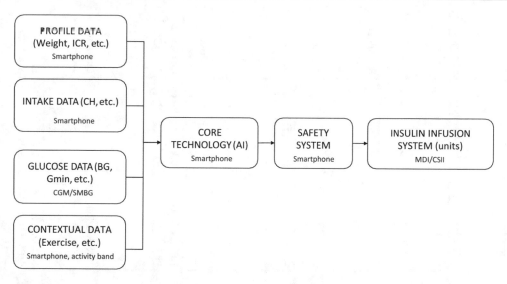

Fig. 1 Generic insulin recommender system

Table 1 Effectiveness of using BC's

Clinical trials	Number of patients	Glucose monitoring system	Insulin infusion system	Mobile application	Improvement	References
NO	–	–	MDI	YES	YES	Bailey and Stone (2017)
NO	81	CGM	MDI/CSII	YES	YES	Clements and Staggs (2017)
YES	63	SMBG	MDI/CSII	YES	YES	Drion et al. (2015)
YES	100	CGM	–	YES	YES	Garg et al. (2017)
YES	73	–	MDI/CSII	YES	YES	Kirwan et al. (2013)
YES	87	SMBG	MDI	YES	YES	Mora et al. (2017)
YES	10	CGM	MDI/CSII	YES	YES	Pesl et al. (2016)
NO	30 VS	CGM	CSII	NO	YES	(Rosales et al. 2018)
YES	51	SMBG	MDI	NO	YES	Schmidt et al. (2012)
YES	70	SMBG	MDI	NO	YES	Vallejo-Mora et al. (2017)

3.2.1 Technology

Criteria in this block include: core technology, number of inputs, name of the attributes, ICR calculation, ISF calculation attribute, and bolus calculation.

The core methodology explains the base technology that lies inside each bolus recommendation system, in order to provide intelligence to the system. In some cases (Al-Taee et al. 2015; Liu et al. 2013; Poerschke 2004), there is more than one methodology. In other cases (Poerschke 2004; Anderson 2009; Glaser et al. 2004; Pańkowska and B-lazik 2010; Pelzer 2006; Rossi et al. 2010), there is a lack of information. Inside parentheses commercial references or references to external publications are included. The major use of the AI inside the system is to personalize recommendations and learn from patients.

Table 2 Insulin recommender system comparative table, part A

Simulator	Clinical trials	Number of patients	Trial/simulation length (days)	Core methodology	Number of attributes	Name of attributes	ICR calculation	ISF calculation	References
NO	NO	–	–	ANN and equations	10	BE, BG, CH, ED, ET, TM, GT, IOB, ICR and ISF	–	–	Al-Taee et al. (2015)
NO	YES	573 (from 1–20 years old)	720	EK (Ezy-BICC)	5	BGL, BGLD, GT, ICR and ISF		–	Anderson (2009)
NO	YES	46	112	BB (Diabetes Insulin Guidance System)	1	TD		–	Bergenstal et al. (2012)
YES (Not defined)	NO	10 VS	3	KF	7	BG, CH, GT, Gmin, ICR, IOB and ISF		–	Boiroux et al. (2015)
YES (UVA-PADOVA)	NO	1 VS	180	Temporal CBR	4	Ex, Gmin, ISF and W		ISF = [1700/ 0.24*W]*0.0555	Brown et al. (2018)
YES (UVAPADOVA)	NO	100 VS	1	ANN	11	BG, CH, GRC, GT, ICR, IPBIR, IISVPC, ISF, IOB, W and Ym		–	Cappon et al. (2018)
NO	YES	36	63	BB (Bolus Wizard)	2	ICR and ISF		–	Carić et al. (2017)
NO	YES	180	180	Equations (Diabeo)	7	BI, GT, CH, Gmin, ISF, Ex and W	BB that depends on GT	BB that depens on BI, CH and W	Charpentier et al. (2011)
YES (UVAPADOVA)	NO	30 VS (10 adults, 10 adolescents and 10 children)	30	RL (Actor-critic method)	3	BG, CH and ICR	–	–	Daskalaki et al. (2013)
YES (UVA-PADOVA)	NO	130 VS (110 adults, 10 adolescents and 10 children)	30	RL (Actor-critic method)	3	BG, CH and ICR	ICR = B/CH		Daskalaki et al. (2016)
YES (Hardware)	NO	1	4	CBR (Herrero2017)	10	BG, Ex, Gmin, IOB, CH, GT, ICR, ISF, TM and W	ICR = [CH +(BG-GT)/(1960/ 2.6*W)]/IB +(1/ISF)*(Gmin-GT) + IOB]	ISF = 1960*ICR/ 2.6*W	El-Sharkawy et al. (2018)

							ICR	ISF	Reference
YES	NO	123	365	BB (ACCUCHEK)	–	–	–	–	Garg et al. (2008)
YES	NO	83 (adolescents)	365	EK (Two metallic disks with tables)	3	BG, CH and ICR	–	–	Glaser et al. (2004)
YES	NO	49	7	BB (Medtronic MiniMed)	6	BG, CH, Gmin, GT, ISF and ICR	–	–	Gross et al (2003)
NO	YES (UVAPADOVA)	22 VS	90	CBR	10	BG, Ex, Gmin, IOB, CH, GT, ICR, ISF, TM and W	$ICR = [CH +(BG-GT)/(1960/2.6*W)]/[B +(1/ISF)*(Gmin-GT) + IOB]$	$ISF = 1960*ICR/2.6*W$	Herrero et al. (2017)
NO	YES (UVAPADOVA)	22 VS	30	CBR	12	AUCk, AUCr, BG, CH, Ex, GT, ICR, ICRO, ISF, K and TM	$ICR = ICRO+K*(AUCr-AUCk)$	$ISF = 4.44*ICR$	Herrero et al. (2014)
YES	NO	168	365	BB (ACCU-CHEK Aviva Expert)	10	BG, CH, EL, GT, HS, ICR, IOB, ISF, ST and HOD	–	–	Hommel et al. (2017)
NO	YES (Meal Simulation model)	10 VS	3	KF	2	BG and CH	–	–	Kirchsteiger and del Re (2014)
YES	NO	7	30	BB (Rapid Calc)	5	AL, BG, CH, Ex and IOB	–	–	Knight et al. (2016)

Table 3 Insulin recommender system comparative table, part B

Bolus calculation	Control measures	Glucose monitoring system	Insulin infusion system	Basal therapy	Initialization methodology	Safety system	System estimated price ($)	Results	Mobile application	References
Bolus = (BG-GT)/ISF + CH/ICR	IOB	CGM	MDI	Optimised	–	NO	640	–	YES	Al-Taee et al. (2015)
Table with BGL, BGLD, GT, ICR and ISF values	NO	SMBG	MDI	Optimised	Using ICR values from old users based on W and ISF new users relations	NO	65	0.6% of HbA1c reduction	NO	Anderson (2009)
BB that depends on TD	NO	SMBG	MDI	Optimised	–	NO	65	0.5% of HbA1c reduction	NO	Bergenstal et al. (2012)
B=CH/ICR + ISF*(BG-GT)-IOB	Gmin	–	–	Optimised	–	NO	–	–	NO	Boiroux et al. (2015)
CBR output	Gmin	–	–	Optimised	–	NO	–	0.05 of LBGI reduction	NO	Brown et al. (2018)
B=CH/ICR+((BG + Ym)-GT)/ISF-IOB	NO	CGM	MDI	Optimised	–	NO	390	0.37 of BGRI reduction	NO	Cappon et al. (2018)
BB that depends on ICR and ISF	NO	SMBG	CSII	Optimised	–	NO	3525	0.05% of HbA1c reduction	NO	Carić et al. (2017)
Bolus = 2.2*ISF	Gmin	CGM	MDI/CSII	Optimised	–	NO	640/4100	0.91% of HbA1c reduction	YES	Charpentier et al. (2011)
BB that depends on BG, CH and ICR	Glycemic profile of the past day	CGM	CSII	Optimised (Using two features describing the glycemic profile of the past day)	Using the ICR's provided by UVA-PADOVA simulator	NO	4100	Percentage in A + B zones of CGVA 100% (adults) and 93% (adolescents +children)	NO	Daskalaki et al. (2013)
BB that depends on BG, CH and ICR	Glycemic profile of the past day	CGM	CSII	Optimised (Using two features describing the glycemic profile of the past day)	Using the ICR's provided by UVA-PADOVA simulator	NO	4100	TIT: 97.18% TAT: 1.03% TBT: 1.78% (30 adults) TIT: 95.66% TAT: 4.07%	NO	Daskalaki et al. (2016)

B = (CH/ICR) + ((BG-GT)/ISF)-IOB	Gmin	CGM (Dexcom G5)	CSII (Tandem t: slim)	Optimised	Using a small set of sub-optimal, but safe, cases	YES	4100	TBT: 0.27% (100 adults) TIT: 86.44% TAT: 11.17% TBT: 2.4% (adolescents) TIT: 74.77% TAT: 4.45% TBT: 20.78% (children)	YES	El-Sharkawy et al. (2018)
BB	NO	SMBG	MDI/CSII	Optimised	Using a questionnaire in order to know the initial ICR's	NO	315/3775	0.4% of HbA1c reduction	YES	Garg et al. (2008)
BB based on EK that depends on BG, CH and ICR	NO	SMBG/ CGM	MDI/CSII	–	Using a questionnaire in order to know the initial ICR's	NO	65/3850	0.9% of HbA1c reduction	NO	Glaser et al. (2004)
B=CH/ICR+(BG-GT)/ISF	Gmin at 2 h after the meal	SMBG	CSII (Medtronic MiniMed)	–	Using patients'logbooks	NO	3775	8.8% of hypoglycaemic events per week reduction	YES	Gross et al. (2003)
B = (CH/ICR) + ((BG-GT)/ISF)-IOB	Gmin	CGM	MDI	Optimised	Using a small set of sub-optimal, but safe, cases	NO	390	TIT: 89.5% TAT: 10.2% TBT: 0.21% (adults) TIT: 77.5% TAT: 19.8% TBT: 2.5% (adolescents)	NO	Herrero et al. (2017)
B=CH/ICR+(BG-GT)/ISF	AUCk	CGM (abbot freestyle navigator)	CSII (Deltec Cozmo)	Optimised (provided by UVA/PADOVA simulator)	Using a clinical experience based on retrospective CGM data and a	YES	3850	MBG from 166 mg/dL to 150 mg/dL (adults) and from 167 mg/	NO	Herrero et al. (2014)

(continued)

Table 3 (continued)

Bolus calculation	Control measures	Glucose monitoring system	Insulin infusion system	Basal therapy	Initialization methodology	Safety system	System estimated price ($)	Results	Mobile application	References
					meal tolerance test data			dL to 162 mg/dL (adolescents)		
BB that depends on BG, CH, GT, IOB, ICR and ISF	NO	CGM (Medtronic iPro2)	MDI	Optimised	Using a questionnaire in order to know the initial ICR's	NO	390	0.5% of HbA1c reduction	NO	Hommel et al. (2017)
BB that depends on BG and CH	NO	CGM	MDI	-	-	NO	390	TIT: 86.5%	NO	Kirchsteiger and del Re (2014)
BB that depends on BG, CH and IOB	NO	SMBG	MDI	Optimised	Using a questionnaire in order to know the initial ICR's	NO	315	-	YES	Knight et al. (2016)

Table 4 Insulin recommender system comparative table, part C

Simulator	Clinical trials	Number of patients	Trial/simulation length (days)	Core methodology	Number of attributes	Name of attributes	ICR calculation	ISF calculation	References
YES (Not defined)	NO	2 VS	10	Fuzzy-Logic-Based and equations	10	BG, CH, Gmin, GT, IOB, ICR, ISF, P1, P2 and PB	$ICR = P1*CH + P2$	$ISF = (1800/500)*ICR$	Liu et al (2013)
NO	NO	–	–	Equations	8	AL, BG, Ex, GT, IOB, ICR, ISF and W	$ICR = 450*W*0.32$	$ISF = 1700/(W*0.23)$	Lloyd et al. (2015)
NO	YES	20	–	1-NN	25	ATSCH, AL, BI, BG, BDICH, BDIFC, CH, Ex, HBGT, LBGT, ICR, DI, ISF, LIISPB, MT, HC, PP, HoS, TM, ST, TD, TCIIS, TBWB, TOI and WS	–	–	Marling et al. (2008)
NO	YES	40	180	BB (Calsulin Thorpe)	4	BG, CH, Ex and ICR	–	–	Maurizi et al. (2011)
NO	NO	–	–	1-NN	25	AG, CT, DT, DFO, HbA1c, HE, HD, HOA, IA, JB, MC, NEP, NUP, NPHI, NOI, OCD, Ex, PI, PR, PS, RI, RT, RTP, SE and WE	–	–	Montani et al. (1998)
YES (based on Hovorka2004)	YES	11 VS	4 real 25 virtual	R2R	8	BG, Gmin, ICR, ICRO, K1, PhiM, PhiO and HOD	$ICR = ICRO + K1*$ (PhiO PhiM)	–	Palerm et al. (2017)

(continued)

Table 4 (continued)

Simulator	Clinical trials	Number of patients	Trial/simulation length (days)	Core methodology	Number of attributes	Name of attributes	ICR calculation	ISF calculation	References
NO	NO	–	–	EK (Database with insulin dose for each meal)	3	CH, MF and MP	–	–	Pańkowska and B-lazik (2010)
NO	YES	8	90	EK (Database of meals, ETS and Ex)	8	AG, BG, ETS, Ex, CH, GN, HE and W	EK (Set by clinician)	EK (Set by clinician)	Pelzer (2006)
NO	YES	10	42	CBR (Herrero2017)	CBR (Herrero2017)	CBR (Herrero2017)	CBR (Herrero2017)	CBR (Herrero2017)	Pesl et al. (2017)
NO	YES	6	–	EK (Hybrid statistical and rule-based expert system based on clinical guideliness)	–	–	–	–	Poerschke (2004)
NO	YES	119	180	EK (Meal database)	2	BG and CH	EK (Set by clinician)	EK (Set by clinician)	Rossi et al. (2010)
NO	YES	51	112	BB	2	BG and CH	EK (Set by clinician according to 500 rule)	EK (Set by clinician according to 1800 rule)	Schmidt et al. (2012)
NO	YES	30	–	KNN	5	BG, CH, IR, Ex and HOD	–	–	Skrøvseth (2015)
YES (UVA PADOVA)	NO	100 VS	98	RL (Actor-critic method)	3	BG, HOA and HEA	RL	–	Sun et al. (2019)
NO	YES	205	–	BB (FreeStyle InsuLinx Blood Glucose Monitoring System)	–	–	–	–	Sussman et al. (2012)

YES (UVAPADOVA)	33 VS (11 adults, 11 adolescents and 11 children)	90	CBR	22	AL, BG, CH, MF, CHO, DI, Ex, FE, Gmin, GT, HAP, HC, TD, HoS, ICR, IOB, ISF, ST, temp, TIR, ToD and W	CBR	ISF=ICR*341.94/W	Torrent-Fontbona and López (2018)
YES (UVAPADOVA)	11 VS (adults)	180	CBR	22	AL, BG, CH, MF, CHO, DI, Ex, FE, Gmin, GT, HAP, HC, TD, HoS, ICR, IOB, ISF, ST, temp, TIR, ToD and W	CBR	ISF=ICR*341.94/W	Torrent-Fontbona et al. (2019)
YES (UVAPADOVA)	10 VS (adults)	50	2nd order R2R	3	BG, CH and ICR	2nd order R2R (3 ICR divide the day)	–	Tuo et a. (2015)
NO	203	180	BB	–	–	–	–	van Niel et al. (2014)
NO	193	180	BB	2	BG and CH	EK (Set by clinician)	EK (Set by clinician)	Ziegler et al. (2013)
NO	104 (children)	180	BB	–	–	EK (Set by clinician)	EK (Set by clinician)	Ziegler et al. (2016)

Table 5 Insulin recommender system comparative table, part D

Bolus calculation	Control measures	Glucose monitoring system	Insulin infusion system	Basal therapy	Initialization methodology	Safety system	System estimated price ($)	Results	Mobile application	References
B=CH/ICR +(BG-110)/ ISF-PB*ISP	Gmin at 3 h after the meal	CGM	MDI/CSII	Optimised (Among 0,5 u/h and 1,5 u/h)	Using a questionnaire in order to know the initial ICR's	NO	390/3850	–	NO	Liu et al. (2013)
B=CH/ICR +(BG-GT)/ ISF-IOB	NO	CGM	CSII	–	–	NO	4100	–	YES	Lloyd et al. (2015)
BB	NO	SMBG/CGM	CSII	–	Using information gathered during clinical trials involving 20 patients	NO	3775	–	NO	Marling et al. (2008)
BB	NO	SMBG/CGM	MDI/CSII	–	–	NO	315	0.85% of HbA1c reduction	NO	Maurizi et al. (2011)
BB	NO	–	MDI/CSII	–	Using 100 cases from 11 patients	NO	265	–	NO	Montani et al. (1998)
BB	Gmin at 3 h after the meal	SMBG	MDI/CSII	Optimised	–	YES (Stability bounds for the ICR)	265	–	NO	Palerm et al. (2007)
BB that depends on CH, MF and MP	NO	CGM	CSII	–	Using meal database	NO	4050	–	NO	Pańkowska and B-lazik (2010)
BB that depends on ETS, Ex and CH	NO	SMBG	MDI	–	Using data from 3-day-long experiments monitoring the patient	NO	265	0.5% of HbA1c reduction	YES	Pelzer (2006)

	CBR (Herrero2017)	CGM	MDI/CSII	Optimised						
CBR (Herrero2017)	NO	–	–	Optimised	–	NO	625	–	YES	Pesl et al. (2017)
BB	NO	–	–	–	–	NO	265	MBI improved from 160 mg/dL to 135 mg/dL	YES	Poerschke (2004)
BB that depends on the CH	NO	SMBG/CGM	MDI/CSII	Optimised (Using an adaptive algorithm based on fasting glucose measures)	Using clinician knowledge	NO	265	0.4% of HbA1c reduction	YES	Rossi e al. (2010)
Bolus that depends on CH	NO	SMBG	MDI	Optimised (By the patient according to some guidelines)	Using patients knowledge	NO	315	0.8% of HbA1c reduction	NO	Schmid et al. (2012)
CBR output	NO	SMBG/CGM	MDI/CSII	–	–	NO	315	0.6% of HbA1c reduction	YES	Skrøvseth (2015)
RL	Hypoglycaemia and hyperglycaemia.	SMBG/CGM	CSII	Optimised (Using reinforcement learning)	Using 1-week CGM values	NO	3775	LBGI from 1.1–2.5 to <1.1	YES	Sun et al. (2019)
BB	NO	SMBG	MDI	–	Using data from training subjects with 2 bolus equations	NO	315	–	NO	Sussman et al. (2012)
B=CH/ICR +(BG-GT)/ISF-IOB	Gmin	SMBG/CGM	MDI/CSII	Optimized (Using iterative algorithm based on KF UVA/PADOVA default therapy)	Using the ICR used by the user	NO	315	TIT: 84% TAT: 13.04% TBT: 2.96% (adults) TIT: 67.83% TAT: 27.34% TBT: 4.84% (adolescents) TIT: 64.31% TAT: 28.39%	YES	Torrent-Fontbona and López (2018)

(continued)

Table 5 (continued)

Bolus calculation	Control measures	Glucose monitoring system	Insulin infusion system	Basal therapy	Initialization methodology	Safety system	System estimated price ($)	Results	Mobile application	References
								TBT: 7.3% (children)		
B=CH/ICR +(BG-GT)/ ISF-IOB	Gmin	SMBG/CGM	MDI/CSII	Optimised (Using iterative algorithm based on KF)	Using the ICR used by the user	NO	315	TIT: 80.14% TAT: 11.63% TBT: 8.23%	YES	Torrent-Fontbona et al. (2019)
B=ICR*CH	BG before and after the meal	CGM	CSII	Optimized (using second order R2R algorithm)	–	NO	4100	TIT: 98.7%, TAT: 0.7%, TBT: 0.6%	NO	(Tuo et al. 2015)
BB	NO	CGM (FreeStyle InsuLinx)	MDI	–	–	NO	640	0.17% of HbA1c reduction	NO	van Niel et al. (2014)
BB	NO	SMBG/CGM	MDI	–	–	NO	315	0.5% of HbA1c reduction	NO	Ziegler et al. (2013)
BB	NO	–	CSII (Accu-Chek Aviva Combo insulin pump)	–	–	NO	3525	–	NO	Ziegler et al. (2016)

The meaning of number of attributes is to quantify the attributes used in the recommender. The attributes can be used in the core or in equations. The names attributes lists the names of the attributes.

ICR calculation indicates the methodology used to determine the ICR. Some systems (Torrent-Fontbona and López 2018; Liu et al. 2013; Daskalaki et al. 2016; El-Sharkawy et al. 2018; Herrero et al. 2014, 2017; Lloyd et al. 2015; Palerm et al. 2007; Pesl et al. 2017; Sun et al. 2019; Torrent-Fontbona et al. 2019; Tuo et al. 2015) provide an equation or method, others (Pelzer 2006; Rossi et al. 2010; Charpentier et al. 2011; Schmidt et al. 2012; Ziegler et al. 2013, 2016) only the variables they depend on. The last ones (Al-Taee et al. 2015; Poerschke 2004; Anderson 2009; Glaser et al. 2004; Pańkowska and B-lazik 2010; Charpentier et al. 2011; Bergenstal et al. 2012; Boiroux et al. 2015; Brown et al. 2018; Cappon et al. 2018; Carić et al. 2017; Daskalaki et al. 2013; Garg et al. 2008; Gross et al. 2003; Hommel et al. 2017; Kirchsteiger and del Re 2014; Knight et al. 2016; Marling et al. 2008; Maurizi et al. 2011; Montani et al. 1998; Skrøvseth 2015; Sussman et al. 2012; van Niel et al. 2014) do not provide any information. ICR explains the relation between insulin and carbohydrates.

So, ICR specifies how much fast-acting insulin is needed per consumed carbohydrate. An example of ICR is 1:12, which means that one unit of insulin is needed for every 12 g of carbohydrate. Thus, a ratio of 1:8 means someone needing more insulin.

The ISF calculation attribute indicates the way in which ISF is calculated by each recommender system. Like in the previous section, some systems (Torrent-Fontbona and López 2018; Liu et al. 2013; El-Sharkawy et al. 2018; Herrero et al. 2014; Herrero et al. 2017; Lloyd et al. 2015; Pesl et al. 2017; Torrent-Fontbona et al. 2019; Brown et al. 2018) provide an equation or method, others (Pelzer 2006; Rossi et al. 2010; Schmidt et al. 2012; Ziegler et al. 2013, 2016) only variables they depend on, and the last ones (Al-Taee et al. 2015; Poerschke 2004; Anderson 2009; Glaser et al. 2004; Pańkowska and B-lazik 2010;

Daskalaki et al. 2016; Palerm et al. 2007; Sun et al. 2019; Tuo et al. 2015; Charpentier et al. 2011; Bergenstal et al. 2012; Boiroux et al. 2015; Cappon et al. 2018; Carić et al. 2017; Daskalaki et al. 2013; Garg et al. 2008; Gross et al. 2003; Hommel et al. 2017; Kirchsteiger and del Re 2014; Knight et al. 2016; Marling et al. 2008; Maurizi et al. 2011; Montani et al. 1998; Skrøvseth 2015; Sussman et al. 2012; van Niel et al. 2014) do not provide any information. ISF explains how sensitive a person is to the effects of insulin. ISF varies from subject to subject. If someone is insulin sensitive, it means they will require smaller amounts of insulin than someone who has low levels of sensitivity.

The bolus calculation attribute specifies how each recommender system calculates the dose of insulin for each intake. In the same way, as stated in two previous cases, some systems (Torrent-Fontbona and López 2018; Al-Taee et al. 2015; Liu et al. 2013; El-Sharkawy et al. 2018; Herrero et al. 2014; Herrero et al. 2017; Lloyd et al. 2015; Pesl et al. 2017; Sun et al. 2019; Torrent-Fontbona et al. 2019; Tuo et al. 2015; Charpentier et al. 2011; Boiroux et al. 2015; Brown et al. 2018; Cappon et al. 2018; Gross et al. 2003; Skrøvseth 2015) provide an equation or method, others (Anderson 2009; Glaser et al. 2004; Pańkowska and B-lazik 2010; Pelzer 2006; Rossi et al. 2010; Daskalaki et al. 2016; Schmidt et al. 2012; Bergenstal et al. 2012; Carić et al. 2017; Daskalaki et al. 2013; Hommel et al. 2017; Kirchsteiger and del Re 2014; Knight et al. 2016) only the variables they depend on, and the last ones (Poerschke 2004; Palerm et al. 2007; Ziegler et al. 2013, 2016; Garg et al. 2008; Marling et al. 2008, 2011; Montani et al. 1998; Sussman et al. 2012; van Niel et al. 2014) do not provide any information.

3.2.2 Control Procedures

Criteria in this block include: control measures and SS.

In some recommender systems (Torrent-Fontbona and López 2018; Al-Taee et al. 2015; Liu et al. 2013; Daskalaki et al. 2016; El-Sharkawy et al. 2018; Herrero et al. 2014, 2017; Palerm et al. 2007; Pesl et al. 2017; Sun et al. 2019; Torrent-Fontbona et al. 2019; Tuo et al. 2015; Charpentier et al. 2011; Boiroux et al. 2015; Brown et al. 2018; Daskalaki et al. 2013; Gross et al. 2003) there is a control process after each intake, usually over BG levels. The goal is to provide information to the recommender in order to take some further corrective actions and learn from subjects. As an example, if a patient takes 30 g of carbohydrates, BG will evolve over the next 3-h according to the carbohydrates (CH) intake, patient features and other external variables such as the amount of administered insulin. Then, depending on the glucose excursion, corrective actions are taken for further insulin recommendations, such as update ICR and ISF parameters.

Some systems (El-Sharkawy et al. 2018; Herrero et al. 2014; Palerm et al. 2007) include another sub-system, working in collaboration with the bolus recommender, in order to provide another layer of control to improve the safety of the whole system. For example, such system may constrain the bolus recommendations, if necessary. Several methodologies are available to do this safety control, but, in general, the reviewed papers do not provide precise information about their systems.

3.2.3 Complementary Processes

Criteria in this block include: basal therapy and initialization procedure. BI acts slowly and it is the cornerstone insulin-based therapy for people with T1DM. The function of BI is to work as background insulin, i.e. correcting glucose levels with a time window from several hours to one day. In the case of MDI, basal insulin is delivered as long-acting or intermediate insulin, injected one or two times a day. The process of adjusting the basal rates (units/hour), for the case of pump users, usually involves a profile of units per hour for a particular user and is usually done by clinicians, in collaboration with the patient. Normally, the better the basal rate adjustment, the lesser fast-acting insulin is needed.

The basal therapy attribute states the method used in the publication in order to set the BI rates. If the paper only describes the process of basal adjustment as an optimization via clinical expert

knowledge (EK), the process is then described as an optimization. If there is more information about how this process was done or something concerning the ratio, then this data is included in brackets.

Stepping on to complementary processes, IRS's provide bolus recommendations based on previous knowledge such as subject-specific ICR, previous insulin recommendations, etc. Thus, the initialization attribute details the methodology used to acquire this information before the first recommendation. In other words, how the cold start problem is solved. The initialization can be made all over the patients or individually. The usual sources of information are from current patients, data-bases or simulations. Not all the publications explain whether they use or not an initialization method.

3.2.4 Hardware

Criteria in this block include: glucose monitoring systems and insulin infusion systems.

IRS's require a device capable of monitoring the subject's BG levels. This process was often performed with a cheap and simple fingerstick that measures the BG concentration, i.e. SMBG. Nowadays, the use of CGM systems is spreading.

Thus, glucose monitoring system attribute specifies the monitoring technology used (SMBG or CGM) and, if the publication includes it, the product name followed by the company name inside parentheses. In some papers (Glaser et al. 2004; Rossi et al. 2010; Ziegler et al. 2013; Marling et al. 2008; Maurizi et al. 2011; Skrøvseth 2015), in which the system was tested conducting clinical trials with real patients, both technologies were used depending on the patient. In other papers (Torrent-Fontbona and López 2018; Liu et al. 2013; Daskalaki et al. 2016; El-Sharkawy et al. 2018; Herrero et al. 2014; Herrero et al. 2017; Palerm et al. 2007; Sun et al. 2019; Torrent-Fontbona et al. 2019; Tuo et al. 2015; Cappon et al. 2018; Daskalaki et al. 2013; Kirchsteiger and del Re 2014), the system was only tested performing simulations, but they decided to indicate for what kind of technology the system was designed. Nevertheless, in some of the previous cases (Poerschke 2004; Ziegler et al. 2016; Boiroux et al. 2015; Brown et al. 2018; Montani et al. 1998), there is no information about the glucose monitoring system.

The insulin infusion system refers to the methodology used by the subject to administer insulin. There are two main trends: (i) MDI, which consists of administering insulin via several injections per day with a syringe or insulin pen, and (ii) CSII, which consists of using a pump to continuously administer a predefined insulin amount according to a time profile.

Besides the insulin infusion technology, if the work includes such information, the product name and the name of the company is provided within brackets. As in the previous case, some papers (Torrent-Fontbona and López 2018; Liu et al. 2013; Glaser et al. 2004; Rossi et al. 2010; Palerm et al. 2007; Pesl et al. 2017; Torrent-Fontbona et al. 2019; Charpentier et al. 2011; Garg et al. 2008; Maurizi et al. 2011; Montani et al. 1998; Skrøvseth 2015) use both, and others (Poerschke 2004; Boiroux et al. 2015; Brown et al. 2018) do not clarify this information.

The mobile application attribute gathers information about the IRS support reported in the publication. Considering that most of the reviewed recommender systems assume that they should be run on a smartphone or require a frequent interaction with the user that can only be reached through a mobile device such a smartphone, the mobile application criterion specifies whether there is a mobile application, or not, that runs the bolus recommender algorithms and is used as user interface. In order to simplify, PDA applications are considered smartphone applications.

3.2.5 Testing and Assessment

Criteria in this block include: simulator, clinical trials, number of patients, and trial/simulation length.

Firstly, simulator criterion explains whether a simulator was used in order to test the IRS, and, if the publication contains this information, which simulator was used. That is, with the aim of testing new BC's, as IRS's are, several companies are offering diabetes simulators. The main features of the simulators are:

- Number and profile (age, weight, etc.) of in silico subjects.
- Basic simulation inputs (meals, exercise, basal, etc.)
- Simulation outputs (BG, insulin doses, meal intakes, etc.)
- Simulation health scores (mean BG, percentage of time in hypoglycaemia, percentage of time in hyperglycaemia, BG risk index, etc.)

The main advantages of using simulators are related to the cost and time requirements to perform clinical trials, and the risk of testing new aggressive BC's on people. The main drawbacks of using simulators are uncertainty in the results and knowledge requirements of some simulators.

Secondly, the clinical trials criterion specifies whether clinical trials were conducted or not to test the IRS of the publication. When clinical trials are designed, clinicians usually search for patients with particular features with the aim of having a significant sample. This is the way to guarantee that the sample follows the distribution of the population they want to study. Subsequently, they split the sample between control group and intervention group. The intervention group is the cohort in which the new therapy, that is new BC, is tested. The other group is used to statistically appreciate the real effects of the intervention. The splitting percentages are frequently close to 50%.

Thirdly, the number of patients refers to the amount of patients used for testing the IRS in the publication and information about their nature: real or virtual patients. In addition, information about the patients age has been considered. The tests done with virtual patients are known as in silico tests.

Finally, the trial/simulation length feature measures the number of days the clinical trials or the simulations last. The longer, the better and more significant the results are.

3.2.6 System Estimated Price

This information is included with the aim of depicting the economic expenditure of each solution. None of the reviewed articles includes this information. In order to provide this information, the commercial prices of several devices were searched, averaged and then rounded off. After doing this process, the following prices were set: (i) mobile unit, either PDA or smartphone, is set to 250 $, (ii) SMBG is set to 25 $, (iii) CGM is set to 350 $, (iv) Insulin pen is set to 40 $, and (v) CSII is set to 3500 $. Other expenditures, such as clinical control, computer technician work, CGM sensors, consumables and server services are not considered.

Doing that, the price of each proposal is approximated on the whole in order to provide a rough idea.

3.2.7   Results

There is a clear lack of consensus regarding to the scores used to provide the results in this topic. The most frequently used metrics are Glycated haemoglobin (bA1c), time in target (TIT) and low blood glucose index (LBGI). HbA1c is the glycosylated haemoglobin test, that is the average of BG during the last 3 months. TIT is referred to the percentage of time within the glycaemic range, which is usually 70–180 mg/dL. LBGI is a metric used to calculate the risk of hypoglycaemia. Other score indicators not directly related with the BG, such as bolus acceptance or degree of happiness using the application, are not considered in this analysis. Generally, the best results in each publication are highlighted.

In summary, the present column provides the metric included in the publication, prioritizing the HbA1c and TIT, if available. If not, the provided score is the one shown in the publication.

4   Discussion

4.1   Technology

The main drawback according to the core of the reviewed insulin recommenders is the high percentage of undocumented systems. Up to 30% of proposals are defined as black boxes (Schmidt et al. 2012; Ziegler et al. 2013, 2016; Bergenstal et al. 2012; Caríc et al. 2017; Garg et al. 2008;

Gross et al. 2003; Hommel et al. 2017; Knight et al. 2016; Maurizi et al. 2011; Sussman et al. 2012; van Niel et al. 2014), but in addition, another 15% are based on an undefined EK (Poerschke 2004; Anderson 2009; Glaser et al. 2004; Pańkowska and B-lazik 2010; Pelzer 2006; Rossi et al. 2010) according to some subjective criteria. Amongst the documented core systems, the most common methodology is case-based reasoning (CBR) (Torrent-Fontbona and López 2018; El-Sharkawy et al. 2018; Herrero et al. 2014, 2017; Torrent-Fontbona et al. 2019; Brown et al. 2018; Pesl et al. 2016), which appears in 18% of the publications. The second and the third are reinforcement learning (RL) (Daskalaki et al. 2016; Sun et al. 2019; Daskalaki et al. 2013) and K-nearest neighbours (KNN) (Marling et al. 2008; Montani et al. 1998; Skrøvseth 2015), used in 8% of the recommender systems each one.

The presented works used 8 attributes on average. There is 1 publication (Bergenstal et al. 2012) in which only 1 attribute is used, whilst 2 proposals (Marling et al. 2008; Montani et al. 1998) that used up to 25 attributes. The fifth most used attributes are: BG, CH, ISF, ICR and GT.

In 58% of the works there is no information about how the process of ICR calculation is performed, that is publications without any information. The same happens in 63% of the publications in ISF case and 25% in bolus equations case. The third most frequently used parameters are BG, CH and W in ICR case, and CH, ICR and W in ISF case. The fifth most used parameters in the bolus calculation process are BG, CH, GT, ICR and ISF. There is a clear homogenization on how the calculation process is currently done, as can be appreciated in the last works.

4.2   Control Measures

Amongst the reviewed systems, 42% use some kind of control measure. And the most popular measure is Gmin, which is used in 43% of cases with control measure. Besides, the use of Gmin is increasing in the last years.

Only 7% of works propose recommender systems with integrated SS's and, among these, only one publication (Palerm et al. 2007) explains how the SS works.

4.3 Complementary Procedures

Near 58% of the works optimized the basal therapy. Only 25% of works explain the technique used to optimize the BI. And, amongst these works, there is a high variability in the proposed methodologies. There are only two methodologies used in more than one work, which are Kalman filtering (KF) and previous glycaemic profile.

A percentage of 40% of publications include an initialization method. There is no apparent tendency in the initialization method used, but patient questionnaires, CGM data and simulator outputs are the most popular used methods.

4.4 Hardware

Regarding the glucose monitoring system, CGM is used in 40% of the works, SMBG in 25% and 22% can use both. There is an obvious preponderance of CGM systems in the most recent works. On the other hand, 33% use MDI as insulin infusion system, 30% CSII and 30% both. There is no clear temporal trend regarding this aspect and both technologies are being used.

With regard to the smartphone applications, 38% of the recommender systems use them, while 62% do not. The more recent publications show a higher amount of IRS's based on smartphone application.

4.5 Testing

Around 35% of works include some kind of simulation, while 52% of them include clinical trials. Roughly 10% of works do not provide any test (simulation or clinical trial), and only Palerm publication (Palerm et al. 2007) provides both.

Among works that use simulation, UVA PADOVA T1DM simulator is the most common, appearing in 67% of the works. Analysing the year of publication, there is a noticeable tendency towards the use of UVA-PADOVA simulator.

Simulation tests are on average performed with 35 virtual subjects (VS's). In particular, Daskalaki work (Daskalaki et al. 2016) is the paper with more VS's, 130, whilst Brown (Brown et al. 2018) only used 1 virtual patient. Besides, 75% of works are based only on adult patient data. Referring to works performing clinical trials, the average number of patients is approximately 102. The publication with less subjects in the clinical trials was performed by El-Sharkawy (El-Sharkawy et al. 2018) with only 1 patient. On the other hand, Anderson (Anderson 2009), which included 573 subjects, is the paper with the greatest number of subjects. The average length of clinical trials and simulations is 186 and 61 days, respectively. The longest clinical trials, 720 days, were performed by Anderson (Anderson 2009), while the longest simulations, 180 days, were run by Brown (Brown et al. 2018) and Torrent-Fontbona (Torrent-Fontbona et al. 2019). The shortest clinical trial, 4 days, was performed by El-Sharkawy et al. (2018), and the shortest simulation, 1 day, by Cappon et al. (2018).

Regarding the length of the clinical trials, the longer the better, since at least 3 months are necessary to see differences in metrics such as the HbA1c. However, longer trials allow the analysis of the systems performance in the long run, when the initially increased motivation of the subject for using the tested system has disappeared. On the other hand, simulations lengths are chosen, usually, according to the convergence time of the system and the VS's, i.e. the time the system needs to learn and becomes optimal plus the time the VS glucose behaviour has been stabilised for the optimised bolus recommender. Systems that want to test how the systems reacts in front of changes of subjects' routine or metabolism may also require long simulations.

4.6   Pricing

About price estimation, 39% of the products have a price lower than 300$. In 34% of the products the price is among 300$ and 1500$. And finally, 27% of them have an estimated price above 1500 $. Providing a general estimation, the idea is to offer a quick view about how the market prices are. The type of insulin infusion system largely determines the price of the IRS.

4.7   Results

No meta-analysis was performed because results comparison is very difficult. This is due to the variability in how they are presented, the metrics used, and how the experiments are performed. Works using simulated results used the TIT as performance score in 54% of the cases. On the other side, 88% of works that conducted clinical trials provided HbA1c as score indicator. In relation with papers with simulated results, the mean HbA1c reduction, considering all the publications, is 0.55%. Looking at all clinical trial publications, the mean TIT is 89.14%, 7.79% in terms of TAT and 2.52% in terms of TBT. Charpentier et al. (2011) achieved the highest performance with 0.91% HbA1c reduction and the highest TIT was achieved by Tuo et al. (2015), attaining a 98.7%. Nevertheless, the conditions of the simulations and trials for the different publications were significantly different, thus it became impossible to fairly compare results provided by the reviewed publications. For the future, it would be also interesting to use the same performance indicators in the topic.

According to the year of publication, 25% of works were done before 2010. The works published between 2010 and 2015 account for the 43% of all. Finally, the papers released after 2015, account for the 32% of the total. It seems that the bolus recommender topic is maintaining its popularity.

5   Conclusions

The current review offers a general perspective of the IRS's after 70 works have been analysed. The main conclusions are provided as follows: There is lack of papers that include both clinical trials and simulations in the experimentation process, while this would be appropriate. In addition, due to the high variability of research works, it would be helpful to establish standards in relation with the minimum number of patients and the experimentation length required to reach statistical significance. As far as core methodology is concerned, there is a clear lack of information and undesired obscurantism in the commercial recommenders. Despite everything, CBR is the most used methodology as core in the AI of IRS's.

About the number of attributes, unexpectedly no apparent trend of increasing them is revealed, even though current devices can afford a higher number. In the latest works, the equations used to calculate ICR, ISF and bolus are slowly converging. Similarly, the most popular control measure is the Gmin. Thus, there is consensus on how to calculate the ICR, ISF, bolus and the measure used to correct or update them.

Addressing the hardware question, CGM and smartphone use is increasing while there is no clear trend about the insulin infusion system. This is the main factor when the price has to be take into consideration as it mainly depends on hardware.

As far as the basal therapy and the initialization methodologies is concerned, there is heterogeneity and widespread use, but no clear future path. This is not the case of SS's, not frequently used and, generally, providing scarce information.

The results cannot be directly compared due to the diversity in the used performance scores and the features of each experimentation such as the length of experiments, the used hardware, the number of patients, etc. The topic requires to set up standards in order to identify and appreciate improvements, ensuring statistical significance of the experiments. However, according to the number of publications, the topic is manifesting effervescence together with the healthcare field.

Acknowledgments This project has received funding from the grant of the University of Girona 20162018 (MPCUdG2016) and the European Union Horizon 2020 research and innovation programme under grant agreement No. 689810, www.pepper.eu.com/, PEPPER. The work has been developed with the support of the research group SITES awarded with distinction by the Generalitat de Catalunya (SGR 20142016). Members of the PEPPER Group are as follows: project leader: C. Martin; project management team: P. Herrero Vinas, L. Nita, J. Massana, J. Masoud, J.M. FernndezReal, B. Lpez; co-investigators: A. Aldea, D. Duce, M. Fernndez-Balsells, P. Georgiou, R. Harrison, B. Innocenti, Y.Leal, N. Oliver, R. Petite, M. Reddy, C. Roman, J. Shapley, M. Waite and M. Wos.

Bibliography

Alberti KGMM, Zimmet PZ (1998) Definition, diagnosis and classification of diabetes mellitus and its complications. Part 1: diagnosis and classification of diabetes mellitus. Provisional report of a WHO consultation. Diabet Med 15(7):539–553

Al-Taee AM, Al-Taee MA, Al-Nuaimy W, Muhsin ZJ, AlZu'bi H (2015) Smart bolus estimation taking into account the amount of insulin on board. In: 2015 IEEE international conference on computer and information technology; ubiquitous computing and communications; Dependable, autonomic and secure computing; pervasive intelligence and computing. IEEE, pp 1051–1056

Anderson DG (2009) Multiple daily injections in young patients using the ezy-BICC bolus insulin calculation card, compared to mixed insulin and CSII. Pediatr Diabetes 10(5):304–309

Atkinson MA, Eisenbarth GS (2001) Type 1 diabetes: new perspectives on disease pathogenesis and treatment. Lancet 358(9277):221–229

Atkinson MA, Eisenbarth GS, Michels AW (2014) Type 1 diabetes. Lancet 383(9911):69–82

Atlas D (2019) International diabetes federation, 9th edn, IDF Diabetes Atlas, Brussels. Available from: http://www.diabetesatlas.org

Bailey TS, Stone JY (2017) A novel pen-based Bluetooth-enabled insulin delivery system with insulin dose tracking and advice. Expert Opin Drug Deliv 14 (5):697–703

Bergenstal RM, Bashan E, McShane M, Johnson M, Hodish I (2012) Can a tool that automates insulin titration be a key to diabetes management? Diabetes Technol Ther 14(8):675–682

Boiroux D, Aradtótir TB, Hagdrup M, Poulsen NK, Madsen H, Jørgensen JB (2015) A bolus calculator based on continuous-discrete unscented Kalman filtering for type 1 diabetics. IFAC-PapersOnLine 48 (20):159–164

Brown D, Aldea A, Harrison R, Martin C, Bayley I (2018) Temporal casebased reasoning for type 1 diabetes mellitus bolus insulin decision support. Artif Intell Med 85:28–42

Cappon G, Vettoretti M, Marturano F, Facchinetti A, Sparacino G (2018) A neural-network-based approach to personalize insulin bolus calculation using continuous glucose monitoring. J Diabetes Sci Technol 12 (2):265–272

Caríc B, Lalíc K, Marin S, Stašíc L, Pejícíc-Popovíc S (2017) The importance of the bolus calculator use for improving glycemic control in patients on the insulin pump therapy. Scr Med 48(1):45–52

Charpentier G, Benhamou PY, Dardari D, Clergeot A, Franc S, Schaepelynck-Belicar P et al (2011) The Diabeo software enabling individ- ualized insulin dose adjustments combined with telemedicine support improves HbA1c in poorly controlled type 1 diabetic patients: a 6-month, randomized, open-label, parallel-group, multicenter trial (TeleDiab 1 study). Diabetes Care 34(3):533–539

Clements MA, Staggs VS (2017) A mobile app for synchronizing glucometer data: impact on adherence and glycemic control among youths with type 1 diabetes in routine care. J Diabetes Sci Technol 11 (3):461–467

Daskalaki E, Diem P, Mougiakakou SG (2013) An Actor–Critic based controller for glucose regulation in type 1 diabetes. Comput Methods Prog Biomed 109 (2):116–125

Daskalaki E, Diem P, Mougiakakou SG (2016) Model-free machine learning in biomedicine: feasibility study in type 1 diabetes. PLoS One 11(7):e0158722

Diabetes AA. Insulin basics (2019). Available from: https://www.diabetes.org/diabetes/medication-management/insulin-other-injectables/insulin-basics

Drion I, Pameijer LR, van Dijk PR, Groenier KH, Kleefstra N, Bilo HJG (2015) The effects of a mobile phone application on quality of life in patients with type 1 diabetes mellitus: a randomized controlled trial. J Diabetes Sci Technol 9(5):1086–1091

El-Sharkawy M, Daniels J, Pesl P, Reddy M, Oliver N, Herrero P et al (2018) A portable low-power platform for ambulatory closed loop control of blood glucose in type 1 diabetes. In: 2018 IEEE International Symposium on Circuits and Systems (ISCAS). IEEE, pp 1–5

Garg SK, Bookout TR, McFann KK, Kelly WC, Beatson C, Ellis SL et al (2008) Improved glycemic control in intensively treated adult subjects with type 1 diabetes using insulin guidance software. Diabetes Technol Ther 10(5):369–375

Garg SK, Shah VN, Akturk HK, Beatson C, Snell-Bergeon JK (2017) Role of mobile technology to improve diabetes care in adults with type 1 diabetes: the remote-T1D study iBGStar§R in type 1 diabetes management. Diabet Ther 8(4):811–819

Glaser NS, Iden SB, Green-Burgeson D, Bennett C, Hood-Johnson K, Styne DM et al (2004) Benefits of an insulin dosage calculation device for adolescents with type 1 diabetes mellitus. J Pediatr Endocrinol Metab 17 (12):1641–1652

Gross TM, Kayne D, King A, Rother C, Juth S (2003) A bolus calcu lator is an effective means of controlling postprandial glycemia in patients on insulin pump therapy. Diabetes Technol Ther 5(3):365–369

Herrero P, Pesl P, Reddy M, Oliver N, Georgiou P, Toumazou C (2014) Advanced insulin bolus advisor based on run-to-run control and case-based reasoning. IEEE J Biomed Health Inform 19(3):1087–1096

Herrero P, Bondia J, Adewuyi O, Pesl P, El-Sharkawy M, Reddy M et al (2017) Enhancing automatic closed-loop glucose control in type 1 diabetes with an adaptive meal bolus calculator–in silico evaluation under intra-day variability. Comput Methods Prog Biomed 146:125–131

Hommel E, Schmidt S, Vistisen D, Neergaard K, Gribhild M, Almdal T et al (2017) Effects of advanced carbohydrate counting guided by an automated bolus calculator in type 1 diabetes mellitus (Steno ABC): a 12-month, randomized clinical trial. Diabet Med 34 (5):708–715

Kirchsteiger H, del Re L (2014) A model based bolus calculator for blood glucose control in type 1 diabetes. In: 2014 American control conference. IEEE, pp 5465–5470

Kirwan M, Vandelanotte C, Fenning A, Duncan MJ (2013) Diabetes selfmanagement smartphone application for adults with type 1 diabetes: randomized controlled trial. J Med Internet Res 15(11):e235

Knight BA, McIntyre HD, Hickman IJ, Noud M (2016) Qualitative assessment of user experiences of a novel smart phone application designed to support flexible intensive insulin therapy in type 1 diabetes. BMC Med Inform Decis Mak 16(1):119

Liu SW, Huang HP, Lin CH, Chien IL (2013) Fuzzy-logic-based supervisor of insulin bolus delivery for patients with type 1 diabetes mellitus. Ind Eng Chem Res 52(4):1678–1690

Lloyd B, Groat D, Cook CB, Kaufman D, Grando A (2015) iDECIDE: a mobile application for insulin dosing using an evidence based equation to account for patient preferences. Stud Health Technol Inform 216:93

Marling C, Shubrook J, Schwartz F (2008) Case-based decision support for patients with type 1 diabetes on insulin pump therapy. In: European conference on case-based reasoning. Springer, Berlin, Germany, pp 325–339

Maurizi AR, Lauria A, Maggi D, Palermo A, Fioriti E, Manfrini S et al (2011) A novel insulin unit calculator for the management of type 1 diabetes. Diabetes Technol Ther 13(4):425–428

Montani S, Bellazzi R, Portinale L, Fiocchi S, Stefanelli M (1998) A casebased retrieval system for diabetic patients therapy. Proc IDAMAP 98:64–70

Mora P, Buskirk A, Lyden M, Parkin CG, Borsa L, Petersen B (2017) Use of a novel, remotely connected diabetes management system is associated with increased treatment satisfaction, reduced diabetes distress, and improved glycemic control in individuals with insulintreated diabetes: first results from the personal diabe. Diabetes Technol Ther 19(12):715–722

Olokoba AB, Obateru OA, Olokoba LB (2012) Type 2 diabetes mellitus: a review of current trends. Oman Med J 27(4):269

Palerm CC, Zisser H, Bevier WC, Jovanovïc L, Doyle FJ (2007) Prandial insulin dosing using run-to-run control: application of clinical data and medical expertise to define a suitable performance metric. Diabetes Care 30(5):1131–1136

Pańkowska E, B-lazik M (2010) Bolus calculator with nutrition database software, a new concept of prandial insulin programming for pump users. J Diabetes Sci Technol 4(3):571–576

Pelzer R (2006) A new approach to improving the control of type 1 diabetes. North-West University, Potchefstroom, South Africa

Pesl P, Herrero P, Reddy M, Xenou M, Oliver N, Johnston D et al (2016) An advanced bolus calculator for type 1 diabetes: system architecture and usability results. IEEE J Biomed Health Inform 20(1):11–17

Pesl P, Herrero P, Reddy M, Oliver N, Johnston DG, Toumazou C et al (2017) Case-based reasoning for insulin Bolus Advice: evaluation of case parameters in a six-week pilot study. J Diabetes Sci Technol 11 (1):37–42

Poerschke C (2004) Development and evaluation of an intelligent handheld insulin dose advisor for patients with Type 1 diabetes. Oxford Brookes University, Oxford, UK

Railton D (2019) How to manage diabetes with basal-bolus in sulin therapy. Medical News Today. Available from: https://www.medicalnewstoday.com/articles/316616.php

Rosales N, De Battista H, Vehí J, Garelli F (2018) Open-loop glucose control: automatic IOB-based super-bolus feature for commercial insulin pumps. Comput Methods Prog Biomed 159:145–158

Rossi MCE, Nicolucci A, Di Bartolo P, Bruttomesso D, Girelli A, Ampudia FJ et al (2010) Diabetes Interactive Diary: a new telemedicine system enabling flexible diet and insulin therapy while improving quality of life: an open-label, international, multicenter, randomized study. Diabetes Care 33(1):109–115

Schmidt S, Meldgaard M, Serifovski N, Storm C, Christensen TM, Gade-Rasmussen B et al (2012) Use of an automated bolus calculator in MDI-treated type 1 diabetes: the BolusCal Study, a randomized controlled pilot study. Diabetes Care 35(5):984–990

Skrøvseth SO (2015) °Arsand E, Godtliebsen F, Joakimsen RM. data- driven personalized feedback to patients with type 1 diabetes: a randomized trial. Diabetes Technol Ther 17(7):482–489

Sun Q, Jankovic M, Budzinski J, Moore B, Diem P, Stettler C et al (2019) A dual mode adaptive basal-bolus advisor based on reinforcement learning. IEEE J Biomed Health Inform 23(6):2633–2641

Sussman A, Taylor EJ, Patel M, Ward J, Alva S, Lawrence A et al (2012) Performance of a glucose meter with a built-in automated bolus calculator versus manual bolus calculation in insulin-using subjects. J Diabetes Sci Technol 6(2):339–344

Torrent-Fontbona F, López B (2018) Personalized adaptive CBR bolus recommender system for type 1 diabetes. IEEE J Biomed Health Inform 23(1):387–394

Torrent-Fontbona F, Massana J, López B (2019) Case-base maintenance of a personalised and adaptive CBR bolus insulin recommender system for type 1 diabetes. Expert Syst Appl 121:338–346

Tuo J, Sun H, Shen D, Wang H, Wang Y (2015) Optimization of insulin pump therapy based on high order run-to-run control scheme. Comput Methods Prog Biomed 120(3):123–134

Vallejo-Mora MR, Carreira-Soler M, Linares-Parrado F, Olveira G, Rojo-Martinez G, Dominguez-López M et al (2017) The Calculating Boluses on Multiple Daily Injections (CBMDI) study: a randomized controlled trial on the effect on metabolic control of adding a bolus calculator to multiple daily injections in people with type 1 diabetes: (CBMDI). J Diabetes 9(1):24–33

van Niel J, Geelhoed-Duijvestijn PH, Group DIS, Others (2014) Use of a smart glucose monitoring system to guide insulin dosing in patients with diabetes in regular clinical practice. J Diabetes Sci Technol 8(1):188

Ziegler R, Cavan DA, Cranston I, Barnard K, Ryder J, Vogel C et al (2013) Use of an insulin bolus advisor improves glycemic control in multiple daily insulin injection (MDI) therapy patients with suboptimal glycemic control: first results from the ABACUS trial. Diabetes Care 36(11):3613–3619

Ziegler R, Rees C, Jacobs N, Parkin CG, Lyden MR, Petersen B et al (2016) Frequent use of an automated bolus advisor improves glycemic control in pediatric patients treated with insulin pump therapy: results of the Bolus Advisor Benefit Evaluation (BABE) study. Pediatr Diabetes 17(5):311–318

Adv Exp Med Biol - Advances in Internal Medicine (2020) 4: 357–373
https://doi.org/10.1007/5584_2020_499
© Springer Nature Switzerland AG 2020
Published online: 13 March 2020

Algorithms for Diagnosis of Diabetic Retinopathy and Diabetic Macula Edema- A Review

Karkuzhali Suriyasekeran,
Senthilkumar Santhanamahalingam,
and Manimegalai Duraisamy

Abstract

Human eye is one of the important organs in human body, with iris, pupil, sclera, cornea, lens, retina and optic nerve. Many important eye diseases as well as systemic diseases manifest themselves in the retina. The most widespread causes of blindness in the industrialized world are glaucoma, Age Related Macular Degeneration (ARMD), Diabetic Retinopathy (DR) and Diabetic Macula Edema (DME). The development of a retinal image analysis system is a demanding research topic for early detection, progression analysis and diagnosis of eye diseases. Early diagnosis and treatment of retinal diseases are essential to prevent vision loss. The huge and growing number of retinal disease affected patients, cost of current hospital-based detection methods (by eye care specialists) and scarcity in the number of ophthalmologists are the barriers to achieve the recommended screening compliance in the patient who is at the risk of retinal diseases. Developing an automated system which uses pattern recognition, computer vision and machine learning to diagnose retinal diseases is a potential solution to this problem. Damage to the tiny blood vessels in the retina in the posterior part of the eye due to diabetes is named as DR. Diabetes is a disease which occurs when the pancreas does not secrete enough insulin or the body does not utilize it properly. This disease slowly affects the circulatory system including that of the retina. As diabetes intensifies, the vision of a patient may start deteriorating and leading to DR. The retinal landmarks like OD and blood vessels, white lesions and red lesions are segmented to develop automated screening system for DR. DME is an advanced symptom of DR that can lead to irreversible vision loss. DME is a general term defined as retinal thickening or exudates present within 2 disk diameter of the fovea center; it can either focal or diffuse DME in distribution. In this paper, review the algorithms used in diagnosis of DR and DME.

K. Suriyasekeran (✉)
Department of Computer Science and Engineering, Kalasalingam Academy of Research and Education, Sriviliputtur, Tamilnadu, India
e-mail: vijikarkuzhali@gmail.com

S. Santhanamahalingam
Department of Chemistry, Ayya Nadar Janaki Ammal College, Sivakasi, Tamilnadu, India
e-mail: santhanasenthil@gmail.com

M. Duraisamy
Department of Information Technology, National Engineering College, Kovilpatti, Tamilnadu, India
e-mail: megalai_nec@yahoo.co.in

Keywords

Diabetic edema · Macula · Blood vessels ·
Classification · Diabetic retinopathy ·
Exudates · Hemorrhages · Macula ·
Microanaurysms · Optic disc · Segmentation

1   Introduction

The human eye is a sensory organ for vision. It is
well designed to collect the significant information about the environment around. The retina is
the light sensitive tissue that lies in the posterior
segment of the eye. It collects light focused from
the lens, converts light signal into neural signal
and then sends these signals to optic nerve. The
optic nerve carries the signal to the brain; it helps
to process the image. The color fundus retinal
photography records interior surface of the eye
that includes retina, retinal vasculature, Optic
Disc (OD), macula and posterior pole. It helps
to interpret retinal landmarks and is used for the
screening of patients suffering from sight-
threatening retinal diseases. The fact from World
Health Organization (WHO) divulges that almost
285 million people are estimated to be visually
impaired, among them 39 million people are blind
and 246 million people have low vision. Approximately, 12.3%, 8.7% and 4.8% of visual
impairment are caused by eye diseases like Glaucoma, Age-Related Macular Degeneration
(ARMD) and Diabetic Retinopathy (DR).

Diabetes or Diabetes mellitus is a disease
which occurs when the pancreas does not secrete
enough insulin or the body is unable to utilize it
properly. The disease slowly affects the circulatory system of the human body and retina. As
diabetes intensifies, the vision of a patient may
start to deteriorate and lead to DR. DR is an
adverse change in retinal Blood Vessels (BVs)
leading to vision loss without any symptoms.
Diabetic Macula Edema (DME) is the advanced
symptom of DR, and leads to irreversible vision
loss. The swelling happens in the macula region
due to leakage of fluid from BVs within the
macula that appears as the presence of Exudates
(EXs) in macula region. Early diagnosis and treatment of retinal diseases are essential to prevent
vision loss, and it would significantly reduce the
workload for the ophthalmologist.

2   Fundus Photography

The drawing of retinal vasculature including OD
and fovea was published by Purkyne in 1823.
Figure 1 shows the first diagrammatic representation of retina.

The first image of the retina was published by
Van Trigt, a Dutch ophthalmologist in 1853 for
analysis of retina. Figure 2 shows the first retinal
image captured by Van Trigt.

After few years, the first photographic image
of the retina was captured by Gerloff, a German
ophthalmologist in 1891 that shows the BVs
clearly (Abràmoff et al. 2010). The fundus camera is a complex optical system used for
illuminating and imaging the retina of the eye.
These devices generally consist of a microscope
attached with a digital camera. It provides an
upright, magnified view of the interior surface of
the eye called fundus. The fundus images are
two-dimensional representations of three-
dimensional retinal tissue projected on to the
imaging plane obtained using reflected light.
The widely used retinal color fundus photographs
have represented the amount of the reflected Red,
Green and Blue wavebands as determined by the
spectral sensitivity of the sensor.

In general to avoid misinterpretations, at least
four images of the fundus are taken during eye

Fig. 1 First diagrammatic
illustration of retina

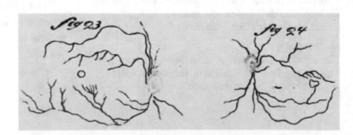

Fig. 2 First image of retina

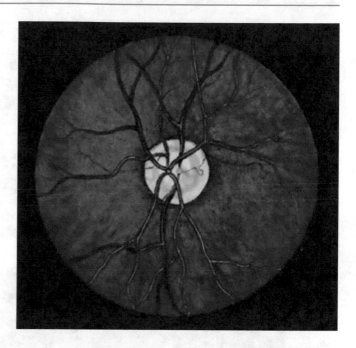

examination for both left and right eye. For each eye, the images will be obtained with centers as Macula and OD separately. The names of the different types of images are given below.

- Macula Centered Left Eye (MCLE)
- Macula Centered Right Eye (MCRE)
- Optic Disc Centered Left Eye (ODCLE)
- Optic Disc Centered Right Eye (ODCRE)

Eye disease like glaucoma is diagnosed using ODCLE and ODCRE, and retinal diseases like ARMD, DR and DME are identified using MCLE and MCRE. A healthy retinal image in MCLE and MCRE may contain clear anatomical structures like macula, OD and BVs, whereas an abnormal image shows the pathological features along with distorted anatomical features. The location of the OD differentiates the left and the right eye images. In the retinal image of a left eye, the macula is located on the right of the OD and in a right eye, macula is located on the left of the OD as shown in Fig. 3a–d.

3 Retinal Image Analysis System

Human retinal images have gained an important role in the detection and diagnosis of many eye diseases. Irrespective of the techniques used, the retinal images are mandatory for this application because the abnormalities are clearly observed in the retina than in any other part of the human eye. Retinal fundus images are easily storable and can be transmitted to anywhere at any time. It can be processed and analyzed to improve image quality and perform objective quantitative analysis. Retinal image analysis system is used for diagnosing and treatment of several diseases related to eye. Examples of such diseases are glaucoma, ARMD, DR, DME and other chronic diseases that may introduce complications in the retina such as cardiovascular and kidney diseases. Early diagnosis and timely treatment of eye diseases are essential to prevent vision loss. It would significantly reduce the workload of the ophthalmologists. The huge and growing number of retinal disease affected patients, cost of current hospital-based detection methods (by eye care specialists) and scarcity in the number of ophthalmologists are the barriers to achieve the recommended screening compliance in the patient who is at the risk of retinal diseases. Developing a retinal image analysis system using image processing, pattern recognition, machine vision and machine learning algorithms to diagnose retinal diseases is a potential solution to this problem. Numerous research efforts have been made in the area of retinal image analysis. Broadly categorized six

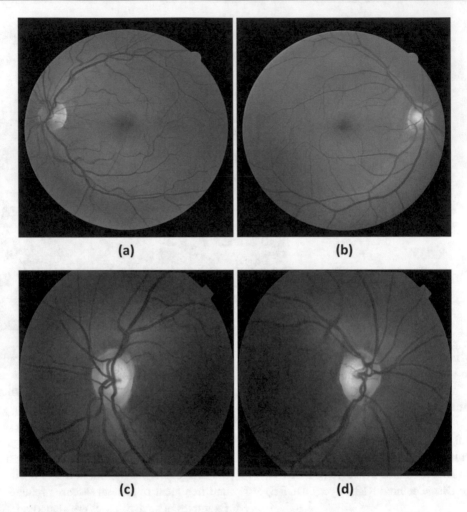

(a)   (b)

(c)   (d)

Fig. 3 Various types of retinal images (**a**) MCLE, (**b**) MCRE (**c**) ODCLE (**d**) ODCRE

components in retinal image analysis systems (RIAS) are explained below.

A. **Image acquisition:** The first stage of RIAS is image acquisition. The color retinal fundus images for processing, implementation and testing of the developed RIAS can be taken from the publicly available retinal image database and in addition to that images are collected from Bejan Singh Eye Hospital, Nagerkoil.

B. **Image preprocessing:** The retinal images, consisting of different sizes and taken from different cameras are used as input data for preprocessing. In retinal image acquisition, the retinal images' quality has been affected by factors such as medical opacities, defocus or presence of artifact. The image quality is improved by image enhancement and restoration techniques. Image enhancement is used to improve or develop the retinal image for further processing and analysis. Techniques like histogram equalization, adaptive histogram equalization and contrast-limited adaptive histogram equalization are used. Image restoration removes salt-and-pepper noise, Gaussian noise and periodic noise by transforming the image to a different structure called Fourier transform, then applying noise filters to transform back to the original image.

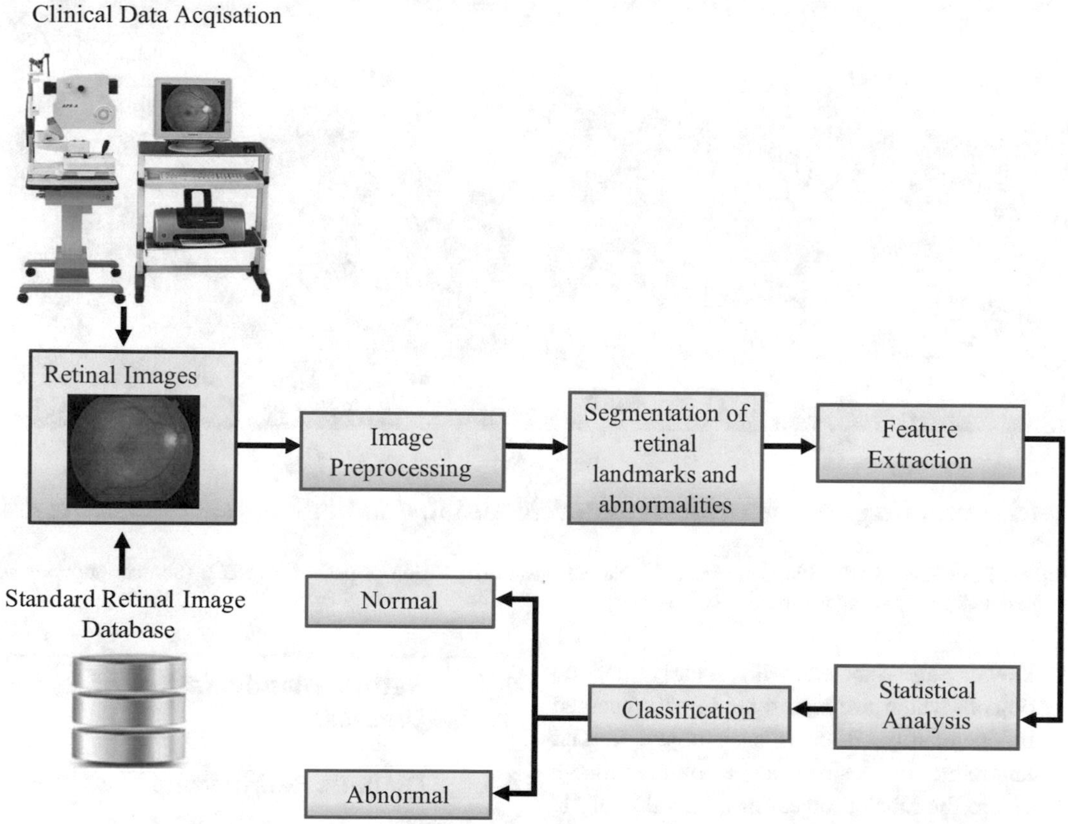

Clinical Data Acqisation

Fig. 4 Flow diagram of RIAS

C. **Image segmentation:** Segmentation subdivides an image into its constituent parts or objects. Segmentation based on discontinuity attempt is to partition the image by detecting abrupt changes in gray level such as point, line, and edge. Segmentation based on similarity attempt is to create uniform regions by grouping together connected pixels that satisfy predefined similarity criteria (Gonzalez and Woods 2002). Therefore, the results of segmentation depend critically on these criteria and on the definition of connectivity. The algorithms are implemented by combining the process like thresholding, edge detection, filters and morphological processing. It has been used for creating mechanism of detecting edges, background removal as well as for finding the specific shape of retinal objects. Segmentation process

in retinal images involves the identification of common retinal landmarks like OD, OC, BVs, macula and fovea or the identification of retinal lesions such as EXs, Cotton Wool Spots (CWSs), Microaneurysms (MAs), Hemorrhages (HAs) and Choroidal NeoVascularizations (CNVs). Many subsequent tasks such as retinal image classification rely on the quality of this segmentation process. Figure 4 shows the retinal image analysis system.

D. **Feature extraction and statistical analysis:** The goal of RIAS is to extract useful retinal landmarks and lesions. The segmentation is used to reduce the amount of image data. Feature extraction technique has been used to extract image-based features which are application-dependent like texture, shape or statistical features. The resultant segmented

(a)   (b)

(Source: http://www.professionaleye.com/faqs/diabetic-eyecare-faq.aspx)

Fig. 5 The effect of DR in human vision. (**a**) Normal vision (**b**) Vision of patient with DR. (Source: http://www.professionaleye.com/faqs/diabetic-eyecare-faq.aspx)

images are labeled with clearly defined objects which are represented and processed independently. It treats each object (retinal landmark or lesion) as a binary image where the labeled object has the value of '1' and everything else is '0'. The features like area, perimeter, mean, variance, and compactness ratio have been extracted, and most suitable features have been selected by performing statistical analysis for image classification.

E. **Image classification:** An ultimate alternative to the conventional disease identification technique is automated disease classification technique. The algorithm is used to analyze retinal images and classify with confirmed diagnosis of the state of (or absence) the disease. The computational techniques employed for medical image analysis can be broadly classified into Artificial Intelligence (AI)-based techniques and conventional-based techniques. Retinal image classification is performed to assign a class for a retinal image such as a normal retina or a retina with lesion, i.e. a retina affected with glaucoma, ARMD, DR or DME.

4 Retinal Manifestation of Eye Diseases

4.1 Diabetic Retinopathy

The diabetes is a chronic condition associated with high blood sugar levels. All patients who have been diagnosed with type-I or type-II diabetes are at risk of developing DR. (Zaki et al. 2016). The consequence of untreated diabetes may lead to progression of DR. DR is an ocular manifestation of diabetes, and diabetics are at risk of vision loss due to DR. (Olson et al. 2015). Figure 5 shows the vision of normal and DR affected patients DR is a progressive disease of the retina that involves pathological changes of BVs and abnormal lesions such as MAs, HAs, EXs and CWSs that appear in fundus images

4.1.1 Signs and Symptoms of DR

The diabetic patients affected by DR have no early symptoms and signs that may cause damage to the retina eventually leading to blindness. The most important abnormal lesions present in fundus images are discussed below.

A. **Micro-aneurysms:** The presence of MAs in retina is the first major symptom of the disease and caused by focal dilations of small BVs. MAs appear as small, round-shaped and red dots. The diameter of the MAs is less than 125 μm and less than the diameter of major optic veins, when crossing the thin BVs leads to smaller circular dots that are similar to MAs (Lazar and Hajdu 2012).

B. **Hemorrhages:** The next major symptom of DR is HAs which are further divided into dot or blot HAs. The wall of thin BVs or MAs is weakened, ruptured, and this gives rise to HAs (Akram et al. 2013). Dot HAs appear as bright small red dots, and blot HAs appear as dark large red lesions. Retinal HAs are caused by pre-proliferative, proliferative DR, retinal ischemia, abnormal BVs in hypertension and malaria. HAs occur in the deeper layers of the retina when blood leaks from BVs (Tang et al. 2012).

C. **Exudates:** The fragile and abnormal BVs leak watery fluid like proteins and lipids that cause EXs to appear. Exudates are yellow-white areas with sharp margins. EXs appear as individual dots, patches or in partial or complete rings surrounding MAs or zones of retinal edema (Mahendran and Dhanasekaran 2015).

D. **Cotton wool spots:** The soft exudates are also known as CWSs. They are round or oval in shape with soft and feathery edges and white or pale yellow in color. CWs occur because of swelling of the surface layer of the retina. The absence of normal blood flow will affect the oxygen supply to retina and cause swelling. Figure 6 shows the physical signs of DR.

4.1.2 Classification of DR

DR is broadly classified into two stages, and they are non-proliferative DR (NPDR) and proliferative DR (PDR). DR is commonly classified into the following stages.

A. **Mild non-proliferative diabetic retinopathy:** At this earliest stage, MAs may occur. MAs are small balloon like swellings in the retina's tiny BVs.

B. **Moderate non-proliferative diabetic retinopathy:** As the disease progresses, some BVs that nourish the retina are blocked.

C. **Severe non-proliferative diabetic retinopathy:** Many more small BVs are blocked, depriving several areas of the retina with their blood supply. These areas of the retina send signals to the body to grow new BVs for nourishment (Yun et al. 2008).

D. **Very Severe non-proliferative diabetic retinopathy:** BVs broken completely in this stage

E. **Proliferative diabetic retinopathy:** At this advanced stage, the signals sent by the retina for nourishment trigger the growth of new BVs. Figure 7 shows the stages of DR.

4.2 Diabetic Macula Edema

Prolonged diabetes can damage the tiny BVs in the retina leading to DR. These damaged BVs leak fluid that gets deposited near macula region leading to distortion in central vision known as DME. The DME may affect up to 10% of people with diabetes. DME can occur at any stage of DR although it is more likely to occur as the DR progresses (Akram et al. 2013).

4.2.1 Types of DME

There are two major types of DME.

A. **Clinically significant macula edema (CSME):** It is the mild form of edema in which EXs are located at a distance from the center of macula region, and the vision is not affected.

B. **Non-Clinically significant macula edema (NCSME):** It is the severe form of edema in which EXs are deposited in and around macula region or on the center of macula (called fovea), and the central vision is affected.

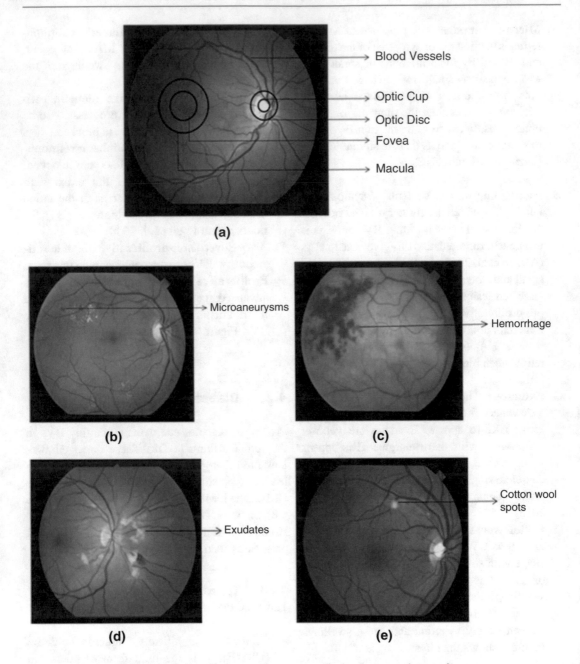

(Source: http://cecas.clemson.edu/~ahoover/stare/)

Fig. 6 The Fundus image (**a**) Normal (**b**) Microaneurysms (**c**) Hemorrhages (**d**) Exudates (**e**) Cotton wool spots. (Source: http://cecas.clemson.edu/~ahoover/stare/)

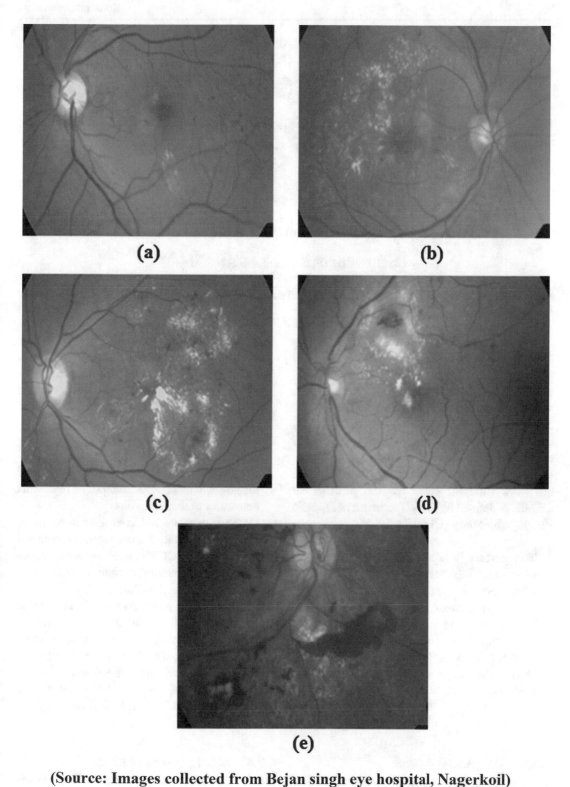

(Source: Images collected from Bejan singh eye hospital, Nagerkoil)

Fig. 7 The Stages of DR (**a**) Mild NPDR (**b**) Moderate NPDR (**c**) Severe NPDR (**d**) Very severe NPDR (**e**) PDR. (Source: Images collected from Bejan singh eye hospital, Nagerkoil)

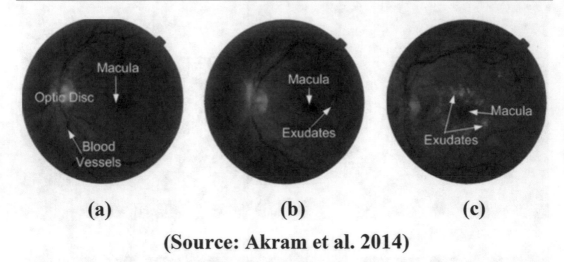

(a)   (b)   (c)

(Source: Akram et al. 2014)

Fig. 8 Retinal fundus image (**a**) Healthy retinal image (**b**) non-CSME retinal image (**c**) CSME retinal image. (Source: Akram et al. 2014)

Figure 8 shows the different stages of edema in digital retinal images.

4.2.2 Signs and Symptoms of DME

The diabetic patients, who are affected with DR, are at risk of developing DME. The major symptoms of DME are blurred vision, double vision and sudden increase in eye floaters. There are three different signs for the presence of CSME, as defined by Early Treatment of Diabetic Retinopathy Study (ETDRS):

1. The increase in retinal thickness ≤ 500 μm of the center of the fovea.
2. EXs present in ≤ 500 μm of the center of the fovea with increased retinal thickness.
3. The increase in retinal thickness ≥ 1 disc diameter with at least within 1 disc diameter at the center of the fovea (Ciulla et al. 2003). Figure 9 shows the diagrammatic representation of CSME

A. **Biomicroscopy:** It is a routine clinical examination tool used to view all ocular structures of the eye. The slit lamps are used in illumination system that delivers bright, sharp and thin light to the focal plane to view ocular structures.
B. **Fluorescein Angiography:** It is performed by injecting fluorescein into the circulatory system. The fundus camera is used to capture image where image intensity represents amount of emitted photons.
C. **OCT:** It captures reflected light from ocular structures to create a cross-sectional image of the retina. The OCT is used to capture three-dimensional structural changes of retina from DME and retinal detachment.
D. **Color fundus photography:** It is used to capture three-dimensional ocular structures into two-dimensional fundus images using fundus camera. The low power microscope provides vertical and magnified view of inner surface of the eye attached on the top of the camera (Mookiah et al. 2015).

4.2.3 Diagnosis of DME

The following examination is performed to provide information for treatment and follow up.

4.2.4 Management of DME

The key part of treating and preventing DME is done by controlling blood glucose level and

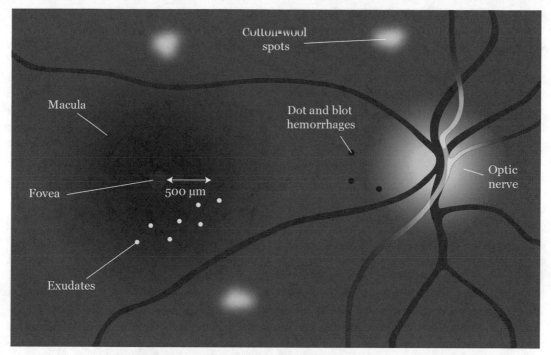

(Source: https://theophthalmologist.com/issues/0815/diabetic-macular-edema-the-crest-of-the-diabetic-tsunami/)

Fig. 9 Diagrammatic illustration of clinically significant macula edema. (Source: https://theophthalmologist.com/issues/0815/diabetic-macular-edema-the-crest-of-the-diabetic-tsunami/)

blood pressure. The main goal of laser surgery used in the treatment of DME is to reduce 50% of the risk of moderate vision loss. The ophthalmologist focuses the tiny laser pulses on areas near the macula where EXs are leaking and seals the leakage of BVs that causes swelling in macula region. The use of anti-VEGF has proven to reducc DME by shutting down the leakage in BVs and stop the growth of abnormal BVs. It automatically reduces the amount of leakage in macula and prevent vision loss (González Cortés 2015).

5   Retinal Image Database

There are several publicly available databases for research purpose which have been provided by research organizations and educational institutes all over the world as described below. DRIVE, STARE, DIARETDB, MESSIDOR, DRISHTI-DS and Indian eye image database have been used for testing and evaluating the developed algorithms.

A. DRIVE database

The DRIVE (Digital Retinal Images for Vessel Extraction) consists of 40 color fundus photographs. It is available for public from their website (http://www.isi.uu.nl/Research/Databases/DRIVE/). The photographs were obtained from a DR screening program in Netherlands. The screening population consisted of 453 subjects between 31 and 86 years of age. The images available in the database are acquired using a Canon CR5 non-mydriatic 3CCD camera with a 45° Field of View (FOV). Each image in the database is captured using 8 bits per plane at 768 × 584 pixel resolution. The FOV of each image is circular with a diameter of approximately 540 pixels. A set of 40 images has been divided into training and testing sets containing 20 images in each set. A single and two manual segmentation of the vasculature are available for training and testing sets. Among the two manual segmentations, one is used as gold standard and the other one is used to compare the computer-generated segmentation.

B. STARE database

The STARE (STructured Analysis of the REtina) consists of a retinal color fundus photograph (Hoover et al. 2000). It is available for educational and research purposes from the website (http://cecas.clemson.edu/~ahoover/stare/). The STARE Project was conceived and initiated in 1975 by Michael Goldbaum, at the University of California, San Diego, and had been funded by the National Institutes of Health, the USA since 1986. During this time, over thirty people have contributed to the project, with backgrounds ranging from medicine to science to engineering. Images and clinical data have been provided by the Shiley Eye Center at the University of California, San Diego, and by the Veterans Administration Medical Center in San Diego. The images available in the database were captured using a TopCon TRV-50 fundus camera at 35° FOV, and subsequently digitized at 605 × 700, 24-bit pixel. The dataset contains 31 normal retina images and 50 diseased retina images with expert annotation of manifestation visible in the retinal images.

C. DIARETDB database

This database consists of two sets, namely DIARETDB0 and DIARETDB1 for DR detection from retinal images (Kauppi et al. 2006, 2007). The DIARETDB0 and DIARETDB1 are available for public in the website (http://www.it.lut.fi/project/imageret/diaretdb0/) and (http://www.it.lut.fi/project/imageret/diaretdb1/). The DIARETDB0 database consists of 130 color retinal images of which 20 are normal and 100 contains signs of DR like EXs, CWSs, MAs, HAs and neovascularization with unknown camera settings. The images in the database were captured using 1500 × 1152 pixels resolution with 50°FOV. The DIARETDB1 database consists of 89 color retinal images of which 5 are normal and 84 with early signs of DR like MAs with varying camera settings. The images in the database were captured using 1500 × 1152 pixels resolutions with 50°FOV.

D. MESSIDOR database

The MESSIDOR (Methods to evaluate segmentation and indexing techniques in the field of retinal ophthalmology) consists of 1200 retinal images (Decencière et al. 2014). The database is available for direct download for research purpose from (http://messidor.crihan.fr.). The images in the database were acquired using 3 ophthalmologic departments using a color video 3CCD camera on a Topcon TRC NW6 non-mydriatic retinograph with a 45°FOV. The images were captured using 8 bits per plane at 1440 × 960, 2240 × 1488 or 2304 × 1536 pixels. The 1200 images are packaged in 3 sets, as per ophthalmologic department. Each set is divided into four zipped subsets containing each 100 images in Tagged Image File Format (TIFF) and an excel file with medical diagnoses of each image.

E. DRISHTI-GS database

Drishti-GS data set consists of 101 retinal images (Sivaswamy et al. 2015). The images in the database are available for public in (http://cvit.iiit.ac.in/projects/mip/drishti-gs/mip-dataset2/Home.php). The images available in the database were collected from Aravind Eye Hospital, Madurai. The patients were between 40 and 80 years of age and OD centered images with dimensions of 2896 × 1944 pixels. Expert annotated images were collected from ophthalmologists with clinical experience of 3, 5, 9 and 20 years.

6 Detection of Diabetic Retinopathy

The growing number of diabetic patients has largely motivated the researchers in developing automated tools to facilitate the screening and evaluation procedures for DR. The literatures related to segmentation of BVs, segmentation of EXs, detection of MAs, identification of HAs and classification of DR are discussed in the following subsections.

6.1 Segmentation of Blood Vessel

BV segmentation is the initial step in the analysis of retina. It is carried out to locate anatomical features like OD and macula. Many researchers have applied various algorithms to segment the BVs successfully. The algorithms and the methodology used for the vessel segmentation are categorized as follows.

6.1.1 Supervised Methods

Aslani and Sarnel (2016) have proposed a pixel classification-based supervised learning for segmentation of blood vessels. The features like contrast enhanced intensity, vesselness measure, intensity of morphological transformed image, multi-scale response of Gabor wavelet and bar-selective combination of shifted filter responses are extracted, and classification of vessel or non-vessel pixels are performed by RF. This algorithm has achieved an SE of 0.7556, SPE of 0.9837, ACC of 0.9605 and area under curve of 0.9789 (Aslani and Sarnel 2016).

Roychowdhury et al. (2014) have developed a three-stage approach to segment BVs. First, the high pass filtering and top hat reconstruction are applied in green plane image to obtain binary image. Second, the major BVs are segmented from regions common to binary image and remaining pixels to construct vessel sub image. The vessel sub images are subjected to two-class classification by Gaussian mixture model. Third, the major vessels are combined with classified pixels in the sub image to get final segmentation output. This methodology has yielded an ACC of 95.2%, 95.15% and 95.3% for DRIVE, STARE and CHASE-DB1 databases with computation time of 3.1 s, 6.7 s and 11.7 s respectively (Roychowdhury et al. 2014).

Wang et al. (2015) have performed hierarchical retinal BV segmentation based on feature and ensemble learning. Histogram equalization and Gaussian filtering are employed as preprocessing steps to remove uneven illumination and improve contrast. The convolutional neural networks are used to extract the required features, and RF is employed to classify based on ensemble method. This methodology has attained an SE of 0.8173, 0.8104, SPE of 0.9733, 09791, ACC of 0.9767, 0.9813 and area under curve of 0.9475, 0.9751 in DRIVE and STARE databases respectively (Wang et al. 2015).

6.1.2 Matched Filters

Kar and Maity (2016a) have employed curvelet transform, matched filter and kernal fuzzy c means to enhance retinal BVs. The enhancement of retinal BVs is done by curvelet transform, and extraction of vessel silhouette is performed by kernel fuzzy c means on matched filter. The step and ramp like signals from vessel structure are distinguished by matched filter and Laplacian of Gaussian (LOG) filter. The proposed method has yielded an SE of 0.7577, 0.7549, false positive rate of 0.0212, 0.0301, SPE of 0.9788, 0.9699 and ACC of 0.9730, 0.9741 respectively in normal/abnormal cases in STARE database (Kar and Maity 2016a).

Kar and Maity (2016b) have employed curvelet transform, matched filter and fuzzy conditional entropy to segment BVs. The removal of noise and improvement of contrast are done by curvelet transform and sequential band pass filter. The different types of vessel silhouettes are extracted by fuzzy conditional entropy on matched filter response. The thin, medium and thick vessels are segmented by multiple thresholding. The proposed method has yielded a true positive rate of 76.32%, 72.82%, false positive rate of 1.99%, 2.6% and an ACC of 96.28%, 96.16% respectively for the DRIVE and STARE databases (Kar and Maity 2016b).

Singh and Srivastava (2016) have proposed Gumble Probability Distribution Function (GPDF)-based matched filter to segment retinal BVs. The PCA-based gray scale conversion and CLAHE are applied as preprocessing steps to enhance retinal BVs. The matched filter response is generated by GPDF-based matched filter. The vessel segmentation is done by entropy-based optimal thresholding, length filtering and removing outer artifacts. The methodology has attained a true positive rate of 0.7594, 0.7939, false positive rate of 0.0292, 0.0624 and an ACC of 0.9522, 0.9270 respectively for DRIVE and STARE databases (Singh and Srivastava 2016).

6.1.3 Thresholding-Based Approach

Christodoulidis et al. (2016) have developed multi-scale tensor voting to segment retinal BVs. The multi-scale line detection is employed to segment small retinal BVs. Adaptive thresholding is used to segment medium and large sized BVs, and multi-scale tensor voting is employed for reconnection of fragmented vessels at variable distances. The proposed method has yielded an improved vessel detection rate of 7.8% against the original multi-scale line detection method (Christodoulidis et al. 2016).

Imani et al. (2015) have employed the Morphological Component Analysis (MCA) to enhance BVs. The lesions and BVs are separated by MCA, and retinal vessels are enhanced by Morlet wavelet transform. Adaptive thresholding is employed to segment final vessel map. The proposed method has yielded an SE of 75.24%, 75.02%, SPE of 97.53%, 97.45% and ACC of 95.23%, 95.90% respectively (Imani et al. 2015).

6.1.4 Other Methods

GeethaRamani and Balasubramanian (2016) have employed image processing and data mining techniques to segment BVs. The image pre-processing, supervised and unsupervised method and image post-processing are the three major steps in the segmentation of BVs. The preprocessing steps used in this algorithm include image cropping, color transformation, color channel extraction, contrast enhancement, Gabor filtering and half wave rectification. Feature extraction is performed by PCA, and classification of vessel or non-vessel cluster is done by k-means clustering. The segmentation of BVs is done by k-means clustering and ensemble classifier. Mathematical morphology and connected component analysis are used as post-processing step. This methodology has yielded an SE of 70.79%, SPE of 97.78%, ACC of 95.36% and Positive Predictive Value (PPV) of 75.76% respectively in DRIVE database (GeethaRamani and Balasubramanian 2016).

Kovács and Hajdu (2016) have performed a self-calibrating approach for segmentation of BVs. The ROI is selected, and binary segmentation is performed by template matching. The edge reconstruction, thin and low contrast vessels are extracted by the trained database. This methodology has yielded an average ACC of 94.94% and 96.10% respectively in DRIVE and STARE databases (Kovács and Hajdu 2016).

Jebaseeli et al. (2019) has used Tandem PCNN model to segment the BVs and deep learning SVM for classification of BVs. The proposed method achieved the SE of 80.61%, SPE of 99.54% and ACC of 99.49% (Jebaseeli et al. 2019). Dash and Senapati (2020) had developed the algorithm to segment BVs by combining discrete wavelet transform (DWT) with Tyler Coye algorithm and achieved the SE of 74.03%, 99.05% and 96.61% (Dash and Senapati 2020). Jiang et al. (2018) used convolution network and transfer learning to segment BVs. The success rate of this algorithm is increased to 1–2% than the existing method (Jiang et al. 2018). Table 1 shows the comparison of performance measures of Blood vessel detection algorithm.

7 Detection of Diabetic Macula Edema

In the literature, numerous examples of the application of digital imaging techniques used in identification of DME can be found. There have been a few research investigations to identify EXs in macula region in the literature.

Table 1 Comparison of retinal BV segmentation algorithm with literature-reviewed methods

Author name	Year	Algorithm	Database	No of images	SE (%)	SPE (%)	ACC (%)	Computation time(s)
Aslani and Sarnel	2016	Hybrid feature vector formation Random forest classifier	DRIVE STARE	40 20	75.45 75.56	98.01 98.37	95.13 96.05	60
Christodoulidis et al.	2016	Multi scale tensor voting approach	Erlangen	45	85.06	95.82	94.79	1200
Kar and Maity	2016a	Matched filtering integrated with curvelet transform and kernel based fuzzy c means	DRIVE STARE DIARETDB1	40 20 89	75.48 75.63	97.92 97.43	96.16 97.35	730.85
GeethaRamani and Balasubramanian	2016	Gabor filtering, half wave rectification, k-means clustering	DRIVE	40	70.79	97.78	95.36	NA
Singh and Srivastava	2016	Gumbel probability distribution function based matched filter	DRIVE STARE	40 20	75.94 79.39	97.08 93.60	95.22 92.70	135.6 144
Kar and Maity	2016b	Band pass filter and fuzzy conditional entropy	DRIVE STARE	40 20	76.32 72.82	98.01 93.38	96.28 96.16	148.48
Kovács and Hajdu	2016	Template matching Contour reconstruction	DRIVE STARE	40 20	72.70 76.65	98.77 98.79	96.48 97.11	225
Roychowdhury et al.	2014	Gaussian mixture model	DRIVE STARE CHASE_DB1	40 20 28	72.5 77.2 72.01	98.3 97.3 98.24	95.2 95.15 95.3	3.1 6.7 11.7
Imani et al.	2015	Morphological component analysis,Shearlet transform	DRIVE STARE	40 20	75.24 75.02	97.53 97.45	95.23 95.90	NA

Akram et al. (2014) have presented a novel method for EX and macula detection by binary map of candidate using filter bank and Gaussian mixture model. A hybrid classifier is an ensemble of Gaussian mixture model and SVM used for classification of images into normal, CSME and non-CSME. This method has yielded an SE, SPE and ACC of 97.3%, 95.9% and 96.8% respectively on HEI-MED and MESSIDOR databases (Akram et al. 2014).

Giancardo et al. (2012) have proposed color, wavelet decomposition and automatic lesion segmentation for diagnosis of DME. The correlation between color/wavelet analysis and EXs probability map is captured by five measures like mean, median, SD, minimum and maximum. The probabilistic, geometric and tree-based techniques are used for classification of DME. The effectiveness of the algorithm is tested with the performance measure of AUC between 0.88 and 0.94 on MESSIDOR database (Giancardo et al. 2012).

Deepak and Sivaswamy (2011) have proposed a supervised learning for DME detection. The exudates are detected by motion pattern generation, and severity of abnormality is calculated by rotational asymmetry of the macular region. This algorithm is evaluated by an ACC of 81% for the moderate case and 100% for the severe case (Deepak and Sivaswamy 2011).

Srinivas et al. (2019) has analyzed the effect of Intravitreal ranibizumab on hard exudates in DME. This work can be extended with large sample size and duration more than 1 year (Srinivas et al. 2019).

8   Conclusion

A survey of techniques for the automatic detection of DR and DME has been presented in this research article. The proposed method involves two phases

(a)   Diagnosis of DR
(b)   Detection of DME.

The intention in this research work is to identify abnormalities from retinal images, locate and segment important structural indicators of eye diseases to extract various features from retinal structures and to classify them based on their severity. Considering the current circumstances in recent developments, the study focuses on, Differentiating bright and dark lesions of varying shapes, size and appearance as appropriate lesion to classify stages of DR and detecting EXs in macula region for diagnosis of DME

References

Abràmoff MD, Garvin MK, Sonka M (2010) Retinal imaging and image analysis. IEEE Rev Biomed Eng 3:169–208

Akram MU, Khalid S, Khan SA (2013) Identification and classification of microaneurysms for early detection of diabetic retinopathy. Pattern Recogn 46(1):107–116

Akram MU, Tariq A, Khan SA, Javed MY (2014) Automated detection of exudates and macula for grading of diabetic macular edema. Comput Methods Prog Biomed 114(2):141–152

Aslani S, Sarnel H (2016) A new supervised retinal vessel segmentation method based on robust hybrid features. Biomed Signal Process Control 30:1–2

Christodoulidis A, Hurtut T, Tahar HB, Cheriet F (2016) A multi-scale tensor voting approach for small retinal vessel segmentation in high resolution fundus images. Comput Med Imaging Graph 52:28–43

Ciulla TA, Amador AG, Zinman B (2003 Sep 1) Diabetic retinopathy and diabetic macular edema: pathophysiology, screening, and novel therapies. Diabetes Care 26 (9):2653–2664

Dash S, Senapati MR (2020) Enhancing detection of retinal blood vessels by combined approach of DWT, Tyler Coye and Gamma correction. Biomed Signal Process Control 57:101740

Decencière E, Zhang X, Cazuguel G, Lay B, Cochener B, Trone C, Gain P, Ordonez R, Massin P, Erginay A, Charton B (2014) Feedback on a publicly distributed image database: the Messidor database. Image Anal Stereol 33(3):231–234

Deepak KS, Sivaswamy J (2011) Automatic assessment of macular edema from color retinal images. IEEE Trans Med Imaging 31(3):766–776

GeethaRamani R, Balasubramanian L (2016) Retinal blood vessel segmentation employing image processing and data mining techniques for computerized retinal image analysis. Biocybern Biomed Eng 36(1):102–118

Giancardo L, Meriaudeau F, Karnowski TP, Li Y, Garg S, Tobin KW Jr, Chaum E (2012) Exudate-based diabetic macular edema detection in fundus images using publicly available datasets. Med Image Anal 16 (1):216–226

González Cortés JH (2015) Treatment of diabetic macular edema (DME): shifting paradigms. Med Univ 17 (69):243–247

Gonzalez RC, Woods RE (eds) (2002) Digital image processing. Prentice Hall Press, Upper Saddle River, p 8. ISBN 0-201-18075

Hoover AD, Kouznetsova V, Goldbaum M (2000) Locating blood vessels in retinal images by piecewise threshold probing of a matched filter response. IEEE Trans Med Imaging 19(3):203–210

Imani E, Javidi M, Pourreza HR (2015) Improvement of retinal blood vessel detection using morphological component analysis. Comput Methods Prog Biomed 118(3):263–279

Jebaseeli TJ, Durai CA, Peter JD (2019) Retinal blood vessel segmentation from diabetic retinopathy images using tandem PCNN model and deep learning based SVM. Optik 199:163328

Jiang Z, Zhang H, Wang Y, Ko SB (2018) Retinal blood vessel segmentation using fully convolutional network with transfer learning. Comput Med Imaging Graph 68:1–5

Kar SS, Maity SP (2016a) Blood vessel extraction and optic disc removal using curvelet transform and kernel fuzzy c-means. Comput Biol Med 70:174–189

Kar SS, Maity SP (2016b) Retinal blood vessel extraction using tunable bandpass filter and fuzzy conditional entropy. Comput Methods Prog Biomed 133:111–132

Kauppi T, Kalesnykiene V, Kamarainen JK, Lensu L, Sorri I, Uusitalo H, Kälviäinen H, Pietilä J (2006) DIARETDB0: Evaluation database and methodology for diabetic retinopathy algorithms. Lappeenranta University of Technology, Finland, Machine Vision and Pattern Recognition Research Group, p 134

Kauppi, T, Kalesnykiene, V, Kamarainen, J.-K, Lensu, L, Sorri, I, Raninen A, Voutilainen R, Uusitalo, H, Kälviäinen, H & Pietilä, J (2007) DIARETDB1 diabetic retinopathy database and evaluation protocol, in proceedings of the eleventh conference on Medical Image Understanding and Analysis, pp 1–10

Kovács G, Hajdu A (2016) A self-calibrating approach for the segmentation of retinal vessels by template matching and contour reconstruction. Med Image Anal 29:24–46

Lazar I, Hajdu A (2012) Retinal microaneurysm detection through local rotating cross-section profile analysis. IEEE Trans Med Imaging 32(2):400–407

Mahendran G, Dhanasekaran R (2015) Investigation of the severity level of diabetic retinopathy using supervised classifier algorithms. Comput Electr Eng 45:312–323

Mookiah MR, Acharya UR, Fujita H, Koh JE, Tan JH, Chua CK, Bhandary SV, Noronha K, Laude A, Tong L (2015) Automated detection of age-related macular degeneration using empirical mode decomposition. Knowl-Based Syst 89:654–668

Olson JL, Asadi-Zeydabadi M, Tagg R (2015) Theoretical estimation of retinal oxygenation in chronic diabetic retinopathy. Comput Biol Med 58:154–162

Roychowdhury S, Koozekanani DD, Parhi KK (2014) Blood vessel segmentation of fundus images by major vessel extraction and subimage classification. IEEE J Biomed Health Inform 19(3):1118–1128

Singh NP, Srivastava R (2016) Retinal blood vessels segmentation by using Gumbel probability distribution function based matched filter. Comput Methods Prog Biomed 129:40–50

Sivaswamy J, Krishnadas S, Chakravarty A, Joshi G, Tabish AS (2015) A comprehensive retinal image dataset for the assessment of glaucoma from the optic nerve head analysis. JSM Biomed Imaging Data Pap 2(1):1004

Srinivas S, Verma A, Nittala MG, Alagorie AR, Nassisi M, Gasperini J, Sadda SR (2019) Effect of intravitreal ranibizumab on intraretinal hard exudates in eyes with diabetic macular edema. Am J Ophthalmol

Tang L, Niemeijer M, Reinhardt JM, Garvin MK, Abramoff MD (2012) Splat feature classification with application to retinal hemorrhage detection in fundus images. IEEE Trans Med Imaging 32(2):364–375

Wang S, Yin Y, Cao G, Wei B, Zheng Y, Yang G (2015) Hierarchical retinal blood vessel segmentation based on feature and ensemble learning. Neurocomputing 149:708–717

Yun WL, Acharya UR, Venkatesh YV, Chee C, Min LC, Ng EY (2008) Identification of different stages of diabetic retinopathy using retinal optical images. Inf Sci 178(1):106–121

Zaki WM, Zulkifley MA, Hussain A, Halim WH, Mustafa NB, Ting LS (2016) Diabetic retinopathy assessment: towards an automated system. Biomed Signal Process Control 24:72–82

Adv Exp Med Biol - Advances in Internal Medicine (2020) 4: 375–389
https://doi.org/10.1007/5584_2020_535
© Springer Nature Switzerland AG 2020
Published online: 3 June 2020

Diabetic Macular Edema: State of Art and Intraocular Pharmacological Approaches

Annalisa Gurreri and Alberto Pazzaglia

Abstract

Diabetic macular edema (DME) is the main cause of vision loss in diabetic retinopathy (DR). Although it is one of the main complications of diabetes, the pathogenesis of DME is not completely understood. The hyperglycemic state promotes the activation of multiple interlinked pathways leading to DME. Different classifications have been proposed: based on clinical features, on pathogenesis or on diagnostic tests (optical coherence tomography – OCT and fluorescin angiography – FA). The multimodal imaging allows a better analysis of the morphological features of the DME. Indeed, new inflammatory biomarkers have been identified on OCT. Also, several studies are evaluating the role of the morphological features, identified on multimodal imaging, to find new prognostic factors. Over the past decade, great progresses have been made in the management of DME. Therapeutic alternatives include intraocular injection of anti-vascular endothelial grow factor agents (anti-VEGF) and steroid molecules, focal/grid laser photocoagulation and vitreoretinal surgery. This review is focused on the description and analysis of the current intravitreal therapeutic pharmacological strategies. Current guidelines recommend anti-VEGF as first line therapy in DME. Corticosteroids are becoming increasingly relevant blocking the inflammatory cascade and indirectly reducing VEGF synthesis.

Keywords

Diabetic macular edema · OCT · OCT-A · Fluorangiography · Pharmacological approaches · Anti-VEGF · Steroids

1 Introduction

Diabetes mellitus (DM) is a serious, long-term condition with a major impact on the lives of individuals, families, and societies worldwide (Caruso et al. 2018).

According to the 9th Edition of International Diabetes Federation (IDF) the worldwide prevalence of diabetes is increasing and it is estimated that 463 million people have diabetes. Also it is predicted that 578 million people will have diabetes in 2030 and the number will increase by 51% (700 million) in 2045. This foreseen increase is probably linked to a raise in obesity and the increase in life expectancy worldwide (Saeedi et al. 2019).

At DM diagnosis half of the patients with DM present with complications, as it is a condition

A. Gurreri (✉)
University of Bologna, Bologna, Italy

Sant' Orsola Malpighi Hospital, Bologna, Italy
e-mail: lisa.gurreri@alice.it

A. Pazzaglia
Sant' Orsola Malpighi Hospital, Bologna, Italy

that could be paucisymptomatic for years (Saeedi et al. 2019).

The most common DM complications include macroangiopathy (myocardial infarction or vasculocerebral stroke) and microangiopathy (diabetic nephropathy, neuropathy, and retinopathy). In this chapter we will discuss diabetic retinopathy (DR) and in particular diabetic macular edema (DME). DME is the leading cause of blindness among working age adults in industrialized countries (Romero-Aroca et al. 2016).

2 Epidemiology and Risk Factors of Diabetic Retinopathy and Diabetic Macular Edema

There are approximately 93 million people with diabetic retinopathy (DR), 17 million with proliferative DR, 21 million with diabetic macular edema (DME), and 28 million with vision-threatening DR (VTDR) worldwide (Yau et al. 2012). DME is a manifestation of DR and the leading cause of central vision loss among patients with DR. Although it is one of the main complications of diabetes, the pathogenesis of DME is not completely understood. A hyperglycemic state promotes the activation of multiple interlinked pathways leading to reactive oxygen species (ROS) formation and augmented concentrations locally surpassing the antioxidant capacity. Free radicals, advanced glycation end products (AGEs), inflammatory processes, and vascular endothelial growth factor (VEGF) have all been implicated in the breakdown of the blood-retinal barrier (BRB) that results in vascular leakage and retinal thickening in DME (Romero-Aroca et al. 2016; Wong et al. 2018).

In a meta-analysis from 20 population-based studies, Yau JW et al., summarized significant risk factors for DR and DME such as: longer duration of diabetes, higher glycosylated hemoglobin (HbA1c) levels and poor glycemic control, and hypertension. Indeed, they had identified a trend toward a higher prevalence of VTDR, but not any DR, in people with cholesterol levels ≥4.0 mmol/L (Yau et al. 2012). Other potential

systemic risk factors for DME include advanced age, presence of sleep apnea, pregnancy, anemia, duration of diabetes, nephropathy/microalbuminuria and systemic fluid retention in congestive heart failure or renal disease (Diep and Tsui 2013; Perkovich and Meyers 1988; Zhang et al. 2017).

The use of the glitazone (thiazolidinedione group) an oral antihyperglycemic drug seems to correlate with higher risk DME (Ryan et al. 2006; Fong and Contreras 2009).

Control of modifiable risk factors could have a positive impact on progression of DR and on DME onset. Kawasaki R et al., observed an association between lipid-lowering medication and a decrease of DR and its complications. Furthermore, an association between statins medication and vitamin C (as an antioxidant agent) supplementation could have a synergistic role in lowering DME onset and DR progression (Kawasaki et al. 2018; Gurreri et al. 2019).

3 Pathogenesis of DME

The pathogenesis of DME is multifactorial and it seems be based on inflammation, vasculopathy, growth factors, angiogenesis and neurodegeneration.

In DME, liquid accumulation can occur both in intracellular or extracellular spaces. Intracellular edema due to liquid in the intracellular space is called cytotoxic, while the accumulation of liquid in the extracellular space is defined as vasogenic edema (Romero-Aroca et al. 2016).

Exudation is due to lesions in vessel walls, combined with increased permeability and changes in blood flow; it determines the onset of edema. The metabolic and hystological disruption of both external-Blood Retinal Barrier (e-BRB) and inner- Blood Retinal Barrier (i-BRB) are due to pro-inflammatory agents in the hyperglycemic environment (Klaassen et al. 2013). In such environment of pro-inflammatory stimuli, the vascular endothelial growth factor (VEGF) display its role on breakdown and increased vessel permeability of BRB (Antonetti et al. 1999).

Another milestone in the pathogenesis of DME are adherent leukocytes. Leukostasis in retinal capillaries in is an early event in the cascade of DME, leading dysfunction of the BRB. Indeed, Joussen AM et al. indicates that the surface expression of Fas L on circulating leukocytes is increased in DR and these leukocytes are capable of inducing Fas-mediated endothelial cell injury, apoptosis, and BRB breakdown (Joussen et al. 2002).

The stimulation and alteration of endothelial cells induces IL-8 production and chemotaxis (for activated leukocytes). So, the breakdown of BRB induces infiltration of activated cells and release of cytokines and free radicals in situ.

The retinal cells suffering related to ischemia activates resident cells (macrophages, microglia, EPR) and increases cellular catabolites. In this proinflammatory state given by the presence of chemokines (MCP-1 and ICAM-1, which increase vascular permeability and white blood cell adhesion) and pro-inflammatory cytokines (IL-1, IL-6, IL-8, which increase vascular permeability) macrophages and microglia cells are activated simultaneously (Romero-Aroca et al. 2016; Klaassen et al. 2013).

4   Classification

For better contextualize DME, it is necessary to comprehend the clinical features of DR. DR is a condition of microvascular abnormalities that are seen in the fundus of diabetic patients on clinical examination.

The first and least severe clinical change observed in DR is the dot-like microaneurysm (MA) which is a localized saccular eversion of the capillary wall. At fundus oculi examination, they appear as well-define red dots on the retina surface. Others frequent features of DR are retinal hemorrhages. They could appear as flame-shaped when located in the nerve fibre layer (RNFL), or dot/spot-like if they originate from middle layers of the retina. Another hallmark of DR are intraretinal microvascular abnormalities (IRMA) that are dilated and tortuous capillaries. Hard exudates could also be found in DR and they are

yellowish spots, expression of chronic localized leakage of lipid from retinal vessels. Hard exudates are usually in proximity of MAs. Conversely, soft exudates, or cotton wall spots, are due to focal ischemia in RNFL for reduction in axoplasm flow (Schmidt-Erfurth et al. 2017).

DME is one of the main complications of DR. To conform the definition of DME, different classifications have been proposed. Considering just the clinical features, DME could be defined as a retinal thickening within two disc diameters from the centre of the macula. DME could be subdivide as follow:

1. Focal edema: well-defined thickness from groups of MAs associated with complete or incomplete hard exudate rings that could assume a circinate distribution.
2. Diffuse edema: results from failure of BRB with leakage from microaneurysms, retinal capillaries and arterioles. Often associated with cystoid macular edema (CME).
3. Clinically significant macular edema (CSME), which was subclassified according the American Academy of ophthalmology of 2001 in:
   - Retina thickening at or within 500 μm from the fovea
   - Hard exudates at or within 500 μm from the fovea, if accompanied by thickening of the adjacent retina.
   - A zone of retinal thickening 1 disc area or larger located 1 disc diameter or less from the fovea (Kanski and Bowling 2011).

Another classification of DME could be based on pathogenesis:

1. Vasogenic edema, secondary to internal BRB breakdown.
2. Edema secondary to external BRB breakdown (rare and/or difficult to evaluate).
3. Tractional edema secondary to epiretinal membrane or to firmly attached posterior hyaloids.
4. Ischemic maculopathy secondary to occlusion of the macular capillary network.

DME may also be classified based on optical coherence tomography (OCT) measurements.

Using OCT, DME could be classified into four main types presented as follows:

1. Type 1: Early diabetic macular edema.
2. Type 2: Simple diabetic macular edema.
3. Type 3: Cystoid diabetic macular edema (mild, intermediate and severe).
4. Type 4: Serous macular detachment.

Fluorescein angiography (FA) is another modality used to classify DME, into three types presented as follows.

1. Focal leakage: localized areas of leakage from MAs or dilated capillaries.
2. Diffuse leakage: leakage involving the entire circumference of the center of the macula.
3. Diffuse cystoid leakage: mainly diffuse leakage, but accumulation of the stain within the cystic areas of the macula during late phase of the angiogram (Mathew et al. 2015).

5   Multimodal Imaging in DME

OCT is the most used imaging method in diagnosis and follow-up of DME. Despite the wide diffusion of OCT, FA is still considered the gold standard. OCT, and the most recent OCT angiography (OCT-A), have reduced the use of FA in clinical practice as diagnostic tool in DME.

In clinical practice, DME treatment is often decided only on OCT images. FA should performed at the beginning of any treatment as it is the only way to distinguish non-leaking from leaking MAs, to highlight the presence of IRMA, and to delineate areas of capillary nonperfusion and the widening of the foveal avascular zone (FAZ) (Bresnick 1986). Emerging data also indicate that the widening of the FAZ is an important prognostic indicator of outcome. However spectral-domain OCT (SD-OCT) cannot identify foveal ischemia and widening of the FAZ, which conversely are clearly visible in FA and OCT-A (Hwang et al. 2016). Nowadays, the implementation of ultra wide-field FA (UWFA) highlighted the importance of peripheral vascularization in

DME. As a matter of fact, various studies have shown that peripheral ischemia is strongly related to presence and severity of DME as well as resistant to medical therapy. This is an additional reason that stresses the usefulness of FA and in particular of UWFA in DME. Wessel MM et al. in their study supported this hypothesis. UWFA permits to unveil larger areas of retinal nonperfusion in DME patients. This study demonstrates a 3.75 times greater chance of having DME in patients with peripheral retinal ischemia. This observation has several important implications. Firstly, it lends support to the growing data suggesting that retinal ischemia and release of VEGF play a role in the pathogenesis of DME. Secondly, it seems to strength the rationale of targeted retinal photocoagulation (TRP) in treating DME. They observed that DME is present despite minimal posterior diabetic pathology and UWFA is able to demonstrate anterior areas of retinal ischemia. They recorded that the area of macular edema is significantly greater in DME associated with peripheral ischemia than those driven by microaneurysms alone (Wessel et al. 2012).

Furthermore, wide-field imaging could be used to calculate the ischemic index. After anti-VEGF treatment the mean decrease in central macular thickness is highest in the lowest ischemic index group and least in the worst ischemic index group (Xue et al. 2017).

OCT has a role both in the diagnosis and follow-up of DME and could verify the efficacy of treatment. It is a fast and noninvasive technique that does not require ionizing radiation. It is useful on providing information on retinal layers structure and morphology. The OCT can investigate the morphological features of the edema and the morphological changes after treatment. It can determine the measure of the central retinal thickness (CRT), the main morphological alteration required for treatment planning.

Morphological signs of de novo, persisting or resolved DME are:

- Sub retinal fluid (SRF): seen as a nonreflective space between the neurosensory retina and the retinal pigment epithelium;

- Intraretinal cystoid fluid (IRC) that appear as minimally reflective, well defined round or oval cystis within the neurosensory retina;
- Disorganization of the inner retinal layers (DRIL) (Sun et al. 2014);
- Microaneurisms (MA);
- hard exudates/hyperreflective foci (Bolz et al. 2009);
- epiretinal membranes;
- changes in choroidal thickness (Gerendas et al. 2014a).

Therefore, a new classification based on OCT findings was proposed in the past years. One of these is the SAVE protocol (Bolz et al. 2014) that, in an acronym, proposes to describe and divide DME according to presence of SRF, Area of affected retina by IRC, Vitreoretinal interface abnormalities and Etiology class of CSME. According to SAVE protocol the following characteristic must be evaluated:

- "S" for Subretinal fluid (SRF): secondary to a breakdown of the outer limiting membrane in CSME leads to subretinal fluid accumulating, typically sub-foveally. Subretinal fluid is evaluated in OCT raster B scans and graded as "1" if present, "0" if absent and "x" if its presence cannot be determined;
- "A" for Area of the retina: As opposed to retinal volume measuring, assessing the planimetric dimension and central retinal thickness allows for a detailed description of the expansion and location of an individual case of CSME. In the SAVE protocol, the number of ETDRS fields that show a local thickening compared with normative database values are counted (ie, 0–9 fields).
- "V" for Vitreo-retinal interface abnormalities: Their presence is important because its presence could indicate the necessity of surgical intervention Their presence (value "1") or absence (physiological findings, value "0") are evaluated in raster B scans.
- "E" for Etiology, divided as:
  1. Focal or multi-focal leakage in FA with a definable leakage source(s) causing exudative edema in OCT.
  2. Non-focal capillary leakage in FA without a definable leakage source causing exudative edema in OCT.
  3. Macular or peripheral ischemia anywhere in FA defined as capillary non-perfusion and graded in a composite (or wide-field) image at minute 3 or 5.
  4. Atrophic edema: retinal cystoid degeneration without Müller cells, usually visible as vertical tissue columns surrounding cysts (pseudo-septa) and/or disruption in the horizontal layer integrity in the central millimetre.

When describing the etiology of the edema, multiple subclass can coexist at the same time. For example a focal macular edema can coexist with a macular ischemia and in the SAVE classification the "E" will assume the value "1" and "3" (Bolz et al. 2014). Instead, in other clinical trials such as RIDE/RISE studies, the only OCT criterion used for the assessment of DME was the CRT (Nguyen et al. 2012).

To date, ophthalmologists agree that treatment regimens must be tailor-made and OCT biomarkers are the key to identify the best PRN treatment scheme for each individual patient. A PRN regimen is the most frequently used treatment regimen for DME in anti-VEGF therapy in everyday clinical practice, but the PRN criteria are not standardized. Alterations at OCT scans and OCT biomarkers are strictly related with treatment response (Schmidt-Erfurth et al. 2017). Several studies have searched for defining relevant morphologic factors, their significance for treatment and prediction of visual outcomes. For example, in the RESTORE study, patients have been treated on a PRN basis after an initial loading phase of three consecutive monthly injections. The authors found that patients with SRF at baseline had higher best corrected visual acuity (BCVA) at 1 year follow up than patients without SRF at baseline (Gerendas et al. 2014b). This protective role of SRF was reconfirmed by an OCT post hoc study analysis of the RIDE/RISE trials (Sophie et al. 2015). When IRC is present patient usually experience worse BCVA letter score at the end of treatment and during follow up (Gerendas et al. 2014b). Furthermore, other patients with vitreomacular adhesion had higher BCVA scores at baseline and maintained a better visual acuity during therapy, while those who had a posterior vitreous detachment experienced lower BCVA letter scores both at baseline and after treatment than patients with vitreomacular adhesion (Gerendas et al. 2014b).

DRIL is another relevant OCT biomarker of visual outcomes. If DRIL extend in more than 50% of the central millimeter was associated with a lower visual acuity both in DME and after edema reabsorption. DRIL leads to a worse visual outcome if it occurred during the first 4 months of treatment (Sun et al. 2015).

Another distinct prognostic marker of DME is DME with association of subfoveal neuroretinal detachment (SND; 15–30% incidence). DME with SND is related to ocular inflammatory evidence such as: higher levels of vitreous IL-6 and multiple hyper-reflective retinal spots (HRS). Recently, SND and HRS have been proposed as new noninvasive OCT-imaging biomarkers of retinal inflammation in DME (Vujosevic and Simó 2017). Decreased retinal sensitivity, increased choroidal thickness and disrupted external limiting membrane (ELM) are specific features of with DME with SND (Sun et al. 2015). Finally, the hyper-reflective retinal spots (HRS) have recently considered as OCT biomarkers of inflammation in DME.

HRS has specific characteristics such as small size (<30 micron), reflectivity similar to nerve fiber layer and no back-shadowing. Increased number of HRS has been documented in DME as well as in early stages of DR and even in diabetes mellitus without clinical signs of DR. Moreover, in DME, a greater number of HRSs were found if SND was present compared to DME without SND (Vujosevic et al. 2017).

In conclusion, the OCT is the most widely used exam in clinical practice in the diagnosis of DME for its rapidity, diagnostic accuracy and non-invasiveness. A careful examination of the OCT images allows, in addition to the measure of CRT, to identify prognostic factors for the visual outcome, inflammatory biomarkers. Moreover, OCT is the milestone exam to decide if it is

necessary a new intravitreal injective treatment. However, the most recent guidelines of EURETINA on DME suggest that FA is still the gold standard for the diagnosis of DME, not only for DR, and it should be performed prior to the initiation of therapy to delineate and stage the DME and DR pathology (Schmidt-Erfurth et al. 2017).

6 Pharmacologic Intravitreal Therapy

Over the past decade, great progresses have been made in the management of DME. Therapeutic alternatives nowadays include intraocular injection of anti-vascular endothelial grow factor agents (anti-VEGF) and steroid molecules, focal/grid laser photocoagulation, subthreshold micropulse yellow laser and vitreo-retinal surgery. This review is focused on the description and analysis of the current therapeutic pharmacological strategies. Currently the only approved pharmacological options for DME treatment are the intravitreal injection of anti-VEGF agents or corticosteroids.

7 Anti-angiogenic Agents

Current guidelines recommend anti-angionetic (anti-VEGF) as first line therapy in DME. Several molecules have been used in past and great attention was given from the scientific community. Bevacizumab, Ranibizumab and Aflibercept are currently commercialized and available for treatment. Recent and on-going trials have been carried out to establish the safety, superiority and optimal treatment regimen.

The rationale of therapy with agents that could block VEGF pathway is based on the fact that high concentration of VEGF have been recorded in DR and DME (Aiello et al. 1994). As previously explained, VEGF has been demonstrated to increase the vessel permeability by increasing the phosphorylation of tight junction proteins (Antonetti et al. 1999). All anti-VEGF agents are products of recombination technology.

Bevacizumab is a full-length antibody and a selective VEGF-A inhibitor. Ranibizumab is a monoclonal antibody fragment that inhibits selectively VEGF-A. Aflibercept is a recombinant fusion protein that blocks VEGF-A, VEGF-B, and placental growth factor (PlGF) (Torres-Costa et al. 2020).

Particularly, Bevacizumab is a full–length, humanized, monoclonal antibody and it is, historically, the first anti-VEGF used in DME. Originally, it was designed as anti-neoplastic agent. Ranibizumab is a recombinant humanized Fab fragment of a monoclonal antibody, designed for intraocular use. Lastly Aflibercept is a recombinant decoy receptor type of inhibitor of VEGF and PIGF. Even though Bevacizumab is the most widely used agent, the indication for DME and neovascular age-related macular degeneration (nAMD), remains off-label (Schmidt-Erfurth et al. 2017).

During last years, great effort was expended in demonstrating the efficacy, safety and superiority between different anti-VEGF agents.

Wells JA et al. conducted a randomized clinical trial (named Protocol T Study) comparing the efficacy and safety of intravitreal Aflibercept, Bevacizumab and Ranibizumab in DME. The authors observed a mean improvement in BCVA letter score at 1 year and 2 years respectively of 13.3 and 12.8 letters for Aflibercept, 9.7 and 10.0 letters for Bevacizumab and 11.2 and 12.3 letters for Ranibizumab. At 1 year follow-up, the authors noted a statistical difference of $P < 0.001$ for aflibercept vs. bevacizumab and $P = 0.03$ for aflibercept vs. ranibizumab. At 2 years follow-up the only statistical difference was reported between Aflibercept vs Bevacizumab ($p = 0.02$), while no difference was noted comparing Ranibizumab with the other two agents. Those results seem to suggest a slight superiority of only Aflibercept and that differences were reduced over time (Diabetic Retinopathy Clinical Research Network et al. 2015).

Another hot spot on anti-VEGF regards the safety of the agent, in particular of bevacizumab as it is usually administered on off-label regimen. It is known that serum levels of bevacizumab and aflibercept are higher than Ranibizumab.

However, the clinical significance is still debated in DME. The RESOLVE study has reported a single cardiovascular adverse event in the Ranibizumab group that seems to be related to the drug. However other studies have not recorded serious cardiovascular event (Schmidt-Erfurth et al. 2014; Massin et al. 2010). A Cochrane database systematic review comparing Bevacizumab and Ranibizumab for the treatment of nAMD did not find any important difference in effectiveness and safety between the two drugs. However, the pathogenic mechanisms of DME and nAMD are different, so patients affected are not completely comparable in terms of cardiovascular risk (Solomon et al. 2016). Further clinical trial and real life data are necessary to evaluate the safety of anti-VEGF in DR.

Other studies compared anti VEGF in monotherapy versus laser therapy or their association. For example the RESTORE trial (randomized double-masked, phase III study) compared three groups: Ranibizumab as monotherapy, only laser therapy and their association (Ranibizumab plus laser therapy). This was the first study that showed the efficacy and superiority of ranibizumab monotherapy over laser photocoagulation at 12 months follow-up. Following the publication of this study the EMA (European Medicine Agency) approved the use of Ranibizumab for DME treatment (Mitchell et al. 2011).

Protocol I study based on DRCR.net seemed to confirm the result of the RESTORE study. The authors noted a superiority of the association of Ranibizumab with laser therapy, prompt or deferred, compared to laser therapy alone (Diabetic Retinopathy Clinical Research Network et al. 2010). The RISE and RIDE studies were two identical clinical trials (double-masked randomized) which demonstrated that a dose of 0.3 mg versus 0.5 mg of Ranibizumab had no statistically significant difference in safety and efficacy. The efficacy of both posology was sustained through 3 years follow-up (Nguyen et al. 2012; Mitchell et al. 2011).

Regarding the efficacy and the security profile of Aflibercept, the VIVID DME and the VISTA DME were designed. In VIVID and VISTA studies two dosing regimens of intravitreal

Aflibercept injection with macular laser photocoagulation for DME were practiced. Eyes were randomized in a 1:1:1 ratio to receive 2 mg of Aflibercept every 4 weeks (2q4), or every 8 weeks after five initial monthly doses (2q8), or macular laser photocoagulation. Proportion of eyes with a ≥2 step improvement in Diabetic Retinopathy Severity Scale (DRSS) score from baseline at week 100 was comparable in both 2q4 and 2q8 groups.

These data suggested that the 8-weekly interval regimen should be preferred over a 4 weekly regimen. Even if the former regimen was more susceptible of oscillating pattern of CRT (mean amplitude of 25–50 μm), no significant differences were noted in terms of BCVA. The drug demonstrated a clear superiority in regards of laser therapy alone (Brown et al. 2015). After these studies were published, both the EMA and the Food and Drug Administration (FDA) approved the use for DME treatment.

These studies seem to suggest that Aflibercept should be the drug of choice in DME, in particular with poor BCVA (<69 letters) as demonstrated in Protocol T study. No consensus exists on administration regimen (PRN-based with monthly revaluation vs fixed bimonthly regimen) following the loading phase with an injection every 4 weeks for 5 months (Diabetic Retinopathy Clinical Research Network et al. 2015). An alternative to Aflibercept is its isomer Ziv-Aflibercept. In 2012 the FDA approved this molecule for the treatment of metastatic colorectal cancer. It seems to guarantee comparable efficacy and safety of Aflibercept but with lower cost effectiveness. It is administered off-label with intravitreal injections and it could be used as a second line therapy in refractory AMD and DME. Also Ziv-Aflibercept demonstrates a high affinity to VEGF-A and inhibits PlGF and VEGF-B, as well (Ebrahimiadib et al. 2020). Further trials are necessary to confirm the safety and efficacy of Ziv-Aflibercept.

8   Steroids

Intravitreal injections of corticosteroids are the other available alternative of pharmacological therapy for DME. Steroids have an important role in DME treatment algorithm even though anti-VEGF agents are considered as first line treatment.

In the light of the recent evidence that underline the importance of the role of inflammation in the development of DME, intravitreal administration of steroids are a fundamental option for both recalcitrant and treatment naive eyes. The role of corticosteroids is mainly evident when anti-VEGF agents are contraindicated or a treatment regimen with fewer intravitreal injections is required, as well as in eyes nonresponding to anti-VEGF agents (Schmidt-Erfurth et al. 2017). The response to anti-VEGF treatment is not predictable and sometimes patients do not improve even after several injections or they develop tachyphylaxis. To note that there is not international consensus on the number of intravitreal injections for declaring treatment failure, before switching to alternatives drugs. The reason to response failure of first line therapy is unclear. It represents a significant challenge in everyday practice and in research fields (Nguyen et al. 2012; Mitchell et al. 2011). As previously mentioned, a hypothesis could be that VEGF is not the unique agent in the DME pathogenesis but there are countless inflammatory cytokines and different signalling pathways (Urias et al. 2017).

In BRB breakdown pathogenesis, multiple factors are involved other than VEGF. As a matter of fact the over-expression of proinflammatory cytokines, such as intracellular adhesion molecule (ICAM)-1, interleukin-6 (IL-6), tumor necrosis factor (TNF), and cyclooxygenase-2, is upregulated, further neutrophils and monocytes are attracted, and vascular permeability deteriorates (McLeod et al. 1995; Funatsu et al. 2009).

The rationale of corticosteroids in DME therapy is based on the fact that these molecules acting downregulating the arachidonic acid pathway and, consequently, reducing the synthesis of thromboxanes, leukotrienes, and prostaglandins. Blocking the inflammatory cascade, corticosteroids could indirectly reduce VEGF synthesis. So, steroids provide powerful anti-inflammatory and anti-edematous effects by targeting not only the synthesis of proinflammatory mediators involved in DME (IL 6, IL 8, MCP 1, ICAM 1, TNF, VEGF, HGF, ANGPT2, and more) but also a decrease in VEGF synthesis (Zur et al. 2019).

Consequently, the BRB can experience a structural improvement, secondary to an increment of density and activity of tight junctions in the retinal capillary endothelium and to an enhancement of retinal oxygenation (Whitcup et al. 2018).

In conclusion corticosterodis act by effecting the early phases of inflammation pathway, inhibiting the inflammatory response and the consequent proinflammatory state.

To date, commercially available corticosteroids for intravitreal use are: Triamcinolone Acetonide (actually on an off-label indication), the Dexamethasone delivery system and the Fluocinolone acetonide insert.

One of most commonly used corticosteroid agents was Triamcinolone Acetonide (TA). One of milestone for the management of DME and for intravitreal TA use is the DRCD.net protocol B. Similarly to the previously mentioned studies, this study was designed as multicentric randomized clinical trial. Patients were randomized in three groups to focal/grid laser, 1 mg intravitreal TA and 4 mg intravitreal TA. Retreatment was administered on a 4 months schedule. The authors reported a better mean BCVA in 4 mg group at 4 months. However no difference in BCVA was noted at 1 year between groups. Extending observational period to 2 years, laser group had a better BCVA outcome than TA groups. Moreover, a majority of patients treated with TA developed cataract in this period. BCVA differences could not be related only to cataract formations. The authors reported that patients developed ocular hypertension in approximately 40% in 4 mg TA group, 20% in 1 mg TA group and 10% laser group. Regarding only phakic eyes at baseline, cataract was performed in 23% vs 13% of patients of TA groups vs laser group respectively (Ip et al. 2008).

Another important study in the history of intravitreal injection of TA is the DRCR.net protocol I study. This trial was built in a randomized, controlled and multicentric fashion. Three arms were defined for different treatment schemes: intravitreal 0.5 mg ranibizumab plus prompt or deferred focal/grid laser; or 4 mg intravitreal TA combined with focal/grid laser compared with focal/grid la ser alone. At a 2 year control, patients treated with ranibizumab had a better BCVA letter score than 4 mg TA and laser groups (Elman et al. 2011). After 2 years follow-up. pseudophakic patients treated with TA faired similarly to the ranibizumab-treated group in terms of visual outcome and anatomical re- sults after 2 years of follow-up. Regarding the pharmacocynetic of TA, it is estimated that mean elimination half-life after intravitreal injection is of 18.6 days in non-vitreoctomized eyes, while it is of 3.2 days in vitreoctimezed eyes (Beer et al. 2003). As every drug with short elimination half life, for TA exists the need to repeat drug administration with higher frequency, in order to maintain the therapeutic effect. This conduct can raise the risk of intravitreal complication, e.g. cataract formation and glaucoma.

In the attempt to improve intravitreal corticosteroids efficacy and feasibility, other drugs and formulation have been studied.

Dexamethasone (DEX) is a potent anti-inflammatory agent; its potency is twice that of FA and five-fold more than that of TA (Ebrahimiadib et al. 2020). DEX is highly water-soluble and it needs to be delivered in a sustained-release system to provide vitreous drug levels over time.

Pharmacokinetic studies of DEX implant showed an initially high rate of DEX release over the first 2 months after injection, followed by a decrease in release until 6 months (Chang-Lin et al. 2011). Intravitreal administration of DEX implant provides a high initial drug concentration with a maximum at 60 days. It is inserted into the vitreous chamber with a 22-G needle via the pars plana and contains 0.7 mg of DEX (Chang-Lin et al. 2011). In contrast to TA, the pharmacokinetics of the DEX implant were not significantly affected in test subjects previously vitrectomized (Shin et al. 2012). This can be explained by the fact that DEX implant does not require the vitreous as a substrate to work (Chang-Lin et al. 2011).

One of the first study that evaluated DEX for DME was the PLACID trial. The study was a randomized, controlled, multicenter clinical trial. Patients were randomized to 0.7 mg DEX implant plus laser therapy and sham implant injection plus laser therapy. All patients were treated with laser at 1 month and retreatment was evaluated after 4 month. The authors noted a significant gain in terms of BCVA letter score during at the beginning, but such difference was lost over time, with no statistic significant difference at 12 months evaluation (Callanan et al. 2013).

The milestone study that assessed the efficacy and safety of DEX intravitreal implant (0.7 and 0.35 mg) has been the MEAD study. These are two 3-years, multicenter, randomized, masked, sham injection-controlled phase III clinical trials. Patients have been randomized in a 1:1:1 ratio to study treatment with DEX implant 0.7 mg, DEX implant 0.35 mg, or sham injection.

Minimum treatment lapses between repeated DEX implants were 6 months, and patients received on average 4–5 treatments over the study period. A substantial part of patients treated with DEX implant achieved a significant improvement in BCVA (22.4 vs. 12.0%, $p < 0.002$) and a statistically significant reduction in central macular thickness (112 vs. 42 μm, $p < 0.001$) compared to patients in the sham group (Boyer et al. 2014).

Patients enrolled were mainly phakic patients (75%) with long-standing DME (mean duration 23 months) who were previously treated by macular laser photocoagulation in 65.8%, other corticosteroids in 16.5%, or anti-VEGF injections in 7.1%. Only 25% of the patients were treatment naive. Common ocular adverse events were cataract and increased intraocular pressure (IOP) related to DEX implant. Regarding the safety, increases in IOP were usually controlled with medication or no therapy; only 2 patients (0.6%) in the DEX implant 0.7 mg group and 1 (0.3%) in the DEX implant 0.35 mg group required trabeculectomy. In the MEAD study it was reported an increase of IOP > 10 mmHg in 27.7% and a IOP > 35 mmHg in 6.6%, so up to 41.5% patients needed to underwent a IOP-lowering treatment. The incidence of

cataract DEX-related increased after the first year of the study, and over three-fourths of the cataract surgeries in the DEX implant groups were performed between 18 and 30 months. However, in phakic eyes, mean BCVA improvement was remarkable until the time of report of cataract, and improvement in vision from baseline was restored after cataract surgery (Boyer et al. 2014).

In a recent article, Callan et al. published the result of a multicentric, randomized, parallel-group, not inferiority study, between DEX implant (0.7 mg) and Ranibizumab. Over the 12 months of observation, DEX implant reached the not inferiority criterion to Ranizumab in improvement of BCVA letter score (respectively 4.34 vs 7,60 letters). Both drugs could reduce CRT and area of fluorescin leakage. Deferred laser treatment (as salvage therapy) was administered in 9,9% patients of DEX group, vs 2.2% of the Ranibizumab group. Average injection needed during study period was 2.85 (median 3) for DEX group vs 8.70 (median 9) in Ranibizumab group. The authors reported a more significant reduction in central macular thickness using DEX implant (122 vs 187 μm, $p = 0.015$). Despite DEX showed a not inferiority to Ranibizumab, an higher rate of complications were recorded in DEX group (65.7% vs 22.7%, $p < 0,001$). In particular, higher rates of cataract and rise of IOP were reported (Callanan et al. 2017).

Actually no consensus exists about administration interval length. Real-life studies are of great importance in case of DEX implant, because intervals can vary respect the MEAD trial. Large-scale studies have been showing the efficacy on BCVA improvements and on decrease of retinal thickness, even in patients non responsive to anti VEGF (Iglicki et al. 2019).

The IRGREL-DEX study, is a recent multicentric observational retrospective study that investigated efficacy and safety of DEX implants in treatment-naïve eyes vs anti-VEGF refractory eyes in real-life environment. Over 24 months, mean BCVA underwent a significant improvement during first 12 months and remained stable until the end of study. These results suggest that DEX implant could improve BCVA in both refractory and naïve eyes, although naïve eyes fared better. Regarding central subfield thickness, was significantly decreased compared with baseline in naive and refractory eyes (P = 0.001), but it was significantly higher in refractory eyes. The mean number of DEX implants received over 24 months was 3.5 ± 1.0 (range: 1–4).

The treatment had a well-acceptable safety profile. In the IRGREL-DEX study recorded cataract surgery rate and intraocular pressure rise, as these are the most frequent side effect of DEX implant. Real-life studies reported a safer profile than the one presented in MEAD study. In IRGREL-DEX study just 14% of all eyes needed IOP-lowering local treatment (7.1% vs 22.8% in naïve and refractory group respectively; p = 0.033), while no glaucoma incision surgery was necessary. Respect to MEAD study, the latter study revealed a lower rate of IOP increase. A possible explanation lays in the fact that in the IRGREL-DEX study the mean number of implants administered was 3.5, while in MEAD trial it was 5 (Iglicki et al. 2019).

Another recent study by Vujosevic S. et al., compared DEX implants and Ranibizumab in 33 treatment naïve DME with subfoveal neuroretinal detachment (SND) eyes. The authors showed evidence in favour of the protective anti-inflammatory effect of DEX. They compared neuroinflammatoru and vascular parameters in DME with SND visible to OCT and OCT-A. In this study DEX implants had a favourable effects in reducing different OCT and OCT-A inflammation biomarker such CMT, hyper-reflective retinal spots (HRS) in inner and full retina, DRIL extension, cysts at deep capillary plexus, circularity index of FAZ at superficial capillary plexus and perfusion density at deep capillary plexus. In particular DEX implants were superior to Ranibizumab in reducing the number of HRS, in particular in case were a high number was calculated at baseline. This observation could indicate that elevated HRS number could be a manifestation of major inflammatory state of the retina, thus requiring steroid treatment (Vujosevic et al. 2017).

Other important findings were a more significant decrease in CMT, retinal cyst at deep capillary plexus and DRIL extension in DEX group than Ranibizumab group. The authors suggested to consider DRIL extension as a surrogate prognostic and predictive marker of visual acuity response in patients with existing or resolved center-involving DME (Vujosevic et al. 2017).

Finally, Fluocinolone Acetonide (FA) is a corticosteroid available in a sustained-release system with a 25 G dispenser, in a not bioerodible implant, and that could last up to 36 months in the vitreous.

The FAME trials were 2 parallel, randomized, multicenter trials. Patients with central DME, BCVA of 20/50 to 20/200 and CRT >250 μm were randomized to receive an intravitreal insert releasing 0.2 or 0.5 μg FA per day or sham injection. 28% of patients treated with a BCVA gain of >15 letters after 2 years vs 16% of the sham group. After stratification, s comparing chronic (defined at least 3 years from diagnosis) with acute DME, the authors observed an significant higher percentage of patients who gained 15 or more letters in chronic DME (34% vs 13% in the sham group) compared to acute DME (22.3 vs 27.8% in the sham group) (Campochiaro et al. 2011).

The FDA approved FA intravitreal implant, 0.19 mg for patients with DME previously treated with a course of corticosteroids and did not developed a clinical significant rise in IOP. In the FAME study almost all phakic patients in the FA groups developed cataract over a 36 months periods. However, visual outcome after cataract surgery was comparable to pseudophakic patients.

The studies cited seem to suggest that corticosteroids are important alternatives in the armamentarium of drugs for treating DME patients, but largely on a second choice level. In a recent article, Pravin et al. used the data from Protocol I study to speculate that patients who had already been treated with ani-VEGF (3–6 injection practiced) could be considered as non-responders, so it should be reasonable to switch to a steroid treatment. As a matter of fact, in this post hoc analysis authors had found that usually the retinal morphologic responses to Ranibizumab plus prompt or deferred laser treatment generally develop rapidly in patients with center-involved DME (at least 20% reduction of CRT within the first 3 months of treatment). In patients who fit in the early-response category, the initial anatomical improvement was maintained also during long-term treatment (83% of patients). By contrast, for the one-third of eyes that showed little or no anatomical improvement after the first three Ranibizumab injections, prospects for future improvement with continued treatment were at best moderate, with 48% of eyes continuing to show not satisfactory CRT reduction at 3 years follow up, despite the intensive treatment and monitoring protocol (Dugel et al. 2019).

Furthermore, according to the latest indications there are special populations in which corticosteroids should be chosen as first-line therapy:

- Patients who have a recent history of a cardio-vascular event
- Pregnancy
- Inability or not willing to adhere to monthly treatments
- DME in eyes undergoing cataract surgery (Zur et al. 2019).

In conclusion, all of three agents are associated with risk of cataract progression and intraocular pressure elevation, but they could be considered to maintain a sufficient safety profile.

Between corticosteroids Dexamethasone shall be used in first line, while FA may be appropriate in select cases of patients with chronic macular edema that is not responsive to other treatments. Since TA is not approved for treatment of DME and it seems to be correlated with increase in IOP and cataract rate, it should be used only in case of shortage of approved agents for this indication. Finally, steroids could be considered as a first line drugs in pseudophakic patients (Schmidt-Erfurth et al. 2017).

9 Conclusion

DME is the main cause of vision loss in diabetic retinopathy and a severely disabling complications of diabetes. It is compulsory screen patients with diabetes for ocular involvement. In case of DR, or suspicious of DME, patients evaluation should include fundus examination and visual acuity measurement in the first instance. However, a multimodal imaging should be mandatory in diabetic retinopathy to detect possible complications as soon as possible. FA is still considered the gold standard for the diagnosis, but also OCT has an important role both in the diagnosis and follow-up of DME. Moreover, OCT is a fast, non-invasive and cost-effective technique to objectified treatment response. The OCT-A and the ultrawidefield FA are becoming increasingly important for a thorough imaging. All these new imaging techniques provide useful information to suggest new classification and to identify prognostic factors or inflammatory biomarkers.

Extensive researches are been conducted with the aim to understand the pathogenetic mechanism and to identify the optimal treatment. Although efforts have been spent, the DME etiology is not completely understood. Angiogenesis and inflammation have been shown to be involved in the pathogenesis. Consequently, the intravitreal injections of anti-VEGF agents or corticosteroids are the pharmacological options for DME treatment. According to EURETINA guidelines, the anti-VEGF agents (Bevacizumab, Ranibizumab and Aflibercept) are usually recommended as first line therapy in DME (Schmidt-Erfurth et al. 2017). Different trials have demonstrated an acceptable efficacy and safety profile. In case of recalcitrant DME, in according to current evidences and literature clinicians should switch between anti-VEGF agents in order to identify a tailor made therapy. In case of anti-VEGF failure, tachyphylaxis or contraindications, corticosteroids are the other available alternative. They are recommended as first-line therapy in pseudophakic patients, patients who have a recent history of a cardiovascular event, during pregnancy or in case of inability or not willing to adhere to monthly treatments. In particular Dexamethasone and Fluocinolone Acetonide are indicated also in vitreoctomized eyes as opposed to anti VEGFs and Triamcinolone Acetonide. Steroids only relevant side effects are the onset of cataract and the increase of intraocular pressure. Non-pharmacological treatments are peripheral laser photocoagulation, focal/grid laser photocoagulation, subthreshold micropulse yellow laser and vitreo-retinal surgery. Both lasertherapy and surgery can be used alone or in combination with intravitreal pharmacological treatment, to achieve a synergistic effect. Optimal therapy in DME has to be tailor-made on patients needs, weighing up the clinical and imaging provided data.

References

Aiello LP, Avery RL, Arrigg PG, Keyt BA, Jampel HD, Shah ST et al (1994) Vascular endothelial growth factor in ocular fluid of patients with diabetic retinopathy and other retinal disorders. N Engl J Med 331 (22):1480–1487

Antonetti DA, Barber AJ, Hollinger LA, Wolpert EB, Gardner TW (1999) Vascular endothelial growth factor induces rapid phosphorylation of tight junction proteins occludin and zonula occluden 1. A potential mechanism for vascular permeability in diabetic retinopathy and tumors. J Biol Chem 274 (33):23463–23467

Beer PM, Bakri SJ, Singh RJ, Liu W, Peters GB, Miller M (2003) Intraocular concentration and pharmacokinetics of triamcinolone acetonide after a single intravitreal injection. Ophthalmology 110(4):681–686

Bolz M, Schmidt-Erfurth U, Deak G, Mylonas G, Kriechbaum K, Scholda C et al (2009) Optical coherence tomographic hyperreflective foci: a morphologic sign of lipid extravasation in diabetic macular edema. Ophthalmology 116(5):914–920

Bolz M, Lammer J, Deak G, Pollreisz A, Mitsch C, Scholda C et al (2014) SAVE: a grading protocol for clinically significant diabetic macular oedema based on optical coherence tomography and fluorescein angiography. Br J Ophthalmol 98(12):1612–1617

Boyer DS, Yoon YH, Belfort R, Bandello F, Maturi RK, Augustin AJ et al (2014) Three-year, randomized, sham-controlled trial of dexamethasone intravitreal implant in patients with diabetic macular edema. Ophthalmology 121(10):1904–1914

Bresnick GH (1986) Diabetic macular edema: a review. Ophthalmology 93(7):989–997

Brown DM, Schmidt-Erfurth U, Do DV, Holz FG, Boyer DS, Midena E et al (2015) Intravitreal aflibercept for diabetic macular edema: 100-week results from the VISTA and VIVID studies. Ophthalmology 122 (10):2044–2052

Callanan DG, Gupta S, Boyer DS, Ciulla TA, Singer MA, Kuppermann BD et al (2013) Dexamethasone intravitreal implant in combination with laser photocoagulation for the treatment of diffuse diabetic macular edema. Ophthalmology 120(9):1843–1851

Callanan DG, Loewenstein A, Patel SS, Massin P, Corcóstegui B, Li X-Y et al (2017) A multicenter, 12-month randomized study comparing dexamethasone intravitreal implant with ranibizumab in patients with diabetic macular edema. Graefes Arch Clin Exp Ophthalmol Albrecht Von Graefes Arch Klin Exp Ophthalmol 255(3):463–473

Campochiaro PA, Brown DM, Pearson A, Ciulla T, Boyer D, Holz FG et al (2011) Long-term benefit of sustained-delivery fluocinolone acetonide vitreous inserts for diabetic macular edema. Ophthalmology 118(4):626–635.e2

Caruso R, Magon A, Baroni I, Dellafiore F, Arrigoni C, Pittella F et al (2018 Jan) Health literacy in type 2 diabetes patients: a systematic review of systematic reviews. Acta Diabetol 55(1):1–12

Chang-Lin J-E, Attar M, Acheampong AA, Robinson MR, Whitcup SM, Kuppermann BD et al (2011) Pharmacokinetics and pharmacodynamics of a sustained-release dexamethasone intravitreal implant. Invest Ophthalmol Vis Sci 52(1):80–86

Diabetic Retinopathy Clinical Research Network, Elman MJ, Aiello LP, Beck RW, Bressler NM, Bressler SB et al (2010) Randomized trial evaluating ranibizumab plus prompt or deferred laser or triamcinolone plus prompt laser for diabetic macular edema. Ophthalmology 117(6):1064–1077.e35

Diabetic Retinopathy Clinical Research Network, Wells JA, Glassman AR, Ayala AR, Jampol LM, Aiello LP et al (2015) Aflibercept, bevacizumab, or ranibizumab for diabetic macular edema. N Engl J Med 372 (13):1193–1203

Diep TM, Tsui I (2013 Jun 1) Risk factors associated with diabetic macular edema. Diabetes Res Clin Pract 100 (3):298–305

Dugel PU, Campbell JH, Kiss S, Loewenstein A, Shih V, Xu X et al (2019) Association between early anatomic response to anti–vascular endothelial growth factor therapy and long term outcome in diabetic macular edema. Retina (Philadelphia, Pa.) 39(1):88–97

Ebrahimiadib N, Lashay A, Riazi-Esfahani H, Jamali S, Khodabandeh A, Zarei M et al (2020) Intravitreal ziv-aflibercept in patients with diabetic macular edema refractory to intravitreal bevacizumab. Ophthalmic Surg Lasers Imaging Retina 51(3):145–151

Elman MJ, Bressler NM, Qin H, Beck RW, Ferris FL, Friedman SM et al (2011) Expanded 2-year follow-up of ranibizumab plus prompt or deferred laser or triamcinolone plus prompt laser for diabetic macular edema. Ophthalmology 118(4):609–614

Fong DS, Contreras R (2009) Glitazone use associated with diabetic macular edema. Am J Ophthalmol 147 (4):583–586.e1

Funatsu H, Noma H, Mimura T, Eguchi S, Hori S (2009) Association of vitreous inflammatory factors with diabetic macular edema. Ophthalmology 116(1):73–79

Gerendas BS, Waldstein SM, Simader C, Deak G, Hajnajeeb B, Zhang L et al (2014a) Three-dimensional automated choroidal volume assessment on standard spectral-domain optical coherence tomography and correlation with the level of diabetic macular edema. Am J Ophthalmol 158(5):1039–1048

Gerendas B, Simader C, Deak GG, Prager SG, Lammer J, Waldstein SM et al (2014b) Morphological parameters relevant for visual and anatomic outcomes during anti-VEGF therapy of diabetic macular edema in the RESTORE trial. Invest Ophthalmol Vis Sci 55 (13):1791–1791

Gurreri A, Pazzaglia A, Schiavi C (2019) Role of statins and ascorbic acid in the natural history of diabetic retinopathy: a new, affordable therapy? Ophthalmic Surg Lasers Imaging Retina 50(5):S23–S27

Hwang TS, Gao SS, Liu L, Lauer AK, Bailey ST, Flaxel CJ et al (2016) Automated quantification of capillary nonperfusion using optical coherence tomography angiography in diabetic retinopathy. JAMA Ophthalmol 134(4):367–373

Iglicki M, Busch C, Zur D, Okada M, Mariussi M, Chhablani JK et al (2019) Dexamethasone implant for diabetic macular edema in naive compared with refractory eyes: the international retina group real-life 24-month multicenter study. The IRGREL-DEX study. Retina (Philadelphia, Pa.) 39(1):44–51

Ip MS, Bressler SB, Antoszyk AN, Flaxel CJ, Kim JE, Friedman SM et al (2008) A randomized trial comparing intravitreal triamcinolone and focal/grid photocoagulation for diabetic macular edema: baseline features. Retina (Philadelphia, Pa.) 28(7):919–930

Joussen AM, Poulaki V, Mitsiades N, Cai W-Y, Suzuma I, Pak J et al (2002) Suppression of Fas-FasL-induced endothelial cell apoptosis prevents diabetic blood–retinal barrier breakdown in a model of streptozotocin-induced diabetes. FASEB J 17(1):76–78

Kanski JJ, Bowling B (2011) Clinical ophthalmology: a systematic approach. Elsevier Health Sciences, 921 p

Kawasaki R, Konta T, Nishida K (2018) Lipid-lowering medication is associated with decreased risk of diabetic retinopathy and the need for treatment in patients with type 2 diabetes: a real-world observational analysis of a health claims database. Diabetes Obes Metab 20 (10):2351–2360

Klaassen I, Van Noorden CJF, Schlingemann RO (2013 May) Molecular basis of the inner blood-retinal barrier and its breakdown in diabetic macular edema and other pathological conditions. Prog Retin Eye Res 34:19–48

Massin P, Bandello F, Garweg JG, Hansen LL, Harding SP, Larsen M et al (2010) Safety and efficacy of ranibizumab in diabetic macular edema (RESOLVE study): a 12-month, randomized, controlled, double-masked, multicenter phase II study. Diabetes Care 33 (11):2399–2405

Mathew C, Yunirakasiwi A, Sanjay S (2015) Updates in the management of diabetic macular edema. J Diabetes Res 2015:794036

McLeod DS, Lefer DJ, Merges C, Lutty GA (1995) Enhanced expression of intracellular adhesion molecule-1 and P-selectin in the diabetic human retina and choroid. Am J Pathol 147(3):642–653

Mitchell P, Bandello F, Schmidt-Erfurth U, Lang GE, Massin P, Schlingemann RO et al (2011) The RESTORE study: ranibizumab monotherapy or combined with laser versus laser monotherapy for diabetic macular edema. Ophthalmology 118(4):615–625

Nguyen QD, Brown DM, Marcus DM, Boyer DS, Patel S, Feiner L et al (2012 Apr) Ranibizumab for diabetic macular edema: results from 2 phase III randomized trials: RISE and RIDE. Ophthalmology 119 (4):789–801

Perkovich BT, Meyers SM (1988 Feb 15) Systemic factors affecting diabetic macular edema. Am J Ophthalmol 105(2):211–212

Romero-Aroca P, Baget-Bernaldiz M, Pareja-Rios A, Lopez-Galvez M, Navarro-Gil R, Verges R (2016) Diabetic macular edema pathophysiology: vasogenic versus inflammatory. J Diabetes Res

Ryan EH, Han DP, Ramsay RC, Cantrill HL, Bennett SR, Dev S et al (2006 Jun) Diabetic macular edema associated with glitazone use. Retina (Philadelphia, Pa.) 26(5):562–570

Saeedi P, Petersohn I, Salpea P, Malanda B, Karuranga S, Unwin N et al (2019) Global and regional diabetes prevalence estimates for 2019 and projections for 2030 and 2045: results from the International Diabetes Federation Diabetes Atlas, 9th edition. Diabetes Res Clin Pract 157:107843

Schmidt-Erfurth U, Lang GE, Holz FG, Schlingemann RO, Lanzetta P, Massin P et al (2014) Three-year outcomes of individualized ranibizumab treatment in patients with diabetic macular edema: the RESTORE extension study. Ophthalmology 121(5):1045–1053

Schmidt-Erfurth U, Garcia-Arumi J, Bandello F, Berg K, Chakravarthy U, Gerendas BS et al (2017) Guidelines for the management of diabetic macular edema by the European Society of Retina Specialists (EURETINA). Ophthalmol J Int Ophthalmol Int J Ophthalmol Z Augenheilkd 237(4):185–222

Shin HJ, Lee SH, Chung H, Kim HC (2012) Association between photoreceptor integrity and visual outcome in diabetic macular edema. Graefes Arch Clin Exp Ophthalmol Albrecht Von Graefes Arch Klin Exp Ophthalmol 250(1):61–70

Solomon SD, Lindsley KB, Krzystolik MG, Vedula SS, Hawkins BS (2016) Intravitreal bevacizumab versus ranibizumab for treatment of neovascular age-related macular degeneration: findings from a cochrane systematic review. Ophthalmology 123(1):70–77.e1

Sophie R, Lu N, Campochiaro PA (2015) Predictors of functional and anatomic outcomes in patients with diabetic macular edema treated with ranibizumab. Ophthalmology 122(7):1395–1401

Sun JK, Lin MM, Lammer J, Prager S, Sarangi R, Silva PS et al (2014) Disorganization of the retinal inner layers as a predictor of visual acuity in eyes with center-involved diabetic macular edema. JAMA Ophthalmol 132(11):1309–1316

Sun JK, Radwan SH, Soliman AZ, Lammer J, Lin MM, Prager SG et al (2015 Jul) Neural retinal disorganization as a robust marker of visual acuity in current and resolved diabetic macular edema. Diabetes 64 (7):2560–2570

Torres-Costa S, Valente MC, Falcão-Reis F, Falcão M (2020) Cytokines and growth factors as predictors of response to medical treatment in diabetic macular edema. J Pharmacol Exp Ther

Urias EA, Urias GA, Monickaraj F, McGuire P, Das A (2017) Novel therapeutic targets in diabetic macular edema: beyond VEGF. Vis Res 139:221–227

Vujosevic S, Simó R (2017) Local and systemic inflammatory biomarkers of diabetic retinopathy: an integrative approach. Invest Ophthalmol Vis Sci 58(6): BIO68–BIO75

Vujosevic S, Torresin T, Berton M, Bini S, Convento E, Midena E (2017) Diabetic macular edema with and without subfoveal neuroretinal detachment: two different morphologic and functional entities. Am J Ophthalmol 181:149–155

Wessel MM, Nair N, Aaker GD, Ehrlich JR, D'Amico DJ, Kiss S (2012) Peripheral retinal ischaemia, as evaluated by ultra-widefield fluorescein angiography, is associated with diabetic macular oedema. Br J Ophthalmol 96(5):694–698

Whitcup SM, Cidlowski JA, Csaky KG, Ambati J (2018) Pharmacology of corticosteroids for diabetic macular edema. Invest Ophthalmol Vis Sci 59(1):1–12

Wong TY, Sun J, Kawasaki R, Ruamviboonsuk P, Gupta N, Lansingh VC et al (2018) Guidelines on diabetic eye care: the International Council of Ophthalmology Recommendations for screening, follow-up, referral, and treatment based on resource settings. Ophthalmology 125(10):1608–1622

Xue K, Yang E, Chong NV (2017) Classification of diabetic macular oedema using ultra-widefield angiography and implications for response to anti-VEGF therapy. Br J Ophthalmol 101(5):559–563

Yau JWY, Rogers SL, Kawasaki R, Lamoureux EL, Kowalski JW, Bek T et al (2012 Mar) Global prevalence and major risk factors of diabetic retinopathy. Diabetes Care 35(3):556–564

Zhang X, Yang J, Zhong Y, Xu L, Wang O, Huang P et al (2017 Oct) Association of bone metabolic markers with diabetic retinopathy and diabetic macular edema in elderly Chinese individuals with type 2 diabetes mellitus. Am J Med Sci 354(4):355–361

Zur D, Iglicki M, Loewenstein A (2019) The role of steroids in the management of diabetic macular edema. Ophthalmic Res 62(4):231–236

Adv Exp Med Biol - Advances in Internal Medicine (2020) 4: 391–415
https://doi.org/10.1007/5584_2020_498
© Springer Nature Switzerland AG 2020
Published online: 3 March 2020

New Concepts in the Management of Charcot Neuroarthropathy in Diabetes

Karakkattu Vijayan Kavitha, Vrishali Swanand Patil,
Carani Balarman Sanjeevi,
and Ambika Gopalakrishnan Unnikrishnan

Abstract

Charcot Neuroarthropathy (CN) is an uncommon, debilitating and often underdiagnosed complication of chronic diabetes mellitus though, it can also occur in other medical conditions resulting from nerve injury. Till date, the etiology of CN remains unknown, but enhanced osteoclastogenesis is believed to play a central role in the pathogenesis of CN, in the presence of neuropathy. CN compromises the overall health and quality of life. Delayed diagnosis can result in a severe deformity that can act as a gateway to ulceration, infection and in the worst case, can lead to limb loss. In an early stage of CN, immobilization with offloading plays a key role to a successful treatment. Medical therapies seem to have limited role in the treatment of CN.

In case of severe deformity, proper footwear or bracing may help prevent further deterioration and development of an ulcer. In individuals with a concomitant ulcer with osteomyelitis, soft tissue infection and severe deformity, where conservative measures fall short, surgical intervention becomes the only choice of treatment. Early diagnosis and proper management at an early stage can help prevent the occurrence of CN and amputation.

Keywords

Bisphosphonates · Charcot · Charcot Neuroarthropathy · Deformity · Diabetes · Immobilization · Neuropathy · Offloading · RANKL antibody · Total contact cast

1 Introduction

Charcot neuroarthropathy (CN) is one of the rare complications of Diabetes Mellitus (DM). It affects the neuromusculoskeletal structures of the foot and ankle, resulting in fractures, resorption, destruction of weight-bearing bones, joints and progressive deformity marked by inflammation in the early stage. DM is the most common etiology of CN though it can also occur in other neurological conditions like tabes dorsalis, syringomyelia etc. (Rogers and Frykberg 2013; Lee et al. 2003; Boulton 2014; Trieb 2016; Papanas and Maltezos 2013a; Gouveri 2011).

CN was first reported in 1831 by American physician *John Kearsley Mitchell* in people with tuberculosis induced spinal damage (Mitchell 1831). In 1868, *Jean-Martin Charcot*, a French neurologist, described denervation-induced joint

K. V. Kavitha, V. S. Patil, and A. G. Unnikrishnan (✉)
Chellaram Diabetes Institute, Pune, Maharashtra, India
e-mail: kkv@cdi.org.in; vsp@cdi.org.in; ceo@cdi.org.in

C. B. Sanjeevi
Department of Medicine, Karolinska Institutet, Head,
Diabetes Immunology Group, Center for Molecular
Medicine, L5:01, Karolinska University Hospital, Solna
Campus, Stockholm, Sweden

destruction as a complication of syphilis (Charcot 1868). However, *William Riely Jordan* was the first to report the association between CN and DM in 1936 (Jayasinghe et al. 2007). Though the etiology of CN is not fully known, it is well accepted that neuropathy precedes the disease (Rogers et al. 2011a). A delay in diagnosis of up to 8 weeks may lead to more rapid progression and increased complications such as joint deformity, ulceration, infection and in the worst scenario, may result in limb loss (Wukich et al. 2011). Early diagnosis and swift care are the keys to reduce amputation risk.

Exploration of existing literature on CN indicates a limited number of studies on the prevalence of CN. Available literature is focused mostly on the western population (Sinha et al. 1972; Leung et al. 2009). Worldwide, the prevalence of CN is reported from 0.08% to 13% in DM and in a high-risk group (Leung et al. 2009; Frykberg and Belczyk 2008; Younis et al. 2015; Milne et al. 2013; Wukich and Sung 2009). In a cross-sectional study done in a specialist diabetes clinic, the prevalence of CN was reported to be 0.4% in individuals with DM (Younis et al. 2015). In another study the prevalence of CN was noted to be 1.4% with radiographic evidence of midfoot CN (Smith et al. 1997). A study done in south India showed a very high prevalence of CN in an individual with DM on radiographs (Viswanathan et al. 2014). A retrospective study conducted at a tertiary health care centre in South India showed a prevalence of 9.8% of CN in Type 2 Diabetes population (Salini et al. 2018). There are limited prospective studies on incidence rate. A study reported an incidence rate ranging from 3 to 11.7/1000 patients per year (Wukich and Sung 2009; Frykberg et al. 2006). It is noted that the incidence of CN is reported to be increasing due to the better diagnostic imaging modalities (Rajbhandari et al. 2002; Foster et al. 1995).

The onset of CN is insidious in nature. It is commonly observed in people of age 50 years or older who had been diagnosed with diabetes for 15 years or more (Petrova et al. 2004). Studies suggest that the prevalence of foot complications in diabetes increases with age and duration of DM, irrespective of its type (Katsilambros et al. 2003). Statistics representing the age for onset of CN is of great significance and is variably stated. Based on a study done in a diabetes centre, individuals with Type 1 DM with an average duration of 24 years of diagnosis present more frequently in their fifth decade, while individuals with type 2 DM with a mean duration of 13 years of diagnosis tend to present in their sixth decade (Sinha et al. 1972; Petrova et al. 2004; Sanders and Frykberg 2001). The rate of occurrence was same in men and women.

2 Pathophysiology

The exact pathogenesis of acute CN remains unclear but neuropathy and inflammation are believed to be the key contributing factors. *Jean-Martin Charcot* proposed the *'French theory'*, also known as 'neurovascular theory', in 1868 which suggests that bony changes result from damage to the central nervous system that directly controls bone nourishment and leads to uncontrolled inflammation. The theory also suggests that joint destruction is secondary to autonomically stimulated vascular reflex that causes hyperemia and periarticular osteopenia with contributory trauma. *Volkman and Virchow* proposed the *'German theory'*, also called as 'neurotraumatic theory', which suggests that multiple sub-clinical trauma in a denervated joint is the initial precipitating factor. It is hypothesized that a minor trauma may trigger the inflammatory response through complex pathways leading to microfracture, subluxation and dislocation of bones. Abnormal joint loading is potentially exacerbated by neuropathy, as partial or complete lack of pain leads to continued weight-bearing and further joint destruction. Continued ambulation on the affected foot accelerates the progression of pathways responsible for osteolysis and osteopenia and weakening of pedal structures, leading to a higher risk of fractures (Varma 2013; Kaynak et al. 2013; Jeffcoate 2015; Wünschel et al. 2012). Nearly 50% of individuals with CN report a precipitating, minor traumatic events such as ankle sprain or a previous foot procedure (Boucho et al. 2016).

Autonomic neuropathy may result in impaired vascular reflexes with arteriovenous (AV) shunting, leading to increased arterial perfusion (Vinik and Ziegler 2007; Gilbey et al. 1989; Edmonds et al. 1982; Watkins 1983) which is clinically manifested as localized increased temperature with redness and dilated dorsal veins (Watkins 1983; Purewal et al. 1995). Increased blood flow to the foot bones, due to the aforementioned factors, has been found responsible for increased bone resorption, reduced bone mineral density, hence predilection for fractures (Edmonds et al. 1985). Increased osteoclastic activity has also been noted (Mabilleau et al. 2008).

Pro-inflammatory cytokines [mainly tumor necrosis factor-alpha (TNF-α) and Interleukin-1 beta (IL-1β)] are produced excessively as a consequence of local inflammation (Gouveri 2011; Rogers et al. 2011a; Jeffcoate 2005, 2008; Jeffcoate et al. 2005). Additionally, the cytokine-driven elevation of the receptor activator of nuclear factor kappa B ligand (RANKL) is accountable for communication between the osteoblasts, and osteoclast. RANKL is secreted by osteocytes, osteoblasts, and certain cells of the immune system. The osteoclast receptor for this protein is referred also as RANK. Activation of RANK by RANKL is a final common path in osteoclasts development and activation. A humoral decoy for RANKL, also secreted by osteoblast is referred to as osteoprotegerin (OPG). RANKL has been shown to mediate osteolysis in CN by stimulating osteoclastic differentiation of monocytes/macrophages. RANKL binds to its target receptor RANK to generate multiple intracellular signals that regulate cell differentiation, function, and survival (Schoppet et al. 2002; Walsh and Choi 2003). The most vital step of the osteoclastogenesis is binding of RANKL to its RANK receptor, anchored in the cell membrane of preosteoclasts. On the other hand, osteoprotegerin (OPG) is a cytokine synthesized and secreted by activated osteoblasts. This acts as a decoy receptor for RANKL and prevents binding of RANKL to RANK. Binding

of RANKL to OPG results in the inhibition of bone resorption and stimulates bone mass building (Fig. 1). RANKL also enhances the synthesis of NF-κB which in turn enhances production of OPG from osteoblast. The dynamic equilibrium between RANKL and OPG concentration is crucial for usual bone metabolism (Boyle et al. 2003).

CGRP, a family of calcitonin produced by healthy neurons is known to reduce the production of RANKL and it contributes to the up keeping of joint integrity. Any reduction of CGRP could be detrimental as it would indirectly enhance the action of RANKL, thereby accelerating the disease process. The role of CGRP is predominantly significant in neuropathy, where the latter contributes to the inflammation-induced osteolysis through reduced secretion of CGRP from affected neurons (Rogers et al. 2011a; Jeffcoate 2008).

In vitro evidence showed the presence of elevated levels of inflammatory cytokines in CN potentially modulating the RANKL/OPG signalling pathway. The evidence also indicated elevated systemic (serum) RANKL levels and RANKL/OPG ratio, driving vascular smooth muscle cells (VSMCs) into an osteoblastic differentiation pathway, causing deposition of a mineralized matrix. Differentiation of VSMCs takes place via RANKL/RANK signaling cascade, setting intracellular mechanisms involving the nuclear translocation of NF-kB. In addition, OPG is shown to prevent nuclear translocation of NF-kB and abrogate differentiation and mineralization of VSMCs induced by RANKL or Charcot serum. Abnormal RANKL/OPG signaling, therefore, could be proposed as the mechanism underpinning the paradoxical osteolysis and MAC seen in CN and similar disease conditions. This clutches promise for the treatment of CN, particularly because new drugs specifically targeting RANKL are already used in the treatment of osteoporosis (Ndip et al. 2011).

Increased non-enzymatic collagen glycation and elevated plantar pressures are observed more commonly in CN (Grant et al. 1997; Armstrong and Lavery 1998). The former is known to increase the forefoot pressure leading

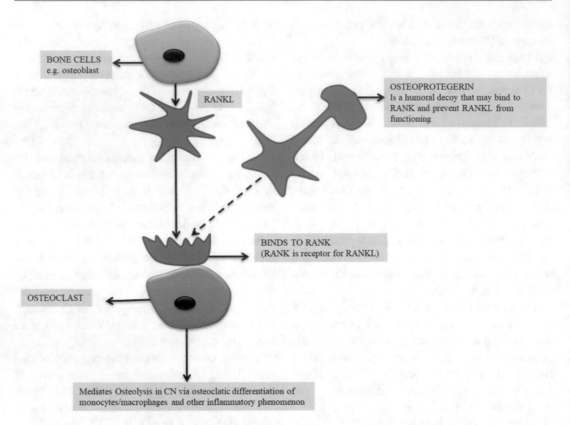

Fig. 1 Mechanism of action of RANK, RANKL and osteoprotegerin. (**Image courtesy**: Chellaram Diabetes Institute, Pune)
Graphical mechanism of action of RANK, RANKL and osteoprotegerin
Note that unopposed RANK- RANKL binding can activate osteolysis via osteoclast development. Osteoprotegerin is a decoy that prevents RANK-RANKL binding and inhibits bone resorption, stimulatory building of bone mass. Denosumab is a therapeutic monoclonal antibody directed at RANK L

to trauma and bone destruction as a result of Achilles tendon shortening (Grant et al. 1997), and the latter would lead to forefoot strain. This would, in turn, lead to increased mechanical stress in the midfoot in the region of Lisfranc's ligament (Armstrong and Lavery 1998). Achilles tendon shortening is caused by increased packing density of collagen fibrils, decreased fibrillar diameter and abnormal fibril morphology (Grant et al. 1997). This abnormal collagen may predispose the individual to CN (Grant et al. 1997; Hough and Sokoloff 1989).

Recently, genetic factors, notably polymorphisms of the gene encoding the beneficial glycopeptide osteoprotegerin, are thought to be potential contributors to pathophysiology

(Grant et al. 1997). Other important factors include the emerging role of genes (Gouveri 2011; Rajbhandari et al. 2002). CN may be further complicated by an ulcer that is often difficult to heal, carrying a high risk of recurrence, infection and even osteomyelitis (Gouveri 2011; Rogers et al. 2011a; Rajbhandari et al. 2002; Petrova and Edmonds 2008).

3 Classifications

Various classification systems have been developed to classify Charcot foot. Eichenholtz classification (Eichenholtz 1966) which was described since 1966 is used in common. This classification

Table 1 Eichenholtz classification of CN (Jeffcoate et al. 2005)

Stage	Radiographic findings	Clinical findings
I Developmental	Osteopenia, osseous fragmentation, joint subluxation or dislocation	Swelling, erythema, warmth, ligamentous laxity
II Coalescence	Absorption of debris, sclerosis, fusion of larger fragments	Decreased warmth, decreased swelling, decreased erythema
III Reconstruction	Consolidation of deformity, fibrous ankylosis, rounding and smoothing of bone fragments	Absence of warmth, absence of swelling, absence of erythema, fixed deformity

Table 2 MRI based Classification of CN (Frykberg and Belczyk 2008; Walsh and Choi 2003; Boyle et al. 2003)

	Severity grade	
Stage	Low severity: grade 0 (without cortical fracture)	High severity: grade 1 (with cortical fracture)
Active arthropathy (acute stage)	Mild inflammation/soft tissue oedema	Severe inflammation/soft tissue oedema
	No skeletal deformity	Severe skeletal deformity
	X-ray: normal	X-ray: abnormal
	MRI: abnormal (bone marrow oedema, microfractures, bone bruise)	MRI: abnormal (bone marrow oedema, macrofractures, bone bruise)
Inactive arthropathy (becalmed stage)	No inflammation	No inflammation
	No skeletal deformity	Severe skeletal deformity
	X-ray: normal	X-ray: abnormal (past macrofractures)
	MRI: no significant bone marrow oedema	MRI: no significant bone marrow oedema

divides the CN into three stages of progression – fragmentation/developmental, coalescence and reconstruction (Table 1). A stage 0 was added later to the aforementioned classification which include localized warmth, redness and swelling with normal radiographic findings (Shibata et al. 1990). However, a more recent study advocated that magnetic resonance imaging (MRI) (Table 2) to be more sensitive and specific than X-rays in detecting bone marrow edema (Milne et al. 2013; Chantelau and Grützner 2014; Johnson 1997; Chantelau and Poll 2006). Sanders and Frykbergs anatomical classification of CN (Sanders and Frykberg 1991; Brodsky 1999) is another classification system which is based on the anatomical location which is divided into five patterns of joint destruction (Table 3). According to this classification, Patterns II and III are associated with the highest complication rates.

Brodsky and Rouse Classification of four types of disease pattern is based on the frequency of joints affected; the commonest being the tarsometatarsal joint (Brodsky 1999). Type 1 includes Lisfranc's joint (tarsometatarsal);

Type 2 – Chopart's joint and/or the subtalar joint; Type 3A – ankle joint, Type 3B – the calcaneum (Brodsky 1999). The existing classifications do not provide a prognostic value or direct treatment. There is no accepted measure to define the transition point. A proposed classification system based on the location of foot involved and associated complications hypothesized a high risk of amputation with rearfoot/ankle involvement in the presence of osteomyelitis in CN (Rogers and Bevilacqua 2008). The same has been confirmed in a study done in a Diabetes centre in South India (Vijay et al. 2012).

4 Clinical Presentation

Numerous authors have asserted the absence of male-female predilections in the presentation of CN (Rajbhandari et al. 2002). In a study conducted on the DM population, no gender-related changes were noted in the presentation of CN (Younis et al. 2015). Various studies reported

Table 3 Sanders and Frykberg's Anatomical classification of CN (Ndip et al. 2011; Grant et al. 1997)

Type of CN	Joint/s involved
I	Metatarsophalangeal, interphalangeal
II	Tarso-metatarsal
III	Tarsal
IV	Sub-talar
V	Calcaneum

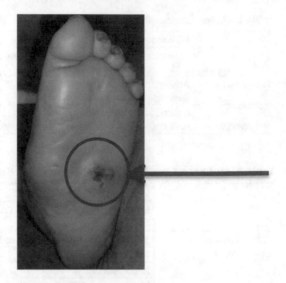

Fig. 2 Chronic CN with infection on the midfoot on the plantar aspect. (**Image courtesy**: Chellaram Diabetes Institute, Pune)

male gender as a risk factor for developing CN. In a study, the higher frequency of CN presentation was noted in male (97.1%) as compared to females (81.2%). The higher rates in males may be partly due to increased physical activity. In another study, presentation of CN in men and women were equally reported (Sohn et al. 2009; Kensarah et al. 2016).

Individuals with CN, especially in diabetes and having peripheral neuropathy, may present with a red, hot swollen foot, with or without any significant history of trauma or surgery (Botek et al. 2010). Sometimes they only present with swelling of the foot, which may put a physician in a clinical dilemma. Thus, a thorough clinical history and physical examination play a crucial role in the diagnosis of CN (Botek et al. 2010). One of the important physical examination findings that may be seen in CN is bounding pedal pulse (Wilson 1991).

Acute CN is defined by a presence of hot, swollen foot, with or without erythema after the exclusion of other conditions resembling CN such as cellulitis, deep vein thrombosis, gout, etc. (Sinha et al. 1972). A significant proportion of subjects with acute CN have a concomitant ulcer, further complicating the diagnosis and raising the possibility of osteomyelitis. Other marked irregularities identified in CN are bony projections, bone formation in soft tissues and sometimes a flabby, distended appearance, resembling a "bag of bones". Usually there is a unilateral involvement of the joint although rare simultaneous bilateral involvement is certainly possible. The most commonly involved sites are tarsus and tarsometatarsal joints, followed by metatarsophalangeal joints and the ankle.

Chronic CN is defined as fracture or dislocation with or without gross deformity of the foot. Continued weight- bearing results in progressive deformity of the foot that is prone to ulceration and amputation (Fig. 2). Common deformities seen are the "rocker bottom foot" (Fig. 3) caused by the collapse of medial arch, medial convexity deformity caused by medial displacement of the talonavicular joint, and tarsometatarsal dislocation. Table 4 explains about the presentation of acute vs chronic CN (Gouveri 2011; Rogers et al. 2011a; Rajbhandari et al. 2002; Eichenholtz 1966; Frykberg et al. 2010a; Papanas and Maltezos 2013b).

5   Diagnosis

The diagnosis of acute CN is primarily clinical, requiring a high index of suspicion. It should be suspected in any individual with diabetes who presents with an inflamed foot, profound neuropathy, bone and joint abnormalities in the absence of fever and elevated erythrocyte sedimentation rate (ESR). In the acute phase, patients present with unilateral erythematous, edematous foot

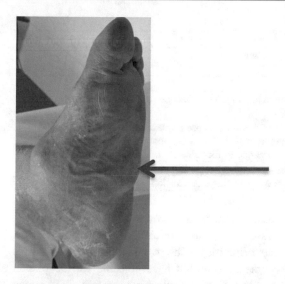

Fig. 3 Representing rocker bottom foot deformity. (**Image courtesy**: Chellaram Diabetes Institute, Pune)

with increased local skin temperature, (Boulton 2014) with a presence or absence of deformity and joint effusion, (Jeffcoate et al. 2005; Jeffcoate 2005) which may or may not be painful (Fig. 4). CN is observed in the contralateral limb in about 25% of people with DM (Milne et al. 2013; Varma 2013; Jeffcoate et al. 2005; Pakarinen et al. 2002; Foltz et al. 2004; Game et al. 2012). Although often unrecalled, CN is probably triggered by a history of trauma. The diagnosis of active CN is primarily based on history and clinical findings but should be confirmed by imaging.

The classical presentation of an acutely inflamed foot may be deceptive as other common lower limb conditions which resemble cellulitis, deep vein thrombosis, or acute gout can be misdiagnosed (Rogers et al. 2011a; Wukich et al. 2011; Jeffcoate 2015; Game et al. 2012). Fig. 5a, b and c shows varying features on radiographs differentiating the clinical presentation. The chances of misdiagnosis are as high as 79% which, eventually, leads to a delay in treatment for an average of 29 weeks (Wukich et al. 2011; Milne et al. 2013; Pakarinen et al. 2002). There is a strong association between the duration of diabetes, elevated HbA1c, and the development of CN (Younis et al. 2015; Milne et al. 2013; Stuck et al. 2008). Often, the triggering

factor for acute CN is repetitive micro-trauma on an insensate foot (Milne et al. 2013; Pakarinen et al. 2002). An individual with CN may remember a precipitating, minor traumatic event, and if no traumatic episode was recollected, then the time frame in which the changes in foot shape and/or gait is noticed should be recorded.

Early diagnosis and appropriate treatment reduce the risk of CN causing permanent debilitating foot deformity or amputation. Subjects with CN have well preserved or exaggerated arterial blood flow in the foot with characteristically bounding pedal pulse unless obscured by concurrent oedema. Laser Doppler shows increased cutaneous blood flow in CN, differentiating from peripheral neuropathy (Krishnan et al. 2004). Other potential risk factors include obesity, advanced age, renal failure, iron deficiency, osteoporosis and rheumatoid arthritis (Trieb 2016; Papanas and Maltezos 2013a; Kaynak et al. 2013; Trieb and Hofstätter 2015).

The initial manifestation of the Charcot foot is mild in nature. The process of CN begins with hyperemia, usually following a trauma to the foot or ankle. Hyperemia may persist for months or years (Jeffcoate et al. 2000) but in some cases the acute phase rapidly progresses to the chronic stage within days, and sometimes in less than 6 months, resulting in permanent deformity (Rajbhandari et al. 2002). The presence of little or no pain may mislead the patient and physician, as peripheral neuropathy is likely to be an essential criterion for the onset of CN (Armstrong et al. 1997a; Jude et al. 2001). There have been no reported cases of CN development in the absence of neuropathy. In patients with red, hot foot, with no ulcers or fever, and a normal or slightly elevated serum C-reactive protein level or ESR, the acute phase of Charcot process should be considered. However, these findings may also be seen in the presence of infection, and in such a condition, the existence of infection cannot be excluded (Loredo et al. 2010). Patients with infection generally have a feeling of sickness, as opposed to the patient with Charcot foot who presents only with a swollen and often painful foot. A valuable clinical finding could be a recent increase in blood

Table 4 Feature of Acute Vs Chronic CN

Acute Charcot neuroarthropathy	Chronic Charcot neuroarthropathy
Acute local inflammation of foot remarkably red, warm and swollen	No longer inflamed
Mild to modest pain or discomfort	Painless
Marked temperature elevation of 2-6° C compared with contralateral limb	Difference in skin temperature between the limb diminishes
No deformity or Mild deformity may be noticed	Chronic deformity with collapse of plantar arch in the midfoot with medial convexity with classic rocker-bottom foot resulting in high pressure points prone to foot ulceration
Foot is stable, with no crepitus or loose bones	Crepitus, palpable loose bodies and large osteophytes are the result of extensive bone and cartilage destruction

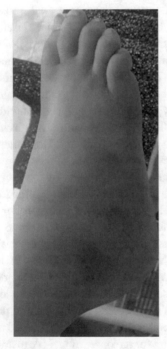

Fig. 4 Acute CN with erythema and edema of the foot. (**Image courtesy**: Chellaram Diabetes Institute, Pune)

glucose levels, increased insulin requirement, or difficulty controlling blood glucose levels that often accompanies infection, but is not true with an acute CN process. A simple physical examination to differentiate between an infectious process and CN is to have the patient lay supine and elevate the affected extremity for 5–10 min. Localized edema will decrease with elevation in CN while it is unlikely in an infectious process (Milne et al. 2013; Ndip et al. 2008; Brodsky 1993). Diagnostic clinical findings and assessment include components of neurological, vascular, musculoskeletal, and radiographic abnormalities. The contraction of the triceps surae muscle that is responsible for the plantar inclination of the calcaneus is relatively common in CN. For further prognosis and management, it is necessary to examine the stability of the foot. Instability of the forefoot can be assessed in the sagittal plane dorsally when the ankle joint is locked in dorsiflexion (Assal and Stern 2009). Often there is a temperature difference of several degrees between the limbs; a temperature difference of 2 °C or more on the infrared thermometer as compared to the contralateral foot indicates the present of CN (Madan and Pai 2013; Khanolkar et al. 2008; Armstrong et al. 1997b). Infrared thermometer allows monitoring the healing (associated with 'foot cooling') and recurrence (associated with 'foot warming').

In chronic CN, symptoms of warmth, swelling is decreased, and inflammation is usually not present (Fig. 6). Classic rocker-bottom foot may be seen in Chronic CN with or without plantar ulceration, representing a severe chronic deformity which begins with the medial column and proceeds to the lateral column in the late stage (Pakarinen et al. 2002; Nielson and Armstrong 2008; Pakariennen et al. 2002). The foot is unstable because of the collapse of longitudinal foot arch (Clouse et al. 1974). The deformity begins with the collapse of naviculocuneiform joint and the perinavicular pattern begins with the osteonecrosis or fracture of the navicular bone. The talus is completely dislocated from the navicular and ulceration of the calcaneocuboid interval begins in the late stage.

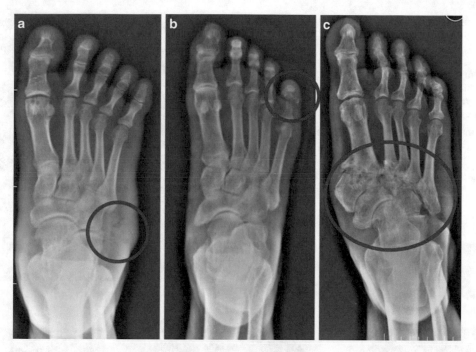

Fig. 5 Shows various features on radiographs. (**a**) Air foci suggestive of infective Etiology and soft tissue edema. (**b**) Bony destruction of 5th toe suggesting osteomyelitis. (**c**) Reduced joint spaces with cortical irregularities at tarsometatarsal joints suggesting CN. (**Image courtesy**: Chellaram Diabetes Institute, Pune)

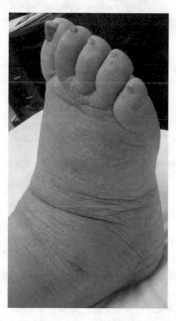

Fig. 6 Chronic Charcot Neuroarthropathy with evidence of deformity. (**Image courtesy**: Chellaram Diabetes Institute, Pune)

Fig. 7 Showing the radiographic differences of normal bony structure and with CN. (**a**) Normal Radiograph. (**b**) Charcot Neuroarthropathy. (**Image courtesy**: Chellaram Diabetes Institute, Pune)

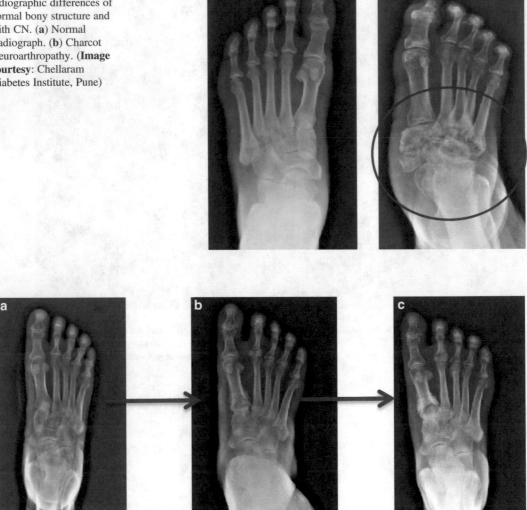

Fig. 8 Shows stages of CN transition as per Eichenholtz classification. (**a**) Developmental stage. (**b**) Coalescence stage. (**c**) Reconstruction. (**Image courtesy**: Chellaram Diabetes Institute, Pune)

5.1 Imaging

Radiographs are the primary imaging method for the initial evaluation of CN which include anteroposterior and lateral weight-bearing views or full series ankle views (anteroposterior, mortise and lateral views) depending on clinical suspicion (Gouveri 2011; Van der Ven et al. 2009; Gold et al. 1995). Both foot should be considered in order to detect and compare metabolic disturbances and subtle changes (Fig. 7a, b). Plain radiographs are useful in the diagnosis of pathology, identification of the area of involvement, evaluation of the bone structure, quality, alignment, mineralization and the process of CN (Fig. 8a, b, c) (Gouveri 2011; Van der Ven et al. 2009; Trepman et al. 2005). The classical presentation of Chronic CN consists of 6 Ds – destruction, debris, joint distention, bone density, dislocation, and deformity (Table 5 and Fig. 9)

Table 5 Classical 6 Ds of CN

Classical 6 Ds of Charcot Neuroarthropathy (CN)
Destruction of articular surface
Debris of bones
Distention of joint spaces
Dense bones
Dislocation (Lisfranc's)
Deformity (loss of foot arches)

Fig. 9 Anteroposterior view of X-ray shows classical 6 Ds of CN. (**Image courtesy**: Chellaram Diabetes Institute, Pune)

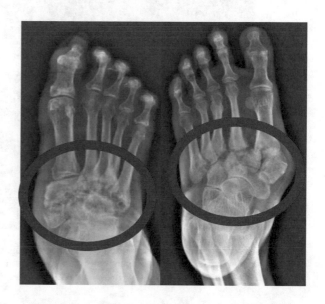

(Rajbhandari et al. 2002; Ledermann and Morrison 2005). Radiographic evidence of demineralization, bone destruction, and periosteal reaction suggest the presence of CN, although these can also be seen in chronic osteomyelitis. The involvement of tarsometatarsal (Lisfranc's) joint leads to the collapse of the longitudinal arch and increased loading on the cuboid thus resulting in rocker-bottom deformity. Talocalcaneal dislocation, talar collapse and atypical calcaneal fractures might be seen in the hindfoot (Rajbhandari et al. 2002). The tarsal bones and proximal metatarsals are typically affected in acute CN (Loredo et al. 2010). The flattening of the metatarsal head is often the first sign of CN.

The plain radiographs can be negative for up to 3 weeks with the only finding being soft tissue swelling. While subtle fractures and dislocations are common in early stages, the reduction in the calcaneal inclination and disruption of the talo-first metatarsal angle has also been noted (Frykberg et al. 2010b). Radiological indicators of Meary's angle and Calcaneal pitch is noted in Chronic CN. Meary's angle of greater than 15° and a calcaneal pitch of less than 17° are predominant in CN. Significant alterations in the Meary's angle and Calcaneal pitch angle can assist in prompt diagnosis and intervention in chronic CN (Gupta et al. 2017). Osteophytes, joint consolidation, and arthrosis can be seen in the radiographic evaluation of chronic CN. Also, dislocation of the tarsometatarsal joint with a break in the talo-first metatarsal line (Fig. 10a, b) and reduced calcaneal inclination angle (Fig. 11a, b) can be seen in the lateral radiograph in a late stage of chronic CN.

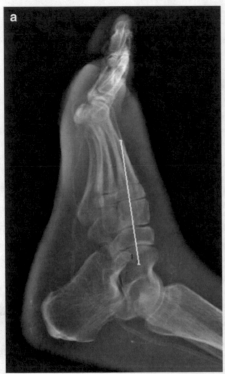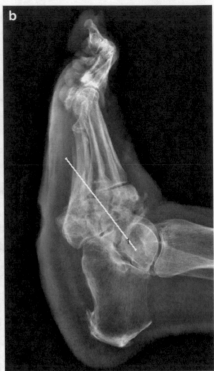

Fig. 10 Shows normal and loss of medial longitudinal arch. (**a**) Normal medial longitudinal arch represented by intact talo-1st metatarsal angle. (**b**) Loss of medial longitudinal arch represented by reversed talo-first metatarsal angle. (**Image courtesy**: Chellaram Diabetes Institute, Pune)

Magnetic resonance imaging (MRI) or nuclear imaging can be considered in case of clinical suspicion and normal-appearing radiographs (Stark et al. 2016). Although plain radiography is the first imaging modality, its sensitivity and specificity in the early stages are low. On the other hand, MRI is regarded as the most sensitive imaging method for detecting early changes in CN and, therefore, a modality of choice to differentiate between CN and infection (Ergen et al. 2013; La Fontaine et al. 2016). MRI is also beneficial in ruling out osteomyelitis, especially in the presence of an ulcer, elevated ESR, C-reactive protein (CRP), or leukocytosis (Trieb 2016; Jeffcoate 2015; Trieb and Hofstätter 2015; Stark et al. 2016). Initial signs of CN in MRI include bone marrow edema, soft tissue edema, joint effusion, and eventually micro fractures (Ledermann and Morrison 2005; Ergen et al. 2013; Schoots et al. 2010; Mautone and Naidoo 2015; Ahmadi et al. 2006).

MRI in middle to late-stage shows joint destruction, cortical fractures, and joint dislocations. Bone marrow edema may or may not present depending on disease activity (Ergen et al. 2013; Mautone and Naidoo 2015; Ahmadi et al. 2006; Rogers et al. 2011b). Noticeable well marginated subchondral cysts is a typical characteristic of chronic CN (Ledermann and Morrison 2005; Ergen et al. 2013). There can be enormous fluid collections surrounding destructed joints. If MRI shows evident bone marrow edema then offloading with total contact cast is continued (Renner et al. 2016). After resolution of bone marrow edema cast is discontinued and customized footwear or orthosis is advised.

High-sensitivity nuclear medicine modalities provide a paradigm shift to the diagnosis (La Fontaine et al. 2016). Radioisotope technetium (Tc-99 m) bone scintigraphy has good sensitivity, but poor specificity for osseous pathology and only shows increased focal uptake during the

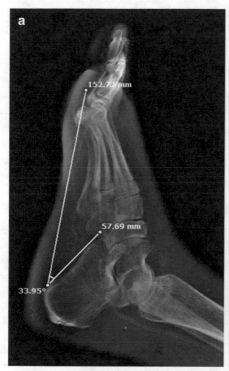

Fig. 11 Showing normal and change in calcaneal inclination angle. (**a**) Normal calcaneal Inclination angle (10–35°). (**b**) Reduced calcaneal inclination angle (3°) suggestive of rocker bottom deformity. (**Image courtesy**: Chellaram Diabetes Institute, Pune)

bony phase. Increased tracer uptake of the involved joint occurs in all three phases of a bone scan in acute CN (Fig. 12). The technetium-99 m methylene diphosphonate (Tc-MDP) scan provides high accuracy in detecting abnormal woven bone but clinical conditions with high bone turnover, such as infection, surgery and trauma, may reduce the specificity rates. Although a four-phased bone scan with delayed image acquisition at 24 h seems to be more specific for detecting abnormal bone, some clinical entities (i.e. tumours, degenerative changes and fractures) may lead to false-positive results (Ergen et al. 2013). Labelled white cell scans (In-WBC) provide high accuracy results in diagnosing osteomyelitis. Accordingly, a combination of In-WBC and Tc- MDP has been shown to increase specificity and sensitivity in diagnosing CN complicated by an ulcer (Ergen et al. 2013; La Fontaine et al. 2016). The role of

FDG-PET and FDG-PET/CT in the diagnosis of acute CN remains a matter of continuing exploration, with favorable outcomes (Pickwell et al. 2011; Höpfner et al. 2004). The use of Computed Tomography (CT) is limited in CN as it cannot detect early findings of acute CN such as bone marrow edema and microfractures (Loredo et al. 2007).

Individuals with acute CN were found to have high levels of bone-specific alkaline phosphatase (a bone formation marker) and urinary deoxypyridinoline (a bone resorption marker) as compared to non-charcot diabetic subjects, indicating continuous bone turnover and remodelling (Selby et al. 1998). There is an increase in the bone resorption marker called pyridinoline crosslinked carboxy-terminal telopeptide domain of type 1 collagen in acute CN. A study has confirmed the presence of the aforementioned marker and its correlation with calcaneal bone density

(Jirkovska et al. 2001). In another study, high levels of cross-linked N telopeptides of type 1 collagen (a urinary marker of bone resorption) were found in CN (Edelson et al. 1996).

5.2   Nerve Conduction

Peripheral neuropathy grounds serious complications like diabetic foot ulcer, gangrene, and Charcot joint, all of which deteriorates the quality of life in an individual with DM (Ogawa et al. 2006). Timely recognition of nerve dysfunction is imperative to deliver appropriate care in an individual with peripheral neuropathy (Chudzik et al. 2007). The diagnosis of diabetic peripheral neuropathy is mainly centred on the characteristic symptoms (Watanabe et al. 2009). But symptoms usually develop at any degree of neuropathic impairment or may not develop at all. This designates the need for nerve conduction studies (NCS) (Asad et al. 2009). The prompt and precise detection can aid in the superior understanding of the pattern of pathophysiological changes of neuropathy as well as helps in controlling crippling ailment (Dobretsov et al. 2007).

5.3   Nerve Biopsy and Histopathological Studies

Histopathology of CN showed a marked loss of sympathetic nerve fibers and inflammation of the bones (Koeck et al. 2009). Histological study demonstrated that CN bone has characteristics of reactive bone with the presence of woven bone that is immature and structurally disorganized. Additionally, the bone marrow spaces were infiltrated by hypervascular, myxoid tissue with spindle fibroblasts with an increase in the number of Howship's lacunae and a decreased number of osteocytes when compared with normal bone and bone in diabetes without CN (La Fontaine et al. 2011). The study also point out that bone in DM is fragile, and the decrease in the cellular component might weaken the reparative process in those with CN (La Fontaine et al. 2011).

6   Management

The management objective is to halt the progression of inflammation, relieve pain, and preserve the architecture of the foot and ankle from the

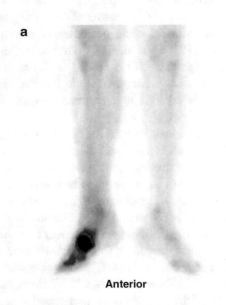

Anterior

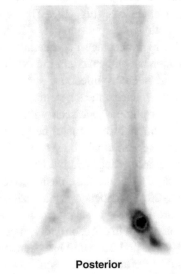

Posterior

Fig. 12 Shows increased tracer uptake in CN. (**a**) shows increased tracer uptake in the anterior and posterior views. (**Image courtesy**: Dr. S V Solav, Spect Lab, Bavdhan, Pune). (**b**) shows increased tracer uptake in the blood pool phase on the dorsal and plantar views. (**Image courtesy**: Dr. S V Solav, Spect Lab, Bavdhan, Pune)

b

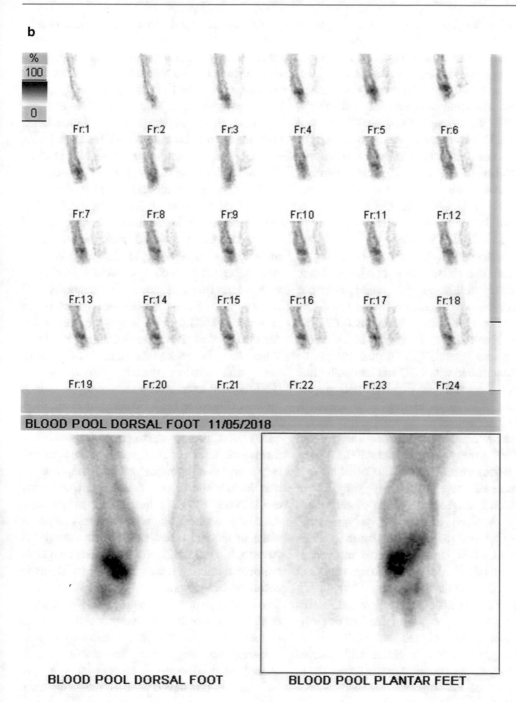

Fig. 12 (continued)

further deformity. The aim is to achieve a planti-grade, stable foot that fits into a shoe and to prevent occurrence or recurrence of ulceration. The primary management of CN remains conservative (non-surgical), comprising offloading techniques including total contact cast (TCC), braces and customized footwear (Ögüt and Yontar 2017). Nevertheless, a few cases of CN eventually need surgical intervention to correct the instability of the foot. Treatment depends upon various factors such as the phase of the disease process, location, deformity, presence or absence of infection, and other comorbidities (Rogers et al. 2011a; Van der Ven et al. 2009). The recommended treatment plan according to the Eichenholtz classification of CN in Stage 0 is a frequent follow-up with serial radiographs to monitor the development of Stage I and foot care education (Petrova and Edmonds 2008; Van der Ven et al. 2009; Jirkovska et al. 2001; Armstrong and Peters 2002). Stage I CN can be successfully treated with TCC and immobilization along with frequent follow-up and radiographic assessment with serial casting until erythema, color, and inflammation is resolved (Jeffcoate 2008; Petrova and Edmonds 2008; Van der Ven et al. 2009; Armstrong and Peters 2002; Young et al. 1995; Pinzur et al. 2006). Stage II (coalescence-subacute phase) is typically treated with protected weight-bearing with a total contact cast or a molded total-contact polypropylene ankle-foot orthosis (Nielson and Armstrong 2008; Pinzur et al. 2006). In Stage III (reconstruction-chronic), an individual is advised to use sandals or shoes with custom molded insoles (Rogers et al. 2011a; Pinzur et al. 2006; Ramanujam and Facaros 2011). In case of a non-plantigrade foot or recurrent history of ulcerations, debridement, exostectomy, correction, or fusion with internal fixation may be an option. Also, in Stage III, in the presence of osteomyelitis, surgical debridement with or without staged reconstruction with internal or external fixation, or amputation is recommended (Eichenholtz 1966; Stuck et al. 2008; Van der Ven et al. 2009; Armstrong and Peters 2002; Pinzur et al. 2006; Ramanujam and Facaros 2011; Perrin et al. 2010).

The use of a total contact cast or instant total contact cast (iTCC) along with a walking frame or crutches or a knee scooter (wherever available) is recommended in the chronic stage. A wheelchair should be prescribed in case of clinical suspicion of non-compliance or bilateral involvement. Frequent cast change as per clinical recommendation and close monitoring is vital in reducing complications as setting can lead to instability and ulceration within the cast (Van der Ven et al. 2009; Armstrong and Peters 2002; Pinzur et al. 2006; Perrin et al. 2010). In the first few weeks, the cast or device should be inspected and changed frequently to avoid "pistoning" as edema reduction is remarkable. In the presence of a wound, regular inspection should be done along with wound measurement and photography, and if required, sharp debridement should be performed (McGill et al. 2000; Sinacore 1998; Armstrong et al. 2005; Armstrong et al. 2002). When the active phase sets in, the patient can be fitted with Charcot restraint orthotic walker (CROW) and later, with a custom-made shoe or orthosis (Sinacore 1998; Armstrong et al. 2002, 2005). The average cast duration in chronic CN with ulceration is 5 weeks, with progression to therapeutic footwear at 12 weeks. Some patients may need a cast for over a year and complications may include simple skin maceration (Armstrong et al. 2002, 2005). Some studies suggest that removable walking braces can successfully treat acute or chronic ulcerated CN (Sinacore 1998; Armstrong et al. 2002, 2005). Disadvantages of removable walking braces include the inability to accommodate severe deformities and possibly limited compliance (Sinacore 1998; Armstrong et al. 2005). Use of assistive modalities such as walker and crutches may allow complete non-weight bearing and are acceptable methods of offloading devices, a three-point gait may increase the pressure on the contralateral limb, leading to repetitive stress, ulceration or neuropathic fractures (Lesko and Maurer 1989). Better alternatives include TCC, CROW, a Scotch cast boot or pneumatic walking braces (Jeffcoate et al. 2005; Jude and Boulton 1999; Morgan et al. 1993). Ambulation in TCC was shown to result in a mean healing time of 86 days, with the most rapid healing occurring in forefoot CN (Sinacore 1998).

Individuals with acute ankle CN have been found to have a better limb survival if treated with non-weight-bearing protective devices, as compared to those who continued weightbearing (Clohisy and Thompson Jr. 1988). Once the inflammation is reduced lifelong protection of foot is important. Recurrence of CN seems to be unusual, but the risk for ulceration due to the presence of severe deformity is always present. Limiting weight-bearing, continued use of customized footwear or braces including a patellar tendon-bearing brace, modified Ankle foot orthosis, a Charcot restraint orthotic walker (CROW) to protect the foot is important (Varma 2013). Foot education, proper foot care practices and use of appropriate orthosis are integral aspects of lifelong foot protection (Milne et al. 2013; Christensen et al. 2012).

6.1 Bisphosphonates

Bisphosphonates are synthetic analogues of inorganic pyrophosphate that decrease bone resorption by hindering the activity of osteoclasts while stimulating osteoblastic activity (Fleisch et al. 2002). Bisphosphonates may shorten the lifespan of osteoclasts and provide pain relief through effects on prostaglandin E2 and other nociceptive substances (Strang 1996). They have also been implicated to interfere with the release of neuropeptides and neuromodulators from afferent nerve endings (Schott 1995). A number of clinical trials assessing bisphosphonates in CN suggest clinical benefit. Pamidronate, alendronate and zolendronic acid have been evaluated in some randomized control studies (RCT) (Jude et al. 2001; Selby et al. 1994; Pitocco et al. 2005; Pakarinen et al. 2011). In one study, the subjects who received an initial infusion of 60 mg of pamidronate followed by a 30 mg infusion fortnightly over 12 weeks showed significant improvement in their symptoms and foot temperature, Also their alkaline phosphatase levels reduced by about 25% by the end of the study. In another randomized double-blind clinical trial, 39 diabetic subjects with active CN were assigned either a placebo (normal saline) or a 90 mg single intravenous infusion of pamidronate (Jude et al. 2001). All participants were recommended immobilization and to avoid weight-bearing. Throughout the study, the pamidronate group had a significant reduction in symptoms compared to the placebo group. Both fasting plasma bone-specific alkaline phosphatase (a bone formation marker) and second-void early morning urinary deoxypyridinoline (a bone resorption marker) showed a significant reduction in the pamidronate treated group compared to the placebo group. Another study reported (Anderson et al. 2004) significantly greater reductions in temperature (measured after 48 h and 2 weeks) and alkaline phosphatase (measured after 2 weeks) among CN patients treated with intravenous pamidronate compared to standard care alone. However, this study was not randomized, and there was bias in the treatment strategy.

An Italian randomized study observed the effect of oral alendronate (70 mg once a week) on 20 subjects with acute CN (Rogers et al. 2011a; Pitocco et al. 2005). All subjects were put on TCC for the first 2 months and a pneumatic walker for the next 4 months, followed by the use of special shoes. Subjects treated with Alendronate had a significant reduction in symptoms, a reduction in the levels of bone resorption markers hydroxyproline and carboxy-terminal telopeptide of type 1 collagen, and an improvement in the bone mineral density of the foot. Another study done in a tertiary care centre in South India showed comparable efficacy with respect to the time taken for attaining complete clinical resolution of acute CN with both zolendronic acid 5 mg as an intravenous infusion and oral alendronate 70 mg once a week along with TCC. The dose of zolendronic acid was reduced to 2.5 mg if the serum creatinine was ≥ 2 mg/dL (Bharath et al. 2013). However, in the absence of conclusive clinical trials, bisphosphonate therapy is not routinely recommended for the management of CN.

6.2 Calcitonin

Intranasal calcitonin has been found effective in arresting excessive bone turnover in subjects with acute CN (Bem et al. 2006). Calcitonin directly affects osteoclasts and interacts with the RANKL pathway (Cornish et al. 2001). In a recent randomized study, 32 subjects were administered with a combination of intranasal calcitonin (200 IU/day) and calcium supplementation (100 mg/day) or calcium supplementation alone (Bem et al. 2006). The disease activity improved in both groups but the calcitonin-treated group had a significant reduction in bone turnover markers. In a follow-up study involving 36 acute CN subjects, the calcitonin-treated group had significantly faster healing compared to the control group (Bem et al. 2006).

6.3 RANKL Antibody

RANKL antibody treatment, which had earlier shown a beneficial effect in osteoporosis-induced fracture, (Cummings et al. 2009) has also been found beneficial in acute CN with respect to fracture resolution time and clinical outcomes (Busch-Westbroek et al. 2018). In another study, subjects were treated with TCC and a single injection of denosumab 60 mg subcutaneously, followed by testing of plasma calcium levels after a week. Fracture resolution was significantly shorter after denosumab administration compared to the usual-care group. Denosumab also called as "bone-modifying agent" is a human monoclonal antibody that works as a RANK ligand (RANKL) inhibitor. Denosumab binds to RANKL, inhibiting its ability to initiate the formation of mature osteoclasts from osteoclasts precursors and to bring mature osteoclasts to the bone surface and initiate bone resorption. It also plays a role in reducing the survival of the osteoclast.

6.4 Surgical Options

With the advancement in surgical corrective techniques and accessibility of instruments, reconstructive procedures beyond the acute phase have gained popularity in recent times (Shen and Wukich 2013). However, because of the increased likelihood of complications associated with poor bone quality such as pseudarthrosis, delayed union or nonunion, potential wound-site complications and infection, surgery in the acute phase has been, and is still a debated matter. Surgery is usually considered in stage 3 when there are no acute inflammatory signs.

There is no single consensus regarding the appropriate timing and the type of surgery in CN. The goals of surgical treatment are to preserve functional ability, restore stability and alignment of the foot so that appropriate footwear or bracing is possible. The main indications for surgical intervention in CN include gross instability, chronic recurrent ulcers, osteomyelitis and progressive joint destruction despite conservative measures (Idusuyi 2015). The surgical techniques described in the literature include simple exostectomy, open reduction and internal fixation of neuropathic fractures, external fixation with Ilizarov, arthrodesis, Achilles tendon lengthening and, eventually, amputation (Lowery et al. 2012; Bari et al. 2018).

The Achilles tendon lengthening is performed to reduce the loading at the midfoot and forefoot by altering its strength; thereby providing more dorsiflexion. Many surgical procedures have been performed in combination with Achilles tendon lengthening or gastrocnemius muscle release (Burns and Wukich 2008; Mueller et al. 2004). According to a systemic review, Achilles tendon lengthening remains a popular method in the management of Charcot foot (Wukich et al. 2016).

7 Summarizing the Clinical Approach to CN

Charcot foot is becoming an increasingly alarming issue as the number of individuals with diabetes is increasing worldwide. Loss of protective sensation secondary to sensory neuropathy in the feet increases the likelihood of trauma, while motor neuropathy could result in altered structure and gait, resulting in abnormal loading. Patients often recall a minor traumatic event as the causative factor for the onset of the condition. Other triggering factors include local inflammation, previous ulceration, or recent foot surgery.

It has been long known that patients who tend to develop CN are in their mid-fifties, with longer duration of diabetes, and are likely to be morbidly obese (Gouveri 2011; Game et al. 2012). Individuals with longstanding and poorly controlled diabetes, neuropathy, history of ulceration, a recent history of trauma, prior neuroarthropathy, or renal transplantation are at high risk and should be closely monitored, for early clinical findings which may be mild. However, the acute phase of CN often goes unnoticed, resulting in a delayed diagnosis and progression to the chronic phase, with irreversible deformation. Unilateral presentation of CN is most common but a significant number of people present with bilateral involvement (ranging from 9% to 75%) (Armstrong et al. 1997a; Clohisy and Thompson 1988).

A thorough clinical evaluation including examination for skin abnormalities and neurological testing should be performed annually to diagnose the problem at an early stage. Patients should be prescribed diabetic footwear, or foot orthosis, and should be advised against wearing normal footwear to prevent the development of CN in high-risk subjects. In clinical practice, the classification systems of Charcot foot often give minimal support to diagnosis and treatment. The two hypotheses – the sensory and the autonomic neuropathy – though differ from the initial cause, agree that peripheral neuropathy together with repeated microtrauma that often go unnoticed, trigger mechanical and vascular alterations that

lead to changes in the bone structure (Vázquez Gutiérrez et al. 2005; Agullera-Cios et al. 2005). The available treatment options for CN aim for the maintenance or recovery of a plantigrade foot, achievement of osseous stability, and prevention of ulceration (Lowery et al. 2012). Regardless of the chosen treatment pathway, treatment protocols should be adjusted according to the patient's lower limb pathology, overall medical status, and ability to comply with the treatment.

Individuals with acute CN have been found to have higher plantar pressures in the metatarsophalangeal joints when compared to people with distal sensorimotor neuropathy or neuropathic ulceration (Armstrong and Lavery 1998). It is proposed that the forefoot acts as a lever and causes the collapse of the midfoot, which is the common site of involvement. Early diagnosis is difficult due to the fact that neuropathic patients are usually pain free (Lowery et al. 2012). Offloading is the most important initial recommendation for the treatment although the gold standard of treatment remains immobilization in a total contact cast. The duration of casting is usually 3–6 months and is continued until signs of inflammation are reduced (Moura-Neto et al. 2012). The Eichenholz and Sanders and Frykberg classification systems are the most commonly used staging systems for CN (Lowery et al. 2012). Current evidence of adjunct therapies such as bisphosphonates or calcitonin in acute CN is inconclusive.

The guidelines available states that it is important to diagnose CN at an early stage to prevent the progression by limiting the deformity. Management includes Immobilization with TCC as the initial treatment. Removable cast walker is another choice but has low compliance. Therapeutic orthotic devices are recommended to prevent complications. Good glucose control is recommended. Reconstructive surgeries may have a role in selected cases. A multidisciplinary approach and long term follow-up for the management of CN is very important (Gooday and Berrington 2015; International Diabetes Federation 2017).

It is also postulated that people with CN also have vitamin D deficiency with subsequent osteoporosis. Peripheral neuropathy, osteoporosis, and morbid obesity appear to be very important predisposing risk factors. People with CN experience a dramatic decrease in their quality of life and have a high risk of losing their foot. Although non-operative methods are considered to be the gold standard treatment in CN, surgical intervention is essential when a conservative approach fails. Surgical treatment is reserved for chronic recurrent ulcerations, unbraceable deformity, acute fracture, dislocation or infection. The common long-term ramifications of failed conservative care of CN include plantar ulceration, superficial and deep tissue infection, osteomyelitis, and amputation. A recent study showed that the presence of ulcer in CN increased the risk of major amputation by 6 – folds when compared to CN without ulcer (Wukich et al. 2011). Patient education, professional foot examination and care remain the cornerstones of lifelong foot protection. Footwear is an important component of long-term management in chronic CN to ensure protection and accommodation of foot. Early recognition and intervention is imperative to avoid rapid progression towards permanent foot deformity, ulceration and limb loss (Wukich et al. 2011; Jeffcoate 2008; Eichenholtz 1966).

Individuals with DM and peripheral neuropathy have a high likelihood of developing CN. The diagnosis of acute CN requires a high index of suspicion. Patient education plays a pivotal role in the long-term management of CN with special focus to appropriate footwear, offloading, regular follow-up, and the risk management of further complications.

Large scale trials would help better understand the benefits of bisphosphonates, calcitonin and RANKL antibodies in the acute setting. A better understanding of the underlying etiopathogenesis, particularly RANKL and osteoprotegerin, may pave the way for new treatment strategies, such as TNF-α antagonists, corticosteroids and nonsteroidal anti-inflammatory agents.

A thorough clinical evaluation including physical, neurological and radiological assessment must be carried out in any patient who presents with erythema, edema, and increased foot temperature. This could prevent any underlying inflammatory process, like Charcot Neuroarthropathy, from going unidentified.

8 Conclusion

CN is a rare, ominously underdiagnosed complication of diabetes. Early diagnosis and management of CN is vital to improve clinical outcomes. Radiology may help to outline the differences between early and late-stage CN and may guide further therapy. Further research may help delineate the benefits and risks of future experimental therapies in the management of CN.

Acknowledgement We Acknowledge Ms. CH. V. Mridula of publication department of Chellaram Diabetes Institute for her contribution towards the manuscript preparation and proof reading.

References

Aguilera-Cros C, Povedano-Gómez J, García-López A (2005) Charcot's neuroarthropathy. Reumatol Clin 1 (4):225–227. https://doi.org/10.1016/S1699-258X(05) 72749-2

Ahmadi ME, Morrison WB, Carrino JA et al (2006) Neuropathic arthropathy of the foot with and without superimposed osteomyelitis: MR imaging characteristics. Radiology 238(2):622–631. https://doi.org/10.1148/radiol.2382041393

Anderson JJ, Woelffer KE, Holtzman JJ, Jacobs AM (2004) Bisphosphonates for the treatment of Charcot neuroarthropathy. J Foot Ankle Surg 43(5):285–289. https://doi.org/10.1053/j.jfas.2004.07.005

Armstrong DG, Lavery LA (1998) Elevated peak plantar pressures in patients who have Charcot arthropathy. J Bone Joint Surg Am 80(3):365–369. https://doi.org/10.2106/00004623-199803000-00009

Armstrong DG, Peters EJ (2002) Charcot's arthropathy of the foot. J Am Podiatr Med Assoc 92:390–394. https://doi.org/10.7547/87507315-92-7-390

Armstrong DG, Todd WF, Lavery LA et al (1997a) The natural history of acute Charcot's arthropathy in a diabetic foot specialty clinic. Diabet Med 14:357–363. https://doi.org/10.1002/(SICI)1096-9136 (199705)14:5<357::AID-DIA341>3.0.CO;2-8

Armstrong DG, Lavery LA, Liswood PJ et al (1997b) Infrared dermal thermometry for the high-risk diabetic foot. Phys Ther 77(2):169–175. https://doi.org/10.1093/ptj/77.2.169

Armstrong DG, Short B, Espensen EH et al (2002) Technique for fabrication for an instant total contact cast for treatment of neuropathic diabetic foot ulcers. J Am Podiatr Med Assoc 92(7):405–408. https://doi.org/10.7547/87507315-92-7-405

Armstrong DG, Lavery LA, Wu S (2005) Evaluation of removable and irremovable cast walkers in the healing of diabetic foot wounds: a randomized controled trial. Diabetes Care 28(3):551–554. https://doi.org/10.2337/diacare.28.3.551

Asad A, Hameed MA, Khan UA et al (2009) Comparison of nerve conduction studies with diabetic neuropathy symptom score and diabetic neuropathy examination score in type-2 diabetics. J Pak Med Assoc 59(9):594–598

Assal M, Stern R (2009) Realignment and extended fusion with use of a medial column screw for midfoot deformities secondary to diabetic neuropathy. J Bone Joint Surg Am 91:812–820

Bari MM, Shahidul I, Shetu NH et al (2018) Charcot's arthropathy in diabetics: treatment by Ilizarov technique. MOJ Orthop Rheumatol 10(1):00388. https://doi.org/10.15406/mojor.2018.10.0038

Bem R, Jirkovská A, Fejfarová V et al (2006) Intranasal calcitonin in the treatment of acute Charcot neuroosteoarthropathy. Diabetes Care 29(6):1392–1394. https://doi.org/10.2337/dc06-0376

Bharath R, Bal A, Shanmuga S et al (2013) A comparative study of zolendronic acid and once weekly Alendronate in the management of acute Charcot arthropathy of foot in patients with diabetes mellitus. Indian J Endocrinol Metab 17(1):110–116. https://doi.org/10.4103/2230-8210.107818

Botek G, Anderson MA, Taylor R (2010) Charcot neuroarthropathy: an often overlooked complication of diabetes. Cleve Clin J Med 77(9):593–599. https://doi.org/10.3949/ccjm.77a.09163

Bouche C, Ratheau L, Gautier JF (2016) Forefoot osteolysis revealing a Charcot Osteoarthropathy. Diabetes Manag 6(3):062–065

Boulton AJ (2014) Diabetic neuropathy and foot complications. Handb Clin Neurol 126:97–107. https://doi.org/10.1016/B978-0-444-53480-4.00008-4

Boyle WJ, Simonet WS, Lacey DL (2003) Osteoclast differentiation and activation. Nature 423(6937):337–342. https://doi.org/10.1038/nature01658

Brodsky JW (1993) The diabetic foot. In: Mann RA, Coughlin MJ (eds) Surgery of the foot and ankle. Mosby, St. Louis, pp 877–958

Brodsky JW (1999) Evaluation of the diabetic foot. Instr Course Lect 48:289–303

Burns PR, Wukich DK (2008) Surgical reconstruction of the Charcot rearfoot and ankle. Clin Podiatr Med Surg 25:95–120. https://doi.org/10.1016/j.cpm.2007.10.008

Busch-Westbroek TE, Delpeut K, Balm R et al (2018) Effect of single dose of RANKL antibody treatment on acute Charcot neuroosteoarthropathy of the foot. Diabetes Care 41:e21–e22. https://doi.org/10.2337/dc17-1517

Chantelau EA, Grützner G (2014) Is the Eichenholtz classification still valid for the diabetic Charcot foot. Swiss Med Wkly 144:w13948. https://doi.org/10.4414/smw.2014.13948

Chantelau D, Poll LW (2006) Evaluation of the diabetic Charcot foot by MR imaging or plain radiography-an observational study. Exp Clin Endocrinol Diabetes 114(8):428–431. https://doi.org/10.1055/s-2006-924229

Charcot JM (1868) Sur quelques arthropathies qui paraissent dépendre d'une lésion du cerveau ou de la moëlle épinière. Arch Physiol Norm Pathol 1:161–178

Christensen TM, Gade-Rasmussen B, Pedersen LW et al (2012) Duration of off-loading and recurrence rate in Charcot osteoarthropathy treated with less restrictive regimen with removable walker. J Diabetes Complicat 26(5):430–434. https://doi.org/10.1016/j.jdiacomp.2012.05.006

Chudzik W, Kaczorowska B, Przybyla M et al (2007) Diabetic neuropathy [in Polish]. Pol Merkur Lekarski 22:66–69

Clohisy DR, Thompson RC Jr (1988) Fractures associated with neuropathic arthropathy in adults who have juvenile-onset diabetes. J Bone Jt Surg Br 70(8):1192–1200

Clouse ME, Gramm HF, Legg M, Flood T (1974) Diabetic osteoarthropathy.Clinical and roentgenographic observations in 90 cases. Am J Roentgenol Radium Therapy Nucl Med 121(1):22–34. https://doi.org/10.2214/ajr.121.1.22

Cornish J, Callon KE, Bava U et al (2001) Effects of calcitonin, amylin, and calcitonin gene-related peptide on osteoclast development. Bone 29(2):162–168. https://doi.org/10.1016/s8756-3282(01)00494-x

Cummings SR, San Martin J, McClung MR et al (2009) Denosumab for prevention of fractures in postmenopausal women with osteoporosis. N Engl J Med 361(8):756–765. https://doi.org/10.1056/NEJMoa0809493

Dobretsov M, Romanovsky D, Stimers JR (2007) Early diabetic neuropathy: triggers and mechanisms. World J Gastroenterol 13:175–191

Edelson GW, Jensen KL, Kaczynski R (1996) Identifying acute Charcot arthropathy through urinary crosslinked N-telopeptides. Diabetes 45(Suppl 2):108

Edmonds ME, Roberts VC, Watkins PJ (1982) Blood flow in the diabetic neuropathic foot. Diabetologia 22(1):9–15. https://doi.org/10.1007/bf00253862

Edmonds ME, Clarke MB, Newton S et al (1985) Increased uptake of bone radiopharmaceutical in diabetic neuropathy. Q J Med 57(224):843–855

Eichenholtz SN (1966) Charcot joints. Charles C Thomas, Springfield, pp 7–8

Ergen FB, Sanverdi SE, Oznur A (2013) Charcot foot in diabetes and an update on imaging. Diab Foot Ankle 20:4. https://doi.org/10.3402/dfa.v4i0.21884

Fleisch H, Reszka A, Rodan G, Rogers M (2002) Bisphosphonates: mechanism of action. In: Bilezikan JP, Raisz LG, Rodan GA (eds) Principles of bone biology, 2nd edn. Academic, San Deigo, pp 1361–1385

Foltz KD, Fallat LM, Schwartz S (2004) Usefulness of a brief assessment battery for early detection of Charcot foot deformity in patients with diabetes. J Foot Ankle Surg 43(2):87–92. https://doi.org/10.1053/j.jfas.2004.01.001

Foster AV, Snowden S, Grenfell A et al (1995) Reduction of gangrene and amputations in diabetic renal transplant patients: the role of a special foot clinic. Diab Med 12:632–635. https://doi.org/10.1111/j.1464-5491.1995.tb00555.x

Frykberg R, Belczyk R (2008) Epidemiology of the Charcot foot. Clin Podiatr Med Surg 25(1):17–28. https://doi.org/10.1016/j.cpm.2007.10.001

Frykberg RG, Zgonis T, Armstrong DG et al (2006) Diabetic foot disorders: a clinical practice guideline. J Foot Ankle Surg 45:S1–S66. https://doi.org/10.1016/S1067-2516(07)60001-5

Frykberg RG, Morrison WB, Shortt CP et al (2010a) Imaging of the Charcot foot. In: Frykberg RG (ed) The diabetic Charcot foot: principles and management. Data Trace Publishing Company, Brooklandville, pp 65–84

Frykberg RG, Morrison WB, Shortt CP et al (2010b) Imaging of the Charcot foot. In: Frykberg RG, Data Trace Publishing Company (eds) The diabetic Charcot Foot: principles and management. Data Trace Publishing Company, Brooklandville, pp 65–84

Game FL, Catlow R, Jones GR et al (2012) Audit of acute Charcot's disease in the UK: the CDUK study. Diabetologia 55(1):32–35. https://doi.org/10.1007/s00125-011-2354-7

Gilbey SG, Walters H, Edmonds ME et al (1989) Vascular calcification, autonomic neuropathy, and peripheral blood flow in patients with diabetic nephropathy. Diabet Med 6:37–42. https://doi.org/10.1111/j.1464-5491.1989.tb01136.x

Gold RH, Tong DJF, Crim JR et al (1995) Imaging the diabetic foot. Skelet Radiol 24:563–571. https://doi.org/10.1007/BF00204853

Gooday C, Berrington R (2015) New NICE guidance on the diabetic foot. Diabetes Prim Care 17:278–284

Gouveri E (2011) Charcot osteoarthropathy in diabetes: a brief review with an emphasis on clinical practice. World J Diabetes 2(5):59–65. https://doi.org/10.4239/wjd.v2.i5.59

Grant WP, Sullivan R, Sonenshine DE et al (1997) Electron microscopic investigation of the effects of diabetes mellitus on the Achilles tendon. J Foot Ankle Surg 36 (4):272–278. https://doi.org/10.1016/s1067-2516(97)80072-5

Gupta RD, Cheema D, Cheema A et al (2017) Clinical characteristics, foot-associated risk factors, offloading practices and radiological assessment in patients with type 2 diabetes mellitus and chronic charcot's neuroarthropathy: a Case-Control Study from India. J Global Diabetes Clin Metab 2(1):011

Höpfner S, Krolak C, Kessler S et al (2004) Preoperative imaging of Charcot neuroarthropathy in diabetic patients: comparison of ring PET, hybrid PET, and magnetic resonance imaging. Foot Ankle Int 25 (12):890–895. https://doi.org/10.1177/107110070402501208

Hough AJ, Sokoloff L (1989) Pathology of osteoarthritis. In: McCarty DJ (ed) Arthritis and allied conditions: a textbook of rheumatology, 11th edn. Lea & Febiger, Philadelphia, p 1571

Idusuyi OB (2015) Surgical management of Charcot neuroarthropathy. Prosthetics Orthot Int 39(1):61–72. https://doi.org/10.1177/0309364614560939

International Diabetes Federation (2017) Clinical practice recommendation on the diabetic foot: a guide for health care professionals. International Diabetes Federation, Brussels

Jayasinghe S, Atukorala I, Gunethilleke B, Siriwardena V, Herath S, De Abrew K (2007) Is walking barefoot a risk factor for diabetic foot disease in developing countries? Rural Remote Health 7. https://doi.org/10.22605/rrh692

Jeffcoate WJ (2005) Theories concerning the pathogenesis of the acute Charcot foot suggest future therapy. Curr Diab Rep 5:430–435

Jeffcoate WJ (2008) Charcot neuro-osteoarthropathy. Diabetes Metab Res Rev 24(Suppl 1):S62–S65. https://doi.org/10.1002/dmrr.837

Jeffcoate WJ (2015) Charcot foot syndrome. Diabet Med 32(6):760–770. https://doi.org/10.1111/dme.12754. Epub 2015 Apr 15

Jeffcoate W, Lima J, Nobrega L (2000) The Charcot foot. Diabet Med 17:253–258. https://doi.org/10.1046/j.1464-5491.2000.00233.x

Jeffcoate WJ, Game F, Cavanagh PR (2005) The role of proinflammatory cytokines in the cause of neuropathic osteoarthropathy (acute Charcot foot) in diabetes. Lancet 366(9502):2058–2061. https://doi.org/10.1016/S0140-6736(05)67029-8

Jirkovska A, Kasalicky P, Boucek P et al (2001) Calcaneal ultrasonometry in patients with Charcot osteoarthropathy and its relationship with densitometry in the lumbar spineand femoral neck and with markers of bone turnover. Diabet Med 18:495–500. https://doi.org/10.1046/j.1464-5491.2001.00511.x

Johnson JE (1997) Surgical reconstruction of the diabetic Charcot foot and ankle. Foot Ankle Clin 2:37–55

Jude EB, Boulton AJM (1999) End stage complications of diabetic neuropathy. Diabetes Rev 7:395–410. https://doi.org/10.1136/bmj.39255.829120.47

Jude EB, Selby PL, Burgess J et al (2001) Bisphosphonates in the treatment of Charcot neuroarthropathy: a double-blind randomized controlled trial. Diabetologia 44(11):2032–2037. https://doi.org/10.1007/s001250100008

Katsilambros N, Dounis E, Tsapogas P et al (2003) Atlas of the diabetic foot. Chichester, Wiley-Blackwell. ISBN: 0-471-48673-6

Kaynak G, Birsel O, Güven MF et al (2013) An overview of the Charcot foot pathophysiology. Diab Foot Ankle 4. https://doi.org/10.3402/dfa.v4i0.21117

Kensarah AMA, Zaidi NH, Noorwali A et al (2016) Evaluation of Charcot Neuroarthropathy in diabetic foot disease patients at tertiary hospital. Surg Sci 7:250–257. https://doi.org/10.4236/ss.2016.76036

Khanolkar MP, Bain SC, Stephens JW (2008) The diabetic foot. QJM 101(9):685–695. https://doi.org/10.1093/qjmed/hcn027

Koeck FX, Bobrik V, Fassold A et al (2009) Marked loss of sympathetic nerve fibers in chronic Charcot foot of diabetic origin compared to ankle joint osteoarthritis. J Orthop Res 27(6):736–741. https://doi.org/10.1002/jor.20807

Krishnan STM, Baker NR, Carrington AL et al (2004) Comparative roles of microvascular and nerve function in foot ulceration in type 2 diabetes. Diabetes Care 27(6):1343–1348. https://doi.org/10.2337/diacare.27.6.1343

La Fontaine J, Shibuya N, Sampson W et al (2011) Trabecular quality and cellular characteristics of normal, diabetic, and Charcot bone. J Foot Ankle Surg 50(6):648–653. https://doi.org/10.1053/j.jfas.2011.05.005]

La Fontaine J, Lavery L, Jude E (2016) Current concepts of Charcot foot in diabetic patients. Foot 26:7–14. https://doi.org/10.1016/j.foot.2015.11.001

Ledermann HP, Morrison WB (2005) Differential diagnosis of pedal osteomyelitis and diabetic neuroarthropathy: MR imaging. Semin Msculoskelet Radiol 9(3):272–283. https://doi.org/10.1055/s-2005-921945

Lee L, Blume P, Sumpio B (2003) Charcot joint disease in diabetes mellitus. Ann Vasc Surg 17(5):571–580. https://doi.org/10.1007/s10016-003-0039-5

Lesko P, Maurer RC (1989) Talonavicular dislocations and midfoot arthropathy in neuropathic diabetic feet. Natural course and principles of treatment. Clin Orthop Relat Res 240:226–231

Leung HB, Ho YC, Wong WC (2009) Charcot foot in a Hong Kong Chinese diabetic population. Hong Kong Med J 15:191–195

Loredo RA, Garcia G, Chhaya S (2007) Medical imaging of the diabetic foot. Clin Podiatr Med Surg 24(3):397–424. https://doi.org/10.1016/j.cpm.2007.03.010

Loredo R, Rahal A, Garcia G et al (2010) Imaging of the diabetic foot diagnostic dilemmas. Foot Ankle Spec 3(5):249–264. https://doi.org/10.1177/1938640010383154

Lowery NJ, Woods JB, Armstrong DG et al (2012) Surgical management of Charcot neuroarthropathy of the foot and ankle: a systematic review. Foot Ankle Int 33(3):113–121. https://doi.org/10.3113/FAI.2012.0113

Mabilleau G, Petrova NL, Edmonds ME et al (2008) Increased osteoclastic activity in acute Charcot's osteoarthropathy: the role of receptor activator of nuclear factor-kappaB ligand. Diabetologia 51(6):1035–1040. https://doi.org/10.1007/s00125-008-0992-1

Madan SS, Pai DR (2013) Charcot neuroarthropathy of the foot and ankle. Orthop Surg 5(2):86–93. https://doi.org/10.1111/os.12032

Mautone M, Naidoo P (2015) What the radiologist needs to know about Charcot foot. J Med Imaging Radiat Oncol 59(4):395–402. https://doi.org/10.1111/1754-9485.12325

McGill M, Molyneaux L, Bolton T et al (2000) Response of Charcot's arthropathy to contact casting:assessment by quantitative techniques. Diabetologia 43(4):481–484. https://doi.org/10.1007/s001250051332

Milne T, Rogers J, Kinnear E, Martin H, Lazzarini P, Quinton T et al (2013) Developing an evidence-based clinical pathway for the assessment, diagnosis and management of acute Charcot Neuro-Arthropathy: a systematic review. J Foot Ankle Res 6(1):30. https://doi.org/10.1186/1757-1146-6-30

Mitchell J (1831) Art. IV. Ona new practice in acute and chronic rheumatism. Am J Med Sci 1(15):55–64. https://doi.org/10.1097/00000441-183108150-00004

Morgan JM, Biehl WC, Wagner FW (1993) Management of neuropathic arthropathy with the Charcot restraint orthotic walker. Clin Orthop 296:58–63

Moura-Neto A, Fernandes TD, Zantut-Wittmann DE, Trevisan RO, Sakaki MH, Santos ALG et al (2012) Charcot foot: skin temperature as a good clinical parameter for predicting disease outcome. Diabetes Res Clin Pract 96:e11–e14

Mueller MJ, Sinacore DR, Hastings MK et al (2004) Impact of Achilles tendon lengthening on functional limitations and perceived disability in people with a neuropathic plantar ulcer. Diabetes Care 27(7):1559–1564. https://doi.org/10.2337/diacare.27.7.1559

Ndip A, Jude EB, Whitehouse R et al (2008) Charcot neuroarthropathy triggered by osteomyelitis and/or surgery. Diabet Med 25(12):1469–1472. https://doi.org/10.1111/j.1464-5491.2008.02587.x

Ndip A, Williams A, Edward B et al (2011) The RANKL/RANK/OPG signaling pathway mediates medial arterial calcification in diabetic Charcot Neuroarthropathy. Diabetes 60(8):2187–2196. https://doi.org/10.2337/db10-1220

Nielson DL, Armstrong DG (2008) The natural history of Charcot's neuroarthropathy. Clin Podiatr Med Surg 25(1):53–62. https://doi.org/10.1016/j.cpm.2007.10.004

Ogawa K, Sasaki H, Yamasaki H et al (2006) Peripheral nerve functions may deteriorate parallel to the progression of microangiopathy in diabetic patients. Nutr Metab Cardiovasc Dis 16(5):313–321. https://doi.org/10.1016/j.numecd.2005.06.003

Ögüt T, Yontar NS (2017) Surgical treatment options for the diabetic Charcot hindfoot and ankle deformity. Clin Podiatr Med Surg 34(1):53–67. https://doi.org/10.1016/j.cpm.2016.07.007

Pakariennen TK, Laine HJ, Honkonen SE et al (2002) Charcot arthropathy of the diabetic foot. Current concepts and review of 36 cases. Scand J Surg 91(2):196–120. https://doi.org/10.1177/145749690209100212

Pakarinen TK, Laine HJ, Honkonen SE, Peltonen J, Oksala H, Lahtela J (2002) Charcot arthropathy of the diabetic foot. Current concepts and review of 36 cases. Scand J Surg 91(2):195–201. https://doi.org/10.1177/145749690209100212

Pakarinen TK, Laine HJ, Mäenpää H et al (2011) The effect of zolendronic acid on the clinical resolution of Charcot neuroarthropathy: a pilot randomized controlled trial. Diabetes Care 34(7):1514–1516. https://doi.org/10.2337/dc11-0396

Papanas N, Maltezos E (2013a) Etiology, pathophysiology and classifications of the diabetic Charcot foot. Diabet Foot Ankle 4(1):20872. https://doi.org/10.3402/dfa.v4i0.20872

Papanas N, Maltezos E (2013b) Etiology, pathophysiology and classifications of the diabetic Charcot foot. Diab Foot Ankle 4:20872. https://doi.org/10.3402/dfa.v4i0.20872

Perrin BM, Gardner MJ, Suhaimi A, Murphy D (2010) Charcot osteoarthropathy of the foot. Aust Fam Physician 39(3):117–119

Petrova NL, Edmonds ME (2008) Charcot neuro-osteoarthropathy-current standards. Diabetes Metab Res Rev 24(Suppl 1):S58–S61. https://doi.org/10.1002/dmrr.846

Petrova NL, Foster AV, Edmonds ME (2004) Difference in presentation of Charcot osteoarthropathy in type 1 compared with type 2 diabetes. Diabetes Care 27:1235–1236. https://doi.org/10.2337/diacare.27.5.1235-a

Pickwell KM, van Kroonenburgh MJ, Weijers RE et al (2011) F-18 FDG PET/CT scanning in Charcot disease: a brief report. Clin Nucl Med 36(1):8–10. https://doi.org/10.1097/RLU.0b013e3181feeb30

Pinzur MS, Lio T, Posner M (2006) Treatment of Eichenholtz stage I Charcot foot arthropathy with a weightbearing total contact cast. Foot Ankle Int 27(5):324–329. https://doi.org/10.1177/107110070602700503

Pitocco D, Ruotolo V, Caputo S et al (2005) Six-month treatment with alendronate in acute Charcot neuroarthropathy: a randomized controlled trial. Diabetes Care 28(5):1214–1215. https://doi.org/10.2337/diacare.28.5.1214

Purewal TS, Goss DE, Watkins PJ et al (1995) Lower limb venous pressure in diabetic neuropathy. Diabetes Care 18(3):377–381. https://doi.org/10.2337/diacare.18.3.377

Rajbhandari SM, Jenkins RC, Davies C et al (2002) Charcot neuroarthropathy in diabetes mellitus. Diabetologia 45:1085–1096. https://doi.org/10.1007/s00125-002-0885-7

Ramanujam CL, Facaros Z (2011) An overview of conservative treatment options for diabetic Charcot foot neuroarthropathy. Diab Foot Ankle 2:6418. https://doi.org/10.3402/dfa.v2i0.6418

Renner N, Wirth SH, Osterhoff G et al (2016) Outcome after protected full weightbearing treatment in an orthopedic device in diabetic neuropathic arthropathy (Charcot arthropathy): a comparison of unilaterally and bilaterally affected patients. BMC Musculoskelet Disord 17:504

Rogers LC, Bevilacqua NJ (2008) The diagnosis of Charcot foot. Clin Podiatr Med Surg 25:43–51. https://doi.org/10.1016/j.cpm.2007.10.006

Rogers L, Frykberg R (2013) The Charcot foot. Med Clin N Am 97(5):847–856. https://doi.org/10.1016/j.mcna.2013.04.003

Rogers L, Frykberg R, Armstrong D, Boulton A, Edmonds M, Van G et al (2011a) The charcot foot in diabetes. Diabetes Care 34(9):2123–2129. https://doi.org/10.2337/dc11-0844

Rogers LC, Frykberg RG, Armstrong DG et al (2011b) The Charcot foot in diabetes. J Am Podiatr Med Assoc 101(5):437–446. https://doi.org/10.7547/1010437

Salini D, Harish K, Minnie P et al (2018) Prevalence of Charcot Arthropathy in type 2 diabetes patients aged over 50 years with severe peripheral neuropathy: a retrospective study in a tertiary care south Indian hospital. Indian J Endocrinol Metab 22(1):107–111. https://doi.org/10.4103/ijem.IJEM_257_17

Sanders LJ, Frykberg RG (1991) Diabetic neuropathic osteoarthropathy: the Charcot foot. In: Frykberg RG (ed) The high risk foot in diabetes mellitus. Churchill Livingstone, New York, pp 297–338

Sanders LJ, Frykberg RG (2001) Charcot neuroarthropathy of the foot. In: Bowker JH, Pfeifer MA (eds) Levin and O'Neal's The Diabetic Foot, 6th edn. Mosby, St. Louis, pp 439–465

Schoots IG, Slim FJ, Busch-Westbroek TE et al (2010) Neuro-osteoarthropathy of the foot-radiologist: friend or foe? Semin Musculoskelet Radiol 14(3):365–376. https://doi.org/10.1055/s-0030-1254525

Schoppet M, Preissner KT, Hofbauer LC (2002) RANK ligand and osteoprotegerin:paracrine regulators of bone metabolism and vascular function. Arterioscler Thromb Vasc Biol 22(4):549–553. https://doi.org/10.1161/01.atv.0000012303.37971.da

Schott GD (1995) An unsympathetic view of pain. Lancet 345:634–636. https://doi.org/10.1016/S0140-6736(95)90528-6

Selby PL, Young MJ, Boulton AJ (1994) Bisphosphonates: a new treatment for diabetic Charcot neuroarthropathy? Diabet Med 11(1):28–31. https://doi.org/10.1111/j.1464-5491.1994.tb00225.x

Selby PL, Jude EB, Burgess J et al (1998) Bone turnover markers in acute Charcot neuroarthropathy. Diabetologia 41(1):A275

Shen W, Wukich D (2013) Orthopaedic surgery and the diabetic Charcot foot. Med Clin North Am 97 (5):873–882. https://doi.org/10.1016/j.mcna.2013.03.013

Shibata T, Tada K, Hashizume C (1990) The results of arthrodesis of the ankle for leprotic neuro-arthropathy. J Bone Joint Surg Am 72(5):749–756

Sinacore DR (1998) Healing times of diabetic ulcers in the presence of fixed deformities of the foot using total contact casting. Foot Ankle Int 19(9):613–618. https://doi.org/10.1177/107110079801900908

Sinha S, Munichoodappa C, Kozak G (1972) Neuroarthropathy (Charcot joints) in diabetes mellitus. Medicine 51(3):191–210. https://doi.org/10.1097/00005792-197205000-00006

Smith DG, Barnes BC, Sands AK et al (1997) Prevalence of radiographic foot abnormalities in patients with diabetes. Foot Ankle Int 18(6):342–346. https://doi.org/10.1177/107110079701800606

Sohn MW, Lee TA, Stuck RM et al (2009) Mortality risk of Charcot arthropathy compared with that of diabetic foot ulcer and diabetes alone. Diabetes Care 32 (5):816–821. https://doi.org/10.2337/dc08-1695

Stark C, Murray T, Gooday C et al (2016) 5 year retrospective follow-up of new cases of Charcot neuroarthropathy-A single centre experience. Foot Ankle Surg 22(3):176–180. https://doi.org/10.1016/j.fas.2015.07.003

Strang P (1996) Analgesic effect of bisphophonates on bone pain in breast cancer patients. Acta Oncol 35 (Suppl):50–54. https://doi.org/10.3109/02841869609083968

Stuck RM, Sohn MW, Budiman-Mak E, Lee TA, Weiss KB (2008) Charcot arthropathy risk elevation in the obese diabetic population. Am J Med 121 (11):1008–1014. https://doi.org/10.1016/j.amjmed.2008.06.038

Trepman E, Nihal A, Pinzur MS (2005) Current topics review: Charcot neuroarthropathy of the foot and ankle. Foot Ankle Int 26(1):46–63. https://doi.org/10.1177/107110070502600109

Trieb K (2016) The Charcot foot. Bone Jt J 98-B (9):1155–1159. https://doi.org/10.1302/0301-620x.98b9.37038

Trieb K, Hofstätter SG (2015) Pathophysiology and etiology of the Charcot foot. Orthopade 44(1):2–7. https://doi.org/10.1007/s00132-014-3049-9

Van der Ven A, Chapman CB, Bowker JH (2009) Charcot neuroarthropathy of the foot and ankle. J Am Acad Orthop Surg 17:562–571

Varma AK (2013) Charcot neuroarthropathy of the foot and ankle: a review. J Foot Ankle Surg 52(6):740–749. https://doi.org/10.1053/j.jfas.2013.07.001. Epub 2013 Aug 18

Vázquez Gutiérrez M, Mangas Cruz MA, Cañas García-Otero E, Astorga JR (2005) Diabetic neuropathy in the acute phase: a diagnostic dilemma. About two cases and review of the bibliography. Span Clin J 205 (11):549–552

Vijay V, Rajesh K, Kavitha KV et al (2012) Evaluation of Roger's Charcot foot classification system in South Indian diabetic subjects with Charcot foot. J Diab Foot Complications 4:67–70

Vinik AI, Ziegler D (2007) Diabetic cardiovascular autonomic neuropathy. Circulation 115(3):387–397. https://doi.org/10.1161/CIRCULATIONAHA.106.634949

Viswanathan V, Kumpatla S, Rao VN (2014) Radiographic abnormalities in the feet of diabetic patients with neuropathy and foot ulceration. J Assoc Physicians India 62:30–33

Walsh MC, Choi YW (2003) Biology of the TRANCE axis. Cytokine Growth Factor Rev 14:251–263. https://doi.org/10.1016/S1359-6101(03)00027-3

Watanabe T, Ito H, Morita A et al (2009) Sonographic evaluation of the median nerve in diabetic patients comparison with nerve conduction studies. J Ultrasound Med 28(6):727–734. https://doi.org/10.7863/jum.2009.28.6.727

Watkins PJ (1983) Foot blood flow in diabetic neuropathy. J R Soc Med 76(12):996. https://doi.org/10.1177/014107688307601203

Wilson M (1991) Charcot foot osteoarthropathy in diabetes mellitus. Mil Med 156:563–569

Wukich D, Sung W (2009) Charcot arthropathy of the foot and ankle: modern concepts and management review. J Diabetes Complicat 23(6):409–426. https://doi.org/10.1016/j.jdiacomp.2008.09.004

Wukich D, Sung W, Wipf S, Armstrong D (2011) The consequences of complacency: managing the effects of unrecognized Charcot feet. Diabet Med 28 (2):195–198. https://doi.org/10.1111/j.1464-5491.2010.03141.x

Wukich DK, Raspovic KM, Hobizal KB et al (2016) Surgical management of Charcot neuroarthropathy of the ankle and hindfoot in patients with diabetes. Diabetes Metab Res Rev 32(suppl 1):292–296. https://doi.org/10.1002/dmrr.2748

Wünschel M, Wülker N, Gesicki M (2012) Charcot arthropathy of the first metatarsophalangeal joint. J Am Podiatr Med Assoc 102(2):161–164. https://doi.org/10.7547/1020161

Young MJ, Marshall A, Adams JE et al (1995) Osteopenia, neurological dysfunction, and the development of Charcot neuroarthropathy. Diabetes Care 18 (1):34–38. https://doi.org/10.2337/diacare.18.1.34

Younis B, Shahid A, Arshad R, Khurshid S, Masood J (2015) Charcot osteoarthropathy in type 2 diabetes persons presenting to specialist diabetes clinic at a tertiary care hospital. BMC Endocr Disord 15(1). https://doi.org/10.1186/s12902-015-0023-4

Adv Exp Med Biol - Advances in Internal Medicine (2020) 4: 417–440
https://doi.org/10.1007/5584_2020_532
© Springer Nature Switzerland AG 2020
Published online: 19 May 2020

Non-alcoholic Fatty Liver Disease and Diabetes Mellitus

Gebran Khneizer, Syed Rizvi, and Samer Gawrieh

Abstract

Nonalcoholic fatty liver disease (NAFLD) has emerged as the leading liver disease globally. NAFLD patients can have a progressive phenotype, non-alcoholic steatohepatitis (NASH) that could lead to cirrhosis, liver failure and cancer. There is a close bi-directional relationship between NAFLD and type 2 diabetes mellitus (T2DM); NAFLD increases the risk for T2DM and its complications whereas T2DM increases the severity of NAFLD and its complications. The large global impact of NAFLD and T2DM on healthcare systems requires a paradigm shift from specialty care to early identification and risk stratification of NAFLD in primary care and diabetes clinics. Approach to diagnosis, risk stratification and management of NAFLD is discussed. In addition to optimizing the control of coexisting cardiometabolic comorbidities, early referral of NAFLD patients at high risk of having NASH or significant fibrosis to hepatology specialist care may improve management and allow access for clinical trials. Lifestyle modifications, vitamin E, pioglitazone and metformin are currently available options that may benefit patients with T2DM and NAFLD. The burst of clinical trials investigating newer therapeutic agents for NAFLD and NASH offer hope for new, effective and safe therapies in the near future.

Keywords

NASH · NAFLD · Diabetes mellitus · vitamin E · Pioglitazone · GLP1 · SGLT2 · Metformin · Diabetes mellitus

G. Khneizer
Department of Medicine, Indiana University School of Medicine, Indianapolis, IN, USA

S. Rizvi
A&M College of Medicine, Round Rock, Austin, TX, USA

S. Gawrieh (✉)
Division of Gastroenterology and Hepatology, Department of Medicine, Indiana University School of Medicine, Indianapolis, IN, USA
e-mail: sgawrieh@iu.edu

1 Introduction

Nonalcoholic fatty liver disease (NAFLD) is the most common liver disease worldwide (Younossi et al. 2016a). It is characterized by increased hepatic fat accumulation in the absence of significant alcohol intake. NAFLD has a wide spectrum that starts with simple accumulation of triglycerides droplets in hepatocytes (macrosteatosis) without associated inflammation, hepatocyte injury or fibrosis, a phenotype called non-alcoholic fatty liver (NAFL) (Chalasani et al. 2018). A more severe NAFLD phenotype is nonalcoholic steatohepatitis (NASH), which in addition to the presence of

A.   B.   C.

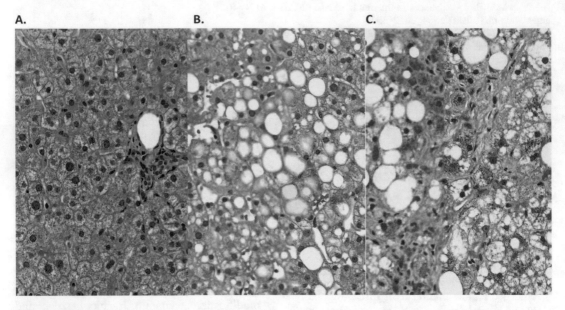

Fig. 1 (**a**) Normal liver histology. (**b**) Non-alcoholic fatty liver where only macrovesicular steatosis is seen. (**c**) Non-alcoholic steatohepatitis where macrovesicular steatosis is seen in addition to hepatocellular ballooning, perisinusoidal fibrosis and lobular inflammation

macrosteatosis is characterized by hepatocyte ballooning, lobular inflammation with or without fibrosis (Fig. 1). While NAFL generally has a benign course, NASH may progress to cause cirrhosis, liver failure and hepatocellular carcinoma (HCC) (Matteoni et al. 1999; Ekstedt et al. 2006). NAFLD is very common in patients with type 2 diabetes mellitus (T2DM), and patients with NAFLD without T2DM at the time of NAFLD diagnosis run a high risk of future T2DM development. In this chapter, we will review clinically relevant state of the art knowledge pertaining to NAFLD in the setting of T2DM. The epidemiology, risk factors and pathophysiology of NAFLD in the context of T2DM will be discussed. This will be followed by review of the approach to diagnosis and utility of screening for NAFLD in T2DM. We discuss the bidirectional effects of NAFLD on development of T2DM and its complication as well as that of T2DM on NAFLD and its complications. Finally, current therapies and ongoing clinical trials for NAFLD including patients who have NAFLD and T2DM are reviewed.

2 Risk Factors for NAFLD

NAFLD can be viewed as the hepatic manifestation of metabolic syndrome (MetS) and therefore other components of the MetS are commonly present in patients with NAFLD (Marchesini et al. 2001; Cusi 2012; Ulitsky et al. 2010). In particular, MetS prevalence is estimated at 43% in NAFLD and 71% in NASH patients (Younossi et al. 2016a). Obesity and T2DM are major risk factors for NAFLD (Tilg et al. 2017). Hypertriglyceridemia is another risk factor for development of both NAFLD and NASH (Jinjuvadia et al. 2017). Elevated very low-density lipoprotein (VLDL) and triglyceride levels along with low HDL are the manifestations of dyslipidemia in NAFLD. Small dense low-density lipoprotein (LDL) and transformation of high-density lipoprotein (HDL) into dysfunctional proatherogenic molecule have also been described (Katsiki et al. 2016). Such manifestations contribute to the elevated cardiovascular disease (CVD) mortality in NAFLD patients (Targher et al. 2010a). Patients that have both NAFLD and T2DM are three times

Table 1 Association and prevalence of common conditions in patients with type 2 diabetics with NAFLD as compared to those without NAFLD

Comorbidities	N	T2DM with NAFLD	T2DM only	OR (95% CI)
MetS	14	77.07 (64.54–86.12)	48.72 (34.94–62.69)	**3.36 (1.82–6.21)**
Hypertension	37	65.03 (56.82–72.44)	58.09 (51.18–64.69)	**1.29 (1.02–1.63)**
Hyperlipidemia	24	51.81 (39.60–63.80)	44.9 (32.66–57.80)	1.30 (0.97–1.76)
CVD	10	24.18 (16.79–33.52)	22.09 (15.44–30.56)	1.03 (0.82–1.29)
PAD	5	9.14 (5.18–15.65)	7.99 (6.14–10.34)	1.25 (0.75–2.07)
CVA	6	7.38 (3.88–13.59)	7.07 (4.31–11.38)	1.00 (0.58–1.69)

Abbreviations: *N* Number of studies, *MetS* Metabolic Syndrome, *CVD* Cardiovascular disease, *PAD* Peripheral arterial disease, *CVA* Cerebrovascular accident, *OR* Odds Ratio, *CI* Confident Interval
From: The global epidemiology of NAFLD and NASH in patients with type 2 diabetes: A systematic review and meta-analysis. J Hepatol, 2019. **71**(4): p. 793–801.**Used with permission**

more likely to have MetS as compared to those with only T2DM (Younossi et al. 2019a). Similarly, hypertension was also more prevalent in NAFLD and T2DM as compared to the patients with T2DM only (Table 1).

3 Epidemiology and Burden of NAFLD on Healthcare Systems

The global prevalence of NAFLD in the general population is estimated at 25% and it increases with age (Younossi et al. 2016a). There is geographic variation in NAFLD prevalence probably due to differences in populations' genetic compositions, dietary habits and environmental aspects. About a quarter of the populations in Europe and North America are estimated to have NAFLD, whereas populations in South America and the Middle East are estimated to have the highest rates of NAFLD globally (range 30–32%) (Younossi et al. 2016a). NASH is estimated to affect 2–6% .of the general population, 7–30% of NAFLD patients who had a liver biopsy without a specific clinical indication (e.g. evaluation for liver organ donation), and 59% in NAFLD patients with clinical indication for liver biopsy (e.g. elevated liver enzymes) (Younossi et al. 2016a). The established rise in the prevalence of obesity is currently at the level of a global epidemic with alarming rates of 18% for men and 21% for women worldwide by year 2025 (Collaboration, N.C.D.R.F 2016). Given the

close correlation between obesity and DM has been established and that excess weight has been closely linked to developing 90% of T2DM cases (Hossain et al. 2007), the number of individuals with impaired glucose tolerance worldwide are projected to exceed 420 million by 2025, and those with T2DM are estimated to increase by 51% in 2045 (Saeedi et al. 2019).

The prevalence and severity of NAFLD increase in T2DM. Worldwide, it is estimated NAFLD and NASH affect 56% and 37% of patients with T2DM, respectively (Younossi et al. 2019a). Just like NAFLD prevalence among the general population, NAFLD prevalence among T2DM also varies geographically (Fig. 2). Europe had the highest prevalence (68%) for NAFLD in patients with T2DM, while Africa had the lowest prevalence at 30% (Younossi et al. 2019a).

Racial difference in NAFLD prevalence and the risk of disease progression have been studied (Anstee and Day 2013). In addition, difference in genetic susceptibility in lipid metabolism is suggested to explain diverse NAFLD prevalence in the United States (US) among African Americans (Foster et al. 2013) and specific gene mutations in those with European descent (Mancina et al. 2016). Socioeconomic and environmental factors such as dietary habits and physical activity levels also have important effects on NAFLD prevalence (Leslie et al. 2014; Oni et al. 2015). In light of that and knowing that prevalence of T2DM is also affected by similar factors (Zheng et al. 2018), population genetics and regional dietary and exercise habits may also

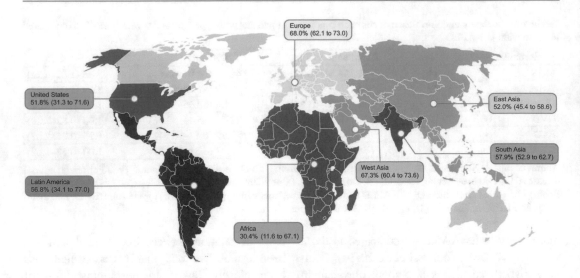

Global prevalence of NAFLD among T2DM patients 55.5% (95% confidence interval: 47.3-63.7)

Fig. 2 Worldwide prevalence of NAFLD in type 2 diabetics
From: The global epidemiology of NAFLD and NASH in patients with type 2 diabetes: A systematic review and meta-analysis. J Hepatol, 2019. **71**(4): p. 793–801. **Used with permission**

influence the prevalence of NAFLD in diabetic patients in different parts in the world.

NAFLD and NASH place a large burden on healthcare systems. In 2013, NASH was the second leading indication for liver transplantation in the US (Wong et al. 2015). With the growing rates of obesity and MetS, NASH-related cirrhosis and hepatocellular carcinoma (HCC) are on the rise. In fact, more recent data demonstrate a rapidly expanding trend of NASH associated HCC in those awaiting liver transplantation in the US (Younossi et al. 2019b). NASH is projected to become the most common indication for liver transplant in the US, surpassing hepatitis C and alcohol-associated liver diseases in the next few years (Charlton et al. 2011; Wong et al. 2014; Gawrieh et al. 2019a; Estes et al. 2018).

With its increasing incidence and prevalence, NAFLD is posing enormous clinical and economic challenges to healthcare systems (Estes et al. 2018; Younossi et al. 2019c). In US Medicare beneficiaries with NAFLD, the median hospital charge was around $36,000 while the median cost of NAFLD care in the outpatient

setting was $9000 in the year 2010 (Sayiner et al. 2017). As expected, NAFLD patients with decompensated cirrhosis had higher costs as compared to their compensated patients. The estimated annual medical costs of NAFLD were $103 billion in the US and € 35 billion in four European countries (Germany, France, Italy, and United Kingdom) in a recent study (Younossi et al. 2016b).

4 NAFLD Awareness in the Primary Care Setting

The primary care or general practice clinic is frequently the first medical access for patients with NAFLD, yet studies suggest that there is low level of awareness of the magnitude and seriousness of NAFLD in the primary care setting (Patel et al. 2018; Wieland et al. 2013). This represents a significant challenge in the care of NAFLD patients as delays of NAFLD diagnosis and assessment of its severity may result in missed opportunities for early specialist care referrals, interventions to halt NASH progression and optimal surveillance for hepatic and extra-hepatic consequences of NAFLD. For example, a survey of primary care providers in Australia with a NAFLD questionnaire revealed that less than half of the practitioners referred the patients to specialist clinics when NAFLD was suspected (Patel et al. 2018). Another study surveyed 64 general practitioners the Netherlands to evaluate their familiarity and approach to NAFLD (van Asten et al. 2017). Surprisingly, 34% and 53% of these practitioners did not know the acronyms NAFLD and NASH, respectively. In addition, 96% of these providers never or rarely screened for NAFLD. In this survey, elevated liver tests were necessary for a provider to consider referral to specialist and over half (53%) of the patients with fatty liver on ultrasound and normal liver tests were never referred to specialist care.

Possible explanations of low awareness of NAFLD in this setting are underestimation of NAFLD prevalence and health consequences, lack of familiarity of risk factors for NAFLD and NASH, limited awareness of available non-invasive tests to risk stratify patients with NAFLD, and overreliance on normal liver biochemistry tests (Patel et al. 2018; Wieland et al. 2013; van Asten et al. 2017). Recent studies demonstrate that a significant proportion of NAFLD patients, including those with T2DM, presents with normal liver tests (Portillo-Sanchez et al. 2015; Gawrieh et al. 2019b). Forty three percent (43%) of NAFLD patients participating in the NASH Clinical Research Network (NASH CRN) studies had normal aminotransferases (<40 U/L), of whom 35% had definite NASH, 20% had advanced fibrosis, and 7% had cirrhosis (Gawrieh et al. 2019b). Severity of NAFLD is therefore not reflected in serum liver enzymes. This is also the case among asymptomatic T2DM patients with normal aminotransferase levels and no prior history of chronic liver disease who are at very high risk for NAFLD and NASH. In a recent study, 50% of patients with T2DM had NAFLD diagnosed by magnetic resonance (MR) spectroscopy, and of those who underwent liver biopsy, 56% were found to have NASH

(Portillo-Sanchez et al. 2015). In that study, NAFLD prevalence was 36% among non-obese study participants, which points to an independent effect of T2DM on NAFLD (Portillo-Sanchez et al. 2015).

Therefore, targeted educational strategies and workshops aimed at increasing awareness and knowledge of NAFLD in the primary care and endocrine practices may help in early recognition of patients with NAFLD in this high-risk population. Further, hepatology, endocrine and primary care societies may also consider coordinating efforts to devise a practical approach for early identification of NAFLD in diabetic patients in the primary care and endocrine outpatient office settings.

5   Guidelines on Screening for NAFLD in High-Risk Groups

Current NAFLD guidelines by the American Association of study of Liver Diseases (AASLD) do not recommend routine screening for NAFLD among patients with diabetes or obesity in the setting of primary care and obesity clinics (Chalasani et al. 2018). This recommendation is based on large gaps in our understanding of the optimal diagnostic and treatment options as well as the current absence of long-term data in terms of the benefits and cost-effectiveness of screening. Such uncertainty is compounded by the low sensitivity of laboratory tests and questionable use of imaging tools for screening purposes. Instead, the guidelines advocate a "high index of suspicion" for NAFLD in T2DM. Use of non-invasive clinical prediction models and detection methods such as vibration-controlled transient elastography to evaluate the presence and severity of fibrosis is proposed (Siddiqui et al. 2019). Several serum-based non-invasive clinical prediction models can be used to screen patients with NAFLD for NASH or advanced fibrosis (Vilar-Gomez and Chalasani 2018; Younossi et al. 2018). For example, the NAFLD fibrosis score (NFS) allows a non-invasive evaluation of the extent of fibrosis using weight, and age and laboratory data which include liver enzymes and serum glucose level (Angulo et al. 2007; Armstrong et al. 2014). A newer model that predicts NASH and fibrosis in patients with NAFLD and T2DM has been devised to assist in guiding clinical decision making and has performed better than NFS in discerning patients with advanced fibrosis (76.6% for the model vs. 71% for NFS) (Bazick et al. 2015). Using NASH CRN studies and including solely biopsy-proven NAFLD, the model used demographic and laboratory data as well as body mass index and waist-to-hip ratio assess fibrosis risk (Bazick et al. 2015). The distinguishing factor of this model is the focus on diabetics which are at an elevated risk for advanced NAFLD. Another useful clinical tool to estimate fibrosis is Fibrosis-4 score which relies solely on laboratory data (Shah et al. 2009). When available, tools such as transient elastography (FibroScan®) with controlled attenuation parameter (CAP) allow simultaneous estimation amount of hepatic fat (using CAP) and fibrosis (using the liver stiffness measurement) (Siddiqui et al. 2019; Vuppalanchi et al. 2018). MR based methods can also offer simultaneous assessment of hepatic fat and fibrosis, albeit it is not widely available and is more expensive (Loomba et al. 2014).

European Association for the Study of Liver (EASL), European Association for the Study of Diabetes (EASD) and European Association for the Study of Obesity (EASO) have issued clinical practice guidelines for NAFLD (European Association for the Study of the Liver (EASL) et al. 2016). The three European societies recommended screening for diabetes in NAFLD patients through fasting or random blood glucose or hemoglobin A1c. Standardized 75 g oral glucose tolerance test can also be used for screening. In addition, EASL and EASD and EASO guidelines advocate a search for NAFLD in patients with T2DM regardless of liver enzymes due to high risk of NAFLD disease progression in diabetics. In particular, presence of metabolic risk

factors including diabetes or obesity should trigger workup to screen for NAFLD according to EASL and EASD and EASO guidelines. In addition, screening for NAFLD is recommended in patients with incidental finding of elevated liver enzymes and metabolic risk factors. A diagnostic algorithm is suggested by EASL and EASD and EASO guidelines which initially stratify patients with metabolic risk factors based on presence of steatosis using ultrasound and steatosis biomarker. Following that, liver enzymes and serum fibrosis markers (such NFS or Fibrosis-4 score) are used to dictate the next diagnostic step such as specialist referral or serial follow up with ultrasound or fibrosis biomarkers.

AASLD guidelines recommend considering liver biopsy in patients at elevated risk for NASH and fibrosis (Chalasani et al. 2018). Those patients are considered for liver biopsy based on the presence of MetS and the use of the clinical prediction models discussed above. Liver biopsy should also be considered in the setting of diagnostic ambiguity when other competing etiologies for liver disease cannot be ruled out with certainty.

6 Diagnostic Approach to NAFLD in Diabetes

The diagnosis of NAFLD requires the presence of hepatic steatosis in the absence of excessive alcohol consumption (defined as average standard drinks per week >21 for men and > 14 for women) or other competing etiologies of liver disease (Chalasani et al. 2018). Abdominal ultrasound is an acceptable initial method to diagnose hepatic steatosis (fatty liver). However, fatty liver is commonly incidentally detected on routine imaging, such as ultrasound, computed tomography or MR imaging, which were performed for other clinical indication (for instance abdominal pain). Identifying the patients with NAFLD at risk for having clinically significant NAFLD (NASH or advanced fibrosis) based on their

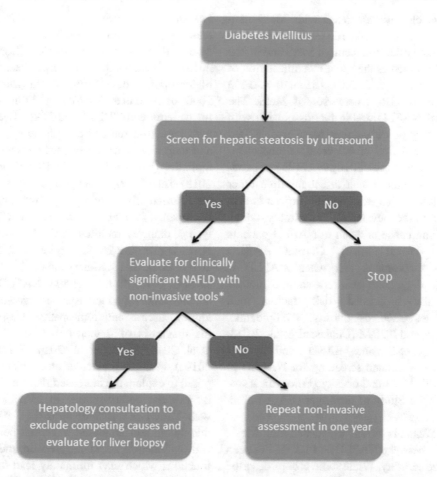

Fig. 3 Diagnostic algorithm to diagnose NAFLD for patients with diabetes mellitus
*Clinically significant NAFLD: NASH or NAFLD with advanced fibrosis

Non-invasive tools: NAFLD fibrosis score, FIB-4, LACSNA (Refer to www.GIHEP.com use hepatology / fibrosis calculators) or Fibroscan®

metabolic profile and using any of the clinical predication models or non-invasive tools discussed above is the next step (Fig. 3). In this setting, T2DM and MetS are considered significant metabolic risk factors for clinically significant NAFLD. Patients identified to be at high risk for having clinically significant NAFLD are referred to hepatology specialist care to exclude coexisting liver disease and consider liver biopsy for confirmation of diagnosis and assessment of the severity of fibrosis. Only histopathological assessment of a liver biopsy can differentiate

NASH from NAFL and give an accurate staging of fibrosis.

7 NAFLD Interaction with Metabolic Syndrome and Diabetes

MetS is comprised of multiple cardiovascular risk factors including elevated fasting blood glucose and waist circumference, dyslipidemia and elevated blood pressure (Nilsson et al. 2019).

NAFLD is closely associated with MetS in a bidirectional fashion (Lonardo et al. 2018). First, NAFLD is regarded as a result of MetS and in that sense, is described as the hepatic manifestation of MetS (Marchesini et al. 2001). Second, NAFLD is interestingly also a precursor of MetS. The presence of NAFLD was in fact associated with increased risk of developing MetS and T2DM over a follow period of 4.5 years, in one meta-analysis (Ballestri et al. 2016). NAFLD has extra-hepatic consequences that affect multiple other organ systems. This extra-hepatic impact is also manifested in the elevated CVD mortality, which is the leading cause of death in NAFLD patients (Chalasani et al. 2018). Current AASLD guidelines recommend that when NAFLD is suspected, diagnostic workup should include probing for associated risk factors and comorbidities such as obesity, dyslipidemia, sleep apnea and T2DM (Chalasani et al. 2018). Similarly, EASL and EASD and EASO guidelines recommend screening for NAFLD in patients with MetS and obesity (European Association for the Study of the Liver (EASL) et al. 2016).

Since T2DM is closely associated with MetS, the relation between NAFLD and T2DM has been under close scrutiny. While some groups describe this interaction as bidirectional similar to that between MetS and NAFLD, a recent study introduced the concept of causality between NAFLD and T2DM (Morrison et al. 2019). Such evidence of causal interplay between NAFLD and T2DM has immense clinical implications particularly on the current effort to devise treatment modalities and reduce the risk of T2DM. The resolution of NAFLD has been shown to decrease the risk of developing T2DM by 27% (Yamazaki et al. 2015). Hence, NAFLD and T2DM interact in a 'cause or consequence' fashion such that presence of one entity increases the predisposition and/or worsens the complications (including CVD events) for the other disease (Lonardo et al. 2018; Loomba et al. 2012; Li et al. 2017; Bril and Cusi 2016; Targher et al. 2007; Fracanzani et al. 2008;

Fukuda et al. 2016; Adams et al. 2005). This interaction between NAFLD and T2DM is also described as a 'vicious circle' (Loria et al. 2013). In fact, NAFLD patients are at elevated lifelong risk of developing T2DM (Ekstedt et al. 2006; Ballestri et al. 2016). In addition, outcomes for patients with T2DM and NAFLD are worse with increased disease complications (Table 1) and mortality, particularly cardiovascular related mortality (Younossi et al. 2019a; Brouha et al. 2018). The microvascular complications of diabetes, particularly nephropathy, retinopathy and neuropathy, are also increased in T2DM and type 1 diabetes mellitus (T1DM) patients with NAFLD (Targher 2014; Targher et al. 2008a, b, 2010b, 2012, 2018; Mantovani et al. 2016, 2017; Li et al. 2014). In this context, NAFLD also has an independently increases the risk of chronic kidney disease and retinopathy irrespective of the diagnosis of diabetes mellitus (Mantovani et al. 2018; Musso et al. 2014; Targher et al. 2010c). Multiple molecular mechanisms are proposed to explain the increased risk for T2DM and its vascular complications in NAFLD patients with T2DM (Targher et al. 2018). Altered gut microbiota and increased intestinal permeability and dysfunction affect the hepatic and vascular functions which may ultimately lead to elevated vascular and renal complications. Lipotoxicity, elevated oxidative stress and inflammation may also contribute to the increased risk of developing vascular complications.

T2DM patients with NAFLD are also more likely to suffer from advanced and progressive liver disease. Prevalence of advanced hepatic fibrosis among patients with both NAFLD and T2DM was shown to be around 17% (Younossi et al. 2019a). Interestingly, with higher stages of fibrosis, all-cause mortality of NAFLD patients increases (Dulai et al. 2017). In addition, liver-related mortality risk exponentially increases with progression of hepatic fibrosis. It has been suggested that fibrosis stage was the strongest predictor of all cause and disease specific mortality among biopsy proven NAFLD patients (Ekstedt et al. 2015; Angulo et al. 2015; Sanyal et al. 2019). With higher rates of advanced fibrosis of patients with NAFLD and T2DM, overall mortality rate of this group is estimated to be 585 per 100,000 (Younossi et al. 2019a). T2DM also accelerates the progression of liver disease in NAFLD patients (Younossi et al. 2019a; European Association for the Study of the Liver (EASL) et al. 2016; Loomba et al. 2012).

It is important to consider T2DM effect on HCC development in the setting of NAFLD. A concerning finding is that HCC associated with NAFLD is on the rise over past 20 years and can even occur in the absence of the background of cirrhosis or significant fibrosis (Gawrieh et al. 2019a; Pais et al. 2017). In fact, HCC with non-cirrhotic background was twice more likely to occur in NAFLD subjects as compared to the cirrhotic HCC counterparts (Gawrieh et al. 2019a). While T2DM is an independent risk factor for HCC, data show that diabetics with NAFLD are also at elevated risk for this malignancy (Yasui et al. 2011; Davila et al. 2005; Polesel et al. 2009). The presence of T2DM was associated with two to threefold increase risk of HCC irrespective of the presence of cirrhosis and other HCC risk factors (such as viral hepatitis) (Davila et al. 2005). In addition, the frequency of T2DM in NASH-related HCC was double that of non-NASH HCC (Davila et al. 2005; Weinmann et al. 2015). Combination of T2DM and other liver injuries such as viral hepatitis and genetic diseases (such as hemochromatosis) is associated with an even higher risk of HCC than that of T2DM alone, suggesting possible synergistic effects in increasing carcinogenesis risk (Davila et al. 2005). Proposed explanation for the higher malignancy in diabetic NAFLD patients include higher likelihood of mutations (particularly of p53 tumor suppressor gene), oxygen free radical production and carcinogenic changes in the milieu of altered lipid metabolism (Hsu et al. 1994; Caldwell et al. 2004; Straub et al. 2010). In an effort to better estimate HCC risks in NAFLD cirrhotic patients, recent models have been proposed and have included T2DM as a predictor of HCC (Ioannou et al. 2019).

8 NAFLD in Type 1 Diabetes Mellitus

While the close relation of T2DM with NAFLD has been well recognized as reflected by extensive relevant literature, it is important to note that T1DM, which accounts for 5–10% of DM cases in adults, is also associated with NAFLD. Prevalence of NAFLD was 44% in a cohort of T1DM in a diabetes clinic in Italy (Targher et al. 2010d). Among that cohort, NAFLD patients with T1DM were more likely to be males and have both higher BMI and blood pressure as compared to those without NAFLD (Targher et al. 2010d). Patients with both T1DM and NAFLD also had a longer duration of diabetes with poor glycemic control. Prevalence of cardiovascular disease was also higher among the NAFLD group (Targher et al. 2010d). On the other hand, an interesting approach was comparing a cohort of T1DM patients with one with T2DM, stratified by use of insulin treatment (Harman et al. 2014). In that study, T1DM patients had lower MetS than T2DM regardless of insulin use. In addition, type 1 diabetics had lower prevalence of NAFLD, steatosis, and NASH as compared to T2DM not on insulin but no difference compared to T2DM patients who are on insulin. Importantly, while no difference in survival existed between the three groups, cirrhosis incidence was lower in T1DM compared to type 2 diabetics on insulin but was not different in the multivariate analysis (Harman et al. 2014). A recent US study reported that 8.8% of patients with T1DM had NAFLD using MRI (Cusi et al. 2017). Although the pathophysiology of NAFLD in this population is not well understood, the mean BMI in T1DM or matched T2DM patients was high in those with increased liver fat content (> 6%) (Cusi et al. 2017). This study also reported a higher need for insulin across all the three groups of T1DM versus T2DM on insulin or T2DM not on insulin. This raises the possibility that insulin resistance (IR) is a shared mechanism between patients with T1DM and T2DM who have NAFLD. This concept is new and challenges the traditional thinking that T1DM is solely characterized by an insulin deficient state. Indeed, the role of IR in T1DM has also been studied in multiple studies. Studies using hyperinsulinemic-euglycemic clamp tests showed total body and hepatic IR (Yki-Jarvinen et al. 1984; DeFronzo et al. 1982a, b). Another study confirmed hepatic and skeletal muscle IR in 25 patients with T1DM with adequate glycemic control (Bergman et al. 2012). Hence, IR in Type 1 diabetics may play a key role in causing steatosis and NASH. Another hypothesis has been proposed emphasizing the role of prolonged hyperglycemia in T1DM patients. Serum glucose is a key source of acetyl-coenzyme A for triglycerides synthesis. Glucose also affects carbohydrate response element-binding protein (ChREBP), which controls hepatocyte glycolytic and lipogenic enzymes, hence, playing a central role in coupling these two pathways (Regnell and Lernmark 2011; Yamashita et al. 2001). Hyperglycemia stimulates transcription of connective tissue growth factor (CTGF), which in part may promote hepatic fibrosis (Paradis et al. 2001). In rat models of hepatic fibrosis, CTGF blockade inhibits hepatic stellate cell activation (Hao et al. 2014; Li et al. 2006).

Since T1DM is predominantly a disease of pediatric patients and constitutes one of the most common chronic diseases in childhood, it is important to examine the association between T1DM and NAFLD in the pediatric population (Simmons and Michels 2015). There are conflicting and small scale data on the prevalence of NAFLD among pediatric T1DM patients. An Egyptian study suggested an increased prevalence of NAFLD in type 1 diabetic children (Elkabbany et al. 2017), whereas a German study of outpatient pediatric patients with T1DM did not reach such conclusion (Kummer et al. 2017). The findings of the German group support the aforementioned finding that NAFLD is more common with longer duration of T1DM. Longer follow up of such patients might explain an increasing prevalence with older age as well. Patients with T1DM and NAFLD have worse clinical outcomes and vascular complications of nephropathy, retinopathy, and polyneuropathy are increased in this population (Targher et al. 2010b; Mantovani et al. 2017).

9   Pathophysiology of NAFLD in T2DM

Multiple factors are implicated in the association between NAFLD and T2DM (Ibrahim et al. 2011; Trauner et al. 2010). IR is a central player (Marchesini et al. 1999). Understanding the pathophysiology behind such interaction is of paramount clinical interest in the current attempts to find effective and early treatment options. On a molecular level, disruption of the insulin signaling pathways affects insulin sensitivity and mediates resistance through activation/inhibition of specific receptors and binding proteins (Tilg et al. 2017). There is increased flux of free fatty acids (FFA) to the liver, which in turn leads to lipotoxicity as manifested by endoplasmic reticulum stress, subsequent hepatocyte injury and eventually apoptosis (Ibrahim et al. 2011; Diehl and Day 2017). Diminished hepatic ability to convert saturated FFA to neutral triglycerides, which are less toxic, is a potential explanation for this lipotoxicity state (Trauner et al. 2010). Importantly, FFA levels closely correlate with NAFLD severity (Nehra et al. 2001). De-novo synthesis of free fatty acids is also promoted by both fatty diets and elevated insulin (Tilg et al. 2017). Impaired hepatic lipid export through very low-density lipoprotein (VLDL) results in increased hepatic fat accumulation and that in turn increases IR further (Perla et al. 2017). Hepatic oxidative stress and mitochondrial dysfunction also promote IR and consequently lead to NAFLD progression and fibrosis (Trauner et al. 2010; Koliaki et al. 2015; Gawrieh et al. 2004). In the described 'multi-hit' hypothesis, with repetitive hepatocyte inflammatory injury and lipotoxicity, wound healing and hepatic regenerative responses become dysfunctional and eventually lead to progressive liver fibrosis (Diehl and Day 2017; Solga and Diehl 2003; Borrelli et al.

2018). The repetitive hits increase the susceptibility of the liver to subsequent insults which in turns worsens the oxidative stress and increase in inflammatory factors. The result of this inflammatory milieu is hepatocyte injury and apoptosis (Solga and Diehl 2003). IR also leads to an increase in hepatic gluconeogenesis and decrease in hepatic glycogen synthesis (Roden 2006). Multiple inflammatory pathways are activated and contribute to propagation of inflammation in NAFLD. Inhibitor of nuclear factor kappa-B kinase subunit beta (IKKβ) pathway is a common mediator of multiple stimuli such as tumor necrosis factor alpha (TNFα) and receptor activator of nuclear factor kappa-B ligand (RANK) that eventually inhibit insulin receptor substrate (Tilg et al. 2017). In addition, adipose tissues contain large amounts of pro-inflammatory cytokines such as interleukin 6 (Perry et al. 2015). In light of this complex pathogenesis, ongoing genetic studies attempt to explain differences in the susceptibility to NAFLD and its severity (Kozlitina et al. 2014; He et al. 2010; Romeo et al. 2008). At the present, the rs738409 variant in *PNPLA3* and rs72613567 variants in *HSD17B13* demonstrate the most convincing effect on modulating the risk for and severity of NAFLD (Romeo et al. 2008; Guichelaar et al. 2013; Sookoian and Pirola 2011; Abul-Husn et al. 2018). Interplay between the environmental factors and gut microbiota through the 'gut-liver axis' is implicated in the pathogenesis of NASH and fibrosis progression (Diehl and Day 2017; Borrelli et al. 2018; Kirpich et al. 2015; Vanni and Bugianesi 2009). Intestinal microbiota acts as a barrier to harmful inflammatory hepatotoxic factors (Solga and Diehl 2003). Therefore, disruption of the intestinal barrier and permeability to endotoxins will adversely affect the body's susceptibility to metabolic injury, and promote inflammatory milieu, liver steatosis, and NASH. Environmental factors such as diet and circadian rhythm also influence the intestinal microbiota (Diehl and Day 2017). To translate such molecular mechanisms into clinical implications, studies are underway in an attempt to devise treatment modalities that potentially target intestinal microbiota (Solga and Diehl 2003).

10 Management of NAFLD

The management of patients with NAFLD is multipronged and aimed to treat not only the liver disease but also the components of MetS. This broadly includes lifestyle and pharmacologic interventions and in select group, bariatric surgery.

10.1 Weight Loss Through Lifestyle Intervention

Lifestyle modifications are a cornerstone to all the treatments prescribed and their importance should be emphasized at any clinical encounter by providers caring for patients with NAFLD. The ultimate goal of lifestyle modifications is to improve all histological features of NAFLD. Lifestyle intervention plan should include (1) dietary modifications; (2) increase in physical activity and possibly; (3) cognitive behavioral therapy to help with overcoming obstacles of weight loss and weight maintenance.

Unfortunately, patients with NAFLD express less readiness to change and less motivation towards healthy lifestyle (Centis et al. 2013). Patients with NAFLD without symptoms do not perceive this condition as a major health issue (Mlynarsky et al. 2016). In addition, physicians are unable to provide sufficient guidance on weight management strategies possibly due to inadequate counseling skills or confidence in providing such guidance (Huang et al. 2004). However, this should not dissuade providers from advising patient and utilizing all available resources and rely on a multidisciplinary approach that includes experts in nutrition, to help patient improve patients dietary habits and health. The value of lifestyle interventions was shown in obese and overweight T2DM patients who despite partially regaining weight achieved a sustained and lower glycohemoglobin levels at 4 years follow up (Look and Wing 2010). Although a multidisciplinary approach involving a team of behavioral therapists, dietician, psychologists and even physical therapists would be ideal, this approach is often met with resistance due to added cost to healthcare, insurance coverage or time constraints. Hence, providers such as primary care physicians, endocrinologists and hepatologists are at the forefront of helping with cognitive therapy. Physician's advice can be a catalyst for lifestyle changes where patients start considering decreased caloric intake, increased physical activity and become more ready to adopt changes to achieve weight loss (Huang et al. 2004; Kreuter et al. 2000; Loureiro and Nayga Jr. 2006; Bellentani et al. 2008).

The most important goal of lifestyle interventions is to achieve weight loss, which has established benefits on improvement in liver histology. The degree of NAFLD histological improvement is proportional to the amount of weight loss achieved. As shown in a meta-analysis of 8 randomized controlled trials (Musso et al. 2012), weight loss by at least 5% of body weight was associated with decline in hepatic steatosis and weight loss of more than 7% was associated with improvement in the NAFLD activity score (NAS) (Elkabbany et al. 2017). In a well-done randomized controlled study of 293 patients with a follow up period of 52 weeks with liver biopsy at start and end of study (Vilar-Gomez et al. 2015), a dose dependent improvement in histology was observed and degree of weight loss was independently associated with NASH histology. Among patients who were able to lose ≥5% of baseline weight, 58% had NASH resolution and 82% had 2 point reduction in NAS score. In patients with ≥10% weight loss, 90% had resolution in NASH and 45% with reduction in fibrosis. Hence, at least 5% weight loss can improve liver histology, however the highest gain and improvement in fibrosis is achieved in those patients able to lose ≥10% of their baseline weight.

10.2 Effects of Different Diets on NAFLD

High fructose corn syrup is almost a staple component of Western diet. Added sugars such as high fructose corn syrup or sucrose increase the risk of obesity, NAFLD and NASH (Jensen et al. 2018). Patients with NAFLD report a twofold increased intake of sugar-sweetened beverages compared to patients without NAFLD (Ouyang et al. 2008).

Observational studies have shown inverse relation between coffee consumption and NASH (Chen et al. 2014; Saab et al. 2014). In a prospective study, coffee consumption was recorded at baseline and after 7 years in patients with NAFLD (Zelber-Sagi et al. 2015). Consumption of more than 3 cups of coffee a day did not affect the risk of hepatic steatosis but was associated with lower risk of significant fibrosis.

Decreasing the calorie intake by 30% improves IR and hepatic steatosis (Kirk et al. 2009; Haufe et al. 2011). Although data is scant on specific macronutrient dietary composition, diet regimens are worth elaborating on as this question is often encountered in clinical practice. Mediterranean diet, which is high in mono-unsaturated fatty acids, resulted in improvement in hepatic steatosis compared to high fat, low carbohydrate diet consumed over a period of 6 weeks (Ryan et al. 2013). Ketogenic diets may help achieve significant weight loss compared to low fat, high carbohydrate diet in adults with obesity (Tobias et al. 2015; Sackner-Bernstein et al. 2015; Mansoor et al. 2016; Gardner et al. 2007). In one study (Vilar-Gomez et al. 2019a), use of carbohydrate restricted ketogenic diet for 1 year resulted in improvement in steatosis and fibrosis Results of recent studies show inconsistent effects of low carbohydrate high fat diet on liver enzymes and intrahepatic lipid content (Ryan et al. 2013; Browning et al. 2011; Volynets et al. 2013; Kani et al. 2014; de Luis et al. 2010; Tendler et al. 2007; Huang et al. 2005). Hence, there is currently insufficient data to make recommendations on benefits of popular diets has high- for NAFLD.

10.3 Effects of Physical Activity on NAFLD

The relationship between sedentary lifestyle, MetS and NAFLD is well established (Byrne and Targher 2015). Several studies demonstrated benefit of exercise as part of the recommended life style modifications for NAFLD patients (Larson-Meyer et al. 2008; Lazo et al. 2010; Hallsworth et al. 2011; Sullivan et al. 2012; de Piano et al. 2012; Scaglioni et al. 2013; Keating et al. 2012; Kantartzis et al. 2009; St George et al. 2009). The degree of physical fitness correlates with the risk of NAFLD; with lower fitness strongly associated with presence of NAFLD and increased fitness associated with fatty liver resolution by MR spectroscopy (Kantartzis et al. 2009). Aerobic and resistance exercise training may improve insulin sensitivity and hepatic steatosis independent of weight loss (Hallsworth et al. 2011; Sullivan et al. 2012; Keating et al. 2012; Johnson et al. 2009). The intensity of exercise rather than its duration seems to be more important in influencing NAFLD risk, and patients exercising more vigorously having lower risk for NASH and advanced fibrosis that those who train for longer duration but at lower intensity (Kistler et al. 2011).

Although it is well established that weight loss achieved by dietary modification and exercise is an effective therapy for patients with NAFLD, long-term compliance with this therapy and ability to maintain it are known practical challenges to these interventions. Because many patients with NAFLD show low enthusiasm and readiness to implement these changes, cognitive therapy and counseling may be necessary to encourage patients to start changing their diet, institute physical activity and address any barriers to compliance with lifestyle modifications (Centis et al. 2013; Fabricatore 2007).

10.4 Bariatric Surgery

Bariatric surgery can produce sustainable and significant weight loss. In a randomized study of patients with T2DM, patients received either Roux-en-Y gastric bypass or sleeve gastrectomy versus medical therapy and were followed for 5 years (Schauer et al. 2017). More patients achieved the primary outcome (a glycated hemoglobin level of 6.0% or less with or without the use of hypoglycemic medications) with bariatric surgery than medical therapy (29% with gastric bypass, 23% with sleeve gastrectomy vs 5% with medical therapy, $p < 0.05$). In addition, more significant weight loss at 5 years was seen with gastric bypass (-23%) and sleeve gastrectomy (-19%) than with medical-therapy (-5%) ($p < 0.05$), Significant improvements in serum insulin, lipoproteins and triglyceride level were observed more commonly in the bariatric surgery vs medical therapy group.

Bariatric surgery has a profound impact on NAFLD. In a meta-analysis of 2374 patients' improvement of steatosis was seen in 88%, in steatohepatitis in 59% of patients and improvement or resolution of fibrosis in 30% of patients (Fakhry et al. 2019). Other metabolic comorbidities including glycemic control improves following bariatric surgery in patients with NAFLD (Mattar et al. 2005). Bariatric surgery may be considered in diabetic patients with NAFLD and NASH and morbid obesity. Limited data currently are available regarding the safety of bariatric surgery in NASH patients with cirrhosis.

Non-surgical endocopic approaches to induce weight loss are under study. Several case series suggest that intragastric balloon may induce weight loss and improve NAFLD histologic and serum based parameters related to NAFLD such as liver enzymes (Ricci et al. 2008; Nguyen et al. 2017; Espinet Coll et al. 2019; Lee et al. 2012).

10.5 Pharmacologic Therapy for NAFLD

10.5.1 Vitamin E

Vitamin E is the major lipid-soluble chain-breaking antioxidant found in the human body. Oxidative stress plays a crucial role in causing hepatocyte injury associated with NAFLD (Gawrieh et al. 2004; Chalasani ct al. 2004). By targeting oxidative stress components, vitamin E demonstrates a convincing therapeutic effect in multiple clinical trials to treat NAFLD or NASH and resulting in improvement in liver biochemistries and histology (Allard et al. 1998; Dufour et al. 2006; Harrison et al. 2003; Hoofnagle et al. 2013; Sanyal et al. 2010). In the PIVENS (Pioglitazone, Vitamin E or Placebo for Nonalcoholic Steatohepatitis) trial, 800 IU/day of Vitamin E was given to non-diabetic, non-cirrhotic NASH patients. The rate of achieving primary outcome of histological improvement was higher in those patients treated with Vitamin E for 96 weeks compared to placebo (43% vs. 19%, p = 0.001), while histological improvement with pioglitazone did not reach statistical significance compared to placebo (34% vs. 19%, P = 0.04). Histological analysis showed an improvement in NASH histological features such as in hepatocyte and lobular inflammation and macrosteatosis with vitamin E (Sanyal et al. 2010). A recent randomized double-blind placebo-controlled trial evaluated whether vitamin E, alone or combined with pioglitazone improved liver histology in patients with T2DM and NASH (Bril et al. 2019). More patients on combination therapy achieved the primary endpoint (2 points reduction in NAS) vs placebo (54% vs. 19%, p = 0.003) but not with vitamin E alone (31% vs. 19%, p = 0.26). However, groups experienced more NASH resolution vs placebo (combination: 43% vs. 12%, p = 0.005; vitamin E alone: 33% vs. 12%, p = 0.04).

Whether chronic use of vitamin E increases overall mortality remains equivocal as different meta-analyses show conflicting results (Miller 3rd et al. 2005; Gerss and Kopcke 2009; Abner et al. 2011). Other safety concerns with chronic vitamin E use pertain to possible association of its use with increased risk of prostate cancer in men (Klein et al. 2011). A meta-analysis suggested an association between vitamin E use and increased risk of hemorrhagic stroke and decreased risk of ischemic stroke (Mattar et al. 2005). In a recent retrospective study of NASH patients with advanced fibrosis (Vilar-Gomez et al. 2018), vitamin E therapy resulted in a significant reduction in the risk of death, liver transplantation and hepatic decompensation, both in patients with and without diabetes. No differences in the adjusted 10-year cumulative probabilities of hepatocellular carcinoma, vascular events, or non-hepatic cancers were noted different between vitamin E treated patients and those who did not receive it. Current practice guidelines recommend use of vitamin E for patients with NASH (Chalasani et al. 2018; European Association for the Study of the Liver (EASL) et al. 2016).

10.6 Insulin Sensitizers

10.6.1 Metformin

Metformin is a first line agent in treatment of T2DM. It suppresses gluconeogenesis, improves insulin sensitivity and may result in weight loss. Although no effects for metformin on liver histology or tests were demonstrated in two meta-analyses (Musso et al. 2012; Rakoski et al. 2010), a recent study in diabetic patients with biopsy-proven NASH and bridging fibrosis or compensated cirrhosis showed that long term ($\geq$ 6 years) metformin use was associated with lower risk of overall mortality, liver transplant and hepatocellular carcinomas independent of the changes in liver biochemistry tests (Vilar-Gomez et al. 2019b). Hence, in patients with established T2DM and NAFLD, metformin is safe and may offer benefit to patients with NAFLD and advanced fibrosis. Although studies suggest survival benefit with metformin even with hepatic impairment (Crowley et al. 2017; Zhang et al. 2014), it should be used with caution in diabetic patients at high risk for lactic acidosis such as those with concomitant hepatic and renal impairment or alcohol use.

10.7 Thiazolidinediones

Thiazolidinediones (TZDs) improve IR by PPAR-γ mediated increases in adiponectin transcription and expression. Pioglitazone improves NASH histology in multiple studies that included diabetics and non-diabetic patients (Sanyal et al. 2010; Aithal et al. 2008; Belfort et al. 2006; Cusi et al. 2013). At doses of 30 mg and 45 mg ranging between 6 months to up to 3 years, pioglitazone has been shown to improve hepatic steatosis, inflammation and fibrosis. The most common side effect with pioglitazone use is weight gain ranging from 2.5 to 4.8 kg (Balas et al. 2007; Boettcher et al. 2012). Peripheral edema is observed in 7–10% of patients especially in those with concomitant use of insulin. Association with increased risk of bladder cancer has been a concern with pioglitazone use but this risk has not been consistently demonstrated (Lewis et al. 2011, 2015; Tuccori et al. 2016; Filipova et al. 2017). Use of pioglitazone is associated with increased risk of bone loss in women (Yau et al. 2013).

10.8 Glucagon-Like Peptide Agonists

Glucagon-like peptide-1 (GLP-1), a hormone secreted by Langerhans cells in response to nutritional intake, exhibits many physiological effects that modulate glucose metabolism: it stimulates glucose-dependent insulin secretion, inhibits glucagon release, induces pancreatic β-cell proliferation, delays gastric emptying and thus resulting in improved insulin sensitivity (Dhir and Cusi 2018; Khan et al. 2019). This class of medications has been shown to result in weight loss by delay of gastric emptying and appetite suppression (Khan et al. 2019). In vitro models have shown GLP-1 receptor plays a key role in decreasing hepatic steatosis. Impaired secretion of GLP-1 has been noted in NAFLD and NASH (Bernsmeier et al. 2014). Based on these effects, GLP-1 agonists have been of great interest as a therapeutic modality for NAFLD and NASH.

10.9 Liraglutide

Liraglutide, a GLP-1 agonist, was tested in a randomized controlled study in 52 patients with biopsy proven NASH (17 with T2DM and 35 non-diabetic) (Armstrong et al. 2016). After 48 weeks of therapy, more patients receiving liraglutide achieved the primary end point (NASH resolution without worsening of fibrosis) compared to placebo (39% vs. 9%; p = 0·019), and worsening of fibrosis was less frequent on liraglutide (9% vs. 36%, p = 0.04). The hepatic beneficial effects of liraglutide are associated with weight loss and reduction in cardiovascular event risk in diabetic patients (Marso et al. 2016).

10.10 Exenatide

Two studies have evaluated exenatide for NAFLD in patients with T2DM. The first study compared the efficacy of exenatide to metformin (Fan et al. 2013). Exenatide showed greater decline in weight and improvement in transaminases. In another study, patients were randomized to exenatide with Insulin Glargine or Insulin Glargine with Insulin aspart for 12 weeks (Shao et al. 2014). Exenatide group showed improvement in transaminases and greater reversal rates in hepatic steatosis. A pilot open label study of 8 patients with diabetes and biopsy proven NASH showed that 28 weeks of exenatide resulted in improvement in NASH histology and fibrosis (Kenny et al. 2010). Randomized controlled trials to study the effects of liver histology are needed to establish the role of this agent in patients with T2DM and NAFLD.

10.11 Semaglutide

Semaglutide is a long-acting GLP-1 analogue, is approved for T2DM. Due to 94% sequence homology to human GLP-1 and a long half-life, it is administered as once weekly dose. Semaglutide has shown promising effects on glucose control and weight loss compared to placebo and other antidiabetics in patients with T2DM in phase III "SUSTAIN" trials (Holst and Madsbad 2017; Madsbad and Holst 2017). It is currently under investigation as a potential as a treatment option for patients with NASH in a randomized, double-blind trial (NCT02970942).

10.12 Sitagliptin and Vildagliptin

Dipeptidyl peptidase 4 (DPP4) deactivates GLP-1. Sitagliptin and vildagliptin are two DDP4 antagonists which have gained interest in patients with T2DM and NASH. Sitagliptin is approved for treatment of T2DM. In two randomized trials in patients with NAFLD, sitagliptin failed to show beneficial effects on liver steatosis, liver enzymes, or IR (Cui et al. 2016; Joy et al. 2017). Recently, in randomized trial in patients with T2DM on Metformin with poor glycemic control, patients were randomized to liraglutide, sitagliptin or Insulin. Groups treated with liraglutide and sitagliptin showed significant weight loss and improvement in intra-hepatic fat content compared to placebo (Yan et al. 2019). Further studies are expected to evaluate the effects of this class of drugs on steatohepatitis and fibrosis in patients with biopsy-proven NASH with and without T2DM.

10.13 Sodium-Glucose Cotransporter-2 Inhibitors

Glucose transport proteins in the kidneys, particularly Sodium-glucose cotransporter-2 (SGLT2), facilitate renal reabsorption of glucose. SGLT2 inhibitors are a novel class of drugs which promotes renal excretion of glucose and have been demonstrated to decrease elevated blood glucose levels in patients with T2DM (Vivian 2014). Currently, three SGLT2 inhibitors, canagliflozin, dapagliflozin, and empagliflozin, are approved by the US Food and Drug Administration (FDA) for monotherapy or combination therapy in patients with T2DM.

SGLT2 inhibitors result in weight loss (around 3–4%) and demonstrate beneficial effects on kidney disease and cardiovascular events in patients with T2DM (Neal et al.

2017). Improvement in alanine aminotransferase in parallel to the decline in weight and improved glycemic control have been reported (Seko et al. 2018; Sattar et al. 2018). Another study reported improvement in liver enzymes and hepatic fat content with in patients with T2DM and NAFLD randomized to empagliflozin plus standard treatment for 20 weeks (Kuchay et al. 2018). Although these results are promising, effects of SGLT2 inhibitors still need to be assessed in randomized studies in patients with NASH to evaluate whether they can improve NASH histology.

10.14 Farsenoid X Receptors Agonists

The farnesoid X receptors (FXR) are nuclear receptors that are highly expressed in tissues that participate in bile acid metabolism such as the liver, intestines, and kidneys (Makishima et al. 1999; Parks et al. 1999). Results from pre-clinical studies indicate that FXR activation results in a reduction in glucose, free fatty acids, and triglycerides and hence FXRs play a critical role in carbohydrate and lipid metabolism and regulation of insulin sensitivity (Mazuy et al. 2014; Neuschwander-Tetri 2012).

Obeticholic acid (OCA,) is a 6α-ethyl derivative of CDCA, and first in its class selective FXR agonist (Adorini et al. 2012). The FLINT (Farnesoid X Receptor Ligand Obeticholic Acid in NASH Treatment) trial was a multicenter, randomized, double-blind, placebo-controlled phase IIb study in which 283 patients received either 25 mg of OCA or placebo for 72 weeks (Neuschwander-Tetri et al. 2014). The study met its pre-determined stopping criteria at interim analysis. Significant improvement in fibrosis stage in the treated group (35% vs. 19%; $P = 0.004$) was noted. No difference in NASH resolution was noted in either arm. Most significant side effects were pruritus in 23% of patients, elevated cholesterol and HOMA-IR. A phase III trial (REGENERATE) is currently ongoing to investigate the use of OCA in patients with

NASH and stage 1–3 fibrosis with a lower dose to determine whether it retains its efficacy with higher tolerability (NCT02548351). Data on from the interim analysis was recently published (Younossi et al. 2019d), and showed fibrosis improvement by ≥1 stage without NASH worsening in 18% in the obeticholic acid 10 mg group (p = 0·045) and 23% in the obeticholic acid 25 mg group (p = 0·0002) compared to 12% in the placebo group. NASH resolution without fibrosis worsening endpoint was not met (11% in the obeticholic acid 10 mg group, 12% in the obeticholic acid 25 mg group vs. 8% in the placebo group, P > 0·05). Pruritus was reported to be 53% with 25 mg of OCA. The final results of this study are awaited.

11 Conclusions

The relationship between NAFLD and T2DM is bidirectional with NAFLD increasing the risk for T2DM and its complications, and T2DM increases the severity of NAFLD and its complications. Given the global magnitude and socioeconomic burden of NAFLD, there needs to be a paradigm shift towards early identification and risk stratification of NAFLD patients in the primary care and diabetes clinics. In addition to optimizing the control of coexisting cardiometabolic comorbidities, early identification and referral of NAFLD patients at high risk of having NASH or significant fibrosis to hepatology specialist care may improve outcomes through earlier diagnosis of NASH, earlier initiation of treatment modalities as well as improved and timely access for clinical trials. Lifestyle modifications, vitamin E, pioglitazone and metformin are currently available options that benefit patients with T2DM and NAFLD. The burst of clinical trials investigating newer therapeutic agents for NAFLD and NASH may offer new effective and safe therapies in the near future.

Acknowledgements None.

Source of Funding None.

Disclosures Dr. Khneizer declares no conflicts. Dr. Rizvi receives speaking fees from Abbvie. Dr. Gawrieh consulting: TransMedics, research grant support: Cirius, Galmed, Viking and Zydus.

Author's Contributions All authors contributed to manuscript preparation.

References

Abner EL et al (2011) Vitamin E and all-cause mortality: a meta-analysis. Curr Aging Sci 4(2):158–170

Abul-Husn NS et al (2018) A protein-truncating HSD17B13 variant and protection from chronic liver disease. N Engl J Med 378(12):1096–1106

Adams LA et al (2005) The histological course of nonalcoholic fatty liver disease: a longitudinal study of 103 patients with sequential liver biopsies. J Hepatol 42(1):132–138

Adorini L, Pruzanski M, Shapiro D (2012) Farnesoid X receptor targeting to treat nonalcoholic steatohepatitis. Drug Discov Today 17(17–18):988–997

Aithal GP et al (2008) Randomized, placebo-controlled trial of pioglitazone in nondiabetic subjects with non-alcoholic steatohepatitis. Gastroenterology 135 (4):1176–1184

Allard JP et al (1998) Effects of vitamin E and C supplementation on oxidative stress and viral load in HIV-infected subjects. AIDS 12(13):1653–1659

Angulo P et al (2007) The NAFLD fibrosis score: a noninvasive system that identifies liver fibrosis in patients with NAFLD. Hepatology 45(4):846–854

Angulo P et al (2015) Liver fibrosis, but no other histologic features, is associated with long-term outcomes of patients with nonalcoholic fatty liver disease. Gastroenterology 149(2):389–397. e10

Anstee QM, Day CP (2013) The genetics of NAFLD. Nat Rev Gastroenterol Hepatol 10(11):645–655

Armstrong MJ et al (2014) Severe asymptomatic non-alcoholic fatty liver disease in routine diabetes care; a multi-disciplinary team approach to diagnosis and management. QJM 107(1):33–41

Armstrong MJ et al (2016) Liraglutide safety and efficacy in patients with non-alcoholic steatohepatitis (LEAN): a multicentre, double-blind, randomised, placebo-controlled phase 2 study. Lancet 387(10019):679–690

Balas B et al (2007) Pioglitazone treatment increases whole body fat but not total body water in patients with non-alcoholic steatohepatitis. J Hepatol 47 (4):565–570

Ballestri S et al (2016) Nonalcoholic fatty liver disease is associated with an almost twofold increased risk of incident type 2 diabetes and metabolic syndrome. Evidence from a systematic review and meta-analysis. J Gastroenterol Hepatol 31(5):936–944

Bazick J et al (2015) Clinical model for NASH and advanced fibrosis in adult patients with diabetes and

NAFLD: guidelines for referral in NAFLD. Diabetes Care 38(7):1347–1355

Belfort R et al (2006) A placebo-controlled trial of pioglitazone in subjects with nonalcoholic steatohepatitis. N Engl J Med 355(22):2297–2307

Bellentani S et al (2008) Behavior therapy for nonalcoholic fatty liver disease: the need for a multidisciplinary approach. Hepatology 47(2):746–754

Bergman BC et al (2012) Features of hepatic and skeletal muscle insulin resistance unique to type 1 diabetes. J Clin Endocrinol Metab 97(5):1663–1672

Bernsmeier C et al (2014) Glucose-induced glucagon-like peptide 1 secretion is deficient in patients with non-alcoholic fatty liver disease. PLoS One 9(1): e87488

Boettcher E et al (2012) Meta-analysis: pioglitazone improves liver histology and fibrosis in patients with non-alcoholic steatohepatitis. Aliment Pharmacol Ther 35(1):66–75

Borrelli A et al (2018) Role of gut microbiota and oxidative stress in the progression of non-alcoholic fatty liver disease to hepatocarcinoma: current and innovative therapeutic approaches. Redox Biol 15:467–479

Bril F, Cusi K (2016) Nonalcoholic fatty liver disease: the new complication of type 2 diabetes mellitus. Endocrinol Metab Clin N Am 45(4):765–781

Bril F et al (2019) Role of vitamin E for nonalcoholic steatohepatitis in patients with type 2 diabetes: a randomized controlled trial. Diabetes Care 42 (8):1481–1488

Brouha SS et al (2018) Increased severity of liver fat content and liver fibrosis in non-alcoholic fatty liver disease correlate with epicardial fat volume in type 2 diabetes: a prospective study. Eur Radiol 28 (4):1345–1355

Browning JD et al (2011) Short-term weight loss and hepatic triglyceride reduction: evidence of a metabolic advantage with dietary carbohydrate restriction. Am J Clin Nutr 93(5):1048–1052

Byrne CD, Targher G (2015) NAFLD: a multisystem disease. J Hepatol 62(1 Suppl):S47–S64

Caldwell SH et al (2004) Obesity and hepatocellular carcinoma. Gastroenterology 127(5 Suppl 1):S97–S103

Centis E et al (2013) Stage of change and motivation to healthier lifestyle in non-alcoholic fatty liver disease. J Hepatol 58(4):771–777

Chalasani N, Deeg MA, Crabb DW (2004) Systemic levels of lipid peroxidation and its metabolic and dietary correlates in patients with nonalcoholic steatohepatitis. Am J Gastroenterol 99(8):1497–1502

Chalasani N et al (2018) The diagnosis and management of nonalcoholic fatty liver disease: practice guidance from the American Association for the Study of Liver Diseases. Hepatology 67(1):328–357

Charlton MR et al (2011) Frequency and outcomes of liver transplantation for nonalcoholic steatohepatitis in the United States. Gastroenterology 141 (4):1249–1253

Chen S et al (2014) Coffee and non-alcoholic fatty liver disease: brewing evidence for hepatoprotection? J Gastroenterol Hepatol 29(3):435–441

Collaboration, N.C.D.R.F (2016) Trends in adult body-mass index in 200 countries from 1975 to 2014: a pooled analysis of 1698 population-based measurement studies with 19.2 million participants. Lancet 387(10026):1377–1396

Crowley MJ et al (2017) Clinical outcomes of metformin use in populations with chronic kidney disease, congestive heart failure, or chronic liver disease: a systematic review. Ann Intern Med 166(3):191–200

Cui J et al (2016) Sitagliptin vs. placebo for non-alcoholic fatty liver disease: a randomized controlled trial. J Hepatol 65(2):369–376

Cusi K (2012) Role of obesity and lipotoxicity in the development of nonalcoholic steatohepatitis: pathophysiology and clinical implications. Gastroenterology 142(4):711–725.e6

Cusi K, Orsak B, Lomonaco R, Bril F, Ortiz-Lopez C, Hecht J, Webb A, Tio F, Darland CM, Hardies J (2013) Extended treatment with pioglitazone improves liver histology in patients with prediabetes or type 2 diabetes mellitus and NASH. Hepatology 58(S1):36A–91A

Cusi K et al (2017) Non-alcoholic fatty liver disease (NAFLD) prevalence and its metabolic associations in patients with type 1 diabetes and type 2 diabetes. Diabetes Obes Metab 19(11):1630–1634

Davila JA et al (2005) Diabetes increases the risk of hepatocellular carcinoma in the United States: a population based case control study. Gut 54(4):533–539

de Luis DA et al (2010) Effect of two different hypocaloric diets in transaminases and insulin resistance in nonalcoholic fatty liver disease and obese patients. Nutr Hosp 25(5):730–735

de Piano A et al (2012) Long-term effects of aerobic plus resistance training on the adipokines and neuropeptides in nonalcoholic fatty liver disease obese adolescents. Eur J Gastroenterol Hepatol 24(11):1313–1324

DeFronzo RA, Simonson D, Ferrannini E (1982a) Hepatic and peripheral insulin resistance: a common feature of type 2 (non-insulin-dependent) and type 1 (insulin-dependent) diabetes mellitus. Diabetologia 23 (4):313–319

DeFronzo RA, Hendler R, Simonson D (1982b) Insulin resistance is a prominent feature of insulin-dependent diabetes. Diabetes 31(9):795–801

Dhir G, Cusi K (2018) Glucagon like peptide-1 receptor agonists for the management of obesity and non-alcoholic fatty liver disease: a novel therapeutic option. J Investig Med 66(1):7–10

Diehl AM, Day C (2017) Cause, pathogenesis, and treatment of nonalcoholic steatohepatitis. N Engl J Med 377(21):2063–2072

Dufour JF et al (2006) Randomized placebo-controlled trial of ursodeoxycholic acid with vitamin e in nonalcoholic steatohepatitis. Clin Gastroenterol Hepatol 4 (12):1537–1543

Dulai PS et al (2017) Increased risk of mortality by fibrosis stage in nonalcoholic fatty liver disease: systematic review and meta-analysis. Hepatology 65 (5):1557–1565

Ekstedt M et al (2006) Long-term follow-up of patients with NAFLD and elevated liver enzymes. Hepatology 44(4):865–873

Ekstedt M et al (2015) Fibrosis stage is the strongest predictor for disease-specific mortality in NAFLD after up to 33 years of follow-up. Hepatology 61 (5):1547–1554

Elkabbany ZA et al (2017) Transient elastography as a noninvasive assessment tool for hepatopathies of different etiology in pediatric type 1 diabetes mellitus. J Diabetes Complicat 31(1):186–194

Espinet Coll E et al (2019) Bariatric and metabolic endoscopy in the handling of fatty liver disease. A new emerging approach? Rev Esp Enferm Dig 111 (4):283–293

Estes C et al (2018) Modeling the epidemic of nonalcoholic fatty liver disease demonstrates an exponential increase in burden of disease. Hepatology 67 (1):123–133

European Association for the Study of the Liver (EASL), European Association for the Study of Diabetes (EASD), European Association for the Study of Obesity (EASO) (2016) EASL-EASD-EASO clinical practice guidelines for the management of non-alcoholic fatty liver disease. J Hepatol 64(6):1388–1402

Fabricatore AN (2007) Behavior therapy and cognitive-behavioral therapy of obesity: is there a difference? J Am Diet Assoc 107(1):92–99

Fakhry TK et al (2019) Bariatric surgery improves nonalcoholic fatty liver disease: a contemporary systematic review and meta-analysis. Surg Obes Relat Dis 15 (3):502–511

Fan H et al (2013) Exenatide improves type 2 diabetes concomitant with non-alcoholic fatty liver disease. Arq Bras Endocrinol Metabol 57(9):702–708

Filipova E et al (2017) Pioglitazone and the risk of bladder cancer: a meta-analysis. Diabetes Ther 8 (4):705–726

Foster T et al (2013) The prevalence and clinical correlates of nonalcoholic fatty liver disease (NAFLD) in African Americans: the multiethnic study of atherosclerosis (MESA). Dig Dis Sci 58(8):2392–2398

Fracanzani AL et al (2008) Risk of severe liver disease in nonalcoholic fatty liver disease with normal aminotransferase levels: a role for insulin resistance and diabetes. Hepatology 48(3):792–798

Fukuda T et al (2016) Transient remission of nonalcoholic fatty liver disease decreases the risk of incident type 2 diabetes mellitus in Japanese men. Eur J Gastroenterol Hepatol 28(12):1443–1449

Gardner CD et al (2007) Comparison of the Atkins, zone, Ornish, and LEARN diets for change in weight and related risk factors among overweight premenopausal women: the a TO Z weight loss study: a randomized trial. JAMA 297(9):969–977

Gawrieh S, Opara EC, Koch TR (2004) Oxidative stress in nonalcoholic fatty liver disease: pathogenesis and antioxidant therapies. J Investig Med 52(8):506–514

Gawrieh S et al (2019a) Characteristics, aetiologies and trends of hepatocellular carcinoma in patients without cirrhosis: a United States multicentre study. Aliment Pharmacol Ther 50(7):809–821

Gawrieh S et al (2019b) Histologic findings of advanced fibrosis and cirrhosis in patients with nonalcoholic fatty liver disease who have normal aminotransferase levels. Am J Gastroenterol 114(10):1626–1635

Gerss J, Kopcke W (2009) The questionable association of vitamin E supplementation and mortality–inconsistent results of different meta-analytic approaches. Cell Mol Biol (Noisy-le-Grand) 55(Suppl):OL1111–OL1120

Guichelaar MM et al (2013) Interactions of allelic variance of PNPLA3 with nongenetic factors in predicting non-alcoholic steatohepatitis and nonhepatic complications of severe obesity. Obesity (Silver Spring) 21 (9):1935–1941

Hallsworth K et al (2011) Resistance exercise reduces liver fat and its mediators in non-alcoholic fatty liver disease independent of weight loss. Gut 60(9):1278–1283

Hao C et al (2014) Inhibition of connective tissue growth factor suppresses hepatic stellate cell activation in vitro and prevents liver fibrosis in vivo. Clin Exp Med 14 (2):141–150

Harman DJ et al (2014) Prevalence and natural history of histologically proven chronic liver disease in a longitudinal cohort of patients with type 1 diabetes. Hepatology 60(1):158–168

Harrison SA et al (2003) Vitamin E and vitamin C treatment improves fibrosis in patients with nonalcoholic steatohepatitis. Am J Gastroenterol 98(11):2485–2490

Haufe S et al (2011) Randomized comparison of reduced fat and reduced carbohydrate hypocaloric diets on intrahepatic fat in overweight and obese human subjects. Hepatology 53(5):1504–1514

He S et al (2010) A sequence variation (I148M) in PNPLA3 associated with nonalcoholic fatty liver disease disrupts triglyceride hydrolysis. J Biol Chem 285 (9):6706–6715

Holst JJ, Madsbad S (2017) Semaglutide seems to be more effective the other GLP-1Ras. Ann Transl Med 5 (24):505

Hoofnagle JH et al (2013) Vitamin E and changes in serum alanine aminotransferase levels in patients with non-alcoholic steatohepatitis. Aliment Pharmacol Ther 38(2):134–143

Hossain P, Kawar B, El Nahas M (2007) Obesity and diabetes in the developing world–a growing challenge. N Engl J Med 356(3):213–215

Hsu HC et al (1994) Allelotype and loss of heterozygosity of p53 in primary and recurrent hepatocellular carcinomas. A study of 150 patients. Cancer 73(1):42–47

Huang J et al (2004) Physicians' weight loss counseling in two public hospital primary care clinics. Acad Med 79 (2):156–161

Huang MA et al (2005) One-year intense nutritional counseling results in histological improvement in patients with non-alcoholic steatohepatitis: a pilot study. Am J Gastroenterol 100(5):1072–1081

Ibrahim SH, Kohli R, Gores GJ (2011) Mechanisms of lipotoxicity in NAFLD and clinical implications. J Pediatr Gastroenterol Nutr 53(2):131–140

Ioannou GN et al (2019) Models estimating risk of hepatocellular carcinoma in patients with alcohol or NAFLD-related cirrhosis for risk stratification. J Hepatol 71(3):523–533

Jensen T et al (2018) Fructose and sugar: a major mediator of non-alcoholic fatty liver disease. J Hepatol 68 (5):1063–1075

Jinjuvadia R et al (2017) The association between nonalcoholic fatty liver disease and metabolic abnormalities in the United States population. J Clin Gastroenterol 51 (2):160–166

Johnson NA et al (2009) Aerobic exercise training reduces hepatic and visceral lipids in obese individuals without weight loss. Hepatology 50(4):1105–1112

Joy TR et al (2017) Sitagliptin in patients with non-alcoholic steatohepatitis: a randomized, placebo-controlled trial. World J Gastroenterol 23(1):141–150

Kani AH et al (2014) Effects of a novel therapeutic diet on liver enzymes and coagulating factors in patients with non-alcoholic fatty liver disease: a parallel randomized trial. Nutrition 30(7–8):814–821

Kantartzis K et al (2009) High cardiorespiratory fitness is an independent predictor of the reduction in liver fat during a lifestyle intervention in non-alcoholic fatty liver disease. Gut 58(9):1281–1288

Katsiki N, Mikhailidis DP, Mantzoros CS (2016) Non-alcoholic fatty liver disease and dyslipidemia: an update. Metabolism 65(8):1109–1123

Keating SE et al (2012) Exercise and non-alcoholic fatty liver disease: a systematic review and meta-analysis. J Hepatol 57(1):157–166

Kenny PR et al (2010) Exenatide in the treatment of diabetic patients with non-alcoholic steatohepatitis: a case series. Am J Gastroenterol 105(12):2707–2709

Khan RS et al (2019) Modulation of insulin resistance in nonalcoholic fatty liver disease. Hepatology 70 (2):711–724

Kirk E et al (2009) Dietary fat and carbohydrates differentially alter insulin sensitivity during caloric restriction. Gastroenterology 136(5):1552–1560

Kirpich IA, Marsano LS, McClain CJ (2015) Gut-liver axis, nutrition, and non-alcoholic fatty liver disease. Clin Biochem 48(13–14):923–930

Kistler KD et al (2011) Physical activity recommendations, exercise intensity, and histological severity of nonalcoholic fatty liver disease. Am J Gastroenterol 106(3):460–468. quiz 469

Klein EA et al (2011) Vitamin E and the risk of prostate cancer: the selenium and vitamin E cancer prevention trial (SELECT). JAMA 306(14):1549–1556

Koliaki C et al (2015) Adaptation of hepatic mitochondrial function in humans with non-alcoholic fatty liver is lost in steatohepatitis. Cell Metab 21 (5):739–746

Kozlitina J et al (2014) Exome-wide association study identifies a TM6SF2 variant that confers susceptibility to nonalcoholic fatty liver disease. Nat Genet 46 (4):352–356

Kreuter MW, Chheda SG, Bull FC (2000) How does physician advice influence patient behavior? Evidence for a priming effect. Arch Fam Med 9(5):426–433

Kuchay MS et al (2018) Effect of empagliflozin on liver fat in patients with type 2 diabetes and nonalcoholic fatty liver disease: a randomized controlled trial (E-LIFT trial). Diabetes Care 41(8):1801–1808

Kummer S et al (2017) Screening for non-alcoholic fatty liver disease in children and adolescents with type 1 diabetes mellitus: a cross-sectional analysis. Eur J Pediatr 176(4):529–536

Larson-Meyer DE et al (2008) Effect of 6-month calorie restriction and exercise on serum and liver lipids and markers of liver function. Obesity (Silver Spring, Md.) 16(6):1355–1362

Lazo M et al (2010) Effect of a 12-month intensive lifestyle intervention on hepatic steatosis in adults with type 2 diabetes. Diabetes Care 33(10):2156–2163

Lee YM et al (2012) Intragastric balloon significantly improves nonalcoholic fatty liver disease activity score in obese patients with nonalcoholic steatohepatitis: a pilot study. Gastrointest Endosc 76(4):756–760

Leslie T et al (2014) Survey of health status, nutrition and geography of food selection of chronic liver disease patients. Ann Hepatol 13(5):533–540

Levin D et al (2015) Pioglitazone and bladder cancer risk: a multipopulation pooled, cumulative exposure analysis. Diabetologia 58(3):493–504

Lewis JD et al (2011) Risk of bladder cancer among diabetic patients treated with pioglitazone: interim report of a longitudinal cohort study. Diabetes Care 34(4):916–922

Li G et al (2006) Inhibition of connective tissue growth factor by siRNA prevents liver fibrosis in rats. J Gene Med 8(7):889–900

Li Y et al (2014) Association between non-alcoholic fatty liver disease and chronic kidney disease in population with prediabetes or diabetes. Int Urol Nephrol 46 (9):1785–1791

Li Y et al (2017) Bidirectional association between non-alcoholic fatty liver disease and type 2 diabetes in Chinese population: evidence from the Dongfeng-Tongji cohort study. PLoS One 12(3):e0174291

Lonardo A et al (2018) Hypertension, diabetes, atherosclerosis and NASH: cause or consequence? J Hepatol 68 (2):335–352

Look ARG, Wing RR (2010) Long-term effects of a life-style intervention on weight and cardiovascular risk factors in individuals with type 2 diabetes mellitus: four-year results of the Look AHEAD trial. Arch Intern Med 170(17):1566–1575

Loomba R et al (2012) Association between diabetes, family history of diabetes, and risk of nonalcoholic steatohepatitis and fibrosis. Hepatology 56(3):943–951

Loomba R et al (2014) Magnetic resonance elastography predicts advanced fibrosis in patients with nonalcoholic fatty liver disease: a prospective study. Hepatology 60(6):1920–1928

Loria P, Lonardo A, Anania F (2013) Liver and diabetes. A vicious circle. Hepatol Res 43(1):51–64

Loureiro ML, Nayga RM Jr (2006) Obesity, weight loss, and physician's advice. Soc Sci Med 62 (10):2458–2468

Madsbad S, Holst JJ (2017) Glycaemic control and weight loss with semaglutide in type 2 diabetes. Lancet Diabetes Endocrinol 5(5):315–317

Makishima M et al (1999) Identification of a nuclear receptor for bile acids. Science 284(5418):1362–1365

Mancina RM et al (2016) The MBOAT7-TMC4 variant rs641738 increases risk of nonalcoholic fatty liver disease in individuals of European descent. Gastroenterology 150(5):1219–+

Mansoor N et al (2016) Effects of low-carbohydrate diets v. low-fat diets on body weight and cardiovascular risk factors: a meta-analysis of randomised controlled trials. Br J Nutr 115(3):466–479

Mantovani A et al (2016) Nonalcoholic fatty liver disease is independently associated with an increased incidence of cardiovascular disease in adult patients with type 1 diabetes. Int J Cardiol 225:387–391

Mantovani A et al (2017) Nonalcoholic fatty liver disease is associated with an increased prevalence of distal symmetric polyneuropathy in adult patients with type 1 diabetes. J Diabetes Complicat 31(6):1021–1026

Mantovani A et al (2018) Nonalcoholic fatty liver disease increases risk of incident chronic kidney disease: a systematic review and meta-analysis. Metabolism 79:64–76

Marchesini G et al (1999) Association of nonalcoholic fatty liver disease with insulin resistance. Am J Med 107(5):450–455

Marchesini G et al (2001) Nonalcoholic fatty liver disease: a feature of the metabolic syndrome. Diabetes 50 (8):1844–1850

Marso SP et al (2016) Liraglutide and cardiovascular outcomes in type 2 diabetes. N Engl J Med 375 (4):311–322

Mattar SG et al (2005) Surgically-induced weight loss significantly improves nonalcoholic fatty liver disease and the metabolic syndrome. Ann Surg 242 (4):610–617. discussion 618-20

Matteoni CA et al (1999) Nonalcoholic fatty liver disease: a spectrum of clinical and pathological severity. Gastroenterology 116(6):1413–1419

Mazuy C et al (2014) Nuclear bile acid signaling through the farnesoid X receptor. Cell Mol Life Sci

Miller ER 3rd et al (2005) Meta-analysis: high-dosage vitamin E supplementation may increase all-cause mortality. Ann Intern Med 142(1):37–46

Mlynarsky L et al (2016) Non-alcoholic fatty liver disease is not associated with a lower health perception. World J Gastroenterol 22(17):4362–4372

Morrison AE et al (2019) Causality between non-alcoholic fatty liver disease and risk of cardiovascular disease and type 2 diabetes: a meta-analysis with bias analysis. Liver Int 39(3):557–567

Musso G et al (2012) Impact of current treatments on liver disease, glucose metabolism and cardiovascular risk in non-alcoholic fatty liver disease (NAFLD): a systematic review and meta-analysis of randomised trials. Diabetologia 55(4):885–904

Musso G et al (2014) Association of non-alcoholic fatty liver disease with chronic kidney disease: a systematic review and meta-analysis. PLoS Med 11(7):e1001680

Neal B, Perkovic V, Matthews DR (2017) Canagliflozin and cardiovascular and renal events in type 2 diabetes. N Engl J Med 377(21):2099

Nehra V et al (2001) Nutritional and metabolic considerations in the etiology of nonalcoholic steatohepatitis. Dig Dis Sci 46(11):2347–2352

Neuschwander-Tetri BA (2012) Farnesoid x receptor agonists: what they are and how they might be used in treating liver disease. Curr Gastroenterol Rep 14 (1):55–62

Neuschwander-Tetri BA et al (2014) Farnesoid X nuclear receptor ligand obeticholic acid for non-cirrhotic, non-alcoholic steatohepatitis (FLINT): a multicentre, randomised, placebo-controlled trial. Lancet

Nguyen V et al (2017) Outcomes following serial intragastric balloon therapy for obesity and nonalcoholic fatty liver disease in a single centre. Can J Gastroenterol Hepatol 2017:4697194

Nilsson PM, Tuomilehto J, Ryden L (2019) The metabolic syndrome – What is it and how should it be managed? Eur J Prev Cardiol 26(2_Suppl):33–46

Oni ET et al (2015) Relation of physical activity to prevalence of nonalcoholic fatty liver disease independent of cardiometabolic risk. Am J Cardiol 115(1):34–39

Ouyang X et al (2008) Fructose consumption as a risk factor for non-alcoholic fatty liver disease. J Hepatol 48(6):993–999

Pais R et al (2017) Temporal trends, clinical patterns and outcomes of NAFLD-related HCC in patients undergoing liver resection over a 20-year period. Aliment Pharmacol Ther 46(9):856–863

Paradis V et al (2001) High glucose and hyperinsulinemia stimulate connective tissue growth factor expression: a potential mechanism involved in progression to fibrosis in nonalcoholic steatohepatitis. Hepatology 34(4 Pt 1):738–744

Parks DJ et al (1999) Bile acids: natural ligands for an orphan nuclear receptor. Science 284(5418):1365–1368

Patel PJ et al (2018) Underappreciation of non-alcoholic fatty liver disease by primary care clinicians: limited awareness of surrogate markers of fibrosis. Intern Med J 48(2):144–151

Perla FM et al (2017) The role of lipid and lipoprotein metabolism in non-alcoholic fatty liver disease. Children (Basel) 4(6)

Perry RJ et al (2015) Hepatic acetyl CoA links adipose tissue inflammation to hepatic insulin resistance and type 2 diabetes. Cell 160(4):745–758

Polesel J et al (2009) The impact of obesity and diabetes mellitus on the risk of hepatocellular carcinoma. Ann Oncol 20(2):353–357

Portillo-Sanchez P et al (2015) High prevalence of nonalcoholic fatty liver disease in patients with type 2 diabetes mellitus and normal plasma aminotransferase levels. J Clin Endocrinol Metab 100(6):2231–2238

Rakoski MO et al (2010) Meta-analysis: insulin sensitizers for the treatment of non-alcoholic steatohepatitis. Aliment Pharmacol Ther 32(10):1211–1221

Regnell SE, Lernmark A (2011) Hepatic steatosis in type 1 diabetes. Rev Diabet Stud 8(4):454–467

Ricci G et al (2008) Bariatric therapy with intragastric balloon improves liver dysfunction and insulin resistance in obese patients. Obes Surg 18(11):1438–1442

Roden M (2006) Mechanisms of disease: hepatic steatosis in type 2 diabetes–pathogenesis and clinical relevance. Nat Clin Pract Endocrinol Metab 2(6):335–348

Romeo S et al (2008) Genetic variation in PNPLA3 confers susceptibility to nonalcoholic fatty liver disease. Nat Genet 40(12):1461–1465

Ryan MC et al (2013) The Mediterranean diet improves hepatic steatosis and insulin sensitivity in individuals with non-alcoholic fatty liver disease. J Hepatol 59 (1):138–143

Saab S et al (2014) Impact of coffee on liver diseases: a systematic review. Liver Int 34(4):495–504

Sackner-Bernstein J, Kanter D, Kaul S (2015) Dietary intervention for overweight and obese adults: comparison of low-carbohydrate and low-fat diets. A meta-analysis. PLoS One 10(10):e0139817

Saeedi P et al (2019) Global and regional diabetes prevalence estimates for 2019 and projections for 2030 and 2045: results from the International Diabetes Federation Diabetes Atlas, 9(th) edition. Diabetes Res Clin Pract 157:107843

Sanyal AJ et al (2010) Pioglitazone, vitamin E, or placebo for nonalcoholic steatohepatitis. N Engl J Med 362 (18):1675–1685

Sanyal AJ et al (2019) The natural history of advanced fibrosis due to nonalcoholic steatohepatitis: data from the simtuzumab trials. Hepatology 70(6):1913–1927

Sattar N et al (2018) Empagliflozin is associated with improvements in liver enzymes potentially consistent with reductions in liver fat: results from randomised trials including the EMPA-REG OUTCOME(R) trial. Diabetologia 61(10):2155–2163

Sayiner M et al (2017) Variables associated with inpatient and outpatient resource utilization among Medicare beneficiaries with nonalcoholic fatty liver disease with or without cirrhosis. J Clin Gastroenterol 51 (3):254–260

Scaglioni F et al (2013) Short-term multidisciplinary non-pharmacological intervention is effective in reducing liver fat content assessed non-invasively in patients with nonalcoholic fatty liver disease (NAFLD). Clin Res Hepatol Gastroenterol 37(4):353–358

Schauer PR et al (2017) Bariatric surgery versus intensive medical therapy for diabetes – 5-year outcomes. N Engl J Med 376(7):641–651

Seko Y et al (2018) Effects of canagliflozin, an SGLT2 inhibitor, on hepatic function in Japanese patients with type 2 diabetes mellitus: pooled and subgroup analyses of clinical trials. J Gastroenterol 53 (1):140–151

Shah AG et al (2009) Comparison of noninvasive markers of fibrosis in patients with nonalcoholic fatty liver disease. Clin Gastroenterol Hepatol 7(10):1104–1112

Shao N et al (2014) Benefits of exenatide on obesity and non-alcoholic fatty liver disease with elevated liver enzymes in patients with type 2 diabetes. Diabetes Metab Res Rev 30(6):521–529

Siddiqui MS et al (2019) Vibration-controlled transient elastography to assess fibrosis and steatosis in patients with nonalcoholic fatty liver disease. Clin Gastroenterol Hepatol 17(1):156–163. e2

Simmons KM, Michels AW (2015) Type 1 diabetes: a predictable disease. World J Diabetes 6(3):380–390

Solga SF, Diehl AM (2003) Non-alcoholic fatty liver disease: lumen–liver interactions and possible role for probiotics. J Hepatol 38(5):681–687

Sookoian S, Pirola CJ (2011) Meta-analysis of the influence of I148M variant of patatin-like phospholipase domain containing 3 gene (PNPLA3) on the susceptibility and histological severity of nonalcoholic fatty liver disease. Hepatology 53(6):1883–1894

St George A et al (2009) Effect of a lifestyle intervention in patients with abnormal liver enzymes and metabolic risk factors. J Gastroenterol Hepatol 24(3):399–407

Straub BK et al (2010) Lipid droplet-associated PAT-proteins show frequent and differential expression in neoplastic steatogenesis. Mod Pathol 23 (3):480–492

Sullivan S et al (2012) Randomized trial of exercise effect on intrahepatic triglyceride content and lipid kinetics in nonalcoholic fatty liver disease. Hepatology 55 (6):1738–1745

Targher GEA (2014) Nonalcoholic fatty liver disease is independently associated with an increased incidence of chronic kidney disease in patients with type 1 diabetes. Diabetes Care 37(6):1729–1736

Targher G et al (2007) Prevalence of nonalcoholic fatty liver disease and its association with cardiovascular disease among type 2 diabetic patients. Diabetes Care 30(5):1212–1218

Targher G et al (2008a) Increased risk of CKD among type 2 diabetics with nonalcoholic fatty liver disease. J Am Soc Nephrol 19(8):1564–1570

Targher G et al (2008b) Non-alcoholic fatty liver disease is independently associated with an increased prevalence of chronic kidney disease and proliferative/laser-treated retinopathy in type 2 diabetic patients. Diabetologia 51(3):444–450

Targher G, Day CP, Bonora E (2010a) Risk of cardiovascular disease in patients with nonalcoholic fatty liver disease. N Engl J Med 363(14):1341–1350

Targher G et al (2010b) Non-alcoholic fatty liver disease is independently associated with an increased prevalence of chronic kidney disease and retinopathy in type 1 diabetic patients. Diabetologia 53(7):1341–1348

Targher G et al (2010c) Relationship between kidney function and liver histology in subjects with nonalcoholic steatohepatitis. Clin J Am Soc Nephrol 5(12):2166–2171

Targher G et al (2010d) Prevalence of non-alcoholic fatty liver disease and its association with cardiovascular disease in patients with type 1 diabetes. J Hepatol 53(4):713–718

Targher G et al (2012) Increased prevalence of chronic kidney disease in patients with type 1 diabetes and non-alcoholic fatty liver. Diabet Med 29(2):220–226

Targher G, Lonardo A, Byrne CD (2018) Nonalcoholic fatty liver disease and chronic vascular complications of diabetes mellitus. Nat Rev Endocrinol 14(2):99–114

Tendler D et al (2007) The effect of a low-carbohydrate, ketogenic diet on nonalcoholic fatty liver disease: a pilot study. Dig Dis Sci 52(2):589–593

Tilg H, Moschen AR, Roden M (2017) NAFLD and diabetes mellitus. Nat Rev Gastroenterol Hepatol 14(1):32–42

Tobias DK et al (2015) Effect of low-fat diet interventions versus other diet interventions on long-term weight change in adults: a systematic review and meta-analysis. Lancet Diabetes Endocrinol 3(12):968–979

Trauner M, Arrese M, Wagner M (2010) Fatty liver and lipotoxicity. Biochim Biophys Acta 1801(3):299–310

Tuccori M et al (2016) Pioglitazone use and risk of bladder cancer: population based cohort study. BMJ 352:i1541

Ulitsky A et al (2010) A noninvasive clinical scoring model predicts risk of nonalcoholic steatohepatitis in morbidly obese patients. Obes Surg 20(6):685–691

van Asten M et al (2017) The increasing burden of NAFLD fibrosis in the general population: time to bridge the gap between hepatologists and primary care. Hepatology 65(3):1078

Vanni E, Bugianesi E (2009) The gut-liver axis in nonalcoholic fatty liver disease: another pathway to insulin resistance? Hepatology 49(6):1790–1792

Vilar-Gomez E, Chalasani N (2018) Non invasive assessment of non-alcoholic fatty liver disease: clinical prediction rules and blood-based biomarkers. J Hepatol 68(2):305–315

Vilar-Gomez E et al (2015) Weight loss through lifestyle modification significantly reduces features of nonalcoholic steatohepatitis. Gastroenterology 149(2):367–378. e5

Vilar-Gomez E et al (2018) Vitamin E improves transplant free survival and hepatic decompensation among patients with nonalcoholic steatohepatitis and advanced fibrosis. Hepatology

Vilar-Gomez E et al (2019a) Post hoc analyses of surrogate markers of non-alcoholic fatty liver disease (NAFLD) and liver fibrosis in patients with type 2 diabetes in a digitally supported continuous care intervention: an open-label, non-randomised controlled study. BMJ Open 9(2):e023597

Vilar-Gomez E et al (2019b) Long-term metformin use may improve clinical outcomes in diabetic patients with non-alcoholic steatohepatitis and bridging fibrosis or compensated cirrhosis. Aliment Pharmacol Ther 50(3):317–328

Vivian EM (2014) Sodium-glucose co-transporter 2 (SGLT2) inhibitors: a growing class of antidiabetic agents. Drugs Context 3:212264

Volynets V et al (2013) A moderate weight reduction through dietary intervention decreases hepatic fat content in patients with non-alcoholic fatty liver disease (NAFLD): a pilot study. Eur J Nutr 52(2):527–535

Vuppalanchi R et al (2018) Performance characteristics of vibration-controlled transient elastography for evaluation of nonalcoholic fatty liver disease. Hepatology 67(1):134–144

Weinmann A et al (2015) Treatment and survival of non-alcoholic steatohepatitis associated hepatocellular carcinoma. BMC Cancer 15:210

Wieland AC et al (2013) Identifying practice gaps to optimize medical care for patients with nonalcoholic fatty liver disease. Dig Dis Sci 58(10):2809–2816

Wong RJ, Cheung R, Ahmed A (2014) Nonalcoholic steatohepatitis is the most rapidly growing indication for liver transplantation in patients with hepatocellular carcinoma in the U.S. Hepatology 59(6):2188–2195

Wong RJ et al (2015) Nonalcoholic steatohepatitis is the second leading etiology of liver disease among adults awaiting liver transplantation in the United States. Gastroenterology 148(3):547–555

Yamashita H et al (2001) A glucose-responsive transcription factor that regulates carbohydrate metabolism in the liver. Proc Natl Acad Sci U S A 98(16):9116–9121

Yamazaki H et al (2015) Independent association between improvement of nonalcoholic fatty liver disease and reduced incidence of type 2 diabetes. Diabetes Care 38(9):1673–1679

Yan J et al (2019) Liraglutide, Sitagliptin, and insulin glargine added to metformin: the effect on body weight and intrahepatic lipid in patients with type 2 diabetes mellitus and nonalcoholic fatty liver disease. Hepatology 69(6):2414–2426

Yasui K et al (2011) Characteristics of patients with non-alcoholic steatohepatitis who develop hepatocellular carcinoma. Clin Gastroenterol Hepatol 9(5):428–433. quiz e50

Yau H et al (2013) The future of thiazolidinedione therapy in the management of type 2 diabetes mellitus. Curr Diab Rep 13(3):329–341

Yki-Jarvinen H et al (1984) Site of insulin resistance in type 1 diabetes: insulin-mediated glucose disposal in vivo in relation to insulin binding and action in adipocytes in vitro. J Clin Endocrinol Metab 59 (6):1183–1192

Younossi ZM et al (2016a) Global epidemiology of nonalcoholic fatty liver disease-meta-analytic assessment of prevalence, incidence, and outcomes. Hepatology 64(1):73–84

Younossi ZM et al (2016b) The economic and clinical burden of nonalcoholic fatty liver disease in the United States and Europe. Hepatology 64(5):1577–1586

Younossi ZM et al (2018) Diagnostic modalities for nonalcoholic fatty liver disease, nonalcoholic steatohepatitis, and associated fibrosis. Hepatology 68 (1):349–360

Younossi ZM et al (2019a) The global epidemiology of NAFLD and NASH in patients with type 2 diabetes: a systematic review and meta-analysis. J Hepatol 71 (4):793–801

Younossi Z et al (2019b) Nonalcoholic Steatohepatitis is the fastest growing cause of hepatocellular carcinoma in liver transplant candidates. Clin Gastroenterol Hepatol 17(4):748–755. e3

Younossi ZM et al (2019c) Burden of illness and economic model for patients with nonalcoholic steatohepatitis in the United States. Hepatology 69 (2):564–572

Younossi ZM et al (2019d) Obeticholic acid for the treatment of non-alcoholic steatohepatitis: interim analysis from a multicentre, randomised, placebo-controlled phase 3 trial. Lancet

Zelber-Sagi S et al (2015) Coffee consumption and nonalcoholic fatty liver onset: a prospective study in the general population. Transl Res 165(3):428–436

Zhang X et al (2014) Continuation of metformin use after a diagnosis of cirrhosis significantly improves survival of patients with diabetes. Hepatology 60 (6):2008–2016

Zheng Y, Ley SH, Hu FB (2018) Global aetiology and epidemiology of type 2 diabetes mellitus and its complications. Nat Rev Endocrinol 14(2):88–98

Adv Exp Med Biol - Advances in Internal Medicine (2020) 4: 441–455
https://doi.org/10.1007/5584_2020_493
© Springer Nature Switzerland AG 2020
Published online: 7 February 2020

Global Experience of Diabetes Registries: A Systematic Review

Roya Naemi and Leila Shahmoradi

Abstract

Introduction

Diabetes mellitus (DM) is as a chronic metabolic disease, and disease registry plays an important role in the care of diabetes. Systematic review of diabetes registry systems in different countries has not been conducted based on evidences. This study conducts a systematic review to determine the goals, data elements, reports, data sources and capabilities of diabetes registry systems.

Method

In this study, searches were conducted in four databases such as PubMed, Scopus, Web of Science and Embase to find available information on diabetes registry systems. Two researchers conducted the search separately to identify related studies based on an input criterion. All controversies were resolved by the consensus.

Results

18,534 studies were identified in the primary search. After reviewing the title and abstract of the articles, 11,344 studies were excluded. Finally, 21 studies were selected for the review. The main characteristics of the diabetes registries have been cited in the study under the categories of country's name with registry, title of the system, data sources and system developers. The information management is considered as the main goal of diabetes registry system. Data elements of diabetes registry were laboratory measurement and chronic complications.

Conclusion

This systematic review provides a global overview of the goals, data elements, reports, data sources and capabilities of the diabetes registries and recommends the use of diabetes registry to increase efficiency of services and quality of care.

R. Naemi
Health Information Management, Department of Health Information Management, School of Allied Medical Sciences, Tehran University of Medical Sciences (TUMS), Tehran, Iran

Department of Health Information Management, School of Paramedical Sciences, Ardabil University of Medical Sciences, Ardabil, Iran

L. Shahmoradi (✉)
Halal Research Center of IRI, FDA, Tehran, Iran

Department of Health Information Management, School of Allied Medical Sciences, Tehran University of Medical Sciences (TUMS), Tehran, Iran
e-mail: Lshahmoradi20@gmail.com;
Lshahmoradi@tums.ac.ir

Keywords

Diabetes · Diabetes mellitus · Diabetes
registry · Disease registry · Registries

Abbreviations

ACC	American College of Cardiology
ADA	American Diabetes Association
CCOPMM	Consultative Council on Obstetric and Paediatric Mortality and Morbidity
CSII	The national register of patients treated with continuous subcutaneous insulin infusion
DCR	Diabetes Collaborative Registry
DCRS	Danish Civil Registration System
DCs	Diabetes Centres
DERI	Diabetes Epidemiology Research International
DM	Diabetes mellitus
DNPaR	Danish National Patient Register
GDM	Gestational Diabetes Mellitus
GIS	Geographic Information System
GP	General Practitioners
HDL	High-Density Lipoprotein
HER	Electronic Health Record
ICDC	Israel Center for Disease Control
IDF	International Diabetes Federation
IOM	Institute of Medicine
IPES	Israel Pediatric Endocrine Society
IT	Information Technology
LDL	Low-Density Lipoprotein
MU	Meaningful Use
NA	Not Available
NDR	National Diabetes Register
NDSS	National Diabetes Support Scheme
NEPDN	The North East Pediatrics Diabetes Network
NGDR	National Gestational Diabetes Register
NHG	National Healthcare Group
PIN	Personal Identification Number
PRISMA	Preferred Reporting Items for Systematic Reviews and Meta-Analyses
RIDI	The Registry for Type 1 Diabetes Mellitus in Italy
SGOT	Serum Glutamic Oxaloacetic Transaminase
SGPT	Serum Glutamic-Pyruvic Transaminase
SNDR	The Saudi National Diabetes Registry
T1DM	Type 1 diabetes mellitus
TSH	Thyroid-Stimulating Hormone
WHO	World Health Organization
YSRCCYP	The Yorkshire Pediatrics Register of Diabetes in Children and Young people

1   Introduction

Diabetes mellitus (DM), is the most common chronic metabolic diseases in the world (Shaw et al. 2010; Guariguata et al. 2014; Sreedharan 2016). Diabetes mellitus has two types type one and type two (Ramadas et al. 2011). Population growth, aging, urbanization, obesity and inactivity are some factors which increase the prevalence of DM (Shaw et al. 2010; Guariguata et al. 2014; Wild et al. 2004). In 1998, the cost of DM care in the USA was estimated to be $ 98.2 billion (Knight et al. 2005). It is estimated that, the prevalence of DM will be 366 million in 2030 (Ramadas et al. 2011). Diabetes can cause severe damage to the body organs, especially the blood vessels and nerves. According to the IOM,[1] diabetes has become a major public health problem in the world because of high prevalence of the disease with its complications and costs of treatment (Kohn, 1999). The information obtained from the disease registry can reduce the rate of the disease and the complications of DM (Leucuţa et al. 2010). Disease registry facilitates the identification of patients with specific conditions and appropriate interventions (Ciemins et al. 2009).

---

[1] Institute of Medicine.

A disease registry is a database of people who have been affected by a specific disease (Viviani et al. 2014). It is considered as a valuable resource for stakeholders to manage chronic diseases by providing the patient list, laboratory information, disease complications and family history (Sreedharan 2016; Cooperberg et al. 2004). Stakeholders may include patients, doctors, researchers, health policymakers and so on. Hospital and population registration are considered as two main types of disease registration. Hospital registration can be either single registration or multi-center registration. Hospital registration focuses on improving patient care, education, information management and clinical research. However, population registration focuses on disease prevention, early diagnosis of diseases, prevalence and trends of the diseases; and research and evaluation of interventions (Types of Registries 2019).

Patient care, medical knowledge management and provision of information are three main goals of diabetes registry. This information is used to monitor and improve the quality of care, health care planning, evaluation, health care interventions and epidemiological researches (Rankin and Best 2014). Management of the outbreaks and complications of chronic diseases such as diabetes can be facilitated by using tools and databases like the national disease registry (Leucuța et al. 2010). IT[2] applications, like registries can increase the quality of health care at the community level (Kohn 1999). In the United States, the disease registry is defined as a vital component of the Meaningful Use (MU) of the Electronic Health Record (EHR) (Han et al. 2015). IT-enabled diabetes management systems make it easier to improve care, prevention of diabetes complications, scientific studies and awareness of the socio-economic effects of diabetes (Al Rubeaan et al. 2014; Bu et al. 2007). Studies and related websites show that, no systematic review has ever been conducted on the diabetes registry systems in the world. The present review is a systematic review of the diabetes registry systems to answer the following questions:

Q1: What are the goals of the diabetes registry systems?
Q2: What are the capabilities and data sources of the diabetes registry systems?
Q3: What are the data elements of the diabetes registry systems?
Q4: What are the reports of the diabetes registry systems?

2 Methods

2.1 Design

This study was a secondary study with narrative synthesis.

2.2 Input and Output Criteria

Input and output criteria for this review are listed in Table 1.

2.3 Search Strategy

In this systematic review, searches were conducted in the databases such as PubMed, Scopus, Web of Science and Embase, because studies of computer science and medical information are found in these databases. The reference list of relevant studies were also investigated; if they had fulfilled the criteria they would have included into the study. Keywords were reviewed in MESH then, evaluated by a specialist librarian. The registry keywords were as follows: Registry, Register, "Registry System", "Electronic Registries", "Registry Database", and "Computerized Registry". The keywords of diabetes were as follows: "Diabetes Mellitus", "Diabetes Mellitus, Type I", "Diabetes Mellitus, Insulin-Dependent", "Diabetes Mellitus, Type 2",

---

[2] Information Technology.

Table 1 Inclusion and exclusion criteria

Input criteria	Output criteria
Published research articles in English with full text and without time limitation which are available at the digital library of Tehran University of Medical Sciences[a] (PubMed, Scopus, Web of Science and Embase databases)	Non- English language articles
	Conference abstracts and letters to the editor, case report, book seasons and Randomized Clinical Trial(RCT)
	Studies with duplicate content
	Studies of another diseases registry system
Studies which evaluated diabetes registry system or any of its components (such as the goal of diabetes registry system, system capabilities, data sources, data elements, reports and outputs of the system)	Studies which addressed the clinical features of diabetes and its complications instead of reporting the goals, outcomes and reports of diabetes registry system

[a]diglib.tums.ac.ir/

"Diabetes Mellitus, Noninsulin-Dependent" and, "Diabetes Mellitus, Type II" and "System", "database", "data repository" and, "databank". The search strategy for PubMed database was as follows:

(diabet∗[tiab] OR T1DM[tiab] OR T2DM[tiab] OR T1D[tiab] OR T2D[tiab] OR MODY [tiab] OR niddm[tiab] OR iddm[tiab] OR ((Noninsulin[tiab] OR Non-insulin[tiab]) AND dependent[tiab]) OR "Diabetes Mellitus"[Mesh]) AND (("Registries" [Mesh] OR Registry[tiab]) AND (System[tiab] OR Systems[tiab] OR database∗[tiab] OR (data [tiab] AND repository[tiab]) OR databank∗ [tiab])). The last search was done on April 28, 2019. Endnote was used to collect articles with the predefined criteria. Duplicate articles were identified and deleted. The article was written according to the PRISMA[3] standard.

2.4   Data Extraction and Synthesis

The articles were independently reviewed by two researchers (RN and LS). At first, the researchers reviewed the title and abstract of the articles. The

article would be omitted, if both researchers were unanimous in not including the article in the review. If there was not enough information to decide at this stage, it was transferred to the next step. In the next step, the researchers decided about accepting or refusing the articles. If the two researchers had a contradicting view, they discussed about their points arguments, and if they did not reach an agreement, a third party made decision on the entry of the article. Two authors extracted the following data independently: goals, capabilities and data sources of the diabetes registry system, data elements and reports of the diabetes registry system.

3   Results

At first, we provide a bibliographic research results; and then, the results of each question are separately provided.

3.1   Bibliographic Research Results

Figure 1 shows the results of selection process. Overall, 18,534 articles were identified in our primary search. 11,400 articles remained after the elimination of duplicates. The two researchers (RN and LS) removed 11,344 articles after a review of their title and abstracts. Finally, the full texts of 56 articles were reviewed by the researchers for more evaluation. In the next step, 18 articles did not meet the inclusion criteria, 15 articles had no full text, 3 articles had duplicate

---

[3] Preferred Reporting Items for Systematic Reviews and Meta-Analyses.

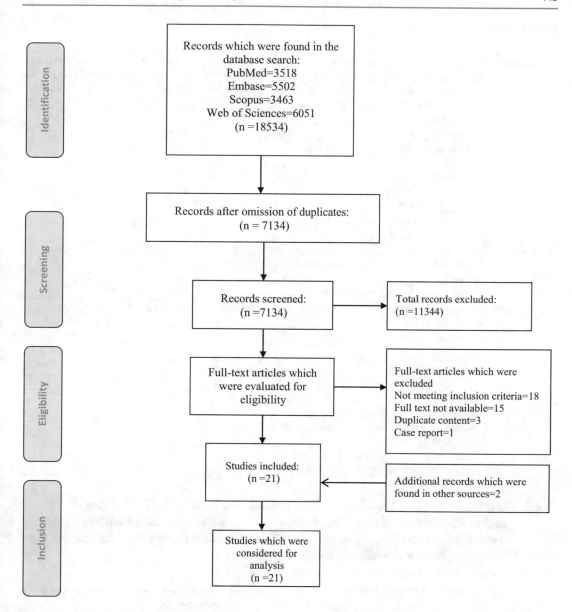

Fig. 1 Shows screening process of articles included in this study

content, and 1 case report was excluded. Therefore, 21 studies were included in the review at the end. Many countries, such as Norway, Mexico, Netherlands, Indonesia, Sri Lanka, India, Uzbekistan, Ireland, etc. were excluded at the time of screening studies because we did not have access to the full text of their article or they did not meet the inclusion criteria.

3.2 Results of Each Question

3.2.1 Q1: What Are the Goals of the Diabetes Registry Systems?

Table 2 provides a list of goals, which are attributed to the diabetes registry systems in 4 categories such as information management, control, follow-up and surveillance.

Table 2 Shows the goals of diabetes registries systems

	Description	References
Information management	To provide reliable and uniform data	Sereday et al. (1994), Green et al. (2015), Bendas et al. (2015), Dedov et al. (2018), Fuziah et al. (2008), Carle et al. (2004), and Thoelke et al. (1990)
	Epidemiological research: Time trends and variations (prevalence, incidence and mortality) by age, gender, demographic and clinical characteristics	
	Monitoring of patients	
	Participate in international collaborative program	
	To promote the establishment of new registries in uncovered areas	
Control	To evaluate DM control	Yang et al. (2006)
	To recognize and control the risk factors	
	Control of complications in patients	
Follow-up	Track and follow-up high-risk patients for early intervention to prevent the progression of diabetes	Boyle et al. (2018) and Yamamoto-Honda et al. (2014)
	Evaluation and management of complication	
Surveillance	To create an instrument to support regional health care policy	Benedetti et al. (2006) and Al-Rubeaan et al. (2013)
	Quality development	
	Shared care	
	Prevention of diabetes	
	To diagnose national diabetic patients	
	Surveillance-monitoring tool for clinical and epidemiology practitioners	
	To evaluate the economic effect (direct and indirect) of DM	
	To use social and cultural variables in planning primary and secondary prevention programs	

3.2.2 Q2: What Are the Capabilities and Data Sources of the Diabetes Registry Systems?

Table 3 shows the main characteristics of diabetes registry systems, which were obtained from the review of 21 documents which include country name, the system title, data sources, capabilities of the system, and system developer/s. Figure 2 shows the diabetes registry in different countries included in this study.

3.2.3 Q3: What Are the Data Elements of the Diabetes Registry Systems?

Table 4 describes the data elements of the diabetes registry systems in five categories of demographic and clinical features, laboratory measurements, social history and chronic complication.

3.2.4 Q4: What Are the Reports of the Diabetes Registry Systems?

Table 5 contains reports about the diabetes registry systems.

4 Discussion

The present systematic review described the goals, capabilities, data elements, reports and data sources of the diabetes registry systems, which can be used to develop any diabetes registry system. The purpose of the present study was to review published articles on diabetes registry systems to answer study questions, not all of the countries with diabetes registry systems. The study of all countries with a diabetes registry system can be done in a separate study. So that it cannot be answered through a systematic

Table 3 Shows the characteristics of diabetes registries systems, which has been included in this study

Registry Country	System title	Data sources	Capabilities of the diabetes registry system	System developers	References
Argentina	Avellaneda Registry	Primary sources: interview at schools, secondary sources: hospitals	The data collection	P. Fiorito Hospital, a reference hospital in Avellaneda	Sereday et al. (1994)
Austria	Austrian Diabetes Register	Hospitals	Network covering all hospitals and wards	Department of Epidemiology, Medical University of Vienna	Wiedemann et al. (2010)
Zhejiang, China	N.A[a]	Local hospitals	Web services and direct network report	Zhejiang Provincial Center	Wu et al. (2017)
Hong Kong, China	Hong Kong Diabetes Registry	Outpatient clinics regional hospitals	Unique identification number	N.A	Yang et al. (2006)
Denmark	National Diabetes Register (NDR)	(DNPaR[b]), the Danish National Prescription Registry, and the Danish National Health Service Register and (DCRS[c])	Personal Identification Number (PIN)	Danish National Board of Health	Green et al. (2015)
Finland	The Finland population-based register	N.A	PIN	Diabetes Epidemiology Research International (DERI)	Akkanen et al. (2009)
Australia	Gestational Diabetes Mellitus (GDM)	(1) Pathology data; (2) birth records from the (CCOPMM[d]); (3) GDM and type 2 diabetes register data from the National Gestational Diabetes Register (NGDR).	Recall system	National Diabetes Support Scheme (NDSS) under the auspices of diabetes Australia	Boyle et al. (2018)
Russia	N.A	N.A	Online access	Federal State Budgetary Institution National Medical Research Center for Endocrinology of the Ministry of Health	Dedov et al. (2018)
Italy	PROMODR (Progressive Model of Diabetes Register) pilot experience	General Practitioners (GPs), Diabetes Centres (DCs)	(a) Acquisition of e-mail, (b) encrypting and decrypting of data (c) generation of patient ID for	Ministry of Health grant	Benedetti et al. (2006)

(continued)

Table 3 (continued)

Registry / Country	System title	Data sources	Capabilities of the diabetes registry system	System developers	References
			longitudinal evaluation		
Italy	(RIDI[e])	Hospital discharges, prescription registries,	N.A	N.A	Carle et al. (2004)
Israel	N.A	N.A	N.A	Israel Pediatric Endocrine Society (IPES) and the Israel Center for Disease Control (ICDC)	Koton (2007)
Kuwait	Registry for IDDM in children as part of the World Health Organization (WHO) Collaborative Multinational Project (DIAMOND)	N.A	N.A	N.A	Shaltout et al. (1995)
Malaysia	Type 1 diabetes mellitus (T1DM)	N.A	N.A	N.A	Fuziah et al. (2008)
Saudi Arabia	The Saudi National Diabetes Registry (SNDR)	N.A	Electronic medical file Geographic Information System (GIS)	N.A	Al-Rubeaan et al. (2013)
Singapore	N.A	Hospitals or clinics	Clinician Decision Support	National Healthcare Group (NHG)	Toh et al. (2009)
Sweden	National Diabetes Register (NDR)	N.A	Web-based quality measure	N.A	Adolfsson and Rosenblad (2011)
USA	Diabetes Collaborative Registry (DCR)	N.A	Quality-oriented registry evaluation of data	American College of Cardiology (ACC), American Diabetes Association (ADA), American College of Physicians, American Association of Clinical Endocrinologists, and the Joslin diabetes center	Fan et al. (2019)
England	The Yorkshire Pediatric Register of Diabetes in	N.A	N.A	Candlelighters Trust	Rankin and Best (2014)

(continued)

Table 3 (continued)

Registry / Country	System title	Data sources	Capabilities of the diabetes registry system	System developers	References
	Children and Young people (YSRCCYP) and The North East Pediatric Diabetes Network (NEPDN)				
Berlin, Germany	N.A	N.A	Recall system	Centre of Diabetes and Metabolic Disorders of Berlin, G.D.R.	Thoelke et al. (1990)
Czech Republic	CSII[f]	N.A	N.A	N.A	Jankovec et al. (2006)
Japan	National Center Diabetes Database	N.A	N.A	National Center for Global Health and Medicine and the Ministry of Health, Labour and Welfare	Yamamoto-Honda et al. (2014)

[a]Not Available
[b]Danish National Patient Register
[c]Danish Civil Registration System
[d]Consultative Council on Obstetric and Pediatric Mortality and Morbidity
[e]The Registry for Type 1 Diabetes Mellitus in Italy
[f]The national register of patients treated with continuous subcutaneous insulin infusion (CSII)

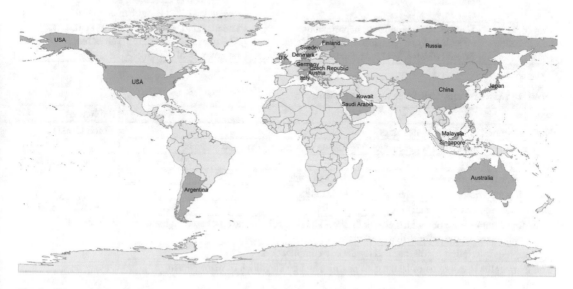

Fig. 2 Shows the diabetes registry in different countries, which has been included in this study

review. Because many countries may have a diabetes registry system, but they have not presented their experiences in the paper.

Generally, 21 diabetes registries were identified according to the inclusion criteria. Included studies in this systematic review came from countries such as Argentina, Austria, China,

Table 4 Shows data elements of the diabetes registries systems

Category	Data elements	References
Demographic features	Address, patient's name, ethnic group, nationality, marital status, gender, family history, insurance type, PIN, complete contact details, identification number of national hospital, complete contact details	Rankin and Best (2014), Bendas et al. (2015), Carle et al. (2004), Thoelke et al. (1990), Yamamoto-Honda et al. (2014), Benedetti et al. (2006), Al-Rubeaan et al. (2013), Wiedemann et al. (2010), Koton (2007), Shaltout et al. (1995), Toh et al. (2009), Adolfsson and Rosenblad (2011), Fan et al. (2019), and Jankovec et al. (2006)
Clinical features	Date of diagnosis, treatment type, diabetes type, date of first insulin injection, referral agent, associated diseases, weight, height, history of diabetes	Rankin and Best (2014), Carle et al. (2004), Thoelke et al. (1990), Benedetti et al. (2006), Al-Rubeaan et al. (2013), Wiedemann et al. (2010), Shaltout et al. (1995), Toh et al. (2009), and Adolfsson and Rosenblad (2011)
Laboratory measurements	Waist circumference, blood pressure, GPT, GOT, fasting plasma glucose, platelets, hba1c, red blood cells, white blood cells, triglyceride, cholesterol, hemoglobin, bilirubin, uric acid, total and, creatinine clearance, High-Density Lipoprotein(HDL), urine analyses, protein, ketones, liver enzymes, Low-Density Lipoprotein (LDL), alkaline phosphate, Serum Glutamic-Pyruvic Transaminase (SGPT), Serum Glutamic Oxaloacetic Transaminase (SGOT), and total protein; thyroid function test, including Thyroid-Stimulating Hormone (TSH), T4, and T3	Rankin and Best (2014), Benedetti et al. (2006), Al-Rubeaan et al. (2013), Toh et al. (2009), Adolfsson and Rosenblad (2011), Fan et al. (2019), and Jankovec et al. (2006)
Social history	Smoking, educational level, occupation, income, physical activity, diet, drinking, history of pregnancy	Rankin and Best (2014), Yamamoto-Honda et al. (2014), Benedetti et al. (2006), Al-Rubeaan et al. (2013), Toh et al. (2009), Adolfsson and Rosenblad (2011), and Fan et al. (2019)
Chronic complications	Neuropathy, retinopathy, nephropathy, vasculopathy, hypertension, hyperlipidemia, thyroid disease, stroke, amputations and others	Thoelke et al. (1990), Benedetti et al. (2006), Al-Rubeaan et al. (2013), Toh et al. (2009), and Jankovec et al. (2006)

Table 5 Shows reports of diabetes registries systems

Reports	References
National trends in DM incidence rates	Koton (2007)
Comparisons of incidence between ages, population groups, genders and seasons	
Time trends	Carle et al. (2004)
Geographical variation with different genetic backgrounds	
Providing the etiology of DM through analytical studies	
Descriptive and analytical statistics	Al-Rubeaan et al. (2013)
Providing economic reports which give both direct and indirect disease cost based on the local and specific cost analyses for different services	
Geographic description for disease and its chronic complications	
The geographic distribution of health care facilities and availabilities of management	

Denmark, Finland, Australia, Russia, Italy, Israel, Kuwait, Malaysia, Saudi Arabia, Singapore, Sweden, USA, England, Germany, Czech Republic and Japan. Information management, control, surveillance, and follow-up were the goals of most of the diabetes registry systems. Using web services, PIN, recall system and GIS were the main capabilities of diabetes registry systems. The main finding of the research will be provided in the following paragraphs.

Determining the registry goals, before the system is launched, which results in the collection of quality data is one of the most important factors in the success of diabetes registry. Mandavia et al., stated that stakeholders' consensus on the goals, data collection and system activities of the registry are essential (Mandavia et al. 2017). Monitoring the health care interventions, investigating the patterns and the causes of the disease, quality improvement, identifying health priorities, planning, and research are known as goals of the registries (Mandavia et al. 2017; Gudbjörnsdottir et al. 2003; Carstensen et al. 2011).

Dombkowski et al., revealed that, using the recall system is one of the important strategies to improve patient's safety (Dombkowski et al. 2017), which is featured in the Australia, Italy and Berlin diabetes registry systems (Thoelke et al. 1990; Boyle et al. 2018; Benedetti et al. 2006). The recall/reminder system can be as phone call, e-mail or message (Dombkowski et al. 2017). Adolfsson et al., found that it was necessary to use a recall system in the diabetes registry for training the patient and to upgrade the quality of care and delivery of effective care (Adolfsson et al. 2010). The results of this research confirm the findings of our study.

Geographic Information System (GIS) is as another characteristic, which has been added to the Saudi Arabia's diabetes registry system (Al-Rubeaan et al. 2013). GIS has developed innovative software for epidemiological planning (Jacquez et al. 2005), decision-making, management and distribution of information in health care (Rytkönen 2004). Naves et al., showed that, GIS includes preparing geographic data, topographic databases and satellite remote sensing to produce maps and make cross-correlate public health registries with the environmental factors. GIS is also useful in disease surveillance and supports planning to improve health care (Naves et al. 2015). The creation of overlapping thematic and disease maps is one of the main tasks of GIS (Jacquez et al. 2005). Therefore, the effective and efficient use of GIS is essential (Ramadan et al. 2017). In Zhejiang, China, Russia and Sweden, using the web is another features of the diabetes registry system (Dedov et al. 2018; Wu et al.

2017; Adolfsson and Rosenblad 2011). The ease of use and collection of information from different locations is the advantage of a web-based registry (Morrow et al. 2013). So, Seneviratne et al.'s, study demonstrated that the use of web-based software is important in rendering up-to-date, valid and comprehensive information to stakeholders (Seneviratne et al. 2017). Kern et al., stated that a diabetes registry can be valuable in combining the web for constant monitoring of patients and study interventions to improve quality (Kern et al. 2008). The results of these studies are consistent with the findings of present study. The PIN has been used in the diabetes registry systems of Hong Kong, China, Finland, and Denmark (Green et al. 2015; Yang et al. 2006; Akkanen et al. 2009). Over time, it is possible to trace the individual histories by a PIN (Wallach Kildemoes et al. 2011). PIN is a key tool for epidemiological research. This national tool is a unique ten-digit number, which is given all persons living in a country, at birth or during emigration. The PIN allows linkage between all registries to be recorded technically, simple, cost-effective, and specific individual-level (Schmidt et al. 2015). Gimsing et al., stated that, we can combine the information of several registries and provide scientific analysis of different health conditions with PIN (Gimsing et al. 2016).

According to studies, data elements of the diabetes registry are divided into demographic and clinical information, laboratory measurements, social history; and complications categories. Mandavia et al., pointed out that a flexible dataset helps to increase the length of registry. It is better to create a dataset with the stakeholders' consensus, and make a balance between comprehensiveness and feasibility. It should be noted that in a registry, the comprehensive datasets are not necessarily lead to provision of complete data; and limited datasets may not be useful for achieving the goals. The datasets must have definite definitions so that national comparisons can be done. The precise design of the registry with the participation of stakeholders during the development will result in the completion of the data (Mandavia et al. 2017). Krysinska et al., stated that data synchronization makes it possible to compare the registries of different countries and regions of the world (Krysinska et al. 2017). The International Diabetes Federation (IDF) and the WHO consider that using standard data are effective in the prevention and treatment of diabetes and cross-national comparisons (Harris et al. 2015). Hanberger et al., acknowledged that standardization and harmonization of data should facilitate collaboration, comparison, exchange experiences and research between nations (Hanberger et al. 2014). Standard data is necessary for the integration of the registry according to Richesson's opinion (Richesson 2011). The use of standard data in the diabetes registry facilitates continuity of care, patient monitoring, global research and quality of care improvement (Richesson 2011). The results of these studies are consistent with the findings of the present study.

The incorrect definition of the system and its data can lead to the production of false data. Some important aspects of the registry design include; setting a specific target, standardization of data, determining the data sources and the inclusion/ exclusion criteria. Also, determining the data collection process and data quality control method lead to the optimal use of data (Dang and Angle 2015; Davids et al. 2015). It is clear that high-quality registry systems affect people's health by providing valuable information from a large population of patients over a long period of time (Liu et al. 2015). According to what was mentioned, an efficient registry can track the long-term incidence trend in the specific geographic regions, etiology of disease, patterns of treatment and also survival variations (Psoter and Rosenfeld 2013; Dabelea et al. 2010).

5   Conclusions

This systematic review provides knowledge about goals, data elements, reports, data sources and capabilities of the diabetes registries in the world, which can be used for creation, and development of diabetes registry systems. It is important to help low and middle-income nations for establishing a diabetes registry system to

overwhelm the rising trend in the rate of diabetes soon. Diabetes registry systems are effective in improving the quality of care, patient follow-up, information management, intervention evaluation, and prevention of diabetes complications. Therefore, the growth of diabetes registry systems can be demanded and should be established by stakeholders. We strongly recommend the use of diabetes registry system and documenting the experiences such as benefits and challenges.

6   Research Limitation

At first, the full texts of some articles were not available, but we solved this problem by contacting the authors of the articles, so we could obtain the full texts of them. Our last study was done in April 2019; we might have missed recently published studies on this topic. Finally, we did not conduct a quality evaluation of the publication included in this review, because of the considerable heterogeneity among types of the selected studies. Two authors answered the study questions separately, and the disagreements were resolved by the consensus. Additional searches were not conducted on grey literature such as conference websites, government websites, and informal sources such as Ph.D. theses.

7   Strengths of Research

This study is the first comprehensive study, which investigated the global experience of diabetes registry with several databases. We used a systematic literature search strategy to identify relevant publications, and focused on the study goals. All steps of the review (i.e., article screening, data extraction) were conducted by two authors to minimize errors using PRISMA guidelines.

Acknowledgments We would like to thank Mrs. Rasha Atlasi; a Ph.D. in Medical Library and Information Science at Tehran University of Medical Sciences who participated in the review of the search strategy for this study.

Financial Support The article is a result of independent research without financial and organizational support.

Intellectual Property The intellectual property of this work belongs to Tehran University of Medical Sciences. The use of information from this study is prohibited without mentioning it.

References

Adolfsson ET, Rosenblad A (2011) Reporting systems, reporting rates and completeness of data reported from primary healthcare to a Swedish quality register—the National Diabetes Register. Int J Med Inform 80 (9):663–668

Adolfsson ET, Rosenblad A, Wikblad K (2010) The Swedish national survey of the quality and organization of diabetes care in primary healthcare—Swed-QOP. Prim Care Diabetes 4(2):91–97

Akkanen MJ, Kivelä S-L, Koistinen V, Sintonen H, Tuomilehto J (2009) Inpatient care of patients with type 1 diabetes mellitus by duration of diabetes and sex: a nationwide population-based longitudinal study. Risk Manag Healthc Policy 2:55

Al-Rubeaan KA, Youssef AM, Subhani SN, Ahmad NA, Al-Sharqawi AH, Ibrahim HM (2013) A Web-based interactive diabetes registry for health care management and planning in Saudi Arabia. J Med Internet Res 15(9):e202

Al-Rubeaan K, Youssef AM, Subhani SN, Ahmad NA, Al-Sharqawi AH, Al-Mutlaq HM et al (2014) Diabetic nephropathy and its risk factors in a society with a type 2 diabetes epidemic: a Saudi National Diabetes Registry-based study. PLoS One 9(2):e88956

Bendas A, Rothe U, Kiess W, Kapellen TM, Stange T, Manuwald U et al (2015) Trends in incidence rates during 1999–2008 and prevalence in 2008 of childhood type 1 diabetes mellitus in Germany—model-based national estimates. PLoS One 10(7):e0132716

Benedetti MM, Carinci F, Federici MO (2006) The Umbria Diabetes Register. Diabetes Res Clin Pract 74:S200–S2S4

Boyle DI, Versace VL, Dunbar JA, Scheil W, Janus E, Oats JJ et al (2018) Results of the first recorded evaluation of a national gestational diabetes mellitus register: challenges in screening, registration, and follow-up for diabetes risk. PLoS One 13(8):e0200832

Bu D, Pan E, Walker J, Adler Milstein J, Kendrick D, Hook JM et al (2007) Benefits of information technology–enabled diabetes management. Diabetes Care 30(5):1137–1142

Carle F, Gesuita R, Bruno G, Coppa GV, Falorni A, Lorini R et al (2004) Diabetes incidence in 0-to 14-year age-group in Italy: a 10-year prospective study. Diabetes Care 27(12):2790–2796

Carstensen B, Kristensen JK, Marcussen MM, Borch-Johnsen K (2011) The national diabetes register. Scand J Publ Health 39(7_suppl):58–61

Ciemins EL, Coon PJ, Fowles JB, Min S-J (2009) Beyond health information technology: critical factors necessary for effective diabetes disease management. J Diabetes Sci Technol 3(3):452–460

Cooperberg MR, Broering JM, Litwin MS, Lubeck DP, Mehta SS, Henning JM et al (2004) The contemporary management of prostate cancer in the United States: lessons from the cancer of the prostate strategic urologic research endeavor (CapSURE), a national disease registry. J Urol 171(4):1393–1401

Dabelea D, Mayer-Davis EJ, Imperatore G (2010) The value of national diabetes registries: SEARCH for Diabetes in Youth Study. Curr Diab Rep 10(5):362–369

Dang A, Angle VS (2015) Utilizing patient registries as health technology assessment (HTA) tool. Syst Rev Pharm 6(1):5

Davids MR, Eastwood JB, Selwood NH, Arogundade FA, Ashuntantang G, Benghanem Gharbi M et al (2015) A renal registry for Africa: first steps. Clin Kidney J 9 (1):162–167

Dedov II, Shestakova MV, Peterkova VA, Vikulova OK, Zheleznyakova AV, Isakov MA et al (2018) Diabetes mellitus in children and adolescents according to the Federal Diabetes Registry in the Russian Federation: dynamics of major epidemiological characteristics for 2013–2016. Diabetes Mellitus 20(6):392–402

Dombkowski KJ, Cowan AE, Reeves SL, Foley MR, Dempsey AF (2017) The impacts of email reminder/recall on adolescent influenza vaccination. Vaccine 35 (23):3089–3095

Fan W, Song Y, Inzucchi SE, Sperling L, Cannon CP, Arnold SV et al (2019) Composite cardiovascular risk factor target achievement and its predictors in US adults with diabetes: the diabetes collaborative registry. Diabetes Obes Metab

Fuziah M, Hong J, Zanariah H, Harun F, Chan S, Rokiah P et al (2008) A national database on children and adolescent with diabetes (e-DiCARE): results from April 2006 to June 2007. Med J Malays 63(Suppl C):37–40

Gimsing P, Holmström MO, Klausen TW, Andersen NF, Gregersen H, Pedersen RS et al (2016) The Danish national multiple myeloma registry. Clin Epidemiol 8:583

Green A, Sortsø C, Jensen PB, Emneus M (2015) Validation of the DANISH NATIONAL DIABETES REGISTER. Clin Epidemiol 7:5

Guariguata L, Whiting DR, Hambleton I, Beagley J, Linnenkamp U, Shaw JE (2014) Global estimates of diabetes prevalence for 2013 and projections for 2035. Diabetes Res Clin Pract 103(2):137–149

Gudbjörnsdottir S, Cederholm J, Nilsson PM, Eliasson B (2003) The National Diabetes Register in Sweden: an implementation of the St. Vincent declaration for quality improvement in diabetes care. Diabetes Care 26 (4):1270–1276

Han W, Sharman R, Heider A, Maloney N, Yang M, Singh R (2015) Impact of electronic diabetes registry 'Meaningful Use'on quality of care and hospital utilization. J Am Med Inform Assoc 23(2):242–247

Hanberger L, Birkebaek N, Bjarnason R, Drivvoll AK, Johansen A, Skrivarhaug T et al (2014) Childhood diabetes in the Nordic countries: a comparison of quality registries. J Diabetes Sci Technol 8(4):738–744

Harris S, Aschner P, Mequanint S, Esler J (2015) Use of diabetes registry data for comparing indices of diabetes management: a comparison of 2 urban sites in Canada and Colombia. Can J Diabetes 39(6):496–501

Jacquez GM, Greiling DA, Kaufmann AM (2005) Design and implementation of a space-time intelligence system for disease surveillance. J Geogr Syst 7(1):7–23

Jankovec Z, Cechurova D, Krcma M, Lacigova S, Zourek M, Rusavy Z (2006) National Register of patients with insulin pump treatment in the Czech Republic. Diabetes Res Clin Pract 74:S135–S1S9

Kern EF, Beischel S, Stalnaker R, Aron DC, Kirsh SR, Watts SA (2008) Building a diabetes registry from the Veterans Health Administration's computerized patient record system. Sage

Knight K, Badamgarav E, Henning JM, Hasselblad V, Gano AD Jr, Ofman JJ et al (2005) A systematic review of diabetes disease management programs. Am J Manag Care 11(4):242–250

Kohn LT (1999) Committee on quality of health care in America, Institute of Medicine. Anyone can make a mistake

Koton S (2007) Group–IIRSG IIRS. Incidence of type 1 diabetes mellitus in the 0-to 17-yr-old Israel population, 1997–2003. Pediatr Diabetes 8(2):60–66

Krysinska K, Sachdev PS, Breitner J, Kivipelto M, Kukull W, Brodaty H (2017) Dementia registries around the globe and their applications: a systematic review. Alzheimers Dement 13(9):1031–1047

Leucuţa DC, Turdeanu C, Cadariu AA (2010) Data sources and information to be collected for a Romanian diabetes register. Appl Med Inform 26(2):101–111

Liu FX, Rutherford P, Smoyer-Tomic K, Prichard S, Laplante S (2015) A global overview of renal registries: a systematic review. BMC Nephrol 16(1):31

Mandavia R, Knight A, Phillips J, Mossialos E, Littlejohns P, Schilder A (2017) What are the essential features of a successful surgical registry? A systematic review. BMJ Open 7(9):e017373

Morrow RW, Fletcher J, Kelly KF, Shea LA, Spence MM, Sullivan JN et al (2013) Improving diabetes outcomes using a web-based registry and interactive education: a multisite collaborative approach. J Contin Educ Health Prof 33(2):136–144

Naves LA, Porto LB, Rosa JWC, Casulari LA, Rosa JWC (2015) Geographical information system (GIS) as a new tool to evaluate epidemiology based on spatial analysis and clinical outcomes in acromegaly. Pituitary 18(1):8–15

Psoter KJ, Rosenfeld M (2013) Opportunities and pitfalls of registry data for clinical research. Paediatr Respir Rev 14(3):141–145

Ramadan AAB, Jackson-Thompson J, Boren SA (2017) Geographic information systems: usability, perception, and preferences of public health professionals. Online J Publ Health Inform 9(2):e191

Ramadas A, Quek K, Chan C, Oldenburg B (2011) Web-based interventions for the management of type 2 diabetes mellitus: a systematic review of recent evidence. Int J Med Inform 80(6):389–405

Rankin J, Best K (2014) Disease registers in England. Paediatr Child Health 24(8):337–342

Richesson RL (2011) Data standards in diabetes patient registries. Sage

Rytkönen MJ (2004) Not all maps are equal: GIS and spatial analysis in epidemiology. Int J Circumpolar Health 63(1):9–24

Schmidt M, Schmidt SAJ, Sandegaard JL, Ehrenstein V, Pedersen L, Sørensen HT (2015) The Danish National Patient Registry: a review of content, data quality, and research potential. Clin Epidemiol 7:449

Seneviratne K, Samaraweera S, Fernando E, Perera S, Hemarathna A (2017) Strengthen cancer surveillance in Sri Lanka by implementing cancer registry informatics to enhance cancer registry data accuracy, completeness, and timeliness. Stud Health Technol Inform 245:1143–1147

Sereday MS, Martí ML, Damiano MM, Moser ME (1994) Establishment of a registry and incidence of IDDM in Avellaneda. Argent Diabetes Care 17(9):1022–1025

Shaltout AA, Qabazard MA, Abdella NA, LaPorte RE, Al Arouj M, Nekhi AB et al (1995) High incidence of childhood-onset IDDM in Kuwait. Diabetes Care 18(7):923–927

Shaw JE, Sicree RA, Zimmet PZ (2010) Global estimates of the prevalence of diabetes for 2010 and 2030. Diabetes Res Clin Pract 87(1):4–14

Sreedharan J (2016) The need to establish local Diabetes Mellitus registries. Nepal J Epidemiol 6(2):551

Thoelke H, Meusel K, Ratzmann K-P (1990) Computer-aided system for diabetes care in Berlin, GDR. Comput Methods Prog Biomed 32(3–4):339–343

Toh MP, Leong HS, Lim BK (2009) Development of a diabetes registry to improve quality of care in the National Healthcare Group in Singapore. Ann Acad Med Singap 38(6):546

Types of Registries (2019) https://training.seer.cancer.gov/registration/types/. Access Date: 13 Jul 2019

Viviani L, Zolin A, Mehta A, Olesen HV (2014) The European Cystic Fibrosis Society Patient Registry: valuable lessons learned on how to sustain a disease registry. Orphanet J Rare Dis 9(1):81

Wallach Kildemoes H, Toft Sørensen H, Hallas J (2011) The Danish national prescription registry. Scand J Publ Health 39(7_suppl):38–41

Wiedemann B, Schober E, Waldhoer T, Koehle J, Flanagan SE, Mackay DJ et al (2010) Incidence of neonatal diabetes in Austria–calculation based on the Austrian Diabetes Register. Pediatr Diabetes 11 (1):18–23

Wild S, Roglic G, Green A, Sicree R, King H (2004) Global prevalence of diabetes: estimates for the year 2000 and projections for 2030. Diabetes Care 27 (5):1047–1053

Wu H, Zhong J, Yu M, Wang H, Gong W, Pan J et al (2017) Incidence and time trends of type 2 diabetes mellitus in youth aged 5–19 years: a population-based registry in Zhejiang, China, 2007 to 2013. BMC Pediatr 17(1):85

Yamamoto-Honda R, Takahashi Y, Yamashita S, Mori Y, Yanai H, Mishima S et al (2014) Constructing the National Center Diabetes Database. Diabetol Int 5 (4):234–243

Yang X, So W, Kong A, Clarke P, Ho C, Lam C et al (2006) End-stage renal disease risk equations for Hong Kong Chinese patients with type 2 diabetes: Hong Kong Diabetes Registry. Diabetologia 49(10):2299

Adv Exp Med Biol - Advances in Internal Medicine (2020) 4: 457–498
https://doi.org/10.1007/5584_2020_518
© Springer Nature Switzerland AG 2020
Published online: 21 April 2020

Diabetes and Genetics: A Relationship Between Genetic Risk Alleles, Clinical Phenotypes and Therapeutic Approaches

Shomoita Sayed and A. H. M. Nurun Nabi

Abstract

Unveiling human genome through successful completion of Human Genome Project and International HapMap Projects with the advent of state of art technologies has shed light on diseases associated genetic determinants. Identification of mutational landscapes such as copy number variation, single nucleotide polymorphisms or variants in different genes and loci have revealed not only genetic risk factors responsible for diseases but also region (s) playing protective roles. Diabetes is a global health concern with two major types – type 1 diabetes (T1D) and type 2 diabetes (T2D). Great progress in understanding the underlying genetic predisposition to T1D and T2D have been made by candidate gene studies, genetic linkage studies, genome wide association studies with substantial number of samples. Genetic information has importance in predicting clinical outcomes. In this review, we focus on recent advancement regarding candidate gene(s) associated with these two traits along with their clinical parameters as well as therapeutic approaches perceived. Understanding genetic architecture of these disease traits relating clinical phenotypes would certainly facilitate population

S. Sayed and A. H. M. N. Nabi (✉)
Department of Biochemistry and Molecular Biology, University of Dhaka, Dhaka, Bangladesh
e-mail: nabi@du.ac.bd

stratification in diagnosing and treating T1D/T2D considering the doses and toxicity of specific drugs.

Keywords

Clinical phenotypes · Gene polymorphisms · Genetics · Genome-wide association study · Precision medicine · Type 1 diabetes · Type 2 diabetes

1 Introduction

Diabetes is one of the biggest public health concerns and leading cause of death world-wide. It is a metabolic disorder caused by insulin deficiency and/or insulin resistance characterized by hyperglycemia, polyphagia, polydipsia and weight loss.

According to the World Health Organization (WHO), prevalence of diabetes in world population is increasing gradually since 1980 and in 2014 the number of people affected by this silent killer was about 422 million while in 2016, 1.6 million people died due to direct cause of diabetes and 500 million prevalent cases of T2D was reported in 2018 (Roglic 2016; Kaiser et al. 2018). Type 1 Diabetes (T1D), a long-standing multicomponent disease with approximate 10% of total diabetic patient burden, stems from selective autoimmune destruction of pancreatic β-cells, resulting in insufficient insulin production

(Onkamo et al. 1999). The β-cell destruction is the amalgamation of innate and adaptive immune entities including CD4⁺ T cells and CD8⁺ T cells, natural killer (NK) cells, B lymphocytes, macrophages, dendritic cells (DC), and antigen-presenting cells (APCs) (Bakay et al. 2013). Evidences confirmed more aggressive insulitis (inflammation in islet cell) in children diagnosed under 7 years of age compared to the children aged 13 and above with reduced β-cells while children aged between 7 and 12 years showed mixture of both phenotypes (Leete et al. 2016). Studies in monozygotic twins regarding development of T1D revealed 50% risk association contributed by HLA genes whereas remaining 50% is the combined outcome of the environmental exposure and epigenetic modifications (Jerram and Leslie 2017; Redondo et al. 2008). Determinants such as diet, vitamin D intake, infections, and gut microbiota contribute to T1D progress via various mechanisms among which modification of gene expression through epigenetic mechanisms resulting in aberrant immune response and islet autoimmunity is a mention-worthy one (Jacobsen and Schatz 2016; Zullo et al. 2017). Moderate methylation variability in genes (MHC, BACH2, INS-IGF2, and CLEC16A) associated strictly with T1D development has been found between discordant monozygotic twins (Elboudwarej et al. 2016). Another epigenetic modification, acetylation especially in lysine 9 of the H3 histone protein (H3K9Ac) in the major T1D susceptibility genes HLA-DRB1 and HLA-DQB1 has been found to be significantly escalated in patients with T1D compared to controls (Miao et al. 2012). Type 2 diabetes, the predominant form of diabetes with 90% of total patient burden is a divergent disorder originating from a blend of genetic loci involved in impaired insulin secretion, insulin resistance and environmental factors (obesity, over eating, lack of exercise, stress and aging) to varying degrees (Kaku 2010; Holt 2004).

Two fundamental defects; impaired insulin secretion in skeletal muscle, liver, and adipocytes, and impaired β-cell function mainly manifests T2D although the corresponding contribution of insulin resistance versus pancreatic β-cell dysfunction to the pathogenesis is still disputed

(Defronzo and Banting Lecture 2009). Aging, physical activity, diets and obesity- all these factors contribute in the epigenetic alterations involved in T2D development (Ling and Rönn 2019).

Simultaneous analysis of DNA methylation in human adipose tissue, liver, and skeletal muscle from subjects with T2D and non-diabetic controls using advanced Illumina arrays discovered numerous CpG sites with altered DNA methylation in target tissues from patients with T2D, which strengthened the input of epigenetics in the pathogenesis of diabetes. (Abderrahmani et al. 2018; Kirchner et al. 2016; Nilsson et al. 2014, 2015; Nitert et al. 2012). Monozygotic twins discordant for T2D; an ideal model to investigate epigenetic contribution also demonstrate such change in methylation in adipose tissues with a noticeable heritable pattern (Ribel-Madsen et al. 2012).

Elucidation of distinctive features of an emerging new entity; lean diabetes (patients with BMI in 17–25) from obesity related T2D and other classical forms of diabetes is also needed to explain the role of defective insulin secretory capacity as opposed to peripheral insulin resistance as the causative factor (George et al. 2015). Pathological attributes of these patients include history of childhood malnutrition with low protein diet resulting in depleted β-cell mass and insulinopenia, fasting C-peptide levels median of type 1 and type 2 diabetics, relatively early age of onset and absence of ketosis on withdrawal of insulin (Huh et al. 1992; Abu-Bakare et al. 1986).

Though it was believed that diabetes is a disease of people from affluent societies and developed countries but people from developing as well as underdeveloped regions are also equally affected. The prevalence of T2D is found to be exponentially greater in developing than in developed countries (69 versus 20%) (Shaw and Sicree 2010). It was estimated that developing country like Bangladesh was the habitat of third largest number of people with diabetes just after China and India (International Diabetes Federation (IDF) 2012).

Using Human Genome Project and the International HapMap project as the cornerstone, GWAS started to thrive in the 2000s assisted by advancement in high-throughput and affordable genotyping technology, investigative tools with abilities of data mining and interpreting large databases (Lander et al. 2001; Venter et al. 2001; The International HapMap Consortium 2003; International HapMap Consortium et al. 2007)

Assaying single nucleotide polymorphisms (SNPs), the most common type of mutation arising almost each 300 base pair emerged as a prevailing way to study the association of genetic variation and disease. Linkage disequilibrium (LD) used for quantifying inheritance pattern can determine the possibility of SNPs staying together in a haplotype during meiosis. Two SNPs in strong LD will be inherited together more frequently than two SNPs in weak LD. So, a minimal set of single nucleotide polymorphisms (SNPs) referred as 'tag' SNPs can be used in genotyping arrays of association analyses to comprehend the majority of remaining common genomic variation (Billings and Florez 2010). The first full-scale GWAS for T1D done on European Ancestry found 392 SNPs; associated strongly to MHCs as well as to KIAA0350 gene (Hakonarson et al. 2007; Wellcome Trust Case Control Consortium 2007).

Investigation led by WTCCC (Wellcome Trust Case Control Consortium) consisting of seven common complex diseases including T1D by genotyping 2000 cases and 3000 controls with ~500,000 SNPs using the Affymetrix GeneChip promulgated a number of novel T1D loci, including the KIAA0350 genomic region (Wellcome Trust Case Control Consortium 2007). This was further validated from a replication study conducted by Todd et al. (2007). 24 SNPs at 23 separate loci within 12q13 region with a combined p-value of 9.13×10^{-10} harbors several genes, including ERBB3, RAB5B, SUOX, RPS26, and CDK2 (Hakonarson et al. 2008). Another significant association between SNP at the UBASH3A locus on 21q22.3 and T1D was discovered by Concannon et al. using SNP genotyping data from a linkage study of affected sib pairs in nearly 2500 multiplex families (Concannon et al. 2008).

With respect to T2D, the first GWAS carried out in a French discovery cohort composed of 661 cases of T2D and 614 non-diabetic controls identified novel and reproducible association signals at SLC30A8 and HHEX, and validated the previously identified association at TCF7L2 (Sladek et al. 2007). Investigators from the Icelandic company deCODE and their collaborators confirmed the association of loci SLC30A8 and HHEX with T2D and identified an additional signal in CDKAL1 (Steinthorsdottir et al. 2007). GWAS have led to the discovery of 38 SNPs associated with T2D (Egefjord et al. 2009). Meta-analysis even in smaller cohorts confirmed the association of T2D with loci such as IRS1, MTNR1B and KCNQ1 (Egefjord et al. 2009; Prokopenko et al. 2009; Lyssenko et al. 2009a; Yasuda et al. 2008). Although GWAS has been successful in discovering risk associated loci in both major types of diabetes; up to 20% of common variations of human genome can be lost in the genotyping assays (Pe'er et al. 2006). Inability of existing genotyping assays to apprehend an even greater percentage of common variation in non-European ancestral populations raises even more doubts regarding the efficacy of GWAS (Gabriel et al. 2002). Discovery of majority of known T2D-associated variants in non-coding intronic or intergenic regions can be construed as proxies in LD with the true causal variants. Fine-mapping and resequencing these regions should identify those causal variants, which should have stronger associations with T2D than their proxies and thus improve genetic prediction (Vassy and Meigs 2012). 'Precision medicine', a rather new concept of personalized treatment based on individual genetic make-up use GWAS as its founding-stone. In this review, we would like to shed light on the genetic aspects in the pathogenesis of both T1D and T2D and would also mention current developments in the realms of precision medicine for diabetes.

2 Type 1 Diabetes and Genetics

Type 1 diabetes is also known as autoimmune diabetes, which accounts for only 5–10% of all cases with diabetes and was previously known as insulin-dependent diabetes, type I diabetes, or juvenile-onset diabetes (American Diabetes

Association 2009). Destruction of pancreatic β-cells leads to insulin deficiency which is the classical feature of T1D. However, as it is an autoimmune disorder thus, measuring the levels of islet cell autoantibodies, autoantibodies to insulin, autoantibodies to glutamic acid decarboxylase, and autoantibodies to the tyrosine phosphatases IA-2 and IA-2β are also other diagnostic criteria to validate or confirm T1D (American Diabetes Association 2009). As a result, the genetic predisposition of this immune mediated diabetes has been largely determined by the polymorphisms or mutations present within multiple genes. Previously, influence of genes outside the major histocompatibility complex (MHC) was ruled out and MHC was thought to be the principal genetic determinant of T1D. Major contribution was attributed to the HLA-DR/DQ along with HLA-DRB1∗04 subtypes and HLA-DP alleles.

Significant association of MHC class I genes such as B∗5701, B∗3906, A∗2402, A∗0201, B∗1801 with T2D was also evident (Noble et al. 2010). Brown et al. hypothesized that a substitution of an amino acid residue at DQα-52 and DQβ-57 positions leads to conformational changes of the antigen-binding site resulting in a modification of the affinity of the class II molecule for the "diabetogenic" peptide(s) (Brown et al. 1993; Morran et al. 2015). Much is still undiscovered regarding class I MHC molecules in T1D pathogenesis.

However, non-MHC genetic factors have also emerged as pivotal causes of risk for developing T1D. Information presented in Table 1 represents that in fact several genes outside MHC area have shown their strong and rational association with T1D. Even, scanning the polymorphic variants available within MHC and MHC-linked loci have contributed to the risk in addition to the HLA-DQ/DR (Lie et al. 1999; Zavattari et al. 2001; Lie and Thorsby 2005). A report from Aly et al. (2008) not only showed involvement of classical HLA loci but also they convincingly demonstrated association of additional locus present in the vicinity of UBD/MASIL genes with T1D, which indicates that T1D-associated polymorphisms/mutations present in chromosome 6 covers almost twice the length as it was thought before. Analyses using linkage disequilibrium has been employed to identify genetic determinants of T1D. Genome-wide association studies have recognized moderate contribution (odds ratio < 2.0) of 60 potent loci associated with the risk of developing T1D (Buniello et al. 2019; Nyaga et al. 2018); although many of the risk-associated SNPs are remained to be defined. Several candidate genes recognized by GWAS are involved in immune system while many of them are expressed in pancreatic β cells. Further, the increasing number of SNPs found to be independent risk factors or tagged SNPs in several risk alleles has led construction of algorithms to calculate T1D genetic risk scores that will ultimately able to predict who is at risk for developing T1D (Eizirik et al. 2012; Brorsson et al. 2016). GWAS studies have uncovered new risk susceptible loci linked to T1D such as CLEC16A (Zoledziewska et al. 2009), PRKCQ (Cooper et al. 2008), KIF5A (Fung et al. 2009), IFIH1 (Liu et al. 2009). Meta-analysis covering enormous data also has enlightened with the knowledge of new loci such as FOSL2, LMO7, EFR3B (Bradfield et al. 2011). Further non-MHC genetics factors showed strong linkage disequilibrium with the MHC alleles implicating their etiological contributions.

The combined effects of candidate loci of T1D contribute to the pathogenesis of T1D in different pathways. For example, a T1D candidate loci, IFIH1 contributes to β-cell destruction by expansion of local production of inflammatory cytokines and chemokines which exacerbate islet immune cell infiltration (Nejentsev et al. 2009; Downes et al. 2010). Protein product of another risk associated loci, TNFAIP3 has anti-apoptotic effects in β-cells which is augmented by Akt survival pathway (Fukaya et al. 2016). T1D risk allele of a SNP (rs2327832) located upstream of TNFAIP3 is associated with poorer residual β-cell function 1 year after T1D diagnosis providing clinical evidence for a role of TNFAIP3 in T1D development (Fukaya et al. 2016). GLIS3; a candidate gene for both T1D and T2D (Barrett et al. 2009; Cho et al. 2011) encodes a transcription factor important for pancreas development and β-cell generation (Senee et al. 2006; Kang et al. 2009) and regulates several key islet

Table 1 Genes and single nucleotide polymorphisms associated with the risk of T1D and/or playing protective role against developing T1D

Candidate genes	Possible associated SNPs	No of Participants	Mean age years ± SD or age range (years)	Study method	Major function	Associated with	Location	References
KIAA0350	rs2903692 rs725613 rs17673553	1146	The median age at onset is 8 year with lower and upper quartiles at 4.6 year and 11 year	GWAS	The gene product of which is predicted to be a sugar-binding, C-type lectin, exclusively expressed in immune cells including dendritic cells, B lymphocytes and natural killer (NK) cells, all of which are pivotal in the pathogenesis of T1D	T1D	16p13.13	Hakonarson et al. (2007)
		2350	Diagnosed below 35 year					
		103	Diagnosed under the age of 18 year					
UBASH3A	rs11203203	2496	Onset at >35 years of age	Illumina human Linkage-12 genotyping Beadchip	This gene encodes one of two family members belonging to the T-cell ubiquitin ligand (TULA) family. Interferes with CBL-mediated down-regulation and degradation of receptor-type tyrosine kinases. Promotes accumulation of activated target receptors, such as T-cell receptors, EGFR and PDGFRB, on the cell surface.	T1D	21q22.3	Concannon et al. (2008)
HLA-DQB1, HLA-DRB1 locus	9.452 Mb of the extended HLA region	257 Sardinian T1DM families	8.15 ± 4.15	Microsatellite markers	An MHC class II protein, displays foreign peptides to immune system to trigger immune response	T1D	6p21.3	Zavattari et al. (2001)
		128 Sardinian T1DM families	8.6 ± 4.0					

(continued)

Table 1 (continued)

Candidate genes	Possible associated SNPs	No of Participants	Mean age years ± SD or age range (years)	Study method	Major function	Associated with	Location	References
CLEC16A	rs725613	584 T1D family trios	7.8 ± 4.1	GWAS	This gene encodes a member of the C-type lectin domain containing family. Regulator of mitophagy through the upstream regulation of the RNF41/NRDP1-PRKN pathway.	T1D	16p13.13	Zoledziewska et al. (2009)
		453 T1D case subjects	8.7 ± 4.4					
BACH2	rs11755527	3561	All were less than 17 years of age at diagnosis	Meta-analysis	Transcriptional regulator that acts as repressor or activator	T1D	6q15	Cooper et al. (2008)
CTSH	rs3825932	3561	All were less than 17 years of age at diagnosis	Meta-analysis	The protein encoded by this gene is a lysosomal cysteine proteinase important in the overall degradation of lysosomal proteins, can act both as an aminopeptidase and as an endopeptidase.	T1D	15q25.1	Cooper et al. (2008)
C1QTNF6	rs229541	3561	All were less than 17 years of age at diagnosis	Meta-analysis	Serve as an endogenous complement regulator, high expression activates the Akt signaling pathway, increase tumor angiogenesis and reduce the necrosis of HepG2 cells	T1D	22q12.3	Cooper et al. (2008)
PRKCQ	rs947474	3561	All were less than 17 years of age at diagnosis	Meta-analysis	The protein is a Ser/Thr Kinase with non-redundant functions in T-cell receptor (TCR) signaling, including T-cells activation, proliferation, differentiation and survival.	T1D	10p15.1	Cooper et al. (2008)

KIF5A	rs1678542	8010		GWAS	Members of this family are part of a multisubunit complex that functions as a microtubule motor in intracellular organelle transport.	T1D	12q13.3	Fung et al. (2009)
TNFAIP3	rs6920220 rs10499194	8010		GWAS	TNF-alpha-induced protein 3 (TNFAIP3), also known as A20, is a cytoplasmic protein which mainly exerts anti-inflammatory function by inhibiting NF-κB activation	T1D	6q23	Fung et al. (2009)
IFIH1	rs3747517 rs1990760	2046	At or before or after the age of 17 years	GWAS	Encoded receptor acts as a cytoplasmic sensor of viral nucleic acids and plays a major role in sensing viral infection and in the activation of a cascade of antiviral responses.	T1D	2q24.2	Liu et al. (2009)
LMO7	rs539514	1120 affected trios		Meta-analysis	An emerin-binding protein that regulates the transcription of emerin and many other muscle-relevant genes	T1D	13q22	Bradfield et al. (2011)
FOSL2	rs6547853	1120 affected trios		Meta-analysis	Encoded leucine zipper protein regulates cell proliferation, differentiation, and transformation.	T1D	2p23.2	Bradfield et al. (2011)
EFR3B	rs478222	1120 affected trios		Meta-analysis	The protein exhibits GTPase activity and plays a role in cell cycle progression	T1D	2p23.3	Bradfield et al. (2011)

(continued)

Table 1 (continued)

Candidate genes	Possible associated SNPs	No of Participants	Mean age years ± SD or age range (years)	Study method	Major function	Associated with	Location	References
CTSH	rs3825932	257	9.1 ± 3.7	qPCR	A ubiquitously expressed lysosomal cysteine protease important in the overall degradation of lysosomal proteins. It is composed of a dimer of disulfide-linked heavy and light chains, both produced from a single protein precursor. CTSH overexpression also up-regulated Ins2 expression and increased insulin secretion	T1D	15q25.1	Floyel et al. (2014)
STAT4	rs7574865 rs8179673 rs10181656	389	7.45 ± 4.05 at diagnosis	TaqMan genotyping	STAT4 gene encodes a transcription factor signal transducer and activator of transcription 4, form homo- or heterodimers through phosphorylation and then, translocate to the cell nucleus where they act as transcription activators.	T1D	2q32.2	Lee et al. (2008)
FoxP3	rs60191426	316	Mean onset age 35.4 years	PCR and gene scan software	Encoded transcription factor appears to be a regulator of the regulatory pathway in the development and function of regulatory T cells.	T1D	Xp11.13	Iwase et al. (2009)

HLA-C	HLA region	283	7.84 = 1.19 to 15.10 ± 1.72, age of onset >0 to >12 years,	PCR-SSOP	Class-I MHC receptors, displays foreign peptides to immune system to trigger immune response	T1D	6p21.33	Valdes et al. (2005)
HLA-A	HLA region	283	7.84 ± 1.19 to 15.10 ± 1.72, age of onset >0 to >12 years,	PCR-SSOP	Class-I MHC receptors, displays foreign peptides to immune system to trigger immune response	T1D	6p21.3	Valdes et al. (2005), Noble et al. (2002)
DPB1*0301 DPB1*0202	HLA region	283	Age of onset >0 to >12 years	PCR-SSOP	Belongs to HLA class II family; presents antigens to T cells	Protective	6p21	Noble et al. (2002)
DPB1*0101DPB1*0402	HLA region	283	Age of onset >0 to >12 years	PCR-SSOP	Belongs to HLA class II family; presents antigens to T cells	Protective	6p21	Noble et al. (2002)
HLA-B	HLA region	4126	Age-at-diagnosis of T1D 0 to >10 years	Dynal RELI SSO assays and Dynal AllSet SSP assay.	Class-I MHC receptors, displays foreign peptides to immune system to trigger immune response	T1D	6p21.33	Nejentsev et al. (2007)
HLA-DQB1 and HLA-DRB1	HLA region	4126	Age-at-diagnosis of T1D 0 to >10 years		Class-II MHC receptors, displays foreign peptides to immune system to trigger immune response	T1D	6p21.33	Nejentsev et al. (2007)
DQB1*0302	HLA region	51	Age-at-diagnosis of T1D <15 years	Sequencing	Belongs to HLA class II family; presents antigens to T cells	T1D	6p21.32	Rajalingan et al. (2004)
HLA-DPB1	HLA region	91 + 84 / 20 / 71	Age 15 years at diagnosis	PCR-SSOP, PCR-RFLP, PCR with primer probe	In multiple ethnic groups but not in Japanese and Sudanese	T1D	6p21	Cruz et al. (2004), Yamagata et al. (1991), Magzoub et al. (1992)
HLA-DQA1	rs7990	151	Age-at-diagnosis of T1D <39 years	PCR followed by sequencing	An MHC class II protein, displays foreign peptides to immune system to trigger immune response	T1D	6p21.32	Raha et al. (2013)

(continued)

Table 1 (continued)

Candidate genes	Possible associated SNPs	No of Participants	Mean age years ± SD or age range (years)	Study method	Major function	Associated with	Location	References
DQA1*0301DQB1*0302 DQA1*0501DQB1*0201	HLA region	645 families	Average age of affected siblings varied from 8.33 ± 5.93 to 12.28 ± 4.99	PCR-based sequence-specific oligonucleotide probes	An MHC class II protein, displays foreign peptides to immune system to trigger immune response	T1D	6p21.32	Erlich et al. (2008)
HLA-DR3 HLA-DR4	Haplotypes	302	Age at onset of T1D patients range 1–55 years (median onset 15 years old) < /=14 years, 15–25 years >/ =26 years	PCR-SSOP	Presents foreign antigen to T-helper cells to elicit immune response	T1D	6p21.31	Noble and Valdes (2011), Noble et al. (1996)
		166 families						
INS	rs1004446 rs7111341	8677	Median follow-up age 57 months	Direct sequencing, oligonucleotide probe hybridization	This gene encodes insulin	IA autoantibodies	11p15.5	Törn et al. (2015)
PTPN22	rs2476601	8677	Median follow-up age 57 months	Direct sequencing, oligonucleotide probe hybridization	Powerful negative regulator of T-cell activation	T1D and IA autoantibodies	1p13.2	Törn et al. (2015), Prasad et al. (2011)
		1434	Onset of T1D before and after 17 years of age					
ERBB3	rs2292239	8677	Median follow-up age 57 months	Direct sequencing, oligonucleotide probe hybridization	This gene encodes a member of the epidermal growth factor receptor (EGFR) family of receptor tyrosine kinases. It does form heterodimers with other EGF receptor family members which do have kinase activity.	IA autoantibodies	12q13.2	Törn et al. (2015)

Gene	SNP	Sample size	Follow-up/age	Method	Function	Association	Locus	Reference
SH2B3	rs3184504	8677	Median follow-up age 57 months	Direct sequencing, oligonucleotide probe hybridization	This gene encodes a member of the SH2B adaptor family of proteins, which are involved in a range of signaling activities by growth factor and cytokine receptors.	IA autoantibodies	12q24.12	Töm et al. (2015)
C14orf64	rs4900384	8677	Median follow-up age 57 months	Direct sequencing, oligonucleotide probe hybridization	Is an RNA gene, and is affiliated with the lncRNA class.	Protective	14q32.2	Töm et al. (2015)
CTSH	rs3825932	8677	Median follow-up age 57 months	Direct sequencing, oligonucleotide probe hybridization	The protein encoded by this gene is a lysosomal cysteine proteinase important in the overall degradation of lysosomal proteins.	Protective	15q25.1	Töm et al. (2015)
CCR5	rs11711054	8677	Median follow-up age 57 months	Direct sequencing, oligonucleotide probe hybridization	The protein regulates G protein-coupled receptor signaling cascades and also inhibits B cell chemotaxis toward CXCL12	T1D	3p21.31	Töm et al. (2015)
CTLA4	rs1427676 rs231727	2298 T1D nuclear families, 5003 affected		Illumina GoldenGate and Sequenom iPlex platforms	Cytotoxic T lymphocyte antigen-4, immune regulation, transmits inhibitory signals to attenuate T-cell activation by competing for the B7 ligands	T1D	2q33	Qu et al. (2009)
PTPN22	rs2476601	316 130	17.3 ± 10.0 11.47 ± 3.72	Taqman genotyping, PCR-RFLP	Powerful negative regulator of T-cell activation,	T1D	1p13.3-p13.1	Santiago et al. (2007), Liu et al. (2015)
PTPN2	rs2542151	419	18.6 ± 11.1	Meta analysis	The protein plays a direct role in B-cell destruction,	T1D	18p11.21	Sharp et al. (2015), Espino-paisan et al. (2011)

(continued)

Table 1 (continued)

Candidate genes	Possible associated SNPs	No of Participants	Mean age years ± SD or age range (years)	Study method	Major function	Associated with	Location	References
IL2RA	rs11594656 rs2104286 rs41295061	10,572		Meta-analysis	Homodimeric alpha chains (IL2RA) result in low-affinity receptor	T1D	10p15.1	Tang et al. (2015)
IL2	rs4505848	6800	>17 years at diagnosis	Taqman genotyping; GWAS meta-analysis	Encoded secretory cytokine is important for the proliferation of T and B lymphocytes; Stimulates the growth and function of T cells, possesses anti-angiogenic activity	T1D	4q27	Todd et al. (2007)
IL7R	rs1445898 rs6897932	301	17.3 ± 10.0	Taqman genotyping	Encoded protein is a receptor for interleukin 7 (IL7).	Protective against T1D	5p13.2	Santiago et al. (2008)
ERBB3	rs2292239	364		Genotyping SNaPshot multiple single-base extension (SBE)	Tyrosine-protein kinase is encoded that plays an essential role as cell surface receptor for neuregulins.	T1D	12q13.2	Sun et al. (2016)
C12orf30 Genomic region	rs17696736	154	22.0 ± 14.3	SNP genotyping assay	NAA25 (N(Alpha)-Acetyltransferase 25, NatB Auxiliary Subunit) is a protein coding gene that non-catalytic subunit of the NatB complex which catalyzes acetylation of the N-terminal methionine residues of peptides beginning with Met-Asp-Glu. May play a role in normal cell-cycle progression.	T1D	12q24.13	Douroudis et al. (2010)

	rs				Function	Disease	Locus	Reference
CD226	rs763361	532	23.9 ± 12.8	PCR-RFLP, Taqman genotyping	The protein is a glycoprotein involved in intercellular adhesion, lymphocyte signaling, cytotoxicity and lymphokine secretion mediated by cytotoxic T-lymphocyte (CTL) and NK cell.	T1D	18q22.2	(El-ella et al. 2018), Mattana e' al. (2014)
IL18RAP	rs917997	8064	7.5 years at diagnosis (range 0.5–16 years)	SNP genotyping	The protein encoded is an accessory subunit of the heterodimeric receptor for interleukin 18 (IL18), a proinflammatory cytokine involved in inducing cell-mediated immunity, in signal transduction, leading to NF-kappa-B and JNK activation.	T1D with celiac disease	2q12.1	Smyth et al. (2008)
EDG7 or LPAR3	rs1983853	563	<18 years at diagnosis	NGS Illumina	Encoded protein functions as a cellular receptor for lysophosphatidic acid and mediates lysophosphatidic acid-evoked calcium mobilization.	T1D	1p22.3	Grant et al (2009)
RASGRP1	rs8035957	563	<18 years at diagnosis	NGS Illumina	*Function* as a diacylglycerol (DAG)-regulated nucleotide exchange factor specifically activating Ras through the exchange of bound GDP for GTP.	T1D	15q14	Grant et al (2009)

(continued)

Table 1 (continued)

Candidate genes	Possible associated SNPs	No of Participants	Mean age years ± SD or age range (years)	Study method	Major function	Associated with	Location	References
BACH2	rs3757247	563	<18 years at diagnosis	GWAS	The protein is a transcription regulator protein which can induce apoptosis in response to oxidative stress through repression of the antiapoptotic factor HMOX1	T1D, relevant to autoimmunity	6q15	Grant et al. (2009)
SH2B3	rs3184504	7514	<18 years	Meta-analysis and Illumina 550 K Infinium platform	Encoded adaptor protein is involved in a range of signaling activities by growth factor and cytokine receptors.	T1D	12q24.12	Nikitin et al. (2010)
GLIS3	rs7020673	113	12-85	RFLP	Act as both a repressor and activator of transcription and is specifically involved in the development of pancreatic β cells	T1D	9p24.2	Bell et al. (1984)
IL2RA/CD25	rs1990760	2134		DNA sequencing	Acts as a co-stimulatory molecule for T cell activation and proliferation.	T1D	10p15.1	Vella et al. (2005)
SKAP2	rs7804356	1434	Onset of T1D before and after 17 years of age	GWAS	An adaptor protein with an essential role in the Src signaling pathway through which it regulates proper activation of the immune system	Protective	7p15.2	Prasad et al. (2011)
ITPR3	rs2077163	1434	Onset of T1D before and after 17 years of age	Taqman genotyping	The protein is a receptor for inositol 1,4,5-trisphosphate, a second messenger that mediates the release of intracellular calcium.	Protective	6p21.31	Prasad et al. (2011)

Gene	SNP	N	Method	Function	Effect	Location	Reference
RNLS	rs10509540	1434	Taqman genotyping	Encoded protein catalyzes the oxidation of the less abundant 1,2-dihydro-beta-NAD(P) and 1,6-dihydro-beta-NAD(P) to form beta-NAD(P)(+)	Protective	10q23.31	Prasad et al. (2011)
CTRB2	rs7202877	1434	Taqman genotyping	Related pathways are matrix Metalloproteinases and degradation of the extracellular matrix	Risk	6q22.3	Prasad et al. (2011)
SIRPG	rs2281808	1434	Taqman genotyping	Probable immunoglobulin-like cell surface receptor. On binding with CD47, mediates cell-cell adhesion.	Protective	20p13	Prasad et al. (2011)
PKD2	rs425105	1434	Taqman genotyping	*Function* through a common signaling pathway that is necessary to maintain the normal, differentiated state of renal tubule cells	Female: protective; Male: Neutral	19q13	Prasad et al. (2011)
HORMAD2	rs5753037	989	Illumina Infinium™ HumanHap550 array	Required for efficient build-up of ATR activity on unsynapsed chromosome regions,	T1D, protective	22q12.2	Wang et al (2010)
IL27	rs4788084	989	Illumina Infinium™ HumanHap550 array	IL27 plays a role in the innate as well as the adaptive immunity	T1D	16p12.3	Wang et al (2010)
UMOD	rs12444268	989	Illumina Infinium™ HumanHap550 array	UMOD gene provides instructions for making a protein called uromodulin. This protein is produced by the kidneys and then excreted from the body in urine	T1D, protective	16p12.3	Wang et al. (2010)

(continued)

Onset of T1D before and after 17 years of age (applies to RNLS, CTRB2, SIRPG, PKD2 rows)

Table 1 (continued)

Candidate genes	Possible associated SNPs	No of Participants	Mean age years ± SD or age range (years)	Study method	Major function	Associated with	Location	References
GSDMB	rs2290400	3109	Diagnosed at <7 years	Genome-wide association study and meta-analysis	Play a role in regulation of epithelial proliferation and has been suggested to act as a tumor suppressor and pyroptotic protein	Affect risk of type 1 diabetes.	17q12	Teo et al. (2007)
		3754	Diagnosed at 7–13 years					
		1724	Diagnosed at ≥13 years					
CCR7	rs7221109	1016	Subjects <12 years Subjects ≥12 years	Illumina Immunochip	C-C chemokine receptor type 7 (CCR7) has been shown to be involved in the recruitment of T cells into inflamed islets.	rs7221109 (CCR7) became significant in the overall model with glucose	17q21.2	Ramesh et al. (2016)
DLK1	rs7111341	10,747		GWAS meta-analysis	The protein is involved in the differentiation of several cell types including adipocytes.	T1D	14q32.2	Nikitin et al. (2010), Wallace et al. (2010)
TYK2	rs2304256	10,747		GWAS meta-analysis	The protein is a member of Janus kinases (JAKs) protein families which can promulgate cytokine signals by phosphorylating receptor subunits	T1D	19p13.2	Wallace et al. (2010)
HTRA1	rs6295	3078	Some were 15 and 34 years while rest were less than 18 years of age at the time of diagnosis	Human660-Quad chip	A protease enzyme which helps break down many other kinds of proteins in the extracellular matrix and also attaches (binds) to proteins in the transforming growth factor-beta (TGF-β) family to slows down their ability to send chemical signal	T1D	10q26.13	Asad et al. (2012)

CUX2	rs1265564		MiniMac for genotype imputation	This is a transcription factor involved in the control of neuronal proliferation and differentiation in the brain.	T1D	12q24	Huang et al. (2012)
NEUROD1	rs1801262	1213	Meta-analysis	This heterodimer protein may be essential for the morphogenesis or differentiation of pancreatic β cells.	T1D	2q32	Kavvoura and Ioannidis (2005)

PCR-SSOP polymerase chain reaction–sequence-specific oligonucleotide probe, *GWAS* Genome-wide association study, *NGS* Next generation sequencing, *RFLP* Restriction Fragment Length Polymorphism

Data are not available for empty cells

transcription factors as well as insulin gene transcription directly and via PDX1, MAFA, and NEUROD1 (Kang et al. 2009; Yang et al. 2009, 2013). Functional studies involving PTPN2 knockdown convey protective and anti-apoptotic role of PTPN2 in β-cells (Colli et al. 2010; Moore et al. 2009; Santin et al. 2011). Risk allele intronic SNP (rs1893217) in PTPN2 can be hypothesized to confer disease susceptibility by sensitizing the β-cells to both immune- and virus-mediated apoptosis. The T1D candidate loci Cathepsin H (CTSH) is suppressed by cytokines in islets and β-cells and also is a positive regulator of insulin transcription (Floyel et al. 2014).

SNP rs3825932 located in intron 1 of CTSH affects the expression level of CTSH in a genotype-dependent manner in multiple tissues (GTEx Portal 2016) and also can be used as critical determinants of β-cell function in children with recent-onset T1D and in healthy adults. STAT4 involved in signaling pathway of cytokines (IL-12, γIFN, and IL-23) partakes in the IL-12 induced differentiation of T cells into the Th1 pathway (Mathur et al. 2007). This loci seemed to be involved in the early onset of autoimmune T1D in Asian population (Lee et al. 2008). Foxp3/Scurfin, plays a dominant role in the development and maintenance of Tregs whose dysfunction manifests in T1D (O'Garra and Vieira 2004; Hori et al. 2003). The (GT)n microsatellite in the FOXP3 gene located on chromosome X seems to be more associated with adult onset of T1D, especially in female in a Japanese study (Iwase et al. 2009).

Apart from impacting immunologically, T1D candidate genes also dominate in islet microenvironment to regulate β-cell apoptosis and immune interactions (Wallet et al. 2017). According to a combination of recent transcriptomic, epigenomic and proteomic studies, β-cell dysfunction is mediated by inflammatory response (IFN-γ and IL-1β) via transcription factor recruitment, DNA looping and chromatin acetylation (Ramos-Rodríguez et al. 2019). Knockdown of candidate gene GLIS3 in human islets resulted in β-cells sensitization to IFN-γ and IL-1β induced death

(Nogueira et al. 2013). This phenomenon is mediated by a depletion in alternative splicing factor SRp55 which negatively regulates apoptosis and endoplasmic reticulum (ER) stress (Juan-Mateu et al. 2018). Thus, crosstalk between T1D candidate genes and splicing regulators may result in increased β-cell susceptibility to immune-mediated as well as endogenous stress related dysfunction.

Based on variability of ethnicity, these candidate genes exert diverse roles in different populations. FOXP3 variant (Zavatari et al. 2004) isn't associated with T1D in the Sardinian population but is associated with adult-onset of T1D in Japanese population Type 2 diabetes. C1858T polymorphism in PTPN22 also is observed to show association with T1DM in a Caucasian population (Bottini et al. 2004) but not in Greek (Giza et al. 2013) or Chinese ethnicity (Taniyama et al. 2010).

3 Type 2 Diabetes and Genetics

Obese individuals with very little physical activity develop insulin resistant, which ultimately leads to T2D. However, this model does not fit in all cases of T2D. Studies demonstrated that majority of the obese individuals do not develop diabetes as their pancreatic β cells compensated the extra insulin demand by adaptive changes (increasing the β cell sizes and numbers). Individuals having a deficit in functional β cell mass (either naturally or due to apoptotic loss of β cells in response to excess metabolic demand) fail to compensate the excess insulin demand and develop diabetes. Thus, T2D is mainly a clinical syndrome defined by hyperglycemia resulting from insufficient insulin secretion in the setting of insulin resistance. T2D has been the forefront of diseases for which enormous efforts regarding genetic association studies have been carried out. Though GWAS have substantially taken our knowledge of understanding regarding the genetics of T2D a step forward, a vast area of genetic heritability is yet to be explained. Till to date, Genome-Wide Association Studies (GWAS) have scanned more than 240 loci associated with type 2 diabetes from modest to largest effect sizes (Grant et al. 2006).

Several approaches such as candidate gene studies, genetic linkage studies, genome wide association studies have identified genes linked with T2D. Candidate gene study approach focuses on association between genetic variations and disease within limited number of pre-specified genes of interest through which a number of genes were recognized to be associated with T2D. However, these studies were performed considering limited number of genes and variations within small number of sample size. Due to lack of sensitivity, these studies were difficult to replicate and did not get enough attention as some genes failed to confirm association which were previously known to have impact on insulin activities. Nonetheless, large scale candidate gene studies identified several genes associated with β cell function by increasing insulin resistance (ABCC8, KCNJ11, SLC2A2, HNF4A and INS) which have been reproduced and thus, confirmed in different ethnic groups (Barroso et al. 2003; Daimon et al. 2003; Fisher et al. 2007).

Genomic linkage studies deal with the genetic variants within the genomic regions that are closely located to each other and inherited together. Genes located on chromosomes 4q, 12q, 22q, 6p, 2p, 13q, 3p at marker D3S2406 have been scanned through linkage studies and found to be associated with T2D that suggested potential functional differences of candidate genes among different ethnic groups (Malhotra et al. 2009; Hunt et al. 2005; Das and Elbein 2006).

Genome wide scanning brought a breakthrough in genetic association studies which investigated the most common variants i.e., the single nucleotide polymorphisms for their probable association with diseases. A notable discovery in recent years demonstrated protective role of the protein truncating variants Arg138* and Lys34Serfs*50 (causes loss of function) within SLC30A8 against developing T2D

(Flannick et al. 2014, 2019) by scanning whole exome of >20,000 patients with T2D and > 24,000 healthy individuals. On the other hand, a nonsense variant Arg684Ter within TBC1D4 gene showed very strong association with developing T2D but only in Greenlanders (Moltke et al. 2014). In addition, several SNPs located at several genes have been identified as risk factors associated with glycemic traits and physiological conditions that confer T2D. A 9 year follow-up study in French population demonstrated that 'A' allele carriers with respect to rs560887 in intron 3 of G6PC2 had decreased fasting plasma glucose and lower risk of developing mild hyperglycemia (Bouatia-Naji et al. 2008). SNP rs563694 in ABCB11 which is in high linkage disequilibrium with 11 kb away from rs560887 was associated with the levels of fasting glucose in non-diabetic individuals from Finland and Sardinia (Chen et al. 2008). Further, a meta-analysis conducted by MAGIC or Meta-Analysis of Glucose and Insulin-related traits Consortium using 10 cohorts consists of 40,735 individuals showed strong association of rs10830963 located within melatonin receptor MT2 encoding gene, MTNR1B, with fasting glucose and T2D (Prokopenko et al. 2009), and validated previous reports regarding G6PC2 and GCK (Bouatia-Naji et al. 2008; Chen et al. 2008; Weedon et al. 2006). Variant rs1387153 near MTNR1B was found to be associated with T2D, fasting glucose, β-cell function in French population while with respect to the same SNP, G allele carriers showed decreased insulin secretion in a study by Lyssenko et al. (2009b). These initial data led MAGIC group to perform profound screening in 76,558 additional individuals from 34 additional cohorts and analyses revealed that SNPs in or near ADCY5, MADD, CRY2, ADRA2A, FADS1, PROX1, SLC2A2, GLIS3, C2CD4B genes and one SNP upstream from IGF1 are associated with fasting insulin and HOMA-IR (Dupuis et al. 2010; Saxena et al. 2010). Assessment of more detail physiological parameters revealed that a death domain-containing adaptor protein encoding gene,

MADD, was strongly associated with proinsulin levels. TCF7L2, SLC30A8, GIPR and C2CD4B were associated with higher proinsulin and lower insulin secretion. Variants in or near MTNR1B, FADS1, DGKB, and GCK impaired early insulin secretion as these were only associated with a lower insulinogenic index while SNPs at GCKR and IGF1 were found to influence insulin sensitivity (Ingelsson et al. 2010). Our recent study on Bangladeshi population revealed association of rs3824662 with in GATA3, a transcription factor, is associated with the risk of T2D (Huda et al. 2018). The candidate genes exert diversified roles in pathogenesis of T2D.

Similar to T1D pathogenesis, T2D risk associated loci contribute to T2D pathogenesis in interlinked pathways. DUSP9; a candidate loci for T2D encodes MKP4 protein with protective effect against the development of insulin resistance due to its ability to inactivate ERK or c-Jun N-terminal kinase (JNK), pathway (Emanuelli et al. 2008). rs5945326 of DUSP9 has been risk associated with several Asian ethnicities (Japanese, Chinese and Pakistani) (Fukuda et al. 2012; Rees et al. 2011). Individuals with rs7903146 TT allele of TCF7L2 are more likely to progress from IGT to T2D than those with CC allele (Florez et al. 2006). Cauchi et al. found that the T allele of rs7903146 also has been linked with hyperglycaemia in a French population (Cauchi et al. 2006). Variations in another T2D candidate gene KCNJ11 (encoding Kir 6.2 protein) E23K gene contributes to the decreased sensitivity of the ion channel to ATP resulting in impaired insulin release. Based on ethnicity this variant also seem to exert different functions in different populations. Meta-analysis in UK consisting of 854 T2D and 1182 control subjects showed E23K allele to be associated with diabetes (Gloyn et al. 2003). However, in Danish and Japanese population, no such association were found (Nielsen et al. 2003; Yokoi et al. 2006). Prediabetes often leads to progression into T2D (American Diabetes Association 2016; Forouhi et al. 2007). Variants of GCK which acts as a gatekeeper for glucose-induced activation or deactivation of biological processes is found to be independently associated with prediabetes in the general population (Gloyn et al. 2003; Calmettes et al. 2015; Choi et al. 2017). Table 2 summarizes involvement of single nucleotide polymorphisms within different genes with T2D along with their genomic location.

4 Diabetes, Genetic Variation and Precision Medicine

A fairly new concept of modern medicine known as 'Precision medicine' is establishing itself as a good treatment option in complex diseases like cancer, diabetes, hypertension and cardiovascular diseases. The underlying foundation of this concept is to customize the treatment of an ailing individual specifically based on his/her distinct phenotypic and genetic features along with individual's data on transcriptomics, epigenomics, proteomics and metabolomics as shown in Fig. 1. The risk probability, progression, treatment options and efficacy of medications of a disease will differ from individual to individual; so it is of paramount importance to optimize a therapeutic strategy based on individualistic behavioral and clinical phenotypic features, standard clinical laboratory findings, gene sequences and molecular markers (de Jong et al. 2003; Malandrino and Smith 2011). T1D and T2D employ personalized medicine on the cornerstone of population genetics paved by human genome sequences which facilitates the chances to design massively parallel, chip-based genotyping arrays (Malandrino and Smith 2011), which can exponentially increase the identification of susceptible risk associated loci (Fitipaldi et al. 2018). Success of precision medicine may depend on identification of susceptible loci contributing to the development of diabetes, prevention of onset of diabetes in high-risk individuals, fixing individualized therapies by establishing drugs with most efficacy and least side effects followed by monitoring the response to the management therapy by measuring the diabetes related

Table 2 Genes and single nucleotide polymorphisms associated with the risk of T2D and/or playing protective role against developing T2D

Name	Associated SNPs	No. of participants	Mean age years ± SD or age range (years)	Study method	Function	Associated with	Location	References
HHEX	rs1111875	686	59.9 ± 10.3	Meta-analysis	A transcription factor involved in Wnt signalling	T2D	10q23.33	Sladek et al. (2007), Cai et al. (2011)
	rs7923837	694	45.0 ± 8.4					
		2617	50.4 ± 11.0					
SLC30A8	rs1995222	1107	51.3	Taqman genotyping	Crucial for insulin processing and secretion	T2D	8q24.11	Sladek et al. (2007)
		115	51.6 ± 13.6					Salem et al. (2014)
								Lara-riegos et al. (2015)
CDKAL1	rs35612982	1399	68.5	Agena MassARRAY	Serine/threonine protein kinase, which contributed to the glucose-dependent regulation of the insulin secretion	T2D	6p22.3	Steinthorsdottir et al. (2007)
	rs4712523	508	59.21 ± 11.9					Tian et al. (2019)
	rs4712524							
	rs10946398							Salem et al. (2014)
TBC1D4	rs61736969	1331	42	Array-based genotyping and exome sequencing	The protein encoded by this gene is a Rab-GTPase-activating protein, s thought to play an important role in glucose homeostasis by regulating the insulin-dependent trafficking of the glucose transporter 4 (GLUT4)	T2D	13q22.2	Moltke et al. (2014)
GIPR	rs10423928	29,084		Meta-analysis	Receptor of gastric inhibitory polypeptide (GIP)	Glucose-raising, defects in insulin processing and insulin secretion	19q13.32	Bouatia-Naji et al. (2008)
DGKB	rs2191349	29,084		Meta-analysis	Regulates the intracellular concentration of DAG	Abnormalities in early insulin secretion	7p21.2	Bouatia-Naji et al. (2008)

(continued)

Table 2 (continued)

Name	Associated SNPs	No. of participants	Mean age years ± SD or age range (years)	Study method	Function	Associated with	Location	References
G6PC2	rs560887	9353		Meta-analysis	Catalyzes the hydrolysis of glucose-6-phosphate	T2D	2q31.1	Bouatia-Naji et al. (2008)
	rs16856187							Shi et al. (2017)
ABCB11	rs563694	5088		Fusion GWAS using Illumina	Encoded protein helps in exportation of bile salt	T2D	2q31.1	Chen et al. (2008)
		18,436						Saxena et al. (2010)
MTNR1B	rs10830963	51, 552		Meta- analysis	The receptor of melatonin which inhibits insulin secretion through its effect on the formation of cGMP	T2D	11q14.3	Prokopenko et al. (2009)
								Xia et al. (2012)
GCK	rs1799884	88, 229		Meta- analysis	Key glucose phosphorylation enzyme responsible for the first rate-limiting step in the glycolysis pathway and regulates glucose-stimulated insulin secretion from pancreatic b-cells a	T2D	7p13	Weedon et al. (2006)
								Fu et al. (2013)
MADD	rs7944584	46,186 76,558		Meta- analysis	Via alternative splicing, regulates cell proliferation, survival and	T2D	11p11.2	Dupuis et al. (2010)
FADS1	rs174550	46,186 76,558		Meta- analysis	Enables desaturation of omega-3 and omega-6 PUFAs at the delta-5 position	T2D	11q12.2	Dupuis et al. (2010)
C2CD4B	rs11071657	29,084		Meta-analysis	Probable role in regulation of cell adhesion and architecture	Glucose-raising, defects in insulin processing and insulin secretion	15q22.2	Bouatia-Naji et al. (2008)
								Dupuis et al. (2010)
								Saxena et al. (2010)

Gene	rs number	n	Age	Method	Function	Disease	Location	Reference
KCNJ11	rs5210 rs5215 rs5218	24	59 ± 10	TaqMan genotyping	Encodes an inward-rectifier potassium ion channel	T2D	11p15.1	Koo et al. (2007)
TCF7L2	rs7901695 rs7903146	763 872	68.5 ± 12.2 50	Taqman genotyping	A transcription factor which modulates MYC expression by binding to its promoter in a sequence-specific manner.	T2D	10q25.2–q25.3	Grant et al. (2006), Ma⁓ans et al. (2007)
IGF2BP2	rs11705701 rs4402960	1161 1215 1470	63.4 60.0 59.9 ± 7.9	Taqman genotyping	Regulates translation of IGF2, a growth factor involved in pancreatic development	T2D	3q27.2	Scott et al. (2007) Chistiakov et al. (2012)
PPARG	rs17036314 rs1152003	479	55	PCR-SSCP (single-strand conformation polymorphism) and TaqMan assay	A key transcription factor in the regulation of adipocyte differentiation.	T2D with overweight individuals having IGT, however with increased physical activity	3p25.2	Stumvoll and Haring (2002) Kilpelainen et al. (2008)
KCNJ11	rs5219	21,464		Meta-analysis	Encodes the subunit protein of KATP (Kir 6.2)	K-allele was relevant to T2DM risk in Caucasian and East Asian, in recessive model rs5219 KK genotype was related to T2DM risk in Caucasian, east Asian, south Asian, and north African, in dominant model rs5219 EE genotype had protective role in Caucasian	11p15.1	Li (2013)
ABCC8	rs1799854	73	56.87 ± 1.19	PCR-RFLP	Encodes sulfonylurea receptor 1 (SUR1) protein	T2D	11p15.1	Engwa et al. (2018)

(continued)

Table 2 (continued)

Name	Associated SNPs	No. of participants	Mean age years ± SD or age range (years)	Study method	Function	Associated with	Location	References
CDKN2A/B	rs10811661	879	21–83	Sequenom MassARRAY® SNP genotyping	Encodes tumor suppressor proteins	TG-HDL family history of T2D	9p21.3	Salem et al. (2014)
FTO	rs7195539, rs1121980, rs1558902	879, 760	21–83, 64.6 ± 11.5	Sequenom MassARRAY® SNP genotyping, Taqman genotyping	RNA demethylase that mediates oxidative demethylation of different RNA species	TG-HDL family history of T2D, increased BMI_{max} with T2D	16q12.2	Salem et al. (2014), Kamura et al. (2016)
HNF1B	rs4430796, rs7501939, rs757210	1000, 1226	30–60, 63 ± 11	Primer extension of multiplex products with detection by matrix-assisted laser desorption ionization–time-of-flight mass spectroscopy using a Sequenom platform, sequence-specific primer–PCR analysis followed by fluorescence correlation spectroscopy, Taqman genotyping	A transcription factor that plays a role in kidney and pancreas development	No or modest association to T2D	17q12	Wang et al. (2014a)
NOTCH2	rs10923931	4516	40.9 ± 10.4	TaqMan allelic discrimination	A trans-membrane protein involved in developmental processes	Impairment of pancreatic β-cell function for diabetes risk alleles	1p12	Sharma et al. (2011)
THADA	rs7578597	4516	40.9 ± 10.4	TaqMan allelic discrimination	Aberration is related to benign thyroid adenomas.	Modestly increased risk of type 2 diabetes	2p21	Sharma et al. (2011)
ADAMTS9	rs4607103	4516	40.9 ± 10.4	TaqMan allelic discrimination	Tumor suppressor	Modestly increased risk of type 2 diabetes	3p14.1	Sharma et al. (2011)
TSPAN8/LGR5	rs7961581	4516	40.9 ± 10.4	TaqMan allelic discrimination	A cell surface protein mediating signal trnsduction	Impairment of pancreatic β-cell function for diabetes risk alleles	12q21.1	Sharma et al. (2011)

Gene	SNP	N	Age	Method	Function	Association	Locus	Reference
JAZF1	rs864745	1496	53.6 ± 12.9	Single Base primer extension reactions using the GenomeLab-SNPstream genotyping system, Sequenom MassARRAY system and TaqMan allelic discrimination	Plays a role in lipid metabolism	Impairment of pancreatic β-cell function for diabetes risk alleles	7p15.1	Langberg et al. (2012)
BCL11A	rs10490072 rs10490072	1496	53.6 ± 12.9	Single base primer extension reactions using the GenomeLab-SNPstream genotyping system, Sequenom MassARRAY system and TaqMan allelic discrimination	Plays a role in the switch from γ- to β-globin expression during the fetal to adult erythropoiesis transition	T2D	2p16.1	Langberg et al. (2012)
KCNQ1	rs2237892 rs2237895	70,577		Meta-analysis	Encodes the pore forming subunit of a voltage-gated K+ channel (KvLQT1) that plays a key role for the repolarization of the cardiac action	With elevated type 2 diabetes susceptibility	11p15.5-p15.4	Sun et al. (2012)
IRS1	rs1801278	236 267 422 268	64.6 ± 10.0 58.7 ± 8.2 64.3 ± 9.2 63.0 ± 9.5	Taqman genotyping	Necessary for transmitting signals from the insulin and insulin-like *growth factor-1 (IGF-1)* receptors	Failure to oral antidiabetic drug among patients with T2D	2q36.3	Prudente et a. (2014)
DUSP9	rs5945326	8318	63.6 ± 11.1	GWAS, multiplex- PCR	Negatively regulates porotein from MAPK family	T2D	X chromosome	Fukuda et al. (2012)

(continued)

Table 2 (continued)

Name	Associated SNPs	No. of participants	Mean age years ± SD or age range (years)	Study method	Function	Associated with	Location	References
PROX1	rs340841	2307 1102	68.0 ± 6.6 65.9 ± 6.6	Affymetrix genome-wide human SNP arrays	A transcription factor involved in developmental processes	Early onset of T2D	1q32.3	Hamet et al. (2017)
ADCY5	rs11708067 rs2877716	21 genome wide associated studies		Meta-analysis	Converts ATP to cAMP	Elevated fasting glucose and increased type 2 diabetes (T2D) risk	3q21.1	Hodson et al. (2014)
SLC2A2	rs5393 rs5394 rs5400 rs5404	522	55	Taqman genotyping	a transmembrane protein which facilitate glucose movement	Onset of diabetes in obese subjects with impaired glucose tolerance	3q26.2	Laukkanen et al. (2005)
WFS1	rs734312 rs10010131	6705		Meta- analysis	Encoded protein is required for Ca regulation in cells		4p16.1	Cheng et al. (2013b)
ZBED3	rs4457053	3132	58.0 ± 7.6	MassARRAY Iplex, Taqman 5' nuclease	An activator of Wnt signaling pathway	Fiber intake on type 2 diabetes incidence	5q13.3	Hindy et al. (2016)
TP53INP1	rs896854	1518 3651	54.0 57.0	GWAS	A tumor suppressor	T2D with elevated waist circumference when age and sex were adjusted	8q22.1	(Qi and Hu 2012)
KLF14	rs972283	6696		Meta- analysis	A transcription factor	T2D	7q32.2	Wang et al. (2014b)
GLIS3	rs7034200	100	35–50	PCR-RFLP	Involved in the development of pancreatic β cells	T2D	9p24.2	Sharif et al. (2018)
ADRA2A	rs10885122	120	61.9 ± 9.8	Taqman genotyping	Involved in regulation of neurotransmitter release	T2D	10q25.2	Welter et al. (2015)
CENTD2	rs1552224	1200 1725	57.4 ± 9.8 58.9 ± 10.3	TaqMan OpenArray genotyping, MassARRAY	Modulates actin cytoskeleton remodeling	T2D	11q13.4	Qian et al. (2015)

ENPP1	rs1044498	110	60.60 ± 10.98	Automated dye terminator sequencing	Mainly breaks down extracellular ATP	T2D	6q23.2	Bouhaha et al. (2008)
ZFAND6	rs11634397	8130 34,412		GWAS meta-analysis	Plays role in insulin secretion	T2D	15q25.1	Voight et al. (2010)
HNF1A	rs7957197	8130 34,412		GWAS meta-analysis	Encoded protein is a transcription factor	T2D	12q24.31	Voight et al. (2010)

GWAS Genome-wide association study, *NGS* Next generation sequencing, *RFLP* Restriction Fragment Length Polymorphism, *T2D* Type 2 diabetes
Data are not available for empty cells

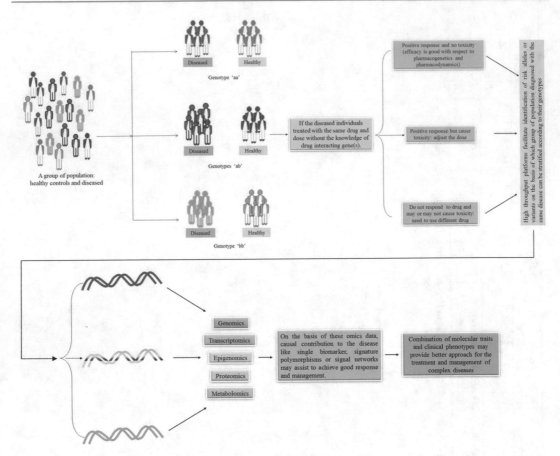

Fig. 1 Key identifiers for characterizing subjects to develop precision medicine. A population can be classified into three groups (wild type homozygous, heterozygous and rare homozygous) based on their genotypic frequencies and this may facilitate identifying risk allele (s) associated with specific disease. Thus, to establish the therapeutic strategy of a group of population diagnosed with the same disease, knowledge of their risk alleles is of great importance. Success of precision medicine may depend on identification of susceptible loci contributing to the development of diabetes, prevention of onset of diabetes in high-risk individuals, fixing individualized therapies by establishing drugs with most efficacy and least side effects followed by monitoring the response to the management therapy by measuring the diabetes related common clinical parameters. However, for characterizing subjects with respect to precision medicine, information of parameters like epigenetics, transcriptomics, metabolomics or proteomics should also play role in identifying susceptible groups and/or detecting biomarkers of a disease in certain population

common clinical parameters (Klonoff 2008; Klein et al. 2001).

Collaborative initiatives like Innovative Medicines Initiative (IMI), UK Biobank, Precision Medicine Initiative (PMI), Nordic Precision Medicine Initiative (NMPI) have been undertaken into different regions with an aim to reserve specific genotyping data in the pursuit of personalizing medicine (Koivula et al. 2014; Rask-andersen et al. 2017; Celis-morales et al. 2017; Kathy et al. 2015). Substantial initiatives in Sweden, Saudi Arabia, and especially in China with enormous budget are going on in full pace (Bergen et al. 2018; Project Team SG 2015; Cyranoski 2016). Table 3 summarizes these aforementioned undertakings.

Table 3 Initiatives taken to develop precision medicine to encounter diabetes

Project	Region	Activities	Budget	References
Innovative Medicines Initiative (IMI)	Europe (established in 2008)	Projects such as Surrogate markers for Micro and Macro-vascular hard endpoints for Innovative diabetes Tools (SUMMIT) for investigation of diabetes complications, Diabetes Research on Patient Stratification (DIRECT) (drug response and glycemic deterioration before and after the onset of type 2 diabetes) with emphasis on type 2 diabetes precision medicine. These projects under IMI prioritize the unification of multiple biomarkers as transcripts, proteins, metabolites and metagenomics sequences	5.6 billion dollar	Koivula et al. (2014)
UK Biobank	UK (established in 2005)	Gathering data regarding non-fasting blood samples and lifestyle, health and well-being concomitant with array-based genotypic and metabolomics assays from a population consisting of approximately 0.5 million. Research findings from this initiative has shown that obesity related genetic variants are expected to influence susceptibility to a range of modifiable lifestyle exposures	NA	Rask-andersen et al. (2017)
Precision Medicine Initiative (PMI)	USA (established in 2015	Cancer research as the short term and all areas of health care with specific emphasis placed on the discovery of predictive biomarkers for type 2 diabetes as the long term focus	US $215 million (initial budget)	Celis-morales et al. (2017)
Nordic Precision Medicine Initiative (NPMI)	Nordic countries	Amassing genetic and other biomedical data from 0.1 million Nordic citizens.	NA	Kathy et al. (2015)
GAPS (Genomic Aggregation Project in Sweden)	Sweden	Assimilation of 160,000 genotyped samples with corresponding phenotype data covering a wide range of diseases including diabetes	NA	Bergen et al. (2018)
Saudi Human Genome Project (SHGP)	Saudi Arabia	Sequencing the genomes of 100,000 Saudi citizens in order to identify the genetic basis of monogenic and complex diseases like type 2 diabetes	NA	Project Team SG (2015)
China		Decoding genomes	US$9 billion	Cyranoski (2016)

NA Not available

5 Use of Personalized Medicine to Treat Monogenic Diabetes

Although the general speculation is that with abundance of information of 'Omics' data in hand, personalized medicine will upgrade the prevention and treatment options of multifactorial disease like T1D and T2D; in reality the practice is still immured to the rare monogenic forms such as neonatal diabetes (Greeley et al. 2011) and maturity-onset diabetes of the young (MODY) (Naylor et al. 2014). Treatment of KCNJ11 Mutations in neonatal diabetes (1,400,000 births) to hinder progression to permanent neonatal

diabetes (1,252,000 in people aged, 20 years) is an exemplary case of precision medicine (Greeley et al. 2011). KCNJ11 encoding the Kir6.2 subunit of the pancreatic ATP dependent potassium (KATP) channel located in β cells is the most common risk associated loci (one-third of all cases) among 24 genetic causes (De Franco et al. 2015). KCNJ11 mutations result in diabetes by blocking the closing of the KATP channels, which prevents β cell depolarization and corresponding insulin secretion (Gloyn et al. 2004). Affected clinically sick babies exhibit insulin deficiency resulting in diabetic ketoacidosis in almost 80% cases which is a frightening scenario (Letourneau et al. 2017). Before the discovery of pathogenic variants of KCNJ11, misconstruing the symptoms in those children as T1D symptoms, they were treated with insulin injections (Gloyn et al. 2004). Sulfonylurea (SU) medication amends this defect in almost 90% of cases by binding to KATP channels and allowing insulin secretion (Pearson et al. 2006). Repurposing an existing oral antidiabetic therapy has become a blessing for majority of KCNJ11 mutant patients. Besides obtaining astounding success in glycemic control, sulphonylurea also show fragmented improvement in some of the neurological features in the first year after switching (Beltrand et al. 2015).

Another mention worthy example of precision medicine is applicable to maturity-onset diabetes of the young (MODY) with different subtypes such as MODY 1, MODY 2 and MODY 3. For MODY 2 patients with hexokinase 4 gene mutation with decreased affinity for glucose, oral antidiabetic agents have miniscule effects (Mark and Andrew 2008). Patients with MODY 1 and MODY 3 with mutation in transcription factors, HNF4A and HFN1A, respectively have markedly increased blood glucose levels and frequent onset of diabetes before 25 which often lead to misdiagnosis as T1D.

Antithetical to T1D, these MODY patients often show higher sensitivity to oral SU drugs and transition from insulin to SU culminates into improvement into blood glucose level (Shepherd et al. 2009). Some studies show MODY 3 to be more responsive to SUs than metformin (Pearson et al. 2003). Combination of pharmacogenomics and pharmacogenetics studies expedited by GWAS can improve precise diagnosis and medication of MODY to change the scenario of inaccurate diagnosis (Shields et al. 2010).

6 Variation in Genetics and their Role in Precision Medicine for Patients with Diabetes

Secretagogue drugs, like SU can rectify the insufficiency of insulin production and secretion by binding to sulphonylurea receptor (SUR) coupled with a KATP leading to the cell depolarization and stimulation of insulin release. Glicklazide modified release drug (MR) and glipizide gastrointestinal transport system (GITS), two once-daily third generation SUs (Schernthaner et al. 2004) possess better safety and lesser weight gain chances compared to second generation glibenclamide, gliquidone and glimepiride. Glucose uptake modulator octameric KATP channel is an assemblage of Kir6.2 and SUR1, encoded by the KCNJ11 and ABCC8 genes, respectively (Mctaggart et al. 2010). Over 150 loss-of- function mutations responsible for congenital hyperinsulinisim are characterized in SUR1 (ABCC8) and additional 24 are found in Kir6.2 (KCNJ11) (Flanagan et al. 2009). A common variant rs5219 (E23K) within KCNJ11 that lead to development of T2D has been found to be associated with an elevated likelihood of SU therapeutic failure (Zhou et al. 2009a). Polymorphic variants in ABCC8, such as rs757110 (Ser1369Ala) isn't found to be linked with the risk for severe SU-induced hypoglycemia in German (Holstein et al. 2012) and Japanese (Sato et al. 2010), while rs1799859 (Arg1273Arg) is reported to be associated with the SU efficacy in European Caucasians (Nikolac et al. 2009). T2D Patients with defective HNF-1a (involved in correction of β-cell development and function) show more sensitivity to SU than matched T2D patients (Pearson et al. 2000). rs7903146 in transcription factor 7-like 2 (TCF7L2), a T2D susceptible loci exhibits depletion in the levels of glycated hemoglobin (HbA1c) and fasting plasma glucose following a combined SU and metformin treatment (Schroner et al. 2011).

Variants in hepatic cytochrome P450 (CYP) isoform CYP2C9; main metabolizer of SUs are associated with variability in patient sensitivity to SU (Becker et al. 2008). CYP2C9∗3 (Ile359Leu) exerts protective role against development of T2D (Ragia et al. 2009), whereas CYP2C9∗2 (Arg144Cys) polymorphism shows no association with diabetes susceptibility (Semiz et al. 2010). CYP2C9∗3 also displays risk of hypoglycemia and depleted SU clearance (Bozkurt et al. 2007). In addition, an astounding finding derived from a GoDARTS (Genetics of Diabetes Audit and Research in Tayside Scotland) study consisting of 1073 incident SU users is that patients with two copies of the *2 or *3 alleles are three times more likely to achieve treatment target with regard to the levels of HbA1c under 7% (53 mmol/mol) than patients with two wild-type CYP2C9*1 alleles (Zhou et al. 2009b). Thiazolidinediones/glitazones improve insulin sensitivity by lessening circulating fatty acid concentrations and lipid availability in liver through activation of PPARs (peroxisome proliferator-activated receptors) (Karalliedde and Buckingham 2007). Severe IR and diabetes mellitus can be the consequence of loss of function mutation in PPARγ (Barroso et al. 1999). Patients with Pro12Ala variant of PPARG show greater reduction of fasting blood glucose and HbA1c levels when treated with rosiglitazone (Kang et al. 2005). Additional study performed on Chinese ethnicity reports on better efficacy of rosiglitazone in T2D treatment associated with Thr394Thr and Gly482Ser SNPs of the peroxisome proliferator activated receptor-γ coactivator-1α (PGC-1α), a transcriptional co-activator of PPARγ (Zhang et al. 2010). Carriers of CYP2C8*3 allele coding for Arg139Lys and Lys399Arg show higher rosiglitazone Clearance (Kirchheiner et al. 2006). There are some overlapping between the distinct types of diabetes such as gestational diabetes can result in T2D. Researches regarding precision medicine are still in nascent phase so whether such a precise drug can be tailored with the ability to treat any ailing individuals even in such transitions is still not known. Another question is if the patients need to be continue non-precision medicines along with the precision ones. Further research is essential to answer such questions.

7   Genetic Variants and Their Influence on Anti-diabetic Drugs

Metformin hinders hepatic gluconeogenesis, increases glucose uptake and utilization, and also improves insulin sensitivity (Kirpichnikov et al. 2002). The postulation regarding the still obscured molecular mechanism is that increasing the AMP/ATP ratio by specific inhibition of mitochondrial respiratory-chain complex 1 results in the activation of adenosine monophosphate-activated protein kinase (AMPK) (Zhou et al. 2001; El-Mir et al. 2000; Owen et al. 2000). Potential mechanism of AMPK phosphorylation by metformin includes the upstream serine-threonine kinase11 (STK11/LKB1) (Shaw et al. 2005). Less than two-thirds of the entire patient population receiving metformin as an initial treatment, achieve desired glycemic control or HB1AC goal of <7% (Kahn et al. 2006). SLC22A1 gene encoding Organic Cation Transporter 1 (OCT1) has polymorphic variants such as R61C (rs12208357), G401S (rs34130495), 420del (rs72552763), and G465R (rs34059508) responsible for lower effects of metformin in the oral glucose tolerance test (Shu et al. 2007). Two variants rs628031 (M408V) and rs36056065 (8 bp insertion) of SLC22A1 are also perceived to be associated with gastrointestinal side effects of metformin (Tarasova et al. 2012). A common 5′-UTR variant, rs12943590 of SLC47A2, is reported to be associated with enhanced promoter activity and weaker response to metformin (Choi et al. 2011).

Meglitinides, a class of short-acting insulin secretagogues acts in a similar way of SUs by binding to two sites of SUR1 and inhibiting KATP channel to stimulate insulin release (Yan et al. 2006). Meglitinides have a lower risk to induce hypoglycemia than SU considering their short action period and can be used as a potent alternative to metformin with lesser side effects. Repaglinide stimulates early insulin secretion in the postprandial period after binding to a distinct site on the β-cell (Fuhlendorff et al. 1998). Diabetic patients with the GA or AA genotype with respect to KCNJ11 Lys23Glu SNP show higher levels of fasting and postprandial glucose and HbA1c following repaglinide treatment compared to that of GG genotype. T2DM patients with the TT genotype of TCF7L rs290487 also exhibit better efficacy with respect to levels of fasting insulin, triglycerides, and low-density lipoprotein cholesterol as compared to carriers of the CC or CT genotype (Yu et al. 2010). SLCO1B1 gene (encodes organic anion transporting polypeptide 1B1 or OATP1B1), exerts different effects on repaglinide pharmacokinetics (Kalliokoski et al. 2008). In healthy Chinese population, presence of 521T>C SNP of SLCO1B1 along with CYP2C9∗3 polymorphism exercises consequential nateglinide pharmacokinetics (Cheng et al. 2013a). Repaglinide is metabolized by CYP2C8 and CYP3A4 (Bidstrup et al. 2003). CYP2C8∗3 carriers have higher drug clearance than carriers having wild-type genotypes (Kirchheiner et al. 2005). rs2237892 (C > T) is found to be associated with the repaglinide efficacy in Chinese with T2D (Dai et al. 2012). T2D patients with rs2237892 T and rs2237895 C alleles are more likely to have a positive effect on postprandial glucose levels than patients with the rs2237892 CC and rs2237895 AA genotype. rs13266634 Arg325Trp) and rs16889462 (Arg325Gln) variants in SLC30A8 are also unearthed to be related to the repaglinide efficacy in Chinese patients with T2D (Huang et al. 2010).

GLP-1, an incretin induces insulin secretion of the β-cells and GLP-1 receptors (GLP-1R) has ability to increase the efficacy T2D treatment (Enigk et al. 2011; Plourde and Matte 2017; Scheen 2015). T allele of rs3765467 and rs761386 were linked to lower and higher standard deviation in plasma glucose in response to exogenous GLP-1, respectively. The rs6923761 variant has shown an increased response from β-cells (Poulsen et al. 2015). Two sodium-dependent glucose transporters, SGLT1 and SGLT2 reabsorb 3 and 90% glucose, respectively (Von Mering 1885). Competitive inhibitor of SGLT1 and SGLT2 enhance glucose excretion via pharmacological inhibition of glucose reabsorption (Crane 1983). Till date, there hasn't been any genetic variants and SNPs found in SLC5A2 gene of SGLT2 protein in treatment of T2D.

8 Conclusion and Future Directions

Researchers are aiming to reclassify types of diabetes with special emphasis on sub-classification on the basis of clinical indices and molecular data for the response to the drug in question, or to develop side effects. Nevertheless, it is going to be a cumbersome work to do so as scanning of whole genome through GWAS are bringing numerous probable region(s), SNPs/SNVs, mutations within genes associated with the risk of T2D. Also, due to their strong connotation with regulatory players, SNPs/SNVs in non-coding sequences are also attracting the interest. Precision medicine is definitely going to vary from region to region owing to the genetic variations present in different ethnicities. Genetic testing can contribute to the clinical management and early prediction followed by prevention of diabetes in future. Thus, we conclude that

(i) robust genome wide association studies with more population should be performed to dissect complex multigenic traits like diabetes which will extend the knowledge of genetic architecture of diabetes for those populations who are inexplicably affected by the disease.

(ii) linkage disequilibrium of tagged SNPs i.e., SNPs that are strongly correlated with each other or randomly selected SNPs, or a combination of both should be taken into considerations by using by deploying major genotyping platforms, which are currently available to get a good coverage.

(iii) though enormous efforts have been made to identify responsible variants for diabetes through GWAS, it represents only a small proportion (~10%) of heritability of diabetes specially T2D. More consortium by blending collaborators from developed, middle, and low-middle income countries should be formed to have a comprehensive panel of genes to overcome the burden of heterogeneity. Big consortiums and working groups comprising clinicians and base line researchers have come in the same platform to address the complex management of diabetes. So far, we have mainly observed the successful applications of such tailored medicine mostly in cases of monogenic forms of diabetes. But high-throughput NGS data and advancements in 'Omics' knowledge will surely go a long way to make this practice of specific medicine more successful.

References

Abderrahmani A, Yengo L, Caiazzo R, Canouil M, Cauchi S, Raverdy V et al (2018) Increased hepatic PDGF-AA signaling mediates liver insulin resistance in obesity-associated type 2 diabetes. Diabetes 67:1310–1321

Abu-Bakare A, Taylor R, Gill GV, Alberti KG (1986) Tropical or malnutrition-related diabetes: a real syndrome. Lancet 1:1135–1138

Aly TA, Baschal EE, Jahromi MM, Fernando MS, Babu SR, Fingerlin TE et al (2008) Analysis of single nucleotide polymorphisms identifies major type 1A diabetes locus telomeric of the major histocompatibility complex. Diabetes 57(3):770–776

American Diabetes Association (2009) Diagnosis and classification of diabetes mellitus. Diabetes Care 32 (Suppl 1):S62–S67

American Diabetes Association (2016) 2. Classification and diagnosis of diabetes. Diabetes Care 39(Suppl 1): S13–S22

Asad S, Nikamo P, Gyllenberg A, Bennet H, Hansson O, Wierup N et al (2012) HTR1A a novel type 1 diabetes susceptibility gene on chromosome 5p13-q13. PLoS One 7(5):e35439

Bakay M, Pandey R, Hakonarson H (2013) Genes involved in type 1 diabetes: an update. Genes (Basel) 4(3):499–521

Barrett JC, Clayton DG, Concannon P, Akolkar B, Cooper JD, Erlich HA, Julier C, Morahan G, Nerup J, Nierras C et al (2009) Genome-wide association study and meta-analysis find that over 40 loci affect risk of type 1 diabetes. Nat Genet 41:703–707

Barroso I, Gurnell M, Crowley VEF, Agostini M (1999) Dominant negative mutations in human PPARgamma associated with severe insulin resistance, diabetes mellitus and hypertension. Nature 402(6764):880–883

Barroso I, Luan J, Middelberg RP, Harding AH, Franks PW, Jakes RW et al (2003) Candidate gene association study in type 2 diabetes indicates a role for genes involved in beta-cell function as well as insulin action. PLoS Biol 1:E20

Becker ML, Visser LE, Trienekens PH, Hofman A, Van Schaik RHN, Stricker B (2008) Cytochrome P 450 2C9 ∗ 2 and ∗ 3 polymorphisms and the dose and effect of sulfonylurea in type II diabetes mellitus. Clin Pharmacol Ther 83(2):288–292

Bell GI, Horita S, Karam JH (1984) A polymorphic locus near the human insulin gene is associated with insulin-dependent diabetes mellitus. Diabetes 33(2):176–183

Beltrand J, Elie C, Busiah K, Fournier E, Boddaert N, Bahi-Buisson N et al (2015) Sulfonylurea therapy benefits neurological and psychomotor functions in patients with neonatal diabetes owing to potassium channel mutations. Diabetes Care 38(11):2033–2041

Bergen SE, Sullivan PF, Carolina N (2018) National-scale precision medicine for psychiatric disorders in Sweden. Am J Med Genet B Neuropsychiatr Genet 177(7):630–634

Bidstrup TB, Bjørnsdottir I, Sidelmann UG, Thomsen MS, Hansen KT (2003) CYP2C8 and CYP3A4 are the principal enzymes involved in the human in vitro biotransformation of the insulin secretagogue repaglinide. Br J Clin Pharmacol 56(3):305–314

Billings LK, Florez JC (2010) The genetics of type 2 diabetes: what have we learned from GWAS? Ann N Y Acad Sci 1212:59–77

Bottini N, Musumeci L, Alonso A, Rahmouni S, Nika K, Rostamkhani M et al (2004) A functional variant of lymphoid tyrosine phosphatase is associated with type I diabetes. Nat Genet 36(4):337–338

Bouatia-Naji N et al (2008) A polymorphism within the G6PC2 gene is associated with fasting plasma glucose levels. Science 320(5879):1085–1088

Bouhaha R, Meyre D, Kamoun HA, Ennafaa H, Vaillant E, Sassi R et al (2008) Effect of ENPP1/ PC-1-K121Q and PPARgamma-Pro12Ala polymorphisms on the genetic susceptibility to T2D in the Tunisian population. Diabetes Res Clin Pract 81 (3):278–283

Bozkurt O, De Boer A, Grobbee DE, Heerdink ER, Burger H, Klungel OH (2007) Pharmacogenetics of glucose-lowering drug treatment: a systematic review. Mol Diagn Ther 11(5):291–302

Bradfield JP, Qu H-Q, Wang K, Zhang H, Sleiman PM, Kim CE et al (2011) A genome-wide Meta-analysis of six type 1 diabetes cohorts identifies multiple associated loci. PLoS Genet 7(9):e1002293

Brorsson CA, Nielsen LB, Andersen ML, Kaur S, Bergholdt R, Hansen L et al (2016) Genetic risk score modelling for disease progression in new-onset type 1 diabetes patients: increased genetic load of islet-expressed and cytokine-regulated candidate genes predicts poorer glycemic control. J Diabetes Res 2016:9570424–9570428

Brown JH, Jardetzky TS, Gorga JC et al (1993) Three-dimensional structure of the human class II histocompatibility antigen HLA-DR1. Nature 364:33–39

Buniello A, MacArthur JAL, Cerezo M, Harris LW, Hayhurst J, Malangone C et al (2019) The NHGRI-EBIGWAS catalog of published genome-wide association studies, targeted arrays and summary statistics 2019. Nucleic Acids Res 47(D1):D1005–D1D12

Cai Y, Yi J, Ma Y, Fu D (2011) Meta-analysis of the effect of HHEX gene polymorphism on the risk of type 2 diabetes. Mutagenesis 26(2):309–314

Calmettes G, Ribalet B, John S, Korge P, Ping P, Weiss JN (2015) Hexokinases and cardioprotection. J Mol Cell Cardiol 78:107–115

Cauchi S, Meyre D, Choquet H et al (2006) TCF7L2 variation predicts hyperglycemia incidence in a French general population: the data from an epidemiological study on the insulin resistance syndrome (DESIR) study. Diabetes 55:3189–3192

Celis-morales CA, Lyall DM, Gray SR, Steell L, Anderson J, Iliodromiti S et al (2017 Dec) Dietary fat and total energy intake modifies the association of genetic profile risk score on obesity: evidence from 48 170 UK Biobank participants. Int J Obes 41 (12):1761–1768

Chen WM, Erdos MR, Jackson AU, Saxena R, Sanna S, Silver KD, Timpson NJ et al (2008) Variations in the G6PC2/ABCB11 genomic region are associated with fasting glucose levels. J Clin Invest 118(7):2620–2628

Cheng Y, Wang G, Zhang W, Fan L, Chen Y, Zhou HH (2013a) Effect of CYP2C9 and SLCO1B1 polymorphisms on the pharmacokinetics and pharmacodynamics of nateglinide in healthy Chinese male volunteers. Eur J Clin Pharmacol 69(3):407–413

Cheng S, Wu Y, Wu W, Zhang D (2013b) Association of rs734312 and rs10010131 polymorphisms in WFS1 gene with type 2 diabetes mellitus: a meta-analysis. Endocr J 60(4):441–447

Chistiakov DA, Nikitin AG, Smetanina SA, Bel LN, Suplotova LA, Shestakova MV et al (2012) The rs11705701 G > a polymorphism of IGF2BP2 is associated with IGF2BP2 mRNA and protein levels in the visceral adipose tissue – a link to type 2 diabetes susceptibility. Rev Diabet Stud 9(2–3):112–122

Cho YS, Chen CH, Hu C, Long J, Ong RT, Sim X, Takeuchi F, Wu Y, Go MJ, Yamauchi T et al (2011) Meta-analysis of genome-wide association studies identifies eight new loci for type 2 diabetes in east Asians. Nat Genet 44:67–72

Choi J, Yee S, Ramirez A, Morrissey K, Jang G, Joski P et al (2011) A common 5′-UTR variant in MATE2-K is associated with poor response to metformin. Clin Pharmacol Ther 90(5):674–684

Choi JW, Moon S, Jang EJ, Lee CH, Park J-S (2017) Association of prediabetes-associated single nucleotide polymorphisms with microalbuminuria. PLoS One 12 (2):e0171367

Colli ML, Moore F, Gurzov EN, Ortis F, Eizirik DL (2010) MDA5 and PTPN2, two candidate genes for type 1 diabetes, modify pancreatic beta-cell responses to the viral by-product double-stranded RNA. Hum Mol Genet 19:135–146. 66

Concannon P, Onengut-Gumuscu S, Todd JA, Smyth DJ, Pociot F, Bergholdt R et al (2008) A human type 1 diabetes susceptibility locus maps to chromosome 21q22.3. Diabetes 57:2858–2861

Cooper JD, Smyth DJ, Smiles AM, Plagnol V, Walker NM, Allen JE et al (2008) Meta-analysis of genome-wide association study data identifies additional type 1 diabetes loci. Nat Genet 40(12):1399–1401

Crane RK (1983) The road to ion-coupled membrane processes. In: Neuberger A, Van Deenen LLM, Semenza G (eds) Comprehensive biochemistry. Elsevier Science, New York, pp 43–69

Cruz TD, Valdes AM, Santiago A, Frazer de Llado T, Raffel LJ, Zeidler A et al (2004) DPB1 alleles are associated with type 1 diabetes susceptibility in multiple ethnic groups. Diabetes 53:2158–2163

Cyranoski D (2016) The sequencing superpower. Nature 534:462–463

Dai X, Huang Q, Yin J, Guo Y, Gong Z, Lei MX et al (2012) KCNQ1 gene polymorphisms are associated with the therapeutic efficacy of repaglinide in Chinese

type 2 diabetic patients. Clin Exp Pharmacol Physiol 39(5):462–468

Daimon M, Ji G, Saitoh T, Oizumi T, Tominaga M, Nakamura T et al (2003) Large-scale search of SNPs for type 2 DM susceptibility genes in a Japanese population. Biochem Biophys Res Commun 302:751–758

Das SK, Elbein SC (2006) The genetic basis of type 2 diabetes. Cell 2:100–131

De Franco E, Flanagan SE, Houghton JAL, Allen HL, Mackay DJG, Temple IK et al (2015) The effect of early, comprehensive genomic testing on clinical care in neonatal diabetes: an international cohort study. Lancet 386(9997):957–963

de Jong PE, Hillege HL, Pinto-Sietsma SJ, de Zeeuw D (2003) Screening for microalbuminuria in the general population: a tool to detect subjects at risk for progressive renal failure in an early phase? Nephrol Dial Transplant 18(1):10–13. PMID: 12480951

Defronzo RA, Banting Lecture (2009) From the triumvirate to the ominous octet: a new paradigm for the treatment of type 2 diabetes mellitus. Diabetes 58:773–795

Douroudis K, Kisand K, Nemvalts V, Rajasalu T, Uibo R (2010) Allelic variants in the PHTF1-PTPN22, C12orf30 and CD226 regions as candidate susceptibility factors for the type 1 diabetes in the Estonian population. BMC Med Genet 11:11

Downes K, Pekalski M, Angus KL, Hardy M, Nutland S, Smyth DJ, Walker NM, Wallace C, Todd JA (2010) Reduced expression of IFIH1 is protective for type 1 diabetes. PLoS One 5:e12646

Dupuis J, Langenberg C, Prokopenko I, Saxena R, Soranzo N, Jackson AU et al (2010) New genetic loci implicated in fasting glucose homeostasis and their impact on type 2 diabetes risk. Nat Genet 42 (2):105–116

Egefjord L, Jensen JL, Bang-Berthelsen CH, Petersen AB, Smidt K, Schmitz O et al (2009) Zinc transporter gene expression is regulated by pro-inflammatory cytokines: a potential role for zinc transporters in beta-cell apoptosis? BMC Endocr Disord 9:7

Eizirik DL, Sammeth M, Bouckenooghe T, Bottu G, Sisino G, Igoillo-Esteve M et al (2012) The human pancreatic islet transcriptome: expression of candidate genes for type 1 diabetes and the impact of pro-inflammatory cytokines. PLoS Genet 8(3): e1002552

Elboudwarej E, Cole M, Briggs FB, Fouts A, Fain PR, Quach H et al (2016) Hypomethylation within gene promoter regions and type 1 diabetes in discordant monozygotic twins. J Autoimmun 68:23–29

El-ella SSA, Khattab ESAEH, El-mekkawy MS, El-shamy AA (2018) CD226 gene polymorphism (rs763361 C > T) is associated with susceptibility to type 1 diabetes mellitus among Egyptian children. Arch Pédiatr 25(6):378–382. 763

El-Mir M-Y, Nogueira V, Fontaine E, Averet N, Rigoulet M, Leverve X (2000) Dimethyl biguanide inhibits cell respiration via an indirect effect targeted on the respiratory chain complex I. J Biol Chem 275 (1):223–228

Emanuelli B, Eberle D, Suzuki R, Kahn R (2008) Overexpression of the dual specificity phosphatase MKP-4/DUSP-9 protects against stress-indused insulin resistance. Proc Natl Acad Sci U S A 105:3545–3550

Engwa GA, Nwalo FN, Chikezie CC, Onyia CO, Ojo OO, Mbacham WF et al (2018) Possible association between ABCC8 C49620T polymorphism and type 2 diabetes in a Nigerian population. BMC Med Genet 19:78

Enigk U, Breitfeld J, Schleinitz D et al (2011) Role of genetic variation in the human sodium-glucose cotransporter 2 gene (SGLT2) in glucose homeostasis. Pharmacogenomics 12:1119–1126. 64

Erlich H, Valdes AM, Noble J, Carlson JA, Varney M, Concannon P et al (2008) HLA DR-DQ haplotypes and genotypes and type 1 diabetes risk: analysis of the type 1 diabetes genetics consortium families. Diabetes 57(4):1084–1092

Espino-paisan L, De Calle H, Fernández-arquero M, Figueredo MÁ, De Concha EG, Urcelay E et al (2011) A polymorphism in PTPN2 gene is associated with an earlier onset of type 1 diabetes. Immunogenetics 63(4):255–258

Fisher E, Nitz I, Lindner I, Rubin D, Boeing H, Mohlig M et al (2007) Candidate gene association study of type 2 diabetes in a nested case-control study of the EPIC-Potsdam cohort - role of fat assimilation. Mol Nutr Food Res 51:185–191

Fitipaldi H, Mccarthy MI, Florez JC, Franks PW (2018) A global overview of precision medicine in type 2 diabetes. Diabetes 67(10):1911–1922

Flanagan SE, Clauin S, Bellanne-Chantelot C, de Lonay P, Harries LW, Gloyn AL et al (2009) Update of mutations in the genes encoding the pancreatic beta-cell K ATP channel subunits Kir6.2 diabetes mellitus and Hyperinsulinism human mutation. Hum Mutat 30 (2):170–585 80

Flannick J, Thorleifsson G, Beer NL, Jacobs SB, Grarup N, Burtt NP et al (2014) Loss-of-function mutations in SLC30A8 protect against type 2 diabetes. Nat Genet 46(4):357–363

Flannick J, Mercader JM, Fuchsberger C, Udler MS, Mahajan A, Wessel J et al (2019) Exome sequencing of 20,791 cases of type 2 diabetes and 24,440 controls. Nature 570(7759):71–76

Florez JC, Jablonski KA, Bayley N et al (2006) TCF7L2 polymorphisms and progression to diabetes in the diabetes prevention program. N Engl J Med 355:241–250

Floyel T, Brorsson C, Nielsen LB, Miani M, Bang-Berthelsen CH, Friedrichsen M et al (2014) CTSH regulates beta-cell function and disease progression in newly diagnosed type 1 diabetes patients. Proc Natl Acad Sci U S A 111:10305–10310

Forouhi NG, Luan J, Hennings S, Wareham NJ (2007) Incidence of type 2 diabetes in England and its association with baseline impaired fasting glucose: the Ely study 1990–2000. Diabet Med 24(2):200–207

Fu D, Cong X, Ma Y, Cai H, Cai M, Li D et al (2013) Genetic polymorphism of Glucokinase on the risk of type 2 diabetes and impaired glucose regulation: evidence based on 298,468 subjects. PLoS One 8(2): e55727

Fuhlendorff J, Rorsman P, Kofod H, Brand CL, Rolin B, Mackay P et al (1998) Stimulation of insulin release by Repaglinide and Glibenclamide involves both common and distinct processes. Diabetes 47(3):345–351

Fukaya M, Brorsson CA, Meyerovich K, Catrysse L, Delaroche D, Vanzela EC, Ortis F, Beyaert R, Nielsen LB, Andersen ML et al (2016) A20 inhibits beta-cell apoptosis by multiple mechanisms and predicts residual beta-cell function in type 1 diabetes. Mol Endocrinol 30:48–61

Fukuda H, Imamura M, Tanaka Y, Iwata M, Hirose H, Kaku K et al (2012) A single nucleotide polymorphism within DUSP9 is associated with susceptibility to type 2 diabetes in a Japanese population. PLoS One 7(9): e46263

Fung E, Smyth DJ, Howson JM, Cooper JD, Walker NM, Stevens H et al (2009) Analysis of 17 autoimmune disease-associated variants in type 1 diabetes identifies 6q23 / TNFAIP3 as a susceptibility locus. Genes Immun 10(2):188–191

Gabriel SB, Schaffner SF, Nguyen H et al (2002) The structure of haplotype blocks in the human genome. Science 296(5576):2225–2229

George AM, Jacob AG, Fogelfeld L (2015) Lean diabetes mellitus: an emerging entity in the era of obesity. World J Diabetes 6(4):613–620

Giza S, Goulas A, Gbandi E, Effraimidou S, Papadopoulou-Alataki E, Eboriadou M et al (2013) The role of PTPN22 C1858T gene polymorphism in diabetes mellitus type 1: first evaluation in Greek children and adolescents. Biomed Res Int 2013:721604

Gloyn AL, Weedon MN, Owen KR, Turner MJ, Knight BA, Hitman G et al (2003) Large-scale association studies of variants in genes encoding the pancreatic β-cell K_{ATP} channel subunits Kir6.2 (KCNJ11) and SUR1 (ABCC8) confirm that the KCNJ11 E23K variant is associated with type 2 diabetes. Diabetes 52 (2):568–572

Gloyn AL, Pearson ER, Antcliff JF, Proks P, Bruining GJ, Slingerland AS et al (2004) Activating mutations in the gene encoding the ATP-sensitive potassium-channel subunit Kir6.2 and permanent neonatal diabetes. N Engl J Med 350(18):1838–1849

Grant SF, Thorleifsson G, Reynisdottir I, Benediktsson R, Manolescu A, Sainz J et al (2006) Variant of transcription factor 7-like 2 (TCF7L2) gene confers risk of type 2 diabetes. Nat Genet 38(3):320–323

Grant SFA, Qu H, Bradfield JP, Marchand L, Kim CE, Glessner JT et al (2009) Follow-up analysis of genome-wide association data identifies novel loci for type 1 diabetes. Diabetes 58(1):290–295

Greeley SA, John PM, Winn AN, Ornelas J, Lipton RB, Philipson LH et al (2011) The cost-effectiveness of personalized genetic medicine. Diabetes Care 34 (3):622–627

GTEx Portal. Available online: https://gtexportal.org. Accessed on 1st Nov 2016

Hakonarson H, Grant SF, Bradfield JP, Marchand L, Kim CE, Glessner J et al (2007) A genome-wide association study identifies kiaa0350 as a type 1 diabetes gene. Nature 448:591–594

Hakonarson H, Qu HQ, Bradfield JP, Marchand L, Kim CE, Glessner JT et al (2008) A novel susceptibility locus for type 1 diabetes on chr12q13 identified by a genome-wide association study. Diabetes 57:1143–1146

Hamet P, Haloui M, Sylvestre M, Tahir M, Simon PHG, Sonja B et al (2017) PROX1gene CC genotype as a major determinant of early onset of type 2 diabetes in slavic study participants from action in diabetes and vascular disease: preterax and diamicron MR controlled evaluation study. J Hypertens 35(Suppl 1): S24–S32

Hindy G, Mollet IG, Rukh G, Ericson U, Orho-melander M (2016) Several type 2 diabetes-associated variants in genes annotated to WNT signaling interact with dietary fiber in relation to incidence of type 2 diabetes. Genes Nutr 11:6

Hodson DJ, Mitchell RK, Marselli L, Pullen TJ, Brias SG, Semplici F et al (2014) ADCY5 couples glucose to insulin secretion in human islets. Diabetes 63 (9):3009–3021

Holstein JD, Kovacs P, Patzer O, Stumvoll M, Holstein A (2012) The Ser1369Ala variant of ABCC8 and the risk for severe sulfonylurea-induced hypoglycemia in German patients with type 2 diabetes. Pharmacogenomics 13(1):5–7

Holt GI (2004) Diagnosis, epidemiology and pathogenesis of diabetes mellitus an update for psychiatrists. Br J Psychiatry 184:s55–s63

Hori S, Nomura T, Sakaguchi S (2003) Control of regulatory T cell development by transcriptional factor foxp3. Science 299:1057–1061

Huang Q, Yin J, Dai X, Wu J, Chen X, Deng CS et al (2010) Association analysis of SLC30A8 rs13266634 and rs16889462 polymorphisms with type 2 diabetes mellitus and repaglinide response in Chinese patients. Eur J Clin Pharmacol 66(12):1207–1215

Huang J, Ellinghaus D, Franke A, Howie B, Li Y (2012) 1000 Genomes-based imputation identifies novel and refined associations for the Wellcome Trust Case Control Consortium phase Data. Eur J Hum Genet 20 (7):801–805

Huda N, Hosen I, Yasmin T, Sarkar PK, Hasan M, Nabi AHMN (2018) Genetic variation of the transcription factor GATA3, not STAT4, is associated with the risk of type 2 diabetes in the Bangladeshi population. PLoS One 13(7):e0198507

Huh KB, Lee HC, Kim HM, Cho YW, Kim YL, Lee KW et al (1992) Immunogenetic and nutritional profile in insulin-using youth-onset diabetics in Korea. Diabetes Res Clin Pract 16:63–70

Hunt KJ, Lehman DM, Arya R, Fowler S, Leach RJ, Goring HH et al (2005) Genome-wide linkage analyses of type 2 diabetes in Mexican Americans: the San Antonio Family Diabetes/Gallbladder Study. Diabetes 54:2655–2662

Ingelsson E, Langenberg C, Hivert MF, Prokopenko I, Lyssenko V, Dupuis J et al (2010) Detailed physiologic characterization reveals diverse mechanisms for novel genetic loci regulating glucose and insulin metabolism in humans. Diabetes 59(5):1266–1275

International Diabetes Federation (IDF) [Internet] (2012) Country estimates table 2011. IDF diabetes atlas, 6th edn. International Diabetes Federation, Brussels

International HapMap Consortium, Frazer KA, Ballinger DG, Cox DR, Hinds DA, Stuve LL et al (2007) A second generation human haplotype map of over 3.1 million SNPs. Nature 449:851–861

Iwase K, Shimada A, Kawai T, Okubo Y, Kanazawa Y, Irie J et al (2009) FOXP3/Scurfin gene polymorphism is associated with adult onset type 1 diabetes in Japanese, especially in women and slowly progressive-type patients. Autoimmunity 42(2):159–167

Jacobsen L, Schatz D (2016) Current and future efforts toward the prevention of type 1 diabetes. Pediatr Diabetes 17(Suppl. 22):78–86

Jerram S, Leslie RD (2017) The genetic architecture of type 1 diabetes. Genes 8(8):pii: E209

Juan-Mateu J, Alvelos MI, Turatsinze JV et al (2018) SRp55 regulates a splicing network that controls human pancreatic beta-cell function and survival. Diabetes 67:423–436

Kahn SE, Haffner SM, Heise MA, Herman WH, Holman RR, Jones NP et al (2006) Glycemic durability of rosiglitazone, metformin, or glyburide monotherapy. N Engl J Med 659(355):2427–2443

Kaiser AB, Zhang N, der Pluijm WVAN (2018) Global prevalence of type 2 diabetes over the next ten years (2018–2028). Diabetes 67(Supplement 1):202

Kaku K (2010) Pathophysiology of type 2 diabetes and its treatment policy. JMAJ 53(1):41–46

Kalliokoski A, Neuvonen M, Neuvonen PJ, Niemi M (2008) The effect of SLCO1B1 polymorphism on repaglinide pharmacokinetics persists over a wide dose range. Br J Clin Pharmacol 66(6):818–825

Kamura Y, Iwata M, Maeda S, Shinmura S, Koshimizu Y, Honoki H et al (2016) FTO gene polymorphism is associated with type 2 diabetes through its effect on increasing the maximum BMI in Japanese men. PLoS One 11(11):e0165523

Kang ES, Park SY, Kim HJ, Kim CS (2005) Effects of Pro12Ala polymorphism of peroxisome proliferator–activated receptor gamma2 gene on rosiglitazone response in type 2 diabetes. Clin Pharmacol Ther 78 (2):202–208

Kang HS, Kim YS, ZeRuth G, Beak JY, Gerrish K, Kilic G, Sosa-Pineda B, Jensen J, Pierreux CE, Lemaigre FP et al (2009) Transcription factor glis3, a novel critical player in the regulation of pancreatic beta cell development and insulin gene expression. Mol Cell Biol 29:6366–6379

Karalliedde J, Buckingham RE (2007) Thiazolidinediones and their fluid-related adverse effects facts, fiction and putative management strategies. Drug Saf 30 (9):741–753

Kathy H, Rick L, Bray P-L, Josh D (2015) The precision medicine initiative cohort program building a research foundation for 21st century Medicine. Precision Medicine Initiative (PMI) Working Group Report to the Advisory Committee to the Director, NIH

Kavvoura FK, Ioannidis JP (2005) Ala45Thr polymorphism of the NEUROD1 gene and diabetes susceptibility: a meta-analysis. Hum Genet 116(3):192–199

Kilpelainen TO, Lakka TA, Laaksonen DE, Lindstrom J, Eriksson JG, Valle TT et al (2008) SNPs in PPARG associate with type 2 diabetes and interact with physical activity. Med Sci Sports Exerc 40(1):25–33

Kirchheiner J, Roots I, Goldammer M, Rosenkranz B, Brockmöller J (2005) Effect of genetic polymorphisms in cytochrome P450 (CYP) 2C9 and CYP2C8 on the pharmacokinetics of Oral antidiabetic drugs clinical relevance. Clin Pharmacokinet 44(12):1209–1225

Kirchheiner J, Thomas S, Bauer S, Jetter A, Stehle S, Tsahuridu M (2006) Pharmacokinetics and pharmacodynamics of rosiglitazone in relation to CYP2C8 genotype. Clin Pharmacol Ther 80(6):657–667

Kirchner H, Sinha I, Gao H, Ruby MA, Schönke M, Lindvall JM, Barrès R et al (2016) Altered DNA methylation of glycolytic and lipogenic genes in liver from obese and type 2 diabetic patients. Mol Metab 5:171–183

Kirpichnikov D, Mcfarlane SI, Sowers JR (2002) Metformin: an update. Ann Intern Med 137(1):25–33

Klein TE, Chang JT, Cho MK, Easton KL, Fergerson R, Hewett M et al (2001) Integrating genotype and phenotype information: an overview of the PharmGKB project. Pharmacogenomics J 1(3):167–170

Klonoff DC (2008) Personalized medicine for diabetes. J Diabetes Sci Technol 2(3):335–341

Koivula RW, Heggie A, Barnett A, Cederberg H, Hansen TH, Siloaho M et al (2014) Discovery of biomarkers for glycaemic deterioration before and after the onset of type 2 diabetes: rationale and design of the epidemiological studies within the IMI DIRECT consortium. Diabetologia 57(6):1132–1142

Koo BK, Cho YM, Park BL, Cheong HS, Shin HD, Jang HC et al (2007) Polymorphisms of KCNJ11 (Kir6.2 gene) are associated with type 2 diabetes and hypertension in the Korean population. Diabet Med 24 (2):178–186

Lander ES, Linton LM, Birren B, Nusbaum C, Zody MC, Baldwin J et al (2001) Initial sequencing and analysis of the human genome. Nature 409:860–921

Langberg KA, Sharma NK, Hanis CL, Elbein SC, Hasstedt SJ, Das SK et al (2012) Single nucleotide polymorphisms in JAZF1 and BCL11A gene are nominally associated with type 2 diabetes in African American Families from the GENNID Study. J Hum Genet 57(1):57–61

Lara-riegos JC, Ortiz-lópez MG, Peña-espinoza BI, Montúfar-robles I, Peña-rico MA (2015) Diabetes susceptibility in Mayas: evidence for the involvement of polymorphisms in HHEX, HNF4α, KCNJ11, PPARγ, CDKN2A/2B, SLC30A8, CDC123/CAMK1D, TCF7L2, ABCA1 and SLC16A11 genes. Gene 565 (1):68–75

Laukkanen O, Lindstro J, Eriksson J, Valle TT, Ha H, Ilanne-parikka P et al (2005) Polymorphisms in the SLC2A2 (GLUT2) gene are associated with the conversion from impaired glucose tolerance to type 2 diabetes the Finnish Diabetes Prevention Study. Diabetes 54(7):2256–2260

Lee H, Park H, Yang S, Kim D, Park Y (2008) STAT4 polymorphism is associated with early- onset type 1 diabetes, but not with late-onset type 1 diabetes. Ann N Y Acad Sci 1150:93–98

Leete P, Willcox A, Krogvold L, Dahl-Jørgensen K, Foulis AK, Richardson SJ et al (2016) Differential insulitic profiles determine the extent of β-cell destruction and the age at onset of type 1 diabetes. Diabetes 65:1362–1369

Letourneau LR, Carmody D, Wroblewski K, Denson AM, Sanyoura M, Naylor RN et al (2017) Diabetes presentation in infancy: high risk of diabetic ketoacidosis. Diabetes Care 40(10):e147–e148

Li Y (2013) The KCNJ11 E23K gene polymorphism and type 2 diabetes mellitus in the Chinese Han population: a meta-analysis. Mol Biol Rep 40(1):141–146

Lie BA, Thorsby E (2005) Several genes in the extended human MHC contribute to predisposition to autoimmune diseases. Curr Opin Immunol 17:526–531

Lie BA, Sollid LM, Ascher H, Ek J, Akselsen HE, Ronningen KS et al (1999) A gene telomeric of the HLA class I region is involved in predisposition to both type 1 diabetes and coeliac disease. Tissue Antigens 54:162–168

Ling C, Rönn T (2019) Epigenetics in human obesity and type 2 diabetes. Cell Metab 29(5):1028–1044

Liu S, Wang H, Jin Y, Podolsky R, Reddy MV, Pedersen J et al (2009) IFIH1 polymorphisms are significantly associated with type 1 diabetes and IFIH1 gene expression in peripheral blood mononuclear cells. Hum Mol Genet 18(2):358–365

Liu HW, Xu RY, Sun RP, Wang Q, Liu JL, Ge W et al (2015) Association of PTPN22 gene polymorphism with type 1 diabetes mellitus in Chinese children and adolescents. Genet Mol Res 14(1):63–68

Lyssenko V, Nagorny CLF, Erdos MR, Wierup N, Jonsson A, Spégel P et al (2009a) A common variant in the melatonin receptor gene (MTNR1B) is associated with increased risk of future type 2 diabetes and impaired early insulin secretion. Nat Genet 41 (1):82–88

Lyssenko V, Nagorny CL, Erdos MR, Wierup N, Jonsson A, Spégel P et al (2009b) Common variant in MTNR1B associated with increased risk of type

2 diabetes and impaired early insulin secretion. Nat Genet 41(1):82–88

Magzoub MM, Stephens HA, Sachs JA, Biro PA, Cutbush S, Wu Z et al (1992) HLA-DP polymorphism in Sudanese controls and patients with insulin-dependent diabetes mellitus. Tissue Antigens 40 (2):64–68

Malandrino N, Smith RJ (2011) Personalized medicine in diabetes. Clin Chem 57(2):231–240

Malhotra A, Igo RP Jr, Thameem F, Kao WH, Abboud HE, Adler SG et al (2009) Genome-wide linkage scans for type 2 diabetes mellitus in four ethnically diverse populations-significant evidence for linkage on chromosome 4q in African Americans: the Family Investigation of Nephropathy and Diabetes Research Group. Diabetes Metab Res Rev 25:740–747

Mark M, Andrew TH (2008) Novel insights arising from the definition of genes for monogenic and type 2 diabetes. Diabetes 57:2889–2898

Mathur AN et al (2007) Stat3 and Stat4 direct development of IL-17-secreting Th cells. J Immunol 178:4901–4907

Mattana TC, Santos AS, Fukui RT, Mainardi-novo DTO, Costa VS, Santos RF et al (2014) CD226 rs763361 is associated with the susceptibility to type 1 diabetes and greater frequency of GAD65 autoantibody in a Brazilian cohort. Mediat Inflamm 2014:694948

Mayans S, Lackovic K, Lindgren P, Ruikka K, Holmberg D (2007) TCF7L2 polymorphisms are associated with type 2 diabetes in northern Sweden. Eur J Hum Genet 15(3):342–346

Mctaggart JS, Clark RH, Ashcroft FM (2010) The role of the KATP channel in glucose homeostasis in health and disease: more than meets the islet. J Physiol 588 (Pt 17):3201–3209

Miao F, Chen Z, Zhang L, Liu Z, Wu X, Yuan YC et al (2012) Profiles of epigenetic histone post-translational modifications at type 1 diabetes susceptible genes. J Biol Chem 287:16335–16345

Moltke I, Grarup N, Jørgensen ME, Bjerregaard P, Treebak JT, Fumagalli M et al (2014) A common Greenlandic TBC1D4 variant confers muscle insulin resistance and type 2 diabetes. Nature 512 (7513):190–193

Moore F, Colli ML, Cnop M, Esteve MI, Cardozo AK, Cunha DA, Bugliani M, Marchetti P, Eizirik DL (2009) PTPN2, a candidate gene for type 1 diabetes, modulates interferon-gamma-induced pancreatic beta-cell apoptosis. Diabetes 58:1283–1291. [CrossRef] [PubMed]

Morran MP, Andrew Vonberg A, Khadra A, Pietropaolo M (2015) Immunogenetics of type 1 diabetes mellitus. Mol Asp Med 42:42–60

Naylor RN, John PM, Winn AN, Carmody D, Greeley SAW, Philipson LH et al (2014) Cost effectiveness of MODY genetic testing: translating genomic advances into practical health applications. Diabetes Care 37 (1):202–209

Nejentsev S, Howson JM, Walker NM et al (2007) Localization of type 1 diabetes susceptibility to the MHC class I genes HLA-B and HLA-A. Nature 450:887–892

Nejentsev S, Walker N, Riches D, Egholm M, Todd JA (2009) Rare variants of IFIH1, a gene implicated in antiviral responses, protect against type 1 diabetes. Science 324:387–389

Nielsen EM, Hansen L, Carstensen B, Echwald SM, Drivsholm T et al (2003) The E23K variant of Kir6.2 associates with impaired post-OGTT serum insulin response and increased risk of type 2 diabetes. Diabetes 52:573–577. 6

Nikitin AG, Lavrikova EY, Seregin YA, Zilberman LI, Tzitlidze NM, Kuraeva TL et al (2010) Association of the Polymorphisms of the ERBB3 and SH2B3 genes with type 1 diabetes. Mol Biol (Mosk) 44(2):257–262

Nikolac N, Simundic A, Katalinic D, Topic E, Cipak A (2009) Metabolic control in type 2 diabetes is associated with sulfonylurea Receptor-1 (SUR-1) but not with KCNJ11 polymorphisms. Arch Med Res 40 (5):387–392

Nilsson E, Jansson PA, Perfilyev A, Volkov P, Pedersen M, Svensson MK et al (2014) Altered DNA methylation and differential expression of genes influencing metabolism and inflammation in adipose tissue from subjects with type 2 diabetes. Diabetes 63:2962–2976

Nilsson E, Matte A, Perfilyev A, de Mello VD, Kakela P, Pihlajamaki J, Ling C (2015) Epigenetic alterations in human liver from subjects with type 2 diabetes in parallel with reduced folate levels. J Clin Endocrinol Metab 100:E1491–E1501

Nitert MD, Dayeh T, Volkov P, Elgzyri T, Hall E, Nilsson E, Yang BT, Lang S, Parikh H, Wessman Y et al (2012) Impact of an exercise intervention on DNA methylation in skeletal muscle from first-degree relatives of patients with type 2 diabetes. Diabetes 61:3322–3332

Noble JA, Valdes AM (2011) Genetics of the HLA region in the prediction of type 1 diabetes. Curr Diab Rep 11 (6):533–542

Noble JA, Valdes AM, Cook M, Klitz W, Thomson G, Erlich HA (1996) The role of HLA class II genes in insulin-dependent diabetes mellitus: molecular analysis of 180 Caucasian, multiplex families. Am J Hum Genet 59(5):1134–1148

Noble JA, Valdes AM, Bugawan TL, Apple RJ, Thomson G, Erlich HA (2002) The HLA class I A locus affects susceptibility to type 1 diabetes. Hum Immunol 63:657–664

Noble JA, Valdes AM, Varney DM, Carlson AJ, Moonsamy P, Fear AL et al (2010) HLA class I and genetic susceptibility to type 1 diabetes: results from the type 1 diabetes genetics consortium. Diabetes 59 (11):2972–2979

Nogueira TC, Paula FM, Villate O et al (2013) GLIS3, a susceptibility gene for type 1 and type 2 diabetes, modulates pancreatic beta cell apoptosis via regulation of a splice variant of the BH3-only protein Bim. PLoS Genet 9:e1003532

Nyaga DM, Vickers MH, Jefferies C, Perry JK, O'Sullivan JM (2018) The genetic architecture of type 1 diabetes mellitus. Mol Cell Endocrinol 477:70–80

O'Garra A, Vieira P (2004) Regulatory T cells and mechanisms of immune system control. Nat Med 10:801–805

Onkamo P, Vaananen S, Karvonen M, Tuomilehto J (1999) Worldwide increase in incidence of type ☐ diabetes—the analysis of the data on published incidence trends. Diabetologia 42:1395–1403

Owen MR, Doran E, Halestrap AP (2000) Evidence that metformin exerts its anti-diabetic effects through inhibition of complex 1 of the mitochondrial respiratory chain. Biochem J 348(Pt 653 3):607–614

Pe'er I, de Bakker PI, Maller J et al (2006) Evaluating and improving power in whole-genome association studies using fixed marker sets. Nat Genet 38(6):663–667

Pearson ER, Liddell WG, Shepherd M, Hattersley AT (2000) Sensitivity to sulphonylureas in patients with hepatocyte nuclear factor-1 a gene mutations: evidence for pharmacogenetics in diabetes. Diabet Med 17 (7):543–545

Pearson ER, Starkey BJ, Powell RJ, Gribble FM, Clark PM, Hattersley AT (2003) Mechanisms of disease genetic cause of hyperglycaemia and response to treatment in diabetes. Lancet 362(9392):1275–1281

Pearson ER, Flechtner I, Njølstad PR, Malecki MT, Flanagan SE, Larkin B et al (2006) Switching from insulin to Oral sulfonylureas in patients with diabetes due to Kir6.2 mutations. N Engl J Med 355 (5):467–477

Plourde G, Matte M-E (2017) Personalised medicine for the treatment of T2DM. Endocrinol Diab Metab J S1 (107):1–7

Poulsen SB, Fenton RA, Rieg T (2015) Sodium-glucose cotransport. Curr Opin Nephrol Hypertens 24:463–469

Prasad MV, Reddy L, Wang H, Liu S, Bode B, Reed JC et al (2011) Association between type 1 diabetes and GWAS SNPs in the southeast US Caucasian population. Genes Immun 12(3):208–212

Project Team SG (2015) The Saudi human genome program: an oasis in the desert of Arab medicine is providing clues to genetic disease. IEEE Pulse 6 (6):22–26

Prokopenko I, Langenberg C, Florez JC, Saxena R, Soranzo N, Thorleifsson G et al (2009) Variants in MTNR1B influence fasting glucose levels. Nat Genet 41(1):77–81

Prudente S, Morini E, Lucchesi D, Lamacchia O, Bailetti D, Mercuri L et al (2014) IRS1 G972R missense polymorphism is associated with failure to oral antidiabetes drugs in white patients with type 2 diabetes from Italy. Diabetes 63(9):3135–3140

Qi Q, Hu FB (2012) Genetics of type 2 diabetes in European populations. J Diabetes 4(3):203–212. 250

Qian Y, Dong M, Lu F, Li H, Jin G, Hu Z (2015) Molecular and cellular endocrinology joint effect of CENTD2 and KCNQ1 polymorphisms on the risk of type 2 diabetes mellitus among Chinese Han population. Mol Cell Endocrinol 407:46–51

Qu H-Q, Bradfield JP, Grant SF, Hakonarson H, Polychronakos C (2009) Remapping the type diabetes association of the CTLA4 locus. Genes Immun 10: 827–832

Ragia G, Petridis I, Tavridou A, Christakidis D (2009) Presence of CYP2C9∗3 allele increases risk of hypoglycemia in Type 2 diabetic patients treated with sulfonylureas 2009;1781–7. Hypoglycemia in Type 2 diabetic patients treated with sulfonylureas. Pharmacogenomics 10(11):1781–1787

Raha O, Sarkar B, Lakkakula BV, Pasumarthy V, Godi S, Chowdhury S et al (2013) HLA class II SNP interactions and the association with type 1 diabetes mellitus in Bengali speaking patients of Eastern India. J Biomed Sci 20(1):12

Rajalingam R, Ge P, Reed EF (2004) A sequencing-based typing method for HLA-DQA1 alleles. Hum Immunol 65(4):373–379

Ramesh R, Munish M, Quang TN, Irma L, Bernhard OB, Flemming P et al (2016) Systematic evaluation of genes and genetic variants associated with type 1 diabetes susceptibility. J Immunol 196(7):3043–3053

Ramos-Rodríguez M, Raurell-Villa H, Colli ML et al (2019) The impact of pro-inflammatory cytokines on the β-cell regulatory landscape provides new insights into the genetics of type 1 diabetes. BioRxiv 560193

Rask-Andersen M, Karlsson T, Ek WE, Johansson Å (2017) Gene-environment interaction study for BMI reveals interactions between genetic factors and physical activity, alcohol consumption and socioeconomic status. PLoS Genet 13(9):e1006977

Redondo MJ, Jeffrey J, Fain PR, Eisenbarth GS, Orban T (2008) Concordance for islet autoimmunity among monozygotic twins. N Engl J Med 359(26):2849–2850

Rees SD, Hydrie MZ, Shera AS, Kumar S, O'Hare JP et al (2011) Replication of 13 genome-wide association (GWA)-validated risk variants for type 2 diabetes in Pakistani populations. Diabetologia 54:1368–1374. 40

Ribel-Madsen R, Fraga MF, Jacobsen S, Bork-Jensen J, Lara E, Calvanese V et al (2012) Genome-wide analysis of DNA methylation differences in muscle and fat from monozygotic twins discordant for type 2 diabetes. PLoS One 7(12):e.51302

Roglic G (2016) WHO global report on diabetes: a summary. Int J NonCommun Dis 1:3–8

Salem SD, Saif-ali R, Ismail IS, Al-hamodi Z, Muniandy S (2014) Contribution of SLC30A8 variants to the risk of type 2 diabetes in a multi-ethnic population: a case control study. BMC Endocr Disord 14:2

Santiago JL, Martínez A, De Calle H, Fernández-arquero M, Figueredo MÁ, De Concha EG et al (2007) Susceptibility to type 1 diabetes conferred by the PTPN22 C1858T polymorphism in the Spanish population. BMC Med Genet 8(1):54

Santiago JL, Alizadeh BZ, Espino L, Figueredo MA, Roep BO, Koeleman BPC et al (2008) Study of the association between the CAPSL-IL7R locus and type 1 diabetes. Diabetologia 51(9):1653–1658

Santin I, Moore F, Colli ML, Gurzov EN, Marselli L, Marchetti P et al (2011) PTPN2, a candidate gene for type 1 diabetes, modulates pancreatic beta-cell apoptosis via regulation of the BH3-only protein bim. Diabetes 60:3279–3288

Sato R, Watanabe H, Genma R, Takeuchi M, Maekawa M, Nakamura H (2010) ABCC8 polymorphism (Ser1369Ala): influence on severe hypoglycemia due to sulfonylureas. Pharmacogenomics 11 (12):1743–1750

Saxena R, Hivert MF, Langenberg C, Tanaka T, Pankow JS, Vollenweider P et al (2010) Genetic variation in *GIPR* influences the glucose and insulin responses to an oral glucose challenge. Nat Genet 42(2):142–148

Scheen AJ (2015) Pharmacodynamics, efficacy and safety of sodium-glucose cotransporter type 2 (SGLT2) inhibitors for the treatment of type 2 diabetes mellitus. Drugs 75:33–59

Schernthaner G, Grimaldi A, Di Mario U, Drzewoski J, Kempler P, Kvapil M et al (2004) GUIDE study: double-blind comparison of once-daily gliclazide MR and glimepiride in type 2 diabetic patients. Eur J Clin Investig 34(8):535–542

Schroner Z, Javorsky M, Tkacova R, Klimcakova L, Dobrikova M, Habalova V et al (2011) Effect of sulphonylurea treatment on glycaemic control is related to TCF7L2 genotype in patients with type 2 diabetes. Diabetes Obes Metab 13(1):89–91

Scott LJ, Mohlke KL, Bonnycastle LL, Willer CJ, Li Y, Duren WL et al (2007) A genome-wide association study of type 2 diabetes in Finns detects multiple susceptibility variants. Science 316(5829):1341–1345

Semiz S, Dujic T, Ostanek B, Prnjavorac B, Bego T, Malenica M et al (2010) Analysis of CYP2C9∗2, CYP2C19∗2, and CYP2D6∗4 polymorphisms in patients with type 2 diabetes mellitus. Bosn J Basic Med Sci 10(4):287–291

Senee V, Chelala C, Duchatelet S, Feng D, Blanc H, Cossec JC, Charon C, Nicolino M, Boileau P, Cavener DR et al (2006) Mutations in glis3 are responsible for a rare syndrome with neonatal diabetes mellitus and congenital hypothyroidism. Nat Genet 38:682–687

Sharif FA, Shubair ME, Zaharna MM, Ashour MJ, Altalalgah IO, Najjar M et al (2018) Genetic polymorphism and risk of having type 2 diabetes in a Palestinian population: a study of 16 gene polymorphisms. Adv Diab Endocrinol 3(1):6

Sharma NK, Langberg KA, Mondal AK, Elbein SC, Das SK (2011) Type 2 diabetes (T2D) associated polymorphisms regulate expression of adjacent transcripts in transformed lymphocytes, adipose, and muscle from Caucasian and African-American subjects. J Clin Endocrinol Metab 96(2):E394–E403

Sharp RC, Abdulrahim M, Naser ES, Naser SA (2015) Genetic variations of PTPN2 and PTPN22: role in the pathogenesis of type 1 diabetes and Crohn's disease. Front Cell Infect Microbiol 5:95

Shaw JE, Sicree RA (2010) Zimmet PZ global estimates of the prevalence of diabetes for 2010 and 2030. Diabetes Res Clin Pract 87:4–14

Shaw RJ, Lamia KA, Vasquez D, Koo S, Depinho RA, Montminy M et al (2005) The kinase LKB1 mediates glucose homeostasis in liver and therapeutic effects of metformin. Science 310(5754):1642–1646

Shepherd M, Shields B, Ellard S, Hattersley AT (2009) Short report a genetic diagnosis of HNF1A diabetes alters treatment and improves glycaemic control in the majority of insulin-treated patients. Diabet Med 26 (4):437–441

Shi Y, Li Y, Wang J, Wang C, Fan J, Zhao J et al (2017) Meta-analyses of the association of G6PC2 allele variants with elevated fasting glucose and type 2 diabetes. PLoS One 12(7):e0181232

Shields BM, Hicks S, Shepherd MH (2010) Maturity-onset diabetes of the young (MODY): how many cases are we missing ? Diabetologia 53 (12):2504–2508

Shu Y, Sheardown SA, Brown C, Owen RP, Zhang S, Castro RA et al (2007) Effect of genetic variation in the organic cation transporter 1 (OCT1) on metformin action. J Clin Invest 117(5):1422–1431

Sladek R, Rocheleau G, Rung J, Dina C, Shen L, Serre D et al (2007) A genome-wide association study identifies novel risk loci for type 2 diabetes. Nature 445:881–885

Smyth DJ, Plagnol V, Walker NM, Cooper JD, Downes K, Yang JHM et al (2008) Shared and distinct genetic variants in type 1 diabetes and celiac disease. N Engl J Med 359(26):2767–2777

Steinthorsdottir V, Thorleifsson G, Reynisdottir I, Benediktsson R, Jonsdottir T, Walters GB et al (2007) A variant in CDKAL1 influences insulin response and risk of type 2 diabetes. Nat Genet 39:770–775

Stumvoll M, Haring H (2002) The peroxisome proliferator-activated receptor-gamma2 Pro12Ala polymorphism. Diabetes 51(8):2341–2347

Sun Q, Song K, Shen X, Cai Y (2012) The association between KCNQ1 gene polymorphism and type 2 diabetes risk: a meta-analysis. PLoS One 7(11):e48578

Sun C, Wei H, Chen X, Zhao Z, Du H, Song W et al (2016) ERBB3 -rs2292239 as primary type 1 diabetes association locus among non- HLA genes in Chinese. Meta Genet 9:120–123

Tang W, Cui D, Jiang L, Zhao L, Qian W, Long SA et al (2015) Association of common polymorphisms in the IL2RA gene with type 1 diabetes: evidence of 32,646 individuals from 10 independent studies characteristics of study. J Cell Mol Med 19(10):2481–2488

Taniyama M, Maruyama T, Tozaki T, Nakano Y, Ban Y (2010) Association of PTPN22 haplotypes with type 1 diabetes in the Japanese population. Hum Immunol 71(8):795–798

Tarasova L, Kalnina I, Geldnere K, Bumbure A, Ritenberga R, Nikitina-zake L et al (2012) Association of genetic variation in the organic cation transporters OCT1, OCT2 and multidrug and toxin extrusion 1 transporter protein genes with the gastrointestinal side effects and lower BMI in metformin-treated type 2 diabetes patients. Pharmacogenet Genomics 22 (9):659–667 66

Teo YY, Inouye M, Small KS, Gwilliam R, Kwiatkowski DP, Clark TG (2007) A genotype calling algorithm for the Illumina BeadArray platform. Bioinformatics 23 (20):2741–2746

The International HapMap Consortium (2003) The international HapMap project. Nature 426:789–796

Tian Y, Xu J, Huang T, Cui J, Zhang W, Song W et al (2019) A novel polymorphism (rs35612982) in CDKAL1 is a risk factor of type 2 diabetes: a case-control study. Kidney Blood Press Res 44:1313–1326

Todd JA, Walker NM, Cooper JD, Smyth DJ, Downes K, Plagnol V et al (2007) Robust associations of four new chromosome regions from genome-wide analyses of type 1 diabetes. Nat Genet 39:857–864

Törn C, Hadley D, Lee H, Hagopian W, Lernmark Å, Simell O et al (2015) Role of type 1 diabetes–associated SNPs on risk of autoantibody positivity in the TEDDY study. Diabetes 64(5):1818–1829

Valdes AM, Erlich HA, Noble JA (2005) Human leukocyte antigen class I B and C loci contribute to type 1 diabetes (T1D) susceptibility and age at T1D onset. Hum Immunol 66:301–313

Vassy JL, Meigs JB (2012) Is genetic testing useful to predict type 2 diabetes? Best Pract Res Clin Endocrinol Metab 26(2):189–201

Vella A, Cooper JD, Lowe CE, Walker N, Nutland S, Widmer B et al (2005) Localization of a type 1 diabetes locus in the IL2RA/CD25 region by use of tag single-nucleotide polymorphisms. Am J Hum Genet 76 (5):773–779

Venter JC, Adams MD, Myers EW, Li PW, Mural RJ, Sutton GG et al (2001) The sequence of the human genome. Science 291:1304–1351

Voight BF, Scott LJ, Steinthorsdottir V, Andrew P, Aulchenko YS, Thorleifsson G et al (2010) Twelve type 2 diabetes susceptibility loci identified through large-scale association analysis. Nat Genet 42 (7):579–589

Von Mering J (1885) Uber Kunstliche diabetes. Centralbl Med Wiss 23:531–532. 5

Wallace C, Smyth DJ, Maisuria-armer M, Walker NM, Todd JA (2010) The imprinted DLK1-MEG3 gene region on chromosome 14q32. 2 alters susceptibility to type 1 diabetes. Nat Genet 42(1):68–71

Wallet MA, Santostefano KE, Terada N, Brusko TM (2017) Isogenic cellular systems model the impact of genetic risk variants in the pathogenesis of type 1 diabetes. Front Endocrinol (Lausanne) 8:276

Wang K, Baldassano R, Zhang H, Qu H, Imielinski M, Kugathasan S et al (2010) Comparative genetic analysis of inflammatory bowel disease and type 1 diabetes implicates multiple loci with opposite effects. Hum Mol Genet 19(10):2059–2067

Wang K, Owusu D, Pan Y, Xu C (2014a) Common genetic variants in the HNF 1 B gene contribute to diabetes and multiple cancers. Austin Biomark Diagn 1(1):5

Wang J, Zhang J, Shen J, Hu D, Yan G, Liu X et al (2014b) Association of KCNQ1 and KLF14 polymorphisms

and risk of type 2 diabetes mellitus: a global meta-analysis. Hum Immunol 75(4):342–347

Weedon MN, Clark VJ, Qian Y, Ben-Shlomo Y, Timpson N, Ebrahim S et al (2006) A common haplotype of the glucokinase gene alters fasting glucose and birth weight: association in six studies and population-genetics analyses. Am J Hum Genet 79(6):991–1001

Wellcome Trust Case Control Consortium (2007) Genome-wide association study of 14,000 cases of seven common diseases and 3,000 shared controls. Nature 447:661–678

Welter M, Frigeri HR, Rea RR, De Souza M, Alberton D, Picheth G et al (2015) The rs10885122 polymorphism of the adrenoceptor alpha 2A (ADRA2A) gene in Euro-Brazilians with type 2 diabetes mellitus. Arch Endocrinol Metab 59(1):29–33

Xia Q, Chen Z, Wang Y, Ma Y, Zhang F, Che W et al (2012) Association between the melatonin receptor 1B gene polymorphism on the risk of type 2 diabetes, impaired glucose regulation: a Meta-analysis. PLoS One 7(11):e50107

Yamagata K, Hanafusa T, Nakajima H, Sada M, Amemiya H, Tomita K et al (1991) HLA-DP and susceptibility to insulin-dependent diabetes mellitus in Japanese. Tissue Antigens 38(3):107–110

Yan F, Casey J, Shyng S (2006) Sulfonylureas correct trafficking defects of disease-causing ATP-sensitive potassium channels by binding to the channel complex. J Biol Chem 281(44):33403–33413

Yang Y, Chang BH, Samson SL, Li MV, Chan L (2009) The kruppel-like zinc finger protein GLIS3 directly and indirectly activates insulin gene transcription. Nucleic Acids Res 37:2529–2538

Yang Y, Chang BH, Chan L (2013) Sustained expression of the transcription factor GLIS3 is required for normal beta cell function in adults. EMBO Mol Med 5:92–104

Yasuda K, Miyake K, Horikawa Y, Hara K, Osawa H, Furuta H et al (2008) Variants in KCNQ1 are associated with susceptibility to type 2 diabetes mellitus. Nat Genet 40(9):1092–1097

Yokoi N, Kanamori M, Horikawa Y, Takeda J, Sanke T et al (2006) Association studies of variants in the genes involved in pancreatic beta-cell function in type 2 diabetes in Japanese subjects. Diabetes 55:2379–2386

Yu M, Xu X, Yin J, Wu J, Chen X, Gong Z et al (2010) KCNJ11 Lys23Glu and TCF7L2 rs290487 (C/T) polymorphisms affect therapeutic efficacy of repaglinide in Chinese patients with type 2 diabetes. Clin Pharmacol Ther 87(3):330–335

Zavatari P, Deidda E, Pitzalis M, Zoa B, Moi L, Lampis R et al (2004) No association between variation of FOXP3 gene and common type 1 diabetes in Sardinian population. Diabetes 53:1911–1914

Zavattari P, Lampis R, Motzo C, Loddo M, Mulargia A, Whalen M et al (2001) Conditional linkage disequilibrium analysis of a complex disease super locus, IDDM1 in the HLA region, reveals the presence of independent modifying gene effects influencing the type 1 diabetes risk encoded by the major

HLA-DQB1, -DRB1 disease loci. Hum Mol Genet 10:881–889

Zhang K, Huang Q, Dai X, Yin J, Zhang W, Zhou G et al (2010) Effects of the peroxisome proliferator activated receptor-γ coactivator-1α (PGC-1α) Thr394Thr and Gly482Ser polymorphisms on rosiglitazone response in Chinese patients with type 2 diabetes mellitus. J Clin Pharmacol 50(9):1022–1030

Zhou G, Myers R, Li Y, Chen Y, Shen X, Fenyk-melody J et al (2001) Role of AMP-activated protein kinase in mechanism of metformin action. J Clin Invest 108 (8):1167–1174

Zhou D, Zhang D, Liu Y, Zhao T, Chen Z, Liu Z et al (2009a) The E23K variation in the KCNJ11 gene is associated with type 2 diabetes in Chinese and east Asian population. J Hum Genet 54(7):433–435

Zhou K, Donnelly L, Burch L, Tavendale R, Doney ASF, Leese G et al (2009b) Loss-of-Function CYP2C9 variants improve therapeutic response to sulfonylureas in type 2 diabetes: a Go- DARTS study. Clin Pharmacol Ther 87(1):52–56

Zoledziewska M, Costa G, Pitzalis M, Cocco E, Melis C, Moi L (2009) Variation within the CLEC16A gene shows consistent disease association with both multiple sclerosis and type 1 diabetes in Sardinia. Genes Immun 10(1):15–17

Zullo A, Sommese L, Nicoletti G, Donatelli F, Mancini FP, Napoli C (2017) Epigenetics and type 1 diabetes: mechanisms and translational applications. Transl Res 185:85–93

Adv Exp Med Biol - Advances in Internal Medicine (2020) 4: 499–519
https://doi.org/10.1007/5584_2020_515
© Springer Nature Switzerland AG 2020
Published online: 20 March 2020

Dietary SCFAs Immunotherapy: Reshaping the Gut Microbiota in Diabetes

Yu Anne Yap and Eliana Mariño

Abstract

Diet-microbiota related inflammatory conditions such as obesity, autoimmune type 1 diabetes (T1D), type 2 diabetes (T2D), cardiovascular disease (CVD) and gut infections have become a stigma in Western societies and developing nations. This book chapter examines the most relevant pre-clinical and clinical studies about diet-gut microbiota approaches as an alternative therapy for diabetes. We also discuss what we and others have extensively investigated- the power of dietary short-chain fatty acids (SCFAs) technology that naturally targets the gut microbiota as an alternative method to prevent and treat diabetes and its related complications.

Keywords

Clinical trials · Diet · Microbiota · SCFAs · T1D · T2D

Diabetes is a chronic immune-metabolic disease in which different mechanisms cause insulin deficiency and impaired insulin action, essential for regulating blood glucose levels. A persistent elevation of glucose concentration in the blood, also known as hyperglycemia, can cause damage to various organs in the body. If left untreated, individuals with hyperglycemia are at a significantly higher risk of developing life-threatening health complications, including CVD, infections, and kidney failure (International Diabetes Federation 2017). Although the exact cause of diabetes remains undetermined, the interactions between genetic and environmental factors cause inflammation, β-cell damage or dysfunction, and hyperglycemia, leading to the diagnosis of diabetes and with it, an increased risk of morbidity and mortality. Many epidemiological studies point to diet as one of the most influential lifestyle factors contributing to the rise of diabetes (Thorburn et al. 2014; Marino 2016). Diet alters the balance of the commensal gut microbiota and the availability and production of microbial metabolites such as SCFAs that can affect many physiological processes. As such, the interplay between the microbiota, metabolism, the immune system, and the nervous system is fundamental in determining the fate of diabetes.

Y. A. Yap and E. Mariño (✉)
Infection and Immunity Program, Biomedicine Discovery Institute, Department of Biochemistry, Monash University, Melbourne, VIC, Australia
e-mail: eliana.marino@monash.edu

1 Factors Influencing the Gut Microbiota and Type 1 Diabetes

Insulin-dependent diabetes, commonly known as type 1 diabetes (T1D), is an organ-specific autoimmune disease that arises from the immune-mediated destruction of pancreatic β-cells that produce insulin (Atkinson et al. 2014). High concentrations of blood glucose induce classic symptoms of T1D such as increased urination (polyuria) and thirst (polydipsia), and uncharacteristic weight loss. Some individuals may also experience fatigue, increased hunger, diminished visual acuity, and numbness in the hands and feet (Harrison 2019). Patients with T1D typically require a lifelong need for exogenous insulin replacement to sufficiently manage the disease (Flier et al. 1986; Atkinson and Eisenbarth 2001; Bluestone et al. 2010). Most commonly diagnosed in young people between the age of 10 and 14 years, it has been estimated that around 85% of T1D patients are 20 years and under (Maahs et al. 2010).

In the past decade, the global incidence of T1D in children <14 years old has increased by 30%, at an average rate of 2.8% per year (Maahs et al. 2010). The development of beta-cell autoimmunity occurs very early in life before the onset of clinical T1D (Krischer et al. 2015). During this early period, the widespread use of antibiotics has been proposed as a cause for the growing rate of T1D (Boursi et al. 2015; Abela and Fava 2013; Livanos et al. 2016). Antibiotics cause profound changes to the gut microbiota (Koren et al. 2012; Tamburini et al. 2016), which play a central role in the development of the infant immune system (Belkaid and Hand 2014). Likewise, mother's nutrition is key to establishing the microbiota in infants (Marino 2016; Koren et al. 2012). During the perinatal stage (pregnancy and breastfeeding), T1D has not yet manifested, but subtle biological changes that contribute to the development of the disease might be already occurring (Tamburini et al. 2016). However, very little is known about the early immune and microbial events that happen during the perinatal stage and how these may impact disease progression.

It has been shown that a drop in the diversity of the gut microbiota occurs in infants well before the onset of TID (Kostic et al. 2015). Similarly, infants present a distinct autoreactive gene signature well before the appearance of beta-cell antigen-specific memory T cells or autoantibodies (Heninger et al.

2017; Mehdi et al. 2018). Thus, strategies to reprogram the immune system to tackle T1D represent an entirely novel approach. Diet shapes the composition of the gut microbiota (Kau et al. 2011) and therefore modulates the production of microbial SCFAs. It has been shown that non-obese diabetic (NOD) mice treated with antibiotics during the perinatal stage cause profound changes in the gut microbiota and accelerate the development of T1D in the offspring (Boursi et al. 2015; Livanos et al. 2016). A single antibiotic treatment not only significantly altered taxonomic and metagenomic composition but also reduced the production of host-signaling microbial SCFAs early in life (Zhang et al. 2018).

Antibiotics are commonly used during pregnancy and delivery (Martinez de Tejada 2014) for treatment of urinary and gastrointestinal (GI) infections, and bacterial vaginosis, as well as for preventative measures during the intrapartum/ peripartum stages to decrease the risk of infection in mother and/or infant after delivery (Tormo-Badia et al. 2014). At least 11 types of broad-spectrum antibiotics can cross the placenta and reach the fetus (Nahum et al. 2006). Antibiotics during fetal development and breastfeeding can induce metabolic changes (Cho et al. 2012) via the gut microbiota (Cox et al. 2014) and increase gut permeability (Kerr et al. 2015), thus accelerating T1D development (Livanos et al. 2016; Rautava et al. 2012). In contrast, findings from a Danish population-based case-control study found no association between overall antibiotic exposure in childhood and the risk of developing T1D (Mikkelsen et al. 2017; Kemppainen et al. 2017). Despite conflicting findings on the impact of antibiotics on T1D risk, there is mounting evidence that perturbations to the gut microbiota may significantly affect disease pathogenesis (Jamshidi et al. 2019).

It is also well understood that diet can shape the gut microbiota (Kau et al. 2011; Makki et al. 2018). The Environmental Determinants of Diabetes in the Young (TEDDY) study found that compared to formula fed babies, breastfeeding was associated with higher levels of Bifidobacterium species (Stewart et al. 2018), a lack of which has been observed in the stool of children with β-cell autoimmunity (de Goffau et al. 2013). Certain dietary components such as protein have been proposed to increase the risk of developing T1D (Serena et al. 2015; Lefebvre et al. 2006). Disease incidence in NOD mice on gluten-free diets were found to be significantly reduced (Marietta et al. 2013; Funda et al. 1999). Furthermore, this protective effect was observed in pups from pregnant NOD mice on a gluten-free diet, even after they had been weaned to a standard diet (Hansen et al. 2014). In humans a gluten-free diet reduced GI symptoms and severe hypoglycemia while significantly increasing the need for exogeneous insulin (Abid et al. 2011; Hansen et al. 2006).

2 Dietary Short-Chain Fatty Acids and Type 1 Diabetes

Yet another component in our diet, dietary fiber, has shown promise in alleviating many inflammatory diseases in recent years (Thorburn et al. 2014; Richards et al. 2016). The gut microbiota ferment undigestible carbohydrates to produce metabolites such as the SCFAs acetate, propionate and butyrate, which are absorbed by the colon epithelium and have downstream effects on metabolic and immune responses (Richards et al. 2016). Research from our lab has demonstrated that the deterioration of immune tolerance in T1D is strongly associated with an altered microbiota and deficiency of acetate and butyrate in mice (Marino et al. 2017). Likewise, T1D individuals have shown reduced bacterial pathways involved in the bio-synthesis of these SCFAs (Vatanen et al. 2018). In a cross-sectional study, similar albeit modest differences have been found in fecal butyrate concentrations in subjects with established T1D, who also exhibited lower intestinal alkaline phosphatase (IAP) activity, immunoglobulin A (IgA) antibody concentrations and elevated fecal calprotectin concentrations (Lassenius et al. 2017). These findings are also supported by reduced SCFA-producing bacteria observed in T1D and T2D individuals (de Goffau et al. 2013, 2014;

Zhao et al. 2018). Thus, one promising therapy appears to be reducing inflammatory responses and inducing immune tolerance through the use of gut microbiota-derived SCFAs.

We have demonstrated that a breakthrough diet intervention protected 90% of NOD mice against T1D, yielding exceptionally high levels of fecal and systemic concentrations of the respective SCFAs acetate and butyrate without any detrimental effects. SCFAs-induced T1D protection happened via changes in gut/immune regulation by expanding regulatory T cells (Tregs) and reducing pathogenic B cells, CD4$^+$, and CD8$^+$ T cells. Diets rich in acetate and butyrate not only reduced the levels of serum LPS and pro-inflammatory interleukin 21 (IL-21) but also increased the concentrations of serum IL-22, an important cytokine that maintains a healthy commensal microbiota, gut epithelial integrity, mucosal immunity and ameliorates metabolic disease (Hasnain et al. 2014; Wang et al. 2014; Dudakov et al. 2015; Sabat et al. 2014).

In T1D, the effects of SCFAs are beyond the gut and the immune system. SCFAs also increase the production of antimicrobial peptides (AMPs) in β-cells (Sun et al. 2015). As previously shown, C-type lectin regenerating islet-derived protein 3 gamma (REGIIIγ) and defensins disrupt surface membranes of bacteria enabling a broad regulation of commensal and pathogenic bacteria in the gut (Gallo and Hooper 2012; Bevins and Salzman 2011; Mukherjee and Hooper 2015). NOD mice present defective production of cathelicidin-related antimicrobial peptide (CRAMP) in insulin-secreting β-cells and administration of soluble butyrate stimulated CRAMP production via G protein-coupled receptors (GPCRs) correlating with the conversion of inflammatory immune cells to a regulatory phenotype (Sun et al. 2015). Likewise, we demonstrated that microbial SCFAs contribute to increased concentrations of serum IL-22 (Marino et al. 2017), which has been shown to be required for β-cell regeneration by up-regulating the expression of regenerating Reg1 and Reg2 genes in the islets (Hill et al. 2013). Altogether, these findings establish an important role of dietary SCFAs in T1D.

There is evidence that mice and humans with T1D display compromised gut integrity and dysbiosis associated with GI inflammation (Alam et al. 2010; Bolla et al. 2017; Bosi et al. 2006; Lee et al. 2010; Leeds et al. 2011; Halling et al. 2017), similar to other inflammatory or autoimmune gut diseases such as infections, celiac disease and inflammatory bowel disease (IBD). Diabetes sufferers present symptoms such as nausea, heartburn, vomiting, diarrhea, abdominal pain, and constipation (Du et al. 2018; Maleki et al. 2000). The gut microbiota and the enteric nervous system (ENS) play a critical role in diabetic gastrointestinal motility disorders. As such, slow GI motility leads to alterations of the gut microbiota that favors pathogenic bacterial overgrowth and subsequently diarrhea (Nguyen et al. 2014; Sellin and Hart 1992). On the other hand, animal studies have suggested that accelerated colonic transit time relative to constipation, could be cause by autonomic neuropathy and diabetes-induced denervation of sympathetic nerve terminals (Du et al. 2018; Rosa-e-Silva et al. 1996). Deficiency of dietary SCFAs can also modulate intestinal motility and survival of enteric neurons by miRNAs, which are involved in energy homeostasis, lipid metabolism and proliferation and development of GI smooth muscles. miRNAs have been vastly studied in organ damage caused by diabetes, and one study in mice have shown that high-fat diets (HFD) delay the GI transit, partly by inducing apoptosis in enteric neuronal cells, an effect mediated by Mir375 associated with reduced levels of 3-phosphoinositide-dependent protein kinase-1 (Pdk1) (Nezami et al. 2014). There is still too much to understand about the intrinsic mechanisms underlying the connection between the gut microbiota and the ENS, and how this impacts the course of T1D. As an example of many beneficial properties ascribed to dietary SCFAs, high-amylose maize starch (HAMS) by itself used as oral rehydration solution decreased diarrhea duration in both adults and children hospitalized for acute infectious diarrhea (Binder et al. 2014).

3 Diet/Gut Microbiota-Interventions in Type 1 Diabetes: Clinical Trials

Treatment of T1D is typically focused on optimizing blood glucose control through different modes of exogenous insulin delivery. Although scientific and technical advances over the past 30 years have resulted in the control of hyperglycemia in those with T1D, methods of prevention and curing the disease remains as elusive as ever. Therefore, recent efforts have been made to identify other targets such diet and the gut microbiota to treat and prevent T1D. Based on 16S rRNA sequencing, these 4 bacterial phyla – Firmicutes, Bacteroidetes, Proteobacteria, and Actinobacteria, are usually dominant in the GI tract (Consortium 2012). Several human studies have shown that T1D could be attributable to an altered microbiota compared to healthy subjects. However, it is particularly challenging to establish a causal relationship between microbiota alterations and T1D in humans due to the complex nature of dietary habits and microbial differences across cultural backgrounds. Taxonomic changes in gut microbiota associated with T1D pathogenesis in humans include a lower Shannon diversity and higher Bacteroides/Firmicutes ratio (Giongo et al. 2011), and a reduction in butyrate producers (i.e., *Faecalibacterium*), lactic acid bacteria (i.e., *Lactobacillus, Bifidobacterium*) and mucin degraders (*Prevotella, Akkermansia*) (Vatanen et al. 2018; de Goffau et al. 2014; Brown et al. 2011; Mejia-Leon and Barca 2015). Alterations in the composition of the gut microbiota in T1D also include changes to the abundance of *Streptococcus spp., Clostridium spp., Staphylococcus spp., Blautia spp.,* and *Roseburia spp.* (de Goffau et al. 2013, 2014; Vatanen et al. 2018; Kostic et al. 2015; Gavin et al. 2018). Moreover, studies have reported correlations between gut microbiota ecology changes and diabetic markers such as HbA1c and inflammatory markers such as TNFα, IL-6, IL-10, IL-13 and IL-1β (Table 1).

Table 1 The gut microbiota: correlations with clinical markers in type 1 diabetes

Study design	Study aims	Outcome/measures	References
Case-control	To evaluate the alteration of gut microbiota between children with newly diagnosed T1D and healthy controls	Gut microbiota was associated with the development of T1DM by affecting autoimmunity. Results suggested modulating the gut microbiota as a potential therapy	Qi et al. (2016)
Case-control	To evaluate the difference in the composition of gut microbiota and glycemic level between children with T1D and healthy controls	The abundance of *Bifidobacterium* and *Lactobacillus*, and the Firmicutes to Bacteroidetes ratio correlated negatively and significantly with plasma glucose levels, while the abundance of *Clostridium* correlated positively and significantly with plasma glucose levels in the diabetic group	Murri et al. (2013)
Case-control	To evaluate the gut inflammatory profile and microbiota in subjects with T1D and healthy controls, and patients with celiac disease as gut inflammatory disease controls	Duodenal mucosa in T1D presents disease-specific abnormalities in the inflammatory profile and microbiota	Pellegrini et al. (2017)
Case-control	To compare the gut microbiota profiles of T1D, MODY2, a monogenic cause of diabetes, and healthy control subjects	Compared with healthy controls, T1D was associated with significantly lower microbiota diversity. Proinflammatory cytokines and LPS, and gut permeability were significantly increased in T1D and MODY2. T1D was also associated with an increment of genes related to lipid and amino acid metabolism, LPS biosynthesis, arachidonic acid metabolism, antigen processing and presentation, and chemokine signaling pathways	Leiva-Gea et al. (2018)
Case-control	To investigate whether intestinal dysbiosis in T1D patients correlate with clinical inflammatory cytokines	IL-6 was significantly increased ($P = 0.017$) in T1D. There was a correlation among patients with poor glycemic control, represented by high levels of HbA1C and Bacteroidetes, Lactobacillales, and *Bacteroides dorei* relative abundances	Higuchi et al. (2018)

T1D type 1 diabetes, *MODY2* type 2 maturity-onset diabetes of the young, *LPS* lipopolysaccharide

Studying the substantial changes in the number of specific species within the makeup of the microbiota may prove an invaluable tool allowing for earlier detection. Beyond the analysis of the gut microbiota composition, detection of microbial SCFAs could be more precise and accurate as biomarkers in autoimmunity. We have already demonstrated through a dietary intervention that the deterioration of immune tolerance in T1D is strongly associated with a deficiency in microbial SCFAs acetate and butyrate. When this deficiency is rectified in the gut by increasing the production of microbial SCFAs, the β-cell damage that leads to T1D ceases (Marino et al. 2017). One of the major barriers to the growth of the T1D market is the high failure rate of trials for disease-modifying (immunomodulatory) therapies. Although the use of diets for the treatment or prevention of T1D is still new, Table 2 provides a summary of selected, currently active clinical trials exploring the role of the microbiota in T1D prevention in high risk or newly diagnosed children and adults.

4 Type 2 Diabetes: A "Gut Origin" Disease

As the most common type of diabetes, T2D accounts for nearly 90% of all cases (Yan et al. 2017). Contrary to T1D, hyperglycemia in T2D is

Table 2 Trials in type 1 diabetes studying the role of diet and gut microbiota

Study design	Intervention	Outcome/measures	References
Single group assignment	To investigate whether a soluble fiber supplement (Benefiber) can improve glycemic control and/or reduce the risk of hypoglycemic events in children with T1D	Decrease in excursion of glucose or hypoglycemia incidence was not observed. However, a strong negative correlation was found between the amount of added fiber and the mean maximum post-prandial blood sugar after the main meals (lunch and breakfast). Researchers suggested that different types of fiber may act differently on regulating the blood glucose level as wheat dextrin showed a higher dampening effect	Nader et al. (2014) NCT01399892
Parallel assignment	A 12-week pilot randomized, double blind, placebo controlled clinical trial testing the effect of prebiotic fiber (1:1 oligofructose: inulin 8 g orally/day) on gut microbiota, intestinal permeability and glycemic control in children (8–17 years of age) with T1D	Serum HbA1c; Gut microbiota composition; Changes in gut permeability	Ho et al. (2016) NCT02442544
Parallel assignment	A randomized placebo controlled clinical trial testing the effect of a gluten-free diet on endogenous insulin production and gut microbiota to reverse or arrest islet destruction in 60 children and adolescents (1–17 years of age) with new onset T1D	AUC of C-peptide on mixed meal tolerance test at baseline, 6 months and one-year post enrolment. Stool sampling to characterize gut microbiota at each time point	NCT02605564
Single group assignment	The ToGeTher trial: specialized fiber supplement in T1D. A single arm, safety, tolerability and feasibility trial of a fiber supplement in 25 adults 18–45 years old with T1D	HbA1c, circulating immune cells numbers, microbiota analysis, C-peptide, fecal and blood SCFAs measurements, proteomics and RNA-seq.	ACTRN12618001391268
Crossover assignment	Experimental: Intervention group. Participants will be instructed to consume HAMS-AB in two divided doses at breakfast and dinner No intervention: Control group	To assess the effect of administering a prebiotic, such as HAMS- AB, on the gut microbiome profile, glycemia and β-cell function in new onset children with T1D in the last 4–24 months	NCT04114357
Sequential assignment	A phase 1b/2a, multi-center study in participants with clinical recent-onset T1D in 2 age groups (18–40 years of age and 12–17 years of age)	A prospective study to assess the safety and tolerability of different doses of AG019 administered alone or in association with Teplizumab in patients with clinical recent-onset T1D. AG019: *Lactococcus lactis*, a naturally occurring gut bacteria genetically modified (GM) to secrete human pro-insulin and IL-10.	NCT03751007

(continued)

Table 2 (continued)

Study design	Intervention	Outcome/measures	References
Single group assignment	Modulation of T1D susceptibility through the use of probiotics (VSL#3). 30 participants (5–17 years old) Child. After 6 weeks of taking VSL#3, they will return for their final visit for stool and blood samples	VSL#3 contains eight probiotic strains: bifidobacteria (*B. longum, B. infantis,* and *B. breve*), lactobacilli (*L. acidophilus, L. casei, L. bulgaricus,* and *L. plantarum*) and *Strepococcus thermophile*	NCT03423589

T1D type 1 diabetes, *AUC* area under curve, *SCFA* short-chain fatty acids, *HAMS* high amylose maize starch, *A* acetate, *B* butyrate

a result of insufficient insulin production and the inability of the body to respond to insulin, also known as insulin resistance (IR). A feedback loop between insulin-secreting β-cells in the pancreas and insulin-sensitive tissues such as liver, muscle and adipose tissue ensures the homeostatic regulation of glucose metabolism (Kahn et al. 2014). A breakdown in this crosstalk between insulin action in tissues and insulin secretion by the pancreas results in abnormal blood glucose levels. The rise in obesity, sedentary lifestyles, energy-dense diets and an ageing population are the main causes of the global T2D epidemic (Chatterjee et al. 2017). Overall, diabetes prevalence has risen steadily in every country since 1980, however the incidence and prevalence of T2D vary across geographical regions and ethnicity. Recent studies suggest that more than 80% of individuals living with T2D reside in low-to-middle-income countries ((NCD-RisC) NRFC 2016). Typically diagnosed in adults over the age of 45, T2D is increasingly seen in younger age groups due to the rising prevalence of childhood obesity (Chen et al. 2011). Indeed, T2D patients have a 15% increased risk of all-cause mortality compared to people without diabetes, especially in young people and those with worse glycemic control (Tancredi et al. 2015).

The variability in T2D prevalence based on ethnicity may also be partially explained by inherent differences in genetic background and phenotype. For example, individuals of Asian descent are generally predisposed to have a higher percentage of total body fat, and visceral and abdominal adiposity (Chan et al. 2009; Kong et al. 2013) compared to their Caucasian counterparts at a

given body mass index (BMI). Furthermore, Asians generally have poorer β-cell function and higher IR compared to Caucasians for a given BMI and waist circumference (Kong et al. 2013). While genetic variants may reveal mechanisms behind T2D development, we have so far been unable to predict disease beyond what is already achieved using current clinical measurements. Therefore, the missing piece of the T2D heritability puzzle may be explained by interactions between genetics, diet and their effects on the gut microbiota.

Although many cases of T2D could be prevented by adopting a healthy lifestyle and maintaining a healthy body weight, some individuals are more susceptible to T2D than others. Genome-wide association studies (GWAS) have implicated up to 250 genomic loci that are significantly associated with T2D (Feero et al. 2010; Anubha et al. 2014; Fuchsberger et al. 2016; Mahajan et al. 2018), primarily affecting insulin secretion and to a lesser degree, insulin action. One such example is the transcription factor 7 like 2 (TCF7L2) gene, which has important biological roles in the pancreas, liver and adipose tissue (Liu and Jin 2008). TCF7L2 polymorphisms can increase T2D susceptibility by decreasing production of glucagon-like peptide 1 (GLP-1), an insulinotropic hormone secreted by intestinal enteroendocrine cells, that works in tandem with insulin to maintain blood glucose homeostasis (Grant et al. 2006). More recently, plasma lipid profiling is an emerging discovery approach that identifies biomarkers in human studies and have been used to predict the development of obesity

(Pietiläinen et al. 2007), T1D (Orešič et al. 2008; La Torre et al. 2013), T2D (Rhee et al. 2011), and CVD (Fernandez et al. 2013). We have identified microbial products and pro-inflammatory molecules that predict the development and progression of diabetes. Thus, if we can effectively and safely target this chronic inflammatory state, it has the potential to reduce body weight gain, halt the progression of diabetes and reduce the risk of other diabetic complications. Microbial SCFAs can act as histone deacetylase (HDAC) inhibitors in immune cells and adipocytes. For example, the protein histone deacetylase 3 (HDAC3) regulates the progression of diet-induced obesity by modulating lipid metabolism of intestinal enterocyte cells in mice (Davalos-Salas et al. 2019). Additionally, we have demonstrated that dietary SCFAs markedly diminished the expression of *Hdac3* transcripts in B cells in T1D (Marino et al. 2017). This would lead to a phenotype similar to that of HDAC3 deficiency in select cells (Davalos-Salas et al. 2019) or of enzymatic inhibition of HDACs with chemical inhibitors such as butyrate, which has anti-inflammatory properties. It will be a novel technology to identify SCFAs-microbiota-inflammation dependent biomarkers involved in disease progression and most importantly to detect early biological clues to stop it.

However, there are currently few if any, real solutions to address the deleterious health consequences from T2D. All efforts to tackle T2D in adults and childhood are worthy of consideration. Diet and lifestyle factors can impact on the gut microbiota, the production of SCFAs and the development of obesity and T2D (Cani et al. 2008a; b; Durack and Lynch 2019; Sanna et al. 2019). As such, the Nutrition Forum from the National Academies of Sciences, Engineering, and Medicine have discussed the potential of utilizing dietary SCFAs during lifespan with an emphasis on healthy aging, beginning in pregnancy and early childhood (National Academies of Sciences, Engineering, and Medicine 2017).

5 Dietary Short-Chain Fatty Acids: Modulators of Meta-Inflammation in Type 2 Diabetes

Our diet is composed of a variety of dietary macronutrients – carbohydrates, proteins, fats, and fibers. Changes in these nutritional components can act as priming triggers for auto-immunity (Funda et al. 2008; Lerner and Matthias 2015), while overconsumption can lead to cell damage and inflammation (Chassaing et al. 2015). The amount of fiber and fat in the diet shapes large bowel microbial ecology (Kau et al. 2011; Makki et al. 2018). The stool samples of African children whose diet consists of cereals, legumes and vegetables are more abundant in bacteria from the phylum Bacteroidetes and less abundant in Firmicutes compared to European children consuming a typical Western diet (De Filippo et al. 2010). Consistent with this result, early studies demonstrated a higher Firmicutes-to-Bacteroidetes ratio in both obese individuals and mouse models of obesity compared to their lean counterparts (Turnbaugh et al. 2008a; Ley et al. 2005). Indeed, diet-associated complications such as CVD are less prevalent in Mediterranean countries where high intakes of fiber from vegetables, fruits and nuts are consumed in preference over highly processed meats and industrialized goods (Estruch et al. 2013). In line with these observations, the global decline in dietary fiber consumption below recommended daily intakes, particularly in Westernized societies, is linked to the rising incidence of inflammatory diseases (Hartley et al. 2016) thus establishing the importance of fiber in affecting the state of health or disease. Foods high in fiber provide many health benefits as it is the source of energy for both our own gut cells and the symbiotic microbial communities that reside within (Bird et al. 2000).

Resistant starches are the preferred energy source for the symbiotic microbiota in our gut. These complex carbohydrates that can be obtained from vegetable, fruits, wheat, corn and nuts, are one such form of dietary fiber (Zaman and Sarbini 2016). They are aptly named due to their strong ability to resist degradation by the body's digestive processes, continue through to the caecum and large intestine where they are fermented by the gut microbiota (Topping and Clifton 2001). This unique property of resistant starches is often utilized in commercial foods to reduce energy density due to the inability of the human body to digest it. In the mammalian gut, primarily the colon, resistant starches are degraded and fermented by the gut microbiota to produce the SCFAs, acetate, propionate and butyrate (Topping and Clifton 2001). These metabolites are produced at varying ratios with acetate being the most abundant in the colon (~60%), followed by propionate (~20%) and butyrate (~20%) (Canfora et al. 2015). In addition, acetate may itself fuel the production of other SCFAs such as butyrate via alternate biochemical pathways. More than 95% of SCFAs are absorbed by the colon, with butyrate being the preferential energy source for colonocytes, as well as having a profound effect on maintaining gut epithelial homeostasis and function (Topping and Clifton 2001). Meanwhile, propionate is metabolized in the liver and thus is only present in small concentrations in the periphery. Acetate, on the other hand, is the most abundant SCFA found in circulation and have been shown to cross the blood-brain barrier (Perry et al. 2016).

Patients with obesity and T2D (and other inflammatory diseases) have reduced levels of beneficial gut bacteria (Zhao et al. 2018). Several studies from our laboratory have demonstrated a remarkable and beneficial role for a SCFAs dietary intervention in the pathogenesis of several inflammatory diseases, such as allergies, asthma, arthritis, IBD, colon cancer, kidney disease, wound healing, hypertension and T1D (Thorburn et al. 2014; Marino et al. 2017; Maslowski and Mackay 2011; Felizardo et al. 2019; Marques et al. 2017). SCFAs are able to lower blood pressure, decrease cardiorenal hypertrophy and fibrosis, and improve cardiorenal function (Marques et al. 2017). These metabolites also have positive effects on appetite regulation and balance of energy intake/expenditure via the nervous system and the brain-gut axis with additional effects inducing lipid oxidation in brown adipose tissue, adipose tissue, liver and intestine (Knauf et al. 2008; den Besten et al. 2015; Kondo et al. 2009; Lu et al. 2016; Canfora et al. 2017; van der Beek et al. 2016). In line with previous findings, SCFAs have beneficial influence on hepatic metabolism preventing progression of non-alcoholic fatty liver disease (NAFLD), T2D and IR in mice and rats (Zhao et al. 2019a; Endo et al. 2013), and in humans (Zhao et al. 2019b; Ding et al. 2019). Similarly, several studies demonstrated that SCFAs are key to reducing and preventing body weight gain and obesity (Cani et al. 2008b; den Besten et al. 2015; Henao-Mejia et al. 2012; Bonfili et al. 2019). In skeletal muscle, SCFAs function in two ways. First, there is a low supply of lipids due to the positive effect of SCFAs in adipose tissue lipid storing capacity and consequent reduction of inflammatory cytokines which in turn prevent IR (Canfora et al. 2015; Chriett et al. 2017; Gao et al. 2009). Secondly, SCFAs directly increases fatty acid oxidation in muscle by stimulating AMPK signaling (Chriett et al. 2017) and inducing expression of metabolic genes like peroxisome proliferator-activated receptor gamma coactivator 1-alpha (PCG1α) and peroxisome proliferator-activated receptor (PPAR) (Gao et al. 2009; Clark and Mach 2017).

SCFAs stimulate the release of the gut hormones GLP-1 and GLP-2 (Tolhurst et al. 2012), which are responsible for modulating gut barrier function and reducing uptake of inflammatory compounds that may trigger chronic low-grade inflammation often linked with obesity and CVD. In addition, the progression of diabetic kidney disease (DKD) is associated with an altered gut microbial ecology and deficiency of SCFAs, which contributes to activation of the Renin-Angiotensin-Aldosterone System (RAAS) (Fernandes et al. 2019; Lu et al. 2018). Two studies using microbiome and metabolomic approaches in DKD patients suggest that dysbiosis in DKD is associated with increased urease-containing bacteria and reduced SCFAs-producing bacteria, resulting in accumulation of toxic metabolites (uremic toxins) (Sharma et al. 2013; Wong et al. 2014). We recently demonstrated that pre-treatment with a high butyrate-yielding diet protected against nephropathy in a mouse model of glomerular disease (Felizardo et al. 2019). Butyrate protected glomerular podocytes from damage thereby reducing proteinuria and glomerulosclerosis, and an overall kidney inflammation. This protective phenotype was associated with maintaining podocyte expression of key functional proteins and a normalized pattern of acetylation and methylation at promoter sites of genes essential for podocyte function (Felizardo et al. 2019). Similarly, oral delivery of sodium butyrate to mice protected against the onset of albuminuria, inflammation, and glomerulosclerosis in a model of type 1 DKD (Dong et al. 2017).

Obesity has been associated with dramatic changes in gut microbial community in both mice and humans (Turnbaugh et al. 2008a, b; Ley et al. 2005; Turnbaugh et al. 2006). Both obesity and T2D in humans have been associated with decreased microbial diversity and reduced abundance of butyrate-producing bacteria (i.e., *Roseburia, Eubacterium halii, Faecalibacterium prausnitzii*) (Qin et al. 2012; Karlsson et al. 2013). However, different to the microbiota profile of T1D patients, the Bacteroides/Firmicutes ratio in T2D subjects appear to be lower (Qin et al. 2012; Karlsson et al. 2013). Having said that, the association between Bacteroides/Firmicutes ratio and diabetes pathogenesis remains controversial and unconfirmed, and therefore may not be a useful biomarker to determine disease causality. A study showed germ free (GF) mice were protected from diet-induced obesity (DIO) and IR (Backhed et al. 2007), elegantly demonstrating the association of the gut microbiota in the development of obesity, inflammation and metabolic dysfunction. Furthermore, treatment of obese mice with antibiotics reduced

inflammation and improved glucose tolerance and insulin sensitivity (Cani et al. 2008a). Moreover, only GF mice transplanted with fecal preparations from obese twins developed increased weight gain and IR (Ridaura et al. 2013), indicating that gut microbiota can transfer and directly induce a metabolic status. Likewise, in humans, the infusion of microbiota from lean donors improved insulin sensitivity in recipients with metabolic syndrome (Vrieze et al. 2012).

6 Harnessing the Potential of the Microbiota for Diabetes Therapy

Obesity-associated deficiency in circulating microbial SCFAs can be due to intestinal permeability and altered microbiota composition. Reduced thickness of the mucosal layer and increased epithelial-cell uptake and translocation all contribute to increased intestinal permeability. Consequently, this allows more bacteria and inflammatory microbial products to enter the circulation. The host innate immune system recognizes the pathogen-associated molecular patterns (PAMPs) such as lipopolysaccharide (LPS) in microbial products (Kumar et al. 2011). LPS subsequently activates toll-like receptor 4 (TLR4) to induce inflammatory responses in immune cells as well as insulin target cells (Shi et al. 2006). Furthermore, LPS have been shown to alter the expression of tight junction proteins thereby increasing intestinal permeability (Guo et al. 2013). Genetically obese mice have increased intestinal permeability and LPS levels in the portal blood, which promote inflammatory liver damage. This is correlated by increased levels of TNF-α and reduced zona occludens 1 mRNA in the proximal colon of obese mice, which correlated with increased macrophage infiltration and levels of inflammatory cytokines TNF-α and IL-6 in the mesenteric fat (Cani et al. 2008a; Cani et al. 2007). Additionally, high-fat feeding in mice have been shown to induce translocation of gram-negative bacteria across the intestinal barrier in a TLR4-dependent manner, to possibly contribute to IR (Amar et al. 2011). Short-term changes in diet such as a single meal high in fat, resulted in an acute elevation of circulating LPS in humans (Erridge et al. 2007). In contrast, the gut anti-inflammatory agent 5-aminosalicyclic acid (5-ASA) was shown to improve metabolic parameters in DIO mice, with associated regulation of gut adaptive immunity and reduced gut permeability (Luck et al. 2015), thus implicating the role of gut leakiness and inflammation in obesity.

Several studies have tried to target the gut microbiota through the use of probiotics and prebiotics to ameliorate or treat T2D. Zhao et al. (2018) showed in a randomized clinical study that diabetic subjects who consumed a high-fiber diet (large amounts of diverse fibers composed of whole grains and traditional Chinese medicinal foods) had better improvement in hemoglobin A1c (HbA1c) levels, partly via increased GLP-1 production. Control and treated group received acarbose (an amylase inhibitor) as the standardized medication. Acarbose transforms part of the starch in the diet into a "fiber" by reducing its digestion and making it more available as fermentable carbohydrate in the colon. Patients exposed to the high fiber diet also showed increased fecal SCFAs and presented diminished bacteria producers of metabolically detrimental compounds such as indole and hydrogen sulfide (Zhao et al. 2018, 2019b). Overall, these findings establish proof-of-principle that a strategy to increase SCFAs may be effective to treat inflammatory and metabolic diseases such as T2D.

7   Diet/Gut Microbiota-Interventions in Type 2 Diabetes

A diet containing high-quality fats (low in transfatty acids and high in polyunsaturated fatty acids) (Ley et al. 2014), and carbohydrates (Bhupathiraju et al. 2014) is significantly more effective in preventing T2D, rather than reducing the relative quantities of these nutrients (Schulze and Hu 2005). Most dietary recommendations for T2D prevention usually promote higher intakes of whole grains, fruits, vegetables, nuts and legumes, and reduced intakes of refined grains, processed meat and sweetened beverages (Ley et al. 2014). The Look AHEAD (Action for Health in Diabetes) trial in the USA (Wadden et al. 2006) is a large multicenter randomized clinical trial spanning 8 years. This clinical trial is one example of how intensive lifestyle intervention through caloric restriction and increased physical activity can impact weight loss and CVD outcomes. However, the study was discontinued as it was determined that the intervention did not reduce the rate of cardiovascular events compared to the control group. Another multicenter study conducted in Spain, the PREDIMED trial, reported that a Mediterranean diet significantly reduced CVD risk by 30% in patients with T2D (Estruch et al. 2013, 2018). Additionally, post hoc analysis of the same trial postulated that a Mediterranean diet supplemented with extra-virgin olive oil or nuts may provide a degree of protection against diabetic retinopathy, but not nephropathy (Diaz-Lopez et al. 2015).

Increased intake of non-digestible fermentable dietary fiber has also been associated with alleviation of T2D phenotypes in various clinical trials (Zhao et al. 2018; Chandalia et al. 2000; Soare et al. 2014; Silva et al. 2015). A 6-week randomized, crossover study found that a high fiber diet composed of 25 g soluble fiber improved glycemic control, decreased hyperinsulinemia and lowered plasma lipid concentrations in patients with T2D (Chandalia et al. 2000). In further support of these findings, participants with T2D who consumed a low-glycemic index and/or high fiber breakfast presented with lower plasma glucose, insulin and ghrelin concentrations (Silva et al. 2015). The MADIAB trial (Soare et al. 2014) investigated the macrobiotic Ma-Pi 2 diet which consisted of whole grains, vegetables and legumes as a potential dietary intervention for the management of

T2D. Short-term intervention with the Ma-Pi 2 diet resulted in significantly better metabolic control compared to the recommended standard diets. As previously discussed, dietary fibers can alter the composition of gut microbiota which may affect glycemic control.

The earliest anti-obesity drugs belong to amphetamine derivatives have undesired effects on the central nervous system, often causing agitation, hallucinations and increased heart rate, and are therefore not optimal for obese patients (Colman 2005). A common drug often used in intervention strategies is metformin, which mainly works to decrease hepatic glucose production through gluconeogenesis inhibition (Aroda et al. 2017). Metformin has been reported to activate 5' AMP-activated protein kinase (AMPK) in hepatocytes, resulting in reduced fatty acid oxidation and suppressed expression of lipogenic enzymes (Zhou et al. 2001). Additionally, metformin has recently been shown to alter macrophage polarization by decreasing proportion of M1, and increasing M2 macrophages in palmitate-stimulated bone marrow-derived macrophages (BMDMs) (Jing et al. 2018). Metformin, prescribed along with lifestyle intervention, has been shown to be effective in preventing diabetes in subjects with impaired fasting glucose (Knowler et al. 2002). Orlistat, which is another commonly used drug for obesity, inhibits pancreatic lipases, leading to reduced fat uptake by the gut (Sternby et al. 2002). Conversely, glucagon-like peptide 1 receptor (GLP-1R) agonists work by suppressing glucagon production, stimulating insulin secretion from the pancreas and promoting satiety (Turton et al. 1996). As such, anti-obesity drugs combined with lifestyle modifications can improve weight loss. However, long-term therapy is required and patients may develop tolerance to the drugs or continue to gain weight during the drug regimen (Martin et al. 2015).

Indeed, therapies combining dietary treatment with immunotherapy are now starting to be tested in various disease models. For example, a very recent study testing the combination of a ketogenic diet and PI3K inhibitors, such as metformin and SGLT2, demonstrated significantly enhanced efficacy/toxicity ratios in various murine models of cancer (Hopkins et al. 2018). If these findings were successful, translation into human clinical trials may revolutionize our current therapeutic strategies to eventually pair dietary or medicinal food supplementation with targeted immunotherapy for the treatment of many autoimmune and inflammatory diseases. Table 3 provides a summary of selected clinical trials exploring the role of the microbiota in human T2D.

8 Conclusions and Future Perspectives

Research into the role of the microbiota and diabetes has grown exponentially in recent years. There is some evidence to suggest the strong influence of diet and gut microbiota in the development of inflammatory diseases such as autoimmune T1D, obesity, and T2D. Although studies in experimental animals have provided us with hints on future strategies and therapies targeting the gut microbiota by dietary intervention to prevent or reverse dysbiosis and reduce the diabetes incidence, current literature in humans is still in early stages. This is the new era of the gut microbiota, and several efforts have been directed to design microbiota-based diets as personalized nutrition extensively discussed in Elinav's review (Kolodziejczyk et al. 2019). Advancing on the power of "superfoods" such as dietary SCFAs, they can act as personalized diets as they target in the same beneficial way the different microbiota of each individual. The mechanism behind this process relies on reprogramming and restoring the microbiota-metabolite signature that has multifactorial effects on many physiological processes. In synergy with each individual's genetics and lifestyle, the gut microbiota is sophisticatedly designed to fight against disease (Fig. 1). Simple alterations in the intestinal microbiota, such as robustly increasing the abundance of SCFA producers and thereby boosting SCFA concentrations by specialized medicinal diets, may represent a novel and attractive therapeutic approach for prevention and treatment of diabetes.

Table 3 The gut microbiota and type 2 diabetes in humans

Study design	Intervention	Outcomes/measures	References
Cohort	Patients diagnosed with clinical T2D: 27 in the W group (acarbose and high fiber diet) and 16 in the U group (acarbose and usual care)	Dietary fiber promoted a select group of acetate- and butyrate-producing bacterial strains and diminished indole and hydrogen sulfide producers	Zhao et al. (2018)
		Participants in W group presented with lower plasma HbA1c partly through increased GLP-1 production	
Case control	In total 100 participants: 65 T2D patients divided into 2 subgroups: 49 with and 16 without chronic complications, and 35 healthy controls	T2D patients had a higher abundance of Proteobacteria and Firmicutes/Bacteroidetes ratio	Zhao et al. (2019b)
		Fecal SCFA concentrations were significantly reduced in T2D patients	
Case control	In total 61 participants with T2D: 31 in the probiotics group and 30 in the placebo group	6-month supplementation with multi-strain probiotics significantly reduced inflammation and HOMA-IR, and improved cardiometabolic profile	Sabico et al. (2019)
Crossover assignment	A randomized trial on 37 with T2D treated for 6 weeks supplementation with a prebiotic fiber mix of inulin and oligofructose compared to maltodextrin as placebo	Additional outcome measures: blood glucose, insulin, GLP-2, ghrelin, PYY, and leptin after a standardized mixed meal test. Also measured changes in microbiota composition and SCFA in feces before and after intervention/placebo periods, and subjective measures of appetite	NCT02569684
Parallel assignment	Role of gastrointestinal microbes on digestion of resistant starch and tryptophan availability to humans. A non-randomized trial on 20 patients	(48 g total/day) suspended in water. 24 g will be consumed 2 times per day. Plasma amino acid levels	NCT02974699
Crossover assignment	Rectal short chain fatty acids combinations and substrate and energy metabolism. A randomized trial on 12 patients obese and T2D	Measure fat oxidation and energy expenditure, hormones that influence energy metabolism and circulating metabolites [Time frame: 4 h total (2 h fasting and 2 h postprandial)	NCT01983046
Parallel assignment	A 12-week pilot randomized controlled trial testing the efficacy of the LoBAG diet (30% carbohydrate low in starch, 30% protein, 40% fat) in 38 participants with T2D	Changes in serum HbA1c, weight, fasting plasma glucose, fasting serum insulin, postprandial plasma and serum insulin, serum fructosamine, fasting serum lipids; Gut microbiota composition at baseline, week 6 and week 12; Urine nitrogen to creatinine ratio	NCT02717078
Single group assignment	A single site, prospective, open label, observational, single arm trial in 30 patients ($\geq$18 years of age) with T2D with GI complaints testing the efficacy of Pendulum Glucose Control formulation (contains 5 human commensal microbial strains including butyrate-producing and mucin-producing strains) to be taken twice daily for 8 weeks with an option of continuing up to 6 months	Decreased GI symptoms in 6 weeks; increased time in glucose range; decreased time in hypoglycemic range; improvement in serum HbA1c at 6, 12 and 24 weeks; improvement in serum fructosamine at 6 weeks	NCT04228003

T2D type 2 diabetes, *HbA1c* hemoglobin A1c, *GLP-1* glucagon-like peptide 1, *SCFA* short-chain fatty acids, *HOMA-IR* homeostatic model assessment of insulin resistance, *PYY* peptide YY, *GI* gastrointestinal

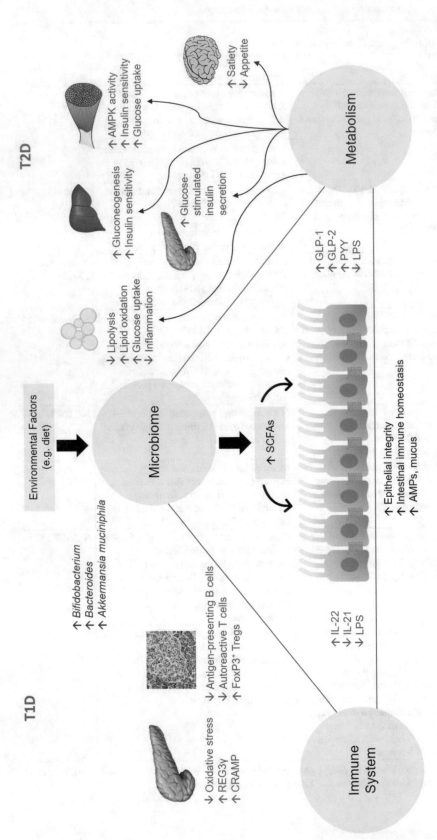

Fig. 1 Schematic representation showing the interactions between dietary SCFAs, the gut microbiota, the immune system, and metabolism in diabetes. The SCFAs acetate, propionate, and butyrate produced from bacterial fermentation of non-digestible carbohydrates are released from the gut lumen into the hepatic portal system and rapidly transported to the liver. In the gut, SCFAs prevent dysbiosis by altering the composition of the gut microbiota, improving epithelial integrity, and regulating immune homeostasis. In T1D, dietary SCFAs reduce serum LPS and modulate the cytokine profile to subvert immune responses in the periphery and prevent the autoimmune destruction of pancreatic β-cells. In T2D, SCFAs prevent insulin resistance and low-grade inflammation by reducing serum LPS and increasing plasma gut hormones that improve insulin secretion, gut motility, and satiety to modulate host metabolism and immune homeostasis in peripheral tissues. *T1D* type 1 diabetes, *T2D* type 2 diabetes, *REG3γ* regenerating islet-derived protein 3 gamma, *CRAMP* cathelicidin-related anti-microbial peptide, *FoxP3* forkhead box P3, *SCFAs* short-chain fatty acids, *LPS* lipopolysaccharide, *GLP* glucagon-like peptide, *PYY* peptide YY, *AMP* antimicrobial peptide

References

(NCD-RisC) NRFC (2016) Worldwide trends in diabetes since 1980: a pooled analysis of 751 population-based studies with 4·4 million participants. Lancet 387 (10027):1513–1530

Abela AG, Fava S (2013) Association of incidence of type 1 diabetes with mortality from infectious disease and with antibiotic susceptibility at a country level. Acta Diabetol 50(6):859–865

Abid N, McGlone O, Cardwell C, McCallion W, Carson D (2011) Clinical and metabolic effects of gluten free diet in children with type 1 diabetes and coeliac disease. Pediatr Diabetes 12(4 Pt 1):322–325

Alam C, Valkonen S, Palagani V, Jalava J, Eerola E, Hänninen A (2010) Inflammatory tendencies and over-production of IL-17 in the colon of young NOD mice are counteracted with diet change. Diabetes 59 (9):2237–2246

Amar J, Chabo C, Waget A, Klopp P, Vachoux C, Bermúdez-Humarán LG et al (2011) Intestinal mucosal adherence and translocation of commensal bacteria at the early onset of type 2 diabetes: molecular mechanisms and probiotic treatment. EMBO Mol Med 3(9):559

Anubha M, Min Jin G, Weihua Z, Jennifer EB, Kyle JG, Teresa F et al (2014) Genome-wide trans-ancestry meta-analysis provides insight into the genetic architecture of type 2 diabetes susceptibility. Nat Genet 46 (3):234

Aroda VR, Knowler WC, Crandall JP, Perreault L, Edelstein SL, Jeffries SL et al (2017) Metformin for diabetes prevention: insights gained from the diabetes prevention program/diabetes prevention program outcomes study. Diabetologia 60(9):1601–1611

Atkinson MA, Eisenbarth GS (2001) Type 1 diabetes: new perspectives on disease pathogenesis and treatment. Lancet 358(9277):221–229

Atkinson MA, Eisenbarth GS, Michels AW (2014) Type 1 diabetes. Lancet (Lond, Engl) 383(9911):69–82

Backhed F, Manchester JK, Semenkovich CF, Gordon JI (2007) Mechanisms underlying the resistance to diet-induced obesity in germ-free mice. Proc Natl Acad Sci U S A 104(3):979–984

Belkaid Y, Hand TW (2014) Role of the microbiota in immunity and inflammation. Cell 157(1):121–141

Bevins CL, Salzman NH (2011) Paneth cells, antimicrobial peptides and maintenance of intestinal homeostasis. Nat Rev Microbiol 9(5):356–368

Bhupathiraju SN, Tobias DK, Malik VS, Pan A, Hruby A, Manson JE et al (2014) Glycemic index, glycemic load, and risk of type 2 diabetes: results from 3 large US cohorts and an updated meta-analysis. Am J Clin Nutr 100(1):218–232

Binder HJ, Brown I, Ramakrishna BS, Young GP (2014) Oral rehydration therapy in the second decade of the twenty-first century. Curr Gastroenterol Rep 16(3):376

Bird AR, Brown IL, Topping DL (2000) Starches, resistant starches, the gut microflora and human health. Curr Issues Intest Microbiol 1(1):25–37

Bluestone JA, Herold K, Eisenbarth G (2010) Genetics, pathogenesis and clinical interventions in type 1 diabetes. Nature 464:1293

Bolla AM, Pellegrini S, Sordi V, Bonfanti R, Bosi E, Piemonti L et al (2017) Duodenal mucosa of patients with type 1 diabetes shows distinctive inflammatory profile and microbiota. J Clin Endocrinol Metab 102 (5):1468–1477

Bonfili L, Cecarini V, Gogoi O, Berardi S, Scarpona S, Angeletti M et al (2019) Gut microbiota manipulation through probiotics oral administration restores glucose homeostasis in a mouse model of Alzheimer's disease. Neurobiol Aging 87:35–43

Bosi E, Molteni L, Radaelli MG, Folini L, Fermo I, Bazzigaluppi E et al (2006) Increased intestinal permeability precedes clinical onset of type 1 diabetes. Diabetologia 49(12):2824–2827

Boursi B, Mamtani R, Haynes K, Yang YX (2015) The effect of past antibiotic exposure on diabetes risk. Eur J Endocrinol 172(6):639–648

Brown CT, Davis-Richardson AG, Giongo A, Gano KA, Crabb DB, Mukherjee N et al (2011) Gut microbiome metagenomics analysis suggests a functional model for the development of autoimmunity for type 1 diabetes. PLoS One 6(10):e25792

Canfora EE, Jocken JW, Blaak EE (2015) Short-chain fatty acids in control of body weight and insulin sensitivity. Nat Rev Endocrinol 11(10):577–591

Canfora EE, van der Beek CM, Jocken JWE, Goossens GH, Holst JJ, Olde Damink SWM et al (2017) Colonic infusions of short-chain fatty acid mixtures promote energy metabolism in overweight/obese men: a randomized crossover trial. Sci Rep 7 (1):2360

Cani PD, Amar J, Iglesias MA, Poggi M, Knauf C, Bastelica D et al (2007) Metabolic endotoxemia initiates obesity and insulin resistance. Diabetes 56 (7):1761–1772

Cani PD, Bibiloni R, Knauf C, Waget A, Neyrinck AM, Delzenne NM et al (2008a) Changes in gut microbiota control metabolic endotoxemia-induced inflammation in high-fat diet–induced obesity and diabetes in mice. Diabetes 57(6):1470

Cani PD, Delzenne NM, Amar J, Burcelin R (2008b) Role of gut microflora in the development of obesity and insulin resistance following high-fat diet feeding. Pathol Biol 56(5):305–309

Chan JCN, Malik V, Jia W, Kadowaki T, Yajnik CS, Yoon K-H et al (2009) Diabetes in Asia: epidemiology, risk factors, and pathophysiology. JAMA 301 (20):2129–2140

Chandalia M, Garg A, Lutjohann D, von Bergmann K, Grundy SM, Brinkley LJ (2000) Beneficial effects of high dietary fiber intake in patients with type 2 diabetes mellitus. N Engl J Med 342(19):1392–1398

Chassaing B, Koren O, Goodrich JK, Poole AC, Srinivasan S, Ley RE et al (2015) Dietary emulsifiers impact the mouse gut microbiota promoting colitis and metabolic syndrome. Nature 519(7541):92–96

Chatterjee S, Khunti K, Davies MJ (2017) Type 2 diabetes. Lancet (Lond, Engl) 389(10085):2239–2251

Chen L, Magliano DJ, Zimmet PZ (2011) The worldwide epidemiology of type 2 diabetes mellitus--present and future perspectives. Nat Rev Endocrinol 8(4):228–236

Cho I, Yamanishi S, Cox L, Methe BA, Zavadil J, Li K et al (2012) Antibiotics in early life alter the murine colonic microbiome and adiposity. Nature 488 (7413):621–626

Chriett S, Zerzaihi O, Vidal H, Pirola L (2017) The histone deacetylase inhibitor sodium butyrate improves insulin signalling in palmitate-induced insulin resistance in L6 rat muscle cells through epigenetically-mediated up-regulation of Irs1. Mol Cell Endocrinol 439:224–232

Clark A, Mach N (2017) The crosstalk between the gut microbiota and mitochondria during exercise. Front Physiol 8:319

Colman E (2005) Anorectics on trial: a half century of federal regulation of prescription appetite suppressants. Ann Intern Med 143(5):380–385

Consortium H (2012) Structure, function and diversity of the healthy human microbiome. Nature 486 (7402):207–214

Cox LM, Yamanishi S, Sohn J, Alekseyenko AV, Leung JM, Cho I et al (2014) Altering the intestinal microbiota during a critical developmental window has lasting metabolic consequences. Cell 158 (4):705–721

Davalos-Salas M, Montgomery MK, Reehorst CM, Nightingale R, Ng I, Anderton H et al (2019) Deletion of intestinal Hdac3 remodels the lipidome of enterocytes and protects mice from diet-induced obesity. Nat Commun 10(1):5291

De Filippo C, Cavalieri D, Di Paola M, Ramazzotti M, Poullet JB, Massart S et al (2010) Impact of diet in shaping gut microbiota revealed by a comparative study in children from Europe and rural Africa. Proc Natl Acad Sci U S A 107(33):14691–14696

de Goffau MC, Luopajarvi K, Knip M, Ilonen J, Ruohtula T, Harkonen T et al (2013) Fecal microbiota composition differs between children with beta-cell autoimmunity and those without. Diabetes 62 (4):1238–1244

de Goffau MC, Fuentes S, van den Bogert B, Honkanen H, de Vos WM, Welling GW et al (2014) Aberrant gut microbiota composition at the onset of type 1 diabetes in young children. Diabetologia 57(8):1569–1577

den Besten G, Bleeker A, Gerding A, van Eunen K, Havinga R, van Dijk TH et al (2015) Short-chain fatty acids protect against high-fat diet–induced obesity via a PPARγ-dependent switch from lipogenesis to fat oxidation. Diabetes 64(7):2398

Diaz-Lopez A, Babio N, Martinez-Gonzalez MA, Corella D, Amor AJ, Fito M et al (2015) Mediterranean diet, retinopathy, nephropathy, and microvascular diabetes complications: a post hoc analysis of a randomized trial. Diabetes Care 38(11):2134–2141

Ding Y, Yanagi K, Cheng C, Alaniz RC, Lee K, Jayaraman A (2019) Interactions between gut microbiota and non-alcoholic liver disease: the role of microbiota-derived metabolites. Pharmacol Res 141:521–529

Dong W, Jia Y, Liu X, Zhang H, Li T, Huang W et al (2017) Sodium butyrate activates NRF2 to ameliorate diabetic nephropathy possibly via inhibition of HDAC. J Endocrinol 232(1):71–83

Du YT, Rayner CK, Jones KL, Talley NJ, Horowitz M (2018) Gastrointestinal symptoms in diabetes: prevalence, assessment, pathogenesis, and management. Diabetes Care 41(3):627–637

Dudakov JA, Hanash AM, van den Brink MR (2015) Interleukin-22: immunobiology and pathology. Annu Rev Immunol 33:747–785

Durack J, Lynch SV (2019) The gut microbiome: relationships with disease and opportunities for therapy. J Exp Med 216(1):20–40

Endo H, Niioka M, Kobayashi N, Tanaka M, Watanabe T (2013) Butyrate-producing probiotics reduce nonalcoholic fatty liver disease progression in rats: new insight into the probiotics for the gut-liver Axis. PLoS One 8 (5):e63388

Erridge C, Attina T, Spickett C, Webb D (2007) A high-fat meal induces low-grade endotoxemia: evidence of a novel mechanism of postprandial inflammation. Am J Clin Nutr 86(5):1286

Estruch R, Ros E, Salas-Salvado J, Covas MI, Corella D, Aros F et al (2013) Primary prevention of cardiovascular disease with a Mediterranean diet. N Engl J Med 368(14):1279–1290

Estruch R, Ros E, Salas-Salvado J, Covas MI, Corella D, Aros F et al (2018) Primary prevention of cardiovascular disease with a Mediterranean diet supplemented with extra-virgin olive oil or nuts. N Engl J Med 378 (25):e34

Feero WG, Guttmacher AE, McCarthy MI (2010) Genomics, type 2 diabetes, and obesity. N Engl J Med 363 (24):2339–2350

Felizardo RJF, de Almeida DC, Pereira RL, Watanabe IKM, Doimo NTS, Ribeiro WR et al (2019) Gut microbial metabolite butyrate protects against proteinuric kidney disease through epigenetic- and GPR109a-mediated mechanisms. FASEB J 33(11):11894–11908

Fernandes R, Viana SD, Nunes S, Reis F (2019) Diabetic gut microbiota dysbiosis as an inflammaging and immunosenescence condition that fosters progression of retinopathy and nephropathy. Biochim Biophys Acta Mol basis Dis 1865(7):1876–1897

Fernandez C, Sandin M, Sampaio JL, Almgren P, Narkiewicz K, Hoffmann M et al (2013) Plasma lipid composition and risk of developing cardiovascular disease. PLoS One 8(8):e71846

Flier JS, Underhill LH, Eisenbarth GS (1986) Type I diabetes mellitus. N Engl J Med 314(21):1360–1368

Fuchsberger C, Flannick J, Teslovich TM, Mahajan A, Agarwala V, Gaulton KI et al (2016) The genetic architecture of type 2 diabetes. Nature 536 (7614):41–47

Funda DP, Kaas A, Bock T, Tlaskalová-Hogenová H, Buschard K (1999) Gluten-free diet prevents diabetes in NOD mice. Diabetes Metab Res Rev 15(5):323–327

Funda DP, Kaas A, Tlaskalova-Hogenova H, Buschard K (2008) Gluten-free but also gluten-enriched (gluten+) diet prevent diabetes in NOD mice; the gluten enigma in type 1 diabetes. Diabetes Metab Res Rev 24 (1):59–63

Gallo RL, Hooper LV (2012) Epithelial antimicrobial defence of the skin and intestine. Nat Rev Immunol 12(7):503–516

Gao Z, Yin J, Zhang J, Ward RE, Martin RJ, Lefevre M et al (2009) Butyrate improves insulin sensitivity and increases energy expenditure in mice. Diabetes 58 (7):1509–1517

Gavin PG, Mullaney JA, Loo D, Cao KL, Gottlieb PA, Hill MM et al (2018) Intestinal metaproteomics reveals host-microbiota interactions in subjects at risk for type 1 diabetes. Diabetes Care 41(10):2178–2186

Giongo A, Gano KA, Crabb DB, Mukherjee N, Novelo LL, Casella G et al (2011) Toward defining the autoimmune microbiome for type 1 diabetes. ISME J 5 (1):82–91

Grant SFA, Thorleifsson G, Reynisdottir I, Benediktsson R, Manolescu A, Sainz J et al (2006) Variant of transcription factor 7-like 2 (TCF7L2) gene confers risk of type 2 diabetes. Nat Genet 38 (3):320

Guo S, Al-Sadi R, Said HM, Ma TY (2013) Lipopolysaccharide causes an increase in intestinal tight junction permeability in vitro and in vivo by inducing enterocyte membrane expression and localization of TLR-4 and CD14. Am J Pathol 182 (2):375–387

Halling ML, Kjeldsen J, Knudsen T, Nielsen J, Hansen LK (2017) Patients with inflammatory bowel disease have increased risk of autoimmune and inflammatory diseases. World J Gastroenterol 23(33):6137–6146

Hansen D, Brock-Jacobsen B, Lund E, Bjorn C, Hansen LP, Nielsen C et al (2006) Clinical benefit of a gluten-free diet in type 1 diabetic children with screening-detected celiac disease: a population-based screening study with 2 years' follow-up. Diabetes Care 29 (11):2452–2456

Hansen CH, Krych L, Buschard K, Metzdorff SB, Nellemann C, Hansen LH et al (2014) A maternal gluten-free diet reduces inflammation and diabetes incidence in the offspring of NOD mice. Diabetes 63 (8):2821–2832

Harrison LC (2019) 71 – type 1 diabetes. In: Rich RR, Fleisher TA, Shearer WT, Schroeder HW, Frew AJ, Weyand CM (eds) Clinical immunology, 5th edn. Elsevier, London, pp 957–966.e1

Hartley L, May MD, Loveman E, Colquitt JL, Rees K (2016) Dietary fibre for the primary prevention of cardiovascular disease. Cochrane Database Syst Rev 1.Cd011472

Hasnain SZ, Borg DJ, Harcourt BE, Tong H, Sheng YH, Ng CP et al (2014) Glycemic control in diabetes is restored by therapeutic manipulation of cytokines that regulate beta cell stress. Nat Med 20(12):1417–1426

Henao-Mejia J, Elinav E, Jin C, Hao L, Mehal WZ, Strowig T et al (2012) Inflammasome-mediated dysbiosis regulates progression of NAFLD and obesity. Nature 482(7384):179–185

Heninger AK, Eugster A, Kuehn D, Buettner F, Kuhn M, Lindner A et al (2017) A divergent population of autoantigen-responsive CD4(+) T cells in infants prior to beta cell autoimmunity. Sci Transl Med 9 (378):eaaf8848

Higuchi BS, Rodrigues N, Gonzaga MI, Paiolo JCC, Stefanutto N, Omori WP et al (2018) Intestinal dysbiosis in autoimmune diabetes is correlated with poor glycemic control and increased Interleukin-6: a pilot study. Front Immunol 9:1689

Hill T, Krougly O, Nikoopour E, Bellemore S, Lee-Chan E, Fouser LA et al (2013) The involvement of interleukin-22 in the expression of pancreatic beta cell regenerative Reg. genes. Cell Regen (Lond) 2(1):2

Ho J, Reimer RA, Doulla M, Huang C (2016) Effect of prebiotic intake on gut microbiota, intestinal permeability and glycemic control in children with type 1 diabetes: study protocol for a randomized controlled trial. Trials 17(1):347

Hopkins BD, Pauli C, Du X, Wang DG, Li X, Wu D et al (2018) Suppression of insulin feedback enhances the efficacy of PI3K inhibitors. Nature 560 (7719):499–503

International Diabetes Federation (2017) IDF diabetes Atlas, 8th edn. ISBN: 978-2-930229-87-4

Jamshidi P, Hasanzadeh S, Tahvildari A, Farsi Y, Arbabi M, Mota JF et al (2019) Is there any association between gut microbiota and type 1 diabetes? A systematic review. Gut Pathog 11(1):49

Jing Y, Wu F, Li D, Yang L, Li Q, Li R (2018) Metformin improves obesity-associated inflammation by altering macrophages polarization. Mol Cell Endocrinol 461:256–264

Kahn SE, Cooper ME, Del Prato S (2014) Pathophysiology and treatment of type 2 diabetes: perspectives on the past, present, and future. Lancet (Lond, Engl) 383 (9922):1068–1083

Karlsson FH, Tremaroli V, Nookaew I, Bergstrom G, Behre CJ, Fagerberg B et al (2013) Gut metagenome in European women with normal, impaired and diabetic glucose control. Nature 498(7452):99–103

Kau AL, Ahern PP, Griffin NW, Goodman AL, Gordon JI (2011) Human nutrition, the gut microbiome and the immune system. Nature 474(7351):327–336

Kemppainen KM, Vehik K, Lynch KF, Larsson HE, Canepa RJ, Simell V et al (2017) Association between early-life antibiotic use and the risk of islet or celiac disease autoimmunity. JAMA Pediatr 171 (12):1217–1225

Kerr CA, Grice DM, Tran CD, Bauer DC, Li D, Hendry P et al (2015) Early life events influence whole-of-life metabolic health via gut microflora and gut permeability. Crit Rev Microbiol 41(3):326–340

Knauf C, Cani PD, Kim D-H, Iglesias MA, Chabo C, Waget A et al (2008) Role of central nervous system glucagon-like peptide-1 receptors in enteric glucose sensing. Diabetes 57(10):2603

Knowler WC, Barrett-Connor E, Fowler SE, Hamman RF, Lachin JM, Walker EA et al (2002) Reduction in the incidence of type 2 diabetes with lifestyle intervention or metformin. N Engl J Med 346(6):393–403

Kolodziejczyk AA, Zheng D, Elinav E (2019) Diet-microbiota interactions and personalized nutrition. Nat Rev Microbiol 17(12):742–753

Kondo T, Kishi M, Fushimi T, Kaga T (2009) Acetic acid upregulates the expression of genes for fatty acid oxidation enzymes in liver to suppress body fat accumulation. J Agric Food Chem 57(13):5982–5986

Kong AP, Xu G, Brown N, So WY, Ma RC, Chan JC (2013) Diabetes and its comorbidities – where East meets West. Nat Rev Endocrinol 9(9):537–547

Koren O, Goodrich JK, Cullender TC, Spor A, Laitinen K, Backhed HK et al (2012) Host remodeling of the gut microbiome and metabolic changes during pregnancy. Cell 150(3):470–480

Kostic AD, Gevers D, Siljander H, Vatanen T, Hyotylainen T, Hamalainen AM et al (2015) The dynamics of the human infant gut microbiome in development and in progression toward type 1 diabetes. Cell Host Microbe 17(2):260–273

Krischer JP, Lynch KF, Schatz DA, Ilonen J, Lernmark A, Hagopian WA et al (2015) The 6 year incidence of diabetes-associated autoantibodies in genetically at-risk children: the TEDDY study. Diabetologia 58 (5):980–987

Kumar H, Kawai T, Akira S (2011) Pathogen recognition by the innate immune system. Int Rev Immunol 30 (1):16–34

La Torre D, Seppänen-Laakso T, Larsson HE, Hyötyläinen T, Ivarsson SA, Lernmark Å et al (2013) Decreased cord-blood phospholipids in young age–at–onset type 1 diabetes. Diabetes 62(11):3951

Lassenius MI, Fogarty CL, Blaut M, Haimila K, Riittinen L, Paju A et al (2017) Intestinal alkaline phosphatase at the crossroad of intestinal health and disease – a putative role in type 1 diabetes. J Intern Med 281(6):586–600

Lee AS, Gibson DL, Zhang Y, Sham HP, Vallance BA, Dutz JP (2010) Gut barrier disruption by an enteric bacterial pathogen accelerates insulitis in NOD mice. Diabetologia 53(4):741–748

Leeds JS, Hopper AD, Hadjivassiliou M, Tesfaye S, Sanders DS (2011) Inflammatory bowel disease is more common in type 1 diabetes mellitus. Gut 60 (Suppl 1):A208-A

Lefebvre DE, Powell KL, Strom A, Scott FW (2006) Dietary proteins as environmental modifiers of type 1 diabetes mellitus. Annu Rev Nutr 26:175–202

Leiva-Gea I, Sanchez-Alcoholado L, Martin-Tejedor B, Castellano-Castillo D, Moreno-Indias I, Urda-Cardona A et al (2018) Gut microbiota differs in composition and functionality between children with type 1 diabetes and MODY2 and healthy control subjects: a case-control study. Diabetes Care 41(11):2385–2395

Lerner A, Matthias T (2015) Changes in intestinal tight junction permeability associated with industrial food additives explain the rising incidence of autoimmune disease. Autoimmun Rev 14(6):479–489

Ley RE, Backhed F, Turnbaugh P, Lozupone CA, Knight RD, Gordon JI (2005) Obesity alters gut microbial ecology. Proc Natl Acad Sci U S A 102 (31):11070–11075

Ley SH, Hamdy O, Mohan V, Hu FB (2014) Prevention and management of type 2 diabetes: dietary components and nutritional strategies. Lancet (Lond, Engl) 383(9933):1999–2007

Liu L, Jin T (2008) Minireview: the Wnt signaling pathway effector TCF7L2 and type 2 diabetes mellitus. Mol Endocrinol 22(11):2383–2392

Livanos AE, Greiner TU, Vangay P, Pathmasiri W, Stewart D, McRitchie S et al (2016) Antibiotic-mediated gut microbiome perturbation accelerates development of type 1 diabetes in mice. Nat Microbiol 1(11):16140

Lu Y, Fan C, Li P, Lu Y, Chang X, Qi K (2016) Short chain fatty acids prevent high-fat-diet-induced obesity in mice by regulating G protein-coupled receptors and gut microbiota. Sci Rep 6:37589

Lu CC, Ma KL, Ruan XZ, Liu BC (2018) Intestinal dysbiosis activates renal renin-angiotensin system contributing to incipient diabetic nephropathy. Int J Med Sci 15(8):816–822

Luck H, Tsai S, Chung J, Clemente-Casares X, Ghazarian M, Revelo XS et al (2015) Regulation of obesity-related insulin resistance with gut anti-inflammatory agents. Cell Metab 21(4):527–542

Maahs DM, West NA, Lawrence JM, Mayer-Davis EJ (2010) Epidemiology of type 1 diabetes. Endocrinol Metab Clin N Am 39(3):481–497

Mahajan A, Taliun D, Thurner M, Robertson NR, Torres JM, Rayner NW et al (2018) Fine-mapping type 2 diabetes loci to single-variant resolution using high-density imputation and islet-specific epigenome maps. Nat Genet 50(11):1505

Makki K, Deehan EC, Walter J, Backhed F (2018) The impact of dietary fiber on gut microbiota in host health and disease. Cell Host Microbe 23(6):705–715

Maleki D, Locke GR 3rd, Camilleri M, Zinsmeister AR, Yawn BP, Leibson C et al (2000) Gastrointestinal tract symptoms among persons with diabetes mellitus in the community. Arch Intern Med 160(18):2808–2816

Marietta EV, Gomez AM, Yeoman C, Tilahun AY, Clark CR, Luckey DH et al (2013) Low incidence of spontaneous type 1 diabetes in non-obese diabetic mice raised on gluten-free diets is associated with changes in the intestinal microbiome. PLoS One 8 (11):e78687

Marino E (2016) The gut microbiota and immune-regulation: the fate of health and disease. Clin Transl Immunol 5(11):e107

Marino E, Richards JL, McLeod KH, Stanley D, Yap YA, Knight J et al (2017) Gut microbial metabolites limit the frequency of autoimmune T cells and protect against type 1 diabetes. Nat Immunol 18(5):552–562

Marques FZ, Nelson E, Chu PY, Horlock D, Fiedler A, Ziemann M et al (2017) High-fiber diet and acetate supplementation change the gut microbiota and prevent the development of hypertension and heart failure in hypertensive mice. Circulation 135(10):964–977

Martin K, Mani M, Mani A (2015) New targets to treat obesity and the metabolic syndrome. Eur J Pharmacol 763(Part A):64–74

Martinez de Tejada B (2014) Antibiotic use and misuse during pregnancy and delivery: benefits and risks. Int J Environ Res Public Health 11(8):7993–8009

Maslowski KM, Mackay CR (2011) Diet, gut microbiota and immune responses. Nat Immunol 12(1):5–9

Mehdi AM, Hamilton-Williams EE, Cristino A, Ziegler A, Bonifacio E, Le Cao K-A, Harris M, Thomas R (2018) A peripheral blood transcriptomic signature predicts autoantibody development in infants at risk of type 1 diabetes. JCI Insight 5:e98212

Mejia-Leon ME, Barca AM (2015) Diet, microbiota and immune system in type 1 diabetes development and evolution. Nutrients 7(11):9171–9184

Mikkelsen KH, Knop FK, Vilsboll T, Frost M, Hallas J, Pottegard A (2017) Use of antibiotics in childhood and risk of type 1 diabetes: a population-based case-control study. Diabet Med 34(2):272–277

Mukherjee S, Hooper LV (2015) Antimicrobial defense of the intestine. Immunity 42(1):28–39

Murri M, Leiva I, Gomez-Zumaquero JM, Tinahones FJ, Cardona F, Soriguer F et al (2013) Gut microbiota in children with type 1 diabetes differs from that in healthy children: a case-control study. BMC Med 11:46

Nader N, Weaver A, Eckert S, Lteif A (2014) Effects of fiber supplementation on glycemic excursions and incidence of hypoglycemia in children with type 1 diabetes. Int J Pediatr Endocrinol 2014(1):13

Nahum GG, Uhl K, Kennedy DL (2006) Antibiotic use in pregnancy and lactation: what is and is not known about teratogenic and toxic risks. Obstet Gynecol 107(5):1120–1138

National Academies of Sciences, Engineering, and Medicine (2017) Nutrition across the lifespan for healthy aging: proceedings of a workshop. The National Academies Press, Washington, DC. https://doi.org/10.17226/24735

Nezami BG, Mwangi SM, Lee JE, Jeppsson S, Anitha M, Yarandi SS et al (2014) MicroRNA 375 mediates palmitate-induced enteric neuronal damage and high-fat diet-induced delayed intestinal transit in mice. Gastroenterology 146(2):473–483.e3

Nguyen NQ, Debreceni TL, Bambrick JE, Bellon M, Wishart J, Standfield S et al (2014) Rapid gastric and intestinal transit is a major determinant of changes in blood glucose, intestinal hormones, glucose absorption and postprandial symptoms after gastric bypass. Obesity (Silver Spring) 22(9):2003–2009

Orešič M, Simell S, Sysi-Aho M, Näntö-Salonen K, Seppänen-Laakso T, Parikka V et al (2008) Dysregulation of lipid and amino acid metabolism precedes islet autoimmunity in children who later progress to type 1 diabetes. J Exp Med 205(13):2975–2984

Pellegrini S, Sordi V, Bolla AM, Saita D, Ferrarese R, Canducci F et al (2017) Duodenal mucosa of patients with type 1 diabetes shows distinctive inflammatory profile and microbiota. J Clin Endocrinol Metab 102(5):1468–1477

Perry RJ, Peng L, Barry NA, Cline GW, Zhang D, Cardone RL et al (2016) Acetate mediates a microbiome-brain-beta-cell axis to promote metabolic syndrome. Nature 534(7606):213–217

Pietiläinen KH, Sysi-Aho M, Rissanen A, Seppänen-Laakso T, Yki-Järvinen H, Kaprio J et al (2007) Acquired obesity is associated with changes in the serum lipidomic profile independent of genetic effects – a monozygotic twin study. PLoS One 2(2):e218

Qi CJ, Zhang Q, Yu M, Xu JP, Zheng J, Wang T et al (2016) Imbalance of fecal microbiota at newly diagnosed type 1 diabetes in Chinese children. Chin Med J 129(11):1298–1304

Qin J, Li Y, Cai Z, Li S, Zhu J, Zhang F et al (2012) A metagenome-wide association study of gut microbiota in type 2 diabetes. Nature 490(7418):55–60

Rautava S, Luoto R, Salminen S, Isolauri E (2012) Microbial contact during pregnancy, intestinal colonization and human disease. Nat Rev Gastroenterol Hepatol 9(10):565–576

Rhee EP, Cheng S, Larson MG, Walford GA, Lewis GD, McCabe E et al (2011) Lipid profiling identifies a triacylglycerol signature of insulin resistance and improves diabetes prediction in humans. J Clin Invest 121(4):1402–1411

Richards JL, Yap YA, McLeod KH, Mackay CR, Marino E (2016) Dietary metabolites and the gut microbiota: an alternative approach to control inflammatory and autoimmune diseases. Clin Transl Immunol 5:e82

Ridaura VK, Faith JJ, Rey FE, Cheng J, Duncan AE, Kau AL et al (2013) Gut microbiota from twins discordant for obesity modulate metabolism in mice. Science 341(6150):1241214

Rosa-e-Silva L, Troncon LE, Oliveira RB, Foss MC, Braga FJ, Gallo JL (1996) Rapid distal small bowel transit associated with sympathetic denervation in type I diabetes mellitus. Gut 39(5):748–756

Sabat R, Ouyang W, Wolk K (2014) Therapeutic opportunities of the IL-22-IL-22R1 system. Nat Rev Drug Discov 13(1):21–38

Sabico S, Al-Mashharawi A, Al-Daghri NM, Wani K, Amer OE, Hussain DS et al (2019) Effects of a 6-month multi-strain probiotics supplementation in endotoxemic, inflammatory and cardiometabolic status of T2DM patients: a randomized, double-blind, placebo-controlled trial. Clin Nutr 38(4):1561–1569

Sanna S, van Zuydam NR, Mahajan A, Kurilshikov A, Vich Vila A, Vosa U et al (2019) Causal relationships among the gut microbiome, short-chain fatty acids and metabolic diseases. Nat Genet 51(4):600–605

Schulze MB, Hu FB (2005) Primary prevention of diabetes: what can be done and how much can be prevented? Annu Rev Public Health 26:445–467

Sellin JH, Hart R (1992) Glucose malabsorption associated with rapid intestinal transit. Am J Gastroenterol 87(5):584–589

Serena G, Camhi S, Sturgeon C, Yan S, Fasano A (2015) The role of gluten in celiac disease and type 1 diabetes. Nutrients 7(9):7143–7162

Sharma K, Karl B, Mathew AV, Gangoiti JA, Wassel CL, Saito R et al (2013) Metabolomics reveals signature of mitochondrial dysfunction in diabetic kidney disease. J Am Soc Nephrol 24(11):1901–1912

Shi H, Kokoeva MV, Inouye K, Tzameli I, Yin H, Flier JS (2006) TLR4 links innate immunity and fatty acid–induced insulin resistance. J Clin Invest 116 (11):3015–3025

Silva FM, Kramer CK, Crispim D, Azevedo MJ (2015) A high-glycemic index, low-fiber breakfast affects the postprandial plasma glucose, insulin, and ghrelin responses of patients with type 2 diabetes in a randomized clinical trial. J Nutr 145(4):736–741

Soare A, Khazrai YM, Del Toro R, Roncella E, Fontana L, Fallucca S et al (2014) The effect of the macrobiotic Ma-Pi 2 diet vs. the recommended diet in the management of type 2 diabetes: the randomized controlled MADIAB trial. Nutr Metab 11:39

Sternby B, Hartmann D, Borgstrom B, Nilsson A (2002) Degree of in vivo inhibition of human gastric and pancreatic lipases by Orlistat (Tetrahydrolipstatin, THL) in the stomach and small intestine. Clin Nutr 21(5):395–402

Stewart CJ, Ajami NJ, O'Brien JL, Hutchinson DS, Smith DP, Wong MC et al (2018) Temporal development of the gut microbiome in early childhood from the TEDDY study. Nature 562(7728):583–588

Sun J, Furio L, Mecheri R, van der Does AM, Lundeberg E, Saveanu L et al (2015) Pancreatic beta-cells limit autoimmune diabetes via an immunoregulatory antimicrobial peptide expressed under the influence of the gut microbiota. Immunity 43(2):304–317

Tamburini S, Shen N, Wu HC, Clemente JC (2016) The microbiome in early life: implications for health outcomes. Nat Med 22(7):713–722

Tancredi M, Rosengren A, Svensson A-M, Kosiborod M, Pivodic A, Gudbjörnsdottir S et al (2015) Excess mortality among persons with type 2 diabetes. N Engl J Med 373(18):1720–1732

Thorburn AN, Macia L, Mackay CR (2014) Diet, metabolites, and "western-lifestyle" inflammatory diseases. Immunity 40(6):833–842

Tolhurst G, Heffron H, Lam YS, Parker HE, Habib AM, Diakogiannaki E et al (2012) Short-chain fatty acids stimulate glucagon-like Peptide-1 secretion via the G-Protein–Coupled receptor FFAR2. Diabetes 61 (2):364

Topping DL, Clifton PM (2001) Short-chain fatty acids and human colonic function: roles of resistant starch and nonstarch polysaccharides. Physiol Rev 81 (3):1031–1064

Tormo-Badia N, Hakansson A, Vasudevan K, Molin G, Ahrne S, Cilio CM (2014) Antibiotic treatment of pregnant non-obese diabetic mice leads to altered gut microbiota and intestinal immunological changes in the offspring. Scand J Immunol 80(4):250–260

Turnbaugh PJ, Ley RE, Mahowald MA, Magrini V, Mardis ER, Gordon JI (2006) An obesity-associated gut microbiome with increased capacity for energy harvest. Nature 444(7122):1027–1031

Turnbaugh PJ, Hamady M, Yatsunenko T, Cantarel BL, Duncan A, Ley RE et al (2008a) A core gut microbiome in obese and lean twins. Nature 457:480

Turnbaugh PJ, Bäckhed F, Fulton L, Gordon JI (2008b) Diet-induced obesity is linked to marked but reversible alterations in the mouse distal gut microbiome. Cell Host Microbe 3(4):213–223

Turton MD, O'Shea D, Gunn I, Beak SA, Edwards CM, Meeran K et al (1996) A role for glucagon-like peptide-1 in the central regulation of feeding. Nature 379 (6560):69–72

van der Beek CM, Canfora Emanuel E, Lenaerts K, Troost Freddy J, Olde Damink SWM, Holst Jens J et al (2016) Distal, not proximal, colonic acetate infusions promote fat oxidation and improve metabolic markers in overweight/obese men. Clin Sci 130(22):2073–2082

Vatanen T, Franzosa EA, Schwager R, Tripathi S, Arthur TD, Vehik K et al (2018) The human gut microbiome in early-onset type 1 diabetes from the TEDDY study. Nature 562(7728):589–594

Vrieze A, Van Nood E, Holleman F, Salojärvi J, Kootte RS, Bartelsman JFWM et al (2012) Transfer of intestinal microbiota from lean donors increases insulin sensitivity in individuals with metabolic syndrome. Gastroenterology 143(4):913–916.e7

Wadden TA, West DS, Delahanty L, Jakicic J, Rejeski J, Williamson D et al (2006) The look AHEAD study: a description of the lifestyle intervention and the evidence supporting it. Obesity (Silver Spring) 14 (5):737–752

Wang X, Ota N, Manzanillo P, Kates L, Zavala-Solorio J, Eidenschenk C et al (2014) Interleukin-22 alleviates metabolic disorders and restores mucosal immunity in diabetes. Nature 514:237–241

Wong J, Piceno YM, DeSantis TZ, Pahl M, Andersen GL, Vaziri ND (2014) Expansion of urease- and uricase-containing, indole- and p-cresol-forming and contraction of short-chain fatty acid-producing intestinal microbiota in ESRD. Am J Nephrol 39(3):230–237

Yan Z, Sylvia HL, Frank BH (2017) Global aetiology and epidemiology of type 2 diabetes mellitus and its complications. Nat Rev Endocrinol 14(2):88

Zaman SA, Sarbini SR (2016) The potential of resistant starch as a prebiotic. Crit Rev Biotechnol 36 (3):578–584

Zhang X-S, Li J, Krautkramer KA, Badri M, Battaglia T, Borbet TC et al (2018) Antibiotic-induced acceleration of type 1 diabetes alters maturation of innate intestinal immunity. eLife 7:e37816

Zhao L, Zhang F, Ding X, Wu G, Lam YY, Wang X et al (2018) Gut bacteria selectively promoted by dietary fibers alleviate type 2 diabetes. Science 359 (6380):1151–1156

Zhao ZH, Lai JK-L, Qiao L, Fan JG (2019a) Role of gut microbial metabolites in nonalcoholic fatty liver disease. J Dig Dis 20(4):181–188

Zhao L, Lou H, Peng Y, Chen S, Zhang Y, Li X (2019b) Comprehensive relationships between gut microbiome and faecal metabolome in individuals with type 2 diabetes and its complications. Endocrine 66(3):526–537

Zhou G, Myers R, Li Y, Chen Y, Shen X, Fenyk-Melody J et al (2001) Role of AMP-activated protein kinase in mechanism of metformin action. J Clin Invest 108 (8):1167–1174

Adv Exp Med Biol - Advances in Internal Medicine (2020) 4: 521–551
https://doi.org/10.1007/5584_2020_527
© Springer Nature Switzerland AG 2020
Published online: 24 April 2020

Animal Models and Renal Biomarkers of Diabetic Nephropathy

Laura Pérez-López, Mauro Boronat, Carlos Melián, Yeray Brito-Casillas, and Ana M. Wägner

Abstract

Diabetes mellitus (DM) is the first cause of end stage chronic kidney disease (CKD). Animal models of the disease can shed light on the pathogenesis of the diabetic nephropathy (DN) and novel and earlier biomarkers of the condition may help to improve diagnosis and prognosis. This review summarizes the most important features of animal models used in the study of DN and updates the most recent progress in biomarker research.

Keywords

Animal models · Chronic kidney disease · Creatinine · Cystatin C · Diabetes mellitus · Diabetic nephropathy · Early markers · Glomerular filtration rate · Kidney injury molecule-1 · Obesity · Symmetric dimethylarginine

L. Pérez-López and Y. Brito-Casillas
Institute of Biomedical and Health Research (IUIBS), University of Las Palmas de Gran Canaria (ULPGC), Las Palmas de Gran Canaria, Spain

M. Boronat and A. M. Wägner (✉)
Institute of Biomedical and Health Research (IUIBS), University of Las Palmas de Gran Canaria (ULPGC), Las Palmas de Gran Canaria, Spain

Department of Endocrinology and Nutrition, Complejo Hospitalario Universitario Insular Materno-Infantil, Las Palmas de Gran Canaria, Spain
e-mail: ana.wagner@ulpgc.es

C. Melián
Institute of Biomedical and Health Research (IUIBS), University of Las Palmas de Gran Canaria (ULPGC), Las Palmas de Gran Canaria, Spain

Department of Animal Pathology, Veterinary Faculty, University of Las Palmas de Gran Canaria, Las Palmas de Gran Canaria, Arucas, Las Palmas, Spain

1 Introduction

Diabetic nephropathy (DN) is a common complication of diabetes mellitus (DM) occurring in 20–40% of people with diabetes (Dronavalli et al. 2008). Although cardiovascular diseases are the first cause of death, approximately 10–20% of people with DM die because of kidney failure, and DM is considered the first cause of end stage chronic kidney disease (CKD) (World Health Organization (WHO) 2019). Improved diagnosis and treatment of renal disease has led to better prognosis (Andrésdóttir et al. 2014). Furthermore, detecting the disease at an earlier stage and building on our understanding of the mechanisms of the disease may help to improve diagnosis further.

Recently, many reports have proposed a wide number of markers of CKD, and it has been shown that many of them reflect damage of one specific part of the nephron (Colhoun and Marcovecchio 2018; Domingos et al. 2016; Kim et al. 2013; Kern et al. 2010; Carlsson et al. 2017;

Colombo et al. 2019a). Albuminuria and microalbuminuria have traditionally been considered as markers of glomerular damage, and they have also been considered the first alterations that can be detected in DN (American Diabetes Association 2004). However, recent studies have shown that some patients with DM have CKD in the absence of microalbuminuria (MacIsaac et al. 2004; Lamacchia et al. 2018; Nauta et al. 2011; Zeni et al. 2017). In addition, the renal tubule could also play an important role in the development of DN (Colombo et al. 2019a). In fact, proteinuria mainly occurs after increased glomerular capillary permeability, but it is also a result of impaired reabsorption by the epithelial cells of the proximal tubule (D'Amico and Bazzi 2003). Thus, the use of tubular markers could also be beneficial for the diagnosis of DN. Indeed, the search for new renal biomarkers could lead to an earlier detection of renal damage. Likewise, animal models are important for improved understanding of the development and progression of DN. This review updates the most recent progress in biomarker research and summarizes the most important features of animal models used in the study of DN.

2 Brief Review of Kidney Anatomy and Physiology

The main functions of the kidney are filtration and excretion of metabolic waste products from the bloodstream, regulation of electrolytes, acidity and blood volume, and contribution to blood cell production (Rayner et al. 2016).

The nephron is the functional unit of the kidney. Each nephron is formed by a glomerulus, a proximal convoluted tubule, loop of Henle, and distal convoluted tubule. The last part of the nephron is the common collecting duct, and is shared by many nephrons (Rayner et al. 2016; National Institute of Diabetes and Digestive and Kidney Disaese (NIDDK) 2019).

The glomerulus is the filtering unit of the nephron. Within the glomerulus, the podocytes are specialized cells lining the outer surfaces of the bed of capillaries, which have interdigitated

foot processes that play an important role in the process of filtration. In fact, podocyte damage leads to proteinuria (Pavenstädt 2000) Podocytes are part of the glomerular barrier, and protein barrier passage of a normal kidney is mainly composed by low molecular weight protein. However, after its structural integrity is affected, high molecular weight proteins can pass through the glomerulus (D'Amico and Bazzi 2003).

In regard to the anatomy of the glomerulus, it has two poles: (1) a vascular pole with the afferent and the efferent arterioles; and (2) a urinary pole with the exit to the proximal tubule. The blood is filtered in a specialized capillary network through the glomerular barrier, which yields the filtrated substances into Bowman's capsule space, and then into the renal tubules. The glomerular barrier is composed by five layers: (1) the inner layer is the glycocalyx covering the surface of the endothelial cells; (2) the fenestrated endothelium, (3) the glomerular basement membrane, (4) the slit diaphragm between the foot-processes of the podocytes; and (5) the sub-podocyte space between the slit diaphragm and the podocyte cell body (Rayner et al. 2016).

Another important structure for the process of filtration and for the regulation of blood pressure is the juxtaglomerular apparatus, which is adjacent to the glomerulus. This structure is formed by the macula densa (cells inside the cortical of the thick ascending limb of Henle), mesangial cells, and the terminal parts of the afferent arteriole that include renin-producing cells. Release of renin is stimulated by decreased sodium concentration in the macula densa, systemic volume loss or reduced blood pressure (Peti-Peterdi and Harris 2010; Castrop and Schießl 2014).

3 Pathophysiology of Diabetic Nephropathy

Hyperglycemia, hypertension and obesity are considered risk factors for DN (Mogensen et al. 1983; Câmara et al. 2017; Kanasaki et al. 2013). Therefore, DN is a heterogeneous syndrome that is common in people with type 1 (T1) and type 2 (T2) DM, although in patients with T2DM and metabolic syndrome, more heterogeneous mechanisms are involved. DN has been classically considered a process with the following sequence of disorders: glomerular hyperfiltration, progression of albuminuria, decline of glomerular filtration rate (GFR) and, finally end stage renal disease (Mogensen et al. 1983); and different pathways that involve hemodynamic, metabolic and inflammatory factors play a role in its development (Table 1).

3.1 Hemodynamic Factors

Glomerular hyperfiltration is considered an alteration of early DN and it has been identified in 10–40% of people with early T1DM, and around 40% of patients with T2DM (Mogensen et al. 1983; Premaratne et al. 2015). However, its role as a leading cause of DN needs further research. In addition, the hyperfiltration mechanism is not well understood yet: it could be the result of a combination of hemodynamic, vasoactive, and tubular factors (Dronavalli et al. 2008; Zeni et al. 2017). Moreover, inhibition of tubuloglomerular feedback could be the main mechanism involved (Zeni et al. 2017). Persistent hyperglycemia produces tubular growth and increased tubular sodium reabsorption, and reduces the delivery of sodium to the macula densa, eliciting the release of renin and the activation of the renin-angiotensin system (RAS). This leads to an inhibition of the tubuloglomerular feedback, causing vasodilation of the afferent arteriole, which increases single-nephron glomerular filtration rate

and consequently leads to hyperfiltration (Premaratne et al. 2015; Vallon and Thomson 2012). An implication of the sodium-glucose co-transporter 2 (SGLT2) in this process has been suggested. SGLT2s are expressed in the proximal tubule and their function is the reuptake of glucose and sodium (ratio 1:1), thus their activity is stimulated by the increased glucose filtration in diabetic subjects (Zeni et al. 2017; Premaratne et al. 2015; Hans-Joachim et al. 2016). Vasodilating substances, such as nitric oxide and cyclo-oxygenase 2 derived prostanoids, also take part in the hyperfiltration process (Wolf et al. 2005). Whether they play a key or secondary role in the pathogenesis of hyperfiltration is not well defined. This process, which is associated with increased intraglomerular pressure, could lead to podocyte stress and nephron loss (Hans-Joachim et al. 2016). Additionally, obesity has been considered as an independent risk factor for CKD, and hyperfiltration has also been detected in non-diabetic obese subjects. In addition, the kidney produces components of the RAS that specifically constrain the efferent rather than the afferent arteriole, increasing GFR and glomerular pressure (Yacoub and Campbell 2015). Several studies have demonstrated local production of RAS components in adipose tissue (Sharma and Engeli 2006; Giacchetti et al. 2002) and it has been suggested that RAS could be overactivated in patients with obesity (Sharma and Engeli 2006; Xu et al. 2017). Indeed, a decrease in RAS activity has been reported in obese women after weight loss (Engeli et al. 2005).

Table 1 Main factors involved in the development of diabetic nephropathy

Mechanisms involved in the pathogenesis of diabetic nephropathy	
Hemodynamic factors	Metabolic and inflammatory factors
Vasodilation of the afferent arteriole of the glomerulus	Increased formation of advanced glycation end-products
Increased glomerular filtration rate	Podocyte stress
Implication of RAS, nitric oxide and cyclo-oxygenase 2	TGF-β1 is a profibrotic cytokine that plays an essential role in inflammatory and fibrotic processes
Increased glomerular pressure	Obesity leads to increased leptin concentration, decreased adiponectin concentration, and increased cytokine production

Abbreviations: *RAS* renin angiotensin system, *TGF-β1* transforming growth factor β1

3.2   Metabolic and Inflammatory Factors

Formation of advanced glycation end-products (AGE), resulting from the reduction of sugars, is increased during chronic hyperglycemia. Accumulation of AGE appears to stimulate production of cytokines and renal fibrosis (Forbes and Cooper 2007). Infiltration by inflammatory cells (monocytes, macrophages and lymphocytes) precedes fibrosis, and these inflammatory cells are responsible for the production of reactive oxygen species, inflammatory cytokines and profibrotic cytokines (Kanasaki et al. 2013). Among the latter, transforming growth factor beta 1 (TGF-β1) plays an essential role in inflammation and fibrosis, and together with other cytokines such as connective tissue growth factor, platelet-derived growth factor and fibroblast growth factor 2; TGF-β1 is involved in fibroblast activation (Kanasaki et al. 2013). This produces an increased deposition of extracellular matrix in the interstitial space, as well as glomerular basement membrane thickening, which could lead to podocyte apoptosis and, as a consequence, to increased vascular permeability in the glomerulus (Wolf et al. 2005).

Additionally, in obese people, high leptin and low adiponectin concentrations result in an increased secretion of several adipokines (i.e. tumor necrosis factor-a, interleukin-6, interleukin-18) that promote an inflammatory state, also leading to extracellular matrix accumulation and renal fibrosis (Câmara et al. 2017; Straczkowski et al. 2007; Kern et al. 2001). In obese mice, reduced plasma adiponectin contributes to albuminuria and podocyte disfunction (Sharma et al. 2008).

3.3   Histologic Changes

The main histologic changes observed in kidneys of people with diabetes are located in the glomerulus, and include glomerular sclerosis (Kimmelstiel-Wilson nodules), basement membrane thickening and mesangial expansion. Tubulo-interstitial and arteriolar lesions have been commonly described as late lesions in patients with T1DM. Nevertheless, some patients with T2DM can show tubulo-interstitial and/or arteriolar lesions with preserved glomerular structure (Fioretto and Mauer 2007).

4   Animal Models

4.1   Rodents

4.1.1   Mouse Models

Rodents are the most studied animal models of human DN. They can show spontaneous or induced diabetes (Kachapati et al. 2012; Song et al. 2009), and DN might develop in the course of the disease, or by induction of renal damage by unilateral nephrectomy, or ischemia and reperfusion (Song et al. 2009; Kitada et al. 2016). However, an animal model that develops all of the features of DN is not available (Betz and Conway 2014). Based on the clinical characteristics of DN in humans, the Animal Models of Diabetic Complications Consortium (AMDCC) has published some criteria to define acceptable models of renal disease in mice with diabetes: (1) 50% decline in GFR; (2) 100 fold increase in proteinuria in comparison to matched controls of the same strain, age and gender; (3) presence of pathologic alterations such as mesangial sclerosis, arterial hyalinosis, 50% thickening of the glomerular basement membrane or tubulointerstitial fibrosis. Nonetheless, to date, no animal model fulfills all these criteria (Diabetes Complications Consortium (DiaComp) 2003).

T1DM rodent models include streptozotocin induced DM or spontaneous models due to genetic mutations, such as AKITA and OVE26 mice. T2DM genetic models are leptin deficient (ob/ob mice) or have inactivating mutations in the leptin receptor (db/db mice) (Kitada et al. 2016; Alpers and Hudkins 2011), but there are also models of induced T2DM, usually through a high fat diet (De Francesco et al. 2019; Ingvorsen et al. 2017). Hypertension plays an important role in the progression of human DN, and the deficiency of endothelial nitric oxide synthase (eNOS) through knockout of eNOS genes, has led to accelerated renal damage in db/db and streptozotocin-treated mice. eNOS$^{-/-}$ C57BL/J^{db} mice can develop T2DM, obesity, hypertension, albuminuria, marked mesangial expansion and mesangiolysis (Nakagawa et al. 2007).

Recently, the black and tan, brachyury (BTBR obese) mouse strain, which is spontaneously insulin resistant, has gained importance in the field of the study of DN. This strain with ob/ob leptin deficiency mutation on BTBR mouse background, has been considered one of the best models of human DN because it rapidly develops pathological changes seen in human DN, such as increased glomerular basement membrane thickness, mesangial sclerosis, focal arteriolar hyalinosis, mesangiolysis, mild interstitial fibrosis and podocyte loss (Hudkins et al. 2010).

4.1.2 Rat Models

The most studied rat models are the Zucker diabetic fatty (ZDF-fa/fa) rat and the Wistar fatty rat (Kitada et al. 2016; Hoshi et al. 2002). Both have an autosomal recessive mutation in the fa gene that encodes the leptin receptor, and both are crossbred with the insulin resistant Wistar Kyoto rats. The ZDF-fa/fa and Wistar fatty rats can develop albuminuria and renal alterations, such as tubular cell damage and tubulointerstitial fibrosis, although nodular glomerular lesions or mesangiolysis have not been observed (Kitada et al. 2016; Hoshi et al. 2002). Another rat strain that is considered a model of DN is The Otsuka Long-Evans Tokushima Fatty (OLETF), in which multiple recessive genes are involved in the development of DM. These rats exhibit mild obesity, hyperinsulinemia with late onset of hyperglycaemia, and can present mesangial matrix expansion, glomerulosclerosis and tubular cell damage (Kawano et al. 1994, 1992).

4.2 Companion Animals

These animals have a great interest as animal models since they develop spontaneous DM and share the human environment (Brito-Casillas et al. 2016). However, it is unclear if dogs and cats with diabetes develop DN.

4.2.1 Cats

Around 80% of cats with DM have T2DM (Nelson and Reusch 2014). As in humans, obesity is a risk factor for DM, which also represents a problem of an epidemic proportion since approximately 35–50% of domestic cats are overweight or obese (Hoenig 2012). Obese cats also show some of the disorders observed in people with the metabolic syndrome; they can present insulin resistance and higher concentrations of very low-density lipoproteins (VLDL), triglycerides and cholesterol (Hoenig 2012; Jordan et al. 2008). Despite these alterations, hypertension has not been linked to feline obesity or DM, and atherosclerosis has not been described either (Jordan et al. 2008; Payne et al. 2017). Lower concentration of angiotensin-converting enzyme 2 has been found in the subcutaneous adipose tissue of overweight or obese cats compared to those with a low body condition score, but whether this could represents a negative mechanism against blood pressure is unknown (Riedel et al. 2006).

Additionally, the link between DM and renal disease is unclear in cats, and discrepancies can be found in the literature. In two retrospective, epidemiological studies, no relationship was found between both diseases (Greene et al. 2014; Barlett et al. 2010). In contrast, higher prevalences of proteinuria and microalbuminuria have been detected in cats with DM compared to age matched controls (Al-Ghazlat et al. 2011) and, recently, in a retrospective study of a population of 561 adult cats, age-adjusted multivariate regression analysis showed an association between both diseases (Pérez-López et al. 2019). Thus, the cat could be a useful model of DN, although prospective studies are still needed for a better understanding and evaluation of the association between feline DM and renal disease.

4.2.2 Dogs

Obesity in dogs is able to induce hyperinsulinemia, to increase blood pressure and to produce glomerular hyperfiltration. Dogs with obesity can develop structural changes in the kidney, such as glomerular basal membrane thickening and mesangial expansion (Henegar et al. 2001). Insulin sensitivity decreases around 35% in obese dogs. However, obesity does not cause T2DM in these animals (Chandler et al. 2017). Indeed, T1DM is the most commonly recognized form of DM in dogs, although they can also develop DM secondarily to dioestrus, Cushing's syndrome or medications (Nelson and Reusch 2014). Regarding the relationship between DM and CKD, it has not been fully investigated in dogs. However, some studies have shown that markers of vascular resistance (ultrasound renal resistive index and pulsatility index), which are associated with progression of kidney disease and hypertension in humans, are also positively correlated with glycemic status in dogs (Priyanka et al. 2018; Novellas et al. 2010). In a case-control and age-matched study, dogs with alloxan-induced DM, uninephrectomized 4 weeks after the induction of DM, were proposed as a valuable model for the study of DN, since they developed greater glomerular basement membrane thickening and mesangial expansion, compared to controls, already 1 year after DM induction (Steffes et al. 1982).

4.3 Production Animals

Mainly two species could represent this group as animal models of DN: rabbits and swine. However, few studies have been reported on rabbits, and they have been mainly focused on the role of obesity in kidney function (Dwyer et al. 2000; Antic et al. 1999). One study showed an increased hyaluronean content in the renal medulla of rabbits with induced high-fat diet induced obesity (Dwyer et al. 2000). Higher renal medullary hyaluronean content has also been reported in humans and rodent models with DM or with renal alterations (Stridh et al. 2012). In regard to the swine, it is considered a more valuable animal model, since it closely resembles human anatomy and physiology (Zhang and Lerman 2016; Li et al. 2011; Rodríguez-Rodríguez et al. 2020; Spurlock and Gabler 2008). Metabolic syndrome can be induced through high fat diet in the Ossabaw and Iberian swine. These pigs exhibit

obesity, insulin resistance, hypertension and dyslipidemia. Regarding renal disease, kidney hypertrophy, increased GFR, renal tubular fibrosis and renal adiposity have all been reported (Zhang and Lerman 2016; Li et al. 2011; Rodríguez-Rodríguez et al. 2020).

Further information in relation to animal models can be found in Table 2.

5 Assessment of Renal Function

Criteria to diagnose CKD are well established in human medicine. However, there are a large number of markers, many still under investigation, that could allow an early detection of impaired renal function. The assortment of markers also represents a variety of mechanisms involved in kidney injury, reflecting damage of different parts of the nephron as well (Fig. 1).

5.1 Routine Evaluation of DN and Gold Standard Methods

In humans, CKD is defined as abnormalities of kidney structure or function, present for more than 3 months, and with implications for health. Patients with altered kidney function present albuminuria and/or a decline in GFR (Levin et al. 2013).

Human DN is diagnosed when urinary albumin to creatinine ratio (ACR) is above 30 mg/g, although some guidelines recommend the use of different ACR thresholds for men (>25 mg/g) and women (>35 mg/g). Despite ACR being the main tool used to diagnose DN, some patients with either T1 or T2DM who have decreased GFR do not show elevated ACR (Gross et al. 2005; Caramori et al. 2003). Therefore, GFR should also be measured in patients with DM to rule out CKD. In fact, estimation of GFR is considered the routine method to evaluate renal function in patients with CKD. The most recent guidelines recommend the interpretation of both albuminuria and GFR for the diagnosis and staging of CKD (Levin et al. 2013).

Table 2 Animal models of diabetic nephropathy

	Animal models of induced diabetes with development of DN			Animal models of spontaneous diabetes with development of DN			Spontaneous animal model of diabetes and unclear development of DN		
	Animal model	Advantages	Disadvantages	Animal model	Advantages	Disadvantages	Animal model	Advantages	Disadvantages
Models of T1DM	Streptozotocin treated C57BL/6 mouse	The most studied strain; Easy to breed and long life span	Less susceptibility to develop renal injury than other strains	AKITA mouse.	Early pathological changes of DN	Strain-dependent susceptibility	Diabetic dog	Gene-environment interaction; Environmental factors shared with humans	Development of DN is still unclear
	Streptozotocin treated DBA/2 mouse	More susceptible to renal injury than C57BL/6 mouse	Tubulointerstitial fibrosis does not occur	OVE26 mouse	To study advanced DN	Poor viability			
Models of T2DM	C57BL/6 mouse on high fat diet	Develops metabolic syndrome (increased SBP and lipids); Renal injury could be due to renal lipid accumulation	Inter-individual phenotypic variability of the amount of food intake after receiving a high fat diet; Phenotype more pronounced in males	eNOS -/-/db/db mouse	Development of glomerular lesions characteristic of advanced DN	Few studies, possible difficulties with breeding	Diabetic cat	Similar mechanisms of human T2DM, with environmental factors and polygenic interactions	Development of DN is still unclear
	Zucker diabetic fatty (ZDF-fa/fa) rat	Development of tubulointerstitial fibrosis	Does not develop advanced features of DN	BTBR ob/ob mice	Rapidly resembles alterations of advanced human DN	Modest interstitial fibrosis			
	Ossabaw pig on high fat diet	Develops metabolic syndrome, increased GFR, renal tubular fibrosis, renal adiposity	Expensive and specialized husbandry						

Abbreviations: *IgG* immunoglobulin G; *kidney injury molecule -1*, *NGAl* neutrophil gelatinase-associated lipocalin, *SDMA* symmetric dimethylarginine, *VEGF* vascular endothelial growth factor, *RBP4* retinol binding protein 4, *suPAR* soluble urokinase type plasminogen activator receptor, *TGFβ1* transforming growth factor β-1

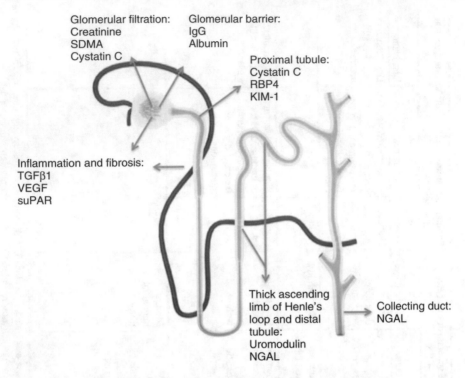

Fig. 1 Classification of markers of renal damage according to part of the nephron involved

The gold standard for assessing GFR is the plasma or urinary clearance of an exogenous filtration marker, such as inulin (Stevens and Levey 2009), Cr-EDTA, ^{125}I-iothalamate or iohexol. However, direct GFR measurement is difficult to perform and is time-consuming; it requires the injection of a suitable marker and several urine sample collections (Stevens and Levey 2009; Levey et al. 2003). Alternatively, equations for estimation of GFR, generally based on serum creatinine levels, are used in clinical practice, and their use is recommended in conjunction with other markers of renal function, such as cystatin C (Levin et al. 2013). However, early alterations of DN are usually associated with hyperfiltration, and these equations are less precise at higher values of GFR, and tend to underestimate GFR in the hyperfiltration state (Levin et al. 2013; Tuttle et al. 2014). Among available formulas, the Chronic Kidney Disease Epidemiology Collaboration (CKD-EPI) has been suggested to be the best one to evaluate early renal impairment in patients with DM,

normoalbuminuria and hyperfiltration (Lovrenčić et al. 2012). Recent studies suggest that the CKD-EPI cystatin C based equations for estimation of GFR, are those that best fit the GFR measurement in patients with DM. However, in general, it is considered that GFR equations show high variability in people with DM (Cheuiche et al. 2019).

As stated above, the development of DN is unclear in companion animals, but diagnosis of CKD is based on serum creatinine or symmetric dimethylarginine (SDMA) concentration together with urinary protein-creatinine ratio (International Renal Interest Society (IRIS) 2015a, 2017), and there is still a lack of standardization in the methods of GFR measurement and their interpretation (International Renal Interest Society (IRIS) 2015b). Only one study proposed a method to estimate GFR in cats. It was based on serum creatinine and it was adjusted for a marker of muscle mass. However, this formula did not prove to be a reliable estimation of GFR (Finch et al. 2018).

5.2 Indirect Markers of Glomerular Filtration Rate

5.2.1 Serum Creatinine

Creatinine is a break-down product of creatine phosphate in muscle tissue (113 Daltons). It is not metabolized and is entirely cleared by the kidneys with minimal reabsorption by the renal tubules (Ferguson and Waikar 2012; Kavarikova 2018). Thus, blood creatinine concentration increases when GFR declines. However, in the early stage of CKD, subtle changes in GFR do not alter creatinine concentration. In humans, other markers seems to correlate better with GFR than creatinine (El-khoury et al. 2016) and, in dogs, its concentrations increase only when around 50% of renal function has been lost (Hokamp and Nabity 2016; Nabity et al. 2015). In addition, both in humans and companion animals, creatinine concentration depends on lean body mass, and it has high inter-individual variability (Delanaye et al. 2017; Hall et al. 2014a, 2015). Other factors, such as hydration and blood volume status, urinary obstructions or urinary infections, can also affect creatinine concentration (Hokamp and Nabity 2016; Blantz 1998).

5.2.2 Symmetric Dimethylarginine

SDMA is a catabolic product of arginine-methylated proteins (202 daltons), which is mainly excreted through the kidneys and is not reabsorbed or secreted by the tubules (Nabity et al. 2015; McDermott 1976). Thus, SDMA concentration is affected by GFR in inverse linear relationship in humans and in companion animals (Pelander et al. 2019; Hall et al. 2014b; Relford et al. 2016). Studies in companion animals have shown that, in contrast to creatinine, SDMA is not affected by muscle mass (International Renal Interest Society (IRIS) 2015a, 2017). On the other hand, human studies have found no correlation or an inverse correlation between SDMA and body mass index (Schepers et al. 2011; Schewedhelm et al. 2011; Potočnjak et al. 2018). Furthermore, discrepancies in the levels of SDMA in patients with DM have been reported among different studies. The offspring cohort

study from the Framingham Heart Study did not find a relationship between SDMA and insulin resistance (Schewedhelm et al. 2011). Moreover, an inverse correlation between SDMA and glycosylated hemoglobin or fructosamine levels has been reported in T2DM patients, i.e. those with poor glycemic control had lower SDMA concentration (Can et al. 2011). In another study, SDMA was increased in patients with T2DM and microalbuminuria, and this marker predicted impaired renal function and cardiovascular disease in this population (Zobel et al. 2017). Additionally, SDMA has been positively associated with proteinuria and inversely associated with GFR in patients with CKD and T2DM, and SDMA to asymmetric dimethylarginine (ADMA) ratio was one of the strongest predictive markers of renal function decline (Looker et al. 2015). ADMA is another byproduct of the proteolytic breakdown of nuclear proteins. It is an inhibitor of NOS and is considered to play a role in endothelial dysfunction (Sibal et al. 2010).

Interestingly, another study showed that young patients with T1DM and microalbuminuria and high GFR, had lower SDMA concentrations compared to normoalbuminuric T1DM patients; and over time, microalbuminuric patients showed an increment in SDMA concentration, probably reflecting a decline in GFR (Marcovecchio et al. 2010). Therefore, the lower levels of SDMA observed in patients with DM might be explained by hyperfiltration occurring in the early stages of DN. However, this requires further investigation, as other explanations have been suggested (Zsuga et al. 2007; Closs et al. 1997; Simmons et al. 1996; Siroen et al. 2005; Nijveldt et al. 2003).

In cats with DM, lower levels of SDMA have also been observed. However, in this species, further research on the relationship between DM and CKD is still needed (Langhorn et al. 2018).

In relation to the inverse association observed between SDMA and GFR, one study in humans, comparing SDMA to the renal clearance of an exogenous molecule (^{125}I – sodium iothalamate), showed that SDMA had a stronger inverse correlation with GFR than creatinine (El-khoury et al. 2016). In contrast, in dogs, the inverse correlation between SDMA and GFR determined through iohexol clearance was similar to the correlation observed between creatinine and GFR. However, SDMA was considered an earlier marker of GFR than creatinine, as SDMA was able to detect a decrease in renal function <20% on average, whereas creatinine increased when renal function was reduced by 50% (Hall et al. 2016). Similarly, in cats, the inverse correlation between SDMA and GFR measured by iohexol clearance was similar to the correlation observed between creatinine and GFR, but the sensitivity to detect impaired renal function was higher using SDMA (100 vs 17%) (Hall et al. 2014b). The established International Renal Interest Society guidelines consider that SDMA persistently >14 ng/ml is consistent with CKD (International Renal Interest Society (IRIS) 2015a). This cut-off (>14 ng/ml) of SDMA is able to detect a decrease of 24% from the median of GFR established for healthy cats, with a sensitivity of 91% (Hall et al. 2014b).

5.3 Markers of Glomerular Damage

5.3.1 Urine Albumin

Albumin is an intermediate molecular weight protein (69 KD) and its urinary concentration could increase with moderate disorders affecting the permeability of the glomerular barrier (D'Amico and Bazzi 2003). The initial alterations of the glomerular barrier usually involve a loss of restriction to passage of negatively charged proteins (especially albumin). Albuminuria can also occur when the tubular cells are damaged and tubular reabsorption is impaired, although intense albuminuria is usually glomerular in origin (Nauta et al. 2011; D'Amico and Bazzi 2003). The evaluation of a single albumin measurement is not recommended because it can vary during the day. Therefore, the measurement of urinary ACR in a random or first morning sample, or through a 24 h urine collection, with a measurement of creatinine, is advised (Basi et al. 2008). In addition, diabetic rodent models that develop albuminuria are considered particularly useful for the study of human DN (Diabetes Complications Consortium (DiaComp) 2003). In

rats, microalbuminuria has been associated with reduced albumin reabsorption in the proximal tubule in early DN (Tojo et al. 2001). Some studies have used ACR to study renal function in dogs (Tvari-jonaviciute 2013). However, in clinical practice of small domestic animals, the protein creatinine ratio is used instead of albumin. Although clinical interpretation of albuminuria is not well established in dogs and cats, borderline protein-creatinine ratio between 0.2 and 0.5 mg/mg in dogs, and between 0.2 and 0.4 mg/mg in cats, are suggestive of microalbuminuria, and its monitorization is recommended (International Renal Interest Society (IRIS) 2017). It should also be highlighted that albuminuria is not specific for kidney function and it could appear in the presence of non-renal diseases (i.e., hyperadrenocorticism, urinary tract infections or neoplasms) (Kivarikova 2015).

5.3.2 Urine Immunoglobulin G

Immunoglobulin (IgG) is a high molecular weight protein (160 kDa) that is involved in antibody-mediated immunity. Due to its size, this protein cannot pass through an intact glomerular barrier. Therefore, urine detection of IgG reflects glomerular damage (D'Amico and Bazzi 2003). IgG has been associated with albuminuria in patients with DM (Carlsson et al. 2017), but few studies have investigated IgG in animals. In dogs, urine detection of IgG has been associated with X-linked hereditary nephropathy even before the onset of proteinuria. The authors hypothesized that in dogs some degree of alteration of the glomerular basement membrane could allow the passage of proteins not detected by the usual assays to measure proteinuria (Nabity et al. 2012). Similarly, in humans, IgG has been detected in normo-albuminuric patients with T2DM, and it has been considered a predictive marker of albuminuria (Narita et al. 2006).

5.4 Markers of Tubular Damage

5.4.1 Serum and Urine Cystatin C

Cystatin C is a low molecular weight protein (13 kDa) that is considered as a marker of both

GFR and proximal tubular damage. Cystatin C is freely filtered by the glomerulus, and it is almost entirely reabsorbed in the proximal tubule by megalin-mediated endocytosis (Mussap and Plebani 2004; Kaseda et al. 2007). Its inverse correlation with GFR is better than that observed between creatinine and GFR (El-khoury et al. 2016). In addition, studies in animal models have suggested that cystatin C could reflect impaired kidney function earlier than creatinine (Song et al. 2009; Togashi and Miyamoto 2013). For example, after ischaemia-reperfusion injury, partial unilateral nephrectomy and bilateral nephrectomy, serum cystatin C concentration increased before creatinine in BALB/c mice models (Song et al. 2009). Another report observed that, in comparison to Zucker diabetic lean rats, Zucker Diabetic Fatty rats showed elevations of urinary cystatin C concentration, along with other renal biomarkers of kidney injury (β2-microglobulin, clusterin, mu-glutathione S-transferase and kidney injury molecule-1 (KIM-1)), but not of serum creatinine, before the appearance of kidney histopathological changes (Togashi and Miyamoto 2013). Additionally, in this study, the authors observed that immunohisto-chemical cystatin C expression was predominantly localized in the proximal tubules of the renal cortex, supporting a role of tubular damage in the development of kidney injury due to obesity (Togashi and Miyamoto 2013).

Furthermore, some advantages of cystatin C as a marker of CKD compared to creatinine should be highlighted. In humans, it is less affected by muscle mass than creatinine and it has lower inter-individual variability (Stevens et al. 2009). However, it should be taken into account that the concentration of cystatin C is subjected to changes in people with thyroid disorders or glucocorticoid treatment (Risch and Huber 2002; Fricker et al. 2003).

In human medicine, clinical practice guidelines for the evaluation and management of CKD recommend the measurement of cystatin C, especially in those patients in whom estimated GFR, based on serum creatinine, might be expected to be less accurate, or in those patients with early stages of CKD (estimated GFR between 45–59 ml/min/1.73 m^2), who do not have other markers of kidney damage (Levin et al. 2013). Additionally, cystatin C might be a predictor of impaired renal function in patients with T2DM and, although its concentration can reach higher levels in macroalbuminuric patients, it is independently associated with GFR, and an increase in serum and urine cystatin C concentration has been observed in subjects with DM, normoalbuminuria and decreased GFR (Kim et al. 2013; Jeon et al. 2011). Thus, cystatin C might predict DN in the early stages of impaired renal function in patients with T2DM (Kim et al. 2013; Jeon et al. 2011).

In contrast, in veterinary medicine, the use of cystatin C does not seem so advantageous. In cats, serum cystatin C was not found to be a reliable marker of GFR, and in regard to urinary cystatin C, one study showed higher concentrations of this marker in cats with CKD compared to healthy cats, although, urinary Cystatin C was below the detection limit of the assay in some of the cats with CKD (Ghys et al. 2016; Williams and Archer 2016). In addition, one study reported lower cystatin C concentration in cats with DM compared to healthy cats (Paepe et al. 2015), although in this species it is not clear whether CKD could occur secondarily to DM (Greene et al. 2014; Barlett et al. 2010; Al-Ghazlat et al. 2011; Pérez-López et al. 2019; Zini et al. 2014).

In a similar manner, in dogs, serum cystatin C does not seem to be superior to creatinine, either, and the sensitivity of cystatin C to detect CKD has been considered similar to SDMA and creatinine, whereas its specificity was lower (Pelander et al. 2019; Almy et al. 2002; Marynissen et al. 2016).

5.4.2 Retinol-Binding Protein 4 (RBP4)

Retinol-binding protein is a low molecular weight protein (21 kDa) that acts as the transport protein for retinol in plasma. It is produced in the liver and circulates in plasma bound to transthyretin (TTR), a protein with a molecular weight of 54 kDa that is too large to pass through the glomerular barrier. However, around 4–5% of serum RBP4 circulates freely and can pass through this barrier, to be then reabsorbed by tubular epithelial cells (Christensen et al. 1999). When tubular damage occurs, reabsorption of retinol is decreased, with subsequent loss of RBP4 into the urine (Zeni et al. 2017). Urinary RBP4 has been considered as a predictive marker of CKD in humans and of microalbuminuria in patients with T2DM (Domingos et al. 2016; Park et al. 2014). However, one study suggested that serum RBP4 does not seem to be a better marker than creatinine or cystatin C to detect GFR impairment (Donadio et al. 2001). On the other hand, serum retinol has also been considered a marker of insulin resistance and cardiovascular risk factors in humans and animal models (Park et al. 2014; Cabré et al. 2007; Mohapatra et al. 2011; Graham et al. 2006; Yang et al. 2005; Akbay et al. 2010), and it could be interesting for the study of early renal alterations in patients with metabolic syndrome. Mouse models have demonstrated that transgenic overexpression of human RBP4, or injection of recombinant RBP4, leads to insulin resistance (Yang et al. 2005), although a few reports disagree on this association (Von Eynatten et al. 2007; Henze et al. 2008).

In dogs and cats, higher urinary RBP4 concentration has been observed in animals with CKD compared to healthy animals (van Hoek et al. 2008; Chakar et al. 2017).

5.4.3 Uromodulin (Tamm-Horsfall Protein)

Uromodulin is a 100 kDa protein synthetized by the epithelial cells of the thick ascending limb of Henle's loop and the distal convoluted tubule (Hokamp and Nabity 2016). Therefore, in healthy individuals it is normal to find uromodulin in urine samples, whereas in patients with tubular damage its urinary concentration is low or even absent (Chakraborty et al. 2004; Steubl et al. 2016). Correlation between uromodulin and GFR has only been assessed through equations of estimated GFR, and conflicting results have been reported. Whereas urinary uromodulin is positively correlated to GFR, the correlation between GFR and serum uromodulin has been reported as positive or negative, depending on the study (Möllsten and Torffvit 2010; Prajczer

et al. 2010; Fedak and Kuźniewski 2016; Wiromrat et al. 2019). Nonetheless, several reports have observed that plasma or serum uromodulin concentrations are lower in patients with nephropathy. Uromodulin could be an indicator of renal damage in people with and without DM (Chakraborty et al. 2004; Steubl et al. 2016; Möllsten and Torffvit 2010). Kidneys of patients with very low levels of serum and urinary uromodulin show more tubular atrophy and decreased concentration is present at the earliest stages of CKD (Prajczer et al. 2010). However, in dogs, uromodulin has been considered a progression marker rather than an early marker of CKD (Chakar et al. 2017).

Additionally, studies in uromodulin knockout mice showed that it has a protective role against urinary tract infections and renal stone formation (Rampoldi et al. 2011; Bates et al. 2004; Liu et al. 2010). However, in humans, mutations in the gene encoding uromodulin lead to a rare autosomal dominant disease, which causes tubulointerstitial damage, but these patients do not show increased rates of urinary tract infections or renal stone formation (Rampoldi et al. 2011).

5.4.4 Neutrophil Gelatinase-Associated Lipocalin (NGAL)

NGAL is a low molecular weight protein of 25 kDa belonging to the lipocalin protein superfamily that is produced in the renal tubules (thick ascending limb of Henle's loop, distal tubule, and collecting duct) after inflammation or tissue injury (Singer et al. 2013). Its concentration may also be increased in case of impaired proximal tubular reabsorption (Singer et al. 2013).

In patients with DM, urinary NGAL shows an inverse correlation with GFR and a positive correlation with albuminuria (Kem et al. 2010; Fu et al. 2012a; Nielsen et al. 2011; Vijay et al. 2018). In addition, NGAL concentrations are higher in normoalbuminuric patients with DM, compared to non-diabetic control subjects, which might suggest that tubular damage could be one of the earliest alterations in patients with DN (Nauta et al. 2011; Vijay et al. 2018). Moreover, higher levels of urinary NAGL have been observed in T2DM diabetic patients with glomerular hyperfiltration compared to T2DM with normal GFR and control subjects (Fu et al. 2012b).

In addition, NGAL has been proposed as a useful indicator of acute kidney injury in human medicine (Singer et al. 2013; Haase et al. 2011) and most studies in animal models focus on its utility to detect acute kidney injury. In dogs, plasma and urinary NGAL are able to distinguish dogs with CKD from those with acute kidney injury (Steinbach et al. 2014). In contrast, in cats with CKD, this marker did not increase until the cats reached an advanced stage of the disease (Wang et al. 2017). Additionally, it should be highlighted that NGAL concentration could be influenced by other conditions, such as urinary tract infections, different types of neoplasms, pre-eclampsia and obstructive pulmonary disease (Giasson et al. 2011; Fjaertoft et al. 2005; Keatings and Barnes 1997; Bolignano et al. 2010).

5.4.5 Kidney Injury Molecule -1 (KIM-1)

KIM-1 is a type 1 transmembrane glycoprotein (90 kDa) located in the proximal tubules and, after tubular injury, its concentration rises before serum creatinine, whereas it is not detected in the urine of humans or other species without kidney damage (Moresco et al. 2018). KIM-1 is a well-known marker of acute kidney injury, and some studies suggest that it might be a good marker for CKD since its concentrations are high in humans with low GFR or albuminuria. Additionally, in humans with T2DM and normoalbuminuria or mild albuminuria, an increment of the urinary concentration of this marker has been observed (De Carvalho et al. 2016). Its ability to predict DN requires further investigation since discrepancies have been observed (Colombo et al. 2019a; Nauta et al. 2011). Two longitudinal studies showed that, the use of KIM-1 in T2DM, together with pro b-type natriuretic peptide or beta 2 microglobulin, GFR and albuminuria, seemed to improve prediction of kidney function decline (Kammer et al. 2019; Colombo et al. 2019b). In contrast, another study reported that KIM-1 was not associated with albuminuria

(Nauta et al. 2011), and, in a longitudinal, multi-center study in T1DM, it did not improve the prediction of progression of DN compared to albuminuria and estimated GFR, either (Panduru et al. 2015). In another large study, from an extensive set of biomarkers, KIM-1 and CD27 antigen combined, were the most important predictors of DN progression, although their predictive power did not improve that of historical estimated GFR and albuminuria. (Colombo et al. 2019a). It should also be highlighted that other conditions such as sepsis and urinary tract disease can increase urinary KIM-1 (Moresco et al. 2018).

In animal models, higher concentrations of this marker have been observed in a rat model of T2DM (*Otsuka Long-Evans Tokushima Fatty rats*) than in healthy rats (*Otsuka Long-Evans Tokushima rats*), and an increase of its concentration was observed prior to the development of hyperfiltration and prior to the increment of serum creatinine concentration (Hosohata et al. 2014).

5.5 Markers of Fibrosis and Inflammation

5.5.1 Transforming Growth Factor-β1 (TGFβ1)

TGFβ1 is a cytokine and pro-fibrotic mediator of kidney damage. It is secreted in an inactivated form associated with the large latent complex (LLC) and it has a biological effect only after it is liberated from the LLC as active TGFβ1, a process which takes part in the extracellular matrix (August and Suthanthiran 2003; Lawson et al. 2016; Sureshbabu et al. 2016; Hinz 2015). In kidneys, increased production of active TGFβ1 leads to interstitial fibrosis, mesangial matrix expansion and glomerular membrane thickening, and, as a consequence, kidney damage and reduction in GFR (August and Suthanthiran 2003; Sureshbabu et al. 2016; Ziyadeh 2004). Animal models have shown that TGFβ1 could also participate in podocyte detachment or apoptosis, stimulating podocyte expression of VEGF that acts in an autocrine loop, leading to increased production of 3(IV) collagen, which probably contributes to the thickening of the glomerular basement membrane (Chen et al. 2004). According to an *in vitro* study, TGFβ1 could also cause oxidative stress in podocytes by itself (Lee et al. 2003). As a consequence of podocyte injury, proteinuria can occur (Nagata 2016).

In vitro studies have also demonstrated that high glucose concentration stimulates TGFβ1 secretion and activation, and this marker has been proposed as an important mediator of DN in animal models (Ziyadeh 2004; Rocco et al. 1992; Hoffman et al. 1998). Also, in humans, high levels of serum and urinary TGFβ1 have been observed in a systematic review that included T2DM patients, showing a positive correlation with albuminuria (Qiao et al. 2017). Treatment with angiotensin-converting enzyme inhibitors has been demonstrated to decrease urinary ACR and plasma TGFβ1, and the latter has been put forward as a mechanism mediating their nephroprotective effects (Andrésdóttir et al. 2014). Moreover, in the db/db mouse, treatment with a neutralizing anti-TGFβ1 antibody avoids the progression of diabetic renal hypertrophy, mesangial matrix expansion, and the development of renal insufficiency, albeit in the absence of a significant reduction in albuminuria (Ziyadeh et al. 2000).

In companion animals, this marker has been studied to investigate DN in cats, where an increment of urinary activated TGFβ1:creatinine ratio precedes the onset of azotemia by 6 months (Lawson et al. 2016).

5.5.2 Vascular Endothelial Growth Factor (VEGF)

VEGF, also named vasopermeability factor, is a homodimeric glycoprotein with different isoforms and heparin-binding properties. VEGF promotes permeability, has mitogenic functions in endothelial cells and is an important angiogenic factor (Khamaisi et al. 2003; Neufeld et al. 1999). In the kidney, reduction of oxygen delivery contributes to inflammation pathways and is the main stimulus for VEGF expression. (Mayer 2011; Ramakrishnan et al. 2014).

In regard to DN, although *in vitro* studies have demonstrated that chronic hyperglycemia is able to increase the production of the VEGF protein and VEGF mRNA expression (Cha et al. 2000; Williams et al. 1997), in humans with DM it is uncertain whether VEGF levels increase or decrease with DN. Down-regulation of VEGF-A mRNA expression has been observed in renal biopsies of patients with DM, and its lower expression was associated with podocyte loss (Baelde et al. 2007). However, two other studies have shown higher concentrations of plasma and urinary VEGF in people with T1DM and T2DM, respectively (Hovind et al. 2000; Kim et al. 2004), and higher urinary VEGF was found in those with advanced DN. Indeed, intervention studies also show conflicting results: both VEGF inhibition and VEGF administration improve kidney function (Schrijvers et al. 2005; Kang et al. 2001).

Additionally, there is scarce information on the role of VEGF in kidney function in companion animals, in which the study of VEGF has been focused on its role in tumor angiogenesis (Millanta et al. 2002; Clifford et al. 2001; Platt et al. 2006). One study detected lower concentrations of urinary VEGF-A creatinine ratio in cats with CKD compared to healthy cats (Habenicht et al. 2013).

5.5.3 Soluble Urokinase Type Plasminogen Activator Receptor (suPAR)

Soluble urokinase-type plasminogen activator receptor (suPAR) is the circulating form of membrane protein urokinase receptor (uPAR), which is a glycosyl-phosphatidylinositol–anchored three-domain membrane protein that is expressed on podocytes and other cells (immunologically active cells and endothelial cells). Both suPAR and uPAR regulate cell adhesion and migration (Salim et al. 2016). Increased levels of plasma or serum suPAR are considered independent risk factors of cardiovascular diseases and CKD. suPAR seems to be a reliable marker of early

CKD since its concentration increases before a decline in GFR is identified (Salim et al. 2016). A cohort study showed that, in patients with T1DM, suPAR predicts cardiovascular events and a decline in GFR, although it was not correlated with albuminuria (Curovic et al. 2019). In contrast, a cross-sectional study showed a positive correlation between suPAR concentration and albuminuria in patients with T1DM (Theilade et al. 2015). Likewise, in patients with T2DM, suPAR concentration showed a positive correlation with albuminuria. Indeed, it was associated with an increased risk of new-onset microalbuminuria in subjects at risk for T2DM (Guthoff et al. 2017). In transgenic mice models, it has also been shown that increased uPAR activity in podocytes leads to proteinuria (Wei et al. 2008).

5.6 Panels of Candidate Biomarkers, Proteomic and Metabolomic Approaches

"Omics" approaches have been used to search for novel biomarkers and different aspects of the pathophysiology of kidney damage (Colhoun and Marcovecchio 2018; Carlsson et al. 2017; Abbiss et al. 2019; Darshi et al. 2016). Each individual biomarker of kidney disease represents a specific pathway. Depending on these pathways, some renal markers can be correlated with others. Therefore, the selection of panels of biomarkers that have low correlation with each other could potentially be beneficial for the prediction of impaired kidney function (Colhoun and Marcovecchio 2018). For example, a set of 297 biomarkers was evaluated in a recent prospective study in two different cohorts of patients with T1DM, and only two biomarkers, CD27, a member of the tumor necrosis factor receptor superfamily, and KIM-1 were considered to give predictive information of DN. This yielded similar predictive power than using historical estimated GFR and albuminuria (Colombo et al. 2019a). Similarly, another study examined 42 biomarkers in 840 serum samples of patients with T2DM and found that prediction of the decline in renal function could be improved by the use of two single markers: KIM-1 and $\beta2$ microglobulin (Colombo et al. 2019b). In contrast, in one study in patients with T1DM, KIM-1 did not seem to predict progression of CKD independently of albuminuria (Panduru et al. 2015) (Table 3).

6 Conclusions

DN is a common complication in patients with DM and research in animal models can be useful to fully understand the mechanisms underlying its development and progression. However, since various renal markers are not equally useful in all species, further studies are required in humans. The differences and similarities between people and animals with DN could also bring the opportunity to investigate new treatments for DN in humans. Although rodent models are the most studied, other animals, such as dogs and cats, could provide important information, since they share the human environment.

Regarding the potential use of markers of early kidney damage, cystatin C in humans, and SDMA in companion animals, have already been incorporated to the guidelines for the Evaluation and Management of CKD and of the International Renal Interest Society, respectively. Nonetheless, many other markers have been proposed, but results are conflicting for most of them. In fact, those biomarkers that have been identified as predictors of CKD or its progression, do not seem to add to established diagnostic tools. It is also important to highlight that most studies performed to date are cross-sectional in their design. More prospective and intervention studies are needed to replicate reported findings and to assess the predictive value of novel biomarkers of DN. Indeed, large consortia such as the SysKid (Systems Biology Towards Novel Chronic

Table 3 Epidemiological studies that provide information for the diagnosis of kidney disease or DN in both humans and animal models (histological, immunochemistry or intervention studies were not included)

Marker	Species	Author (reference)	N	Study design	Main results	Potential clinical application
Plasma SDMA and cystatin C	Humans	El-khoury et al. (2016)	40 patients who had clinical indication for measuring GFR	Cross-sectional	SDMA and cystatin C are highly and inversely correlated with GFR, more than creatinine	SDMA and cystatin C could detect an earlier decline in GFR
Plasma SDMA and ADMA	Humans	Zobel et al. (2017)	200 patients with T2DM	Longitudinal	Higher SDMA was associated with incident cardiovascular disease, and deterioration in renal function	SDMA could be a marker of DN and cardiovascular disease in patients with T2DM
Serum SDMA	Dogs	Hall et al. (2016)	19 dogs with CKD and 20 control dogs	Retrospective	SDMA detected CKD earlier than creatinine	SDMA should also be used to evaluate kidney function in dogs
Serum SDMA	Cats	Hall et al. (2014)	21 cats with CKD and 21 healthy control cats	Retrospective	SDMA detects CKD earlier than creatinine	SDMA should also be used to evaluate kidney function in cats
Serum SDMA	Cats	Langhorn et al. (2018)	17 with CKD, 40 with HCM, 17 with DM, and 20 healthy controls	Cross-sectional	Cats with DM had significantly lower SDMA concentrations than controls	SDMA is probabli not a useful marker of DN in cats; and further research about DN is still needed in cats
Serum cystatin C	BALB/c mice	Song et al. (2009)	23 partial nephrectomy 6 ischaemia reperfusion injury model / 8 controls	Cross-sectional	Cystatin C increases before creatinine in mice with kidney damage. CysC levels show an earlier and sharper increase than creatinine after bilateral nephrectomy	Cystatin C could be a more precise marker compared to creatinine
Urinary cystatin C	Humans	Jeon et al. (2011)	335 T2DM patients with normoalbuminuria (n = 210), those with microalbuminuria (n = 83) and those with macroalbuminuria (n = 42)	Retrospective	Cystatin C was independently associated with GFR, and was increased in people with diabetes, normoalbuminuria and decreased GFR	Might predict early stages of DN in T2DM patients
Serum and urinary Cystatin C	Cats	Ghys et al. (2016)	49 cats with CKD and 41 healthy cats	Cross-sectional	Sensitivity and specificity to detect CKD were 22 and 100% for Cystatin C and 83 and 93% for creatinine	Cystatin C should not be used to evaluated kidney function in cats

Biomarker	Reference	Species	Population	Study design	Findings	Conclusion
Serum SDMA and Cystatin C	Pelander et al. (2019)	Dogs	30 healthy dogs and 67 dogs with diagnosis or suspicion of CKD	Cross-sectional	Creatinine and SDMA were similar to detect reduced GFR, whereas cystatin C was inferior	SDMA should be measured together with creatinine as it might add information of kidney function in dogs
Urinary Cystatin C- creatinine ratio and nonalbumin protein – creatinine ratio	Kim et al. (2013)	Humans	237 T2DM patients	Longitudinal	After adjusting for several clinical factors, both urinary Cystatin C- creatinine ratio and non albuminuric protein – creatinine ratio had significant associations with the decline of the estimated glomerular filtration rate (eGFR)	Cystatin C could be a useful marker of renal function decline in patients with T2DM
$\beta2$-microglobulin, calbindin, clusterin, EGF, GST-α, GST-μ, KIM-1, NGAL, osteopontin, TIMP-1, and VEGF	Togashi and Miyamoto (2013)	Zucker diabetic fatty rats	5 Male Zucker diabetic fatty rats (ZDF/CrlCrlj-Leptfa/fa) and 5 Male nondiabetic lean rats (ZDF/CrlCrlj-Lept?/+	Cross-sectional	Urinary levels of cystatin C, $\beta2$-microglobulin, clusterin, GST-μ, KIM-1 were increased before the development of histophatological changes consistent with DN	Cystatin C, $\beta2$-microglobu in, clusterin, GST-μ, KIM-1 could be used as markers of DN in mice models of DN
Urinary RBP4	Domingos et al. (2016)	Humans	454 participants with stages 3 and 4 CKD	Cross-sectional	A logistic regression model showed an inverse association between CKD-EPI eGFR and urinary retinol binding protein	Urinary retinol binding protein might be a promising marker of chronic kidney disease progression
Urinary RBP4	Park et al. (2014)	Humans	471 type 2 diabetes patients, 143 with impaired glucose tolerance and 75 controls	Cross-sectional	Urinary RBP4 concentration was higher in insulin resistant patients, and it was highly associated with microalbuminuria (odds ratio 2.6, 95% CI 1.6–4.2),	Kidney function of diabetic patients with higher levels of urinary RBP4 should be evaluated closely, and it could be useful in the management and stratifications of insulin resistant patients. Further investigation is needed
Serum RBP4, CysC, b2M	Donadio et al. (2001)	Humans	110 patients with various kidney diseases	Cross-sectional	Serum concentrations of CysC, b2M and RBP4 increase with the reduction of GFR ROC analysis showed that diagnostic accuracy of CysC and b2M was similar to	Cys, b2M and RBP4 do not seems more reliable markers to detect a decline in kidney function compared to creatinine

(continued)

Table 3 (continued)

Marker	Species	Author (reference)	N	Study design	Main results	Potential clinical application
Plasma RBP4	Humans	Cabré et al. (2007)	165 T2DM patients	Cross-sectional	creatinine and better than RBP4. Patients with moderate renal dysfunction (MDRD-GFR <60 mL min^{-1} 1.73 m^{-2} had higher plasma RBP4 than those with normal renal function. Albuminuria was not associated with RBP4	RBP4 might predict early decline in GFR in patients with DN
Serum RBP4	Humans	Akbay et al. (2010)	53 T2DM patients and 30 controls	Cross-sectional	Logistic regression analysis showed that microalbuminuria is associated with increased serum RBP4 concentration	RBP4 might predict early DN
Serum RBP4	Cats	van Hoek et al. (2008)	10 cats with CKD, 10 cats with hyperthyroidism and 10 healthy cats	Cross-sectional	Cats with CKD and hyperthyroidism had higher concentration of RBP4 than healthy cats	RBP4 should be investigated as a marker of impaired kidney function in cats
Urinary and serum uromodulin	Humans	Prajczer et al. (2010)	77 patients with CKD and 14 healthy subjects	Cross-sectional	Urinary uromodulin was positively correlated with GFR and negatively correlated with serum creatinine. Patients with lower uromodulin values showed higher degree of tubular atrophy (assessed through biopsy)	Urinary uromodulin might be a early marker of tubular damage
Serum uromodulin	Humans	Wiromrat et al. (2019)	179 T1DM adolescents patients and 61 control subjects	Cross-sectional	Lower levels of serum uromodulin are associated with albumin excretion	Serum uromodulin should be assessed as a marker of DN
Urinary uromodulin	Humans	Möllsten and Torffvit (2010)	301 patients with T1DM, 164 with normoalbuminuria, 91 with microalbuminuria and 46 with macroalbuminuria	Cross-sectional	Patients with albuminuria had lower uromodulin concentrations	Urinary uromodulin might reflect tubular damage in patients with T1DM

Biomarker	Species	Reference	Subjects	Study type	Results	Conclusions
Serum uromodulin	Humans	Fedak and Kuźniewski (2016)	170 patients with CKD and 30 healthy subjects	Cross-sectional	Serum uromodulin was inversely correlated with other renal markers and positively correlated with estimated GFR	Serum uromodulin might be assessed as an early marker of CKD
Urinary albumin excretion ratio (AER), N-acetyl-β-D-glucosaminidase, advanced glycosylation end-products (AGEs) pentosidine and AGE-fluorescence	Humans	Kern et al. (2010)	55 T1DM patietns with and 110 without macroalbuminuria 91 T1DM patients with and 178 without microalbuminuria	Retrospective case-control	N-acetyl-β-D-glucosaminidase independently is associated with both macroalbuminuria and microalbuminuria. Other markers did not independently predict macro or microalbuminuria.	Tubular damage occurs in patients with T1DM and measurement of urinary albumin excretion ratio and urinary N-acetyl-β-D-glucosaminidase might predict early impaired kidney function
Plasma, urinary NGAL, and urinary NGAL-to-creatinine ratio (UNCR)	Cats	Wang et al. (2017)	80 cats with CKD and 18 healthy cats	Longitudinal	NGAL values were statistically different between healthy cats and cats with stage 3 or 4 CKD, however, no statistical differents were found between healthy cats and cats with those with stage 2 CKD	Plasma NGAL cannot distinguish CKD in cats. Urinary NGAL does not detect early stages of CKD in cats
Plasma NGAL and UNCR	Dogs	Steinbach et al. (2014)	17 dogs with CKD 48 dogs with AKI and 18 controls subjects	Cross-sectional	Plasma NGAL concentration and UNCR was significantly higher in dogs with AKI or CKD compared to healthy dogs. In addition, these markers were higher in dogs with AKI compared with dogs with CKD	Although NGAL is an established marker for AKI it might also be useful to distinguish dogs with CKD from healthy dogs, although further research is needed
Urinary NAGL, NAG and KIM-1	Human	Fu et al. (2012b)	101 T2DM patients 28 control subjects	Cross-sectional	All marker showed higher levels in patients with DM. NGAL and NAG were positively correlated with albuminuria. NGAL showed significant differences between micro and macroalbuminuric patients	NAGL and KIM-1 could be early markers of DN. Those patients with glomerular hyperfiltration and higher levels of urinary NAGL or KIM-1, should be closely monitored
					KIM-1 was not associated with albuminuria	

(continued)

Table 3 (continued)

Marker	Species	Author (reference)	N	Study design	Main results	Potential clinical application
Urinary KIM-1 and NGAL	Humans	de Carvalho et al. (2016)	117 T2DM patients	Cross-sectional	Both markers were observed in patients with T2DM with normal or mild albuminuria, and they were independently associated with albuminuria	Tubular markers could help in the early detection of DN
Urinary KIM-1, NGAL and vanin-1	Rats	Hosohata et al. (2014)	8 male spontaneous type 2 diabetic OLETF rats and 8 male non-diabetic Long-Evans Tokushima Otsuka (LETO) rats	Cross-sectional	Urinary KIM-1 was more sensitive than albumin to detect DN	KIM-1 detect early tubular damage
Urinary KIM-1	Humans	Panduru et al. (2015)	1573 T1DM patients	Longitudinal multicenter study	Mendelian randomization (MR) approach suggested a causal link between increased urinary KIM-1 and decreased GFR. KIM-1 did not predict progression of albuminuria	Further studies to evaluate KIM-1 as an early marker of DN could be interesting
Urinary active TGFβ1: creatinine ratio	Cats	Lawson et al. (2016)	6 non-azotaemic cats that developed azotaemia within 24 months; 6 cats and with renal azotaemia at baseline; and 6 non-azotemic cats	Longitudinal	Increased active TGFβ1: creatinine ratio was observed 6 month before the development of azotaemia	Urinary active TGFβ1: creatinine ratio might be an early marker of CKD in cats, this marker should be assessed in a larger sample.
Serum and urinary TGFβ1	Humans	Qiao et al. (2017)	63 case-control studies (364 T2DM patients, 1604 T2DM and DN patients, and 2100 healthy controls)	Systematic review	TGFβ1 levels were higher in T2DM patients and were positively correlated with albuminuria	TGFβ1 could be a promising marker of DN, although future research is needed
Plasma uromodulin and cystatin C	Humans	Steubl et al. (2016)	426 individuals of whom 71 healthy subjects and 355 had CKD (stages I-V)	Cross-sectional	Multiple linear regression modeling showed significant association between uromodulin and eGFR (coefficient estimate b.0.696, 95% confidence interval [CI] 0.603–0.719, $P < 0.001$) Uromodulin was able to differentiate between patients in stage 0 and I	Uromodulin could be an earlier marker of CKD compared to creatinine and CysC

Biomarker	Species	Reference	Study population	Study type	Findings	Conclusion
Urinary Ig G, KIM-1, NAGL, NAG	Humans	Nauta et al. (2011)	94 T1DM and T2DM, and 45 control subjects	Cross-sectional	Neither cystatin C nor creatinine distinguished between stages 0 and I. Glomerular and tubular markers were associated with albuminuria independently of GFR (except KIM-1)	Markers of glomerular and tubular damage should be evaluated in-patients with DN
Urinary RBP4, b2-microglobulin, NAGL, NAG, and IgG creatinine ratios	Dogs	Nabity et al. (2015)	20–25 dogs with X-linked hereditary nephropathy and 10–19 dogs control subjects	Retrospective	Urinary RBP4 was the most strongly correlated with serum creatinine and GFR. Although logistic regression analysis showed serum creatinine, uIgG/c, and uB2M, but not uRBP4/c, as significant independent predictors of GFR	b2-microglobulin, NAGL/c, NAG/c, IgG/ could allow early detection of CKD, and RBP4/c is a marker of CKD progression in dogs
Urinary albumin vitamin D-binding protein, RBP4 uromodulin,	Dogs	Chakar et al. (2017)	40 dogs with CKD and 9 control subjects	Cross-sectional	Increased vitamin D-binding protein and RBP4 were detected in early stages of CKD with or without albuminuria. Dogs with CKD had undetectable or lower RBP4 than controls, although among dogs in early stages of CKD there were no differences	Vitamin D binding protein and RBP4 should be studied as an early marker of CKD, whereas uromodulin could be a marker of progression of CKD in dogs
Plasma and urinary VEGF	Humans	Kim et al. (2004)	147 patients with T2DM and 47 healthy controls	Cross-sectional	VEGF concentration was higher in T2DM patients and was associated with albuminuria	VEGF might be a useful marker of DN in patients with T2DM
Plasma VEGF	Humans	Hovind et al. (2000)	199 patients with T1DM and DN, and 188 patients with T1DM and normoalbuminuria	Cross-sectional	Men with DN had higher concentration of VEGF than normoalbuminuric patients	Sex differences might affect VEGF concentration
Urinary cytokine (IL-8, MCP-1. TGF-β1, VEGF): urine creatinine ratios	Cats	Habenicht et al. (2013)	26 cats with CKD and 18 healthy cats	Cross-sectional	Cats with CKD had a significantly lower urinary levels of VEGF and higher urinary levels of IL-8 and	Further research is needed to evaluate the utility of measure urinary cytokines in cats with CKD, but it markers might

(continued)

Table 3 (continued)

Marker	Species	Author (reference)	N	Study design	Main results	Potential clinical application
					TGF-β1 compared to healthy cats	reflect kidney inflammaniton and fibrosis
Plasma suPar	Humans	Salim et al. (2016)	2292 patients from Emory cardiovascular biobank whose renal function was sequentially evaluated	Longitudinal	suPAR concentration increased before a decline in estimated GFR was observed	Kidney function of patients with elevated suPAR should be monitored closely
Plasma suPar	Humans	Curovic et al. (2019)	667 patients with T1DM and different levels of albuminuria	Longitudinal	suPAR predicted cardiovascular events and a decline in GFR but was not associated with albuminuria	suPAR is useful to detect early DN
Plasma suPAR	Humans	Theilade et al. (2015)	667 patients with T1DM and 51 control subjects	Cross-sectional	suPAR levels were higher in patients with cardiovascular disease and in patients with albuminuria. Multivariate logistic regression analysis showed an association between suPAR and albuminuria in patients with DM	suPAR and its relations with albuminuria should be investigated
Plasma suPAR	Humans	Guthoff et al. (2017)	258 patients at risk of T2DM	Longitudinal	Higher suPAR levels are associated with an increased risk of new-onset microalbuminuria in subjects at risk for type 2 diabetes	suPAR might be a useful marker of early DN in T2DM patients
Panel of 42 serum biomarkers	Humans	Colombo et al. (2019b)	840 patients with T2DM	Longitudinal multicenter study	Kim-1 and β2-microglobulin improve prediction of renal function decline	Until further validation Kim-1 and β2-microglobulin could be useful in clinical trials to select those patients with higher risk of DN
Panel of 297 serum biomarkers	Humans	Colombo et al. (2019a)	1174 patients with T1DM	Longitudinal multicenter study	Predictive information can be obtained using just two biomarkers (CD27 and KIM-1)	Few biomarkers are necessary to gain prediction of DN diagnosis.
Panel of 207 serum biomarkers	Humans	Looker et al. (2015)	154 cases (40% reduction in GFR) and 153 controls	Longitudinal multicenter study	14 biomarkers were associated with CKD progression (SDMA, SDMA/ADMA ratio,	These panel detected some novel biomarkers that require further investigation

| Panel of 402 plasma biomarkers | Humans | Kammer et al. (2019) | 481 T2DM patients with incident or early CKD (comparing patients with stable GFR and patients with a rapid decline of GFR) | Longitudinal multicenter study | Kim-1, creatinine, β2-Microglobulin, α1 Antitrypsin, Uracil, N-terminal prohormone of brain natriuretic peptide, C16-acylcarnitine, Hydroxyproline, Fibroblast growth factor-21, Fatty acid-binding protein heart, Creatine, Adrenomedullin) | KIM-1 was the most important predictor of GFR decline | KIM-1 might be a useful marker of DN in patients with T2DM |

Abbreviations: *eGFR* estimated glomerular filtration rate, *DN* diabetic nephropathy, *T1DM* patients with type 1 diabetes, *T2DM* patients with type 2 diabetes, *SDMA* symmetric dimethylarginine, *NGAl* neutrophil gelatinase-associated lipocalin, *AER* Albumin excretion ratio, *AGEs* advanced glycosylation end-products, *AKI* acute kidney injury, *CKD* chronic kidney disease, *LN* diabetic nephropathy, *UNCR* urinary NGAL-to-creatinine ratio, *KIM-1*, kidney injury molecule −1, *IgG* immunoglobulin G, *MCP-1* urinary monocyte chemoattractant protein-1, *NAG* N-acetyl-β-D-glucosaminidase, *GST-α* alpha glutation S transferasa, *GST-μ* mu glutation S transferasa, *IL-8* interleukin 8, *TIMP-1* tissue inhibitor of metalloprotease-1, *VEGF* vascular endothelial growth factor, *RBP4* retinol binding protein 4, *suPAR* soluble urokinase type plasminogen activator receptor, *TGFβ1* transforming growth factor β-1

Kidney Disease Diagnosis and Treatment), SUM-MIT (Surrogate markers for micro and macrovascular hard endpoints for innovative diabetes tools) and BEAt-DKD (Biomarker Enterprise to Attack Diabetic Kidney Disease) are providing and will provide important results in the near future.

References

Abbiss H, Maker GL, Trengove RD (2019) Metabolomics approaches for the diagnosis and understanding of kidney diseases. Meta 9(2):E34

Akbay E, Muslu N, Nayir E, Ozhan O, Kiykim A (2010) Serum retinol binding protein 4 level is related with renal functions in type 2 diabetes. J Endocrinol Investig 33(10):725–729

Al-Ghazlat SA, Cathy B, Langston E, Greco DS, Reine NJ, May SN et al (2011) The prevalence of microalbuminuria and proteinuria in cats with diabetes mellitus. Top Companion Anim Med 26(3):154–157

Almy FS, Christopher MM, King DP, Brown SA (2002) Evaluation of cystatin C as an endogenous marker of glomerular filtration rate in dogs. J Vet Intern Med 16 (1):45–51

Alpers CE, Hudkins KL (2011) Mouse models of diabetic nephropathy. Curr Opin Nephrol Hypertens 20 (3):278–284

American Diabetes Association (2004) Nephropathy in diabetes. Diabetes Care 27(Suppl 1):S79–S83

Andrésdóttir G, Jensen ML, Carstensen B, Parving HH, Rossing K, Hansen TW et al (2014) Improved survival and renal prognosis of patients with type 2 diabetes and nephropathy with improved control of risk factors. Diabetes Care 37(6):1660–1667

Antic V, Tempini A, Montani JP (1999) Serial changes in cardiovascular and renal function of rabbits ingesting a high-fat, high-calorie diet. Am J Hypertens 12(8 Pt 1):826–829

August P, Suthanthiran M (2003) Transforming growth factor beta and progression of renal disease. Kidney Int Suppl 87:S99–S104

Baelde HJ, Eikmans M, Lappin DW, Doran PP, Hohenadel D, Brinkkoetter PT et al (2007) Reduction of VEGF-A and CTFG expression in diabetic nephropathy is associated with podocyte loss. Kidney Int 71 (7):637–645

Barlett PC, Van Buren JW, Barlett AD, Zhou C (2010) Case-control study of risk factors associated with canine and feline chronic kidney disease. Vet Med Int 2010:957570

Basi S, Fesler P, Mimran A, Lewis LB (2008) Microalbuminuria in type 2 diabetes and hypertension. Diabetes care 2008. Diabetes Care 31(Suppl 2):S194–S201

Bates JM, Raffi HM, Prasadan K, Mascarenhas R, Laszik Z, Maeda N (2004) Tamm-Horsfall protein knockout mice are more prone to urinary tract infection: rapid communication. Kidney Int 65:791–797

Betz B, Conway BR (2014) Recent advances in animal models of diabetic nephropathy. Nephron Exp Nephrol 126:191–195

Blantz RC (1998) Pathophysiology of pre-renal azotemia. Kidney Int 53:512–523

Bolignano D, Donato V, Lacquaniti A, Fazio MR, Bono C, Coppolino G et al (2010) Neutrophil gelatinase-associated lipocalin (NGAL) in human neoplasias: a new protein enters the scene. Cancer Lett 288:10–16

Brito-Casillas Y, Melián C, Wägner AM (2016) Study of the pathogenesis and treatment of diabetes mellitus through animal models. Endocrinol Nutr 63 (7):345–353

Cabré A, Lázaro I, Girona J, Manzanares J, Marimón F, Plana N et al (2007) Retinol-binding protein 4 as a plasma biomarker of renal dysfunction and cardiovascular disease in type 2 diabetes. J Intern Med 262 (4):496–503

Câmara NO, Iseki K, Kramer H, Liu ZH, Sharma K (2017) Kidney disease and obesity: epidemiology, mechanisms and treatment. Nat Rev Nephrol 13 (3):181–190

Can A, Bekpinar S, Gurdol F, Tutuncu Y, Unlucerci Y, Dinccag N (2011) Dimethylarginines in patients with type 2 diabetes mellitus: relation with the glycaemic control. Diabetes Res Clin Pract 94(3):e61–e64

Caramori ML, Fioretto P, Mauer M (2003) Low glomerular filtration rate in normoalbuminuric type 1 diabetic patients: an indicator of more advanced glomerular lesions. Diabetes 52(4):1036–1040

Carlsson AC, Ingelsson E, Sundström J, Carrero JJ, Gustafsson S, Feldreich T et al (2017) Use of proteomics to investigate kidney function decline over 5 years. Clin J Am Soc Nephrol 12(8):1226–1235

Castrop H, Schießl IM (2014) Physiology and pathophysiology of the renal Na-K-2Cl cotransporter (NKCC2). Am J Physiol Renal Physiol 307(9):F991–F1002

Cha DR, Kim NH, Yoon JW, Jo SK, Cho WY, Kim HK et al (2000) Role of vascular endothelial growth factor in diabetic nephropathy. Kidney Int Suppl 77:S104–S1012

Chakar F, Kogika M, Sanches TR (2017) Urinary Tamm-Horsfall protein, albumin, vitamin D-binding protein, and retinol-binding protein as early biomarkers of chronic kidney disease in dogs. Phys Rep 5(11): e13262

Chakraborty J, Below AA, Solaiman D (2004) Tamm-Horsfall protein in patients with kidney damage and diabetes. Urol Res 32(2):79–83

Chandler M, Cunningham S, Lund EM, Khanna C, Naramore R et al (2017) Obesity and associated comorbidities in people and companion animals- a one health perspective. J Comp Pathol 156(4):296–309

Chen S, Kasarma Y, Lee JS, Jim B, Marin M, Ziyadeh FN (2004) Podocyte-derived vascular endothelial growth factor mediates the stimulation of 3(IV) collagen production by transforming growth factor 1 in mouse podocytes. Diabetes 52(11):2939–2949

Cheuiche AV, Queiroz M, Azeredo-da-Silva ALF, Silveiro SP (2019) Performance of cystatin C-based equations for estimation of glomerular filtration rate in diabetes patients: a Prisma-compliant systematic review and Meta-analysis. Sci Rep 9:1418

Christensen EI, Moskaug JO, Vorum H (1999) Evidence for an essential role of megalin in transepithelial transport of retinol. J Am Soc Nephrol 10(4):685–695

Clifford CA, Hughes D, Beal MW, Mackin AJ, Henry CJ, Shofer FS et al (2001) Plasma vascular endothelial growth factor concentrations in healthy dogs and dogs with hemangiosarcoma. J Vet Intern Med 15 (2):131–135

Closs EI, Basha FZ, Habermeier A, Förstermann U (1997) Interference of L-arginine analogues with L-arginine transport mediated by the y/carrier hCAT-2B. Nitric Oxide 1(1):65–73

Colhoun HM, Marcovecchio ML (2018) Biomarkers of diabetic kidney disease. Diabetologia 61(5):996–1011

Colombo M, Valo E, McGurnaghan SJ, Sandholm N, Blackbourn LAK, Dalton RN et al (2019a) Biomarker panels associated with progression of renal disease in type 1 diabetes. Diabetologia 62(9):1616–1627

Colombo M, Looker HC, Farran B, Hess S, Groop L, Palmer CAN et al (2019b) Serum kidney injury molecule 1 and β2-microglobulin perform as well as larger biomarker panels for prediction of rapid decline in renal function in type 2 diabetes. Diabetologia 62 (1):156–158

Curovic VR, Theilade S, Winther SA, Tofte N, Eugen-Olsen J, Persson F et al (2019) Soluble Urokinase plasminogen activator receptor predicts cardiovascular events, kidney function decline, and mortality in patients with type 1 diabetes. Diabetes Care 42 (6):1112–1119

D'Amico G, Bazzi C (2003) Pathophysiology of proteinuria. Kidney Int 63(3):809–825

Darshi M, Van Espen B, Sharma K (2016) Metabolomics in diabetic kidney disease: unraveling the biochemistry of a silent killer. Am J Nephrol 44(2):92–103

De Carvalho JAM, Tatsch E, Hausen BS, Bollick Y, Moretto MB, Duarte T et al (2016) Urinary kidney injury molecule-1 and neutrophil gelatinase-associated lipocalin as indicators of tubular damage in normoalbuminuric patients with type 2 diabetes. Clin Biochem 49(3):232–236

De Francesco PN, Cornejo MP, Barrile F, García Romero G, Valdivia S, Andreoli MF et al (2019) Inter-individual variability for high fat diet consumption in inbred C57BL/6 mice. Front Nutr 6:67

Delanaye P, Cavalier E, Pottel H (2017) Serum creatinine-not so simple! Nephron 136(4):302–308

Diabetes Complications Consortium (DiaComp) (2003) Validation of mouse models of diabetic nephropathy [Internet]. EEUU: Diabetes Complications Consortium (DiaComp); [Cited 2019 July 24]. Available from: https://www.diacomp.org/shared/document.aspx? Id=23&docType=Protocol

Domingos MA, Moreira SR, Gomez L, Goulart A, Lotufo PA, Benseñor I et al (2016) Urinary retinol-binding protein: relationship to renal function and cardiovascular risk factors in chronic kidney disease. PLoS One 11 (9):e0162782

Donadio C, Lucchesi A, Ardini M, Giordani R (2001) Cystatin C, b2-microglobulin, and retinol-binding protein as indicators of glomerular filtration rate- comparison with plasma creatinine. J Pharm Biomed Anal 24 (5–6):835–842

Dronavalli S, Duka I, Bakris GL (2008) The pathogenesis of diabetic nephropathy. Nat Clin Pract Endocrinol Metab 4(8):444–452

Dwyer TM, Banks SA, Alonso-Galicia M (2000) Distribution of renal medullary hyaluronan in lean and obese rabbits. Kidney Int 58(2):721–729

El-khoury JM, Bunch DR, Hu B, Payto D, Reineks EZ, Wang S (2016) Comparison of symmetric dimethylarginine with creatinine, cystatin C and their eGFR equations as markers of kidney function. Clin Biochem 49(15):1140–1143

Engeli S, Böhnke J, Gorzelniak K, Janke J, Schiling P, Bader M et al (2005) Weight loss and the renin-angiotensin-aldosterone system. Hypertension 45 (3):356–362

Fedak D, Kuźniewski M, Fugiel A (2016) Serum uromodulin concentrations correlate with glomerular filtration rate in patients with chronic kidney disease. Pol Arch Med Wewn 126(12):995–1004

Ferguson MA, Waikar SS (2012) Established and emerging markers of kidney function. Clin Chem 58 (4):680–689

Finch NC, Syme HM, Elliott J (2018) Development of an estimated glomerular filtration rate formula in cats. J Vet Intern Med 32(6):1970–1976

Fioretto P, Mauer M (2007) Histopathology of diabetic nephropathy. Semin Nephrol 27(2):195–207

Fjaertoft G, Foucard T, Xu S, Venge P (2005) Human neutrophil lipocalin (HNL) as a diagnostic tool in children with acute infections: a study of the kinetics. Acta Paediatr 94:661–666

Forbes JM, Cooper ME (2007) Diabetic nephropathy: where hemodynamics meets metabolism. Exp Clin Endocrinol Diabetes 115:69–84

Fricker M, Wiesli P, Brändle M, Schwegler B, Schmind C (2003) Impact of thyroid dysfunction on serum cystatin C. Kidney Int 63(5):1944–1947

Fu WJ, Xiong SL, Fang YG, Wen S, Chen ML, Deng RT et al (2012a) Urinary tubular biomarkers in short-term type 2 diabetes mellitus patients: a cross-sectional study. Endocrine 41(1):82–88

Fu WJ, Li BL, Wang SB, Chen ML, Deng RT, Ye CQ et al (2012b) Changes of the tubular markers in type 2 diabetes mellitus with glomerular hyperfiltration. Diabetes Res Clin Pract 95(1):105–109

Ghys LFE, Paepe D, Lefebvre HP, Reynolds BS, Croubels S, Meyer E et al (2016) Evaluation of

Cystatin C for the detection of chronic kidney disease in cats. J Vet Intern Med 30:1074–1082

Giacchetti G, Faloia E, Mariniello B, Sardu C, Gatti C, Camilloni MA et al (2002) Overexpression of the renin-angiotensin system in human visceral adipose tissue in normal and overweight subjects. Am J Hypertens 15(5):381–382

Giasson J, Li GH, Chen Y (2011) Neutrophil gelatinase-associated lipocalin (NGAL) as a new biomarker for non-acute kidney injury (AKI) diseases. Inflamm Alerrgy Drug Targets 10(4):272–282

Graham TE, Yang Q, Blüher M et al (2006) Retinol-binding protein 4 and insulin resistance in lean, obese, and diabetic subjects. N Engl J Med 354(24):2552–2563

Greene JP, Lefebvre SL, Wang M, Wang M, Yang M, Lund EM et al (2014) Risk factors associated with the development of chronic kidney disease in cats evaluated at primary care veterinary hospitals. J Am Vet Med Assoc 244(3):320–327

Gross JL, de Azevedo MJ, Silveiro SP, Canani LH, Caramori ML, Zelmanovitz T (2005) Diabetic nephropathy: diagnosis, prevention, and treatment. Diabetes Care 28(1):164–176

Guthoff M, Wagner R, Randrianarisoa E, Hatziagelaki E, Peter A, Häring HU et al (2017) Soluble urokinase receptor (suPAR) predicts microalbuminuria in patients at risk for type 2 diabetes mellitus. Sci Rep 7:40627

Haase M, Haase-Fielitz A, Bellomo R, Mertens PR (2011) Neutrophil gelatinase-associated lipocalin as a marker of acute renal disease. Curr Opin Hematol 18:11–18

Habenicht LM, Webb TL, Clauss LA, Dow SW, Quimby JM (2013) Urinary cytokine levels in apparently healthy cats and cats with chronic kidney disease. J Feline Med Surg 15(2):99–104

Hall JA, Yerramilli M, Obare E, Yerramilli M, Yu S, Jewell DE (2014a) Comparison of serum concentrations of symmetric dimethylarginine and creatinine as kidney function biomarkers in healthy geriatric cats fed reduced protein foods enriched with fish oil, L-carnitine, and medium-chain triglycerides. Vet J 202(3):588–596

Hall JA, Yerramilli M, Obare E, Yerramilli M, Jewell DE (2014b) Comparison of serum concentrations of symmetric dimethylarginine and creatinine as kidney function biomarkers in cats with chronic kidney disease. J Vet Intern Med 28(6):1676–1683

Hall JA, Yerramilli M, Obare E, Yerramilli M, Melendez LD, Jewell DE (2015) Relationship between lean body mass and serum renal biomarkers in healthy dog. J Vet Intern Med 29(3):808–814

Hall JA, Yerramilli M, Obare E, Yerramilli M, Almes K, Jewell DE (2016) Serum concentrations of symmetric dimethylarginine and creatinine in dogs with naturally occurring chronic kidney disease. J Vet Intern Med 30(3):794–802

Hans-Joachim A, Davis JM, Thurau K (2016) Nephron protection in diabetic kidney disease. N Engl J Med 375:2096–2098

Henegar JR, Bigler SA, Henegar LK, Tyagi SC, Hall JE (2001) Functional and structural changes in the kidney in the early stages of obesity. J Am Soc Nephrol 12(6):1211–1217

Henze A, Frey SK, Raila J, Tepel M, Scholze A, Pfeiffer AF et al (2008) Evidence that kidney function but not type 2 diabetes determines retinol-binding protein 4 serum levels. Diabetes 57(12):3323–3326

Hinz B (2015) The extracellular matrix and transforming growth factor-β1 tale of a strained relationship. Matrix Biol 47:54–65

Hoenig M (2012) The cat as a model for human obesity and diabetes. J Diabetes Sci Technol 6(3):525–533

Hoffman BB, Sharma K, Zhu Y, Ziyadeh FN (1998) Transcriptional activation of transforming growth factor-beta1 in mesangial cell culture by high glucose concentration. Kidney Int 54(4):1107–1116

Hokamp JA, Nabity MB (2016) Renal biomarkers in domestic species. Vet Clin Pathol 45(1):28–56

Hoshi S, Shu Y, Yoshida F, Inagaki T, Sonoda J, Watanabe T et al (2002) Podocyte injury promotes progressive nephropathy in Zucker diabetic fatty rats. Lab Investig 82(1):25–35

Hosohata K, Ando H, Takeshita Y, Misu H, Takamura T, Kaneko S et al (2014) Urinary Kim-1 is a sensitive biomarker for the early stage of diabetic nephropathy in Otsuka Long-Evans Tokushima Fatty rats. Diab Vasc Dis Res 11(4):243–250

Hovind P, Tarnow L, Oestergaard PB, Parving HH (2000) Elevated vascular endothelial growth factor in type 1 diabetic patients with diabetic nephropathy. Kidney Int Suppl 75:S56–S61

Hudkins KL, Pichaiwong W, Wietecha T, Kowalewska J, Banas MC, Spencer MW et al (2010) BTBR Ob/Ob mutant mice model progressive diabetic nephropathy. J Am Soc Nephrol 21(9):1533–1542

Ingvorsen C, Karp NA, Lelliott CJ (2017) The role of sex and body weight on the metabolic effects of high-fat diet in C57BL/6N mice. Nutr Diabetes 7(4):e261

International Renal Interest Society (IRIS) (2015a) GFR in practice: urine specific gravity. [Internet]. International Renal Interest Society (IRIS), Vienna; [Cited 2019 July 25]. Available from: http://www.iris-kidney.com/education/urine_specific_gravity.html

International Renal Interest Society (IRIS) (2015b) GFR in practice: assessment of glomerular filtration rate in dogs. [Internet]. International Renal Interest Society (IRIS), Vienna; [Cited 2019 July 25]. Available from: http://www.iris-kidney.com/education/gfr_in_practice.html

International Renal Interest Society (IRIS) (2017) IRIS staging of CKD [Internet]. International Renal Interest Society (IRIS), Vienna; [Cited 2019 May 22]. Available from: http://www.iris-kidney.com/guidelines/staging.html

Jeon YK, Kim MR, Huh JE, Mok JY, Song SH, Kim SS et al (2011) Cystatin C as an early biomarker of nephropathy in patients with type 2 diabetes. J Korean Med Sci 26:258–263

Jordan E, Kley S, Le NA, Waldron M, Hoenig M (2008) Dyslipidemia in obese cats. Domest Anim Endocrinol 35(3):290–299

Kachapati K, Adams D, Bednar K, Ridgway WM (2012) The Non-Obese Diabetic (NOD) mouse as a model of human type 1 diabetes. Methods Mol Biol 933:3–16

Kammer M, Heinzel A, Willency JA, Duffin KL, Mayer G, Simons K et al (2019) Integrative analysis of prognostic biomarkers derived from multiomics panels helps discrimination of chronic kidney disease trajectories in people with type 2 diabetes. Kidney Int 96(6):1381–1388

Kanasaki K, Gangadhar T, Koya D (2013) Diabetic nephropathy: the role of inflammation in fibroblast activation and kidney fibrosis. Front Endocrinol (Lausanne) 4:7

Kang DH, Hughes J, Mazzali M, Schreiner GF, Johnson RJ (2001) Impaired angiogenesis in the remnant kidney model: II. Vascular endothelial growth factor administration reduces renal fibrosis and stabilizes renal function. J Am Soc Nephrol 12(7):1446–1457

Kaseda R, Iino N, Hosojima M, Takea T, Hosaka K, Kobayashi A (2007) Megalin-mediated endocytosis of cystatin C in proximal tubule cells. Biochem Biophys Res Commun 357(4):1130–1134

Kavarikova S (2018) Indirect markers of glomerular filtration rate in dogs and cats: a review. Vet Med 63(9):395–412

Kawano K, Hirashima T, Mori S, Saitho Y, Kurosumi M, Natori T (1992) Spontaneous long-term hyperglycemic rat with diabetic complications. Diabetes 41(11):1422–1428

Kawano K, Hirashima T, Mori S, Natori T (1994) OLETF (Otsuka Long-Evans Tokushima Fatty) rat: a new NIDDM rat strain. Diabetes Res Clin Pract 24(Suppl):S317–S320

Keatings VM, Barnes PJ (1997) Granulocyte activation markers in induced sputum: comparison between chronic obstructive pulmonary disease, asthma, and normal subjects. Am J Respir Crit Care Med 155:449–453

Kem EF, Erhard P, Sun W, Genuth S, Weiss MF (2010) Early urinary markers of diabetic kidney disease: a nested case-control study from the Diabetes Control and Complications Trial (DCCT). Am J Kidney Dis 55(5):824–834

Kern PA, Ranganathan S, Li C, Li C, Wood L, Ranganathan G (2001) Adipose tissue tumor necrosis factor and interleukin-6 expression in human obesity and insulin resistance. Am J Physiol Endocrinol Metab 280(5):E745–E751

Khamaisi M, Schrijivers BF, De Vriese AS, Itamar R, Flyvbjerg A et al (2003) The emerging role of VEGF in diabetic kidney disease. Nephrol Dial Transplant 18(8):1427–1430

Kim NH, Kim KB, Kim DL, Kim SG, Choi KM, Baik SH et al (2004) Plasma and urinary vascular endothelial growth factor and diabetic nephropathy in type 2 diabetes. Diabet Med 21(6):545–551

Kim SS, Song SH, Kim IJ, Jeon YK, Kim BH, Kwak IS et al (2013) Urinary cystatin C and tubularproteinuria predict progression of diabetic nephropathy. Diabetes Care 36:656–661

Kitada M, Ogura Y, Koya D (2016) Rodent models of diabetic nephropathy: their utility and limitations. Int J Nephrol Renov Dis 9:279–290

Kivarikova S (2015) Urinary biomarkers of renal function in dogs and cats: a review. Vet Med 60(11):589–602

Lamacchia O, Viazzi F, Fioretto P et al (2018) Normoalbuminuric kidney impairment in T1DM patients. Diabetol Metab Syndr 10:60

Langhorn R, Kieler IN, Koch J, Christiansen LB, Jessen LR (2018) Symmetric dimethylarginine in cats with hypertrophic cardiomyopathy and diabetes mellitus. J Vet Intern Med 32:57–63

Lawson JS, Syme HM, Wheeler-Jones CP, Elliot J (2016) Urinary active transforming growth factor β in feline chronic kidney disease. Vet J 214:1–6

Lee HB, Yu MR, Yang Y, Jiang Z, Ha H (2003) Transforming growth factor β1-induced apoptosis in podocytes via the extracellular signal-regulated kinase-mammalian target of rapamycin complex 1-NADPH oxidase 4 Axis. J Am Soc Nephrol 14(8 Suppl 3):S241–S245

Levey AS, Coresh J, Balk E et al (2003) National kidney foundation practice guidelines for chronic kidney disease: evaluation, classification, and stratification. Ann Intern Med 139(2):137–147

Levin A, Steven PE, Bilous RW, Coresh J, De Francisco ALM, De Jong PE et al (2013) Kidney disease: improving global outcomes (KDIGO) CKD work group. KDIGO 2012 clinical practice guideline for the evaluation and management of chronic kidney disease. Kidney Int Suppl 3(1):1–150

Li Z, Woollard JR, Wang S, Korsmo MJ, Ebrahimi B, Grande JP et al (2011) Increased glomerular filtration rate in early metabolic syndrome is associated with renal adiposity and microvascular proliferation. Am J Physiol Renal Physiol 301(5):F1078–F1087

Liu Y, Mo L, Goldfarb DS, Evan AP, Liang F, Khan SR et al (2010) Progressive renal papillary calcification and ureteral stone formation in mice deficient for Tamm-Horsfall protein. Am J Physiol Renal Physiol 299:F469–F478

Looker HC, Colombo M, Hess S, Brosnan MJ, Farran B, Dalton RN et al (2015) Biomarkers of rapid chronic kidney disease progression in type 2 diabetes. Kidney Int 88(4):888–896

Lovrenčić MV, Biljak ZR, Božičević S, Prašek M, Pavković P, Knotek M (2012) Estimating glomerular filtration rate (GFR) in diabetes: the performance of MDRD and CKD-EPI equations in patients with various degrees of albuminuria. Clin Biochem 45(18):1694–1696

MacIsaac RJ, Tsalamandris C, Panagiotopoulos S, Smith TJ, McNeil KJ, Jerums G (2004) Non-albuminuric renal insufficiency in T2DM. Diabetes Care 27(1):195–200

Marcovecchio ML, Dalton RN, Turner C, Prevost AT, Widmer B, Amin R et al (2010) Symmetric dimethylarginine, an endogenous marker of glomerular filtration rate, and the risk for microalbuminuria in young people with type 1 diabetes. Arch Dis Child 95(2):119–124

Marynissen SJJ, Pascale MYS, Ghys LFE, Paepe D, Delanghe J, Galac S et al (2016) Long-term follow-up of renal function assessing serum cystatin C in dogs with diabetes mellitus or hyperadrenocorticism. Vet Clin Pathol 452(2):320–329

Mayer G (2011) Capillary rarefaction, hypoxia, VEGF and angiogenesis in chronic renal disease. Nephrol Dial Transplant 26(4):1132–1137

McDermott JR (1976) Studies on the catabolism of NG-methylargire, NG, NlG-dimtylarginie and; NG, NG dimethylarginine in the rabbit. Biochem J 154(1):179–184

Millanta F, Lazzeri G, Vannozzi I, Viacava P, Poli A (2002) Correlation of vascular endothelial growth factor expression to overall survival in feline invasive mammary carcinomas. Vet Pathol 39(6):690–696

Mogensen CE, Christensen CK, Vittinghus E (1983) The stages in diabetic renal disease with emphasis on the stage of incipient diabetic nephropathy. Diabetes 32 (Suppl. 2):64–78

Mohapatra J, Sharma M, Acharya A, Pandya G, Chatterjee A, Balaraman R et al (2011) Retinol-binding protein 4: a possible role in cardiovascular complications. Br J Pharmacol 164(8):1939–1948

Möllsten A, Torffvit O (2010) Tamm-Horsfall protein gene is associated with distal tubular dysfunction in patients with type 1 diabetes. Scand J Urol Nephrol 44(6):438–434

Moresco RN, Bochi GV, Stein CS, De Carvalho JAM, Cembranel BM, Bollick YS (2018) Urinary kidney injury molecule-1 in renal disease. Clin Chim Acta 487:15–21

Mussap M, Plebani M (2004) Increased biochemistry and clinical role of human cystatin C. Crit Rev Clin Lab Sci 41(5–6):467–550

Nabity MB, Lees GE, Cianciolo R, Boggess MM, Steiner JM, Suchodolski JS (2012) Urinary biomarkers of renal disease in dogs with X-linked hereditary nephropathy. J Vet Intern Med 26(2):282–293

Nabity MB, Lees GE, Boggess MM, Yerramilli M, Obare E, Yerramilli M et al (2015) Symmetric dimethylarginine assay validation, stability, and evaluation as a marker for the early detection of chronic kidney disease in dogs. J Vet Intern Med 29(4):1036–1044

Nagata M (2016) Podocyte injury and its consequences. Kidney Int 89(6):1221–1230

Nakagawa T, Sato W, Glushakova O, Heinig M, Clarke T, Campbell-Thompson M et al (2007) Diabetic endothelial nitric oxide synthase knockout mice develop advanced diabetic nephropathy. J Am Soc Nephrol 18(2):539–550

Narita T, Hosoba M, Kakei M (2006) ItoS. Increased urinary excretions of immunoglobulin G, ceruloplasmin, and transferrin predict development of microalbuminuria in patients with type 2. Diabetes Care 29(1):142–144

National Institute of Diabetes and Digestive and Kidney Disaese (NIDDK) (2019) Glomerular disease primer: the normal kidney (Kidney anatomy and physiology) [Internet]. National Institute of Diabetes and Digestive and Kidney Disaese (NIDDK), Phoenix; [Cited 2020 March 13]. Available from: https://www.niddk.nih.gov/research-funding/at-niddk/labs-branches/kidney-diseases-branch/kidney-disease-section/glomerular-disease-primer/normal-kidney

Nauta FL, Wendy EB, Bakker SJL et al (2011) Glomerular and tubular damage markers are elevated in patients with diabetes. Diabetes Care 34(4):975–981

Nelson RW, Reusch CE (2014) Animal models of disease: classification and etiology of diabetes in dogs and cats. J Endocrinol 222(3):T1–T9

Neufeld G, Cohen T, Gengrinovithch S, Poltorak Z (1999) Vascular endothelial growth factor (VEGF) and its receptors. FASEB J 13(1):9–22

Nielsen SE, Andersen S, Zdunek D, Andersen S, Zdunek D, Hess G et al (2011) Tubular markers do not predict the decline in glomerular filtration rate in type 1 diabetic patients with overt nephropathy. Kidney Int 79(10):1113–1118

Nijveldt RJ, Teerlink T, Siroen MP, van Lambalgen AA, Rauwerda JA, van Leeuwen PA (2003) The liver is an important organ in the metabolism of asymmetrical dimethylarginine (ADMA). Clin Nutr 22(1):17–22

Novellas R, Ruiz de Gopegui R, Espada Y (2010) Assessment of renal vascular resistance and blood pressure in dogs and cats with renal disease. Vet Rec 166(20):618–623

Paepe D, Ghys LF, Smets P, Lefebvre HP, Croubels S, Daminet S (2015) Routine kidney variables, glomerular filtration rate and urinary cystatin C in cats with diabetes mellitus, cats with chronic kidney disease and healthy cats. J Feline Med Surg 17(10):880–888

Panduru NM, Sandholm N, Forsblom C, Saraheimo M, Dahlström EH, Thorn LM et al (2015) Kidney injury molecule-1 and the loss of kidney function in diabetic nephropathy: a likely causal link in patients with type 1 diabetes. Diabetes Care 38(6):1130–1137

Park SE, Lee NS, Park JW, Rhee EJ, Lee WY, Oh KW et al (2014) Association of urinary RBP4 with insulin resistance, inflammation, and microalbuminuria. Eur J Endocrinol 171(4):443–449

Pavenstädt H (2000) Roles of the podocyte in glomerular function. Am J Physiol Renal Physiol 278(2):F173–F179

Payne JR, Brodbelt DC, Fuentes VL (2017) Blood pressure measurements in 780 apparently healthy cats. J Vet Intern Med 31(1):15–21

Pelander L, Häggström J, Larsson A (2019) Comparison of the diagnostic value of symmetric dimethylarginine, cystatin C, and creatinine for detection of decreased glomerular filtration rate in dogs. J Vet Intern Med 33 (2):630–639

Pérez-López L, Boronat M, Melián C, Saavedra P, Brito-Casillas Y, Wägner AM (2019) Assessment of the association between diabetes mellitus and chronic kidney disease in adult cats. J Vet Intern Med 33 (5):1921–1925

Peti-Peterdi J, Harris RC (2010) Macula Densa sensing and signaling mechanisms of renin release. J Am Soc Nephrol 21(7):1093–1096

Platt SR, Scase TJ, Adams V, Adams V, Wieczorek L, Miller J et al (2006) Vascular endothelial growth factor expression in canine intracranial meningiomas and association with patient survival. J Vet Intern Med 20 (3):663–668

Potočnjak I, Radulović B, Degoricija V, Trbušić M, Pregarner G, Berghold A et al (2018) Serum concentrations of asymmetric and symmetric dimethylarginine are associated with mortality in acute heart failure patients. Int J Cardiol 261:109–113

Prajczer S, Heidenreich U, Pfaller W, Kotanko P, Lhotta K, Jennings P (2010) Evidence for a role of uromodulin in chronic kidney disease progression. Nephrol Dial Transplant 25(6):1986–1903

Premaratne E, Verma S, Ekinci EI, Theverkalam G, Jerums G, Maclsaac RJ (2015) The impact of hyperfiltration on the diabetic kidney. Diabetes Metab 41(1):5–17

Priyanka M, Jeyaraja K, Thirunavakkarasu PS (2018) Abnormal renovascular resistance in dogs with diabetes mellitus: correlation with glycemic status and proteinuria. Iran J Vet Res 19(4):304–309

Qiao YC, Chen YL, Pan YH, Ling W, Tian F, Zhan XX et al (2017) Changes of transforming growth factor beta 1 in patients with type 2 diabetes and diabetic nephropathy: a PRISMA-compliant systematic review and meta-analysis. Medicine (Baltimore) 96(15):e6583

Ramakrishnan S, Anand V, Roy S (2014) Vascular endothelial growth factor signaling in hypoxia and inflammation. J NeuroImmune Pharmacol 9(2): 142–160

Rampoldi L, Scolari F, Amoroso A, Ghiggeri G, Devuyst O (2011) The rediscovery of uromodulin (Tamm–Horsfall protein) from tubulointerstitial nephropathy to chronic kidney disease. Kidney Int 80 (4):338–347

Rayner H, Thomas M, Milford D (2016) Kidney anatomy and physiology, In: Understanding kidney diseases, 1st edn. Springer, Cham, pp 1–10

Relford R, Robertson J, Clements C (2016) Symmetric dimethylarginine improving the diagnosis and staging of chronic kidney disease in small animals. Vet Clin North Am Small Anim Pract 46(6):941–960

Riedel J, Badewien-Rentzsch B, Kohn B, Hoeke L, Einspanier R (2006) Characterization of key genes of the renin-angiotensin system in mature feline adipocytes and during invitro adipogenesis. J Anim Physiol Anim Nutr 100:1139–1148

Risch L, Huber AR (2002) Glucocorticoids and increased serum cystatin C concentrations. Clin Chim Acta 320:133–134

Rocco MV, Chen Y, Goldfarb S, Ziyadeh FN (1992) Elevated glucose stimulates TGF-beta gene expression and bioactivity in proximal tubule. Kidney Int 41 (1):107–114

Rodríguez-Rodríguez R, González-Bulnes A, Garcia-Contreras C et al (2020) The Iberian pig fed with high-fat diet: a model of renal disease in obesity and metabolic syndrome. Int J Obes 44:457–465

Salim S, Hayek SS, Sever S, Ko Y-A, Tratchman H, Awad M et al (2016) Soluble urokinase receptor and chronic kidney disease. N Engl J Med 374(9):891

Schepers E, Barreto DV, Liabeuf S, Glorieux G, Eloot S, Barreto FC et al (2011) Symmetric dimethylarginine as a proinflammatory agent in chronic kidney disease. Clin J Am Soc Nephrol 6(10):2374–2383

Schewedhelm E, Xanthakis V, Maas R, Sullivan LM, Atzler D, Lüneburg N et al (2011) Plasma symmetric dimethylarginine reference limits from the Framingham offspring cohort. Clin Chem Lab Med 49(11):1907–1910

Schrijvers BF, Flyvbjerg A, Tilton RG (2005) Pathophysiological role of vascular endothelial growth factor in the remnant kidney. Nephron Exp Nephrol 101(1):e9–e15

Sharma AM, Engeli S (2006) Obesity and the renin–angiotensin–aldosterone system. Expert Rev Endocrinol Metab 1(2):255–264

Sharma K, Ramachandrarao S, Qiu G, Usui HK, Zhu Y, Dunn SR et al (2008) Adiponectin regulates albuminuria and podocyte function in mice. J Clin Invest 118 (5):1645–1656

Sibal L, Agarwal SC, Home PD, Boger RH (2010) The role of asymmetric dimethylarginine (ADMA) in endothelial dysfunction and cardiovascular disease. Curr Cardiol Rev 6(2):82–90

Simmons WW, Closs EI, Cunningham JM, Smith TW, Kelly RA (1996) Cytokines and insulin induce cationic amino acid transporter (CAT) expression in cardiac myocytes. Regulation of L-arginine transport and no production by CAT-1, CAT-2A, and CAT-2B. J Biol Chem 271(2):11694–11702

Singer E, Markó L, Paragas N, Barasch J, Dragun D, Müller DN et al (2013) Neutrophil gelatinase-associated lipocalin: pathophysiologyand clinical applications. Acta Physiol 207(4):663–672

Siroen MP, van der Sijp JR, Teerlink T, van Schaik C, Nijveldt RJ, van Leeuwen PA (2005) The human liver clears both asymmetric and symmetric dimethylarginine. Hepatology 41(3):559–565

Song S, Meyer M, Türk TR, Wild B, Feldkamp T, Assert R et al (2009) Serum cystatin C in mouse models: a reliable and precise marker for renal function and superior to serum creatinine. Nephrol Dial Transplant 24(4):1157–1161

Spurlock ME, Gabler NK (2008) The development of porcine models of obesity and the metabolic syndrome. J Nutr 138(2):397–402

Steffes MW, Buchwald H, Wigness BD (1982) Diabetic nephropathy in the uninephrectomized dog: microscopic lesions after one year. Kidney Int 21 (5):721–724

Steinbach S, Weis J, Schweighauser A, Francey T, Neiger R (2014) Plasma and urine neutrophil gelatinase–associated Lipocalin (NGAL) in dogs with acute kidney injury or chronic kidney disease. J Vet Intern Med 28(2):264–269

Steubl D, Block M, Herbst V (2016) Plasma uromodulin correlates with kidney function and identifies early stages in chronic kidney disease patients. Medicine (Baltimore) 95(10):e3011

Stevens LA, Levey AS (2009) Measured GFR as a confirmatory test for estimated GFR. J Am Soc Nephrol 20 (11):2305–2313

Stevens LA, Schmid C, Greene T, Li L, Beck GJ, Joffe M et al (2009) Factors other than GFR affecting serum cystatin C levels. Kidney Int 75(6):652–660

Straczkowski M, Kowalska I, Nikolajuk A et al (2007) Increased serum interleukin-18 concentration is associated with hypoadiponectinemia in obesity, independently of insulin resistance. Int J Obes 31 (2):221–225

Stridh S, Palm F, Hansell P (2012) Renal interstitial hyaluronan- functional aspects during normal and pathological conditions. Am J Phys Regul Integr Comp Phys 302(11):R1235–R1249

Sureshbabu A, Muhsin SA, Choi ME (2016) TGF-β signaling in the kidney: profibrotic and protective effects. Am J Phys 310(7):F596–F606

Theilade S, Lyngbaek S, Hansen TW, Eugen-Olsen J, Fenger M, Rossing P et al (2015) Soluble urokinase plasminogen activator receptor levels are elevated and associated with complications in patients with type 1 diabetes. J Intern Med 277(3):362–371

Togashi Y, Miyamoto Y (2013) Urinary cystatin C as a biomarker for diabetic nephropathy and its immuno-histochemical localization in kidney in Zucker diabetic fatty (ZDF) rats. Exp Toxicol Pathol 65(5):615–622

Tojo A, Onozato ML, Ha H, Kurihara H, Sakai T, Goto A et al (2001) Reduced albumin reabsorption in the proximal tubule of early-stage diabetic rats. Histochem Cell Biol 116(3):269–276

Tuttle KR, Bakris GL, Bilous RW, Coresh J, Balk E, Kausz AT et al (2014) Diabetic kidney disease: a report from ADA consensus conference. Diabetes Care 37 (10):2864–2883

Tvari-jonaviciute (2013) Effect of weight loss in obese dogs on indicators of renal function or disease. J Vet Intern Med 27:31–38

Vallon V, Thomson SC (2012) Renal function in diabetic disease models: the tubular system in the pathophysiology of the diabetic kidney. Annu Rev Physiol 74:351–375

van Hoek I, Daminet S, Notebaert S, Janssens I, Meyer E (2008) Immunoassay of urinary retinol binding protein as a putative renal marker in cats. J Immnunol Methods 329(1–2):208–213

Vijay S, Hamide A, Senthilkumar GP, Mehalingam V (2018) Urinary biomarkers for early diabetic nephropathy in type 2 diabetic patients. Diabetes Metab Syndr 12(5):649–652

Von Eynatten M, Lepper PM, Liu D, Lang K, Baumann M, Nawroth PP et al (2007) Retinol-binding protein 4 is associated with components of the metabolic syndrome, but not with insulin resistance, in men with type 2 diabetes or coronary artery disease. Diabetologia 50(9):1939–1937

Wang IC, Hsu WL, Wu PH, Yin HY, Tsai HJ, Lee YJ (2017) Neutrophil gelatinase-associated lipocalin in cats with naturally occurring chronic kidney disease. J Vet Intern Med 31(1):102–108

Wei C, Möller CC, Altintas MM, Li J, Schwarz K, Zacchigna S et al (2008) Modification of kidney barrier function by the urokinase receptor. Nat Med 14 (1):55–63

Williams TL, Archer J (2016) Evaluation of urinary biomarkers for azotemic chronic disease in cats. J Small Anim Pract 57(3):122–129

Williams B, Gallacher B, Patel H, Orme C (1997) Glucose-induced protein kinase C activation regulates vascular permeability factor mRNA expression and peptide production by human vascular smooth muscle cells in vitro. Diabetes 46(9):1497–1503

Wiromrat P, Bjornstad P, Roncal C, Pyle L, Johnson RJ, Cherney DZ et al (2019) Serum uromodulin is associated with urinary albumin excretion in adolescents with type 1 diabetes. J Diabetes Complicat 33(9):648–650

Wolf G, Chen S, Zidayeh FN (2005) From the periphery of the glomerular capillary wall toward the center of disease: podocyte injury comes of age in diabetic nephropathy. Diabetes 54(6):1626–1634

World Health Organization (WHO) (2019) Diabetes: data and statistics. World Health Organization, Copenhagen. [Cited 2019 May 12]. Available from: http://www.euro.who.int/en/health-topics/noncommunicable-diseases/diabetes/data-and-statistics

Xu T, Sheng Z, Yao L (2017) Obesity-related glomerulopathy- pathogenesis, pathologic, clinical characteristics and treatment. Front Med 11(3):340–348

Yacoub R, Campbell KN (2015) Inhibition of RAS in diabetic nephropathy. Int J Nephrol Renovasc Dis 8:29–40

Yang Q, Graham TE, Mody N, Preitner F, Peroni OD, Zabolotny JM et al (2005) Serum retinol binding protein 4 contributes to insulin resistance in obesity and type 2 diabetes. Nature 436(7049):356–362

Zeni L, Norden AGW, Cancarini G, Unwin RJ (2017) A more tubulocentric view of diabetic kidney disease. J Nephrol 30(6):701–717

Zhang X, Lerman LO (2016) Investigating the metabolic syndrome: contributions of swine models, Toxicol Pathol 44(3):338–366

Zini E, Benali S, Coppola L, Guscetti F, Ackermann M, Lutz TA et al (2014) Renal morphology in cats with diabetes mellitus. Vet Pathol 51(6):1143–1150

Ziyadeh FN (2004) Mediators of diabetic renal disease: the case for tgf-beta as the major mediator. J Am Soc Nephrol 15(Suppl1):S55–S57

Ziyadeh FN, Hoffman BB, Han DC, Iglesias-De La Cruz MC, Hong SW, Isono M et al (2000) Long-term prevention of renal insufficiency, excess matrix gene expression, and glomerular mesangial matrix expansion by treatment with monoclonal antitransforming growth factor-beta antibody in db/db diabetic mice. Proc Natl Acad Sci U S A 97(14):8015–8020

Zobel EH, von Scholten BJ, Reinhard H (2017) Symmetric and asymmetric dimethylarginine as risk markers of cardiovascular disease, all-cause mortality and deterioration in kidney function in persons with type 2 diabetes and microalbuminuria. Cardiovasc Diabetol 16 (4):888–896

Zsuga J, Török J, Magyar MT, Valikovics A, Gesztelyi R, Lenkei A et al (2007) Dimethylarginines at the crossroad of insulin resistance and aterosclerosis. Metabolism 56(3):394–300

Adv Exp Med Biol - Advances in Internal Medicine (2020) 4: 553–576
https://doi.org/10.1007/5584_2020_536
© Springer Nature Switzerland AG 2020
Published online: 6 June 2020

In Vivo and In Vitro Models of Diabetes: A Focus on Pregnancy

Joaquín Lilao-Garzón, Carmen Valverde-Tercedor, Silvia Muñoz-Descalzo, Yeray Brito-Casillas, and Ana M. Wägner

Abstract

Diabetes in pregnancy is associated with an increased risk of poor outcomes, both for the mother and her offspring. Although clinical and epidemiological studies are invaluable to assess these outcomes and the effectiveness of potential treatments, there are certain ethical and practical limitations to what can be assessed in human studies.

Thus, both *in vivo* and *in vitro* models can aid us in the understanding of the mechanisms behind these complications and, in the long run, towards their prevention and treatment. This review summarizes the existing animal and cell models used to mimic diabetes, with a specific focus on the intrauterine environment.

Keywords

Animal model · Embryo culture · Fertility · Gestation · Gestational diabetes · *In vitro* model · Intrauterine programing · Organoids · Pregestational diabetes · Streptozotocin

J. Lilao-Garzón, C. Valverde-Tercedor, S. Muñoz-Descalzo, and Y. Brito-Casillas (✉)
Instituto Universitario de Investigaciones Biomédicas y Sanitarias (IUIBS), Universidad de Las Palmas de Gran Canaria (ULPGC), Las Palmas de Gran Canaria, Islas Canarias, Spain
e-mail: yeray.brito@ulpgc.es

A. M. Wägner
Instituto Universitario de Investigaciones Biomédicas y Sanitarias (IUIBS), Universidad de Las Palmas de Gran Canaria (ULPGC), Las Palmas de Gran Canaria, Islas Canarias, Spain

Servicio de Endocrinología y Nutrición, Complejo Hospitalario Universitario Insular Materno-Infantil de Gran Canaria, Las Palmas de Gran Canaria, Spain

Diabetes Mellitus (DM) is a group of metabolic diseases associated with defects in insulin secretion, insulin action or both, inducing chronic hyperglycaemia, which leads to organ damage (American Diabetes Association 2013).

According to the current classification of DM, type 1 diabetes (T1DM) is an autoimmune disorder leading to the destruction of β-cells, ultimately causing total insulin deficiency. Type 2 diabetes (T2DM), produced by progressive loss of insulin secretion by the β-cells, is usually associated with peripheral insulin resistance and obesity (American Diabetes Association 2019). Other specific types of diabetes include monogenic DM, pancreatic diseases and drug-induced DM (American Diabetes Association 2019). Monogenic DM include Maturity-onset diabetes of the young (MODY) and neonatal diabetes, caused by a single mutation that leads to deficient insulin secretion (Vaxillaire and Froguel 2006).

DM first diagnosed in the second or third trimester of pregnancy is defined as Gestational Diabetes (GDM) (American Diabetes Association 2019). From the second trimester, maternal insulin sensitivity decreases due to the effect of placental hormones (i.e. oestrogen, progesterone, leptin, cortisol, placental lactogen, and growth hormone), which facilitate glucose transport across the placenta to the growing foetus (Plows et al. 2018a). In physiological conditions, these glycaemic changes are compensated by pancreatic β-cells hypertrophy and hyperplasia, and the secretion of glucose-stimulated insulin. However, upon failure of these compensatory mechanisms, hyperglycaemia is triggered and GDM occurs (Barbe et al. 2019; Plows et al. 2018b).

DM in pregnancy can be present before conception (pregestational diabetes) or develop during pregnancy (GDM). Moreover, mild hyperglycaemia, not fulfilling the criteria for DM diagnosis, may be present before pregnancy, too. When working with DM models, it is important to define the type of DM targeted in each study, since model choice and appropriateness will depend on this. Indeed, model characterisation is also challenging, and may lead to discrepancies in the nomenclature used (e.g. GDM vs mild DM).

(Huang et al. 2009). Other models, such as diabetic cats and dogs, combine genetic alterations with environmental factors (Moshref et al. 2019).

Induced models are easier to generate and have the advantage that they phenotypically match hyperglycaemia, obesity or the metabolic syndrome. However, it is necessary to be cautious when generating them, to ensure that they represent the targeted disease phenotype. There are surgically induced models (SIM); chemically, toxic or drug induced models (CIM); and diet induced models (DIM). In the SIM, a part or all of the pancreas is surgically removed. In CIM several drugs with affinity and toxicity for the β-cells are used, though streptozotocin (STZ) and alloxan are most commonly employed. On the other hand, DIM are widely used in T2DM studies, especially in rodents, but also in other species that develop insulin resistance and glucose impairment upon high fat diets (Brito-Casillas et al. 2016). A combination of these methods can also be used in order to mimic different types of DM.

1 Animal Models of Diabetes

Animal models are a valuable tool to study DM, although none is comprehensive enough to fully reproduce any form of the disease. The most extensively used species are mice (Hanafusa et al. 1994) and rats (Phillips et al. 1996), but also rabbits (Mage et al. 2019), cats (Pérez-López et al. 2019), dogs (Moshref et al. 2019) or bigger animals like sheep (Dickinson et al. 1990) and pigs (Ezekwe et al. 1984) are used.

These animal models can be classified according to whether their diabetes is spontaneous or induced (Brito-Casillas et al. 2016). The mechanisms of disease in spontaneous DM models are presumed to be similar to human DM, especially in polygenic models. The latter include Non-Obese-Diabetic mouse (NOD) (Hanafusa et al. 1994), the BioBreeding Diabetes-Prone rat (BB) (Jacob et al. 1992a) and the Zucker fatty rat (Phillips et al. 1996). Spontaneous mouse models for monogenic DM include ob/ob, Lep^{ob} (Kaufmann et al. 1981) and Lepr^{db}

1.1 Animal Models of Gestational and Pregestational Diabetes

1.1.1 Surgically Induced Models (SIM)

The first models developed for GDM were based on partial pancreatectomy in dogs, during different stages of pregnancy (Pasek and Gannon 2013; Markowitz and Soskin 1927; Carlson and Drennan 1911). Due to ethical, economic and practical reasons, rodents are the most currently used species to develop SIMs (Sharma et al. 1999; Jawerbaum et al. 1993). They allow for the study of both utero-placental defects and foetal alterations, as well as differences in intra-uterine metabolic patterns (Jawerbaum et al. 1993). One of the biggest disadvantages of these models is their lack of specificity since both the endocrine and exocrine tissues are removed, resulting in other, non-DM related symptoms (Pasek and Gannon 2013). Furthermore, the diabetic phenotype is caused by the lack of endocrine pancreas. Therefore, these models only mimic certain aspects of GDM, caused by insulin deficiency, but do not account for insulin resistance.

1.1.2 Chemically, Toxic or Drug Induced Models (CIM)

CIM are mainly generated in rodents and are commonly used to develop T1DM. Since this approach allows to control the time-point of the induction of DM, authors use this strategy to develop models of GDM (Türk et al. 2018). Induction is usually done in early pregnancy, before the embryonic pancreas develops (Martin et al. 1995; Sugimura et al. 2009), in order to avoid foetal β-cell destruction by the chemical agent utilized (Pasek and Gannon 2013; López-Soldado and Herrera 2003; Aerts et al. 1997).

As described for SIM, this strategy induces DM within days, with mild to severe hyperglycaemia, depending on the doses and the regime applied (López-Soldado and Herrera 2003; Lee et al. 2019). Due to the fast induction, STZ is commonly used to study a variety of pregnancy-related DM complications, such as preimplantation embryo defects, congenital malformations, utero-placental defects, foetal alterations or offspring defects (Jawerbaum and White 2010).

The main disadvantage of CIM models is the variability in the response to the drug. Even within the same species, many variables (strain, age, sex. . .) can change the outcome. This has led to a lack of consensus in the scientific community for a dosing strategy to generate the DM models (Pasek and Gannon 2013). Table 1 summarises the different STZ regimes used in studies published between 2012 and 2018. Indeed, even with the same regime, the type of DM obtained is classified differently.

1.1.3 Diet Induced Models (DIM)

Dietary modification is another strategy to generate DM. Different approaches are used, such as continuous glucose infusions (Gauguier et al. 1990) or a diet enriched in fat (High Fat Diet, HFD) that will induce obesity, a known risk factor for DM. These approaches are mostly used to generate T2DM (Surwit et al. 1988), but also GDM models (Holemans et al. 2004). Although diet alteration is one of the most complete strategies to mimic DM, there are some factors that need to be taken into account, since they may have an impact on the response to the diet. These include dietary fat origin and animal housing and number, as density and hierarchical behaviours can affect feeding (Feige et al. 2008). Combination of HFD with high fructose diet (HFHF) has been used to mimic fast food diets that usually combine fat enriched food with carbonated beverages enriched in sugars (Lozano et al. 2016). HFHF fed rats develop not only T2DM but also associated complications like hepatic fibrosis, inflammation and oxidative stress (Lozano et al. 2016).

The main advantages of this strategy are that it can be used to study both pre-existing DM during pregnancy and GDM, both in maternal and embryonic scenarios. Another advantage is that it provides an alternative to study DM in animals where genetic modification is not possible either due to the lack of genetic engineering availability or because of long generation times. The main disadvantages are that it does not consider genetic variability and other possible collateral disorders, such as obesity and hypertension, often associated to T2DM but not necessarily present in GDM (Feige et al. 2008).

1.1.4 Genetic Models

Genetic models have an important role in understanding the general influence of DM on health. However, as their diabetic phenotype is usually severe already before pregnancy, it has been difficult to generate specific models of GDM using this approach (Pasek and Gannon 2013).

The NOD mouse (Hanafusa et al. 1994; Pearson et al. 2016) and BB rat (Jacob et al. 1992b) that develop spontaneous DM, similar to human T1DM, are the main genetic models for this disease. Both have been used as models to study diabetes complications before and during pregnancy, such as intrauterine growth restriction, neural tube defects, premature birth or macrosomia (Moley et al. 1991; Burke et al. 2007; Eriksson et al. 1989).

Several polygenic strategies have been found to study T2DM, such as inbred mouse models

Table 1 Examples of streptozotocin regimes to generate diabetic mouse models

STZ dose (mg/Kg)	Number of IP injections	Mouse Strain	Type of DM induced according to the reference	Developmental Stage	References
50	1	ICR	DM induced on neonatal mice	Neonate	Kataoka et al. (2013)
50	5	C57BL/6	T1DM	Adult	Chaudhry et al. (2013)
50	5	C57BL/6	Hyperglycemia defined by non-fasting BG >300 mg/dL	Adult	Kim et al. (2016a)
50	5	C57BL/6	Defined as subdiabetogenic doses of STZ	Adult	Kurlawalla-Martinez et al. (2005)
150	1	C57BL/6	Combined with HFD to model T2DM	Adult	Song et al. (2008)
180	1	C57BL/6	T1DM effect on preimplantation embryos	Adult	Brown et al. (2018)
60	1	C57BL/6	DM induction on neonates	E16.0-E18.0	Yang et al. (2013)
35	5	ICR	T1DM triggered by gradual insulitis	Adult	Priel et al. (2007)
240	1	Kunming mice	T1DM defined by fasting BG >11.1 mmol/L	Adult	Liao et al. (2017)
100	2	C57BL/6 and ICR	Defined as mild DM	Adult	Shimizu et al. (2012)
100 + 80	1 + 1	C57BL/6	Combination with genetic and HFD model for T2DM	Adult	Chen et al. (2015)
75	3	Kunming mice	DM characterized by BG >16 mmol/L	Adult	Wang et al. (2018)
190	1	C57BL/6	Pregestational DM defined by non-fasting BG >13.3 mmol/L	Immature (20–24 days)	Chang et al. (2005)
230	1	CD1	Pregestational DM defined by non-fasting BG >17.0 mmol/L	Adult	Ge et al. (2014)
50	5	C57BL/6	Pregestational T1DM defined by fasting BG >11 mmol/L	Adult	Dowling et al. (2014)
75	3				

STZ streptozotocin, *BG* blood glucose, *DM* Diabetes mellitus, *HFD* high fat diet, *IP* intraperitoneal, *T1DM* type 1 diabetes mellitus, *T2DM* type 2 diabetes mellitus

like the TALLYHO/Jng (Dontas et al. 2012) and the New Zealand Obese mouse (Herberg and Coleman 1977), or the sand rat (*Psammomys obesus*). In laboratory conditions, these animals develop T2DM within weeks (Kaiser et al. 2005).

The mutant mouse for the leptin receptor *db/db* is a classic monogenic model to study T2DM. Lack of leptin receptor causes an increase in appetite, obesity and insulin resistance. This model has specific characteristics to be considered: the homozygous *db/db* females are sterile, and heterozygous animals do not show a specific phenotype until pregnancy, when they develop insulin resistance and GDM (Yamashita et al. 2001, 2003).

Something similar happens in mutant mice for the prolactin receptor (*PrlR*), responsible for increasing the proliferation of β-cells during pregnancy, in response to prolactin. This β-cell increment is a compensatory strategy to counterbalance insulin resistance in pregnancy. The complete lack of this receptor renders females infertile but, in the heterozygous state, pregnancy is possible and is followed by moderate glucose intolerance, similar to GDM (Huang et al. 2009).

The main advantages of genetic models are the wide range of studied phenotypes and the possibility of generating new ones through genetic engineering. The main disadvantage is their monogenic character, and the limited number of species where genetic modifications are possible. In spontaneous genetic models, the main limitation is long generation time, as in bigger animals like sheep or pigs, where gestation lasts between 152 and 114 days (Pasek and Gannon 2013).

1.2   Animal Studies on Gestational and Pregestational Diabetes

1.2.1   Fertility

Ovulation is the point in time when female fertility can be first affected by DM. Animal models used to study ovulation in DM are usually models with mild hyperglycaemia. Hence, the oestrus cycle still occurs, and neither oocyte maturation nor fertilization are affected (Novaro et al. 1998). Folliculogenesis, oogenesis, and preimplantation embryogenesis are impaired in mouse models with severe hyperglycaemia, independently of the induction strategy (Lee et al. 2019).

Preimplantation embryos from diabetic BB rats show difficulties to reach the expanded blastocyst state and around 30% are abnormal (Vercheval et al. 1990). In fact, experiments on cultured embryos from CIM diabetic mice, show that even brief exposure to maternal hyperglycaemia can negatively affect embryonic development (Wang and Moley 2010). This was also confirmed by one-cell stage embryo transfer from diabetic mice to non-diabetic mice (Wyman et al. 2008).

The observed damage in preimplantation embryos has been related to a decrease in the expression of glucose transporters like *GLUT1* in the embryo caused by maternal hyperglycaemia. Indeed, a decrease in these transporters can lead to apoptosis in early embryos (Wang and Moley 2010; Chi et al. 2000).

Implantation is the next developmental step that can be affected in DM to cause impaired fertility. Different studies, both with NOD and STZ treated mice, show more resorption sites in these models than in non-diabetic or prediabetic animals (Brown et al. 2018; Burke et al. 2007). These resorption sites represent places in the uterus where implantation is initiated, but the embryo fails to develop.

The reason behind these implantation defects is unknown. However, one possibility is that hyperglycaemia induces inflammatory cytokines in the female reproductive tract, leading to DNA damage that results in suboptimal embryos, which start implantation but whose development does not progress and are therefore reabsorbed (Brown et al. 2018).

1.2.2   Placentation

STZ administration in mice or rats is the main strategy to develop *in vivo* DM for placenta studies. These models show not only structural, functional and developmental abnormalities (Capobianco et al. 2005; Acar et al. 2008; Suwaki et al. 2007; Favaro et al. 2013), but also abnormal gene expression (Yu et al. 2008). An upregulation in placental nutrient transporters, such as GLUT1 for glucose or *SystemA* for amino acids, has been observed in T1DM and GDM models, which may result in an increased flux of nutrients to the foetus (Jansson et al. 1999, 2006).

Oxidative stress seems to have an important role in the aberrant placenta observed in diabetic pregnancies. The diabetic placenta shows a limited ability to respond to oxidative stress as well as an enhanced production of mitochondrial ROS (Lappas et al. 2011). It has been proposed that this stress is caused by extracellular matrix metalloproteinases, since there is a correlation between blood glucose concentrations and the placental expression of these proteinases (Ding et al. 2018).

Furthermore, during diabetic pregnancy, nutrient transporters increase and promote a more nutritious foetal environment causing macrosomia. However, there are also placental disorders related to reduced blood flow and increased placental glucose consumption, that lead to intrauterine growth restriction (see Fig. 1) (Zamudio et al. 2010; Illsley and Baumann 1866).

1.2.3   Perinatal Outcomes

Human epidemiological studies show negative effects of maternal DM on their offspring.

Maternal DM can cause macrosomia at birth that, in turn, increases the risk of shoulder dystocia, insulin resistance, caesarean section delivery, and neonatal hypoglycaemia (Araujo Júnior et al. 2017; Schwartz et al. 1994; Esakoff et al. 2009). Such disorders cause high neonatal morbidity and mortality.

Most of the research on this matter is done through human epidemiology. However, some studies in rodents are available. To evaluate the intrauterine environment effect on foetuses, CIMs are mainly used (Gonzalez and Jawerbaum 2006; Martínez et al. 2008), but also inbred genetic models like BB rats. In the latter, the diabetic environment during organogenesis decreases embryonic survival and growth rates, and can alter skeletal and internal organ development (Eriksson et al. 1989).

An interesting strategy to study the intrauterine environment in these genetic models while avoiding the embryonic genetic background is to transfer the embryos, one-cell mouse zygotes or blastocysts, from healthy to diabetic females. Also, embryo transfer from diabetic to healthy females is being used to study more acute, short-term effects on the embryos (Wyman et al. 2008; Rousseau-Ralliard et al. 2019).

1.2.4 Intrauterine Programming

High glucose levels during pregnancy promote malformations and inadequate organization of the hypothalamus. These changes affect food intake and energy balance increasing offspring's susceptibility to body weight dysregulation (Šeda et al. 2018). In the long term, babies from diabetic mothers show an increased risk of obesity, T2DM and cardiovascular disease (Plows et al. 2018b).

On the other hand, several studies show lower body weight in offspring from rats where GDM is chemically induced by STZ. Furthermore, other authors show that in male offspring from diabetic mothers, the weight of the reproductive organs is significantly lower. It is suggested that this reduction is due to a decrease in testosterone levels (Türk et al. 2018; Amorim et al. 2011). Finally, low body weight can persist into adulthood (Van

Assche et al. 2001), which in rats is explained by deficiencies in milk production and quality during lactation (Lau et al. 1993). However, a direct toxic effect of STZ on the offspring cannot be ruled out. Indeed, there is discrepancy between these findings and what is found in human studies, where macrosomia is the predominant complication of GDM (see Fig. 1).

Animal models are a valuable tool to study the mechanisms behind these effects. A recent report shows how the oxidative stress present in the maternal diabetic environment programmes adult hypertension in the offspring by hyperphosphorylation of the renal dopamin1 (D1) receptor. In this case, hypertension is caused by dysfunction of the D1 receptor that is responsible for ≈60% of sodium excretion during increased sodium intake (Luo et al. 2018).

Finally, *Socs2* null mutant mice have been recently proposed as a model for GDM associated to aging, as older pregnant females show lower birth and higher mortality rates, associated with a high frequency of foetal macrosomia (Brito-Casillas 2019).

2   *In Vitro* Models

In vitro models are complementary to *in vivo* models. They do not represent complex physiological systems as faithfully as *in vivo* models do. However, they allow researchers to study numerous and precise experimental conditions simultaneously, investing fewer resources than with animal studies. They have certain advantages, as they can be used in parallel to test the reproducibility of results, be genetically modified by transfection to investigate the role of several genes and be used to screen for drugs. Furthermore, the use of *in vitro* models allows to reduce the number of animals used in research, in compliance with the 3 R's principles. These *in vitro* models include cell-lines, islets, human stem cells and organoids. Their specific advantages and disadvantages are summarized in Table 2.

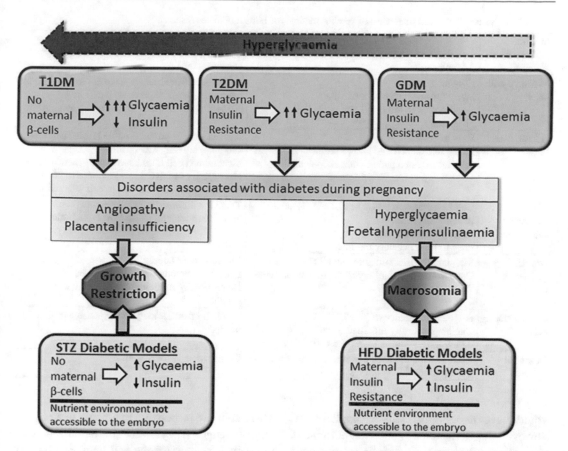

Fig. 1 Overview of the mechanisms involved in diabetes and its complications during pregnancy and the animal models used to mimic them. Both type 2 and gestational diabetes are associated with insulin resistance and mild-to-moderate hyperglycaemia, whereas type 1 diabetes is associated with insulin deficiency and more severe hyperglycaemia. The most frequent complications of diabetes in pregnancy are associated with hyperglycaemia itself, compensated by the foetus with an increased insulin secretion that leads to macrosomia. Vascular complications, on the other hand, are associated with placental insufficiency, and induce complications such as preeclampsia and intrauterine growth restriction. Animal models do not always truthfully reflect the human disease. *T1DM* Type 1 Diabetes Mellitus, *T2DM* Type 2 Diabetes Mellitus, *GDM* Gestational diabetes Mellitus, *STZ* Streptozotocin, *HFD*: High Fat Diet

2.1 *In Vitro* Models of Diabetes

2.1.1 Cell Models of Diabetes

Until recently, rodents have been the main source of β-cell lines. The latter are obtained from radiation induced insulinoma, such as RIN cell lines (Chick et al. 1977); virally induced insulinoma, like the cell line In-111 (Uchida et al. 1979); oncogenic transfection of primary β-cells using the simian-vacuolating virus 40 (SV40) antigen oncogene from the insulin gene promoter, like MIN-6 (Miyazaki et al. 1990); and electrofusion of primary pancreatic β-cells. The latter technique was used to generate the BRIN-BD11 cell line from the RINm5F clone (McClenaghan et al. 1996). This cell line is widely used in β-cell biology and drug target research since it behaves like β-cells, with an almost normal response to glucose and to other modulators of insulin secretion. A high number of researchers use INS-1 cells, obtained from rat insulinoma, which proliferate easily and fast in culture, though their response to glucose is lower than that of primary islets (Chick et al. 1977; Asfari et al. 1992).

While rodent cell lines allow great advances in β-cell research, the use of a stable human cell line

Table 2 Advantages and disadvantages of *in vitro* models for human diabetes research

In vitro models of diabetes	Advantages	Disadvantages
Murine β-cell lines	Easy to culture Many types of cells are available Good option to test drugs and study cell physiology	Differences from human make difficult to choose the most appropriate murine β-cell line No vascularization and cell to cell interaction
Human β-cell lines	Easy to culture Stable human cell line allows progress in human diabetes research and in clinical applicability	It is not easy to produce stable human cell lines, and there are only few of them Most human cell lines have some genetic defects, grow slowly or have a low response to glucose No vascularization and cell to cell interaction
Murine pancreatic islets	Can be isolated in less time and with less cost than human islets Short generational times Easy to be genetically modified	There are differences in islet architecture, vascularization and blood flow from human islets
Human pancreatic islets	Maintain the islet structure and all cell types Used to study the biology of the human pancreas	Limited donor supply Do not allow long functional studies Heterogeneity in their characteristics: size, genetics...
Human stem cells	A renewable source of β-cells Can be genetically modified Allow longer studies than pancreatic islets	To obtain them, a long and expensive process is needed
Organoid cultures	Resemble the diseased organ architecture better than traditional 2D cultures	Do not have vascularization

would have more direct clinical applicability. However, their generation is not straight forward. Most of them have genetic defects, grow slowly or have an insufficient response to glucose (Green et al. 2015). Nevertheless, there are also well characterized cell lines, such as NAKT-15, created by transfection of human islet cells with SV40 T-antigen, that restore normoglycemia in diabetic mice (Narushima et al. 2005). The 1.1 B4 cell line, produced by electrofusion of isolated cadaveric human β-cells with immortal PANC-1 epithelial cells, expresses most of the genes and proteins related with insulin production. Furthermore it shows similar characteristics to isolated human β-cell lines, although insulin content is lower than in rodent cells. 1.1 B4 cells are used to study human β-cell biology and function (Ota et al. 2013; Krause et al. 2014; Elumalai et al. 2017) and have been used to demonstrate glucotoxic effects on gene expression and insulin secretion (Vasu et al. 2013).

Another, well characterized cell line is EndoC-βH1, considered a stable human β-cell line with many applications (Scharfmann et al. 2016).

These cells are responsive to glucose at a physiological range. They were immortalized with transgenes to generate the cell lines EndoC-βH2 and EndoC-βH3 which are very similar to primary β-cells (Scharfmann et al. 2014; Benazra et al. 2015). The EndoC-βH1 cell line is widely used to study human β-cell physiology and to screen and identify glucose-lowering drugs (Grieco et al. 2014; Andersson et al. 2015; Gurgul-Convey et al. 2015; Krishnan et al. 2015; Tsonkova et al. 2018).

2.1.2 Pancreatic Islets

The endocrine pancreas is organized in islets with five major cell types, each one responsible for the secretion of different hormones: α-cells (glucagon), β-cells (insulin), PP-cells (pancreatic polypeptide), δ-cells (somatostatin) and ε-cells (ghrelin) (Zhou and Melton 2018).

The advantages of using whole pancreatic islets are that they maintain their tridimensional structure and contain all cell types and, thus, would be expected to reproduce what happens *in vivo* more faithfully. Pancreatic islets have been used to study the biology of the pancreas, test drugs and study potentially preventive interventions. However, due to the vascular damage after their isolation, they cannot be used in long studies (Schmied et al. 2000).

Rodent islets predominantly employed in DM research are currently being replaced by human islets. Thanks to the increased availability of the latter, several advances have been accomplished: knowledge of the islet structure, gene expression, insulin secretion, cell proliferation and their response to stress situations (Hart and Powers 2019). Indeed, comparisons between islets from healthy donors and those with DM give information on differential β-cell identity, gene expression, etc., that can be useful to elucidate mechanisms of disease (Marchetti et al. 2019).

However, universally accepted criteria to define human islet quality are not yet available. Some authors use islet insulin production to measure quality (Hart and Powers 2019) and propose a list of features which should be reported on each human islet preparation. This list includes information about the origin/source, isolation centre, estimated purity and viability, total culture time, functional measurement, and description of islet use.

The organization of networks and consortia, such as the Islet Cell Resource Center Consortium (ICRCC), have increased human pancreatic islet availability (Kaddis et al. 2009; Niland et al. 2010).

2.1.3 Human Stem Cells

Human pluripotent stem cells (hPSCs), human embryonic stem cells (hESCs) and human induced pluripotent stem cells (hiPSCs) have been differentiated into pancreatic β-like cells. To this end, it is necessary to collect fibroblasts or peripheral blood mononuclear cells (PBMCs) from patients and then reprogramme them into hiPSCs (Amirruddin et al. 2019).

Human pluripotent stem cells have some advantages for human DM research, as they are a renewable source of β-cells, and can be modified genetically (Balboa et al. 2019). Hence, these cells represent a new model to

study pathogenic mechanisms and test drugs (Amirruddin et al. 2019).

Moreover, they can be used as part of newer technologies like organ-on-a-chip devices, single cell RNA-sequencing, development of hypoimmunogenic cells and promotion of vascularization of the β-like cells (Amirruddin et al. 2019).

On the other hand, they also present some disadvantages, since long and cumbersome procedures are necessary to obtain them. However, recent progress has been made in the protocols to differentiate hPSCs into pancreatic β-like cells.

Many authors have developed protocols leading to insulin producing, β-like cells. Sometimes these cells are polyhormonal β-like cells, which do not express key β cell transcription factors (Bruin et al. 2014; Guo et al. 2013), and show limited *in vitro* insulin secretion in response to glucose. Furthermore, these cells resemble transient endocrine cells found in human foetal pancreases (Bruin et al. 2014; D'Amour et al. 2006).

In 2014, Rezania et al. (2014) generated insulin expressing β-like cells from hPSCs, cells capable of secreting insulin in response to glucose and reversing diabetes *in vivo* restoring the blood glucose levels of STZ-induced diabetic mice. According the authors (Rezania et al. 2014), the differentiation process occurs in seven stages which are defined by specific markers. The key stages are: development of the definitive endoderm (stage 1), primitive gut tube (stage 2), posterior foregut (stage 3), pancreatic progenitor (stage 4), endocrine progenitor (stage 5), immature β-cells (stage 6) and mature cells (stage 7). In stage 4, PDX1 and NKX6.1 are co-expressed and are needed for the generation of glucose-responsive and mono-hormonal β-cells (Rezania et al. 2014). From this stage the cells can differentiate *in vitro* Into mature β-cells and can be transplanted into mice for further *in vivo* maturation. Pagliuca et al. also developed a differentiation protocol to generate β-like cells from hPSC. The cells were responsive to glucose *in vitro* and functionally resembled adult cells (Pagliuca et al. 2014). In 2015, Russ et al. published a differentiation protocol without using bone morphogenic protein (BMP) inhibitors during pancreatic specification to avoid formation of non-functional polyhormonal cells. This represented an improvement in the protocol, generating pancreatic progenitors that were then differentiated into glucose responsive cells *in vitro* (Russ et al. 2015).

More recently, Aigha et al. developed a protocol to generate a population of pancreatic progenitor population expressing NKX6.1 in the absence of PDX1 from hPSCs. These cells are potentially able to generate endocrine islet cells (Aigha et al. 2018). Some authors have used hPSC-derived pancreatic progenitor cells and have observed that when transplanted into mice with T1DM, the cells are able to differentiate and mature into glucose-sensitive insulin cells and may even reverse DM (Rezania et al. 2012, 2014; Bruin et al. 2013).

Thus, optimizing protocols to generate β-like cells from hPSC is a challenge, though great progress has been made in recent years and many research groups are capable of producing pancreatic progenitor and immature islet-like cells. Nevertheless, work still needs to be done in the process of obtaining mature cells capable of secreting insulin, *in vivo,* in response to glucose (Balboa et al. 2019; Shahjalal et al. 2018; Abdelalim and Memon 2020).

2.1.4 Organoid Cultures

In recent years, modelling diseases *in vitro* has bloomed with the generation of three-dimensional (3D) cell culture systems, which resemble the diseased organ architecture better than traditional 2D cultures. Among the 3D cell structures, cell aggregates or spheroids, cysts and true organoids can be distinguished. Spheroids are solid cell aggregates without a central lumen, cysts are spheres of polarised epithelial cells with a central lumen, while true organoids are complex, polarised structures with a central lumen (Bakhti et al. 2019). An organoid can be defined as a 3D structure derived from either pluripotent stem cells (PSCs), neonatal tissue stem cells or adult-derived stem/progenitor cells (AdSCs), in which cells spontaneously self-organise into structures

that resemble the *in vivo* tissue in terms of cellular composition and tissue function (Hindley and Cordero-Espinoza 2016). These organoids can be used to model diseases *in vitro* and as a source for cell therapy or organ transplant.

The ultimate goal of organoid biology for regenerative medicine is to faithfully generate *in vitro* functional organs using the patient's own cells for cell therapy, organ transplant or replacement. To this end, the best option is to generate the organoid using induced pluripotent stem cells (iPSCs) derived from the patient's fibroblasts or peripheral blood mononuclear cells (see above). To accomplish this, a detailed knowledge of the organ embryonic development, function and cell composition is required. The organ development will inform about signals and gene activity that will be required to efficiently direct iPSCs differentiation into the cells present in the organ. The detailed organ function is needed to test how faithfully the organoid recapitulates the organ. Finally, organ cell composition knowledge is required to ensure that all the needed cell types for organ function are present in the organoid. Ideally, the progenitor population that allows for the homeostasis of the organ should also be identified to facilitate the organoid generation from the adult organ.

Pancreatic cysts and organoids can be derived from mouse foetal tissues (Greggio et al. 2013; Sugiyama et al. 2013; Bonfanti et al. 2015). These organoids and cysts are composed by acinar, ductal, endocrine and exocrine cells, hence faithfully representing the pancreatic cell composition. Given the poor regenerative potential of the pancreas, it is controversial whether an established progenitor population (AdSCs) or rather transdifferentiation from one pancreatic cell type into another is responsible for pancreas homeostasis (Kopp et al. 2016). Regardless, pancreatic cysts and organoids have been generated from adult mice using different starting cell populations, indicating that a clear pancreatic progenitor population is elusive (Huch et al. 2013; Jin et al. 2013, 2014). If Lgr5+ cells are used, organoids composed by duct cells are obtained, that only upon transplantation *in vivo* can differentiate into endocrine cells (exocrine and acinar fates were not tested) (Huch et al. 2013). However, if CD133 + Sox9+, are used, cysts composed by endocrine and acinar fates are obtained (no transplantation tests were done) (Jin et al. 2013, 2014). In summary, all these studies using mouse samples help to understand the mechanisms of pancreas development. Hence, they can be used to model pancreatic diseases like DM, though their suitability for cell therapy is still to be tested.

The generation of human pancreatic spheres from adult samples was initially achieved upon cell reprogramming with islet developmental regulators (Lee et al. 2013). These spheres can differentiate into endocrine cells *in vitro*. More recently, applying adult mouse organoid generation conditions to human foetus samples has allowed to generate human pancreatic cysts with limited differentiation success (Bonfanti et al. 2015; Loomans et al. 2018).

Several groups have targeted their efforts into generating islet-like spheroids using cell lines, hESCs or hiPSCs rather than pancreatic organoids. The cells EndoC-βH1 (Ravassard et al. 2011), EndoC-βH2 (Scharfmann et al. 2014) and EndoC-βH3 can be cultured as 3D pseudo-islets *in vitro* (Benazra et al. 2015). Other cells, such as 1.1 B4, form pseudo-islets spontaneously and show higher resistance to toxicity and DNA damage if grown as pseudo-islets than in monolayers (Green et al. 2015). The first advantage of using pseudo-islets instead of pancreatic islets is their availability. Furthermore they are similar to primary islets and they present an increase in insulin production in response to glucose (Rogal et al. 2019). Indeed, interaction between cells increases the response to glucose (Kojima 2014). Furthermore, pseudo-islets can be modified genetically, accordingly to the research goal. However, they neither have vascularization, nor an intrinsic structure. Given the scarcity of human donor, islet-like spheroids generated from hESCs or hiPSCs could be a potential source for transplantation into diabetic patients. The first protocols using hESCs or hIPSCs only produced immature β-cells (D'Amour et al. 2006; Hrvatin et al. 2014). Later protocols have achieved mature

β-cells from hESCs or hiPSCs (Rezania et al. 2014; Pagliuca et al. 2014; Russ et al. 2015; Kim et al. 2016b; Shim et al. 2015). Importantly, the islet-like spheroids improve hyperglycaemia in diabetic mice. The improvement and efficiency vary depending on the protocol used for islet-like spheroid generation. Only 2 protocols report an improvement after 3–4 days that only lasts for a short period of time (Russ et al. 2015; Kim et al. 2016b). The other protocols only show improvement weeks after transplantation (Rezania et al. 2014; Pagliuca et al. 2014; Shim et al. 2015). These limited results in reversing hyperglycaemia are likely due to their limited functionality, which might include poor tissue engraftment, lack of revascularization of the transplanted tissue or absence of a niche.

The improvement of organoid functionality can be achieved by increasing its complexity applying bioengineering approaches like co-cultures and biomimetic scaffolds (Yin et al. 2016). Co-cultures of islet-like cells with mesenchymal stem cells (Shin et al. 2015; Takebe et al. 2015) and human amniotic epithelial cells (Lebreton et al. 2019) enhance engraftment, viability and function in transplant experiments by increasing vascularisation. The co-cultures can also be done using cell lines like MIN6 (Takahashi et al. 2018). The aim of using biomimetic scaffolds is to provide an extrinsic microenvironment that supplies biochemical and biophysical signals, including structural support, to the growing organoid. For example, islet-like organoids from hESCs using Matrigel and collagen as scaffold secrete insulin in response to glucose (Wang et al. 2017). Finally, both co-cultures together with biomimetic scaffolds can be used (Candiello et al. 2018). In this case hESCs together with endothelial cells and a novel engineered hydrogel are used, which also secrete insulin in response to glucose.

In summary, multiple protocols and approaches have been developed to generate pancreatic or islet-like organoids that harbour the potential to be used for transplantation in diabetic patients. However, their efficiency is not always comparable among them as not all test their *in vivo* functionality in diabetic mice.

2.2   *In Vitro* Models of Diabetes in Pregnancy

The human placenta is the interface between the mother and the foetus. During pregnancy, the placenta is exposed to morphological and functional changes at the cellular and tissue level (Desoye and Mouzon 2007). The foetal-maternal interface is formed by a heterogeneous group of trophoblast cells with a wide range of functions.

There are several models and cell lines available to investigate the effect of DM on the placenta.

2.2.1   Primary Cell Lines

These cell lines are of placental origin from early gestation or at term. Some have been used to study different aspects of trophoblast function, such as invasion, migration and immunology.

Primary term trophoblast cells have been used to study the association between maternal lipid metabolism and GDM (Stirm et al. 2018). The human first trimester trophoblast cell line Sw.71 has been used to study the effects of high concentrations of glucose on trophoblasts and the effect of metformin (Han et al. 2015) and to understand how hyperglycaemia increases the risk of preeclampsia (Heim et al. 2018).

Primary human villous trophoblasts have been also used in culture to study the effect of free fatty acids (palmitate and oleate) on cell viability and function (Colvin et al. 2017). Specifically, to understand the effect of maternal obesity before pregnancy on GDM, and the latter on preeclampsia and complications in the foetus. Similar studies investigate the effect of IL-6 and TNF-α (elevated in GDM and obesity) on fatty acid accumulation in human primary trophoblast cells culture (Lager et al. 2011).

Primary cells have some limitations, such as relatively low number of isolated cells, short lifespan and lack of proliferation *in vitro*. To overcome these limitations, several trophoblastic cell lines have been established, using two methods: establishment of human choriocarcinoma cell lines and introduction of the gene encoding simian virus 40 large T (SV40T) antigen (Graham et al. 1993).

2.2.2   Choriocarcinoma Cells

Despite being a cancer cell model, choriocarcinoma cells are widely used to study the function of the placenta. These cells are easy to culture and propagate. They present barrier capacity, are able to express glucose transporters (GLUT-1, GLUT-3) and release hormones (Schmitz et al. 1999; Nusrat et al. 2001; Brown et al. 2011; Saleh et al. 2007). Nevertheless, only some clones are available, including: BeWo (Heaton et al. 2008), JAR (Azizkhan et al. 1979) and Jeg-3 (Frank et al. 2000). In addition, ACH-3P (Hiden et al. 2007; Fröhlich et al. 2015; Weiss et al. 2014) and AC1-M59 are AC-1 choriocarcinoma cells fused with first and third trimester trophoblast cells.

These cells have been broadly used as *in vitro* models. However, they show relevant differences with respect to primary trophoblasts (Poaty et al. 2012). For example, the HLA expression of these cells is distinct. Furthermore, studies in DNA methylation show high variability producing differential expression profiles (Apps et al. 2011). Hence, to interpret the results obtained using choriocarcinoma cells it is necessary to compare them with those obtained using primary cells.

In DM research, the BeWo choriocarcinoma cell line can be exposed to high glucose levels to identify changes in transcripts and metabolites (Hulme et al. 2018). It has also been used to assess the effect of choline and betaine on the placental transport, the role of miR-130b-3p in regulating oxidative stress during GDM and to test the effects of hyperglycaemia on endoplasmic reticulum stress (Nanobashvili et al. 2018; Jiang et al. 2017; Yung et al. 2016).

2.2.3   Immortalized Cells

The human, first trimester, extravillous trophoblast cell line (HTR-8/SVneo) was developed using first trimester extravillous trophoblast infected with SV40 T-antigen. It contains two populations of cells suggesting the presence of trophoblast and stromal/mesenchymal cells (Abou-Kheir et al. 2017). HTR-8/SVneo cells are largely used to study trophoblast functions including cell fusion, migration and invasion during hyperglycaemia (Wu et al. 2018).

HTR-8/SVneo cells are also used to study microRNA expression profiles under high glucose conditions. They have been used, for example, to assess the influence of miR-137 on trophoblast viability and migration under high glucose conditions, to evaluate the combined effect of glucose and fatty acids on early placentation and to study the effects of hypoxia on placentation and the role of glucose and its transporters (Peng et al. 2018; Basak et al. 2015, 2019; Bermejo-Alvarez et al. 2012a).

2.2.4 Human Trophoblast Stem Cells

There are three main trophoblast populations in the human placenta: cytotrophoblast (CT), extravillous cytotrophoblast and syncytiotrophoblast (Okae et al. 2018). Recently, a culture system has been developed to generate trophoblast stem cells (Okae et al. 2018). The authors derived human syncytiotrophoblast from cytotrophoblast and from blastocysts. These cells are genetically stable and can be potentially used to study growth, trophoblast defects and preeclampsia (Okae et al. 2018). They seem promising to evaluate the influence of DM on trophoblast development and function. However, further studies are needed to fully characterise the model.

2.2.5 Placental Organoids

Recently, trophoblast organoids that model maternal-foetal interactions have been developed (Haider et al. 2018; Turco et al. 2018). The authors from both studies successfully generate the trophoblast organoids using first trimester placentas, from either 6th–9th (Turco et al. 2018) or 6th–7th weeks, but not from 10th–11th weeks of pregnancy (Haider et al. 2018). Moreover, decidual glandular organoids can also be obtained to model the placenta (Turco et al. 2018). Using similar culture conditions, both studies rely on the long-term culture capabilities of the organoids (over a year versus 5 months) and the starting material (individual patient versus pooled samples) in (Turco et al. 2018) versus (Haider et al. 2018). Moreover, the main focus of both studies varies, being in (Haider et al. 2018) to model the development of the placenta

understanding the signalling pathways involved. While in (Turco et al. 2018), the authors focus on fully characterising the organoids, including cell origin (foetal), transcriptomic and methylation profile, structural, secretome and invasive capabilities.

It is also possible to generate extravillous trophoblast spheroids (Nandi et al. 2018), which can be obtained by culturing HTR8/SVneo EVT cells in ultralow attachment plates and be used as a model for placental extravillous trophoblast invasion (Wong et al. 2019).

The new trophoblast organoids derived from human samples opens new avenues to model maternal-foetal interactions that will be of interest to further progress in how placentas develop in women with DM. However, these models present with some limitations, as they only represent the trophoblast component of the placenta (Turco and Moffett 2019). In this sense, new organoid models mimicking diabetic microenvironments are being developed, and by the use co-culture techniques, this may provide more complex models also under diabetic conditions (Tsakmaki et al. 2020). Further studies, maybe including bioengineering approaches, will be necessary before true placental organoids can be generated.

2.2.6 Embryo Cultures

Maternal DM is associated with retarded preimplantation embryo development in mouse models (Moley et al. 1991; Beebe and Kaye 1991), due to apoptosis (Moley et al. 1998) and changes in their metabolic state (Moley et al. 1996). Preimplantation development comprises the stage of development from fertilization till embryo implantation in the uterus. During this period, two sequential cell fate decisions occur that result in three cell populations. Upon the first decision, cells become either trophectoderm (TE) or inner cell mass (ICM) cells. Descendants of TE cells form the foetal portion of the placenta, while the ICM cells make a further decision: they differentiate either into Epiblast (Epi) or into Primitive Endoderm (PrE). Epi cells predominantly give rise to the embryo proper while PrE cell descendants generate the endodermal part of the yolk sac (Bassalert et al. 2018).

To assess the effect of intrauterine conditions in diabetic mothers during preimplantation development, embryos can be cultured under high glucose concentrations. This approach allows investigating early developmental defects without inducing DM in the mothers. Therefore, this approach can be considered a refinement method following the 3R's principles. Under these culture conditions, embryos exhibit higher intracellular concentration of glucose (Moley et al. 1996). There is controversy about the glucose concentration present in the reproductive tract milieu where embryos develop and the concentration that provides optimal development when embryos are cultured *in vitro* (Biggers and McGinnis 2001). Hence, different studies use increasing concentrations of glucose in their embryo culture medium to understand the underlying cause of poor embryo development in diabetic mothers. While control conditions use a 0.2 mM concentration of glucose, experimental conditions use up to 52 mM. However most studies use 15–27 mM, that is 2–3 times that of serum of non-diabetic mice (Bermejo-Alvarez et al. 2012b). Summarising, *in vitro* culture of mouse embryos in the presence of glucose reduces total and TE cell numbers at the blastocyst stage (Bermejo-Alvarez et al. 2012b) and also cell allocation to the ICM (Fraser et al. 2007).

An alternative model to study the effect of high glucose during early embryonic development is the use of blastoids and ICM organoids (Rivron et al. 2018; Mathew et al. 2019). Blastoids are blastocyst-like structures obtained from mouse trophoblast and embryonic stem cells, which morphologically and transcriptionally resemble blastocysts. ICM organoids are three-dimensional spheroids obtained from modified mouse embryonic stem cells that recapitulate Epi and PrE differentiation. In this context, these models could serve as an approach to replace the use of mice.

Altogether, the use of cultured embryos (blastoids or ICM organoids) in the presence of high glucose levels allows characterising the defects and causes of poor embryo quality obtained by diabetic mothers without inducing DM in the females. However, the results are limited and more studies to address the molecular mechanisms responsible of the embryonic defects will be needed in the future.

2.3 Recreation of the Diabetic *Milieu*

In vitro models are composed, not only by the cells themselves, but also by the conditions under which they are grown. To study the effect of DM on the β-cells, pancreatic islets etc. and even the effect on the development and physiology of the placenta, it is crucial to select a good cell model, but also to imitate the diabetic environment as closely as possible.

DM is defined by fasting plasma glucose concentrations above 7.0 mmol/L (126 mg/dL) or random glucose concentrations above 11.1 mmol/L (200 mg/dl) (American Diabetes Association 2013). For GDM, the cut-offs are even lower. However, concentrations used to mimic diabetes *in vitro* are not always consistent with this.

To recreate a non-diabetic milieu *in vitro* most authors agree with the concentrations to mimic normoglycemia, using 5.5 mM of glucose (Han et al. 2015; Heim et al. 2018; Yung et al. 2016; Peng et al. 2018; Basak et al. 2015), whereas other authors use 7 mM (Li et al. 2019), 10 mM (Jiang et al. 2017) or even 11 mM of glucose (Basak et al. 2019).

To simulate diabetic conditions, different concentrations are used: 20 mM (Yung et al. 2016), 22 mM (Triñanes et al. 2017), but 25 mM is the concentration of glucose most frequently used (Han et al. 2015; Hulme et al. 2018; Peng et al. 2018; Basak et al. 2015, 2019; Li et al. 2019), although 35.5 mM is also employed (Nanobashvili et al. 2018). Regarding prediabetic conditions, 10 mM (Basak et al. 2015) and 11 mM of glucose are the concentrations chosen (Han et al. 2015; Heim et al. 2018; Yung et al. 2016).

On the other hand, to simulate high fatty acids concentrations, authors employ the two most common dietary fatty acids, palmitate and oleate. Since it has been shown that palmitate is toxic to human trophoblast but oleate is not toxic (Colvin et al. 2017), the first is mainly used to understand the effect of fatty acids on the developing placenta and foetus. It is also the most frequently employed to study the effects of fatty acids on the β-cells and pancreas physiology. Most of the authors apply a concentration of 100 µM (Stirm et al. 2018; Colvin et al. 2017; Triñanes et al. 2017), but 500 µM palmitate has also been used (Hong et al. 2018).

To summarize, in recent years there has been important progress in the development of *in vitro* models to study DM. The *in vitro* models derived from animal cells have allowed to study DM physiology and test drugs. However, the advance in human *in vitro* models should lead to important progress in the understanding of the disease and the applications in medicine. Significant advances have been made in the field. For example, in the differentiation of human pluripotent stem cells into pancreatic β-like cells, in pancreatic islet research and in the development of 3D cultures (Tsakmaki et al. 2020). Development of molecular techniques have also improved that allow the modification of the *in vitro* models according to the goal. Regarding the placenta, it is still not very a well-known organ and much research is needed to understand its development and functions in humans. However, recent progress such as new human cell lines, human trophoblast stem cells, human embryo culture and trophoblast organoids, should allow improved understanding of the effect of DM and other diseases on placental and foetal development.

3 Potential Translation, Research Gaps and Future Perspectives

The translation of research from the experimental frame to the patient, usually lacks of direct transmission of the result success from the first scenario to the latter. This fact has driven research organizations to claim and establish common criteria for adequate methodology and reporting of the results, as this could be one of the main factors affecting replication and translation

(Jastreboff 2014; Starling 2019; Nakao 2019; Kilkenny et al. 2010).

In this regard, several aspects can be highlighted regarding the DM models described in the present review. Criteria for animal model definition and selection are not uniform, and methods to induce and test the models are also varied, even when similar aims are pursued. The type of DM targeted in the animal model is commonly not well established, or disagrees with the actual definitions of the disease. Regarding *in vitro* modelling, there are also important variations amongst authors in the definition of hyper- and normo-glycaemia. Therefore, starting a new study can be challenging if the protocol is to be based on published evidence. Common criteria need to be established for methodological aspects, both *in vivo* and *in vitro*. This will improve reproducibility of the results, as well as potential translation. The recent organoid and stem cell based strategies, and initiatives like the standards established by the ICRCC, will probably improve research translation.

Furthermore, models that adequately match the specific phenotype of GDM and pre-gestational DM or their complications are also scarce and necessary for this translation. In this sense, genetic models may also help, and recent advances in genetic engineering will boost the available options, given the growing list of candidate genes for GDM (Moon et al. 2019; Li et al. 2020; Lin et al. 2020; Thong et al. 2020).

Nevertheless, an immediate interesting strategy to improve these options could be a thorough evaluation of DM around pregnancy, in previously known diabetic models, and especially in the genetically modified. A previous example is available, where a macrosomic phenotype associated with impaired glucose metabolism was identified during routine rodent management, in a well-characterized *Socs2* null mouse model (Lau et al. 1993). Therefore, approaches to evaluate DM in pregnancy in previously known genetic models should be considered, as they are especially necessary in the assessment of such important areas like intrauterine programming, placentation and perinatal complications.

4 Summary and Recommendations

Maternal diabetes is associated with reduced fertility (Fraser et al. 2007) and an increased risk of poor pregnancy outcomes, both for the mother and her offspring (Hjort et al. 2019). Although clinical and epidemiological studies are invaluable to assess these outcomes and the effectiveness of potential treatments, there are certain ethical and practical limitations to what can be assessed in human studies. Thus, both *in vivo* and *in vitro* models can aid us in the understanding of the mechanisms behind these complications and, in the long run, towards their prevention and treatment. However, some important issues should be considered in the design and performance of the studies.

1. The aim of the study should be clearly defined: what type of DM are we focusing on? Is it pregestational or gestational? In the former case, is it T1DM or T2DM we want to assess?
2. The model should be selected in accordance with the aim of the study. In the case of animal studies, this includes the species, strain and mode and timing of DM induction.
3. The experimental (and control) milieu should mimic DM (and healthy control) as closely as possible, i.e. mildly hyperglycaemic for GDM and T2DM and more overtly so for T1DM.

Increasing progress in the development of *in vitro* models, as well as standardisation of methods and reporting should improve translation of experimental research in diabetic pregnancy to human health.

5 Funding

JLG is supported by the pre-doctoral program from the Universidad de Las Palmas de Gran Canaria (ULPGC) (2017), CVT is supported by the Postdoctoral Program from the ULPGC (2018). SMD is supported by the 'Viera y Clavijo' Program from the Agencia Canaria de Investigación Innovación y Sociedad de la Información (ACIISI)CIISI and the ULPGC. This manuscript is linked to the project PI16/00587 from the National Institute of Health Carlos III from the Spanish Government (ISCIII), cofounded by the European Funding for Regional Development (FEDER).

References

Abdelalim E, Memon B (2020) Stem cell therapy for diabetes: beta cells versus pancreatic progenitors. Cell 9(2):pii: E283

Abou-Kheir W, Barrak J, Hadadeh O, Daoud G (2017) HTR-8/SVneo cell line contains a mixed population of cells. Placenta 50:1–7

Acar N, Korgun ET, Cayli S, Sahin Z, Demir R, Ustunel I (2008) Is there a relationship between PCNA expression and diabetic placental development during pregnancy? Acta Histochem [Internet] 110(5):408–417. [cited 2019 Nov 6]. Available from: http://www.ncbi.nlm.nih.gov/pubmed/18377963

Aerts L, Vercruysse L, Van Assche FA (1997) The endocrine pancreas in virgin and pregnant offspring of diabetic pregnant rats. Diabetes Res Clin Pract [Internet] 38(1):9–19. [cited 2019 Nov 5]. Available from: http://www.ncbi.nlm.nih.gov/pubmed/9347241

Aigha II, Memon B, Elsayed AK, Abdelalim EM (2018) Differentiation of human pluripotent stem cells into two distinct NKX6.1 populations of pancreatic progenitors. Stem Cell Res Ther 9(1):83

American Diabetes Association (2013) Diagnosis and classification of diabetes mellitus, ADA clinical practice recommendations. Diabetes Care 36(Suppl 1): S67–S74

American Diabetes Association (2019) 2.Classification and diagnosis of diabetes: standards of medical care in diabetes 2019. Diabetes Care 42:S13–S28

Amirruddin NS, Low BSJ, Lee KO, Tai ES, Teo AKK (2019) New insights into human beta cell biology using human pluripotent stem cells. Semin Cell Dev Biol. pii: S1084-9521(18)30308-2

Amorim EMP, Damasceno DC, Perobelli JE, Spadotto R, Fernandez CDB, Volpato GT et al (2011) Short- and long-term reproductive effects of prenatal and lactational growth restriction caused by maternal diabetes in male rats. Reprod Biol Endocrinol 9:154

Andersson LE, Valtat B, Bagge A, Sharoyko VV, Nicholls DG, Ravassard P et al (2015) Characterization of stimulus-secretion coupling in the human pancreatic EndoC-βH1 beta cell line. PLoS One 10(3):e0120879

Apps R, Sharkey A, Gardner L, Male V, Trotter M, Miller N et al (2011) Genome-wide expression profile of first trimester villous and extravillous human trophoblast cells. Placenta 32(1):33–43

Araujo Júnior E, Peixoto AB, Zamarian ACP, Elito Júnior J, Tonni G (2017) Macrosomia. In: Best practice and research: clinical obstetrics and gynaecology, vol 38. Bailliere Tindall Ltd, pp 83–96

Asfari M, Janjic D, Meda P, Li G, Halban PA, Wollheim CB (1992) Establishment of ?-mercaptoethanol-dependent differentiated insulin secreting cell lines. Endocrinology 130:167–178

Azizkhan JC, Speeg KV, Stromberg K, Goode D (1979) Stimulation of human chorionic gonadotropin by JAr line choriocarcinoma after inhibition of DNA synthesis. Cancer Res 39(6 Pt 1):1952–1959

Bakhti M, Böttcher A, Lickert H (2019) Modelling the endocrine pancreas in health and disease. Nat Rev Endocrinol 15(3):155–171

Balboa D, Saarimäki-Vire J, Otonkoski T (2019) Concise review: human pluripotent stem cells for the modeling of pancreatic β-cell pathology. Stem Cells 37(1):33–41

Barbe A, Bongrani A, Mellouk N, Estienne A, Kurowska P, Grandhaye J et al (2019) Mechanisms of adiponectin action in fertility: An overview from gametogenesis to gestation in humans and animal models in normal and pathological conditions. Int J Mol Sci [Internet] 20(7). [cited 2019 Oct 29]. Available from: http://www.ncbi.nlm.nih.gov/pubmed/30934676

Basak S, Das MK, Srinivas V, Duttaroy AK (2015) The interplay between glucose and fatty acids on tube formation and fatty acid uptake in the first trimester trophoblast cells, HTR8/SVneo. Mol Cell Biochem 401 (1–2):11–19

Basak S, Vilasagaram S, Naidu K, Duttaroy AK (2019) Insulin-dependent, glucose transporter 1 mediated glucose uptake and tube formation in the human placental first trimester trophoblast cells. Mol Cell Biochem 451 (1–2):91–106

Bassalert C, Valverde-Estrella L, Chazaud C (2018) Primitive endoderm differentiation: from specification to epithelialization. Curr Top Dev Biol [Internet] 128:81–104. Available from: http://linkinghub. elsevier.com/retrieve/pii/S0070215317300716

Beebe LF, Kaye PL (1991) Maternal diabetes and retarded preimplantation development of mice. Diabetes 40 (4):457–461

Benazra M, Lecomte M-J, Colace C, Müller A, Machado C, Pechberty S et al (2015) A human beta cell line with drug inducible excision of immortalizing transgenes. Mol Metab 4(12):916–925

Bermejo-Alvarez P, Rosenfeld CS, Roberts RM (2012a) Effect of maternal obesity on estrous cyclicity, embryo development and blastocyst gene expression in a mouse model. Hum Reprod [Internet] 27 (12):3513–3522. Available from: https://academic. oup.com/humrep/article-lookup/doi/10.1093/humrep/des327

Bermejo-Alvarez P, Roberts RM, Rosenfeld CS (2012b) Effect of glucose concentration during in vitro culture of mouse embryos on development to blastocyst, success of embryo transfer, and litter sex ratio. Mol Reprod Dev 79(5):329–336

Biggers JD, McGinnis LK (2001) Evidence that glucose is not always an inhibitor of mouse preimplantation development in vitro. Hum Reprod 16(1):153–163

Bonfanti P, Nobecourt E, Oshima M, Albagli-Curiel O, Laurysens V, Stangé G et al (2015) Ex vivo expansion and differentiation of human and mouse fetal pancreatic progenitors are modulated by epidermal growth factor. Stem Cells Dev 24(15):1766–1778

Brito-Casillas Y, Melián C, Wägner AM (2016) Study of the pathogenesis and treatment of diabetes mellitus through animal models. Endocrinol Nutr [Internet] 63 (7):345–353. Available from: http://linkinghub. elsevier.com/retrieve/pii/S1575092216300481

Brito-Casillas Y, Aranda-Tavío H, Rodrigo-González L, Expósito-Montesdeoca AB, Martín-Rodríguez P, Guerra B, Wägner AM, Fernández-Pérez L (2019) SOCS2$^{-/-}$ mouse as a potential model of macrosomia and gestational diabetes. Eur Med J [Internet]. [cited 2019 Dec 20]. Available from: https://www. emjreviews.com/diabetes/abstract/socs2-mouse-as-a-potential-model-of-macrosomia-and-gestational-diabetes/

Brown K, Heller DS, Zamudio S, Illsley NP (2011) Glucose transporter 3 (GLUT3) protein expression in human placenta across gestation. Placenta 32 (12):1041–1049

Brown HM, Green ES, Tan TCY, Gonzalez MB, Rumbold AR, Hull ML et al (2018) Periconception onset diabetes is associated with embryopathy and fetal growth retardation, reproductive tract hyperglycosylation and impaired immune adaptation to pregnancy. Sci Rep 8 (1):2114

Bruin JE, Rezania A, Xu J, Narayan K, Fox JK, O'Neil JJ et al (2013) Maturation and function of human embryonic stem cell-derived pancreatic progenitors in macroencapsulation devices following transplant into mice. Diabetologia 56(9):1987–1998

Bruin JE, Erener S, Vela J, Hu X, Johnson JD, Kurata HT et al (2014) Characterization of polyhormonal insulin-producing cells derived in vitro from human embryonic stem cells. Stem Cell Res 12(1):194–208

Burke SD, Dong H, Hazan AD, Croy BA (2007) Aberrant endometrial features of pregnancy in diabetic NOD mice. Diabetes 56(12):2919–2926

Candiello J, Grandhi TSP, Goh SK, Vaidya V, Lemmon-Kishi M, Eliato KR et al (2018) 3D heterogeneous islet organoid generation from human embryonic stem cells using a novel engineered hydrogel platform. Biomaterials 177:27–39

Capobianco E, Jawerbaum A, Romanini MC, White V, Pustovrh C, Higa R et al (2005) 15-Deoxy-Delta12,14-prostaglandin J2 and peroxisome proliferator-activated receptor gamma (PPARgamma) levels in term placental tissues from control and diabetic rats: modulatory effects of a PPARgamma agonist on nitridergic and lipid placental metabolism. Reprod Fertil Dev [Internet] 17(4):423–433. [cited 2019 Nov 6]. Available from: http://www.ncbi.nlm.nih.gov/pubmed/15899154

Carlson AJ, Drennan FM (1911) The control of pancreatic diabetes in pregnancy by the passage of the internal secretion of the pancreas of the fetus to the blood of the mother. Am J Physiol Content 28(7):391–395

Chang AS, Dale AN, Moley KH (2005) Maternal diabetes adversely affects Preovulatory oocyte maturation, development, and granulosa cell apoptosis. Endocrinology [Internet] 146(5):2445–2453. [cited 2019 Nov 29]. Available from: https://academic.oup.com/endo/article-lookup/doi/10.1210/en.2004-1472

Chaudhry ZZ, Morris DL, Moss DR, Sims EK, Chiong Y, Kono T et al (2013) Streptozotocin is equally diabetogenic whether administered to fed or fasted mice. Lab Anim 47(4):257–265

Chen Z, Canet MJ, Sheng L, Jiang L, Xiong Y, Yin L et al (2015) Hepatocyte TRAF3 promotes insulin resistance and type 2 diabetes in mice with obesity. Mol Metab 4(12):951–960

Chi MMY, Pingsterhaus J, Carayannopoulos M, Moley KH (2000) Decreased glucose transporter expression triggers BAX-dependent apoptosis in the murine blastocyst. J Biol Chem 275(51):40252–40257

Chick WL, Warren S, Chute RN, Like AA, Lauris V, Kitchen KC (1977) A transplantable insulinoma in the rat. Proc Natl Acad Sci U S A 74:628–632

Colvin BN, Longtine MS, Chen B, Costa ML, Nelson DM (2017) Oleate attenuates palmitate-induced endoplasmic reticulum stress and apoptosis in placental trophoblasts. Reproduction 153(4):369–380

D'Amour KA, Bang AG, Eliazer S, Kelly OG, Agulnick AD, Smart NG et al (2006) Production of pancreatic hormone–expressing endocrine cells from human embryonic stem cells. Nat Biotechnol 24(11):1392–1401

Desoye G, Mouzon SH (2007) The human placenta in gestational diabetes mellitus. Diabetes Care 30(Supplement 2):S120–S126

Dickinson JE, Meyer BA, Brath PC, Chmielowiec S, Walsh SW, Parisi VM et al (1990) Placental thromboxane and prostacyclin production in an ovine diabetic model. Am J Obstet Gynecol [Internet] 163(6 Pt 1):1831–1835. [cited 2019 Oct 30]. Available from: http://www.ncbi.nlm.nih.gov/pubmed/2147814

Ding R, Liu XM, Xiang YQ, Zhang Y, Zhang JY, Guo F et al (2018) Altered matrix metalloproteinases expression in placenta from patients with gestational diabetes mellitus. Chin Med J 131:1255–1258

Dontas IA, Marinou KA, Karatzas T (2012) Research in diabetes using animal models. Br J Pharmacol 166(3):877–894

Dowling D, Corrigan N, Horgan S, Watson CJ, Baugh J, Downey P et al (2014) Cardiomyopathy in offspring of pregestational diabetic mouse pregnancy. J Diabetes Res 2014:624939

Elumalai S, Karunakaran U, Lee IK, Moon JS, Won KC (2017) Rac1-NADPH oxidase signaling promotes CD36 activation under glucotoxic conditions in pancreatic beta cells. Redox Biol 11:126–134

Eriksson UJ, Bone AJ, Turnbull DM, Baird JD (1989) Timed interruption of insulin therapy in diabetic BB/E rat pregnancy: effect on maternal metabolism and fetal outcome. Acta Endocrinol (Copenh) [Internet] 120(6):800–810. [cited 2019 Nov 5]. Available from: http://www.ncbi.nlm.nih.gov/pubmed/2658457

Esakoff TF, Cheng YW, Sparks TN, Caughey AB (2009) The association between birthweight 4000 g or greater and perinatal outcomes in patients with and without gestational diabetes mellitus. Am J Obstet Gynecol 200(6):672.e1–672.e4

Ezekwe MO, Ezekwe EI, Sen DK, Ogolla F (1984) Effects of maternal streptozotocin-diabetes on fetal growth, energy reserves and body composition of newborn pigs. J Anim Sci [Internet] 59(4):974–980. [cited 2019 Oct 30]. Available from: http://www.ncbi.nlm.nih.gov/pubmed/6239852

Favaro RR, Salgado RM, Covarrubias AC, Bruni F, Lima C, Fortes ZB et al (2013) Long-term type 1 diabetes impairs decidualization and extracellular matrix remodeling during early embryonic development in mice on occasion of the 30th anniversary of the Laboratory of Reproductive and Extracellular Matrix Biology we dedicate this article to its founder, Professor Paulo Abrahamsohn. Placenta 34(12):1128–1135

Feige JN, Lagouge M, Auwerx J, Feige JN, Lagouge M, Auwerx J (2008) Dietary manipulation of mouse metabolism. In: Current protocols in molecular biology [Internet]. Wiley, Hoboken, pp 29B.5.1–29B.5.12. Available from: http://doi.wiley.com/10.1002/0471142727.mb29b05s84

Frank HG, Gunawan B, Ebeling-Stark I, Schulten HJ, Funayama H, Cremer U et al (2000) Cytogenetic and DNA-fingerprint characterization of choriocarcinoma cell lines and a trophoblast/choriocarcinoma cell hybrid. Cancer Genet Cytogenet 116(1):16–22

Fraser RB, Waite SL, Wood KA, Martin KL (2007) Impact of hyperglycemia on early embryo development and embryopathy: in vitro experiments using a mouse model. Hum Reprod 22(12):3059–3068

Fröhlich JD, Desoye G, König J, Huppertz B (2015) Oxygen and glucose dependent viability of HLA-G positive and negative trophoblasts using ACH-3P cells as first trimester trophoblast-derived cell model. J Reprod Heal Med 1(1):4–9

Gauguier D, Bihoreau MT, Ktorza A, Berthault MF, Picon L (1990) Inheritance of diabetes mellitus as consequence of gestational hyperglycemia in rats. Diabetes [Internet] 39(6):734–739. [cited 2019 Oct 30]. Available from: http://www.ncbi.nlm.nih.gov/pubmed/2189765

Ge ZJ, Liang QX, Hou Y, Han ZM, Schatten H, Sun QY et al (2014) Maternal obesity and diabetes may cause DNA methylation alteration in the spermatozoa of offspring in mice. Reprod Biol Endocrinol 12(1):29

Gonzalez E, Jawerbaum A (2006) Diabetic pregnancies: the challenge of developing in a pro-inflammatory environment. Curr Med Chem 13(18):2127–2138

Graham CH, Hawley TS, Hawley RC, MacDougall JR, Kerbel RS, Khoo N et al (1993) Establishment and characterization of first trimester human trophoblast cells with extended lifespan. Exp Cell Res 206 (2):204–211

Green AD, Vasu S, McClenaghan NH, Flatt PR (2015) Pseudoislet formation enhances gene expression, insulin secretion and cytoprotective mechanisms of clonal human insulin-secreting 1.1B4 cells. Pflügers Arch – Eur J Physiol 467(10):2219–2228

Greggio C, De Franceschi F, Figueiredo-Larsen M, Gobaa S, Ranga A, Semb H et al (2013) Artificial three-dimensional niches deconstruct pancreas development in vitro. Development 140(21):4452–4462

Grieco FA, Moore F, Vigneron F, Santin I, Villate O, Marselli L et al (2014) IL-17A increases the expression of proinflammatory chemokines in human pancreatic islets. Diabetologia 57(3):502–511

Guo S, Dai C, Guo M, Taylor B, Harmon JS, Sander M et al (2013) Inactivation of specific β cell transcription factors in type 2 diabetes. J Clin Invest 123 (8):3305–3316

Gurgul-Convey E, Kaminski MT, Lenzen S (2015) Physiological characterization of the human EndoC-βH1 β-cell line. Biochem Biophys Res Commun 464 (1):13–19

Haider S, Meinhardt G, Saleh L, Kunihs V, Gamperl M, Kaindl U et al (2018) Self-renewing trophoblast organoids recapitulate the developmental program of the early human placenta. Stem Cell Rep 11 (2):537–551

Han CS, Herrin MA, Pitruzzello MC, Mulla MJ, Werner EF, Pettker CM et al (2015) Glucose and metformin modulate human first trimester trophoblast function: a model and potential therapy for diabetes-associated Uteroplacental insufficiency. Am J Reprod Immunol 73(4):362–371

Hanafusa T, Miyagawa J, Nakajima H, Tomita K, Kuwajima M, Matsuzawa Y et al (1994) The NOD mouse. Diabetes Res Clin Pract [Internet] 24:S307–S311. [cited 2019 Nov 15]. Available from: https://linkinghub.elsevier.com/retrieve/pii/0168822794902674

Hart NJ, Powers AC (2019) Use of human islets to understand islet biology and diabetes: progress, challenges and suggestions. Diabetologia 62(2):212–222

Heaton SJ, Eady JJ, Parker ML, Gotts KL, Dainty JR, Fairweather-Tait SJ et al (2008) The use of BeWo cells as an in vitro model for placental iron transport. Am J Physiol Cell Physiol 295(5):C1445–C1453

Heim KR, Mulla MJ, Potter JA, Han CS, Guller S, Abrahams VM (2018) Excess glucose induce trophoblast inflammation and limit cell migration through HMGB1 activation of Toll-Like receptor 4. Am J Reprod Immunol 80(5):e13044

Herberg L, Coleman DL (1977) Laboratory animals exhibiting obesity and diabetes syndromes. Metabolism 26(1):59–99

Hiden U, Wadsack C, Prutsch N, Gauster M, Weiss U, Frank H-G et al (2007) The first trimester human trophoblast cell line ACH-3P: a novel tool to study autocrine/paracrine regulatory loops of human trophoblast subpopulations – TNF-α stimulates MMP15 expression. BMC Dev Biol 7(1):137

Hindley CJ, Cordero-Espinoza L (2016) Organoids from adult liver and pancreas: stem cell biology and biomedical utility. Dev Biol 420(2):251–261

Hjort L, Novakovic B, Grunnet LG, Maple-Brown L, Damm P, Desoye G et al (2019) Diabetes in pregnancy and epigenetic mechanisms—how the first 9 months from conception might affect the child's epigenome and later risk of disease. Lancet Diabetes Endocrinol (Lancet Publishing Group) 7:796–806

Holemans K, Caluwaerts S, Poston L, Van Assche FA (2004) Diet-induced obesity in the rat: a model for gestational diabetes mellitus. Am J Obstet Gynecol 190(3):858–865

Hong Y, Ahn H-J, Shin J, Lee JH, Kim J-H, Park H-W et al (2018) Unsaturated fatty acids protect trophoblast cells from saturated fatty acid-induced autophagy defects. J Reprod Immunol 125:56–63

Hrvatin S, O'Donnell CW, Deng F, Millman JR, Pagliuca FW, DiIorio P et al (2014) Differentiated human stem cells resemble fetal, not adult, β cells. Proc Natl Acad Sci U S A 111(8):3038–3043

Huang C, Snider F, Cross JC (2009) Prolactin receptor is required for normal glucose homeostasis and modulation of beta-cell mass during pregnancy. Endocrinology [Internet] 150(4):1618–1626. [cited 2019 Nov 5]. Available from: http://www.ncbi.nlm.nih.gov/pubmed/19036882

Huch M, Bonfanti P, Boj SF, Sato T, Loomans CJM, van de Wetering M et al (2013) Unlimited in vitro expansion of adult bi-potent pancreas progenitors through the Lgr5/R-spondin axis. EMBO J 32(20):2708–2721

Hulme CH, Stevens A, Dunn W, Heazell AEP, Hollywood K, Begley P et al (2018) Identification of the functional pathways altered by placental cell exposure to high glucose: lessons from the transcript and metabolite interactome. Sci Rep 8(1):5270

Illsley NP, Baumann MU (1866) Human placental glucose transport in fetoplacental growth and metabolism. Biochim Biophys Acta Mol basis Dis 2020(2):165359

Jacob HJ, Pettersson A, Wilson D, Mao Y, Lernmark Å, Lander ES (1992a) Genetic dissection of autoimmune type I diabetes in the BB rat. Nat Genet 2(1):56–60

Jacob HJ, Pettersson A, Wilson D, Mao Y, Lernmark Å, Lander ES (1992b) Differential effects of fat and sucrose on the development of obesity and diabetes in C57BL/6J and A/J mice. Nat Genet 2(1):56–60

Jansson T, Wennergren M, Powell TL (1999) Placental glucose transport and GLUT 1 expression in insulin-dependent diabetes. Am J Obstet Gynecol 180 (1 I):163–168

Jansson T, Cetin I, Powell TL, Desoye G, Radaelli T, Ericsson A et al (2006) Placental transport and metabolism in fetal overgrowth – a workshop report. Placenta 27(SUPPL):109–113

Jastreboff AM (2014) Giving leptin a second chance. Sci Transl Med 6(220):220ec14

Jawerbaum A, White V (2010) Animal models in diabetes and pregnancy. Endocr Rev [Internet] 31(5):680–701. [cited 2019 Nov 1].. Available from: http://www.ncbi.nlm.nih.gov/pubmed/20534704

Jawerbaum A, Gonzalez ET, Catafau JR, Rodriguez RR, Gomez G, Gimeno AL et al (1993) Glucose, glycogen and triglyceride metabolism, as well as prostaglandin production in uterine strips and in embryos from diabetic pregnant rats. Influences of the presence of substrate in the incubation medium. Prostaglandins [Internet] 46(5):417–431. [cited 2019 Oct 31] Available from: http://www.ncbi.nlm.nih.gov/pubmed/8278619

Jiang S, Teague AM, Tryggestad JB, Chernausek SD (2017) Role of microRNA-130b in placental PGC-1α/TFAM mitochondrial biogenesis pathway. Biochem Biophys Res Commun 487(3):607–612

Jin L, Feng T, Shih HP, Zerda R, Luo A, Hsu J et al (2013) Colony-forming cells in the adult mouse pancreas are expandable in Matrigel and form endocrine/acinar colonies in laminin hydrogel. Proc Natl Acad Sci U S A 110(10):3907–3912

Jin L, Feng T, Zerda R, Chen C-C, Riggs AD, Ku HT (2014) In vitro multilineage differentiation and self-renewal of single pancreatic colony-forming cells from adult C57BL/6 mice. Stem Cells Dev 23 (8):899–909

Kaddis JS, Olack BJ, Sowinski J, Cravens J, Contreras JL, Niland JC (2009) Human pancreatic islets and diabetes research. JAMA 301(15):1580–1587

Kaiser N, Nesher R, Donath MY, Fraenkel M, Behar V, Magnan C et al (2005) Psammomys obesus, a model for environment-gene interactions in type 2 diabetes. Diabetes 54(SUPPL. 2):S137–S144

Kataoka M, Kawamuro Y, Shiraki N, Miki R, Sakano D, Yoshida T et al (2013) Recovery from diabetes in neonatal mice after a low-dose streptozotocin treatment. Biochem Biophys Res Commun 430 (3):1103–1108

Kaufmann RC, Amankwah KS, Dunaway G, Maroun L, Arbuthnot J, Roddick JW (1981) An animal model of gestational diabetes. Am J Obstet Gynecol 141 (5):479–482

Kilkenny C, Browne WJ, Cuthill IC, Emerson M, Altman DG (2010) Improving bioscience research reporting: the ARRIVE guidelines for reporting animal research. PLoS Biol [Internet] 8(6):e1000412. Available from: https://dx.plos.org/10.1371/journal.pbio.1000412

Kim JH, Pan JH, Cho HT, Kim YJ (2016a) Black ginseng extract counteracts Streptozotocin-induced diabetes in mice. Irwin N, editor. PLoS One [Internet] 11(1): e0146843. [cited 2019 Nov 29]. Available from: http://dx.plos.org/10.1371/journal.pone.0146843

Kim Y, Kim H, Ko UH, Oh Y, Lim A, Sohn J-W et al (2016b) Islet-like organoids derived from human pluripotent stem cells efficiently function in the glucose responsiveness in vitro and in vivo. Sci Rep 6 (1):35145

Kojima N (2014) In vitro reconstitution of pancreatic islets. Organogenesis 10(2):225–230

Kopp JL, Grompe M, Sander M (2016) Stem cells versus plasticity in liver and pancreas regeneration. Nat Cell Biol 18(3):238–245

Krause M, Keane K, Rodrigues-Krause J, Crognale D, Egan B, De Vito G et al (2014) Elevated levels of extracellular heat-shock protein 72 (eHSP72) are positively correlated with insulin resistance in vivo and cause pancreatic β-cell dysfunction and death in vitro. Clin Sci 126(10):739–752

Krishnan K, Ma Z, Björklund A, Islam MS (2015) Calcium signaling in a genetically engineered human pancreatic β-cell line. Pancreas 44(5):773–777

Kurlawalla-Martinez C, Stiles B, Wang Y, Devaskar SU, Kahn BB, Wu H (2005) Insulin hypersensitivity and resistance to streptozotocin-induced diabetes in mice lacking PTEN in adipose tissue. Mol Cell Biol [Internet] 25(6):2498–2510. [cited 2019 Nov 29]. Available from: http://www.ncbi.nlm.nih.gov/pubmed/15743841

Lager S, Jansson N, Olsson AL, Wennergren M, Jansson T, Powell TL (2011) Effect of IL-6 and TNF-α on fatty acid uptake in cultured human primary trophoblast cells. Placenta 32(2):121–127

Lappas M, Hiden U, Desoye G, Froehlich J, De Mouzon SH, Jawerbaum A (2011) The role of oxidative stress in the pathophysiology of gestational diabetes mellitus. Antioxid Redox Signal 15:3061–3100

Lau C, Sullivan MK, Hazelwood RL (1993) Effects of diabetes mellitus on lactation in the rat. Proc Soc Exp Biol Med 204(1):81–89

Lebreton F, Lavallard V, Bellofatto K, Bonnet R, Wassmer CH, Perez L et al (2019) Insulin-producing organoids engineered from islet and amniotic epithelial cells to treat diabetes. Nat Commun 10(1):4491

Lee J, Sugiyama T, Liu Y, Wang J, Gu X, Lei J et al (2013) Expansion and conversion of human pancreatic ductal cells into insulin-secreting endocrine cells. elife 2: e00940

Lee J, Lee HC, Kim SY, Cho GJ, Woodruff TK (2019) Poorly-controlled type 1 diabetes mellitus impairs LH-LHCGR signaling in the ovaries and decreases female fertility in mice. Yonsei Med J 60(7):667–678

Li G, Lin L, Wang Y, Yang H (2019) 1,25(OH)2D3 protects trophoblasts against insulin resistance and inflammation via suppressing mTOR signaling. Reprod Sci 26(2):223–232

Li M, Rahman ML, Wu J, DIng M, Chavarro JE, Lin Y et al (2020) Genetic factors and risk of type 2 diabetes among women with a history of gestational diabetes: findings from two independent populations. BMJ Open Diabetes Res Care 8(1):e000850

Liao Z, Wang J, Tan H, Wei L (2017) Cinnamon extracts exert intrapancreatic cytoprotection against streptozotocin in vivo. Gene 627:519–523

Lin R, Yuan Z, Zhang C, Ju H, Sun Y, Huang N et al (2020) Common genetic variants in ADCY5 and gestational glycemic traits. Petry CI, editor. PLoS One [Internet] 15(3):e0230032. [cited 2020 Apr 12]. Available from: https://dx.plos.org/10.1371/journal.pone.0230032

Loomans CJM, Williams Giuliani N, Balak J, Ringnalda F, van Gurp L, Huch M et al (2018) Expansion of adult human pancreatic tissue yields organoids harboring progenitor cells with endocrine differentiation potential. Stem Cell Rep 10(3):712–724

López-Soldado I, Herrera E (2003) Different diabetogenic response to moderate doses of streptozotocin in pregnant rats, and its long-term consequences in the offspring. Exp Diabesity Res 4(2):107–118

Lozano I, Van Der Werf R, Bietiger W, Seyfritz E, Peronet C, Pinget M et al (2016) High-fructose and high-fat diet-induced disorders in rats: impact on diabetes risk, hepatic and vascular complications. Nutr Metab 13:15

Luo H, Chen C, Guo L, Xu Z, Peng X, Wang X et al (2018) Exposure to maternal diabetes mellitus causes renal dopamine D1 receptor dysfunction and hypertension in adult rat offspring. Hypertension 72(4):962–970

Mage RG, Esteves PJ, Rader C (2019) Rabbit models of human diseases for diagnostics and therapeutics development. Dev Comp Immunol (Elsevier Ltd) 92:99–104

Marchetti P, Schulte AM, Marselli L, Schoniger E, Bugliani M, Kramer W et al (2019) Fostering improved human islet research: a European perspective. Diabetologia 62(8):1514–1516

Markowitz J, Soskin S (1927) Pancreatic diabetes and pregnancy. Am J Physiol Content 79(3):553–558

Martin ME, Garcia AM, Blanco L, Herrera E, Salinas M (1995) Effect of streptozotocin diabetes on polysomal aggregation and protein synthesis rate in the liver of pregnant rats and their offspring. Biosci Rep 15(1):15–20

Martínez N, Capobianco E, White V, Pustovrh MC, Higa R, Jawerbaum A (2008) Peroxisome proliferator-activated receptor α activation regulates lipid metabolism in the feto-placental unit from diabetic rats. Reproduction 136(1):95–103

Mathew B, Muñoz-Descalzo S, Corujo-Simon E, Schröter C, Stelzer EHK, Fischer SC (2019) Mouse ICM organoids reveal three-dimensional cell fate clustering. Biophys J 116(1):127–141

McClenaghan NH, Barnett CR, Ah-Sing E, Abdel-Wahab YHA, O'Harte FPM, Yoon T-W et al (1996) Characterization of a novel glucose-responsive insulin-secreting cell line, BRIN-BD11, produced by electrofusion. Diabetes 45(8):1132–1140

Miyazaki J-I, Araki K, Yamato E, Ikegami H, ASANO T, Shibasaki Y et al (1990) Establishment of a pancreatic β cell line that retains glucose-inducible insulin secretion: special reference to expression of glucose transporter isoforms*. Endocrinology 127(1):126–132

Moley KH, Vaughn WK, DeCherney AH, Diamond MP (1991) Effect of diabetes mellitus on mouse pre-implantation embryo development. J Reprod Fertil [Internet] 93(2):325–332. Available from: http://www.ncbi.nlm.nih.gov/pubmed/1787451

Moley KH, M-Y Chi M, Manchester JK, McDougal DB, Lowry OH (1996) Alterations of intraembryonic metabolites in preimplantation mouse embryos exposed to elevated concentrations of glucose: a metabolic explanation for the developmental retardation seen in preimplantation embryos from diabetic Animals1. Biol Reprod 54(6):1209–1216

Moley KH, Chi MM-Y, Knudson CM, Korsmeyer SJ, Mueckler MM (1998) Hyperglycemia induces apoptosis in pre-implantation embryos through cell death effector pathways. Nat Med [Internet] 4(12):1421–1424. Available from: http://www.nature.com/articles/nm1298_1421

Moon S, Bin KDY, Ko JH, Kim YS (2019) Recent advances in the CRISPR genome editing tool set. Exp Mol Med (NLM Medline) 51:130

Moshref M, Tangey B, Gilor C, Papas KK, Williamson P, Loomba-Albrecht L et al (2019) Concise review: canine diabetes mellitus as a translational model for innovative regenerative medicine approaches. Stem Cells Transl Med (Wiley) 8:450–455

Nakao K (2019) Translational science: newly emerging science in biology and medicine – lessons from translational research on the natriuretic peptide family and leptin. Proc Jpn Acad Ser B Phys Biol Sci 95(9):538–567

Nandi P, Lim H, Torres-Garcia EJ, Lala PK (2018) Human trophoblast stem cell self-renewal and differentiation: role of decorin. Sci Rep 8(1):8977

Nanobashvili K, Jack-Roberts C, Bretter R, Jones N, Axen K, Saxena A et al (2018) Maternal choline and betaine supplementation modifies the placental response to hyperglycemia in mice and human trophoblasts. Nutrients 10(10):1507

Narushima M, Kobayashi N, Okitsu T, Tanaka Y, Li S-A, Chen Y et al (2005) A human β-cell line for transplantation therapy to control type 1 diabetes. Nat Biotechnol 23(10):1274–1282

Niland JC, Stiller T, Cravens J, Sowinski J, Kaddis J, Qian D (2010) Effectiveness of a web-based automated cell distribution system. Cell Transplant 19(9):1133–1142

Novaro V, Jawerbaum A, Faletti A, Gimeno MA, González ET (1998) Uterine nitric oxide and prostaglandin E during embryonic implantation in non-insulin-dependent diabetic rats. Reprod Fertil Dev [Internet] 10(3):217–223.[cited 2019 Nov 7]. Available from: http://www.ncbi.nlm.nih.gov/pubmed/11596867

Nusrat A, von Eichel-Streiber C, Turner JR, Verkade P, Madara JL, Parkos CA (2001) Clostridium difficile toxins disrupt epithelial barrier function by altering membrane microdomain localization of tight junction proteins. Infect Immun 69(3):1329–1336

Okae H, Toh H, Sato T, Hiura H, Takahashi S, Shirane K et al (2018) Derivation of human trophoblast stem cells. Cell Stem Cell 22(1):50–63.e6

Ota H, Itaya-Hironaka A, Yamauchi A, Sakuramoto-Tsuchida S, Miyaoka T, Fujimura T et al (2013) Pancreatic β cell proliferation by intermittent hypoxia via up-regulation of Reg family genes and HGF gene. Life Sci 93(18–19):664–672

Pagliuca FW, Millman JR, Gürtler M, Segel M, Van Dervort A, Ryu JH et al (2014) Generation of functional human pancreatic β cells in vitro. Cell 159 (2):428–439

Pasek RC, Gannon M (2013) Advancements and challenges in generating accurate animal models of gestational diabetes mellitus. Am J Physiol Endocrinol Metab 305:1327–1338

Pearson JA, Wong FS, Wen L (2016) The importance of the Non Obese Diabetic (NOD) mouse model in autoimmune diabetes. J Autoimmun (Academic Press) 66:76–88

Peng HY, Li MQ, Li HP (2018) High glucose suppresses the viability and proliferation of HTR-8/SVneo cells through regulation of the miR-137/PRKAA1/IL-6 axis. Int J Mol Med 42(2):799–810

Pérez-López L, Boronat M, Melián C, Saavedra P, Brito-Casillas Y, Wägner AM (2019) Assessment of the association between diabetes mellitus and chronic kidney disease in adult cats. J Vet Intern Med 33 (5):1921–1925

Phillips MS, Liu Q, Hammond HA, Dugan V, Hey PJ, Caskey CT et al (1996) Leptin receptor missense mutation in the fatty Zucker rat. Nat Genet 13(1):18–19

Plows JF, Stanley JL, Baker PN, Reynolds CM, Vickers MH (2018a) Molecular sciences the pathophysiology of gestational diabetes mellitus. Int J Mol Sci 19(11): pii: E3342. [cited 2019 Oct 29] Available from: www.mdpi.com/journal/ijms

Plows J, Stanley J, Baker P, Reynolds C, Vickers M (2018b) The pathophysiology of gestational diabetes mellitus. Int J Mol Sci [Internet] 19(11):3342. [cited 2019 Nov 11]. Available from: http://www.mdpi.com/1422-0067/19/11/3342

Poaty H, Coullin P, Peko JF, Dessen P, Diatta AL, Valent A et al (2012) Genome-wide high-resolution aCGH analysis of gestational Choriocarcinomas. Krahe R, editor. PLoS One 7(1):e29426

Priel T, Aricha-Tamir B, Sekler I (2007) Clioquinol attenuates zinc-dependent β-cell death and the onset of insulitis and hyperglycemia associated with experimental type I diabetes in mice. Eur J Pharmacol 565 (1–3):232–239

Ravassard P, Hazhouz Y, Pechberty S, Bricout-Neveu E, Armanet M, Czernichow P et al (2011) A genetically engineered human pancreatic β cell line exhibiting glucose-inducible insulin secretion. J Clin Invest 121 (9):3589–3597

Rezania A, Bruin JE, Riedel MJ, Mojibian M, Asadi A, Xu J et al (2012) Maturation of human embryonic stem cell-derived pancreatic progenitors into functional islets capable of treating pre-existing diabetes in mice. Diabetes 61(554722):1–14

Rezania A, Bruin JE, Arora P, Rubin A, Batushansky I, Asadi A et al (2014) Reversal of diabetes with insulin-producing cells derived in vitro from human pluripotent stem cells. Nat Biotechnol 32(11):1121–1133

Rivron NC, Frias-Aldeguer J, Vrij EJ, Boisset J-C, Korving J, Vivié J et al (2018) Blastocyst-like structures generated solely from stem cells. Nature [Internet] 557(7703):106–111. Available from: http://www.nature.com/articles/s41586-018-0051-0

Rogal J, Zbinden A, Schenke-Layland K, Loskill P (2019) Stem-cell based organ-on-a-chip models for diabetes research. Adv Drug Deliv Rev 140:101–128

Rousseau-Ralliard D, Couturier-Tarrade A, Thieme R, Brat R, Rolland A, Boileau P et al (2019) A short periconceptional exposure to maternal type-1 diabetes is sufficient to disrupt the feto-placental phenotype in a rabbit model. Mol Cell Endocrinol [Internet] 480:42–53. [cited 2019 Mar 29]. Available from: https://www.sciencedirect.com/science/article/pii/S0303720718302910?via%3Dihub#

Russ HA, Parent AV, Ringler JJ, Hennings TG, Nair GG, Shveygert M et al (2015) Controlled induction of human pancreatic progenitors produces functional beta-like cells in vitro. EMBO J 34(13):1759–1772

Saleh L, Prast J, Haslinger P, Husslein P, Helmer H, Knöfler M (2007) Effects of different human chorionic gonadotrophin preparations on trophoblast differentiation. Placenta 28(2–3):199–203

Scharfmann R, Pechberty S, Hazhouz Y, von Bülow M, Bricout-Neveu E, Grenier-Godard M et al (2014) Development of a conditionally immortalized human pancreatic β cell line. J Clin Invest 124(5):2087–2098

Scharfmann R, Didiesheim M, Richards P, Chandra V, Oshima M, Albagli O (2016) Mass production of functional human pancreatic β-cells: why and how? Diabetes Obes Metab 18(Suppl 1):128–136

Schmied BM, Ulrich A, Matsuzaki H, Batra SK, Pour PM, Schmied BM et al (2000) Maintenance of human islets in long term culture. Differentiation 66(4–5):173–180

Schmitz H, Barmeyer C, Fromm M, Runkel N, Foss H-D, Bentzel CJ et al (1999) Altered tight junction structure contributes to the impaired epithelial barrier function in ulcerative colitis. Gastroenterology 116(2):301–309

Schwartz R, Gruppuso PA, Petzold K, Brambilla D, Hiilesmaa V, Teramo KA (1994) Hyperinsulinemia and macrosomia in the fetus of the diabetic mother. Diabetes Care 17(7):640–648

Šeda O, Vieira AR, Proshchina A, Molina-Hernández A, Márquez-Valadez B, Valle-Bautista R et al (2018) Maternal diabetes and fetal programming toward neurological diseases: beyond neural tube defects. Front Endocrinol [Internet] 9:664. [cited 2019 Nov 28]. Available from: www.frontiersin.org

Shahjalal HM, Abdal Dayem A, Lim KM, Jeon T-l, Cho SG (2018) Generation of pancreatic β cells for treatment of diabetes: advances and challenges. Stem Cell Res Ther 9(1):1–19

Sharma A, Zangen DH, Reitz P, Taneja M, Lissauer ME, Miller CP et al (1999) The homeodomain protein IDX-1 increases after an early burst of proliferation during pancreatic regeneration. Diabetes [Internet] 48 (3):507–513. [cited 2019 Oct 30]. Available from: http://www.ncbi.nlm.nih.gov/pubmed/10078550

Shim J-H, Kim J, Han J, An SY, Jang YJ, Son J et al (2015) Pancreatic islet-Like three-dimensional aggregates derived from human embryonic stem cells ameliorate hyperglycemia in Streptozotocin-induced diabetic mice. Cell Transplant 24(10):2155–2168

Shimizu R, Sakazaki F, Okuno T, Nakamuro K, Ueno H (2012) Difference in glucose intolerance between C57BL/6J and ICR strain mice with streptozotocin/nicotinamide-induced diabetes. Biomed Res 33:63–66

Shin J-Y, Jeong J-H, Han J, Bhang SH, Jeong G-J, Haque MR et al (2015) Transplantation of Heterospheroids of islet cells and mesenchymal stem cells for effective angiogenesis and Antiapoptosis. Tissue Eng Part A 21(5–6):1024

Song B, Scheuner D, Ron D, Pennathur S, Kaufman RJ (2008) Chop deletion reduces oxidative stress, improves β cell function, and promotes cell survival in multiple mouse models of diabetes. J Clin Invest 118 (10):3378–3389

Starling S (2019) New therapeutic promise for leptin. Nat Rev Endocrinol (Nature Publishing Group) 15:625

Stirm L, Kovářová M, Perschbacher S, Michlmaier R, Fritsche L, Siegel-Axel D et al (2018) BMI-independent effects of gestational diabetes on human placenta. J Clin Endocrinol Metab 103 (9):3299–3309

Sugimura Y, Murase T, Oyama K, Uchida A, Sato N, Hayasaka S et al (2009) Prevention of neural tube defects by loss of function of inducible nitric oxide synthase in fetuses of a mouse model of streptozotocin-induced diabetes. Diabetologia 52(5):962–971

Sugiyama T, Benitez CM, Ghodasara A, Liu L, McLean GW, Lee J et al (2013) Reconstituting pancreas development from purified progenitor cells reveals genes essential for islet differentiation. Proc Natl Acad Sci 110(31):12691–12696

Surwit RS, Kuhn CM, Cochrane C, McCubbin JA, Feinglos MN (1988) Diet-induced type II diabetes in C57BL/6J mice. Diabetes [Internet] 37(9):1163–1167. Available from: http://www.ncbi.nlm.nih.gov/pubmed/3044882

Suwaki N, Masuyama H, Masumoto A, Takamoto N, Hiramatsu Y (2007) Expression and potential role of peroxisome proliferator-activated receptor gamma in the placenta of diabetic pregnancy. Placenta [Internet] 28(4):315–323. [cited 2019 Nov 6]. Available from: http://www.ncbi.nlm.nih.gov/pubmed/16753211

Takahashi Y, Takebe T, Taniguchi H (2018) Methods for generating vascularized islet-Like organoids via self-condensation. Curr Protoc Stem Cell Biol 45(1):e49

Takebe T, Enomura M, Yoshizawa E, Kimura M, Koike H, Ueno Y et al (2015) Vascularized and complex organ buds from diverse tissues via mesenchymal cell-driven condensation. Cell Stem Cell 16 (5):556–565

Thong EP, Codner E, Laven JSE, Teede H (2020) Diabetes: a metabolic and reproductive disorder in women. Lancet Diabetes Endocrinol 8(2):134–149

Triñanes J, Rodriguez-Rodriguez AE, Brito-Casillas Y, Wagner A, De Vries APJ, Cuesto G et al (2017) Deciphering tacrolimus-induced toxicity in pancreatic β cells. Am J Transplant [Internet] 17(11):2829–2840. Available from: http://www.ncbi.nlm.nih.gov/pubmed/28432716

Tsakmaki A, Fonseca Pedro P, Bewick GA (2020) Diabetes through a 3D lens: organoid models. Diabetologia 27:1–10

Tsonkova VG, Sand FW, Wolf XA, Grunnet LG, Kirstine Ringgaard A, Ingvorsen C et al (2018) The EndoC-β H1 cell line is a valid model of human beta cells and applicable for screenings to identify novel drug target candidates. Mol Metab 8:144–157

Turco MY, Moffett A (2019) Development of the human placenta. Development 146(22):pii: dev163428

Turco MY, Gardner L, Kay RG, Hamilton RS, Prater M, Hollinshead MS et al (2018) Trophoblast organoids as a model for maternal–fetal interactions during human placentation. Nature 564(7735):263–267

Türk G, Rişvanlı A, Çeribaşı AO, Sönmez M, Yüce A, Güvenç M et al (2018) Effect of gestational diabetes mellitus on testis and pancreatic tissues of male offspring. Andrologia [Internet] 50(4):e12976. [cited 2019 Nov 29] Available from: http://doi.wiley.com/10.1111/and.12976

Uchida S, Watanabe S, Aizawa T, Furuno A, Muto T (1979) Polyoncogenicity and insulinoma-inducing ability of BK Virus, a human Papovavirus, in Syrian golden hamsters. J Natl Cancer Inst 63(1):119–126

Van Assche FA, Holemans K, Aerts L (2001) Long-term consequences for offspring of diabetes during pregnancy. Br Med Bull 60:173–182

Vasu S, McClenaghan NH, McCluskey JT, Flatt PR (2013) Cellular responses of novel human pancreatic β-cell line, 1.1B4 to hyperglycemia. Islets 5 (4):170–177

Vaxillaire M, Froguel P (2006) Genetic basis of maturity-onset diabetes of the young [Internet]. Endocrinol Metab Clin N Am 35:371–384. Available from: https://linkinghub.elsevier.com/retrieve/pii/S0889852906000107. [cited 2020 Mar 24]

Vercheval M, De Hertogh R, Pampfer S, Vanderheyden I, Michiels B, De Bernardi P et al (1990) Experimental diabetes impairs rat embryo development during the preimplantation period. Diabetologia [Internet] 33 (4):187–191. [cited 2019 Nov 1]. Available from: http://www.ncbi.nlm.nih.gov/pubmed/2347432

Wang Q, Moley KH (2010) Maternal diabetes and oocyte quality. Mitochondrion 10:403–410

Wang W, Jin S, Ye K (2017) Development of islet organoids from H9 human embryonic stem cells in biomimetic 3D scaffolds. Stem Cells Dev 26 (6):394–404

Wang G, Liang J, Gao LR, Si ZP, Zhang XT, Liang G et al (2018) Baicalin administration attenuates hyperglycemia-induced malformation of cardiovascular system article. Cell Death Dis 9(2):234

Weiss G, Huppertz B, Lang I, Siwetz M, Moser G (2014) First trimester trophoblast cell line ACH-3P as model to study invasion into arteries vs. veins. Placenta 35(9): A99–A100

Wong MK, Wahed M, Shawky SA, Dvorkin-Gheva A, Raha S (2019) Transcriptomic and functional analyses of 3D placental extravillous trophoblast spheroids. Sci Rep 9(1):1–13

Wu L, Song W, Xie Y, Hu L, Hou X, Wang R et al (2018) miR-181a-5p suppresses invasion and migration of HTR-8/SVneo cells by directly targeting IGF2BP2. Cell Death Dis 9(2):16

Wyman A, Pinto AB, Sheridan R, Moley KH (2008) One-cell zygote transfer from diabetic to nondiabetic mouse results in congenital malformations and growth retardation in offspring. Endocrinology 149 (2):466–469

Yamashita H, Shao J, Ishizuka T, Klepcyk PJ, Muhlenkamp P, Qiao L et al (2001) Leptin administration prevents spontaneous gestational diabetes in heterozygous Lepr(db/+) mice: effects on placental leptin and fetal growth. Endocrinology [Internet] 142 (7):2888–2897. [cited 2019 Nov 5]. Available from: http://www.ncbi.nlm.nih.gov/pubmed/11416008

Yamashita H, Shao J, Qiao L, Pagliassotti M, Friedman JE (2003) Effect of spontaneous gestational diabetes on fetal and postnatal hepatic insulin resistance in Leprdb/ + mice. Pediatr Res 53(3):411–418

Yang S-C, Tseng H-L, Shieh K-R (2013) Circadian-clock system in mouse liver affected by insulin resistance. Chronobiol Int [Internet] 30(6):796–810. [cited 2019 Nov 29]. Available from: http://www.tandfonline.com/doi/full/10.3109/07420528.2013.766204

Yin X, Mead BE, Safaee H, Langer R, Karp JM, Levy O (2016) Engineering stem cell organoids. Cell Stem Cell 18(1):25–38

Yu Y, Singh U, Shi W, Konno T, Soares MJ, Geyer R et al (2008) Influence of murine maternal diabetes on placental morphology, gene expression, and function. Arch Physiol Biochem 114(2):99–110

Yung H, Alnæs-Katjavivi P, Jones CJP, El-Bacha T, Golic M, Staff A-C et al (2016) Placental endoplasmic reticulum stress in gestational diabetes: the potential for therapeutic intervention with chemical chaperones and antioxidants. Diabetologia 59(10):2240–2250

Zamudio S, Torricos T, Fik E, Oyala M, Echalar L, Pullockaran J et al (2010) Hypoglycemia and the origin of hypoxia-induced reduction in human fetal growth. PLoS One 5(1):e8551

Zhou Q, Melton DA (2018) Pancreas regeneration. Nature (Nature Publishing Group) 557:351–358

Adv Exp Med Biol - Advances in Internal Medicine (2020) 4: 577
https://doi.org/10.1007/5584_2020_559
© Springer Nature Switzerland AG 2020
Published online: 8 July 2020

Correction to: Hypoglycemia, Malnutrition and Body Composition

I. Khanimov, M. Shimonov, J. Wainstein, and Eyal Leibovitz

Correction to:
Adv Exp Med Biol - Advances in Internal Medicine
https://doi.org/10.1007/5584_2020_526

The original version of this chapter was inadvertently published with a subtitle which was a duplication of the chapter title. The chapter subtitle has been removed.

The original article was corrected.

The updated online version of this protocol can be found at
https://doi.org/10.1007/5584_2020_559

Adv Exp Med Biol – Advances in Internal Medicine (2020) 4: 579–582
https://doi.org/10.1007/978-3-030-51089-3
© Springer Nature Switzerland AG 2021

Index

Printed in the United States
by Baker & Taylor Publisher Services